Resources for Success

MyEngineeringLab® for **Thinking Like an Engineer**
(access code required)

MyEngineeringLab is an online homework, tutorial, and assessment program designed to work with your book to engage students and improve results. Students practice what they learn, test their understanding, and pursue a personalized study plan that helps them better absorb course material and understand difficult concepts.

Book-Specific Exercises

MyEngineeringLab's varied homework and practice questions are correlated to the textbook and many regenerate algorithmically to give students unlimited opportunity for practice and mastery.

Question features include:

- Learning aids and immediate feedback
- Show My Work functionality to view all student work
- Downloadable templates in Microsoft® PowerPoint, Word, and Excel

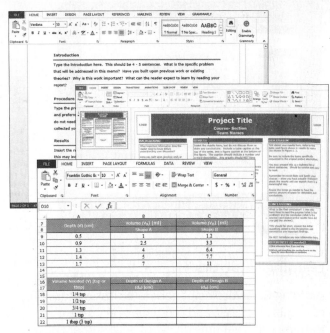

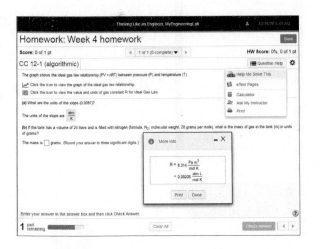

Question Help and Support

Many exercises provide step-by-step instruction, input-specific feedback, hints, and videos. They also provide links to spreadsheets, example problems, and contextually appropriate links to the eText.

For more information, visit **myengineeringlab.com**

Pearson

Student Videos

Short videos are available to help explain concepts and skills that may be difficult to explain on paper.

Topics include:

- Engineering Ethics
- Functions in Excel
- MATLAB Introduction
- Conditional Statements

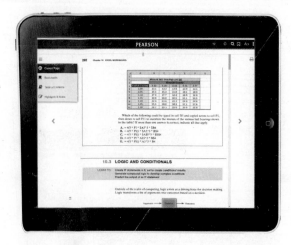

eText

Pearson eText 2.0 is optimized for mobile and offers:

- offline access and downloading for most iOS and Android devices;
- seamlessly integrated videos and other rich media;
- accessibility (screen-reader ready); and
- note-taking, highlighting, bookmarking, and search for instructors and students.

Learning Catalytics™

Learning Catalytics helps generate class discussion, customize lectures, and promote peer-to-peer learning with real-time analytics. As a student response tool, it uses students' own mobile devices to engage them in more interactive tasks and thinking.

- Help students develop critical thinking skills.
- Monitor responses to find out where students are struggling.
- Rely on real-time data to adjust your teaching strategy.
- Automatically group students for discussion, teamwork, and peer-to-peer learning.

for more information, visit **myengineeringlab.com**

Pearson

Thinking Like An Engineer

An Active Learning Approach

Fourth Edition

Elizabeth A. Stephan
Clemson University

David R. Bowman
Boeing

William J. Park
Clemson University

Benjamin L. Sill
Clemson University

Matthew W. Ohland
Purdue University

330 Hudson Street, NY NY 10013

Senior Vice President, Portfolio Management, Engineering and Computer Science: *Marcia J. Horton*
Director, Portfolio Management: *Julian Partridge*
Executive Portfolio Manager: *Holly Stark*
Portfolio Management Assistant: *Amanda Brands*
Field Marketing Manager: *Demetrius Hall*
Product Marketing Manager: *Yvonne Vannatta*
Marketing Assistant: *Jon Bryant*
Managing Producer: *Scott Disanno*
Content Producer: *Erin Ault*
Manager, Rights and Permissions: *Ben Ferinni*

Operations Specialist: *Maura Zaldivar-Garcia*
Cover Designer: *Black Horse Designs*
Cover Photos: *Red guitar: Evgeny Guityaev/ Shutterstock; X-ray of guitar: Gustoimages/ Science Photo Library*
Composition/Full-Service Project Management: *Cenveo Publisher Services*
Full-Service Project Management: *Louise Capulli*
Cover Printer: *Phoenix Color*
Printer/Binder: *LSC Communications*
Typeface: *10/12 Times Ten LT STD Roman*

Credits and acknowledgments borrowed from other sources and reproduced, with permission, in this textbook appear on appropriate page within text.

MATLAB is a registered trademark of The MathWorks, Inc., 3 Apple Hill Drive, Natick, MA 01760-2098.

Library of Congress Cataloging-in-Publication Data

Names: Stephan, Elizabeth A., author. | Bowman, D. R. (David Richard) author.
  | Park, W. J. (William John), author. | Sill, Ben L., author. |
  Ohland, Matthew W., author.
Title: Thinking like an engineer : an active learning approach / Elizabeth A.
  Stephan, Clemson University, David R. Bowman, Boeing, William J. Park,
  Clemson University, Benjamin L. Sill, Clemson University, Matthew W.
  Ohland, Purdue University.
Description: Fourth edition. | Includes index.
Identifiers: LCCN 2016046504 | ISBN 9780134639673 (alk. paper)
Subjects: LCSH: Engineering—Study and teaching (Higher) | Active learning.
Classification: LCC TA147 .T45 2017 | DDC 620.0071/1—dc23 LC record available at
https://lccn.loc.gov/2016046504

Pearson

4 18

ISBN-10: 0-13-463967-7
ISBN-13: 978-0-13-463967-3

Contents

Preface   ix

Acknowledgments   xvii

Part 1

Engineering Essentials   2

Engineering is an . . . Itch!   6

Chapter 1

Everyday Engineering   8

1.1 Choosing a Career   8

1.2 Choosing Engineering as a Career   9

1.3 NAE Grand Challenges for Engineering   11

1.4 Choosing a Specific Engineering Field   14

1.5 Engineering Technology—A Related Field   22

1.6 Gathering Information   24

1.7 Pursuing Student Opportunities   26

Review Questions   40

Chapter 2

Ethics   45

2.1 Ethical Decision Making   46

2.2 Plagiarism   51

2.3 Engineering Creed   52

2.4 Social Responsibility   53

In-Class Activities   54

Review Questions   58

Chapter 3

Design, Teamwork, and Project Management   61

3.1 Design Processes   61

3.2 Defining the Problem or Need   62

3.3 Criteria: Defining What Is Important   64

3.4 Generating Ideas   67

3.5 Comparing Designs and Making Decisions   68

3.6 Prototyping and Testing   70

3.7 Sustainability—A Special Design Criterion   70

3.8 Working in Teams   73

3.9 Experimental Design: PERIOD Analysis   79

3.10 Project Timeline   82

3.11 Modern Project Management   84

In-Class Activities   85

Review Questions   86

Mini Design Projects   87

Chapter 4

Engineering Communication   91

4.1 Basic Presentation Skills   91

4.2 Sample Presentations   94

4.3 Basic Technical Writing Skills   98

4.4 Common Technical Communication Formats   102

In-Class Activities   111

Review Questions   118

Chapter 5

Estimation   124

5.1 General Hints for Estimation   127

5.2 Estimation by Analogy   129

5.3 Estimation by Aggregation   129

5.4 Estimation by Upper and Lower Bounds   130

5.5 Estimation Using Modeling   130

5.6 Significant Figures   131

5.7 Reasonableness   135

5.8 Notation   139

In-Class Activities   142

Review Questions   145

Chapter 6
Solving Problems   146
6.1 Problem Types   146

6.2 SOLVEM—One Approach to Solving Problems   149

6.3 Representing Final Results   155

6.4 Avoiding Common Mistakes   155

6.5 Examples of SOLVEM   156

In-Class Activities   159

Review Questions   163

Part 2
Ubiquitous Units   164

Chapter 7
Fundamental Dimensions and Base Units   167
7.1 The Metric System   168

7.2 Other Unit Systems   171

7.3 Conversion Procedure for Units   171

7.4 Conversions Involving Multiple Steps   174

7.5 Conversions Involving "New" Units   178

7.6 Derived Dimensions and Units   179

7.7 Equation Laws   183

7.8 Conversion Involving Equations   187

In-Class Activities   190

Review Questions   196

Chapter 8
Universal Units   202
8.1 Force   202

8.2 Weight   205

8.3 Density   206

8.4 Amount   211

8.5 Temperature   214

8.6 Pressure   218

8.7 Gas Pressure   223

8.8 Energy   226

8.9 Power   229

8.10 Efficiency   231

8.11 Electrical Concepts   235

In-Class Activities   245

Review Questions   259

Chapter 9
Dimensionless Numbers   269
9.1 Constants with Units   269

9.2 Common Dimensionless Numbers   272

9.3 Dimensional Analysis   275

9.4 Rayleigh's Method   278

In-Class Activities   286

Review Questions   290

Part 3
Scrupulous Worksheets   294

Chapter 10
Excel Workbooks   297
10.1 Cell References   298

10.2 Functions in Excel   302

10.3 Logic and Conditionals   310

10.4 Lookup and Data Validation   319

10.5 Conditional Formatting   324

10.6 Sorting and Filters   327

In-Class Activities   334

Review Questions   349

Chapter 11
Graphical Solutions   361
11.1 Graphing Terminology   361

11.2 Proper Plots   362

11.3 Available Graph Types in Excel   369

11.4 Graph Interpretation   372

11.5 Meaning of Line Shapes   376

11.6 Graphical Solutions   382

In-Class Activities   389

Review Questions   400

Chapter 12
Models and Systems   412
12.1 Proper Plot Rules for Trendlines   413

12.2 Linear Functions   414

12.3 Linear Relationships   417

12.4 Combinations of Linear Relationships   422

12.5 Power Functions   432

12.6 Exponential Functions   435

In-Class Activities   441

Review Questions   452

Chapter 13
Mathematical Models   465
13.1 Selecting a Trendline Type   466

13.2 Interpreting Logarithmic Graphs   474

13.3 Proper Plot Rules for Log Plots   479

13.4 Converting Scales to Log in Excel   480

13.5 Dealing with Limitations of Excel   482

In-Class Activities   487

Review Questions   495

Chapter 14
Statistics   502

14.1 Histograms   503

14.2 Statistical Behavior   506

14.3 Distributions   509

14.4 Cumulative Distribution Functions   515

14.5 Statistical Process Control (SPC)   518

14.6 Statistics in Excel   523

14.7 Statistics in MATLAB   528

In-Class Activities   535

Review Questions   544

Part 4
Programming Prowess   546

Chapter 15
MATLAB Basics   550

15.1 Variable Basics   553

15.2 Numeric Types and Scalars   556

15.3 Vectors   560

15.4 Matrices   572

15.5 Character Strings   585

15.6 Cell Arrays   587

15.7 Structure Arrays   596

In-Class Activities   602

Review Questions   613

Chapter 16
Algorithms, Programs, and Functions   619

16.1 Algorithms   619

16.2 Programs   627

16.3 Functions   637

16.4 Deriving Mathematical Models   646

16.5 Debugging MATLAB Code   650

In-Class Activities   653

Review Questions   661

Chapter 17
Input/Output in MATLAB   669

17.1 Input   669

17.2 Output   680

17.3 Plotting   684

17.4 Trendlines   695

17.5 Microsoft Excel I/O   700

In-Class Activities   706

Review Questions   719

Chapter 18
Logic and Conditionals   731

18.1 Algorithms Revisited—Representing Decisions   732

18.2 Relational and Logical Operators   736

18.3 Logical Variables   739

18.4 Conditional Statements in MATLAB   746

18.5 Application: Classification Diagrams   750

18.6 switch Statements   755

18.7 Errors and Warnings   758

In-Class Activities   762

Review Questions   774

Chapter 19
Looping Structures   788

19.1 Algorithms Revisited—Loops   788

19.2 while Loops   794

19.3 for Loops   801

In-Class Activities   814

Review Questions   827

Comprehension Check Answers   855

Index   885

Equation Tables   889

MATLAB Graphing Properties   891

Appendix A: Basic Engineering Math—Online

Appendix B: Basic Workbooks—Online

Appendix C: Basic Excel Graphs—Online

Appendix D: Basic Excel Trendlines—Online

Appendix E: PROCESS Problem Solving Approach–Online

Preface

At Clemson University, all students who wish to major in engineering begin in the General Engineering Program, and after completing a core set of classes, they can declare a specific engineering major. Within this core set of classes, students are required to take math, physics, chemistry, and a two-semester engineering sequence. Our courses have evolved to address not only the changing qualities of our students, but also the changing needs of our customers. The material taught in our courses is the foundation upon which the upper-level courses depend for the skills necessary to master more advanced material. It was for these freshman courses that this text was created.

We didn't set out to write a textbook: we simply set out to find a better way to teach our students. Our philosophy was to help students move from a mode of learning, where everything was neatly presented as lecture and handouts where the instructor was looking for the "right" answer, to a mode of learning driven by self-guided inquiry. We wanted students to advance beyond "plug-and-chug" and memorization of problem-solving methods—to ask themselves if their approaches and answers make sense in the physical world. We couldn't settle on any textbooks we liked without patching materials together—one chapter from this text, four chapters from another—so we wrote our own notes. Through them, we tried to convey that engineering isn't always about having the answer—sometimes it's about asking the right questions, and we want students to learn how to ask those sorts of questions. Real-world problems rarely come with all of the information required for their solutions. Problems presented to engineers typically can't be solved by looking at how someone else solved the exact same problem. Part of the fun of engineering is that every problem presents a unique challenge and requires a unique solution. Engineering is also about arriving at an answer and being able to justify the "why" behind your choice, and equally important, the "why not" of the other choices.

We realized quickly, however, that some students are not able to learn without sufficient scaffolding. Structure and flexibility must be managed carefully. Too much structure results in rigidity and unnecessary uniformity of solutions. On the other hand, too much flexibility provides insufficient guidance, and students flounder down many blind alleys, thus making it more difficult to acquire new knowledge. The tension between these two must be managed constantly. We are a large public institution, and our student body is very diverse. Our hope is to provide each student with the amount of scaffolding they need to be successful. Some students will require more background work than others. Some students will need to work five problems, and others may need to work 50. We talk a great deal to our students about how each learner is unique. Some students need to listen to a lecture; some need to read the text over three times, and others just need to try a skill and make mistakes to discover what they still don't understand. We have tried to provide enough variety for each type of learner throughout.

Over the years, we have made difficult decisions on exactly what topics, and how much of each topic, to teach. We have refined our current text to focus on mastering four areas, each of which is introduced below.

Part 1: Engineering Essentials

There are three threads that bind the first six chapters in *Engineering Essentials* together. The first is expressed in the part title: all are essential for a successful career in engineering. The second is communications. The third and final thread is an introduction to a problem-solving methodology.

First, as aspiring engineers, students should try to verify that engineering is not only a career that suits their abilities but also one in which they will find personal reward and satisfaction.

Second, practicing engineers often make decisions that will affect not only the lives of people but also the viability of the planetary ecosystem that affects all life on Earth. Without a firm grounding in making decisions based on ethical principles, there is an increased probability that undesirable or even disastrous consequences may occur.

Third, most engineering projects are too large for one person to accomplish alone; thus, practicing engineers must learn to function effectively on teams, putting aside their personal agendas and combining their unique talents, perspectives, and ideas to achieve the goal.

Finally, communications bind it all together. Communication, whether written, graphical, or spoken is essential to success in engineering.

This part ends where all good problem solving should begin—with estimation and a methodology. It's always best to have a good guess at any problem before trying to solve it more precisely. SOLVEM provides an example of a framework for solving problems that encourages creative observation as well as methodological rigor.

Part 2: Ubiquitous Units

The world can be described using relatively few dimensions. We need to know what these are and how to use them to analyze engineering situations. Dimensions, however, are worthless in allowing engineers to find the numeric solution to a problem. Understanding units is essential to determine the correct numeric answers to problems. Different disciplines use different units to describe phenomena (particularly with respect to the properties of materials such as viscosity, thermal conductivity, density and so on). Engineers must know how to convert from one unit system to another. Knowledge of dimensions allows engineers to improve their problem-solving abilities by revealing the interplay of various parameters.

Part 3: Scrupulous Worksheets

When choosing an analysis tool to teach students, our first pick is Excel. Students enter college with varying levels of experience with Excel. To allow students who are novice users to learn the basics without hindering more advanced users, we have placed the basics of Excel in the appendix material, which is available online. To help students determine if they need to review the appendix material, an activity has been included in the introductions to Chapter 10 (Excel Workbooks), Chapter 11 (Graphical Solutions), and Chapter 12 (Models and Systems) to direct students to Appendices B, C, and D, respectively.

Once students have mastered the basics, each chapter in this part provides a deeper usage of Excel in each category. Some of this material extends beyond a simple introduction to Excel, and often, we teach the material in this unit by jumping around, covering half of each chapter in the first semester, and the rest of the material in the second semester course.

Chapter 12 introduces students to the idea of similarities among the disciplines, and how understanding a theory in one application can often aid in understanding a similar theory in a different application. We also emphasize the understanding of models (trendlines) as possessing physical meaning. Chapter 13 discusses a process for determining a mathematical model when presented with experimental data and some advanced material on dealing with limitations of Excel.

Univariate statistics and statistical process control wrap up this part of the book by providing a way for engineering students to describe both distributions and trends.

Part 4: Programming Prowess

Part 4 (Programming Prowess) covers a variety of topics common to any introductory programming textbook. In contrast to a traditional programming textbook, this part approaches each topic from the perspective of how each can be used in unison with the others as a powerful engineering problem-solving tool. The topics presented in Part 4 are introduced as if the student has no prior programming ability and are continually reiterated throughout the remaining chapters.

For this textbook we chose MATLAB as the programming language because it is commonly used in many engineering curricula. The topics covered provide a solid foundation for using computers as problem-solving tools, and they provide enough scaffolding for transfer of programming knowledge into other languages commonly used by engineers (such as C, C++, and Java).

The "Other" Stuff We've Included . . .

Throughout the book, we have included sections on surviving engineering, time management, goal setting, and study skills. We did not group them into a single chapter but have scattered them throughout the part introductions to assist students on a topic when they are most likely to need it. For example, we find students are much more open to discussing time management in the middle of the semester rather than the beginning.

In addition, we have called upon many practicing and aspiring engineers to help us explain the "why" and "what" behind engineering. They offer their "Wise Words" throughout this text. We have included our own set of "Wise Words" as the introduction to each topic here as a glimpse of what inspired us to include certain topics.

New to This Edition

The Fourth Edition of *Thinking Like an Engineer: An Active Learning Approach* contains new material and revisions based on the comments from faculty who teach with our textbook, reviewer recommendations, and most importantly, the feedback from our students. We continue to strive to include the latest software releases; in this edition, we have upgraded to Microsoft Office (Excel) 2016 and MATLAB 2016. We have added approximately 30% new questions and modified 25% of questions previously appearing. We have also added new material that reflects the constant changing face of engineering education because many of our upperclassman teaching assistants often comment to us "I wish I had ____when I took this class."

New to this edition, by chapter:

- Chapter 1: Everyday Engineering

  - New specific engineering fields show the diversity of engineering.
  - Improved representation of women in list of engineers in Review Questions.

- Chapter 3: Design, Teamwork, and Project Management

  - Expanded discussion of design, particularly a more general discussion of idea generation.
  - Expanded discussion of teamwork, including discussion of common team dynamics.
  - New discussion of project management approaches.

- Chapter 4: Engineering Communication

  - Improved examples of bad practice in presentations along with improved versions.
  - Extensive do's and don't's list for designing technical posters.

- Chapter 6: Solving Problems

  - More general treatment of problem solving added preceding SOLVEM as a specific process.

- Chapter 8: Universal Units

  - Improved discussion of mass versus weight.
  - Improved discussion of pressure.
  - New examples throughout chapter.

- Part 3: Scrupulous Worksheets

  - Revised to be consistent with the appearance and operation of Excel 2016.

- Part 4: Programming Prowess

  - Substantial revision to integrate algorithms (previously Chapter 15) into the other chapters where appropriate.
  - Revised to be consistent with the appearance and operation of MATLAB 2016.

- Online appendix materials

  - Umbrella Projects have all been moved online to allow for easier customizing of the projects for each class.
  - Curriculum materials for alternative problem solving approaches.

How to Use

This text contains many different types of instruction to address different types of learners. There are two main components to this text: hardcopy and online.

In the hardcopy, the text is presented topically rather than sequentially, but hopefully with enough autonomy for each piece to stand alone. For example, we routinely discuss only part of the Excel material in our first-semester course and leave the rest to the second semester. We hope this will give you the flexibility to choose how deeply into any given topic you wish to dive, depending on the time you have, the starting abilities of your students, and the outcomes of your course. More information about topic sequence options can be found in the instructor's manual.

Within the text, there are several checkpoints for students to see if they understand the material. Within the reading are **Comprehension Checks**, with the answers provided in the back of the book. Our motivation for including Comprehension Checks within the text rather than including them as end-of-part questions is to maintain the active spirit of the classroom within the reading, which allows students to evaluate their own understanding of the material in preparation for class—to encourage students to be self-directed learners, we must encourage them to self-evaluate regularly. At the end of each chapter, **In-Class Activities** are given to reinforce the material in each chapter. In-Class Activities exist to stimulate active conversation within pairs and groups of students working through the material. We generally keep the focus on student effort and ask them to keep working the problem until they arrive at the right answer. This provides them with a set of worked-out problems, using their own logic, before they are asked to tackle more difficult problems. The **Review Questions** sections provide additional questions, often combining skills learned in the current chapter with previous concepts to help students climb to the next level of understanding. By providing these three types of practice, students are encouraged to reflect on their understanding in preparing for class, during class, and at the end of each chapter as they prepare to transfer their knowledge to other areas. Finally we have provided a series of online **Umbrella Projects** to allow students to apply skills that they have mastered to larger-scope problems. We have found the use of these problems extremely helpful in providing context for the skills that they learn throughout a unit.

Understanding that every student learns differently, we have included several media components in addition to traditional text. Each section within each chapter has an accompanying set of **video lecture slides** . Within these slides, the examples presented are unique from those in the text to provide another set of sample solutions. The slides are presented with **voiceover**, which has allowed us to move away from traditional in-class lecture. We expect the students to listen to the slides outside of class, and then in class we typically spend time working problems, reviewing assigned problems, and providing **"wrap-up" lectures**, which are mini-versions of the full lectures to summarize what they should have gotten from the assignment. We expect the students to come to class with questions from the reading and lecture that we can then help clarify. We find with this method, the students pay more attention, as the terms and problems are already familiar to them, and they are more able to verbalize what they don't know. Furthermore, they can always go back and listen to the lectures again to reinforce their knowledge as many times as they need.

Some sections of this text are difficult to lecture, and students will learn this material best by **working through examples**. This is especially true with Excel and MATLAB, so you will notice that many of the lectures in these sections are shorter than previous material. The examples are scripted the first time a skill is presented, and students are expected to have their laptops open and to work through the examples (not just read them). When students ask us questions in this section, we often start the answer by asking them to "show us your work from Chapter XX." If the student has not actually worked the examples in that chapter, we tell them to do so first; often, this will answer their questions.

After the first few basic problems, in many cases where we are discussing more advanced skills than data entry, we have **provided starting worksheets and code** . Students can access the starting data in the MyEngineeringLab. In some cases, though, it is difficult to explain a skill on paper, or even with slides, so for these instances we have included **videos** .

Finally, for the communication section, we have provided **templates** 🅿 🆆 for several types of reports and presentations. These are available on the Instructor Resource Center at www.pearsonhighered.com/irc and with the adoption of MyEngineeringLab. Visit www.myengineeringlab.com for more information.

MyEngineeringLab™

Thinking Like an Engineer, Fourth Edition, together with MyEngineeringLab provides an engaging in-class experience that will inspire your students to stay in engineering, while also giving them the practice and scaffolding they need to keep up and be successful in the course. It's a complete digital solution featuring:

- **Book-Specific Exercises**—MyEngineeringLab's varied homework and practice questions are correlated to the textbook and many regenerate algorithmically to give students unlimited opportunity for practice and mastery. Question features include: Learning aids and immediate feedback; Show My Work functionality to view all student work; and downloadable templates in Microsoft® PowerPoint, Word, and Excel.
- **Question Help and Support**—Many exercises provide step-by-step instruction, input-specific feedback, hints, and videos. They also provide links to spreadsheets, example problems, and contextually appropriate links to the eText.
- **Student Videos**—Short videos are available to help explain concepts and skills that may be difficult to explain on paper. Topics include: Engineering Ethics, Functions in Excel MATLAB Introduction, and Conditional Statements.
- **eText**—Pearson eText 2.0 is optimized for mobile and offers: offline access and downloading for most iOS and Android devices; seamlessly integrated videos and other rich media; accessibility (screen-reader ready); and note-taking, highlighting, bookmarking, and search for instructors and students.

If adopted, access to MyEngineeringLab can be bundled with the book or purchased separately. For a fully digital offering, learn more at www.myengineeringlab.com.

Resources for Instructors

Instructor's Manual—Available to all adopters, this provides a complete set of solutions for all activities and review exercises. For the In-Class Activities, suggested guided inquiry questions along with time frame guidelines are included. Suggested content sequencing and descriptions of how to couple assignments to the Umbrella Projects are also provided.

PowerPoints—A complete set of lecture PowerPoint slides make course planning as easy as possible.

Sample Exams—Available to all adopters, these will assist in creating tests and quizzes for student assessment.

MyEngineeringLab—Provides web-based assessment, tutorial, homework and course management. www.myengineeringlab.com

Learning Catalytics™—Learning Catalytics helps generate class discussion, customize lectures, and promote peer-to-peer learning with real-time analytics. As a student response tool, it uses students' own mobile devices to engage them in more interactive tasks and thinking. Help students develop critical thinking skills. Monitor responses to find out where students are struggling. Rely on real-time data to adjust your teaching strategy. Automatically group students for discussion, teamwork, and peer-to-peer learning.

All requests for instructor resources are verified against our customer database and/or through contacting the requestor's institution. Contact your local Pearson representative for additional information.

What Does Thinking Like an Engineer Mean?

We are often asked about the title of the book. We thought we'd take a minute and explain what this means, to each of us. Our responses are included in alphabetical order.

For me, thinking like an engineer is about creatively finding a solution to some problem. In my pre-college days, I was very excited about music. I began my musical pursuits by learning the fundamentals of music theory by playing in middle school band and eventually worked my way into different bands in high school (orchestra, marching and, jazz band) and branching off into teaching myself how to play guitar. I love playing and listening to music because it gives me an outlet to create and discover art. I pursued engineering for the same reason; as an engineer, you work in a field that creates or improves designs or processes. For me, thinking like an engineer is exactly like thinking like a musician—through my fundamentals, I'm able to be creative, yet methodical, in my solutions to problems.

D. Bowman, Computer Engineer

Thinking like an engineer is about solving problems with whatever resources are most available—or fixing something that has broken with materials that are just lying around. Sometimes, it's about thinking ahead and realizing what's going to happen before something breaks or someone gets hurt—particularly in thinking about what it means to fail safe—to design how something will fail when it fails. Thinking like an engineer is figuring out how to communicate technical issues in a way that anyone can understand. It's about developing an instinct to protect the public trust—an integrity that emerges automatically.

M. Ohland, Civil Engineer

To me, understanding the way things work is the foundation on which all engineering is based. Although most engineers focus on technical topics related to their specific discipline, this understanding is not restricted to any specific field, but applies to everything! One never knows when some seemingly random bit of knowledge, or some pattern discerned in a completely disparate field of inquiry, may prove critical in solving an engineering problem. Whether the field of investigation is Fourier analysis, orbital mechanics, Hebert boxes, personality types, the Chinese language, the life cycle of mycetozoans, or the evolution of the music of Western civilization, the more you understand about things, the more effective an engineer you can be. Thus, for me, thinking like an engineer is intimately, inextricably, and inexorably intertwined with the Quest for Knowledge. Besides, the world is a truly fascinating place if one bothers to take the time to investigate it.

W. Park, Electrical Engineer

Engineering is a bit like the game of golf. No two shots are ever exactly the same. In engineering, no two problems or designs are ever exactly the same. To be successful, engineers need a bag of clubs (math, chemistry, physics, English, social studies) and then need to have the training to be able to select the right combination of clubs to move from the tee to the green and make a par (or if we are lucky, a birdie). In short, engineers need to be taught to THINK.

B. Sill, Aerospace Engineer

I like to refer to engineering as the color grey. Many students enter engineering because they are "good at math and science." I like to refer to these disciplines as black and white—there is one way to integrate an equation and one way to balance a chemical reaction. Engineering is grey, a blend of math and science that does not necessarily have one clear answer. The answer can change depending on the criteria of the problem. Thinking like an engineer is about training your mind to conduct the methodical process of problem solving. It is examining a problem from many different angles, considering the good, the bad and the ugly in every process or product. It is thinking creatively to discover ways of solving problems, or preventing issues from becoming problems. It's about finding a solution in the grey and presenting it in black and white.

E. Stephan, Chemical Engineer

Lead author note: When writing this preface, I asked each of my coauthors to answer this question. As usual, I got a wide variety of interpretations and answers. This is typical of the way we approach everything we do, except that I usually try and mesh the responses into one voice. In this instance, I let each response remain unique. As you progress throughout this text, you will (hopefully) see glimpses of each of us interwoven with the one voice. We hope that through our uniqueness, we can each reach a different group of students and present a balanced approach to problem solving, and, hopefully, every student can identify with at least one of us.

—Beth Stephan
Clemson University
Clemson, SC

Acknowledgments

When we set out to formalize our instructional work, we wanted to portray engineering as a reality, not the typical flashy fantasy portrayed by most media forums. We called on many of our professional and personal relationships to help us present engineering in everyday terms. During a lecture to our freshmen, Dr. Ed Sutt [PopSci's 2006 Inventor of the Year for the HurriQuake Nail] gave the following advice: *A good engineer can reach an answer in two calls: the first, to find out who the expert is; the second, to talk to the expert.* Realizing we are not experts, we have called on many folks to contribute articles. To our experts who contributed articles for this text, we thank Dr. Lisa Benson, Dr. Neil Burton, Jan Comfort, Jessica (Pelfrey) Creel, Jason Huggins, Leidy Klotz, and Troy Nunmaker.

To Dr. Lisa Benson, thank you for allowing us to use "Science as Art" for the basis of many photos that we have chosen for this text. To explain "Science as Art": *Sometimes, science and art meet in the middle. For example, when a visual representation of science or technology has an unexpected aesthetic appeal, it becomes a connection for scientists, artists and the general public. In celebration of this connection, Clemson University faculty and students are challenged to share powerful and inspiring visual images produced in laboratories and workspaces for the "Science as Art" exhibit.* For more information, please visit www.scienceasart.org. To the creators of the art, thank you for letting us showcase your work in this text: Martin Beagley, Dr. Caye Drapcho, Eric Fenimore, Dr. Scott Husson, Dr. Jaishankar Kutty, Dr. Kathleen Richardson, and Dr. Ken Webb. A special thanks to Russ Werneth for getting us the great Hubble teamwork photo.

To the Rutland Institute for Ethics at Clemson University: The four-step procedure outlined in Chapter 2 on Ethics is based on the toolbox approach presented in the Ethics Across the Curriculum Seminar. Our thanks to Dr. Daniel Wueste, director, and the other Rutlanders (Kelly Smith, Stephen Satris, and Charlie Starkey) for their input into this chapter.

To Jonathan Feinberg and all the contributors to the Wordle (http://www.wordle .net) project, thank you for the tools to create the Wordle images in the introduction sections. We hope our readers enjoy this unique way of presenting information and are inspired to create their own Wordles!

To our friends and former students who contributed their Wise Words: Tyler Andrews, Corey Balon, Ed Basta, Sergey Belous, Brittany Brubaker, Tim Burns, Ashley Childers, Jeremy Comardelle, Matt Cuica, Jeff Dabling, Christina Darling, Ed D'Avignon, Brian Dieringer, Lauren Edwards, Andrew Flowerday, Stacey Forkner, Victor Gallas Cervo, Lisa Gascoigne, Khadijah Glast, Tad Hardy, Colleen Hill, Tom Hill, Becky Holcomb, Beth Holloway, Selden Houghton, Allison Hu, Ryan Izard, Lindy Johnson, Darryl Jones, Maria Koon, Rob Kriener, Jim Kronberg, Rachel Lanoie, Mai Lauer, Jack Meena, Alan Passman, Mike Peterson, Candace Pringle, Derek Rollend,

Eric Roper, Jake Sadie, Janna Sandel, Ellen Styles, Adam Thompson, Kaycie (Smith) Timmons, Devin Walford, Russ Werneth, and Aynsley Zollinger.

To the faculty members at Clemson, for providing inspiration, ideas, and helping us find countless mistakes: Dr. Steve Brandon, Dr. Ashley Childers, Michael Giebner, Dr. Sarah Grigg, Dr. Mariah Magagnotti, Dr. Jonathan Maier, Dr. William Martin, Matthew Miler, John Minor, Dr. Andrew Neptune, and Dr. Joe Watkins. You guys are the other half of the team that makes Clemson such a great place to work! We could not have done this without you.

To the staff of the GE program, we thank you for your support of us and our students: Kelli Blankenship, Lib Crockett, Chris Porter, and all of our terrific advising staff both past and present. To the administration at Clemson, we thank you for your continued support of our program, especially Dean Dr. Anand Gramopadhye.

To the thousands of students who used this text in various forms over the years—thanks for your patience, your suggestions, and your criticism. You have each contributed not only to the book, but to our personal inspiration to keep doing what we do.

To all the reviewers who provided such valuable feedback to help us improve: We appreciate the time and energy needed to review this material, and your thoughtful comments have helped push us to become better.

Emily Ohland's project management skills kept this edition on schedule. Working with this team has been more like herding squirrels than herding cats. We are forever grateful.

To the great folks at Pearson—this project would not be a reality without all your hard work. To Eric Hakanson, without that chance meeting this project would not have begun! Thanks to Holly Stark for her belief in this project and in us! Thanks to Erin Ault for keeping us on track and having such a great vision to display our hard work. You have put in countless hours on this edition—thanks for making us look great! Thanks to Tim Galligan, Demetrius Hall, and the fabulous Pearson sales team all over the country for promoting our book to other schools and helping us allow so many students to start "Thinking Like Engineers"! We would not have made it through this without all of the Pearson team efforts and encouragement!

Finally, on a Personal Note

DRB: Thanks to my parents and sister for supporting my creative endeavors with nothing but encouragement and enthusiasm. To my grandparents, who value science, engineering, and education to be the most important fields of study. To my coauthors, who continue to teach me to think like an engineer. To my employers and coworkers past present and future, thank you for encouraging my growth and service as an engineer. To Dana, you are the glue that keeps me from falling to pieces. Thank you for your support, love, laughter, inspiration, and determination, among many other things. You have the hardest job on earth—living with an engineer—but do know that I love you exponentially, iteratively, and infinitely more every day.

MWO: My wife Emily has my love, admiration, and gratitude for all she does, including holding the family together. For my children, who share me with my students—Everett, whose "old soul" and courage touches so many; Carson, who models how to be inspiring without drawing attention to yourself; and Anders, who is discovering who he wants to be. My father Theodor inspired me to be an educator, my mother Nancy helped me understand people, my sister Karen lit a pathway for me in engineering, my brother Erik showed me that one doesn't need to be loud to be a leader, and my mother-in-law Nancy Winfrey shared the wisdom of a long career. I recognize those who helped me create an engineering education career path, most prominently: Fred Orthlieb, Marc Hoit, Tim Anderson, Ben Sill, Sue Kemnitzer, and Purdue's School of Engineering Education.

WJP: Choosing only a few folks to include in an acknowledgment is a seriously difficult task, but I have managed to reduce it to five. First, Beth Stephan has been the guiding force behind this project, without whom it would never have come to fruition. In addition, she has shown amazing patience in putting up with my shenanigans and my weird perspectives. Next, although we exist in totally different realities, my parents have always supported me, particularly when I was a newly married, destitute graduate student fresh off the farm. Third, my son Isaac, who has the admirable quality of being willing to confront me with the truth when I am behaving badly, and for this I am grateful. Finally, and certainly most importantly, to Lila, my partner of more than four decades, I owe a debt beyond anything I could put into words. Although life with her has seldom been easy, her influence has made me a dramatically better person.

BLS: To my amazing family, who always picked up the slack when I was off doing "creative" things, goes all my gratitude. To Anna and Allison, you are wonderful daughters who both endured and "experienced" the development of many "in class, hands on" activities—know that I love you and thank you. To Lois who has always been there with her support and without whining for over 40 years, all my love. Finally, to my coauthors who have tolerated my eccentricities and occasional tardiness with only minimum grumbling, you make great teammates.

EAS: To my coauthors, for hanging in there to the Fourth Edition and always keeping things interesting! To my mom, Kay and Denny—thanks for your love and support. To Khadijah, Steven and Miles, wishes for you to continue to conquer the world! To Brock and Katie, I love you both a bushel and a peck. It is amazing to me to look at how much you both have grown since the first time we did this, and how much you now give me insight into what I design for my students—who are now your peers. Finally, to Sean . . . one more time! I love you more than I can say—and know that even when I forget to say it, I still believe in us. "Show a little faith, there's magic in the night . . ."

PART 1

Engineering Essentials

Chapter 1

Everyday Engineering

1.1 Choosing a Career

1.2 Choosing Engineering As a Career

1.3 NAE Grand Challenges for Engineering

1.4 Choosing a Specific Engineering Field

1.5 Engineering Technology—A Related Field

1.6 Gathering Information

1.7 Pursuing Student Opportunities

Chapter 2

Ethics

2.1 Ethical Decision Making

2.2 Plagiarism

2.3 Engineering Creed

2.4 Social Responsibility

Chapter 3

Design, Teamwork, and Project Management

3.1 Design Processes

3.2 Defining the Problem or Need

3.3 Criteria: Defining What is Important

3.4 Generating Ideas

3.5 Comparing Designs and Making Decisions

3.6 Prototyping and Testing

3.7 Sustainability—A Special Design Criterion

3.8 Working in Teams

3.9 Experimental Design: Period Analysis

3.10 Project Timeline

3.11 Modern Project Management

Chapter 4

Engineering Communication

4.1 Basic Presentation Skills

4.2 Sample Presentations

4.3 Basic Technical Writing Skills

4.4 Common Technical Communication Formats

Chapter 5

Estimation

5.1 General Hints for Estimation

5.2 Estimation by Analogy

5.3 Estimation by Aggregation

5.4 Estimation by Upper and Lower Bounds

5.5 Estimation Using Modeling

5.6 Significant Figures

5.7 Reasonableness

5.8 Notation

Chapter 6

Solving Problems

6.1 Problem Types

6.2 SOLVEM—One Approach to Solving Problems

6.3 Representing Final Results

6.4 Avoiding Common Mistakes

6.5 Examples of SOLVEM

You are no doubt in a situation where you have an idea you want to be an engineer. Someone or something put into your head this crazy notion—that you might have a happy and successful life working in the engineering profession. Maybe you are good at math or science, or you want a job where creativity is as important as technical skill. Maybe someone you admire works as an engineer. Maybe you are looking for a career that will challenge you intellectually, or maybe you like to solve problems.

You may recognize yourself in one of these statements from practicing engineers on why they chose to pursue an engineering degree.

I chose to pursue engineering because I enjoyed math and science in school, and always had a love for tinkering with electronic and mechanical gadgets since I was old enough to hold a screwdriver.

S. Houghton, Computer Engineer

I chose to pursue engineering because I always excelled in science and math and I really enjoy problem solving. I like doing hands-on activities and working on "tangible" projects.

M. Koon, Mechanical Engineer

Learning Objectives

The overall learning objectives for this part include the following:

Chapter 1:

- Explore the variety of collegiate and career opportunities of an engineering discipline.

Chapter 2:

- Conduct research on ethical issues related to engineering; formulate and justify positions on these issues.

Chapter 3:

- Demonstrate an ability to design a system, component, or process to meet desired needs.
- Demonstrate an ability to function on multidisciplinary teams.

Chapter 4:

- Communicate technical information effectively by composing clear and concise oral presentations and written descriptions of experiments and projects.

Chapter 5:

- Identify process variability and measurement uncertainty associated with an experimental procedure, and interpret the validity of experimental results.
- Use "practical" skills, such as visualizing common units and conducting simple measurements, calculations, and comparisons to make estimations.

Chapter 6:

- Classify types of problems and how to approach them.

I wanted to pursue engineering to make some kind of positive and (hopefully) enduring mark on the world.

J. Kronberg, Electrical Engineer

I was good at science and math, and I loved the environment; I didn't realize how much I liked stream and groundwater movement until I looked at BioSystems Engineering.

C. Darling, Biosystem Engineer

My parents instilled a responsibility to our community in us kids. As an engineer, I can serve my community through efficient and responsible construction while still satisfying my need to solve challenging problems.

J. Meena, Civil Engineer

I asked many different majors one common question: "What can I do with this degree?" The engineering department was the only one that could specifically answer my question. The other departments often had broad answers that did not satisfy my need for a secure job upon graduating.

L. Johnson, Civil Engineer

Engineering is a highly regarded and often highly compensated profession that many skilled high school students choose to enter for the challenge, engagement, and ultimately the reward of joining the ranks of the esteemed engineers of the world. But what, exactly, does an engineer do? This is one of the most difficult questions to answer because of the breadth and depth of the engineering field. So, how do the experts define engineering?

The National Academy of Engineering (NAE) says:

Engineering has been defined in many ways. It is often referred to as the "application of science" because engineers take abstract ideas and build tangible products from them. Another definition is "design under constraint," because to "engineer" a product means to construct it in such a way that it will do exactly what you want it to, without any unexpected consequences.

I am a first-generation college student, and I wanted to have a strong foundation when I graduated from college.

C. Pringle, Industrial Engineer

Since I knew I wanted to design computers, I had a choice between electrical and computer engineering. I chose computer engineering, so I could learn about both the hardware and software. It was my interests in computers and my high school teachers that were the biggest influence in my decision.

E. D'Avignon, CpE

My first choice in majors was mechanical engineering. I changed majors after taking a drafting class in which I did well enough to get a job teaching the lab portion, but I did not enjoy the work. After changing to electrical and computer engineering, I took a Statics and Dynamics course as part of my required coursework and that further confirmed my move as I struggled with that material.

A. Flowerday, EE

Some people come into college knowing exactly what they want their major and career to be. I, on the other hand, was not one of those people. I realized that I had a wide spectrum of interests, and college allows you to explore all those options. I wanted a major that was innovative and would literally change the future of how we live. After looking through what I loved and wanted to do, my choice was computer engineering.

S. Belous, CpE

WISE WORDS: WHAT WAS THE HARDEST ADJUSTMENT FROM HIGH SCHOOL TO COLLEGE?

The biggest adjustment was the overwhelming amount of responsibility that I had to take on. There was no longer anybody there to tell me what to do or when to do it. I had to rely on myself to get everything done. All the things I took for granted when I was at home—not having to do my own laundry, not preparing all of my meals, not having to rely on my alarm clock to wake me up, etc.— quickly became quite apparent to me after coming to college. I had to start managing my time better so that I would have time to get all of those things done.

T. Andrews, CE

For me, the most difficult adjustment from high school to college has been unlearning some of the study habits adopted early on. In high school, you can easily get by one semester at a time and just forget what you "learned" when you move into a new semester or a new chapter of your text.

The National Academy of Engineering (NAE) is an independent, nonprofit institution that serves as an adviser to government and the public on issues in engineering and technology. Its members consist of the nation's premier engineers, who are elected by their peers for their distinguished achievements. Established in 1964, NAE operates under the congressional charter granted to the National Academy of Sciences. http://www.nae.edu/About.aspx

According to the Merriam-Webster online dictionary:

> *Engineering is the application of science and mathematics by which the properties of matter and the sources of energy in nature are made useful to people.*

Engineering is a broad, hard-to-define field requiring knowledge of science and mathematics and other fields to turn ideas into reality. The ideas and

College is just a little bit different. To succeed, you have to really make an effort to keep up with your studies—even the classes you have finished already. If you do not, chances are that a topic mentioned in a prerequisite course is going to reappear in a later class, which requires mastery of the previous material in order to excel.

R. Izard, CpE

The hardest adjustment was learning how to study. I could no longer feel prepared for tests by simply paying attention in class. I had to learn to form study groups and begin studying for tests well in advance. You can't cram for engineering tests.

M. Koon, ME

The hardest adjustment was taking full personal responsibility for everything from school work, to social life, and to finances. Life becomes a lot more focused when you realize that you are paying for your education and that your decisions will greatly impact your future. The key is to manage your time between classes, studying, having fun, and sleeping.

S. Belous, CpE

Studying, networking, talking to my professors about my strengths and weaknesses, taking responsibility for my actions, just the whole growing up into an adult was tough.

C. Pringle, IE

The hardest adjustment I had to make going from high school to college was realizing that I was on my own—and not just for academics, either. I was responsible for making sure I remembered to eat dinner, for not eating candy bars for lunch everyday, for balancing my social life with my studies, for managing my money . . . for everything.

J. Sandel, ME

The hardest adjustment from high school to college was changing my study habits. In high school, teachers coordinated their tests so we wouldn't have several on the same day or even in the same week. I had to learn how to manage my time more efficiently. Moreover, it was difficult to find a balance between both the social and academic aspects of college.

D. Walford, BioE

Since the tests cover more material and have more weight in college, I had to alter my study habits to make myself start studying more than a day in advance. It was overwhelming my first semester because there was always something that I could be studying for or working on.

A. Zollinger, CE

problems posed to engineers often do not require a mastery-level knowledge of any particular scientific field, but instead require the ability to put together all of the pieces learned in those fields.

Because engineers solve real-life problems, their ultimate motivation is to work toward making life better for everyone. In "The Heroic Engineer" (*Journal of Engineering Education*, January 1997) by Taft H. Broome (Howard University) and Jeff Peirce (Duke University), those authors claimed:

> *Engineers who would deem it their professional responsibility to transcend self-interests to help non-experts advance their own interests may well prove indispensable to free societies in the twenty-first century.*

Broome and Peirce go on to explain that the traits and behaviors of engineers can be compared to those of a hero. The motivation of any hero is to save someone's life; engineers create products, devices, and methods to help save lives. Heroes intervene to protect from danger; engineers devise procedures, create machines, and improve processes to protect people and the planet from danger. While learning an engineering discipline can be challenging, the everyday engineer does not see it as an obstacle: It is merely an opportunity to be a hero.

You will see many quotes from practicing engineers. As a good engineering team would, we recognize that we (the authors) are not experts at all things, and request input and advice when needed. We asked engineers we know who work at "everyday engineering" jobs to reflect on the choices they made in school and during their careers. We hope you benefit from their collective knowledge. When asked for advice to give to an incoming freshman, one gave the following reply, summing up this section better than we ever could have imagined.

> *[A career in engineering] is rewarding both financially and personally. It's nice to go to work and see some new piece of technology—to be on the cutting edge. It's also a great feeling to know that you are helping improve the lives of other people. Wherever there has been a great discovery, an engineer is to thank. That engineer can be you.*
>
> A. Thompson, Electrical Engineer

Engineering is an . . . Itch!

Contributed by Dr. Lisa Benson, Assistant Professor of Engineering and Science Education, Clemson University

There are a lot of reasons why you are majoring in engineering. Maybe your goal is to impress someone, like your parents, or to defy all those who said you would never make it, or simply to prove to yourself that you have it in you. Maybe your goal is to work with your hands as well as your mind. Maybe you have no idea why you are here, but you know you like cars. There are about as many goals as there are students, and they all motivate students to learn. Some goals are better motivators than others.

Lots of experts have studied goals and how they affect what students do in school. Not surprisingly, there are as many ideas and theories about goals as there are experts. But most experts agree on the idea that there are mastery goals and performance goals.

> *Students who are **mastery oriented** try to do things well because they want to do their best. They are not driven by external factors like grades or praise, but instead*

they seek to learn things because they want to really understand them and not just get the correct answer.

Students who are **performance oriented** *seek to earn good grades to reflect how hard they've worked. They study because they know it will get them something—a scholarship, an above-average grade, or praise from their parents. Since grades are tied to their sense of achievement, students who are performance oriented tend to feel discouraged and anxious when they earn low grades. They tend to want to memorize and pattern-match to solve problems, rather than learn the underlying concepts and methods.*

Most students have been performance oriented throughout high school. In college, you will be more successful if you start thinking in terms of mastery. If you seek to really understand what you are learning in your classes, performance (i.e., good grades) will follow. But performance (a grade, an award, or praise) is not everything, and it will not be enough to keep you motivated when projects and coursework are challenging.

There is nothing like the feeling when you finally understand something that you did not get before. Sometimes it is an "aha!" moment, and sometimes it is a gradual dawning. The feeling is like an itch you can't scratch—you will want to keep at it once you get it. *When you are motivated to understand and master something, you're taking pride in your achievement of conquering the material, not just getting a good grade on an exam.* And you are going to keep scratching that itch. Keep "scratching" at the material—working, practicing, drilling skills if you need to, whatever it takes—to master that . . . itch!

1

Everyday Engineering

Most students who start off in a technical major know very little about their chosen field. This is particularly true in engineering, which may not be explicitly present in the high-school curriculum. Students commonly choose engineering and science majors because someone suggested them. In this section, we help you ask the right questions about your interests, skills, and abilities; we then show you how to combine the answers with what you learn about engineering and science in order to make the right career decision.

1.1 Choosing a Career

LEARN TO: Think about the kind of career you want and training you need

In today's society, the careers available to you upon your graduation are numerous and diverse. It is often difficult as a young adult to determine exactly what occupation you want to work at for the rest of your life because you have so many options. As you move through the process, there are questions that are appropriate to ask. You cannot make a good decision without accurate information. No one can (or should) make the decision for you: not your relatives, professors, advisors, or friends. Only you know what feels right and what does not. You may not know all the answers to your questions right away. That means you will have to get them by gathering more information from outside resources and through your personal experience. Keep in mind that choosing your major and ultimately your career is a process. You constantly evaluate and reevaluate what you learn and experience. A key component is whether you feel challenged or overwhelmed. True success in a profession is not measured in monetary terms; it is measured in job satisfaction . . . enjoying what you do, doing what you enjoy. As you find the answers, you can choose a major that leads you into a successful career path that you enjoy.

Before you decide, answer the following questions about your tentative major choice. Start thinking about the questions you cannot answer and look for ways or resources to get the information you need. It may take a long time before you know, and that is okay!

- What do I already know about this major?
- What courses will I take to earn a degree in this major?
- Do I have the appropriate academic preparation to complete this major? If not, what will I have to do to acquire it?
- Am I enjoying my courses? Do I feel challenged or stressed?
- What time demands are involved? Am I willing to spend the time it takes to complete this major?

- What kinds of jobs will this major prepare me for? Which sounds most interesting?
- What kinds of skills will I need to do the job I want? Where can I get them?

This process will take time. Once you have the information, you can make a choice. Keep in mind, nothing is set in stone—you can always change your mind!

1.2 Choosing Engineering as a Career

LEARN TO: Understand the relationship between an engineering major and a technical industry
Think about different technical industries that might interest you
Think about different engineering majors that might interest you

In the previous section, we gave several examples of why practicing engineers wanted to pursue a career in engineering. Here are a few more:

> I was always into tinkering with things and I enjoyed working with computers from a young age. Math, science, and physics came very natural to me in high school. For me it was an easy choice.
>
> *J. Comardelle, Computer Engineer*

> My initial instinct for a career path was to become an engineer. I was the son of a mechanical engineer, performed well in science and mathematics during primary education, and was always "tinkering" with mechanical assemblies.
>
> *M. Ciuca, Mechanical Engineer*

> I chose engineering for a lot of the same reasons that the "typical" entering first-year does—I was good at math and science. I definitely did not know that there were so many types of engineering and, to be honest, I was a little overwhelmed by the decision I needed to make of what type of engineering was for me.
>
> *L. Edwards, Civil Engineer*

> I wasn't really sure what I wanted to do. My parents were not college graduates so there was not a lot of guidance from them, so my high school teachers influenced me a lot. I was taking advanced math and science classes and doing well in them. They suggested that I look into engineering, and I did.
>
> *S. Forkner, Chemical Engineer*

> I was a nighttime/part-time student while I worked full time as a metallurgical technician. I was proficient in math and science and fortunate to have a mentor who stressed the need for a bachelor's degree.
>
> *E. Basta, Materials Engineer*

> Coming into college, I knew I wanted to pursue a career in medicine after graduation. I also knew that I did not want to major in chemistry, biology, etc. Therefore, bioengineering was a perfect fit. It provides a challenging curriculum while preparing me for medical school at the same time. In addition, if pursuing a career in medicine does not go according to plan, I know that I will also enjoy a career as a bioengineer.
>
> *D. Walford, Bioengineer*

Table 1-1 Sample career paths and possible majors. (Shaded boxes indicate a good starting point for further exploration.)

Careers	Engineering										Science				
	Aerospace	Biomedical	BioSystems	Civil	Chemical	Materials	Electrical/computer	Environmental	Industrial	Mechanical	Chemistry	Computer Science	Geology	Mathematics	Physics
GENERAL															
Energy industry				■			■	■		■	■			■	
Machines		■					■		■	■					
Manufacturing					■				■	■					
Materials	■			■	■	■				■	■				
Structures	■			■		■				■					
Technical sales		■	■		■		■					■			
SPECIFIC															
Rocket/airplane	■									■					
Coastal engineering			■	■				■					■		
Computing							■					■		■	
Cryptography												■		■	
Defense	■	■		■	■		■								
Environment			■	■				■					■		
Fiber optics							■					■			■
Forensics		■									■	■			
Groundwater			■	■				■					■		
Healthcare		■							■			■			
Human factors	■								■			■			
Industrial sensors	■				■		■								
Intelligent systems		■					■					■		■	
Management	■	■		■	■		■		■						
Operations research									■			■		■	
Outdoor work			■	■				■					■		
Pharmaceutical		■	■		■						■				
Plastics					■	■				■					
Robotics	■						■			■		■			
Semiconductors	■					■	■								■
Telecommunications	■						■					■		■	
Transportation				■				■		■					
Waste management			■	■				■							

Table 1-1 describes the authors' perspective on how various engineering and science disciplines might contribute to different industries or innovations. This table is only an interpretation by a few engineers and does not handle every single possibility of how an engineer might contribute toward innovation. For example, an industrial engineer might be called in to work on an energy product to share a different perspective on energy efficiency. The broad goal of any engineering discipline is to solve problems, so there is often a need for a different perspective to possibly shed new light toward an innovative solution.

1.3 NAE Grand Challenges for Engineering

LEARN TO: Learn about the challenges facing the engineer of the future
Consider the NAE Grand Challenges and think about your own interests

History (and prehistory) is replete with examples of technological innovations that forever changed the course of human society: the mastery of fire, the development of agriculture, the wheel, metallurgy, mathematics of many flavors, the printing press, the harnessing of electricity, powered flight, nuclear power, and many others. The National Academy of Engineering (NAE) has established a list of 14 challenges for the twenty-first century, each of which has the potential to transform the way we live, work, and play. Your interest in one or more of the Grand Challenges for Engineering may help you select your engineering major. For more information, visit the NAE website at http://www.engineeringchallenges.org/. In case this address changes after we go to press, you can also type "NAE Grand Challenges for Engineering" into your favorite search engine.

A burgeoning planetary population and the technological advances of the last century are exacerbating many current problems, as well as engendering a variety of new ones. For example:

- Relatively inexpensive and rapid global travel make it possible for diseases to quickly span the globe, whereas a century ago, they could spread, but much more slowly.
- The reliance of the developed world on computers and the Internet makes the fabric of commerce and government vulnerable to cyberterrorism.
- Increased demand for limited resources not only drives up prices for those commodities, but also fosters strain among the nations competing for them.

These same factors can also be a force for positive change in the world:

- Relatively inexpensive and rapid global travel allows even people of modest means to experience different cultures and hopefully promote a more tolerant attitude toward those who live by different sets of social norms.
- Modern communications systems—cell phones, the Internet, etc.—make it essentially impossible for a government to control the flow of information to isolate the members of a population or to isolate that population from the political realities in other parts of the world. An excellent example was the rapid spread of rebellion in the Middle East and Africa in early 2011 against autocratic leaders who had been in power for decades.
- Increased demand for and rising prices of limited resources is driving increased innovation in alternatives, particularly in meeting the world's energy needs.

As should be obvious from these few examples, technology both solves problems and creates them. A significant portion of the difficulty in the challenges put forth by the NAE to solve critical problems in the world lies in finding solutions that do not create other problems. Let us consider a couple of the stated challenges in a little more detail. You probably already have some familiarity with several of them, such as "make solar energy economical," "provide energy from fusion," "secure cyberspace," and "enhance virtual reality," so we will begin with one of the NAE Grand Challenges for Engineering that is perhaps less well known.

Manage the Nitrogen Cycle

Nitrogen is an element required for all known forms of life, being part of every one of the 20 amino acids that are combined in various ways to form proteins, all five bases used to construct RNA and DNA, and numerous other common biological molecules, such as chlorophyll and hemoglobin. Fortunately, the supply of nitrogen is—for all practical purposes—inexhaustible, constituting over 75% of the Earth's atmosphere. However, nitrogen is mostly in the molecular form N_2, which is chemically unavailable for uptake in biological systems, since the two nitrogen atoms are held together by a very strong triple bond.

For atmospheric nitrogen to be available to biological organisms, it must be converted, or fixed, by the addition of hydrogen, into ammonia, NH_3, that may then be used directly or converted by other microorganisms into other reactive nitrogen compounds for uptake by microorganisms and plants. The term "nitrogen fixation" includes conversion of N_2 into both ammonia and these other reactive compounds, such as the many oxides of nitrogen. Eventually, the cycle is completed when these more readily available forms of nitrogen are converted back to N_2 by microorganisms, a process called denitrification.

Prior to the development of human technology, essentially all nitrogen fixation was performed by bacteria possessing an enzyme capable of splitting N_2 and adding hydrogen to form ammonia, although small amounts of fixed nitrogen are produced by lightning and other high-energy processes. In the early twentieth century, a process called the Haber-Bosch process was developed that would allow conversion of atmospheric nitrogen into ammonia and related compounds on an industrial scale. Today, slightly more than a century later, approximately one-third of all fixed nitrogen is produced using this process.

The ready availability of relatively inexpensive nitrogen fertilizers has revolutionized agriculture, allowing people to increase yields dramatically and to grow crops on previously unproductive lands. However, the widespread use of synthetic nitrogen has caused many problems, including water pollution, air pollution, numerous human health problems, and disruption of marine and terrestrial ecosystems to the extent that entire populations of some organisms have died off.

Deliberate nitrogen fixation is only one part of the nitrogen cycle problem, however. Many human activities, especially those involving the combustion of fossil fuels, pump huge quantities of various nitrogen compounds into the atmosphere. Nitrous oxide (N_2O), commonly known as "laughing gas," is particularly problematic since it is about 200 times more effective than carbon dioxide as a greenhouse gas, and it persists in the atmosphere for over a century.

Altogether, human-caused conversion of nitrogen into more reactive forms now accounts for about half of all nitrogen fixation, meaning that there is twice as much nitrogen fixed today than there was a little more than a century ago. However, we have done little to augment the natural denitrification process, so the deleterious effects of excessive fixed nitrogen continue to increase. We have overwhelmed the natural nitrogen cycle. If we are to continue along this path, we must learn to manage the use of these products more efficiently and plan strategies for denitrification to bring the cycle back into balance.

Reverse-Engineer the Brain

The development of true artificial intelligence (AI) holds possibly the most overall potential for positive change in the human race, as well as the most horrendous possible negative effects. This is reflected in science fiction, where the concept of thinking machines is a common plot device, ranging from Isaac Asimov's benevolent R. Daneel Olivaw to the malevolent Skynet in the *Terminator* movies. If history is any guide, however, the potential for disastrous consequences seldom deters technological advances, so let us consider what is involved in the development of AI.

Although great strides have been made in creating machines that seem to possess "intelligence," almost all such systems that have come to the public notice either rely on brute-force calculations, such as the chess-playing computer, Deep Blue, that defeated world champion Garry Kasparov in 1997, or reliance on incredibly fast access to massive databases, such as the Jeopardy-playing computer, Watson, that defeated both the highest money winner, Brad Rutter, and the record holder for longest winning streak, Ken Jennings, in 2011. Perhaps, needless to say, these are oversimplifications, and there are many more aspects to both of these systems. However, one would be hard-pressed to argue that these computers are truly intelligent—that they are self-aware and contain the unexplainable spark of creativity, which is the hallmark of humans, and arguably other highly intelligent creatures on Earth.

Today's robots perform many routine tasks, from welding and painting vehicles to vacuuming our homes and cutting our grass. However, all of these systems are programmed to perform within certain restrictions and have serious limitations when confronted with unexpected situations. For example, if your school utilized vacuuming robots to clean the floors in the classrooms, they would probably be unable to handle the situation effectively if someone became nauseous and regurgitated on the carpet. If we could endow such robots with more human-like intelligence, the range of tasks that they could successfully accomplish would increase by orders of magnitude, thus increasing their utility tremendously.

To date, we have almost exclusively tried merely to construct intelligent systems that mimic behavior and thought, not design systems that actually store and process information in a manner analogous to that of a biologically based computer (a brain). The human brain uses a network of interconnections between specialized subsections that makes even the most advanced computers look like a set of children's building blocks. Although some understanding has been gained, the means of encoding information and its transfer in the brain is almost completely a mystery.

Gaining even a basic understanding of brain function might allow us to develop prosthetic limbs that actually function as well as the originals, restore sight to the blind, repair brain damage, or even enhance human intelligence.

1.4 Choosing a Specific Engineering Field

LEARN TO: Compare and contrast various engineering majors
Think about engineering majors you have never considered before

The following paragraphs briefly introduce several different types of engineering majors. By no means is this list completely inclusive.

Agricultural Engineering

Agricultural engineering (AgE) focuses on producing food and fiber in environmentally sound ways. This includes the design of machines and systems supporting agricultural operations, the modeling and management of soil and water, and the development and operation of large-scale manufacture of food products.

Agricultural engineers work in such areas as irrigation systems, identification and elimination of bacteria that cause food poisoning, and grain transportation. Agricultural engineers are employed in industry, universities, research facilities, and government. In industry, they are commonly part of a team serving in food production. In government agencies, they are involved in regulation and environmental conservation.

Bioengineering or Biomedical Engineering

Bioengineering (BioE) and biomedical engineering (BME) apply engineering principles to the understanding and solution of medical problems. Bioengineers are involved in research and development in all areas of medicine, from investigating the physiological behavior of single cells to designing implants for the replacement of diseased or traumatized body tissues. Bioengineers design new instruments, devices, and software; assemble knowledge from many scientific sources to develop new procedures; and conduct research to solve medical problems.

Typical bioengineers work in such areas as artificial organs, automated patient monitoring, blood chemistry sensors, advanced therapeutic and surgical devices, clinical laboratory design, medical imaging systems, biomaterials, and sports medicine.

Bioengineers are employed in universities, industry, hospitals, research facilities, and government. In industry, they may be part of a team serving as a liaison between engineers and clinicians. In hospitals, they select appropriate equipment and supervise equipment performance, testing, and maintenance. In government agencies, they are involved in safety standards and testing.

Biosystems Engineering

Biosystems engineering (BE) is the field of engineering most closely allied with advances in biology. BE emphasizes two main areas: (1) bioprocess engineering, with its basis in microbiology, and (2) ecological engineering, with its basis in ecology. The field focuses on the sustainable production of biorefinery compounds (biofuels, bioactive molecules, and biomaterials) using metabolic pathways found in nature and green processing

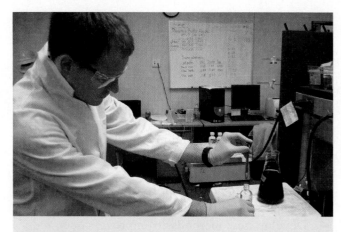

My research is part of a Water Research Foundation project, which is investigating the formation of emerging disinfection byproducts (DBPs) in drinking water treatment.

DBPs are undesirable, toxic compounds that are formed when water is chlorinated. I am investigating the effects of pH, bromide, and iodide concentrations, and preoxidants on the formation of a specific family of DBPs.

D. Jones, BE

technologies. Further, BE encompasses the design of sustainable communities using low-impact development strategies (bioretention basins, rainwater harvesting) for stormwater retention and treatment—and ecologically sound food and energy-crop production. Scientific emphasis is shifting toward the biosciences. Biosystems engineers apply engineering design and analysis to biological systems and incorporate fundamental biological principles to engineering designs to achieve ecological balance.

Here are some activities of biosystems engineers:

- Design bioprocesses and systems for biofuels (biodiesel, hydrogen, ethanol), biopharmaceuticals, bioplastics, and food processing industries
- Develop ecological designs (permeable pavement, bioswales, green infrastructure) to integrate water management into the landscape
- Integrate biological sustainability concepts into energy, water, and food systems
- Provide engineering expertise for agriculture, food processing, and manufacturing
- Pursue medical or veterinary school or graduate school in the fields of BE, BME, or ecological engineering

I am a project manager for new product development.

I oversee and coordinate the various activities that need to be completed in order to get a new product approved and manufactured, and ultimately in the hands of our consumers.

S. Forkner, ChE

Chemical Engineering

Chemical engineering (ChE) incorporates a strong emphasis on three sciences: chemistry, physics, and mathematics. Chemical engineers are involved in the research and development, manufacture, sales, and use of chemicals, pharmaceuticals, electronic components, food and consumer goods, petroleum products, synthetic fibers and films, pulp and paper, and many other products. They work on environmental remediation and pollution prevention, as well as in medical and health-related fields. Chemical engineers:

- Conduct research and develop new products
- Develop and design new manufacturing processes
- Earn additional degrees to practice medicine or patent, environmental, or corporate law
- Sell and provide technical support for sophisticated chemical products to customers
- Solve environmental problems; work in biotechnology
- Troubleshoot and solve problems in chemical manufacturing facilities

Civil Engineering

Civil engineering (CE) involves the planning, design, construction, maintenance, and operation of facilities and systems to control and improve the environment for modern

My team is responsible for implementing the engineered design in the field.

We install, tune, test, and accept into operations all of the electronics that allow customers to use our state-of-the-art fiber optic network to run voice, video, and data for their residential needs.

L. Gascoigne, CE

civilizations. This includes projects of major importance, such as bridges, transportation systems, buildings, ports, water distribution systems, and disaster planning.

Here are just a few of many opportunities available for civil engineers:

- Design and analyze structures ranging from small buildings to skyscrapers to off-shore oil platforms
- Design dams and building foundations
- Develop new materials for pavements, buildings, and bridges
- Design improved transportation systems
- Design water distribution and removal systems
- Develop new methods to improve safety, reduce cost, speed construction, and reduce environmental impact
- Provide construction and project management services for large engineered projects throughout the world

My current responsibilities include:

- Analysis of traffic signal operations and safety for municipal and private clients;
- Preparation of traffic impact studies;
- Review of plans and traffic studies for municipalities and counties;
- Design of traffic signal installations and traffic signing projects.

C. Hill, CE

I am responsible for assisting in the management of commercial and healthcare projects for Brasfield & Gorrie. Working closely with the owner and architect, I maintain open lines of communication and aim to provide exceptional service to the entire project team from the preconstruction phase of the project through construction. I assist in establishing and monitoring procedures for controlling the cost, schedule, and quality of the work in accordance with the construction contract.

L. Edwards, CE

I develop, manage, and support all software systems. I also deal with system scalability, customer satisfaction, and data management.

J. Comardelle, CpE

Computer Engineering

Computer engineering (CpE) spans the fields of computer science and engineering, giving a balanced view of hardware, software, hardware-software trade-offs, and basic modeling techniques that represent the computing process involving the following technologies:

- Communication system design
- Computer interface design
- Computer networking
- Digital signal processing applications
- Digital system design
- Embedded computer design
- Process instrumentation and control
- Software design

I am a digital designer and work on the read channel for hard-disk drives. The read channel is the portion of the controller SOC (system on a chip) that decodes the analog signal read from the hard disk and converts it to digital data. I am responsible for writing Verilog RTL code, verification, synthesis into gates, and meeting timing requirements of my blocks.

E. D'Avignon, CpE

Electrical Engineering

Electrical engineering (EE) ranges from the generation and delivery of electrical power to the use of electricity in integrated circuits. The rapid development of technology, based on integrated circuit devices, has enabled the pervasive use of computers in command, control, communication, and computer-aided design. Some systems electrical engineers work on include the following:

As a radiation effects engineer, I test the performance of electronic components in a specific application exposed to different types of radiation. Responsibilities include interfacing with design and system engineers, creating test plans, performing testing and data analysis, and authoring test reports.

A. Passman, EE

I manage global programs that help develop leadership capabilities and skills of our current and future leaders. I am a consultant, a coach, a mentor, and a guide. If leaders are interested in improving how they lead and the impact they have on their employees and on company results, we work with them to identify the best ways for them to continue their development.

A. Hu, EE

- Communication system design
- Control systems—from aircraft to automotive
- Electrical power generation and distribution
- Electromagnetic waves
- Integrated circuit design
- Process instrumentation and control
- Robotic systems design
- Telecommunications

My group supports [a major automotive manufacturer's] decisions pertaining to where to put new plants around the world, what products to build in them, and at what volumes.

In particular, my work involves understanding what the other auto manufacturers are planning for the future (footprint, capacity, technology, processes, etc.), so that information can be used to affect decisions about how to compete around the globe.

M. Peterson, EE

Environmental Engineering

I work with scientists and engineers to protect their innovations by writing patent applications describing their inventions and presenting the applications before the U.S. Patent & Trademark Office. I also assist clients in determining whether another party is infringing their patents and help my clients to avoid infringing other's patents.

M. Lauer, EnvE

Environmental engineering (EnvE) is an interdisciplinary field of engineering that is focused on cleaning up environmental contamination, as well as designing sustainable approaches to prevent future contamination. Environmental engineers apply concepts from basic sciences (including chemistry, biology, mathematics, and physics) to develop engineered solutions to complex environmental problems.

Environmental engineers design, operate, and manage both engineered and natural systems to protect the public from exposure to environmental contamination and to develop a more sustainable use of our natural resources. These activities include the following:

- Production of safe, potable drinking water
- Treatment of wastewater so that it is safe to discharge to surface water or reuse in such applications as landscape irrigation
- Treatment of air pollutants from mobile (e.g., automobiles) and stationary (e.g., power plants) sources
- Characterization and remediation of sites contaminated with hazardous wastes (e.g., polychlorinated biphenyls, or PCBs)
- Disposal of municipal solid wastes
- Management of radioactive wastes, including characterization of how radioactive materials move through the environment and the risks they pose to human health
- Evaluation of methods to minimize or prevent waste production and inefficient use of energy by manufacturing facilities
- Reduce human health risks by tracking contaminants as they move through the environment
- Design a more sustainable future by understanding our use of resources

WISE WORDS: WHAT DID YOU DO YESTERDAY MORNING AT WORK?

I worked on completing a failure analysis report for an industrial client.

E. Basta, Materials Engineer

I reviewed the results of the overnight simulation runs. There were several failures, so I analyzed the failures and devised fixes for the problems.

E. D'Avignon, CpE

On any given day, my morning might be spent this way: in meetings, at my computer (e-mail, drafting documents/reports), making phone calls, talking to other project members, running a test on the manufacturing lines. Not glamorous, but necessary to solve problems and keep the project moving forward.

S. Forkner, ChE

I continued to design a warehouse/office building on a nuclear expansion project.

T. Hill, CE

Yesterday, I designed a spreadsheet to assist in more precisely forecasting monthly expenditures.

R. Holcomb, IE

With a BS degree in EnvE, students will find employment with consulting engineering firms, government agencies involved in environmental protection, and manufacturing industries.

Industrial Engineering

Industrial engineering (IE) deals with the design and improvement of systems, rather than with the objects and artifacts that other engineers design. A second aspect of IE is the involvement of people in these systems—from the people involved in the design and production to the people who are ultimate end users. A common theme is the testing and evaluation of alternatives that may depend on random events. Industrial engineers use mathematical, physical, social sciences, and engineering combined with the analytical and design methods to design, install, and improve complex systems that provide goods and services to our society. Industrial engineers are called upon to:

- Analyze and model complex work processes to evaluate potential system improvements
- Analyze how combinations of people and machines work together
- Analyze how the surroundings affect the worker, and design to reduce the negative effects of this environment
- Develop mathematical and computer models of how systems operate and interact
- Improve production and service processes from the perspectives of quality, productivity, and cost
- Work on teams with other professionals in manufacturing, service industries, and government agencies

As the Business Leader for Central Florida at a major power company, I develop and manage a $42 million budget. I ensure that our engineering project schedule and budget match, and report on variances monthly. I also conduct internal audits and coach employees on Sarbanes–Oxley compliance requirements.

R. Holcomb, IE

I attended the plant morning meeting and the Boardmill leadership team meeting, followed by the plant budget meeting. In between meetings, I returned e-mails and project-related phone calls. Typically, I spend about 50% of my time in meetings. I use the information I receive at these meetings to direct and focus the efforts of the engineering staff.

J. Huggins, ME

Testing some failed prototype biostimulators returned by a trial user, to determine why and how they failed and how to prevent it from happening in coming production versions.

J. Kronberg, EE

Yesterday, I worked on a patent infringement opinion involving agricultural seeding implements, a Chinese patent office response for a component placement and inspection machine used in circuit board manufacturing, and a U.S. Patent Office response for database navigation software.

M. Lauer, EnvE

In my current position, I spend much of my time reading technical manuals and interface control documents. I attended a meeting detailing lightning protection for the Ares rocket.

E. Styles, EE

My primary job responsibilities include maintaining, upgrading, and designing all the computer systems and IT infrastructure for the Vermont Railroad. I handle all the servers and take care of network equipment. When needed, I also program customized applications and websites for customers or our own internal use. I also serve as a spare conductor and locomotive engineer when business needs demand.

S. Houghton, CpE

Currently I am working on a project to determine patient priorities for evacuations from healthcare facilities during emergencies. The assumption of an evacuation is that there will be enough time to transfer all of the patients, but in the event of limited resources, there may not be enough time to move all of the patients to safety. Further—and depending on the emergency type—it may be an increased risk to transport some patient types. Based on certain objectives, we are developing guidelines to most ethically determine a schedule for choosing patients for emergency evacuations.

A. Childers, IE

Materials Engineering

As a metallurgical engineer, my duties include

- Consulting firm management/administration
- Failure analysis
- Subcontracted metals testing services
- Metallurgical quality systems design/auditing
- Metallurgical expert in litigation cases
- Materials selection and design consultant, in-process and final inspection and testing services

E. Basta, Materials Engineer

Materials engineering or metallurgical engineering focuses on the properties and production of materials. Nature supplied only 92 naturally occurring elements to serve as building blocks to construct all modern conveniences. A materials engineer works to unlock the relationship between atomic, molecular, and larger-scale structures and the resultant properties. This category includes such majors as ceramic engineering, metallurgical engineering, and polymer science and engineering.

Here is a partial list of products designed and manufactured by material engineers:

- Brick, tile, and whitewares research and manufacturing for the home and workplace
- Ceramic spark plugs, oxygen sensors, and catalytic converters that optimize engine performance
- Metal and ceramic materials that enable biomedical implants and prosthetics
- Microwave responsive ceramics that stabilize and filter cellular phone reception
- Nanotechnology, including silver nanoparticles used as antibacterial agents in socks and t-shirts and carbon nanotubes used to reinforce the forks of racing bicycles
- Plastics found in bulletproof vests, replacement heart valves, and high tension wires on bridges
- Superconducting metals that are used in medical imaging devices, such as magnetic resonance imaging (MRI) equipment
- Ultrapure glass optical fibers that carry telephone conversations and Internet communications

I implement technology to protect national assets against adversaries.

J. Dabling, ME

Mechanical Engineering

Mechanical engineering (ME) involves areas related to machine design, manufacturing, energy production and control, materials, and transportation. Areas supported by mechanical engineers include:

- Construction
- Energy production and control
- Environmental systems
- Food production
- Management
- Materials processing
- Medicine
- Military service
- Propulsion and transportation systems
- Technical sales

As plant engineering manager I report directly to the plant manager. My primary responsibilities are managing all the capital investments; providing technical support and expertise to the plant leadership team; and mentoring and developing the plant's engineering staff and technical resources.

J. Huggins, ME

I am a salesman, so at the end of the day I'm looking to grow my market share while trying to protect the market share I already have. I help companies maintain a safe, reliable, and efficient steam and condensate system by utilizing the many products and services that we have to offer. This is mostly done by designing and installing upgrades and improving my customer's existing steam systems.

T. Burns, ME

I am an aerospace engineering manager responsible for developing unique astronaut tools and spacewalk procedures and for testing and training for NASA's Hubble Space Telescope servicing missions. My job ranges from tool and procedure design and development to underwater scuba testing to real-time, on-console support of Space Shuttle missions.

R. Werneth, ME

In my job as a management consultant, I address CEO-level management decisions as part of a project team by helping clients identify, analyze, and solve business-related problems. My responsibilities include generating hypotheses, gathering, and analyzing data, conducting benchmarking and best practices assessments, recommending actions, and working with clients to develop implementation plans.

M. Ciuca, ME

Nuclear Engineering

Nuclear engineering (NucE) focuses on the application of the breakdown (fission) and fusion of atomic nuclei. It relies heavily on principles from nuclear physics. While the most commonly recognized applications of nuclear processes are power-generating nuclear reactors and nuclear weapons, a broad range of opportunities is available. Nuclear engineers do the following:

- Provide electrical energy needs
- Assist in space exploration through nuclear propulsion systems, radiation power sources, and detector systems
- Apply radiation-based methods to sterilize medical instruments and food processing systems and preserve food
- Design radiation-based systems to diagnose and eradicate cancer and other diseases
- Apply plasma technology in material processing and other technologies-
- Contribute to national security through the stewardship of nuclear weapons and engineering safeguards against nuclear proliferation
- Design systems to protect people and equipment from radiation and for the safe storage of nuclear waste.

1.5 Engineering Technology — A Related Field

LEARN TO: Understand the difference between engineering and engineering technology
Understand differences in curricula and in career paths for the two disciplines

As its name suggests, engineering technology is related to engineering. In a formal sense, the two fields use different requirements for accreditation and are accredited by different commissions. While it is possible to earn an associate's degree in engineering technology, it is clearer to compare the bachelor's degrees in engineering technology and engineering to observe the formal differences. The student outcomes required of accredited engineering technology programs are shown in the table below alongside those required of accredited engineering programs. Note that the criteria do not appear in the same order in the two sets of criteria. The notable differences between the two are noted. In some cases, whereas the wording may be very different between the two criteria, the sense is very similar. For example, engineering student outcome "(g), an ability to communicate effectively," is similar to engineering technology student outcome "g. an ability to apply written, oral, and graphical communication in both technical and nontechnical environments; and an ability to identify and use appropriate technical literature." Whereas the engineering technology outcome provides much greater detail, there is no reason to suggest that these are different outcomes.

Engineering Technology (1)	Engineering (2)
a. an ability to select and apply the knowledge, techniques, skills, and modern tools of the discipline to broadly defined engineering technology activities;	(k) an ability to use the techniques, skills, and modern engineering tools necessary for engineering practice.
b. an ability to select and apply a knowledge of mathematics, science, engineering, and technology to engineering technology problems that require the application of principles and applied procedures or methodologies;	(a) an ability to apply knowledge of mathematics, science, and engineering
c. an ability to conduct standard tests and measurements; to conduct, analyze, and interpret experiments; and to apply experimental results to improve processes;	(b) an ability to design and conduct experiments, as well as to analyze and interpret data
d. an ability to design systems, components, or processes for broadly defined engineering technology problems appropriate to program educational objectives;	(c) an ability to design a system, component, or process to meet desired needs within realistic constraints, such as economic, environmental, social, political, ethical, health and safety, manufacturability, and sustainability
e. an ability to function effectively as a member or leader on a technical team;	(d) an ability to function on multidisciplinary teams
f. an ability to identify, analyze, and solve broadly defined engineering technology problems;	(e) an ability to identify, formulate, and solve engineering problems
g. an ability to apply written, oral, and graphical communication in both technical and nontechnical environments; and an ability to identify and use appropriate technical literature;	(g) an ability to communicate effectively
j. a knowledge of the impact of engineering technology solutions in a societal and global context;	(h) the broad education necessary to understand the impact of engineering solutions in a global, economic, environmental, and societal context
i. an understanding of and a commitment to address professional and ethical responsibilities including a respect for diversity;	(f) an understanding of professional and ethical responsibility
	(j) a knowledge of contemporary issues
h. an understanding of the need for and an ability to engage in self-directed continuing professional development;	(i) a recognition of the need for, and an ability to engage in lifelong learning
k. a commitment to quality, timeliness, and continuous improvement.	

Sources: (1) http://www.abet.org/accreditation/accreditation-criteria/criteria-for-accrediting-engineering-technology-programs-2016-2017/#studentoutcomes
(2) http://www.abet.org/accreditation/accreditation-criteria/criteria-for-accrediting-engineering-programs-2016-2017/#outcomes

Differences in Academic Curricula

Generally, engineering program curricula are more academic, focusing more on theory and concepts, whereas engineering technology program curricula are more practical, focusing on applications and skills. This difference can be seen in the table above, whereas E(b) requires that engineering graduates are able to design and conduct experiments, ET(c.) does not require engineering technology graduates to be able to design experiments. Engineering outcome E(e) requires the more theory-oriented ability to "formulate" problems, an outcome that is missing from ET(f.).

This difference is often noted in a more general treatment in engineering curricula compared to a more specific treatment in engineering technology curricula.

Whereas engineering graduates must learn to face a wide variety of design constraints in E(c), engineering technology graduates have the more application-oriented option in ET(d.) of focusing on a narrower set of constraints appropriate to a particular context. Similarly, while ET(j.) addresses societal and global impact, E(h) additionally includes economic and environmental impact. Whereas engineering graduates must function on "multidisciplinary" teams per E(d), engineering technology graduates are required in ET(e.) to function on "technical" teams that need not be multidisciplinary. In some cases, the application-oriented focus of engineering technology appears easier—conducting experiments that others design, solving problems that others formulate. Yet even in the criteria above, it is clear that engineering technology students must figure out how to act on things that engineering students must only understand—whereas E(f) requires that engineering graduates understand professional and ethical responsibility, ET(i.) also requires that engineering technology graduates have a commitment to address those issues, including a respect for diversity; whereas engineering students are required by E(b) to be able to design and conduct experiments, engineering technology graduates are expected to be able to apply the results of experiments.

Differences in Typical Career Pathways

Engineering graduates have wide-ranging jobs ranging from design to analysis, office work to field work, from companies that make things to companies that design things that are made by others. Graduates of four-year engineering technology programs are more often found in jobs where things are made or sold and are more often engaged in field work. Four-year technology graduates are called technologists—the term "technician" is appropriate for two-year engineering technology graduates. Generally, engineering careers are more flexible, whereas engineering technology careers tend to result in more tangible accomplishments (rather than accomplishments on paper).

1.6 Gathering Information

LEARN TO: Research different professional organizations for engineering disciplines

You will need to gather a lot of information in order to answer your questions about engineering or any other major. Many resources are available on your campus and online.

The Career Center

Most universities have a centralized campus career center. The staff specializes in helping students explore various occupations and make decisions. They offer testing and up-to-date information on many career fields. Professional counselors are available by appointment to assist students with job and major selection decisions.

Table 1-2 Website research starting points

Society	Abbreviation
American Ceramic Society	ACerS
American Indian Science and Engineering Society	AISES
American Institute of Aeronautics and Astronautics	AIAA
American Institute of Chemical Engineers	AIChE
American Nuclear Society	ANS
American Society of Agricultural and Biological Engineers	ASABE
American Society of Civil Engineers	ASCE
American Society for Engineering Education	ASEE
American Society of Mechanical Engineers	ASME
American Society of Metals International	ASM Int'l.
Association for Computing Machinery	ACM
Audio Engineering Society	AES
Biomedical Engineering Society	BMES
Engineers Without Borders	EWB
Institute of Biological Engineering	IBE
Institute of Electrical and Electronics Engineers	IEEE
Institute of Industrial Engineers	IIE
Institute of Transportation Engineers	ITE
Materials Research Society	MRS
National Academy of Engineering	NAE
National Society of Black Engineers	NSBE
National Society of Professional Engineers	NSPE
Society of Automotive Engineers International	SAE Int'l.
Society of Hispanic Professional Engineers	SHPE
Society of Petroleum Engineers	SPE Int'l.
Society of Plastics Engineers	SPE
Society of Women Engineers	SWE
Tau Beta Pi, The Engineering Honor Society	TBP

Career Websites

To learn more about engineering and the various engineering fields, you can find a wealth of information from engineering professional societies. Each engineering field has a professional society dedicated to promoting and disseminating knowledge about that particular discipline. Table 1-2 provides a list of most major engineering professional socities in the United States. In some cases, more than one society is connected with different subdisciplines. Other regions of the world may have their own professional societies.

Perusing the various societies' websites can provide you with information invaluable in helping you decide on a future career. We have not given URLs for the societies, since these sometimes change. To find the current address, simply use an online search engine with the name of the society.

In addition, a few engineering societies are not specific to a discipline, but to their membership:

- American Indian Science and Engineering Society (AISES)
- National Society of Black Engineers (NSBE)
- National Society of Professional Engineers (NSPE)
- Society of Hispanic Professional Engineers (SHPE)
- Society of Women Engineers (SWE)
- Tau Beta Pi, The Engineering Honor Society (TBP)

Most engineering schools have student chapters of the relevant organizations on campus. These organizations provide an excellent opportunity for you to learn more about your chosen discipline or the ones you are considering, and they also help you meet other students with similar interests. Student membership fees are usually nominal, and the benefits of membership far outweigh the small cost.

Active participation in these societies while in school not only gives you valuable information and experience, but also helps you begin networking with professionals in your field and enhances your résumé.

1.7 Pursuing Student Opportunities

LEARN TO: Understand what a cooperative experience entails
Understand what an internship experience entails
Understand what a study abroad experience entails

In addition to the traditional educational experience, many students seek experience outside of the classroom. Many engineering colleges and universities have special departments that help place students in programs to gain real engineering work experience or provide them with a culturally rich study environment. Ask a professor or advisor if your university provides experiences similar to those described in this section.

Cooperative Education

Contributed by Dr. Neil Burton, Executive Director of Career Services,
Michelin® Career Center, Clemson University

People learn things in many different ways. Some people learn best by reading; others by listening to others; and still others by participating in a group discussion. One very effective form of learning is called **experiential learning**, also referred to as engaged learning in some places. As the name suggests, experiential learning means learning through experience, and there is a very good chance you used this method to learn how to ride a bike, bake a cake, change a flat tire, or perform any other complex process that took some practice to perfect. The basic assumption behind experiential learning is that you learn more by doing than by simply listening or watching.

Becoming a good engineer is a pretty challenging process, so it seems only natural that experiential learning would be especially useful to an engineering student. In 1906, Herman Schneider, the Dean of Engineering at the University of Cincinnati, developed

WISE WORDS: ADVICE ABOUT SOCIETY PARTICIPATION

Get involved! It is so much fun! Plus, you're going to meet a ton of cool people doing it!

T. Andrews, CE

I am a very involved person, and I love it. I definitely recommend participating in professional societies because not only do they look good on a résumé but they also provide you with useful information for your professional life. It also allows you to network with others in your field which can be helpful down the road. Also, do something fun as it is a nice stress relief and distraction when life seems to become really busy.

C. Darling, BE

My advice to students willing to participate on student activities is for them to not be shy when going to a student organization for the first couple of times. It takes time to get well known and feel comfortable around new people, but don't let that prevent you from being part of a student organization that can bring many benefits to you. Always have a positive attitude, be humble, and learn to listen to others; these are traits which you will use in your professional life.

V. Gallas Cervo, ME

My advice to first-year students is to not get involved with too many organizations all at once. It is easy to get distracted from your class work with all the activities on campus. Focus on a couple and be a dedicated officer in one of them. This way you have something to talk about when employers see it on your résumé. I would recommend that you are involved with one organization that you enjoy as a hobby and one organization that is a professional organization.

D. Jones, BE

The most important thing to do when joining any group is to make sure you like the people in it. This is probably even more important than anything the group even does. Also make sure that if the group you're joining has a lot of events they expect you to be at, you have the time to be at those events.

R. Kriener, EE

The field of engineering is a collaborative project; therefore, it is important to develop friendships within your major.

S. Belous, CpE

Make the time to participate in student activities. If possible, try to get a leadership position in one of the activities because it will be useful in interviews to talk about your involvement. While employers and grad schools may not be impressed with how you attended meetings occasionally as a general member, they will be interested to hear about the projects that you worked on and the challenges that you faced in a leadership position.

K. Smith, ChE

Get involved! College is about more than just academics. Participating in student activities is a lot of fun and makes your college experience more memorable. I've made so many friends not just at my school, but all over the country by getting involved. It is also a great way to develop leadership and interpersonal skills that will become beneficial in any career.

A. Zollinger, CE

On my co-op, I worked on the hazard analysis for the new Ares I launch vehicle that NASA designed to replace the Space Shuttles when they retired in 2010.

Basically, we looked at the design and asked: What happens if this part breaks, how likely is it that this part will break, and how can we either make it less likely for the part to break or give it fault tolerance so the system can withstand a failure?

J. Sandel, ME

an experiential learning program for engineering students because he felt students would understand the material in their engineering classes much better if they had a chance to put that classroom knowledge into practice in the workplace. Schneider called this program **Cooperative Education**, and more than 100 years later, colleges and universities all over the world offer cooperative (co-op) education assignments to

WISE WORDS: WHAT DID YOU GAIN FROM YOUR CO-OP OR INTERNSHIP EXPERIENCE?

I was able to learn how to practically apply the knowledge I was gaining from college. Also, the pay allowed me to fund my schooling.

B. Dieringer, ME

The best part about that experience was how well it meshed with my courses at the time. My ability to apply what I was learning in school every day as well as to take skills and techniques I was learning from experienced engineers and use them toward the projects I was working on in school was invaluable.

A. Flowerday, CpE

My internships were a great introduction to the professional workplace — the skills and responsibilities that are expected; the relationships and networks that are needed.

S. Forkner, ChE

I decided to pursue a co-op because I had trouble adapting to the school environment in my first years and taking some time off to work at a company seemed a good way to rethink and reorganize myself. I made a good decision taking some time off, since it allowed me to learn a lot more about myself, how I work, how I learn, and how I operate. I learned that the biggest challenges were only in my mind and believing in me was, and still is, the hardest thing.

V. Gallas Cervo, ME

My internship taught me that I could be an engineer — and a good one too. I had a lot of self-doubt before that experience, and I learned that I was better than I thought I was.

B. Holloway, ME

The value of the cooperative education was to apply the classroom material to real-world applications, develop an understanding of the expectations post-grad, and provide the opportunity for a trial run for a future career path in a low-risk environment.

M. Ciuca, ME

students in just about every major, although engineering students remain the primary focus of most co-op programs.

There are many different kinds of co-op programs, but all of them offer engineering students the chance to tackle real-world projects with the help and guidance of experienced engineers. One common model of cooperative education allows students to alternate semesters of co-op with semesters of school. In this model, students who accept co-op assignments spend a semester working full-time with a company, return to school for the following academic term, go back out for a second co-op **rotation**, return to school the following term, and continue this pattern until they have spent enough time on assignment to complete the co-op program.

Students learn a lot about engineering during their co-op assignments, but there are many other benefits as well. Engineering is a tough discipline, and a co-op assignment can often help a student determine if he or she is in the right major. It is a lot better to figure out that you do not want to be an engineer before you have to take thermodynamics or heat transfer! Students who participate in co-op programs also have a chance to develop some great professional contacts, and these contacts are very handy when it comes time to find a permanent job after graduation. The experience students receive while on a co-op assignment is also highly valued by employers who want to know if a student can handle the challenges and responsibilities of a certain position. In fact, many students receive full-time job offers from their co-op employers upon graduation.

Being able to immediately apply the things I learned in school to real-world applications helped reinforce a lot of the concepts and theories. It also resulted in two job offers after graduation, one of which I accepted.

J. Huggins, ME

This was an amazing opportunity to get a real taste for what I was going to be doing once I graduated. I began to realize all the different types of jobs I could have when I graduated, all working in the same field. In addition, my work experience made my résumé look 100 times more appealing to potential employers. The experience proved that I could be a team player and that I could hit the ground running without excessive training.

L. Johnson, CE

It gave me a chance to see how people work together in the "real world" so that I could learn how to interact with other people with confidence, and also so that I could learn what kind of worker or manager I wanted to be when I "grew up."

M. Peterson, ME

Without a doubt, the best professional decision of my life. After my co-op rotation finished, I approached school as more like a job. Furthermore, co-oping makes school easier! Imagine approaching something in class that you have already seen at work!

A. Thompson, EE

I did my research, and it really made sense to pursue a co-op—you get to apply the skills you learn in class, which allows you to retain the information much better, as well as gain an increased understanding of the material.

As for choosing a co-op over an internship, working for a single company for an extended period of time allows students to learn the ropes and then progress to more intellectually challenging projects later in the co-op. And, if you really put forth your best effort for the duration of your co-op, you could very well end up with a job offer before you graduate!

R. Izard, ME

I wanted some practical experience, and I wasn't exactly sure what career I wanted to pursue when I graduated. My experience at a co-op set me on a completely different career path than I had been on previously.

K. Smith, ChE

Perhaps the most important benefit a co-op assignment can provide is improved performance in the classroom. By putting into practice the theories you learn about in class, you gain a much better understanding of those theories. You may also see something on your co-op assignment that you will cover in class the following semester, putting you a step ahead of everyone else in the class. You will also develop time management skills while on a co-op assignment, and these skills should help you complete your school assignments more efficiently and effectively when you return to the classroom.

Companies that employ engineers often have cooperative education programs because co-op provides a number of benefits to employers as well as students. While the money companies pay co-op students may be double or even triple than what those students would earn from a typical summer job, it is still much less than companies would pay full-time engineers to perform similar work. Many employers also view cooperative education as a recruiting tool—what better way to identify really good employees than to bring aboard promising students and see how they perform on co-op assignments!

Internship

Contributed by Mr. Troy Nunamaker, Director of Graduate and Internship Programs, Michelin Career Center, Clemson University

Internships offer the unique opportunity to gain career-related experience in a variety of settings. Now, more than ever, employers look to hire college graduates with internship experience in their field.

Employers indicate that good grades and participation in student activities are not always enough to help students land a good, full-time job. In today's competitive job market, the students with career-related work experience are the students getting the best interviews and job offers. As an added bonus, many companies report that over 70% of full-time hires come directly from their internship program.

Searching for an Internship

Although a number of students will engage in an internship experience during their first and sophomore years, most students pursue an internship during their junior and senior years. Some students might participate in more than one internship during their college career. Allow plenty of time for the search process to take place, and be sure to keep good records of all your applications and correspondences.

- **Figure out what you are looking for.** You should not start looking for an internship before you have answered the following questions:
  - What are my interests, abilities, and values?
  - What type of organization or work environment am I looking for?
  - Are there any geographical constraints, or am I willing to travel anywhere?
- **Start researching internship opportunities.** Start looking one to two semesters before your desired start date. Many students find that the search process can take anywhere from 3 to 4 weeks up to 5 to 6 months before securing an internship. You should utilize as many resources as possible in order to have the broadest range of options.
  - Visit your campus's career center office to do the following: meet with a career counselor; attend a workshop on internships; find out what positions and resources are available; and look for internship postings through the career center's recruiting system and website resources.
  - Attend a career fair on your campus or in your area. Career fairs typically are not just for full-time jobs, but are open to internship applicants as well. In addition, if there are specific companies where you would like to work, contact them directly and find out if they offer internships.
  - Network. Network. Network. Only about a quarter of internship opportunities are actually posted. Talk to friends, family, and professors and let them know that you are interested in an internship. Networking sites like LinkedIn and Facebook are also beginning to see more use by employers and students. However, be conscious what images and text are associated with your profile.
- **Narrow down the results and apply for internships.** Look for resources on your campus to help with developing a résumé and cover letter. *Each résumé and cover letter should then be tailored for specific applications.* As part of the application

The biggest project I worked on was the Athena model. The Athena is one of NASA's launch platforms.

Before I came here to the contract, several other interns had taken and made SolidWorks parts measured from the actual Athena. My task was to take their parts, make them dimensionally correct and put it together in a large assembly.

I took each individual part (about 300 of them!) and made them dimensionally correct, then put them together into an assembly. After I finished the assembly, I animated the launcher and made it move and articulate.

E. Roper, ME

process, do not be surprised if a company requests additional documents, such as references, transcripts, writing samples, and formal application packets.
- **Wait for responses.** It may take up to a month to receive any responses to your applications. One to two weeks after you have submitted your application, call the organization to make sure they received all the required documents from you.
- **Interview for positions.** Once you have your interviews scheduled, stop by the career center to see what resources they have available to help you prepare for the interview. Do not forget to send a thank you note within 24 hours of the interview, restating your interest in the position.

Why Choose An Internship?
- Bridge classroom applications to the professional world
- Build a better résumé
- Possibly receive higher full-time salary offers upon graduation
- Gain experience and exposure to an occupation or industry
- Network and increase marketability
- Potentially fulfill academic requirements and earn money

Accepting an Internship

Once you have secured an internship, look to see if academic internship coursework is available on your campus so that the experience shows up on your transcripts. If you were rejected from any organizations, take it as a learning experience and try to determine what might have made your application stronger.

WISE WORDS: DESCRIBE A PROJECT YOU WORKED ON DURING YOUR CO-OP OR INTERNSHIP

The large project I worked on was an upgrade of an insulin production facility. A small project I worked on for two weeks was the design of a pressure relief valve for a heat exchanger in the plant.

D. Jones, BE

I have been working on a series of projects, all designed to make the production of electric power meters more efficient. I am rewriting all the machine vision programs to make the process more efficient and to provide a more sophisticated graphical user interface for the operators. These projects have challenged me by requiring that I master a new "machine vision" programming language, as well as think in terms of efficiency rather than simply getting the job done.

R. Izard, ME

A transmission fluid additive was not working correctly and producing harmful emissions, so I conducted series of reactions adding different amounts of materials in a bioreactor. I determined the best fluid composition by assessing the activation energy and how clean it burned.

C. Darling, BE

POINTS TO PONDER

Am I eligible for an internship?
Most companies look to hire rising juniors and seniors, but a rising sophomore or even a first-year student with relevant experience and good grades can be a strong candidate.

Will I be paid for my internship?
The pay rate will depend on your experience, position, and the individual company. However, most engineering interns receive competitive compensation; averages are $14–$20 per hour.

When should I complete an internship?
Contrary to some popular myths, an internship can be completed not only during summers, but also during fall and spring semesters. Be sure to check with your campus on how to maintain your student enrollment status while interning.

Will I be provided housing for an internship?
Do not let the location of a company deter you. Some employers will provide housing, while others will help connect you with resources and fellow interns to find an apartment in the area.

At Boeing, I worked with Liaison Engineering. Liaison engineers provide engineering solutions to discrepancies on the aircraft that have deviated from original engineering plans. As one example, I worked closely with other engineers to determine how grain properties in titanium provide a sound margin or safety in the seat tracks.

J. Compton, ME

The site I worked at designs and manufactures radar systems (among others). During my internship I wrote C code that tests the computer systems in a certain radar model. The code will eventually be run by an operator on the production floor before the new radars are sent out to customers.

D. Rollend, EE

One of my last projects I worked on during my first term was building a new encoder generator box used to test the generator encoder on a wind turbine. The goal was to make a sturdier box that was organized inside so that if something had broken, somebody who has no electrical skills could fix it. I enjoyed this project because it allowed me to use my skills I have learned both from school and my last internship.

C. Balon, EE

Study Abroad

Contributed by Mrs. J. P. Creel, (previous) International Programs
Coordinator, College of Engineering and Science, Clemson University

In today's global economy, it is important for engineering students to recognize the importance of studying abroad. A few reasons to study abroad include the following:

- Taking undergraduate courses abroad is an exciting way to set your résumé apart from those of your peers. Prospective employers will generally inquire about your international experiences during the interview process, giving you the chance to make a lasting impression that could be beneficial.
- Studying abroad will give you a deeper, more meaningful understanding of a different culture. These types of learning experiences are not created in traditional classrooms in the United States and cannot be duplicated by traveling abroad on vacation.
- Students who study abroad generally experience milestones in their personal development as a result of stepping outside of their comfort zone.
- There is no better time to study abroad than now! Students often think that they will have the opportunity to spend significant amounts of time traveling the world after graduation. In reality, entering the workforce typically becomes top priority.

I think that first-year students in engineering should do a co-op or internship. It was extremely valuable to my education. Now that I am back in the classroom, I know what to focus on and why what I am learning is important.

Before the experience I did not know what I wanted to do with my major, and I didn't fully understand word problems that were presented in a manner that applied to manufacturing or real life. Being in industry and working in a number of different departments, I figured out that I liked one area more than any other area, and that is where I am focusing my emphasis area studies during my senior year.

K. Glast, IE

I studied abroad twice. The first time I spent a semester in the Netherlands, experiencing a full immersion in the Dutch culture and exploring my own heritage, and the second time I spent a summer in Austria taking one of my core chemical engineering classes. Both countries are beautiful and unique places and they will always hold a special place in my heart.

Through studying abroad, I was able to expand my own comfort zone by encountering novel situations and become a more confident individual. Although the experiences were amazing and the memories are truly priceless, the biggest thing I gained from studying abroad was that I was able to abandon many perceptions about other cultures and embrace new perspectives.

R. Lanoie, studying abroad in the Netherlands and Austria

I think study abroad is a wonderful learning experience everyone should have. In going abroad you get to meet people you would never meet otherwise and experience things you never thought you would see.

I learned about independently navigating a new country, as well as adjusting to foreign ways of doing things. It is interesting to take classes that you would have taken at home and experience them a completely different way.

Studying abroad teaches you how to easily adjust to new situations and allows you to learn about cultures you had no idea of before.

B. Brubaker, studying abroad in Scotland

- Large engineering companies tend to operate on a global scale. For instance, a company's headquarters may be in the United States, but that company may also control factories in Sweden and have parts shipped to them from Taiwan. Having an international experience under your belt will give you a competitive edge in your career because you will have global knowledge that your coworkers may not possess.

    While programs differ between universities, many offer a variety of choices:

- **Exchange Programs:** Several institutions are part of the Global Engineering Education Exchange (GE³). This consortium connects students from top engineering schools in the United States with foreign institutions in any of 18 countries. In an ideal situation, the number of international students on exchange in the United States would be equal to the number of U.S. students studying abroad during the same time frame.

    Generally, students participating in these types of exchange programs will continue to pay tuition at their home institution; however, you should check your university's website for more information. Also, by entering "Global E³" into a web browser, you will be able to access the website to determine if your institution currently participates, which foreign institutions offer courses taught in English, and which schools offer courses applicable to your particular major. If your institution is not a member of GE³, consult with someone in the international, or study abroad, office at your institution to find out about other exchange opportunities.

- **Faculty-Led Programs:** It is not unusual for faculty members to connect with institutions or other professionals abroad to establish discipline-specific study-abroad programs. These programs typically allow students to enroll in summer classes at their home institution, then travel abroad to complete the coursework. Faculty-led programs offer organized travel, lodging, administration, and excursions, making the overall experience hassle-free. Consult with professors in your department to find out if they are aware of any programs that are already in place at your institution.

- **Third-Party Providers:** Many universities screen and recommend providers of programs for students. If there's a place you want to go for study and your university does not have an established program at that particular location, you will probably find a study program of interest to you by discussing these options with the study-abroad office at your university.

- **Direct Enrollment:** If you are interested in a particular overseas institution with which your home university does not have an established program, there is always the option of direct enrollment. This process is basically the same as applying to a university in the United States. The school will likely require an admissions application, a purpose statement, transcripts, and a letter of reference. Be cautious if you choose this route for study abroad, as it can be difficult to get credits transferred back to your home institution. It is a good idea to get your international courses preapproved prior to your departure. On a positive note, it can be cheaper to directly enroll in a foreign institution than it is to attend at your home university. This option is best suited for students who want to go abroad for a semester or full year.

Typically, students go abroad during their junior year, though recently there has been an increase in the number of second-year students participating. The timing of your experience should be agreed upon by you and your academic advisor. Each program offers different international incentives, and some are geared for better opportunities later in your academic career, whereas others may be better toward the beginning.

India is quite an eye-opening country. All of your senses work on overload just so you can take everything in at once. My time studying in India was simply phenomenal. I had an excellent opportunity placed before me to travel to India to earn credit for electrical engineering courses, and I would have been a fool to pass up that chance. I regret nothing, I would do it again, and I would urge everyone with the slightest inkling of studying abroad to put their worries aside and have the experience of a lifetime.

J. Sadie, studying abroad in India

If you would like to go abroad for a semester or year but do not feel entirely comfortable with the idea, why not get your feet wet by enrolling in a summer program first? This will give you a better idea of what it is that you are looking for without overwhelming you. Then, you can plan for a semester or year abroad later in your academic career. In fact, more students are electing to spend a semester or year of study abroad, and increasingly more opt for an academic semester plus a semester internship combination.

There are three basic principles to follow when deciding on a location:

- **Personal Preference:** Some people are more interested in Asia than Europe, or maybe Australia instead of Latin America. There are excellent opportunities worldwide, regardless of the location.
- **Program Opportunities:** Certain countries may be stronger or have more options in certain fields. For example, Germany is well known for innovations in mechanical engineering, while the Japanese tend to be more widely recognized for their efforts in computer engineering. Listen to your professors and weigh your options.
- **Language:** You may not feel like you are ready to study in a foreign country, speaking and reading in a foreign language. There are numerous institutions that offer courses taught in English. However, if you have taken at least two years of the same foreign language, you should be knowledgeable enough to succeed in courses taught in that language. Do not let your fears restrict you!

WISE WORDS: HOW DID YOUR ENGINEERING STUDIES PREPARE YOU FOR YOUR CAREER . . . ?

In sales: *Though sales work is rather different from the engineering that I studied in school, the education has proven to be remarkably valuable. Engineering is the type of discipline that teaches discipline. The problem solving necessary to complete tasks in both an individual and a team environment has been extremely useful in my sales career. I remember one first-year project, designing the pulley system for a Sky Cam over a sports stadium, required teaming with two other students and often times the willingness to sit back and let the discussion unfold. No one had the answers immediately and the collective mind of the three proved indispensable in solving the problem. Though I contributed my part to the project, I might still be working on it had it not been a team assignment! There was much creativity necessary in engineering study. This was especially true when it came to the Senior Project. This type of creativity I find is similar to that in an entrepreneurial role such as business-to-business sales. Engineering is also a discipline that, by its nature, pulls from many other disciplines (math, physics, art, biology, sociology, economics, etc.) much in the same way that sales people have to be results-oriented and pull from many other resources (marketing, finance, operations, IT).*

Tom Lee, Vice President of Sales, Transworld Systems

As an engineering society administrator: *While sitting in a classroom and studying thermodynamics and applied energy systems, I was thinking of a long career in engineering. Getting involved in association management was never even a thought. But obtaining that engineering degree taught me two vital things that have helped me in both engineering and nonengineering positions. Engineering taught me how to be a linear thinker and that has helped me in so many ways from writing to problem solving to project management. The other important aspect was to really look and understand problems before trying to solve them. Seeing this big picture view fostered my ability to think and to be creative in problem solving, again something that helps you succeed in any occupation*

Burt Dicht, Director, IEEE University Programs

As a chef: *Easy answer: motion and time study! In the kitchen, and I am a chef, motion and time is the name of the game. My background in engineering at NCSU was in the Furniture Manufacturing and Management department of the Industrial Engineering School. The kitchen is the heart of the "food manufacturing facilities" we commonly call restaurants. Designing maximum potential work spaces is the most important aspect of laying out a kitchen with respect to job performance, execution, presentation, and of course, labor and product cost controls. As a chef, I always tried to lay out kitchens with a minimum of wasted steps and motions, thus saving precious time, maximizing workplace enjoyment, and reducing fatigue. It is a tough business, so anything we can do to increase workplace pleasure is huge—which also puts us into design of HVAC and exhaust as well. Looking back over it, I believe designing a furniture plant may be easier.*

Chef Jim Noble

In politics: *Most Senators and staff are lawyers. Engineers are taught to approach decision making and problem solving very differently than lawyers. Most of the time, my solutions to problems sounded more pragmatic and less political or ideological. Of course, it also helped to not be intimidated by problems that required math or science to resolve.*

Senator Ted Kaufman

As a member of the clergy: *As a second-career minister, I benefit enormously from my engineering education that emphasized structured and disciplined thinking and taught me to focus on solving particular problems despite many intrusions and distractions. During my theological education, I found the emphasis on disciplined thinking surprisingly helpful to me when I was studying church doctrine—now called systematic theology, and when I was reading the teachings of some of the greatest minds in history: Augustine, Aquinas, and Luther. In my work as a parish pastor, I constantly need to solve problems by identifying the essential elements of the situation, creating a solution that is a series of steps, all the while identifying and ignoring distractions. I am an engineering faculty member as well. I am an associate professor of engineering and computer science.*

Pastor Charles Stevenson, St. John Lutheran Church, Peabody, MA

As a high school teacher: *As a high school teacher, I constantly faced challenges related to what I taught, managing my classroom, and engaging my students. Engineering taught me a general framework for how to develop solutions for the problems I encountered in my classroom given various constraints involved with working at my school. My knowledge of engineering also helped me to provide useful examples of how the mathematics and science that my students were learning in the classroom applied to real-world situations.*

Noah Salzman, Assistant Professor of Electrical and Computer Engineering, Boise State University and former high school teacher

REVIEW QUESTIONS

1. Which of these is closest to the definition of engineering put forward by the National Academy of Engineering (NAE)?
   (A) The practice of using mathematical equations
   (B) The application of science
   (C) The discovery of technological inventions
   (D) The supervision of project design

2. Which sentence best describes someone who is motivated by mastery-oriented goals?
   (A) I want to understand things.
   (B) I want to get excellent grades.
   (C) I want my parents to be proud of me.
   (D) I want to earn a lot of money.

3. Which sentence best describes someone who is motivated by performance-oriented goals?
   (A) I want to understand things.
   (B) I want to get excellent grades.
   (C) I want to help people succeed.
   (D) I want to solve problems.

4. How do you know you have chosen the "right" career?
   (A) Your family and friends feel it is a good fit.
   (B) You are satisfied and challenged.
   (C) Your boss is happy with your work.
   (D) You are appropriately compensated.

5. What is the goal of the Nitrogen Cycle Grand Challenge?
   (A) Discover new uses for fixed nitrogen compounds.
   (B) Reduce the amount of nitrogen being used in fertilization.
   (C) Discover ways to augment natural denitrification.
   (D) Reduce the rate of nitrogen uptake in plants.

6. Synthetic (human caused) conversion of nitrogen into more reactive forms now accounts for what percent of all nitrogen fixation?
   (A) 1%
   (B) 10%
   (C) 50%
   (D) 90%
   (E) 99%

7. What type of engineer is likely to work on implantable devices, such as artificial kidneys?
   (A) A biosystems engineer
   (B) An electrical engineer
   (C) A mechanical engineer
   (D) A biomedical engineer

8. Which engineering field is primarily concerned with the planning, construction, and operation of facilities, such as offshore oil platforms or solar panel factories?
   (A) Civil engineering
   (B) Mechanical engineering
   (C) Bioengineering
   (D) Chemical engineering

9. What type of engineers are most likely to design products that provide communication and/ or control, such as a household thermostat or robotic microcontrollers?
   (A) Computer engineers
   (B) Industrial engineers
   (C) Electrical engineers
   (D) Materials engineers

10. What engineer would be most likely to work on identifying and reducing the number of defective car engines built on an assembly line?
    (A) A materials engineer
    (B) An industrial engineer
    (C) A mechanical engineer
    (D) A civil engineer

11. What type of engineer is most likely to work on designing the fiber orientation of a composite for a new carbon fiber airplane?
    (A) An industrial engineer
    (B) A mechanical engineer
    (C) A materials engineer
    (D) An electrical engineer

Writing Assignments

For each question, write a response according to the directions given by your instructor. Each response should contain correct grammar, spelling, and punctuation. Be sure to answer the question completely, but choose your words carefully so as to not exceed the word limit if one is given. There is no right or wrong answer; your score will be based upon the strength of the argument you make to defend your position.

12. On a separate sheet, write a one-page résumé for one of the engineers from the following list. Include information such as education, job experience, primary accomplishments (inventions, publications, etc.), and references. If you want, you can include a photo or likeness no larger than 2 in. × 3 in. You can add some "made up" material, such as current address and references, but do not overdo this.

- Ammann, Othmar
- Ampere, Andre Marie
- Arafat, Yasser
- Archimedes
- Avogadro, Armedeo
- Bernoulli, Daniel
- Bessemer, Henry
- Bezos, Jeffrey P.
- Birdseye, William
- Blanchard, Helena Augusta
- Bloomberg, Michael
- Bohr, Niels
- Boyle, Robert
- Brezhnev, Leonid
- Brill, Yvonne
- Brown, Robert
- Brunel, Isambard
- Calder, Alexander
- Capra, Frank
- Carnot, Nicolas
- Carrier, Willis Haviland
- Carter, Jimmy

- Cauchy, Augustin Louis
- Cavendish, Henry
- Celsius, Anders
- Clausius, Rudolf
- Clarke, Edith
- Coston, Martha
- Coulomb, Charles
- Cray, Seymour
- Crosby, Philip
- Curie, Marie
- Dalton, John
- Darcy, Henri
- de Coriolis, Gaspard
- de Mestral, George
- Deming, W. Edwards
- Dennis, Olive
- Diesel, Rudolf
- Dunbar, Bonnie
- Eaves, Elsie
- Eiffel, Gustave
- Euler, Leonhard
- Fahrenheit, Gabriel
- Faraday, Michael

- Fitzroy, Nancy
- Fleming, Sandford
- Flugge-Lotz, Irmgard
- Ford, Henry
- Fourier, Joseph
- Fung, Yuan-Cheng
- Gantt, Henry
- Gauss, Carl Friedrich
- Gibbs, Josiah Willard
- Gilbert, William
- Gilbreth, Lillian
- Gleason, Kate
- Goizueta, Robert
- Grove, Andrew
- Hancock, Herbie
- Henry, Beulah
- Hertz, Heinrich Rudolf
- Hicks, Beatrice
- Hitchcock, Alfred
- Hooke, Robert
- Hoover, Herbert
- Hopper, Grace

- Iacocca, Lee
- Joule, James Prescott
- Juran, Joseph Moses
- Kelvin, Lord
- Kraft, Christopher, Jr.
- Kwolek, Stephanie
- Lamarr, Hedy
- Landry, Tom
- Laplace, Pierre-Simon
- Leibniz, Gottfried
- LeMessurier, William
- MacCready, Paul
- Mach, Ernst
- Mayer, Marissa
- McDonald, Capers
- Midgley, Thomas Jr.
- Millikan, Robert
- Navier, Claude-Louis
- Newton, Isaac

- Nielsen, Arthur
- Ochoa, Ellen
- Ohm, Georg
- Pascal, Blaise
- Poiseuille, Jean Loius
- Porsche, Ferdinand
- Prandtl, Ludwig
- Rankine, William
- Rayleigh, Lord
- Reece, Marilyn Jorgensen
- Resnik, Judith
- Reynolds, Osborne
- Roebling, Emily
- Rømer, Ole
- Sikorsky, Igor Ivanovich
- Stanton, Nora

- Stinson, Katherine
- Stokes, George
- Sununu, John
- Taguchi, Gen'ichi
- Taylor, Fredrick
- Teller, Edward
- van der Waals, Johannes
- Venturi, Giovanni
- Volta, Count Alessandro
- von Braun, Wernher
- von Kármán, Theodore
- Walton, Mary
- Watt, James
- Welch, Jack
- Wyeth, Nathaniel
- Yeltsin, Boris

13. Please address the following questions in approximately one page. You may write, type, draw, sketch your answers . . . whatever form you would like to use.

  (a) Where are you from?
  (b) What type of engineering are you interested in? Why?
  (c) Are there any engineers in the circle of your friends and family?
  (d) How would you classify yourself as a student (new first-year, transfer, or upperclassman)?
  (e) What are your activities and hobbies?
  (f) What are you most proud of?
  (g) What are you fantastic at?
  (h) What are you passionate about?
  (i) What has been the most difficult aspect of college so far?
  (j) What do you expect to be your biggest challenge this term?
  (k) Do you have any concerns about this class?
  (l) Anything else you would like to add?

14. In 2008, experts convened by the National Academy of Engineering (NAE) met and proposed a set of Grand Challenges for Engineering, a list of the 14 key goals for engineers to work toward during the twenty-first century. For any aspiring engineer, reading this list should feel like reading a description of the challenges you will face throughout your career. To read a description of each Challenge, as well as a description of some connected areas within each, visit the NAE's Grand Challenges for Engineering website at http://www.engineeringchallenges.org.

   After reading through the list, write a job description of your dream job. The job description should include the standard components of a job description: title, responsibilities (overall and specific), a description of the work hierarchy (whom you report to, whom you work with), as well as any necessary qualifications. Do not include any salary requirements, as this is a completely fictional position.

   After writing the description of your dream job, identify which of the 14 Grand Challenges for Engineering this position is instrumental in working toward solving. Cite specific examples of the types of projects your fictional job would require you to do, and discuss the impact of those projects on the Grand Challenges you have identified.

15. Choose and explain your choice of major and the type of job you envision yourself doing in 15 years. Consider the following:

  (a) What skills or talents do you possess that will help you succeed in your field of interest?
  (b) How passionate are you about pursuing a career in engineering? If you do not plan on being an engineer, what changed your mind?

(c) How confident are you in your choice of major?

(d) How long will it take you to complete your degree?

(e) Will you obtain a minor?

(f) Will you pursue study abroad, co-op, or internship?

(g) Do you plan to pursue an advanced degree or become a professional engineer (PE)?

(h) What type of work (industry, research, academic, medical, etc.) will you pursue?

16. In 2008, the University of Memphis made national headlines when the Memphis Tigers played in the NCAA basketball national championship game. When interviewed by a local newspaper about how he helped his team "own" the tournament, Coach Calipari revealed that before each game, he had his star player write an essay about how the game would play out. This particular player was prone to nervousness, so to help him focus, the coach told him to mentally envision the type of plays and how he himself would react to them.

In this assignment, write a short essay on how you will prepare for and take the final exam. Consider the following:

(a) What will you do to study? What materials will you gather, and how will you use them? Where will you study? Will it be quiet? Will you play music?

(b) What kinds of things could go wrong on the day of the exam, and how would you avoid them? (List at least three.)

(c) What will the exam look like, and how will you work through it?

Thank you to Dr. Lisa Benson for contributing this assignment.

17. Read the essay "Engineering Is an . . . Itch!" in the Engineering Essentials introduction. Reflect on what it means to have performance-focused versus mastery-focused learning goals.

(a) Describe in your own words what it means to be a performance-based learner compared with a mastery-based learner.

(b) What learning goals do you have? Are these goals performance based or mastery based?

(c) Is it important to you to become more mastery focused?

(d) Do you have different kinds of learning goals than you had in the past, and do you think you will have different learning goals in the future?

18. An article in *Science News* addressed the topics of nature and technology. In our electronic world, we are in constant contact with others, through our cell phones, iPods, Facebook, e-mail, etc. Researchers at the University of Washington have determined that this effect may create long-term problems in our stress levels and in our creativity.

Scenes of Nature Trump Technology in Reducing Low-Level Stress

Technology can send a man to the moon, help unlock the secrets of DNA, and let people around the world easily communicate through the Internet. But can it substitute for nature?

"Technology is good and it can help our lives, but let's not be fooled into thinking we can live without nature," said Peter Kahn, a University of Washington associate professor of psychology who led the research team.

We are losing direct experiences with nature. Instead, more and more we're experiencing nature represented technologically through television and other media. Children grow up watching Discovery Channel and Animal Planet. That's probably better than nothing. But as a species we need interaction with actual nature for our physical and psychological well-being.

Part of this loss comes from what the researchers call environmental generational amnesia. This is the idea that across generations the amount of environmental degradation increases, but each generation views conditions it grew up with as largely non-degraded and normal. Children growing up today in the cities with the worst air pollution often, for example, don't believe that their communities are particularly polluted.

"This problem of environmental generational amnesia is particularly important for children coming of age with current technologies," said Rachel Severson, a co-author of the study and a University of Washington psychology doctoral student. "Children may not realize they are not getting the benefits of actual nature when interacting with what we're calling technological nature."

Source: Joel Schwartz. *"Scenes of Nature Trump Technology in Reducing Low-Level Stress." Science Daily, June 16, 2008.*

Go someplace quiet and spend at least 10 minutes clearing your head. Only after these 10 minutes, get out a piece of paper and sketch something you see; are thinking of; or want to create or invent or imagine. On the same sheet of paper, write a poem about your sketch; something you are thinking of; or about the quiet, wonder of the universe, lack of technology, etc.

Rules:

- Must be done all by hand.
- Must be original work (no copied poems or artwork).
- Ability does not count—draw like you are 5 years old!
- Draw in any medium you want: use pencil, pen, colored pencils, markers, crayons, watercolors.
- Poetic form does not matter: use rhyme, no rhyme, haiku, whatever you want it to be.
- It does not need to be elaborate; simple is fine.

Ethics

Every day, we make many ethical decisions, although most are so minor that we do not even view them as such:

- When you drive your car, do you knowingly violate the posted speed limit?
- When you unload the supermarket cart at your car, do you leave it in the middle of the parking lot, or do you spend the extra time to return it to the cart corral?
- You know that another student has plagiarized an assignment; do you rat him or her out?
- A person with a mental disability tries to converse with you while waiting in a public queue. Do you treat him or her with respect or pretend he or she does not exist?
- In the grocery, a teenager's mother tells her to put back the package of ice cream she brought to the shopping cart. The teenager walks around the corner and places the ice cream on the shelf with the soft drinks and returns to the shopping cart. Do you ignore this or approach the teenager and politely explain that leaving a package of ice cream in that location will cause it to melt, thus increasing the cost of groceries for everyone else, or do you replace it in the freezer yourself?
- When going through a public door, do you make a habit of looking back to see if releasing the door will cause it to slam in someone's face?
- You notice a highway patrol officer lying in wait for speeders. Do you flash your lights at other cars to warn them?
- A cashier gives you too much change for a purchase. Do you correct the cashier?
- You are on the lake in your boat and notice a person on a JetSki chasing a great blue heron across the lake. The skier stops at a nearby pier. Do you pilot your craft over to the dock and reprimand him for harassing the wildlife?

On a grand scale, none of these decisions is particularly important, although some might lead to undesirable consequences. However, as an aspiring engineer, you may face many decisions in your career that could affect the lives and well-being of thousands of people. Just like almost everything else, practice makes perfect, or at least better. The more you practice analyzing day-to-day decisions from an ethical standpoint, the easier it will be for you to make good decisions when the results of a poor choice may be catastrophic.

In very general terms, there are two reasons people try to make ethical decisions:

- They wish to make the world a better place for everyone—in a single word, altruism.
- They wish to avoid unpleasant consequences, such as fines, incarceration, or loss of job.

In an ideal society, the second reason would not exist. However, history is replete with examples of people, and even nations, who do not base their decisions solely on whether or not they are acting ethically. Because of the common occurrence of unethical behavior and the negative impact it has on others, almost all societies have developed rules, codes, and laws to specify what is and is not acceptable behavior, and the punishments that will be meted out when violations occur.

The major religions all have fairly brief codes summarizing how one should conduct their life. Some examples follow; other examples exist as well:

> Good people do not need laws to tell them to act responsibly, while bad people will find a way around the laws.
>
> *Plato*

- Judaism, Christianity, and derivatives thereof have the Decalogue, or Ten Commandments.
- Islam has the Five Pillars in addition to a slightly modified and reorganized form of the Decalogue.
- Buddhism has the Noble Eightfold Path.
- Bahá'í has 12 social principles.
- In Hinduism, Grihastha Dharma has four goals.

Secular codes of conduct go back more than four millennia to the Code of Ur-Nammu. Although by today's standards, some of the punishments in the earliest codes seem harsh or even barbaric, it was one of the earliest known attempts to codify crimes and corresponding punishments.

Admittedly, although not specifically religious in nature, these codes are usually firmly rooted in the prevailing religious thought of the time and location. Through the centuries, such codes and laws have been expanded, modified, and refined so that most forms of serious antisocial behavior are addressed and consequences for violations specified. These codes exist from a local to a global level. Several examples follow:

- Most countries purport to abide by the Geneva Conventions, which govern certain types of conduct on an international scale.
- Most countries have national laws concerning murder, rape, theft, etc.
- In the United States of America, it is illegal to purchase alcohol unless you are 21 years of age. In England, the legal age is 18.
- In North and South Dakota, you can obtain a driver's license at age $14\frac{1}{2}$. In most other states, the legal age is 16.
- It is illegal to say "Oh boy!" in Jonesboro, Georgia.
- Nearly all states in the U.S.A. (46 of them) have ordinances prohibiting text messaging while driving.

2.1 Ethical Decision Making

LEARN TO: Apply the four-step procedure for making ethical decisions
Consider different perspectives in ethical decisions
Determine different stakeholders involved in ethical decisions

Some ethical decisions are clear-cut. For example, essentially everyone (excluding psychopaths) would agree that it is unethical to kill someone because you do not like his or her hat. Unfortunately, many real-world decisions that we must make are far from black and white issues, instead having many subtle nuances that must be considered to arrive at what one believes is the best decision.

There is no proven algorithm or set of rules that one can follow to guarantee that the most ethical decision possible is being made in any particular situation. However, many people have developed procedures that can guide us in considering questions with ethical ramifications. A four-step procedure is discussed here, although there are various other approaches:

Step 1: Determine *What* the issues are and *Who* might be affected by the various alternative courses of action that might be implemented.

We will refer to the *Who* as **stakeholders**. Note that at this point, we are not trying to determine how the stakeholders will be affected by any particular plan of action:

- The issues (What) can refer to a wide variety of things, including, for example, personal freedom, national security, quality of life, economic issues, fairness, and equality.

- The term "stakeholders" (Who) does not necessarily refer to people but might be an individual, a group of people, an institution, or a natural system, among other things.

EXAMPLE 2-1

Consider the question of whether to allow further drilling for oil in the Alaska National Wildlife Refuge (ANWR). List several issues and stakeholders.

Issues:

- *Oil independence*
- *The price of gasoline*
- *Possible impacts on the ecosystem*

Stakeholders:

- *Oil companies*
- *The general population of the United States*
- *Other countries from whom we purchase oil*
- *The flora and fauna in ANWR*
- *The native people in Alaska*

Step 2: Consider the effects of alternative courses of action from different perspectives.

Here, we look at three perspectives: consequences, intent, and character:

Perspective 1: Consequences

When considering this perspective, ask how the various stakeholders will be affected by each alternative plan being contemplated. In addition, attempt to assign a relative level of importance (weight) to each effect on each stakeholder. For instance, an action that might affect millions of people adversely is almost always more important than an action that would cause an equivalent level of harm to a dozen people.

Fingerprint technology has advanced in recent years with the implementation of computer recognition for identification. Originally in the United States, the Henry Classification System was used to manually match fingerprints based on three main patterns: arches, loops, and whorls (shown below from left to right).

Sources: Andrey Kuzmin/Shutterstock, BeRad/Shutterstock, and Janaka Dharmasena/Shutterstock.

Today, the Automated Fingerprint Identification System (AFIS) uses algorithmic matching to compare images. Future work of AFIS systems is in the adoption and creation of secure multitouch devices like mobile computers and tablets, which can identify different security levels for the operator of the device. For example, a multitouch computer owner might be able to issue permissions to an administrator that might not be available to a 5 year old, all without providing a single password!

EXAMPLE   2-2

Should all U.S. children be fingerprinted when entering kindergarten and again each third year of grade school (3, 6, 9, 12)? Identify the stakeholders and consequences.

Stakeholders:

- *All U.S. children*
- *All U.S. citizens*
- *Law enforcement*
- *The judicial system*
- *The U.S. Constitution*

Consequences:

- *Provides a record to help identify or trace missing children (not common, but possibly very important in some cases)*
- *Affords an opportunity for malicious use of the fingerprint records for false accusation of crime or for identity theft (probability unknown, but potentially devastating to affected individuals)*
- *Could help identify perpetrators of crimes, thus improving the safety of law-abiding citizens (importance varies with type of crime)*
- *Raises serious questions concerning personal freedoms, possibly unconstitutional (importance, as well as constitutionality, largely dependent on the philosophy of the person doing the analysis)*

This list could easily be continued.

Perspective 2: Intent

The intentions of the person doing the acting or deciding are considered in this perspective, sometimes called the rights perspective. Since actions based on good intentions can sometimes yield bad results, and vice versa, the intent perspective avoids this possible pitfall by not considering the outcome at all, only the intentions.

It may be helpful when considering this perspective to recall Immanuel Kant's Categorical Imperative: "Act only according to that maxim whereby you can at the same time will that it should become a universal law." To pull this out of the eighteenth century, ask yourself the following questions:

(a) Is the action I am taking something that I believe everyone should do?
(b) Do I believe that this sort of behavior should be codified in law?
(c) Would I like to be on the receiving end (the victim) of this action?

EXAMPLE   2-3

Should you download music illegally over the Internet?

Rephrasing this question using the suggestions above yields:

(a) Should everyone illegally download the music they want if it is there for the taking?
(b) Should the laws be changed so that anyone who obtains a song by any means can post it on the web for everyone to get for free?
(c) If you were a struggling musician trying to pay the bills, would you like your revenue stream to dry up because everyone who wanted your music got it for free?

Perspective 3: Character

Character is the inherent complex of attributes that determines a person's moral and ethical actions and reactions. This perspective considers the character of a person who takes the action under consideration. There are different ways of thinking about this. One is to simply ask: Would a person of good character do this? Another is to ask: If I do this, does it enhance or degrade my character? Yet another way is to ask yourself if a person you revere as a person of unimpeachable character (whoever that might be) would take this action.

EXAMPLE   2-4

Your friends are deriding another student behind her back because she comes from a poor family and does not have good clothes.
    Do you:

(a) Join in the criticism?
(b) Ignore it, pretend it is not happening, or simply walk away?
(c) Tell your friends that they are behaving badly and insist that they desist?

- *Which of these actions would a person of good character take?*
- *Which of these actions would enhance your character and which would damage it?*
- *What would the founder of your religion do? (Moses or Jesus or Buddha or Mohammed or Bahá'u'lláh or Vishnu or whoever.) If you are not religious, what would the person who, in your opinion, has the highest moral character do?*

Step 3: Correlate perspectives.

Now look back at the results of considering the issues from the three perspectives. In many cases, all three perspectives will lead to the same or a similar conclusion. When this occurs, you have a high level of confidence that the indicated action is the best choice from an ethical standpoint.

    If the three perspectives do not agree, you may wish to reconsider the question. It may be helpful to discuss the issue with people whom you have not previously consulted in this matter. Did you omit any factors? For complicated issues, it is difficult to make sure you have included all possible stakeholders and consequences. Did you properly assign weights to the various aspects? Upon reconsideration, all three perspectives may converge.

    If you cannot obtain convergence of all three perspectives, no matter how hard you try to make sure you left nothing out, then go with two out of three.

Step 4: Act.

This is often the hardest step of all to take, since ethical action often requires courage. The whistle-blower who risks losing his or her job, Harriet Tubman repeatedly risking her life to lead slaves to freedom via the Underground Railroad, the elected official standing up for what she knows to be right even though it will probably cost her the next election, or even something as mundane as risking the ridicule of your friends because you refuse to go along with whatever questionable activities they are engaging in for "fun." Ask yourself the question: "Do *I* have the courage to do what I know is right?"

EXAMPLE	2-5

Your company has been granted a contract to develop the next generation of electronic cigarette, also known as a "nicotine delivery system," and you have been assigned to the design team. Can you in good conscience contribute your expertise to this project?

Step 1: Identify the issues (What) and the stakeholders (Who).

Issues:

- *Nicotine is poisonous and addictive.*
- *These devices eliminate many of the harmful components of tobacco smoke.*
- *Laws concerning these devices range from completely legal, to classification as a medical device, to banned, depending on country.*
- *There are claims that such devices can help wean tobacco addicts off nicotine.*
- *The World Health Organization does not consider this an effective means to stop smoking.*
- *Whether an individual chooses to use nicotine should be a personal decision, since its use does not generally degrade a person's function in society.*
- *The carrier of the nicotine (80–90% of the total inhaled product) is propylene glycol, which is relatively safe, but can cause skin and eye irritation, as well as other adverse effects in doses much larger than would be obtained from this device.*
- *A profit can be made from nicotine products or antismoking devices.*

Note

In the interest of brevity, this is not an exhaustive analysis but shows the general procedure.

Stakeholders:

- *You (your job and promotions)*
- *Your company and stockholders (profit)*
- *Cigarette manufacturers and their employees and stockholders (lost revenue)*
- *Tobacco farmers (less demand)*
- *The public (less second-hand smoke)*
- *The user (various health effects, possibly positive or negative)*

Step 2: Analyze alternative courses of action from different perspectives.

1. *Consequences*
   - *You may lose your job or promotion if you refuse.*
   - *If you convince management to abandon the project, the company may lose money.*
   - *If you succeed brilliantly, your company may make a lot of money, and you may receive a promotion.*
   - *If the project goes ahead, the possibility of future lawsuits exists.*
   - *Users' health may be damaged.*
   - *Users' dependence on nicotine may either increase or decrease.*

2. *Intent*
   - *Should everyone use electronic cigarettes, or at least condone their use?*
   - *Should use of electronic cigarettes be unrestricted by law?*
   - *Would I like to risk nicotine addiction because of using these devices?*
   - *Would I be able to kick my tobacco habit by using these devices?*

3. *Character*
   - *Would a person of good character develop this device, use it, or condone its use?*
   - *Would work on this project (thus implicitly condoning its use) or use of the device itself enhance or degrade my character?*
   - *Would my personal spiritual leader, or other person I revere, condone development or use of this product?*

Step 3: *Correlate perspectives.*

Here we enter the realm of subjective judgment. The individual author responsible for this example has a definite personal answer, but it is in the nature of ethical decision making that different people will often arrive at different results in good conscience. You would have to weigh the various factors (including any that have been overlooked or knowingly omitted) to arrive at your own conclusion. We refuse to dictate a decision to you.

Step 4: *Act on your decision.*

If your decision was that working on this project poses no threat to your soul (if you happen to believe in such), probably little courage is required to follow through, since your career may blossom, or at least not be curtailed.

On the other hand, if you believe that the project is unethical, you need to have the intestinal fortitude to either attempt to change the minds of management or refuse to work on the project, both of which may put your career at risk.

2.2 Plagiarism

LEARN TO: Understand what plagiarism is and how it can be avoided
Recognize similarities between plagiarism and copyright infringement
Understand that not properly attributing work is plagiarism, intentional or not

Did you know? There are Internet services available that will accept a document and search the web for exact or similar content. Also, there are programs that will scan multiple documents and search for exact or similar content.

Did you know? Prior to the romantic movement of the eighteenth century, European writers were encouraged not to be inventive without good reason and to carefully imitate the work of the great masters of previous centuries.

You probably know what plagiarism is—claiming someone else's work as your own. This is most often used in reference to written words, but may be extended to other media as well. From a legal standpoint, plagiarism per se is not illegal, although it is widely considered unethical. However, if the plagiarism also involves copyright infringement, then this would be a violation of the law. Certainly, in the context of your role as a student, plagiarism is almost universally regarded as academic dishonesty and subject to whatever punitive actions your school deems appropriate.

In some cases, plagiarism is obvious, as when an essay submitted by a student is almost identical to one found on the Internet or is the same as that submitted by another student. It is amazing how frequently students are caught cheating because they copied verbatim from another student's work, complete with strange mistakes and bizarre phrasing that grab the grader's attention like an 18-wheeler loaded with live pigs locking its brakes at 80 miles per hour. (*Thanks to Gilbert Shelton for that image.*)

In other cases, things are far less clear. For example, if you were writing a short story for your English class and used the simile "her lips were like faded tulips, dull and wrinkled," can you (or the professor) really be sure whether that was an original phrase or if you had read it at some time in the past, and your brain dragged it up from your subconscious memory as though it were your own?

We all hear or read things during our lives that hang around in our brains whether we are consciously aware of them or not. We cannot go through life in fear of being accused of plagiarism because our brain might drag up old data masquerading as our own original thought, or even worrying about whether our own original thoughts have ever been concocted by another person completely independently.

Any reasonable person (although admittedly, there is a surfeit of unreasonable people) will take the work as a whole into account. If there is simply a single phrase or a couple of instances of wordings that are similar to another source, this is most likely an innocent coincidence. On the other hand, if a work has many such occurrences, the probability that the infractions are innocent is quite low.

We arrive here at intent. Did you knowingly copy part of someone else's work and submit it as your own without giving proper credit? If you did not, stop worrying about it. If you did, Big Brother, also known as your professor, is watching, possibly with the assistance of high-tech plagiarism detection tools. (*A tip of the hat to George Orwell.*)

2.3 Engineering Creed

LEARN TO: Recognize the importance of considering the ethical aspects of engineering problems
Understand the Engineer's Creed
Understand the Fundamental Canons of the Engineer's Creed

Ethical decisions in engineering have, in general, a narrow focus specific to the problems that arise when designing and producing products or services of a technical nature. Engineers and scientists have, by the very nature of their profession, a body of specialized knowledge that is understood only vaguely, if at all, by most of the population. This knowledge can be used for tremendous good in society, but can also cause untold mischief when used by unscrupulous practitioners. Various engineering organizations have thus developed codes of conduct specific to the profession. Perhaps the most well known is the Code of Ethics for Engineers developed by the National Society of Professional Engineers (NSPE). The entire NSPE Code of Ethics is rather long, so we list only the Engineer's Creed and the Fundamental Canons of the Code here.[1]

Engineer's Creed

As a Professional Engineer, I dedicate my professional knowledge and skill to the advancement and betterment of human welfare. I pledge:

- To give the utmost of performance
- To participate in none but honest enterprise
- To live and work according to the laws of man and the highest standards of professional conduct
- To place service before profit, the honor and standing of the profession before personal advantage, and the public welfare above all other considerations

In humility and with need for Divine Guidance, I make this pledge.

Fundamental Canons

Engineers, in the fulfillment of their professional duties, shall:

- Hold paramount the safety, health, and welfare of the public
- Perform services only in areas of their competence
- Issue public statements only in an objective and truthful manner
- Act for each employer or client as faithful agents or trustees
- Avoid deceptive acts
- Conduct themselves honorably, responsibly, ethically, and lawfully so as to enhance the honor, reputation, and usefulness of the profession

---

[1]"NSPE Code of Ethics." National Society of Professional Engineers (NSPE). Revised July 2007. www.nspe.org. Used with permission.

The complete code can easily be found online at a variety of sites. At the time of this writing the URL for the Code of Ethics on the NSPE site was http://www.nspe.org/Ethics/CodeofEthics/index.html.

2.4 Social Responsibility

Contributed by Jason Huggins, P.E., Executive Councilor for Tau Beta Pi,
the National Engineering Honor Society, 2006–2014

> **LEARN TO:**   Recognize the need for professional conduct within engineering
> Understand the reach of the work done by engineers
> Understand the drive to help solve problems by engineers

NOTE

Social responsibility is the ideology that an individual has an obligation to act to benefit society at large.

As a starting engineering student, you are just beginning your journey to join the engineering profession. Have you thought about what it will mean to be a part of a profession? Being a professional means we hold the public's trust and confidence in our training, skills, and knowledge of engineering. As a profession, we recognize the importance of this trust in the Engineering Canons and the Engineering Creed that define our standards for ethics, integrity, and regard for public welfare. Does adherence to the Engineering Canons and the Engineering Creed fulfill our social responsibilities as engineers?

Traditionally, professions have always been held in very high regard by society, largely due to the extensive amount of training, education, and dedication required for membership. With this come high expectations of how the members of a profession conduct themselves both in their professional and private lives: doctors save lives, lawyers protect people's rights, and engineers make people's lives better. I did not really make this connection or understand what it meant until I was initiated into Tau Beta Pi, the National Engineering Honor Society. The Tau Beta Pi initiation ceremony has remained largely unchanged for over 100 years and emphasizes the obligation as engineers and members of Tau Beta Pi to society that extends beyond the services we offer to our employers and our clients.

Over the years, I have taken these obligations to mean that as a profession we are not elevated above anyone else in society. We are affected by the same problems as the general public, and we must have an equal part in addressing them. In your lifetime, you will be affected by such issues as the strength of the economy, the effectiveness of the public educational system, unemployment, the increasing national debt, national security, and environmental sustainability. You cannot focus your talents as an engineer on solving only technical issues and assume the rest of society will address the nontechnical issues. The same skill sets you are currently developing to solve technical issues can be applied to solve issues outside the field of engineering. Your ability as an engineer to effectively examine and organize facts and information in a logical manner and then present your conclusions in an unbiased fashion allows others to more fully understand complex issues and, in turn, help develop better solutions.

This does not mean that as an engineering profession, we are going to solve all of the world's problems. It simply means that it is our responsibility to use our skills and talents as engineers in helping to solve them. It is our obligation to actively use our skills and talents to act upon issues impacting our local, national, and global communities, not merely watching as passive observers.

I challenge you to pick one issue or problem facing society that you feel passionate about and get involved. Once you do, you will be surprised at the impact you can have, even if on a small scale. By adhering to the Engineering Canon and Creed in your professional life and getting actively involved trying to solve societal issues in your personal life, you will be fulfilling your social responsibility.

IN-CLASS ACTIVITIES

ICA 2-1

For each of the following situations, indicate whether you think the action is ethical or unethical or you are unsure. Do not read ahead; do not go back and change your answers.

Situation	Ethical	Unethical	Unsure
1. Not leaving a tip after a meal because your steak was not cooked to your liking			
2. Speeding 5 miles per hour over the limit			
3. Killing a roach			
4. Speeding 15 miles per hour over the limit			
5. Having plastic surgery after an accident			
6. Killing a mouse			
7. Driving 90 miles per hour			
8. Using Botox			
9. Not leaving a tip after a meal because the waiter was inattentive			
10. Killing a healthy cat			
11. Driving 90 miles per hour taking an injured child to the hospital			
12. Killing a healthy horse			
13. Dyeing your hair			
14. Killing a person			
15. Having liposuction			

ICA 2-2

For each of the following situations, indicate whether you think the action is ethical or unethical or you are unsure. Do not read ahead; do not go back and change your answers.

Situation	Ethical	Unethical	Unsure
1. Using time at work to IM your roommate			
2. Accepting a pen and pad of paper from a company trying to sell a new computer system to your company			
3. Obtaining a fake ID to purchase alcohol			
4. Using time at work to plan your friend's surprise party			
5. Accepting a wedge of cheese from a company trying to sell a new computer system to your company			
6. Taking a company pen home from work			
7. Taking extra time at lunch once a month to run a personal errand			
8. Accepting a set of golf clubs from a company trying to sell a new computer system to your company			
9. Drinking a beer while underage at a party in your dorm			
10. Using the company copier to copy your tax return			
11. Drinking a beer when underage at a party, knowing you will need to drive yourself and your roommate home			
12. Taking extra time at lunch once a week to run a personal errand			
13. Borrowing company tools			
14. Going to an NC-17 rated movie when underage			
15. Accepting a Hawaiian vacation from a company trying to sell a new computer system to your company			

ICA 2-3

For each of the following situations, indicate whether you think the action is ethical or unethical or you are unsure. Do not read ahead; do not go back and change your answers.

Situation	Ethical	Unethical	Unsure
1. Acting happy to see an acquaintance who is spreading rumors about you			
2. Letting a friend who has been sick copy your homework			
3. Shortcutting by walking across the grass on campus			
4. "Mooning" your friends as you drive by their apartment			
5. Registering as a Democrat even though you are a Republican			
6. Cheating on a test			
7. Shortcutting by walking across the grass behind a house			
8. Saying that you lunched with a coworker, rather than your high school sweetheart, when your spouse asks who you ate lunch with			
9. Helping people with their homework			
10. Shortcutting by walking through a building on campus			
11. Not telling your professor that you accidentally saw several of the final exam problems when you visited his or her office			
12. Suppressing derogatory comments about the college because the dean has asked you not to say anything negative when he or she invited you to meet with an external board evaluating the college			
13. Letting somebody copy your homework			
14. Shortcutting by walking through a house			
15. Not telling your professor that your score on a test was incorrectly totaled as 78 instead of the correct 58			

ICA 2-4

For each of the following situations, indicate how great you feel the need is in the world to solve the problem listed. Do not read ahead; do not go back and change your answers.

Situation	Urgent	Great	Somewhat	Little	None
1. Teaching those who cannot read or write					
2. Helping starving children in poor nations					
3. Helping people locked in prisons					
4. Helping to slow population growth					
5. Helping to reduce dependence on foreign oil					
6. Helping to reduce greenhouse gas emissions					
7. Helping people persecuted for sexual orientation					
8. Helping to reduce gun ownership					
9. Helping those who are mentally disabled					
10. Helping to supply laptops to poor children					
11. Helping prevent prosecution of "victimless" crimes					
12. Helping to end bigotry					
13. Helping to prevent development of WMD (weapons of mass destruction)					
14. Helping prosecute "hate" crimes					
15. Helping to eliminate violence in movies					
16. Helping homeless people in your community					
17. Helping people with AIDS					
18. Helping people in warring countries					
19. Helping endangered species					

CHAPTER 2

REVIEW QUESTIONS

1. Which statement about ethical codes is NOT true?
   (A) Ethical codes have been around for thousands of years.
   (B) Ethical codes exist at local and global levels.
   (C) Ethical codes are often rooted in religious thought.
   (D) Ethical codes prevent people from acting unethically.

2. Who is primarily responsible for ethical decisions?
   (A) The company that employs the decision makers
   (B) Everyone involved in the decision-making process
   (C) The company ethical/legal compliance officers
   (D) All of the above

3. When considering alternative solutions, the book suggests viewing the outcomes from three perspectives. Which one was NOT a perspective listed in the book?
   (A) Fairness
   (B) Consequences
   (C) Intent
   (D) Character

4. In the (fictitious) story of George Washington and the cherry tree, a young George Washington tells the truth even when the consequence would likely result in punishment. This story is an appeal to ethical action from which perspective?
   (A) Consequences
   (B) Intent
   (C) Fairness
   (D) Character

5. In a common ethical riddle, a person is forced to make a decision that will result in injury to either their child or 100 strangers. This riddle is an appeal to which perspective?
   (A) Intent
   (B) Consequences
   (C) Character
   (D) Fairness

6. Your lab report is due soon and you're still a long way away from finishing. Your friend lets you borrow a few lines from his conclusion section. If you accept, you are
   (A) Stealing your friend's ideas.
   (B) Claiming another's work as your own.
   (C) Honoring your friend's efforts.
   (D) Being a good steward of your time.

7. Why do you have an obligation to act to benefit society at large?
   (A) You are a member of a respected profession.
   (B) You are passionate about helping others.
   (C) You have skills that can be useful to others.
   (D) You joined an engineering society.

8. Discuss the possible actions, if any, that you would take in each of the following situations. In each case, use the four-step analysis procedure presented in Section 2.1 to help determine an appropriate answer.
   (a) Your roommate purchased a theme over the Internet and submitted it as his or her own work in English class.
   (b) Your project team has been trying to get your design to work reliably for 2 weeks, but it still fails about 20% of the time. Your teammate notices another team's design that

is much simpler, that is easy to build, and that works almost every time. Your teammate wants your group to build a replica of the other team's project at the last minute.

(c) You notice that your professor forgot to log off the computer in lab. You are the only person left in the room.

(d) The best student in the class, who consistently wrecks the curve by making 15 to 20 points higher than anyone else on every test, accidentally left her notes for the course in the classroom.

(e) You have already accepted and signed the paperwork for a position as an intern at ENGR-R-US. You then get an invitation to interview for an intern position (all expenses paid) at another company in a city you have always wanted to visit. What would you do? Would you behave differently if the agreement was verbal, but the papers had not been signed?

(f) One of your professors has posted a political cartoon with which both you and your friend vehemently disagree. The friend removes the cartoon from the bulletin board and tears it up.

9. Discuss the possible actions, if any, that you would take in each of the following situations. In each case, use the four-step analysis procedure presented in Section 2.1 to help determine an appropriate answer.

(a) You witness several students eating lunch on a bench on campus. When they finish, they leave their trash on the ground.

(b) You see a student carving his initials in one of the largest beech trees on campus.

(c) You see a student writing graffiti on a trash dumpster.

(d) There is a squirrel in the road ahead of a car you are driving. You know that a squirrel's instinct is to dart back and forth rather than run in a straight line away from a predator (in this case a vehicle) making it quite likely it will dart back into the road at the last instant.

(e) You find a wallet containing twenty-three $100 bills. The owner's contact information is quite clear. Does your answer change if the wallet contained three $1 bills?

10. Read the Engineer's Creed section of this chapter.

If you are planning to pursue a career in engineering, type the creed word for word, then write a paragraph (100–200 words) on what the creed means to you, in your own words, and how the creed makes you feel about your chosen profession (engineering).

If you are planning to pursue a career other than engineering, does your future discipline have such a creed? If so, look this up and type it, then write a paragraph (100–200 words) on what the creed means to you, in your own words, and how the creed makes you feel about your chosen profession. If not, write a paragraph (100–200 words) on what items should be included in a creed if your profession had one and how the lack of a creed makes you feel about your chosen profession.

11. Engineers often face workplace situations in which the ethical aspects of the job should be considered.

Table 2-1 lists a variety of types of organizations that hire engineers and one or more possibly ethical issues that might arise.

Pick several of the organizations from the table that interest you (or those assigned by your professor) and answer the following:

(a) Can you think of other ethical problems that might arise at each of these organizations?

(b) Apply the four-step ethical decision-making procedure to gain insight into the nature of the decision to be made. In some cases, you may decide that an ethical issue is not really involved, but you should be able to justify why it is not.

(c) List 10 other types of organizations at which engineers would confront ethical problems, and explain the nature of the ethical decisions to be made.

(d) How does one find a balance between profit and environmental concerns?

(e) Under what circumstances should an engineer be held liable for personal injury or property damage caused by the products of his or her labor?

(f) Under what situations would you blow the whistle on your superior or your company?

(g) Should attorneys specializing in personal injury and property damage litigation be allowed to advertise, and if so, in what venues?

Table 2-1 Industries and issues

Organization/Occupation	Possible Issues
Alternative energy providers	Use of heavy metals in photovoltaic systems
	Effect of wind generators on bird populations
	Aesthetic considerations (e.g., NIMBY)
	Environmental concerns (e.g., Three Gorges project)
Environmental projects	Fertile floodplains inundated by dams/lakes
	Safety compromised for cost (e.g., New Orleans levees)
	Habitat destruction by projects
	Habitat renovation versus cost (e.g., Everglades)
	Environmental impact of fossil fuels
Chemical processing	Toxic effluents from manufacturing process
	Pesticide effect on ecosystem (e.g., artificial estrogens)
	Insufficient longitudinal studies of pharmaceuticals
	Nonbiodegradable products (e.g., plastics)
Transportation and building industry	Runoff/erosion at large projects
	Disruption of migration routes (freeways)
	Quality of urban environments
	Failure modes of structures
	Automotive safety versus cost
Computers	Vulnerability of software to malware
	Intellectual property rights (e.g., illegal downloads)
	Safety issues (e.g., programmed medical devices, computer-controlled transportation)
Electric industry	Toxic materials in batteries
	Cell phone safety concerns
	Power grid safety and quick restoration in crises
	Possible use to break the law (e.g., radar detectors)
	Shipment of high sulfur coal to China
	Disposal of nuclear waste
	Environmental issues (e.g., spraying power-line corridors)
Food processing industry	Health possibly compromised by high fat/sugar/salt products
	Use of genetically engineered organisms
	Sanitation (e.g., *Escherichia coli*, *Salmonella*)
	Use of artificial preservatives
Manufacturing companies	Manufacturing in countries with poor labor practices
	Lax safety standards in some countries
	Domestic jobs lost
	Environmental pollution due to shipping distances
	Trade imbalance
	Quality/safety compromised by cost
	Efficiency versus quality of working environment
	Management of dangerous tools and materials

Design, Teamwork, and Project Management

Due to the complexity of many analysis and design projects, it is necessary for all engineers to operate effectively on a team, regardless of your selected engineering discipline or career path. Because design projects must be managed to ensure a robust process, project management issues are interwoven with those of design and teamwork. This chapter introduces design, teamwork, and project management to emphasize the integration of all three.

3.1 Design Processes

LEARN TO: Define design
List and describe common design processes

Design is a creative process that requires problem definition, idea generation and selection, solution implementation and testing, and evaluation. Design is inherently multifaceted, so any problem addressed will have multiple solutions. While a particular solution might address some objectives well, other objectives might not be met at all. The goal is to identify a design that meets the most important objectives. To evaluate ideas and communicate them to others, engineers commonly sketch possible solutions and even build models of their work, so those are commonly a part of any design process.

There are many different versions of "design processes," so it is better to discuss "design" or "a design process" rather than "THE design process." When designers or textbook authors describe design, the processes they describe have a lot of similarities, but those features can have a variety of names. To introduce design concepts, we use various examples so that you can see how the design process is applied in different contexts.

It is common to draw a diagram of the design process to help others understand how process steps are connected. Two example design processes for engineering programs are shown in Figure 3-1, but many more images of design processes are easily found using your favorite Internet browser. These two are chosen as examples because of their differences and their similarities. While Figure 3-1a is cyclic and Figure 3-1b is a block diagram sequence, they each communicate the iterative nature of design.

The two processes in Figure 3-1 have a rough correspondence. The "Define Problem" step roughly corresponds to "Ask" and has also been called "Specification." "Generate Concepts" is similar to "Imagine." Other names for that process are "ideation," "innovating," and "identifying possible solutions." "Develop a Solution" or "Plan" adds details to some of concepts generated in the previous step. The process of evaluating the ideas before doing detailed planning is not pictured in either process, but it is an important one—it is not practical to follow all imagined solution pathways,

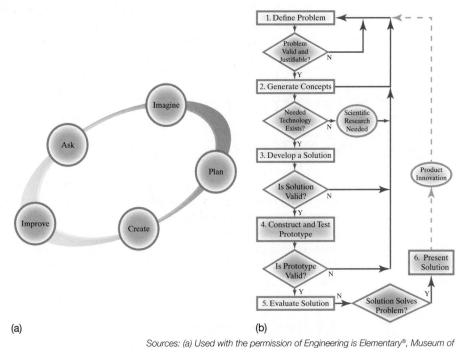

(a)   (b)

Sources: (a) Used with the permission of Engineering is Elementary®, Museum of
Science, Boston. (b) Courtesy of Project Lead The Way. Used with permission.

Figure 3-1 Two models of the design process: (*a*) the processes used in the Boston Museum of Science's
Engineering is Elementary curriculum and (*b*) in Project Lead The Way's Introduction to Engineering Design.

so it is critical to systematically choose those most likely to succeed. While the Project
Lead the Way curriculum includes "Identifying Criteria" as part of their "Define
Problem" step and "Evaluate Possible Solutions" as part of their "Generate Concepts"
step, we will discuss these processes separately. "Construct and Test Prototype" is sim-
ilar to "Create" and may be called "Implement" in other models of the design process.
"Evaluate Solution" coordinates with "Improve," known as "Refine," "Reflect and
Redesign," "Product Testing," and "Communicate Results" elsewhere.

What follows describes the tools engineers use in this process, providing detail
about each of these processes.

3.2 Defining the Problem or Need

LEARN TO: Describe the need to define problems
Give at least one example of how a problem can actually be an opportunity
Determine the stakeholders in a design

The process of problem definition is critical because it shapes all the others. Problem
definition includes getting input from stakeholders—users, clients, and others who will
be affected by your design. As with other parts of the design process, problem definition is iterative because it is important to confirm that the constituencies are pleased
with the results. You may also need to adapt to changing needs.

In education, during college and earlier, the problem or need is often identified
and described by a teacher or professor. The "problem" might be something to be
calculated, a topic for a term paper, a position to debate, or a subject to paint—but it

is commonly chosen in advance and provided to the student. Similarly, in the workplace, many "problems" are assigned to employees—to make sales calls to a list of phone numbers, to stock shelves, to gather vital health data from a set of patients, to make a series of deliveries, or to serve vegetarian lasagna to the traveling engineer at table 8 after a long day of field work. In engineering, a problem or need may be provided by a client or by an employer—where certain basic decisions about the design pathway have been made before the engineer makes any decisions. At other times, engineers have much more latitude in defining (or redefining) the problem. If an engineer is asked to "design a process by which the asbestos in a school boiler room can be removed," the problem is already well defined, even though there are still decisions to make. On the other hand, if that same engineer is asked to "design a process by which the school boiler room can be made safe from airborne asbestos," the engineer could choose from multiple ways to define the problem—removing the asbestos, containing it, increasing airflow and filtration to ensure that any loose particulate is swept away and captured, and others.

There are countless examples of how redefining a problem has been commercially profitable. When Dum Dum Pops® are made, the process is continuous, which causes the manufacture of some lollipops that combine two of the flavors. This might have been considered a problem, and some companies would have designed their manufacturing process so as to stop and clean the machinery between flavor runs. Instead, the Dum Dum company wraps these "combination" flavor lollipops and labels them "Mystery" flavor. By their defining the problem a different way, a creative solution emerged, saving money and providing a market attraction. In all designs, the way in which the problem is defined will affect the set of solutions explored.

EXAMPLE   3-1

An engineering professor is using a computer projector to share information with the class, but students are having difficulty reading the screen. After talking with a large number of students, you conclude that it's not likely that so many students have uncorrected vision problems. The students seem to agree that the problem is related to sunlight coming through the windows and hitting the screen. What are some other possible problem definitions, and how would those affect the sorts of solutions you would consider?

There are a variety of possibilities to choose from:

- The image isn't bright enough. Design a projector with a brighter bulb.
- There is glare on the screen. Design an antiglare coating for the screen.
- There is sunlight coming in the windows. Cover the windows temporarily with shades, blinds, awnings, or tinting or permanently with bricks.
- There is sunlight hitting the projector screen. Move the screen or its angle, adjusting classroom seating as needed so students can see the screen.
- The sun shines through the windows during class. Design the building so that classrooms don't have direct sunlight during class times.
- Class is being held during sunlight hours. Reschedule class to when it is dark, and reserve daylight hours for outside activities.

Each of the possibilities is only the definition of a problem, leaving much of the design process to go.

3.3 Criteria: Defining What Is Important

LEARN TO: Describe the importance of criteria in design
Contrast "must" and "should" design criteria
Manage a large number of criteria in design

Starting with a problem definition, we can begin to generate an appropriate set of criteria. At first, those criteria will provide some basic direction as we consider all the possible solutions. Later, those same criteria will be used to compare design options and narrow our choices objectively. We might be particularly concerned about cost, so certain approaches become less attractive. Other proposed solutions might be hazardous, thus further narrowing our options. The criteria for evaluating potential solutions should be discussed before you start thinking of solutions to avoid choosing criteria that favor a popular solution. Criteria must be identified before evaluation begins so that all ideas can be considered fairly—this helps avoid arguments based on hidden criteria—such as criteria that favor giving the contract to someone's sister-in-law.

Criteria range from **must criteria** that any successful solution has to have to **should criteria**, which are qualities that are generally agreed upon as desirable and that help distinguish one solution as better than another. Some solution parameters are preferences or options where there is no agreement on what is better. Must criteria are commonly called constraints and may be established legally at a regional level, such as California's automotive emission laws and Florida's building code provisions regarding wind-borne debris and at the national level, such as the corporate average fuel economy (CAFE) standards, by an independent certifying body, such as the American National Standards Institute (ANSI) or the International Organization for Standardization (ISO), or by an international agreement, such as the Basel Convention on the Control of Transboundary Movements of Hazardous Wastes and Their Disposal, a treaty that has 182 United Nations (UN) member states and the European Union as parties (Haiti and the United States have signed, but not ratified the Convention).

To be valid, criteria must be clearly understood and measurable. For criteria to be considered should criteria, usually called simply criteria as opposed to constraints, it must also be clear what is better. While the temperature of an office work area is important to productivity, there is no general agreement on the best temperature. On the other hand, it is universally agreed that a building floor system should be stiff—that the floor should not sag beneath our feet while we walk on it. While we will face limits based on cost, material selection, and other factors, the stiffness of a floor system is clearly a should criterion. Similarly, the energy consumed by an appliance is a should criterion—we may accept higher energy usage to meet other needs, but we can all agree that appliances that use less energy to accomplish the same task are better.

Other criteria include preferences or options. These are features that distinguish designs, but that different people, groups, or applications indicate a different choice for what is better. No matter how comfortable a couch is, if it is too large to fit in your living room, it's not the right couch for you. An elderly couple living in an efficiency apartment might want a small refrigerator that uses less space, but if two parents, two children, and two grandparents live in the same house, a larger unit is probably desirable.

The same criterion can vary from a constraint to a preference based on the context of the problem. While color is an important consideration in painting a house, it is clearly a preference—there is no one "best" color to paint a house. Color is a constraint in some design applications, however, such as National School Bus Glossy Yellow, fire engine red, and safety orange.

EXAMPLE **3-2**

In selecting a cordless screwdriver, what criteria are important?

While a particular person might have a variety of criteria, they do not all meet the test of good criteria:

- Battery life is measurable, and longer is clearly better.
- Charging time is measurable, and shorter is clearly better.
- Versatility might be measured by the range of attachments the device can use. This is measurable and more is better, but this fails to distinguish options because the ¼-inch socket is standard and attachments can be purchased separately.
- Torque is measurable, and more seems better, but some users could be injured by excessive torque.
- Weight is measurable, and less is clearly better.
- Comfort is qualitative, but it can be measured. Yet is it is unclear whether users could agree on what is better.

Each possibility listed is only the definition of a problem, leaving much of the design process incomplete.

Comprehension Check **3-1**

We often express criteria in terms that are not clear and measurable. Write a clear criterion to replace each of these vague criteria for the products.

Product	Computer	Automobile	Bookshelf
Inexpensive	Less than $300		
Small			
Easy to assemble			Requires only a screwdriver
Aesthetically pleasing			
Lightweight			
Safe			
Durable			
Environmentally friendly		Has an estimated MPG of at least 50	

In general, using fewer criteria keeps things simple. A rule of thumb is that if you can think of 10 criteria that are meaningful, you should consider only the two most important criteria to compare solutions. In this way, you can be sure that the important criteria maintain their importance in your decision making. Seek consensus on what the most important criteria are. Experts, specialists, managers, customers, and research can help you focus on the most important criteria. In certain situations, many criteria are used. For example, when a magazine rates consumer products, extra criteria should be included to be sure that all the criteria important to the magazine's readers have been considered.

Another situation, in which a large number of criteria are included, occurs when very complex decisions are being evaluated. The U.S. Green Building Council has long lists of criteria used for certifying various types of projects according to the Leadership in Energy and Environmental Design (LEED) Green Building Rating System. A subset of items from the LEED for Homes project checklist follows.

LEED is an internationally recognized green building certification system, providing third-party verification that a building or community was designed and built using strategies aimed at improving performance across all the metrics that matter most: energy savings, water efficiency, CO_2 emissions reduction, improved indoor environmental quality, and stewardship of resources and sensitivity to their impacts.

From the U.S. Green Building Council website: http://www.usgbc.org

These checklist items are neither clear nor measurable, however, as shown below. The full LEED for Homes rating system and the reference guide associated with the rating system provide the additional detail needed to make the insulation criteria both clear and measurable. Note that these criteria build from existing standards, meaning that even greater levels of detail can be found elsewhere (e.g., Chapter 4 of the 2004 International Energy Conservation Code and The National Home Energy Rating Standards).

LEED for Homes Simplified Project Checklist
Addendum: Prescriptive Approach for Energy and Atmosphere (EA) Credits

Points cannot be earned in both the Prescriptive (below) and the Performance Approach (pg 2) of the EA section

Energy and Atmosphere (EA)	(No minimum Points Required)	OR	Max Points	Preliminary			Final
				Y/Pts	Maybe	No	Y/Pts
2. Insulation	2.1 Basic Insulation 2.2 Enhanced Insulation		Prereq 2	0	0		0

PREREQUISITES

2.1 Basic Insulation. Meet all the following requirements:

(a) Install insulation that meets or exceeds the R-value requirements listed in Chapter 4 of the 2004 International Energy Conservation Code. Alternative wall and insulation systems, such as structural insulated panels (SIPs) and insulated concrete forms (ICFs), must demonstrate a comparable R-value, but thermal mass or infiltration effects cannot be included in the R-value calculation.

(b) Install insulation to meet the Grade II specifications set by the National Home Energy Rating Standards (Table 16). Installation must be verified by an energy rater or Green Rater conducting a predrywall thermal bypass inspection, as summarized in Figure 3.

Note: For any portion of the home constructed with SIPs or ICFs, the rater must conduct a modified visual inspection using the ENERGY STAR Structural Insulated Panel Visual Inspection Form.

Note that the LEED criteria specify only those characteristics that are important to the overall goal of green building. This leaves the choice of the type of insulation (fiberglass, cellulose) up to the designer and the consumer. Because homes, outdoor facilities, and factories face different challenges, different LEED criteria apply. It is important to recognize which choices are options rather than criteria—even the most energy-efficient factory cannot substitute for a house. Further, the LEED criteria allow different thresholds for different sizes of houses, so as not to penalize multifamily households—even though multifamily households will require more material to build and more energy to operate, the marginal cost is not as high as building another home.

When designing a product or a process, you must first listen to what a client wants and needs. The client might be from a department, such as manufacturing, marketing, or accounting, a customer, your boss, or the government. Subsequently, you will need to translate those wants and needs into engineering specifications. Pay particular attention to separating the wants (criteria) from the needs (requirements). Requirements can be imposed legally (e.g., Corporate Average Fuel Economy regulations, building codes), they can be from within your company (e.g., the finished product must not cost more than $50 to manufacture), or they may come from your client (e.g., the product must be made from at least 90% recycled material).

Comprehension Check	3-2

You have been asked to improve the fuel efficiency of an automobile by 20%. Convert this request into engineering criteria. What changes might be made to the automobile to achieve this objective?

3.4 Generating Ideas

LEARN TO:	List various ways of generating ideas
	Recognize the need for diverse perspectives in design

Most high school graduates have heard the term **brainstorming,** and many have participated in a process called by that name. This term refers to one particular process by which ideas are generated. Idea generation follows three common rules: encourage a lot of ideas, encourage a wide variety of ideas, and do not criticize. The third rule is important if the first two are to be achieved. It has been said: "The best way to get a good idea is to get lots of ideas!" Brainstorming may be useful if individuals are stuck and hearing the ideas of others helps get them thinking differently, but individuals thinking separately and then pooling their ideas typically outperform the results of brainstorming as a group. Sometimes, brainstorming fails because participants don't follow the rules well, but there are other processes that limit the productivity of brainstorming. In brainstorming, only one person can contribute at a time without interrupting, so ideas may be lost or idea generation might be slowed while participants wait their turn. Further, some participants will limit their participation out of laziness, introversion, or social pressure to avoid dominating discussion. More effective than brainstorming is generating ideas in a group where any idea is welcome, but critical reflection is permitted to spend more time on productive directions as long as the group doesn't run out of ideas. It's still important that everyone in the group feels safe to contribute, but critical reflection makes better use of everyone's time. Formal brainstorming may be useful in groups where criticism is likely to prevent some people from contributing.

Defining criteria before generating ideas seems as if it might threaten the "do not criticize" objective—proposed solutions could be steered to match the expectations established by the criteria and thus limit the range of ideas. While defining criteria first might constrain idea generation, generating solutions first might cause participants to develop an opinion of a "best solution" and then force their criteria choices to fit their preferred idea. The process of generating solutions and the process of generating criteria to evaluate those solutions are linked. The second interaction is more damaging to the quality of the design process, so we recommend identifying constraints and criteria first.

While brainstorming is well known, it is far from the only approach to generating ideas. The most common approach to solving a problem is to find out if it has been solved before. The biblical proverb, "there is nothing new under the sun" shows that it has long been known that all of our ideas have roots in our previous experience—even if we cannot explicitly identify where an idea came from. This approach can range from "phone a friend" to a comprehensive search of published material, including journals, patents, and product catalogs.

Contacting a colleague to ask if any similar problem has been solved previously is most effective for engineers who have large, diverse networks to draw on. An experienced engineer with a robust network may know just the person to go to when certain kinds of questions arise:

- "Javon, I've got to put some electronics where they will be exposed to gamma radiation—what are my options for shielding?"
- "Isabela, I've run into a problem—how can I damp the vibration in this equipment so that the operator doesn't feel uncomfortable?

A more formal study of the way a problem has been solved before is called research on **prior art**, particularly when discussing existing solutions in a legal sense related to a patent application. More generally, this sort of research will extend to related products. Printers, fax machines, and copiers have many common technologies, so an engineer designing a new copy machine would be expected to be familiar with recent advances in the other two technologies. A special subset of prior art is known as **biomimetics**— exploring how a particular problem has been solved in nature. Birds were studied for many years in the pursuit of a technological means of human flight, and more recently, the structure of the wings of an owl that permit silent flight have been studied in hopes of reducing the noise made by aircraft.

There are various systems of questions that encourage the exploration of a larger design space from a more diverse perspective. Examples are Edward de Bono's Six Thinking Hats and Alex Osborn's checklist, and more recently Tom Kelley's Ten Faces of Innovation.

3.5 Comparing Designs and Making Decisions

> **LEARN TO:**   Use pairwise comparison to narrow a set of design choices
> Conduct and analyze a weighted benefit analysis

The criteria identified earlier guided our design and must also guide the process of choosing a solution from a list of possible solutions. Simply identifying criteria is not enough information for a decision—each proposed solution must be evaluated against those criteria. The first step will be to eliminate any solutions that do not meet the minimum requirements—those solutions are out-of-bounds and need not be considered further. Those unreasonable solutions may have been important in the process of generating ideas, because some features of those unreasonable solutions may have been incorporated in other solutions, or maybe because hearing those ideas helped others think differently, leading to better solutions—but we don't need to feel guilty about throwing them away now—they have served their purpose.

Once the minimum requirements have been satisfied, there are many ways of applying the remaining criteria to select a solution. Voting is a quick way to reduce a large number of choices to a smaller number of choices, because one of many voting processes (weighted voting, multivoting, ranking, and more) can quickly eliminate many noncontroversial options. In spite of the use of voting to select leaders in many countries, voting is not the best approach to make a final selection, because it is likely to disenfranchise a large fraction of voters, particularly in a two-party system.

Other approaches are typically used to further narrow the set of choices, such as making **pairwise comparisons**. In this approach, you use a table for each criterion to summarize how each of the solutions compares with others. An example is shown in Table 3-1 for the criterion "safety." Among the table entries, 0 indicates that that option in that column is worse than the option in that row; 1 indicates that both options rank equally; and 2 indicates that that option in that column is better than the option in that row.

Table 3-1 Comparing options

Safety	Option 1	Option 2	Option 3
Option 1		0	1
Option 2	0		0
Option 3	1	2	
Total	1	2	1

Since a high level of safety is preferred, "better" means "safer." In the example, the first column indicates that option 1 is less safe than option 2, but the same as option 3. The resulting totals indicate that option 2 ranks best in terms of safety. To complete the process, generate similar tables for each of the criteria being considered and then sum the totals to identify the best solution.

A disadvantage of the pairwise comparisons approach is that all criteria have equal weight, whereas some criteria are likely to be more important than others. An alternative approach is to use **weighted benefit analysis**, shown in Table 3-2. In this approach, each option is scored against each of the criteria.

Table 3-2 Options with a weighted benefit analysis

	Weights	Option 1	Option 2	Option 3
Cost	2	2	6	10
Safety	8	10	4	6
Weight	10	7	7	2
Wow	5	2	4	6
Totals		4 + 80 + 70 + 10 = 164	12 + 32 + 70 + 20 = 134	20 + 48 + 20 + 30 = 118

Table 3-3 Sample scoring rubric

Score	Meaning
0	Not satisfactory
1	Barely applicable
2	Fairly good
3	Good
4	Very good; ideal

In Table 3-2, the weights are in the first column and each option has been assigned a score from 0 to 10, indicating how well that option meets each criterion. This approach may be inconsistent in that a 7 for one rater may be different from a 7 for another rater, so it can help to better define the scale. For example, the options may be scored on the scale shown in Table 3-3 as to how well the option fits each criterion.

An important aspect of such scoring methods is that they are not a strictly mathematical determination, but rather provide input to the decision-making process. Based on the scores that result, judgment is required to ensure that all criteria have been properly considered and that there is agreement on the final decision.

3.6 Prototyping and Testing

LEARN TO: Define prototyping
Define design testing
Understand how the process of reevaluation fits into prototyping and testing

When a solution has been identified as the best fit to the criteria, it is best to stop for a reality check before moving forward. After the evaluation process, some ideas will be left on the cutting room floor. Do any of these merit further consideration? Are there important elements of those ideas that can be incorporated into the chosen design? If the reality check reveals that an idea really should not have been eliminated, then a change in the selection criteria may be appropriate.

An important reality check is to make sure a particular design will work in practice—to actually try it out—to build a sample, a scale model, or a **prototype** and find out if it performs the way we think it will. Historically, generating prototypes has been resource-intensive—both in terms of materials and construction time—so it has been critical to reduce the number of design options before considering making a prototype. In many applications, a prototype was only made to ensure that a chosen design would function correctly, and no further design options were considered unless the prototype did not function as expected. More recently, advances in rapid prototyping, including 3D printing and digital fabrication have reduced the cost of materials and labor need-ed to generate a prototype, so it has become more feasible to use prototypes to explore a larger range of the design space.

Even when the selection of a particular design is considered final, it will undergo testing. Sometimes, this is intentional testing in laboratories and in user studies. This may include the introduction of the design into a pilot installation or a test market, which yields information both about the performance of the design and about its marketability.

After deciding, you will implement your chosen solution. Undoubtedly, both carrying out the design and using the design once it is complete will provide new information about how the design might be improved. In this way, design tends to be iterative—design, build, test, redesign, and so on. Even when a design performs as expected, it may be important to test a model extensively or even build multiple models to be sure that the design is reliable.

3.7 Sustainability—A Special Design Criterion

LEARN TO: Define sustainability
Understand how sustainability fits within the design process
Recognize how sustainability can also apply to topics aside from design

Contributed by Dr. Leidy Klotz, S.E. Liles Family Associate Professor
of Civil Engineering, Clemson University

The most common definition of sustainability is "meeting the needs of the present without compromising the ability of future generations to do the same."[1] Notice that

---

[1]From the United Nations' Brundtland report.

this definition is fundamentally about people. There is no mention of hippies or saving trees just for the sake of saving trees (of course, we are dependent on the ecosystem services trees provide). Notice also that the definition includes future generations as well as present ones. Sustainability is not just an issue for our children and grandchildren, it is an issue that is affecting all of us right now.

The terms "environmental" and "sustainability" are often used interchangeably; however, sustainability also has social and economic dimensions. All three of these dimensions must be balanced for truly sustainable engineering solutions. The figures show the relationships between these dimensions. Our society would not exist if our environment did not support human life. Our economy would not exist if we did not have a stable society (most people do not want to start businesses in failed states). These relationships seem quite obvious but can be overlooked if we just focus on one dimension of sustainability. As business leader Peter Senge points out: "the economy is the wholly owned subsidiary of nature, not the other way around."

You can apply a basic understanding of sustainability to your own engineering solutions. Sustainability is not a stand-alone topic. It cannot be bolted onto an engineering design at the end of the project. For example, in a new building project, one of the first sustainability considerations should be whether this project is even necessary. Perhaps similar goals could be achieved by more efficient use of existing facilities. This is quite an ethical dilemma! Imagine telling a potential client they do not need to hire and pay your engineering firm to design a new building. Assuming the building project is necessary, some of the best sustainability opportunities occur early on in the project, during project planning and design. This is where engineers play a key role. Teamwork and communication in the process are vital because sustainable solutions require consideration of multiple issues. We must be able to work with engineers from different disciplines and with non-engineers, such as architects, contractors, and lawmakers. We must be able to communicate with the end users who will occupy and operate the building. After all, the end user is the recipient of your design. These basic ideas apply across disciplines, whether you are designing a building, an engine, or a new material.

You may be wondering how much of humanity currently considers sustainability. Maybe we are already on a sustainable path? Unfortunately, this is not the case. Our use of critical resources, such as energy and water, and less critical resources, such as tequila and chocolate, cannot be sustained at current rates. Increasing population and affluence will stress these resources even more. Allocation of resources for the present and future is a huge ethical question engineers must consider. Should you build a reservoir that will provide water for an impoverished area, but restrict availability downstream? Do the risks associated with nuclear energy outweigh the fact that it is a carbon-free source of energy?

Creating solutions for sustainability issues requires expanding the boundaries of single-discipline thinking, being able to recognize relationships between systems and the associated problems and opportunities. For example, our fossil-fuel-based energy system has increased standards of living all over the world, but this same system also contributes to climate change, which is already having significant negative impacts, with more predicted for the future. In addition, the system contributes to inequalities between those who have energy and those who do not, which is a major source of poverty and conflict. These complex relationships can make problems seem overwhelming; however, these relationships also offer opportunities. Engineers creating sustainable energy solutions will have positive impacts in multiple areas, such as helping to curb climate changing emissions, while reducing energy poverty and resource conflict. You can make a conscious effort to build your skills in the broad systems thinking needed to identify these opportunities.

PERSONAL REFLECTION ON SUSTAINABILITY

Dr. Leidy Klotz

I see sustainability issues as challenges, but also as incredible opportunities for engineers. Is your goal to save the world? Here is your chance. Is your goal to make as much money as possible? Engineering solutions that address sustainability issues offer huge opportunities for profit. Those who figure out ways to make solar energy more economical or provide greater access to clean water will be the Bill Gates and Steve Jobs of their time.

In particular, young engineers must play a key role. It is unlikely that an engineer who graduated before 2005 was exposed to sustainability topics during college. This is an area where older engineers need your help, where you can be a leader right away. History shows us that groundbreaking advances, like those needed in engineering for sustainability, are often made by young people. Albert Einstein had his most groundbreaking year at age 26, the same age at which Martin Luther King Jr. led the Montgomery bus boycott. Thomas Jefferson wrote the declaration of independence at 33, and Harriet Tubman started the Underground Railroad at 28. We need similar innovative ideas and bold actions in all areas of engineering for sustainability. I think young people are our best shot.

Please work hard to learn more about sustainability in engineering. In your area of engineering, learn as much as possible about the fundamentals and how they are related to sustainability. Develop your broad, systems-thinking skills. Take classes that provide additional information on engineering for sustainability. Pursue opportunities for hands-on practice with engineering for sustainability on your campus and beyond. I think you will truly enjoy the collaborative process and unique design challenges associated with creating sustainable solutions!

WISE WORDS: IN YOUR JOB, DO YOU WORK ALONE OR ON A TEAM?

I often work alone on projects and analyses, but I do need to interact with client company teams, sometimes leading the team as an outside expert, but still a "member" of the company I am trying to help. Some of the teams I will work with also have mechanical, chemical, industrial, and process engineers involved.

E. Basta, Materials Engineer

I work on a team made up of all electrical and computer engineers.

E. D'Avignon, CpE

Team: chemist, package engineer, process development engineer (chemical engineer), industrial designer, line engineer (mechanical engineer or chemical engineer), process engineer (chemical engineer), planner, quality engineer, marketing, and market research.

S. Forkner, ChE

I work mostly alone on my assignments. I am given tasks and I have to find solutions for my problems on my own. Only when needed I consult someone for questions and guidance on my task.

V. Gallas Cervo, ME

I worked the first 5 years of my career as a "sole contributor" in an engineering role. Since then, I have worked in a team setting managing technical employees.

L. Gascoigne, CE

3.8 Working in Teams

LEARN TO:	Define ground rules for working in teams
	Resolve issues in communication or roles in teams
	Recognize the importance of personal participation within a team

Group: *A number of people who come together at the same place, at the same time.*

Team: *Individuals cooperating to accomplish a common goal.*

The ability to work in a team is one of the most critical traits an engineer needs. Even if you're the greatest engineer in the world, you will not know all of the answers or have all of the right ideas. We can always learn something from our peers.

A. Thompson, EE

As a student and in the workplace, you will complete some assignments individually and complete some as part of a team. When you work independently, you are mostly free to choose when and how you will work. When you work as part of a team, make sure the team has ground rules for how it will operate. Any time several people are asked to work closely together, there is a potential for much good from a diversity of ideas and skills, but there is also a potential for conflict. Because conflict can be both productive and unproductive, you need to manage it.

Team Behavior

The most critical task for a team, particularly a new team, is to establish its purpose, its process (its way of doing things), and a means of measuring team progress. Here are several topics regarding team behavior that you may wish to consider:

- **Ground rules:** Each team needs to come to a consensus about acceptable and unacceptable individual behaviors as well as team behavior.

I work with a chemical engineer, an instrumentation engineer, a mechanical engineer, and a piping engineer.

D. Jones, BE

Every project involves a team. My typical team includes surveyors, structural engineers, environmental scientists (wetlands, endangered species), permit specialists, electrical engineers, land appraisers, archaeologists, and architects.

J. Meena, CE

I work alone, but seek advice from my management team.

C. Pringle, IE

I work on a team including mechanical, electrical, aeronautical and systems hardware designers, software designers, integrators, power engineers, and other liaisons like myself.

E. Styles, EE

The Hubble Space Telescope servicing mission project is made up of a team of mechanical design engineers, human factors engineers, environmental test engineers, thermal engineers, underwater test engineers, safety engineers, mechanical technicians, documentation specialists, space scientists, and systems engineers.

R. Werneth, ME

- **Decision making:** Teams by necessity make decisions. Each team needs to decide how these decisions will be made. For example, will they be done through consensus, majority vote (either secret or show of hands), or by other methods?
- **Communication:** This is often one of the hardest parts of working effectively as a team. Team members need to recognize the value of real listening and constructive feedback. During the course of team meetings, *every* team member needs to participate *and* listen.
- **Roles:** You may adopt various roles on your team. In a long-term project, roles should rotate so that everyone has a chance to learn each role. Some typical roles are a meeting organizer who plans the meeting, including reviewing what tasks team members should complete before the meeting; a recorder who keeps track of decisions made and assignments agreed upon; a timekeeper who alerts the team if a topic is taking too long; and an encourager-gatekeeper-facilitator who helps ensure that everyone contributes and is heard during team meetings.
- **Participation:** Decide as a team how work will be distributed. Your team should also consider how to handle shifts in workload when a team member is sick or otherwise unavailable.
- **Values:** The team as a whole needs to acknowledge and accept the unique insights that each team member can contribute.
- **Outcomes:** Discuss and agree on what types of measures will be used to determine that the team has reached its final goal.

> We trained hard but it seemed that every time we were beginning to form teams we would be reorganized. I was to learn later in life that we tend to meet every situation by reorganizing, and a wonderful method it can be for creating the illusion of progress while producing confusion, inefficiency, and demoralization.
>
> *Petronius*

Teammate Evaluation: Practicing Accountability

Engineering is a self-governing profession. ABET (pronounced with a long A) is charged with accrediting engineering programs in the United States through membership from the profession. Among a set of required outcomes of engineering graduates, ABET requires that engineering students graduate with an ability to function on multidisciplinary teams. To ensure that each student achieves this outcome, individual accountability is needed. Realizing that much of the activity of a team happens when the team is meeting privately (without a professor), an effective and increasingly common way of addressing this tenet is to have team members provide feedback to each other and to the professor.

It is important to learn how to be an effective team member now, because most engineering work is done in teams, and studies show that most engineering graduates will have supervisory responsibility (at least project management) within five years of graduation. You have worked in teams before, so you have probably noticed that some team members are more effective than others. Consider these three snapshots of interactions with engineering students. All three are true stories:

- Three team members approach the professor, concerned that they have not seen the fourth team member yet. The professor speaks with the student, who quickly becomes despondent, alerting the professor to a number of serious burdens the student is bearing. The professor alerts an advisor and the student gets needed help.
- A student, acting as the team spokesperson, tells the professor that one team member never comes to meetings. The professor speaks to the nonparticipating student, who expects to be contacted about meetings by cell phone, not e-mail. The professor explains that the student's expectations are unrealistic.
- A student is insecure about being able to contribute during team activities. After the team's ratings of that student are in, the professor talks to the class about the importance of participating and the different ways students can contribute to a team. In the next evaluation, the student receives the highest rating on the team.

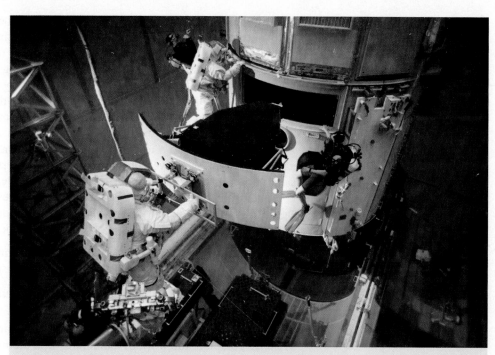

Source: NASA

Underwater Training for a Hubble Space Telescope Servicing Mission. NASA engineers (on SCUBA) and astronauts (in modified space suits) take advantage of the effects of neutral buoyancy to practice replacing a Wide Field Camera in preparation for a Hubble Space Telescope (HST) servicing mission. Extensive teamwork is required in a 6.2-million gallon pool in Houston, TX, to develop, refine, and practice the procedures to be used on spacewalks in orbit. Engineers from Goddard Space Flight Center in Greenbelt, MD, and Johnson Space Center in Houston work together to perfect the procedures and hardware, including specialized astronaut tools. The team uses models representing the flight items for conducting end-to-end tasks in the neutral buoyancy facility. This unique example of engineering teamwork has resulted in five very successful HST servicing missions involving complex astronaut spacewalks. The challenge to the team is to develop the nominal and contingency procedures and tools on the ground (and in the water!) to be used for mission success with HST in orbit 300 miles up.

R. Werneth, ME

Peer evaluations are a useful way for team members to communicate to one another and to their professor about how the members of a team are performing. Reviewing and evaluating job performance is a marketable skill and is as useful to the employee seeking a job or a job advancement as it is to the supervisor.

Focus on What Your Teammates Do Rather Than What You Think of Them

It is challenging to give a team member a single rating on their effectiveness as a teammate because some team members will be helpful to the team in some ways, but engage in some behaviors that hinder the team. Another difficulty is that each team member is likely to consider some ways of contributing more valuable than others, so the evaluation of a particular teammate will be overly influenced by that teammate's performance in certain areas. The only way to be fair is to focus on behaviors—what your teammates do—rather than opinions, such as how you feel about them. One way to focus on behaviors would be to ask you to take an inventory of what behaviors your teammates demonstrate and how often. The result would be that you might need to answer 50 or more questions about each member of the team. It is difficult to stay focused on answering accurately when completing such a long survey. A better way to focus on behaviors is by using sample behaviors to anchor each point of a rating scale. A peer evaluation instrument that is widely used in engineering education (and in other fields) is the Comprehensive Assessment of Team-Member Effectiveness (CATME, see http://info.catme.org/). CATME measures five different types of contributions to a team using such a behaviorally anchored rating scale. Each scale includes representative behaviors describing exceptionally good, acceptable, and deficient performance in each area. Each team should determine what the level of expected behavior should be. Recognizing that an individual team member may exhibit a combination of behaviors, the CATME instrument also includes in-between ratings. The five types of contributions are described below the associate behaviors.

Contributing to the Team's Work describes a team member's commitment to the effort, quality, and timeliness of completing the team's assigned tasks.

- A student who is exceptional at contributing to the team's work
  - Does more or higher-quality work than expected
  - Makes important contributions that improve the team's work
  - Helps to complete the work of teammates who are having difficulty

- A student who does an acceptable job at contributing to the team's work
  - Completes a fair share of the team's work with acceptable quality
  - Keeps commitments and completes assignments on time
  - Fills in for teammates when it is easy or important

- A student who is deficient at contributing to the team's work
  - Does not do a fair share of the team's work. Delivers sloppy or incomplete work
  - Misses deadlines. Is late, unprepared, or absent for team meetings
  - Does not assist teammates. Quits if the work becomes difficult

Interacting with Teammates measures how a team member values and seeks contributions from other team members.

- A student who is exceptional at interacting with teammates
  - Asks for and shows an interest in teammates' ideas and contributions
  - Improves communication among teammates. Provides encouragement or enthusiasm to the team
  - Asks teammates for feedback and uses their suggestions to improve

- A student who does an acceptable job at interacting with teammates
  - Listens to teammates and respects their contributions
  - Communicates clearly. Shares information with teammates. Participates fully in team activities
  - Respects and responds to feedback from teammates

- A student who is deficient at interacting with teammates
  - Interrupts, ignores, bosses, or makes fun of teammates
  - Takes actions that affect teammates without their input. Does not share information
  - Complains, makes excuses, or does not interact with teammates. Accepts no help or advice

Keeping the Team on Track describes how a team member monitors conditions that affect the team's progress and acts on that information as needed.

- A student who is exceptional at keeping the team on track
  - Watches conditions affecting the team and monitors the team's progress
  - Makes sure teammates are making appropriate progress
  - Gives teammates specific, timely, and constructive feedback

- A student who does an acceptable job at keeping the team on track
  - Notices changes that influence the team's success
  - Knows what everyone on the team should be doing and notices problems
  - Alerts teammates or suggests solutions when the team's success is threatened

- A student who is deficient at keeping the team on track
  - Is unaware of whether the team is meeting its goals
  - Does not pay attention to teammates' progress
  - Avoids discussing team problems, even when they are obvious

Expecting Quality is about voicing expectations that the team can and should do high-quality work.

- A student who is exceptional at expecting quality
  - Motivates the team to do excellent work
  - Cares that the team does outstanding work, even if there is no additional reward
  - Believes that the team can do excellent work

- A student who does an acceptable job at expecting quality
  - Encourages the team to do good work that meets all requirements
  - Wants the team to perform well enough to earn all available rewards
  - Believes that the team can fully meet its responsibilities

- A student who is deficient at expecting quality
  - Is satisfied even if the team does not meet assigned standards
  - Wants the team to avoid work, even if it hurts the team
  - Doubts that the team can meet its requirements

Having Relevant Knowledge, Skills, and Abilities accounts for both the talents a member brings to the team and those talents a member develops for the team's benefit.

- A student who has exceptional knowledge, skills, and abilities
  - Demonstrates the knowledge, skills, and abilities to do excellent work
  - Acquires new knowledge or skills to improve the team's performance
  - Is able to perform the role of any team member if necessary

- A student who has an acceptable level of knowledge, skills, and abilities
  - Has sufficient knowledge, skills, and abilities to contribute to the team's work
  - Acquires knowledge or skills needed to meet requirements
  - Is able to perform some of the tasks normally done by other team members

- A student who has deficient knowledge, skills, and abilities is
  - Missing basic qualifications needed to be a member of the team
  - Unable or unwilling to develop knowledge or skills to contribute to the team
  - Unable to perform any of the duties of other team members

Comprehension Check   3-3

Research shows that team performance can be enhanced if team members reflect on their own and their teammates' performance and give each other high-quality feedback. High-quality ratings are consistent with observed behavior, which may or may not be high ratings. We also know that rating quality (again, consistency with observed behavior) improves with practice. Guided practice in giving and receiving feedback and in practicing self- and peer evaluations using behavioral criteria will help you improve. Please take your time in evaluating the members of the following fictitious team.

Pat   Friendly and very well-liked, makes working fun, and keeps everyone excited about working together. Relies on teammates to make sure everything is going okay. Pays attention to keeping the team upbeat but does not seem to notice if the team's work is getting done. Struggles to keep up with the rest of the team and often asks teammates for explanations. The team has to assign Pat the least difficult jobs because Pat does not have the skills to do more complex work. Offers ideas when able, but does not make suggestions that add anything unique or important to the final product. Always shows up for meetings, prepares beforehand, and does everything promised. Is confident that the team can do everything that is essential. Agrees that the team should meet all explicit task requirements.

Chris   Okay as a person and does not interfere with the contributions of others but rubs teammates the wrong way by frequently griping about the work and making excuses for not following through on promises to the team. Chris has the brains and experience to make a unique and valuable contribution, but does not try. The fact that Chris is so smart frustrates some teammates who have to try hard to accomplish tasks that would be easy for Chris. Ignores assigned tasks or does a sloppy job because "Robin will redo the work anyway." Misses meetings or shows up without assigned work. Contributes very little during meetings. Was late to one meeting because "no one told me the meeting time." Missed another meeting because "the alarm clock did not go off." After missing meetings, he asks lots of questions to make sure that everyone is making progress and the team's work is being accomplished. Spends more time checking that everyone else is doing their work than getting the job done. Chris always seems sure that the team will do fine and says that the team should do good work that fully meets the standards for acceptable performance. In response to a teammate's question about Chris' failure to deliver a promised piece of work, Chris said, "Why should I bother? Robin won't let the team fail."

Robin   Very bright. Has far greater knowledge of the subject than any of the other team members. Extremely skilled in problem solving. Robin has very high standards and wants the team's work to be impressive, but Robin worries whether the team's work will be good enough to stand out. Robin completes a big chunk of the team's work and takes on a lot of the really difficult work. Does the work that Chris leaves unfinished. The quality of Robin's work is consistently outstanding. Tends to just work out the solutions and discourages teammates' attempts to contribute. Reluctant to spend time explaining things to others. Does not like to explain "obvious" things. Is particularly impatient with Pat's questions and once told Pat "You are not smart enough to be on this team." Complains that Chris is a "lazy freeloader." Sometimes gets obsessed with grand plans and ignores new information that would call for changes. Does not pay attention to warning signs that the current plan might not be effective until the problems are obvious. Then handles the situation as a crisis and takes over without getting team input. Robin is reluctant to acknowledge or discuss problems in the team until they affect his work.

Terry   Not nearly as bright as Robin, but works to develop enough knowledge and skills to do the assigned tasks. Terry can usually fill in for other team members if given specific directions, but does not understand most of the tasks that other team members normally perform. Does more grunt work than any of the other team members, but does not do as good a job as Robin and does not take on difficult tasks. Sometimes makes mistakes on the more complex work. Super responsible, spends a lot of time giving one-on-one help to Pat. Always on time to meetings. Often calls to remind everyone (especially Chris) about meetings and usually makes some nice comment about one of the teammate's strengths or a valuable contribution that the teammate has recently made. Terry is outgoing and highly supportive of teammates when well-rested, but is sometimes too tired to get excited about teammates' ideas. Is not defensive when teammates offer feedback, but does not ask for teammates' suggestions, even when teammates' input could help Terry to do better work. Terry thinks that the team can do great work and encourages teammates to do their best. When Robin expresses doubts if the team can do superior work, Terry reassures everyone that the team is capable of outstanding work. When the team is headed in the wrong direction, Terry is quick to notice and say something, but usually does not suggest a way to fix it. Terry reviews the team's objectives and alerts the team to anything that comes up that would affect the team. Terry was reluctant to press the issue when Robin's plan ignored one of the guidelines specified for the project.

To test your ability to focus on individual behaviors, go to https://www.catme.org/login/survey_instructions and rate each team member on each type of contribution to the team. On the Scenario Results page, a green arrow indicates that your rating matches the expected rating. If your rating does not match the expert rating, the blue arrow shows your rating and the red arrow indicates the rating experts would have assigned. If you count one point for every level separating your rating from the expert rating on the five different types of contribution, a low score is best, indicating the greatest agreement with the expert ratings. You can mouse over the red arrows to read the rationales underlying the expert ratings.

3.9 Experimental Design: PERIOD Analysis

LEARN TO: Discuss why experimental design is important
Define the steps in the PERIOD analysis method
Understand how many measurements might be necessary in an analysis

Experiments enable engineers to come up with a creative solution to a problem and test the validity of the proposed idea. An experiment is a test of a proposed explanation of a problem. A good design of an experiment is a critical part of the scientific method.

What Constitutes the Scientific Method?

1. Observation: Observe the problem and note items of interest.
2. Hypothesis: Search for a known explanation of the phenomenon or attempt to formulate a new explanation.
3. Prediction: Create a model or prediction of behavior based on that hypothesis.
4. Experiment: Test your predictions. If necessary, modify your hypothesis and retest.

Why Is Experimental Design Important?

As you move through your college career, you will be inundated with many equations and theories. These are useful in solving a wide variety of problems. However, as you will see, often the equations are only really useful in solving the most basic type of problems.

As an example, suppose you are interested in the speed of a ball as it rolls across the floor after rolling down a ramp. In physics, you will learn the equations of motion for bodies moving under the influence of gravity. If you are good, you can use these to examine rolling balls. What you will quickly find, however, is that numerous complicating factors make it difficult to apply the basic equations to obtain an adequate answer. Let us suppose you are interested in smooth balls (such as racquetballs), rough balls (tennis balls), heavy balls (bowling balls), and lightweight balls (ping-pong balls). The simplified equations of motion predict that all these will behave in essentially the same way. You will discover, however, that the drag of the air affects the ping-pong ball, the fuzz affects the tennis ball, and the flexible nature of the racquetball will allow it to bounce at steep ramp angles. It is difficult to predict the behavior analytically. Often, one of the quickest ways to learn about the performance of such complex situations is to conduct experiments.

What Are Experimental Measurements?

Most scientific experiments involve measuring the effect of variability of an attribute of an object. In an experiment, the **independent variable** is the variable that is controlled. The **dependent variable** is a variable that reacts to a change in the independent variable. A **control variable** is part of the experiment that can vary but is held constant to let the experimenter observe the influence of the independent variable on the dependent variable. Keeping control variables constant throughout an experiment eliminates any confounding effects resulting from excess variability.

Any measurement acquired in an experiment contains two important pieces of information. First, the measurement contains the actual value measured from the instrument. In general, a measurement is some physical dimension that is acquired with some man-made data-collection instrument. As with any man-made device, there may be some imperfection that can cause adverse effects during data collection. Thus, the second piece of information that goes along with any measurement is the level of uncertainty.

Any uncertainty in measurement is not strictly by instrumentation error. Systematic error is any error resulting from human or instrumentation malfunction. Random error is caused by the limits of the precision of the data-collection device. It is possible to minimize the systematic error in an experiment, but random error cannot be completely eliminated.

What Measurements Do You Need to Make?

You need to develop a coherent experimental program. You should make enough measurements to answer any anticipated questions, but you do not usually have the time or money to test every possible condition. Points to consider:

- What are the parameters of interest?
- What is the range of these parameters—minimum values, maximum values?
- What increments are reasonable for testing (every 10 degrees, every 30 seconds, etc.)?
- What order is best to vary the parameters? Which should be tested first, next, etc.?

The acronym PERIOD can help you to remember these important steps. As an example, it is applied to the problem of the ramp and rolling balls previously described.

P—Parameters of interest determined

- Parameter 1 is the ramp angle.

- Parameter 2 is the distance up the ramp that we release the ball.
- Parameter 3 is the type of ball.

E—Establish the range of parameters

- Ramp angle can vary between 0 and 90 degrees in theory, but in reality can only vary between 10 degrees (if too shallow, the ball would not move) and 45 degrees (if too steep, the ball will bounce).
- The distance we release the ball up the ramp can vary between 0 and 3 feet in theory, assuming that the ramp is 3 feet long. We cannot release the ball too close to the bottom of the ramp or it would not move. In reality, we can only vary between 0.5 and 3 feet.
- We will test as many types of balls as we have interest in.

R—Repetition of each test specified

- The ramp angle will be set according to the height of the ramp from the floor, so there is not much room for error in this measurement; only one measurement is needed for such geometry.
- Each placement of the ball before release will vary slightly and may cause the ball to roll slightly differently down the ramp; this is probably the most important factor in determining the speed, so three measurements at each location are needed.
- We will assume that every ball is the same, and the actual ball used will not change the outcome of the experiment; only one ball of each type is needed.

I—Increments of each parameter specified

- We will test every 10 degrees of ramp angle, starting at 10 degrees and ending at 40 degrees.
- We will release the balls at a height of 0.5, 1, 1.5, 2, 2.5, and 3 feet up the ramp.
- We will test five types of balls: racquetball, baseball, tennis ball, ping-pong ball, and bowling ball.

O—Order to vary the parameters determined

- We will set the ramp angle and then test one ball type by releasing it at each of the four different distances up the ramp.
- We will repeat this process three times for each ball.
- We will then repeat this process for each type of ball.
- We will then change the ramp angle by 10 degrees and repeat the process.
- This process is repeated until all conditions have been tested.

D—Determine number of measurements needed and Do the experiment

It is always important to determine before you start how many measurements you need to make. Sometimes you can be too ambitious and end up developing an experimental program that will take too much effort or cost too much money. If this is the case, then you need to decide which increments can be relaxed, to reduce the number of overall measurements.

   The number of measurements (N) you will need to make can be easily calculated by the following equation for a total of n parameters:

N = (# increments parameter 1 $\star$ number of repetitions for parameter 1)

(# increments parameter 2 $\star$ number of repetitions for parameter 2) $\star \ldots$

(# increments parameter n $\star$ number of repetitions for parameter n)

Continuing the preceding examples, the number of actual measurements that we need to make is calculated as

N = (4 angles) $*$ (6 distances $*$ 3 repetitions) $*$ (5 types of balls) = 360 measurements

In this example, 360 measurements may be extreme. If we examine our plan, we can probably make the following changes without losing experimental information:

- We decide to test every 10 degrees of ramp angle, starting at 20 degrees and ending at 40 degrees. This will lower the angle testing from four to three angles.
- We will release the balls at a height of 1, 2, 2.5, and 3 feet up the ramp. This will lower the distances from six to four.
- We will test three types of balls: racquetball, ping-pong ball, and bowling ball. This will lower the type of balls from five to three.

The number of actual measurements that we now need to make is calculated as

N = (3 angles) $*$ (4 distances $*$ 3 repetitions) $*$ (3 types of balls) = 108 measurements

This result seems much more manageable to complete than 360!

3.10 Project Timeline

LEARN TO: Create a project timeline
Define and use a responsibility matrix
Recognize the importance of team dynamics within a project timeline

To complete a project successfully, on schedule, and satisfying all constraints, careful planning is required. The following steps should help your team plan the completion of project work.

Step 1: Create a Project Timeline.

The first consideration is the project's due date. All team members need to note this on a calendar. Examine the due date within the context of other assignments and classes. For example, is there a calculus test in the fourth week? When is the first English paper due?

Next, look at the project itself and break into individual tasks and subtasks. Create a list, making it as specific, detailed, and thorough as possible. Your list should include:

- All tasks needed to complete the project
- Decisions that need to be made at various times
- Any supplies/equipment that will need to be obtained

Carefully consider the order in which the tasks should be completed. Does one task depend on the results of another? *Then, working backward from the project due date, assign each task, decision, or purchase its own due date on the calendar.*

Finally, your team should consider meeting at least once a week at a consistent time and location for the duration of the semester. More meetings will be necessary, but there should be at least once per week when the entire team can get together and review the project status. A standing meeting time will prevent issues of "I did not know we were going to meet" or "I did not get the message."

Step 2: Create a Responsibility Matrix.

List the project's tasks and subtasks one by one down the left side of the paper. Then, create columns beneath each team member's name, written side by side across the top. Put a check mark in the column beneath the name of the member who agrees to perform each task. It then becomes the responsibility of that team member to successfully perform the task by the due date that was agreed upon in Step 1. An alternate grid is shown in Table 3-4.

Table 3-4 Sample responsibility matrix

Task	Completed by	Checked by
Purchase supplies	Pat and Chris by 9/15 team mtg	
Write initial proposal	Terry—Email to Robin by 9/22	Robin by 9/25 team mtg
Conduct preliminary calculations on height	Robin—Email to Pat by 9/22	Pat by 9/25 team mtg
Build prototype in lab	All—Lab: 7–9 p.m., 9/28	

In assigning the tasks, consider the complexity and time required for the job. One team member may have five small tasks while another may have one major task, with the goal being an equal distribution of effort. A second team member should be assigned to each task to assist or check the work completed by the first team member. Be sure all team members are comfortable with the assignments.

Step 3: Consider Team Dynamics.

Communication: The success of any project depends to a great extent on how well the team members communicate. Do not hesitate to share ideas and suggestions with the group and consider each member's input carefully. Do not be afraid to admit that you are having difficulties with a task or that the task is taking longer than expected. Be ready and willing to help one another.

Trust and respect: Remember the team is working toward a shared objective. Therefore, you must choose to trust and respect one another. Treat fellow team members with simple courtesy and consideration. Follow through with promises of completed tasks, remembering the team is counting on your individual contributions. Try to deal honestly and openly with disagreements. However, do not hesitate to ask for help from faculty if problems begin to escalate.

Nothing is carved in stone: It is important to plan the project as carefully as possible; however, unforeseen problems can still occur. Treat both the Project Timeline and Responsibility Matrix as working documents. Realize they were created to serve as guides, not as inflexible standards. Watch the project progress relative to the timeline, and do not hesitate to redesign, reallocate, or reschedule should the need arise. Review your matrix each week and adjust as needed.

Finally, do not forget to have fun!

3.11 Modern Project Management

LEARN TO: List and define modern project management roles
Use modern project management tools

Modeled after the way that a rugby team works as a team to move the ball down the field toward the goal, scrum is a project management process that was popularized as a method of software development. The scrum method is distinguished from earlier methods in that it is holistic rather than linear and flexible rather than sequential. This flexibility is particularly with respect to user requirements, which may change during the development process. Another reason for the rugby metaphor is that control of the project is passed back and forth among the team members, which makes the scrum method particularly appropriate in today's engineering workforce, where it is common for engineers to manage certain projects and participate in projects that others manage without any hierarchical management structure. As a result of this more shared leadership structure, the term **scrum master** is used rather than project manager. The scrum master is primarily responsible for ensuring that nothing hinders the team's progress—in a facilitative role. Another role in the scrum process is the **product owner**, who is the primary liaison to customers and clients and represents their voice in the design process. The remaining team members (as many as are needed) comprise the **development team**. The development team focuses on project deliverables—sharing their progress publicly as often as possible in completed, if minor, advancements.

IN-CLASS ACTIVITIES

ICA 3-1

With your team, compose a plan to build the longest bridge possible using the K'Nex™ pieces provided by your instructor. The longest part of your bridge will be defined as the longest stretch of K'Nex pieces that are not touching another surface (table, floor, chair, etc.).

During the planning phase (15 minutes), your team will only be allowed to connect eight K'Nex pieces together at any one moment. As soon as your team has finished the planning phase, disconnect all K'Nex pieces and place them in the provided container. The container will be shaken before you begin, so do not bother attempting to order the pieces in any way.

During the building phase (60 seconds), the restriction on the number of connected pieces goes away, but your team will not be allowed to talk. Your instructor will say go, and then stop after 60 seconds has elapsed. At the end of the building phase, your team must step away from the bridge and remove all hands and other body parts from the K'Nex bridge. If any pieces fall after time is called, you are not allowed to stabilize the structure in any way.

ICA 3-2

With your team, come up with a plan to build the tallest tower possible using the K'Nex™ pieces provided by your instructor. The tallest portion of your tower will be defined as the longest stretch of K'Nex pieces that are not touching another surface (table, floor, chair, etc.).

During the planning phase (15 minutes), your team will only be allowed to connect eight K'Nex pieces together at any one moment. As soon as your team has finished the planning phase, disconnect all K'Nex pieces and place them in the provided container. The container will be shaken before you begin, so do not bother attempting to order the pieces in any way.

During the building phase (60 seconds), the restriction on the number of connected pieces goes away, but your team will not be allowed to talk. Your instructor will say go, and then stop after 60 seconds has elapsed. At the end of the building phase, your team must step away from the tower and remove all hands and other body parts from the K'Nex tower. If any pieces fall after time is called, you are not allowed to stabilize the structure in any way.

ICA 3-3

Following the rules for brainstorming (encourage a lot of ideas, encourage a wide variety of ideas, and do not criticize), develop ideas for the following with your team.

- **(a)** A better kitty litter box
- **(b)** A new computer interface device
- **(c)** A new kind of personal transportation device
- **(d)** Reducing noise pollution
- **(e)** Reducing light pollution
- **(f)** A new board game
- **(g)** A squirrel-proof bird feeder
- **(h)** A no-kill mole trap
- **(i)** A tub toy for children of 4 years or younger
- **(j)** A jelly bean dispenser
- **(k)** A new musical instrument
- **(l)** A self-cleaning bird bath
- **(m)** A new smart phone application

REVIEW QUESTIONS

1. Which of the following statements about design is true?
   (A) The best designs are the result of a singular creative genius.
   (B) The goal of any designer is to find the single best solution.
   (C) Designs must satisfy many objectives.
   (D) The best design will always satisfy all of the objectives.
   (E) None of the above statements is true.

2. Which of the following statements about design is NOT true?
   (A) Design is an iterative process.
   (B) Customer feedback is part of the design process.
   (C) Designs must result in tangible products.
   (D) Failure is an important part of the design process.

3. Which of the following do NOT describe a possible design criteria?
   (A) Design criteria should be specific.
   (B) Design criteria may include a must-have feature in the design.
   (C) Design criteria may include a should-have feature in the design.
   (D) Design criteria may include preferences.
   (E) Design criteria may include potential solutions.

4. Which of the following is NOT a good criterion for a new vehicle?
   (A) The car should be a silver color.
   (B) The car should be spacious.
   (C) The car should have at least 15 cubic feet of cargo space.
   (D) The car must have a backup camera.

5. What is the best way to choose a final design?
   (A) Let each engineer argue for the merits of their design ideas.
   (B) Use the design criteria to compare designs.
   (C) Build prototypes of each design idea and test them.
   (D) Ask a customer focus group to choose the best design.

6. Which of the following definitions for sustainability is used in the text?
   (A) Sustainability is about finding long-term solutions that do not pollute the environment.
   (B) Sustainability is finding solutions that use renewable resources or provide the maximum benefit from nonrenewable resources.
   (C) Sustainability is meeting our needs without compromising the ability of future generations to do the same.
   (D) Sustainability means generating products that can be recycled or making use of recycled material.

7. Which of the following is NOT a good way to build teamwork?
   (A) Focus on personal relationships.
   (B) Set rules for team behavior.
   (C) Agree on the purpose of the team.
   (D) Clearly define a means of measuring team progress.

8. Which of the following describes random error?
   (A) Random error is the uncertainty caused by instrumentation malfunction.
   (B) Random error is the uncertainty inherently present in the instrumentation.
   (C) Random error is the uncertainty that we try to eliminate from our experiment.
   (D) Random error is the uncertainty cause by human mistakes.

9. Using the PERIOD method, design an experiment that compares how four different tires perform. You decide to run an experiment that tests all four tires on dry, damp, and wet roads for decelerations of 1, 3, 5, 7, and 9 m/s². If each experiment is repeated two times, how many measurements are needed?
   (A) 240 measurements
   (B) 120 measurements
   (C) 60 measurements
   (D) 30 measurements

10. What should be the first consideration when creating a project timeline?
    (A) Determine the project's due date.
    (B) Determine which tasks depend on other tasks.
    (C) Create a matrix of responsibilities for each team member.
    (D) Elect a team leader.

MINI DESIGN PROJECTS

This section provides a wide range of design projects, varying in difficulty and time commitment. Your instructor may assign projects required for your specific course and provide more details.

Demonstrate a Physical Law or Measure a Material Property

1. Prove the law of the lever.

2. Demonstrate conservation of energy (potential energy + kinetic energy = constant).

3. Determine the coefficient of static and sliding friction for a piece of wood.

4. Prove that the angle of incidence is equal to the angle of reflection.

5. Demonstrate momentum conservation (force = mass * acceleration).

6. Demonstrate the ideal gas law.

7. Obtain a series of data points from an experiment you conduct that, when plotted, exhibit a normal distribution.

8. Show that forces can be resolved into horizontal and vertical components.

9. Find the center of gravity of an irregular piece of plywood.

10. Show that for circular motion, force = mass * velocity squared / radius.

11. Show that for circular motion, velocity = angular velocity * radius.

12. Measure the effective porosity of a sand sample.

13. Prove the law of the pendulum.

14. Prove Hooke's law for a spring.

15. Prove Hooke's law for a metal rod (in deflection).

16. Measure the coefficient of thermal expansion for a solid rod or bar.

17. Estimate the heat capacity for several objects; compare with published results.

18. Prove Archimedes' law of buoyancy.

19. Determine the value of pi experimentally.

20. Prove the hydrostatic pressure distribution.

21. Relate the magnetic strength to the radius.

22. Determine the density and specific gravity of a rock.

Solve a Problem

23. What is the volume of a straight pin?

24. Determine the thickness of a specified coin or a piece of paper.

25. How many pennies are needed to sink a paper cup in water?

26. Determine the specific gravity of your body.

27. What is the volumetric flow rate from your shower?

28. Use a coat hanger to make a direct reading scale for weight.

Design a Solution

29. THE GREAT EGG DROP

You have no doubt seen the "Odyssey of the Mind" type of assignment in which you are to design protection for an egg that is to be dropped from some height without being broken. This assignment is to have the same end product (i.e., an unbroken egg) but in a different way.

You cannot protect the egg in any way but are allowed to design something for it to land on. You will be allowed three drops per team and will be assigned at random the heights from which you will drop the egg. The egg must freefall after release. If the egg breaks open, the height will be taken as zero. If the eggshell cracks, the height from which you dropped it will be divided by 2. If you drop the egg and miss the catching apparatus, that is your tough luck (and a zero height will be used)—suggesting that you need to devise a way to always hit the target.

Your grade will be determined by the following equation:

Ranking = (height from which egg is dropped)/(weight of catching apparatus)

After demonstrations, the average ranking number for each team will be calculated and the value truncated to an integer. The teams will then be ranked from highest to lowest value. The heights and weights will be measured in class. The actual grade corresponding to your class ranking will be determined by your instructor.

30. TREE HEIGHT

We have been contacted by a power company to conduct a study of tree height and interference with high-voltage wires. Among other requirements, the company is looking for a quick, easy, and inexpensive method to measure the height of a tree.

Your project is to develop different methods of measuring the height of a tree. As a test case, use a tree designated by your instructor. You may *not* climb the tree as one of the methods! In addition, you may not harm the trees or leave any trace of your project behind.

To sell your methods to the customer, you must create a poster. The poster should showcase your measurement methods, including instructions on how to conduct the experiment, any important calculations, graphs or photos, and your resulting measurements. The poster will be graded on neatness, organization, spelling and grammar, formatting, and strength of conclusions. You may use any piece of poster board commercially available, or a "science fair" board. The poster can be handwritten, or typed and attached, or … here is a chance to use your creativity! If you present more than one method, you must indicate and justify your "best" choice.

Your grade will be determined as follows:

- Method 1:   30 points
- Method 2:   20 points
- Method 3:   10 points
- Presentation Board: 40 points

31. FIRE EXTINGUISHER

Make a fire extinguisher for a candle. The candle will be lit, and the extinguisher will put out the flame at a predetermined time after the candle is lit (say, 20 seconds). The only thing that the participant can do to start the time is to light the candle. The candle can be mounted anywhere you like. The results will be scored as follows:

You will be allowed three trials. You will be allowed to use your best trial for grading. You must extinguish the candle between 19 and 21 seconds; for every second or fraction thereof outside this range, you will lose five points.

32. CLEPSYDRA

Construct a clepsydra (water clock). When you bring it to class to demonstrate its performance, the following test will be used:

You will have three times to measure: a short time, a medium time, and an extended period. The actual times to be measured will be given to you at the time of demonstration, and you will have 2 minutes to set up your apparatus.

- Short: Between 10 and 30 seconds (in 2-second increments)
- Medium: Between 1 and 4 minutes (in 30-second increments)
- Extended: Between 5 and 10 minutes (in 1-minute increments)

You will start the clock and tell the timekeeper to begin. You will then call out the times for each of the three intervals and the timekeeper will record the actual times. Your clock must "run" for the total time.

Your grade will be determined by the average percent error of your timings.

For example, if the specified "medium" time was 2 minutes and 30 seconds and you said "mark" at an actual time of 2 minutes and 50 seconds, the absolute value of the percent error would be (20 s) / (150 s) = 13%. The absolute values of the three errors will be summed, divided by 3, and subtracted from 100% to get a numerical grade.

33. ON TARGET

Each team will design, build, and test a device that will allow you to successfully hit a target with a ping-pong ball. The target will be a flat sheet of poster paper placed on the floor with a bull's-eye and two other rings around it for scores of 100, 90, and 80 with a score of 60 for hitting the paper. The target will be placed at a location of 15 feet from the point at which you release the ball. Once the ball is released, you cannot touch it again, and it must be airborne before it hits the target. When demonstrating your device, you will not be allowed any trial run.

Your grade will be determined as follows:

Average numerical score + bonus (10, 8, or 6 points for creativity and simplicity)

The class (each person) will be given a slip of paper on which they will rank their top three teams with respect to creativity and also with respect to simplicity. These will be tallied and the top three teams in each category will receive 10, 8, or 6 bonus points.

34. KEEPING TIME

Each team is to build a "clock." When you say "go," a stopwatch will be started, and you are to tell the timekeeper when 10, 30, and 60 seconds have elapsed. Differences between the actual times and the predicted times at the three checkpoints will be noted and the percent error calculated. Average the absolute values of these three errors and subtract from 100 to obtain your final score. You may not use any store-bought device that is designed to measure time. No electronic devices may be used.

Your final grade will be determined as follows:

Average numerical score + bonus points (10, 8, or 6 for creativity and simplicity)

The class (each person) will be given a slip of paper on which they will rank their top three teams with respect to creativity and simplicity. These will be tallied and the top three teams in each category will receive 10, 8, or 6 bonus points.

Additional Projects

35. Develop a device that can be placed into a container of water and used to measure the pressure as a function of depth. Take measurements and plot them against theory for a hydrostatic pressure distribution.

36. A two-liter soft-drink container, nearly full of water and open to the atmosphere, is placed on the floor. Where could you locate an orifice in the side of the bottle so that the jet of water that squirts out will have the maximum range? Keep the bottle filled by continuously pouring water in the container as the tests are conducted. Justify your answer with theory by discussion rather than equations.

37. Design and build a device that will allow a ping-pong ball to hit a target (small circle) between 5 and 15 feet away from the point at which you release the ball with the device on the floor. Points will be given for accuracy (distance from the target center). From a hat, you will draw two slips of paper: a short distance (3–7 feet) and a long distance (8–15 feet). The slips of paper will have values in 1-foot increments on them (3, 4, 5, 6, or 7 feet for the short distances, and similar for the long distances). When you set up your device, you will be given 2 minutes to set the device for the first test and 2 minutes to set the device for the second (long distance) test.

38. Build a vehicle that will travel over a flat surface (hallway in the building). Points will be awarded for the distance traveled divided by the total (initial) weight of the vehicle. The vehicle must move under its own internal "engine"—the team can only release the stationary vehicle when the test begins. You will have two attempts, and the best value will be recorded. Your instructor may impose an allowable maximum weight. No batteries or electricity can be used.

39. Build a thermometer. You are to be able to measure the temperature of cold water in a bucket, room temperature, and the temperature of hot water in a container (degrees Celsius). The total percent errors will be summed (absolute values), averaged, and subtracted from 100.

40. Without moving more than 10 feet from your initial location, position a person a distance of 50 feet (or 100 feet) away from your initial location; bonus points for doing it several ways; the person must initially start beside you. You cannot be connected to the other person in any way (e.g., string, rope). Points are given for accuracy.

41. Roll an object of your design down an inclined plane (provided and the same for all participants). The object is to knock over a small piece of wood placed at a distance of 5 feet from the base of the incline and 5 feet to one side.

42. We are interested in rolling plastic drink bottles across the floor. Rather than hold a race to see which bottle will roll the fastest, we want to determine which will roll the farthest. It is important that each bottle be given a fair chance, so the starting conditions must be the same for each. Each team will use the same ramp (18 inches wide by 24 inches long) and supported by the 4-inch dimension of a 2 × 4-inch board. The bottles to be used will be clear-plastic soft-drink bottles. The two-liter size is probably the best, but smaller bottles could also be used. The objective is to determine the answer to questions. You must develop a defendable test program, carry out the tests, present your results in an easy-to-understand manner, and defend your conclusions.
  (a) How much water should the bottle contain in order to roll the farthest (until it stops)?
  (b) How much water should the bottle contain in order to roll the shortest distance (until it stops)?
  (c) As a part of your test program, you will release two bottles simultaneously on the ramp (with differing amounts of water in them, including one empty). Do not let them roll all the way until they stop, but catch them about one foot after they leave the bottom of the ramp. Which moves the fastest, which the slowest, and why?

Engineering Communication

It is a common joke that most engineers cannot construct a grammatically correct sentence, and there is all too much truth in this anecdote. In reality, the most successful engineers have developed good communication skills, not only oral and written, but also those involving multimedia formats. You might have the best idea in the known universe, but at some point you are going to have to convince someone to supply the $200 million needed to develop it. You *must* be able to communicate effectively not only that you have this great idea for a practical antigravity device, but also that it will actually work and that you are the person to lead the team developing it.

Our intent here is not to make you expert communicators, but to at least make you aware of the importance of good communication skills in engineering, as well as to give you a bit of guidance and practice developing these skills.

4.1 Basic Presentation Skills

LEARN TO: Use Microsoft PowerPoint to create a presentation
Understand the audience of a presentation
Define the 4-S Formula

Since most students consider giving an oral presentation a more daunting task than submitting written documents, we focus on live presentations first, although many of the suggestions apply to all forms of engineering communication.

Many years ago, one of us was responsible for a program to recruit high school students into engineering. Each engineering department made a short presentation to the visiting students, extolling the glories of its particular discipline. One department sent its most personable and able communicator about half the time and Professor X came the other half. Both used the same set of PowerPoint slides, but when Professor X showed up, every single student seemed to be completely brain-dead within three minutes. It was *awful*! The other professor maintained their rapt attention for the entire 15 minutes, with supposedly the same presentation.

With this in mind, you need to focus on several factors when planning a presentation. Note that the first item in our list is *who* the audience is, although the other factors mentioned are equally important.

Preplanning

- **Who is my audience?** Know the age group, demographics, prior knowledge about the topic, and what positions or opinions they may hold.
- **What is my purpose?** What do I hope to accomplish? What response do I expect? What will the audience get out of my speech?

NOTE

5 Ws and 1 H:
- Who
- What
- Where
- When
- Why
- How

- **Where is all the equipment I need?** Where will the talk be held?
- **When am I on the program agenda?** Will I be the first presenter (when audience is most alert) or the last one before lunch (when they are becoming restless) or after lunch (when they are sleepy)? What will I need to do to keep my listeners attentive?
- **Why am I giving this talk?** Why is the audience here?
- **How long should I talk?** Remember that only a few people can focus for more than 20 minutes. Trim your talk so that people will ask for more information rather than thinking "When will he sit down?"

Preparing the Verbal Elements

The preceding list focused primarily on logistics. In addition to these considerations, the structure of your presentation is vital. As a simple example, which of the following two sentences is easier to understand?

> Sentence A: While perambulating in the antithesis of the metropolis to evade the intemperate brouhaha thereof, my visual cortex perceived an ophidian.

> Sentence B: I saw a snake while taking a relaxing walk in the woods.

Although sentence A may be phrased in a more intriguing manner, it tends to obscure the underlying meaning. This is perhaps desirable in poetry or fancy fiction but generally is detrimental to a professional engineering presentation.

To help you avoid such pitfalls, we offer the 4-S formula for structuring presentations.

- **Short:** Use short sentences, avoid too many details, and do not talk too long.
- **Simple:** Avoid wordy, lengthy phrases.
- **Strong:** Use active voice and action verbs, not passive voice and "to be" verbs.
- **Sincere:** Convey empathy, understanding, and respect for the audience.

WISE WORDS: HOW IMPORTANT ARE COMMUNICATION SKILLS AT YOUR JOB?

When working with clients, we deliver our approach, analysis structure, status, findings, and final deliverable by presentation.

M. Ciuca, ME

Absolutely essential. While I thankfully have a job that allows me to dig into the math and analysis, I still deal with a lot of people. Being able to communicate effectively, where two people (or more) really understand what each is saying is very important on a complex project, and is often more difficult than one would think; or at least it involves more active participation than many are inclined to put into it.

J. Dabling, ME

I have to write design documentation, edit customer specs, produce design review presentations, and sometimes present to our customers.

E. D'Avignon, CpE

Being able to connect with customers, and internal team members, is the means to develop relationships and win new business.

B. Dieringer, ME

Three Structural Parts

Keep in mind the purpose of the following three discrete elements of a speech:

- **Introduction:** Purpose: to capture the interest of the audience. Your first task is to *hook your audience*. What is it about your subject that *they* (and not necessarily you) would find most interesting and relevant?
- **Body:** Purpose: to keep your audience interested. They will continue to pay attention if you keep the material interesting and relevant to them.
  - Divide the presentation into two or three main points.
  - Use one or more simple examples to illustrate each major point.
- **Conclusion:** Purpose: to pull it all together.
  - Summarize major points.
  - Show appreciation for your audience's attention.
  - Allow for a few questions, but be sensitive to your audience and the other speakers.

Preparing Visual Aids

Most of you have seen presentations that used slides with unreadable text, incomprehensible graphics, or annoying special effects. Well-designed graphics can greatly enhance your presentation, not only making it easier for the audience to understand, but also keeping their attention focused. A picture really is worth about 2^{10} words! Although our focus is on PowerPoint presentations, these suggestions apply to other media as well.

Helpful Hints

- Keep each slide simple, with one concept per slide. As a rule, use no more than *six lines per slide*. Each slide should correspond to an average of *60 seconds* of speech.
- If possible, make slides in landscape format.

Absolutely critical. I can think of people that I have worked with who do not have these skills, and they are not easy or fun to work with.

S. Forkner, ChE

Social skills are crucial to success when working at a company. Many times the person who is most successful is not necessarily the one who has the best or brightest ideas, but rather the individual who has solid ideas and is able to communicate in a manner that allows others to easily understand the vision and path to goal achievement.

L. Gascoigne, CE

Good social skills are very important because they improve productivity, teamwork, and goodwill. I have also found socials skills can be a deciding factor in job advancement.

R. Holcomb, IE

- Present data in simple graphs rather than in lists or tables. Avoid excessively complex graphs with extensive data. If you must present tables, divide them among several slides.
- Pictures, diagrams, and video simulations all may enhance your presentation. Be sure that all are large enough to be seen by the audience, and choose color schemes that do not appear washed out when projected. Often, such items are designed for viewing on a small screen and do not project well. Be sure to test them prior to your presentation.
- Use bullet points with important phrases to convey ideas. Avoid complete sentences.
- Large-size text is best. A font size of at least 18 points and preferably no less than 24 points should be used. This includes *all* objects, such as axis and legend captions, table headings, figure symbols, and subscripts.
- Use high-contrast colors. Avoid fancy fonts, such as cursive, or light colors, such as yellow or other pastels. Avoid using all capitals.
- Use a light background and dark print to keep the room brighter.
- Keep background styles simple and minimize animation to avoid distracting from the presentation. Keep all the slide backgrounds the same throughout a single presentation.

4.2 Sample Presentations

To illustrate the visual aids caveats, we critiqued three student presentations.

EXAMPLE   4-1

Sample Student Presentation 1

Original Presentation: Critique

- Slide 1 font is difficult to read, poor choice of abbreviation for approximately.
- Slide 2 dates and text appear disjointed due to text size and graphic; the graphic is too large. The graph explanation is better summarized using an assertion as the title on the graph slide and a small amount of text highlighting the important points about the graph.
- Slide 3 graph is difficult to read. The graph is now included with the explanation of the important point on Slide 2. In case the absolute growth (in the number of flights) is important, this is included on the conclusions slide.
- Slide 4 has too many words, and the graphic is too large. Removing rambling text has made room for an additional meaningful conclusion.

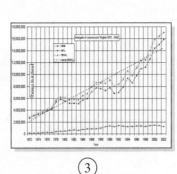

Improved Presentation:

(1)

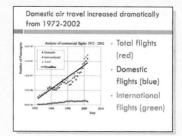

(2)

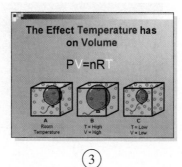

(3)

EXAMPLE 4-2

Sample Student Presentation 2

Original Presentation: Critique

- Title slide has a less distracting graphic element, but it still evokes gas bubble imagery.
- Different graphics on every page is distracting.
- White color is hard to project over graphics.
- Slide 4 graphic is difficult to read; yellow highlights make it worse.
- Titles of Slides 3 and 4 now contain meaningful assertions evoking the point of the slide.

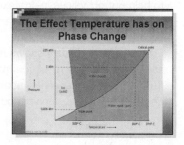

(1)

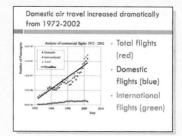

Source: Topseller/Shutterstock

(2)

(3)

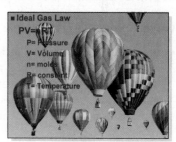

(4)

Improved Presentation:

1

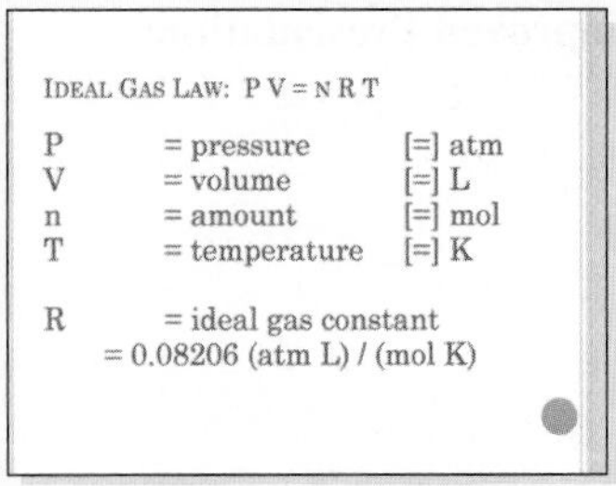

2

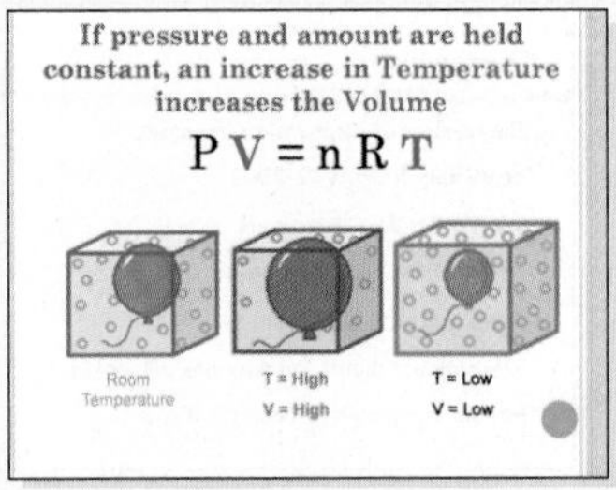

3

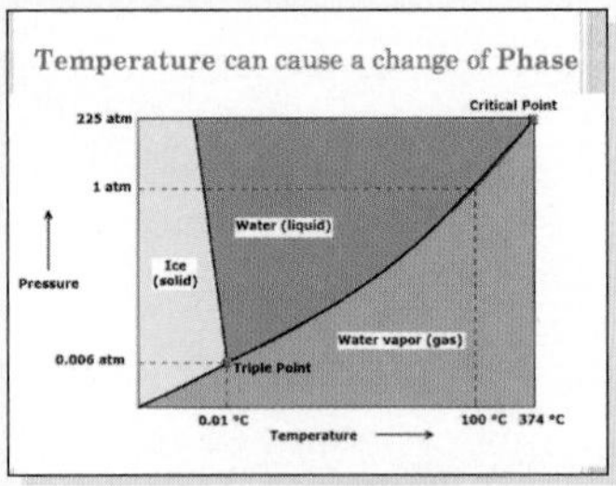

4

EXAMPLE   4-3

Sample Student Presentation 3

Original Presentation: Critique

- Green backgrounds with white text do not project well; blue and red text is especially hard to read. Added subtitle explains "Arrhenius" and discourages a U.S.-centric view of science.
- Slides 2 and 4 have too many words; should use bullets, not sentences. Full range of pH added to help put pH values in context.
- No graphics; some pictures of acid rain damage would be helpful.
- Graph added to Slide 4 shows relative occurrence of most common free radicals. Time series shows decrease during this time period.

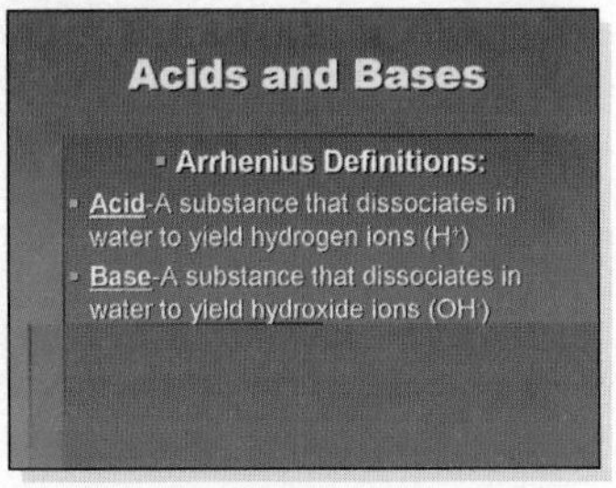

1

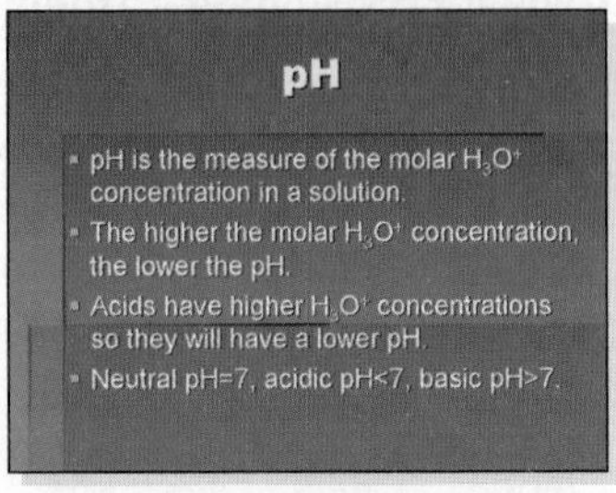

2

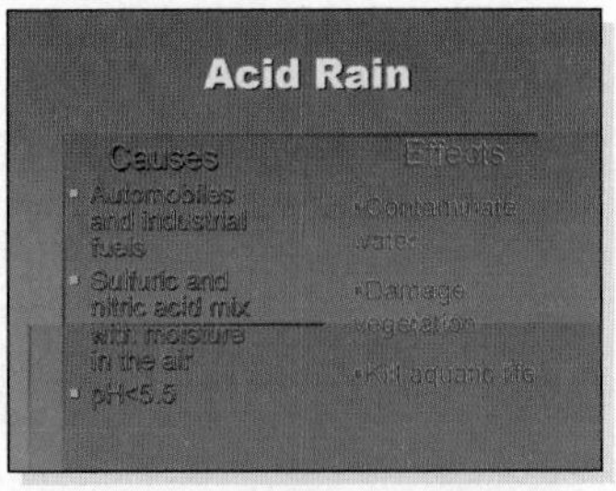

3

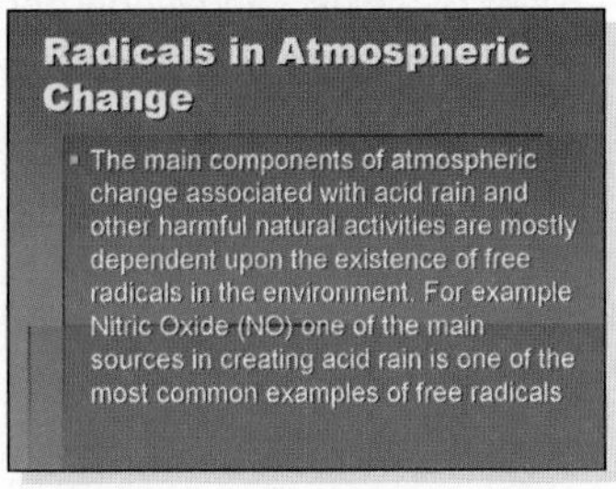

4

Improved Presentation:

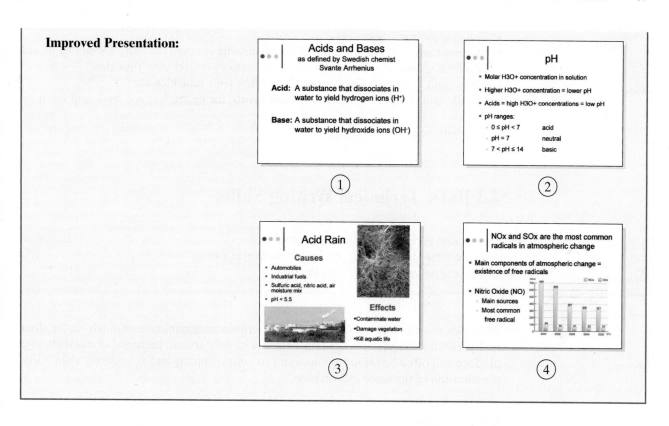

Making the Presentation

Oral presentations present several challenges for effective communication. How many of you have had an instructor who simply reads the contents of the slides with no embellishment? How many have had an instructor who seems to be terrified of the audience, cowering in fear and trying to disappear into the wall? How many have tried desperately to read the contents of a slide containing hundreds of words in a minuscule font? If you have not yet suffered through one or more such painful presentations—you will! Do not inflict such things on your own audiences.

Presentation Dos and Don'ts

When delivering a presentation, do:

- Relax!
- Speak slowly and clearly, making good eye contact.
- When your hands are not busy, drop them to your sides.
- Rehearse your presentation *out loud* multiple times. If possible, have a friend critique.
- Arrive early enough to make sure that all technology is present and working, and resolve any problems you may discover.

When delivering a presentation, do not:

- Lean on your surroundings, turn your back to the audience, or cover your mouth while speaking.
- Read your presentation from a prepared text.

NOTE

The key to improving presentation skills is practice, practice, practice!

- Tell inappropriate jokes.
- Stammer, overuse the pronoun "I," or repeatedly say "um" or "uh." Do not be afraid of a little silence if you need to glance at notes or collect your thoughts.
- Chew gum, remove coins from pockets, crack your knuckles, etc.
- Shuffle your feet or slouch; move repetitively, for example, pace back and forth or sway.
- Play with your notes.

4.3 Basic Technical Writing Skills

LEARN TO: Utilize basic principles for technical writing
Recognize the importance of editing and revising in writing
Use proper references in technical documents

Although most of you probably consider written communications much easier than oral presentations because the fear factor is largely absent, technical documents you produce will often be far more important to your company and your career than a live presentation of the same information.

General Guidelines

In addition to many of the points made earlier, effective technical writing requires its own set of guidelines:

- **Be clear**; use precise language. Keep wording efficient without losing meaning. Do not exaggerate.
- Ensure that the finished copy logically and smoothly *flows* toward a conclusion. Beware of choppiness or discontinuity. Avoid extremely long sentences because they may confuse the reader.
- If possible, use **10-point font size and 1.5 line spacing**.
- **Generally, prefer past tense verbs.** Keep verb tenses in agreement within a paragraph.
- **Define any terms** that might be unfamiliar to the reader, including acronyms and symbols within equations.
- **Present facts or inferences** rather than personal feeling.
- **Maintain a professional tone.** Do not be emotional or facetious.
- **Number and caption all tables, figures, and appendices.** Refer to each from within the body of the text, numbering them in order of appearance within the text.
  - Tables are numbered and captioned *above* the table (Table 4-1).
  - Figures are numbered and captioned *below* the figure (Figure 4-1).

Table 4-1 Example of a properly formatted table

Current (I) [A]	2	6	10	14	16
Energy of Inductor #1 (E1) [J]	0.002	0.016	0.050	0.095	0.125
Energy of Inductor #2 (E2) [J]	0.010	0.085	0.250	0.510	0.675
Energy of Inductor #3 (E3) [J]	0.005	0.045	0.125	0.250	0.310

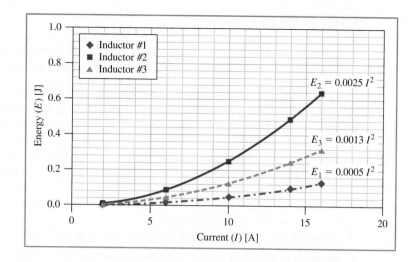

Figure 4-1
Example of a properly formatted figure.

- **Proofread and edit several times.**
  - **Remember to include** headings, figures, tables, captions, and references.
  - Do not assume that the spell check on the computer will catch everything! It will not distinguish between such words as "whether" and "weather," or "was" and "as."
- **Read it twice:** once for technical content and once for flow.
  - As you proofread, look for and *remove* the following: unnecessary words; sentences that do not add to the message; superfluous paragraphs.
  - Do one proofreading aloud. When you encounter commas, semicolons, colons, or periods, pause. Read a comma as a brief pause. Read a colon or semicolon as a longer pause. Read a period as a complete stop before the next sentence. Read what is actually written, not what you *think* it should say. If the text sounds stilted or blurred when read, you probably need to reconsider your use of these punctuation marks.
  - If possible, have someone not associated with the project (a roommate, a friend, or a mentor) read it, and ask that person for suggestions.
- **Spell out a number that starts a sentence.** If the number is large (e.g., a date), reword the sentence.
  - 23 points were outliers. (Unacceptable)
  - Twenty-three points were outliers. (OK)
- **Keep the leading zero with a decimal.**
  - The bridge cost .23 dollars per gram. (Unacceptable)
  - The bridge cost $0.23 per gram. (OK)
- **For long numbers, do not spell out.**
  - The average was one thousand, two hundred fifty-five grams. (Unacceptable)
  - The average was 1,255 grams. (OK)
- **Use the dollar symbol.**
  - The bridge cost four thousand dollars. (Unacceptable)
  - The bridge cost $4,000. (OK)
- **Watch for significant figures. Keep it reasonable!**
  - The bolt is 2.5029 inches long. (Unacceptable)
  - The bolt is 2.5 inches long. (OK)

Proper Use of References

Contributed by Ms. Jan Comfort—Engineering Reference Librarian, Clemson University Libraries

In today's wired age, most students immediately go to the Internet to find information. Although this can be an excellent source, particularly for preliminary research, there are definite risks associated with using online sources, since essentially anyone can put anything they want on the web. For example, type "flat earth society" into your favorite search engine and check out the "truth" concerning our home planet, or explore how you can save the endangered Pacific Northwest Tree Octopus.

When making presentations or writing reports, it is important to verify the veracity of any sources you consult. These guidelines will help you avoid egregious errors in your own technical communications.

The **ABCs of evaluating information** offer a useful start:

- **Authority:** Is it clear who is responsible for the site? What are the author's credentials? Is the author an expert in the field? Is it a .com or .gov or .edu site?
- **Bias:** What is the purpose of the article? Is it free of obvious bias? Is the author presenting an objective view of the subject matter?
- **Currency:** When was the information created or last updated?

But there is more to evaluating resources than that. Good students take it to the next level. Here is how you can, too.

- **Use sources that have been reviewed by experts.** Instead of searching for hours trying to find websites that meet stringent requirements, try using library sources to identify good quality sources that have already been through a review process.

WISE WORDS: HOW IMPORTANT ARE COMMUNICATION SKILLS AT YOUR JOB?

Projects are successful when the people who work together on them are able to communicate clearly with each other and work together to achieve a common goal. Misunderstanding and miscommunication leads to delays, poor quality, and frustration.

A. Hu, EE

As an engineer you communicate at all levels from the least senior production employee on the factory floor to the president of the company. Tailoring the message to the audience is the difference in acceptance and rejection.

J. Huggins, ME

Writing clear and concise specifications for construction can make the difference between an under-budget, on-time project and an over-budget, late, and unsafe final product.

L. Johnson, CE

As a consultant, extremely. If clients can't get along with you, they won't hire you. Every job requires a proposal and an interview.

J. Meena, CE

- **Secure a peer review:** An expert in the appropriate field evaluates something proposed (as for research or publication).[1] Academic Search Premier and Expanded Academic ASAP are the names of two very good multisubject databases that contain scholarly (peer-reviewed) as well as popular articles. One or both of them should be available at your library.
- **Compare the information found in your article or website with content from other websites, or from reviewed sources.** Comparing sources can also alert you to controversial information or bias that will need further study. Are facts from one website the same as those of another? How about depth of coverage? Maybe one site has better-quality information. Does the site have photos or other unique features that make it a good choice? Or perhaps a journal article from a library database is a better source. Until you compare several sources, you will not know what you are missing!
- **Corroborate the information.** Verify the facts from your source—regardless of where you found it—against one or more different sources. Do not take the word of one person or organization. A simple rule might be: "Do not use information unless you have corroborated it. Corroboration with varied and reviewed sources increases the probability of success."[2]

---

[1]Peer review. (2009). In *Merriam-Webster Online Dictionary*. Retrieved May 13, 2009, from http://www.merriam-webster.com/dictionary/peerreview

[2]Meola, Marc. (2004). "Chucking the checklist: A contextual approach to teaching undergraduates web-site evaluation." Portal: *Libraries and the Academy*, 4(3), 331–344.

Much of my work in my current job involves researching what is going on in the world, and then putting that information into a format that makes sense to people and helps other people draw conclusions from it. Good written communication skills are essential for what I do every day.

M. Peterson, EE

All the social skills are extremely *important because of the different functions and technical levels that I interface with.*

E. Styles, EE

Communication—along with teamwork—really separates bad engineers from good ones. Someone could have the best idea in the world, but if he isn't able to describe the invention or provide reasons as to why it should be developed, the idea is useless. Plus, engineers are trained to be rational and thus perfect for managerial positions. If you have good communication skills, one can easily expect you to climb quickly up the corporate ladder.

A. Thompson, EE

4.4 Common Technical Communication Formats

Technical communications can take on a variety of formats. Here, we will specifically address e-mails, memos, and short technical reports. Other, usually longer, formats will probably be addressed later in your engineering career, but the same general guidelines apply regardless of form or length of content.

E-mail

Many students believe that the rules they use for instant messaging (IM) and Twitter apply to e-mail also. When using e-mail in a professional context (including e-mail to professors!), more formal rules should be followed. The following suggestions will help you write e-mail that is clear, concise, and appropriate for the recipient.

After you have composed your e-mail, ask yourself if you would mind the president of the university, the CEO of your company, or your parents reading it. If the answer to any of these is no, then you probably should reword it.

E-mail Etiquette

NOTE

Keep two e-mail accounts: one for professional use and the other for personal use.

Choose e-mail names carefully. Some of our favorite "bad email addresses" follow:

- sugarbritches
- guitarfreak
- jessiethestudent
- Bombom bombombo
- YoltsPreston
- fatmarauder

- Be sure to correctly address the recipient. If you are unsure of a person's proper title (Dr., Mrs., Prof.), look it up!
- Use an appropriate subject line. Avoid silly subjects (Hey—Read this!) or omitting the subject line—this may cause the e-mail to end up in the junk mail folder.
- Sign your *full* name and include contact information for e-mail, phone, or mailing address if appropriate. When sending e-mail about a class, including your course number and section or day/time is often helpful. (Your professor may teach more than one section of the class).
- Change your sending name (who your e-mail program says the e-mail is from) to your full name (such as Elizabeth Stephan) or an appropriate nickname (Beth Stephan). Do not leave your account as Student or the computer default setting (such as Noname Stephan).
- Keep it brief. Do not use one continuous paragraph—make it easy to read.
- If you expect a response, be sure that action items are clearly defined.
- Use correct capitalization and punctuation. Spelling does count—even in e-mail! Avoid IM speak (e.g., LOL, IMHO, IIRC).
- Avoid putting anything in e-mail you would not say in person, and remember that an e-mail does not convey your tone of voice. Do not use e-mail to "vent" or write anything that can be easily misinterpreted by the reader.
- To avoid sending an e-mail before you have a chance to check over your work, fill in the To: and CC: lines last.
- When waiting for a reply, allow a grace period of 48 to 72 hours. If you have not received a reply after 48 hours and a deadline is approaching, you can resend your message, inquiring politely if it was received. Items do sometimes get lost in cyberspace! If the matter is critical, try the phone or request a face-to-face meeting if the first contact does not elicit a response. If you need a faster response, put that in the message or even the subject.

Sample E-mail

To…	R. Swarthmore, Ph.D. [swart@reactorsealsrus.com]
cc…	C. Ohland [carson@reactorsealsrus.com];
	K. Stephan [katie@reactorsealsrus.com]
Subject:	Leaky gel reactor seal

Dr. Swarthmore:

The gel reactor seals in B4L3 are leaking and causing production losses (over 200K for FY 2001). The Materials Engineering Lab was asked to test other seal materials. Laboratory tests identified six material couples that produced better wear resistance than the current seal. A prototype seal was made with a new material, self-mated cemented carbide, but the carbide on the seal cracked during fabrication.

The purpose of this e-mail is to request an additional $40,000 and four months' project time to fabricate and test another new seal configuration.

Your approval of this program before Friday noon will allow us to proceed with the project as quickly as possible without any delay. If you have further questions or would like more information, please contact me.

Sincerely,
J. Brock
Design Team Manager, Reactor Seals R Us
(123) 456-7890 x 1234
jbrock@reactorsealsrus.com

EXAMPLE 4-4

I won't go into detail but my partner didn't really give me enough heads up to let me know he needed to do the analysis. Approximately 30 minutes before 5 when it was due so I am going to submit an answer for the final question into the course management system where we would turn in the workbook. Weather you choose to accept it or not is up to you.

Better solution:

I have been having a difficult time communicating with my partner (insert name), and we had some confusion about who was responsible for submitting the analysis portion of Project 2. Just before 5 p.m., when the project was due, he informed me that the analysis was not complete. I will finish the project analysis tonight to include the additional questions I did not submit previously, and bring you the completed document in the morning at 10 a.m. during your office hours. I would appreciate the opportunity to discuss this situation with you further at that time.

I dont know what going on with the computer but I have been trying to upload the assignement for an hour and it will not work. I have to drive to (insert state) tonight so I wont be able to fix this problem later. I can not files on monday so you can see when they were last saved and show you them then. I really need this one point. Please consider this.

Better solution:

I have been trying to upload the assignment since 6 p.m. and will continue to try; however, I wanted to e-mail you this assignment since the deadline of 10 p.m. is approaching. The system will let me browse and select a document, but it will not do anything when I hit Submit. I have tried to use a wired connection instead of wireless, with no luck. Do you have any suggestions for fixing this problem? Thank you in advance for your assistance.

I've been throwing up all morning and did not make it to class. i did not think it would be a good idea to possibly get anyone else sick or disrupt class with me running out. I hope you can excuse my absence if I'm not there. i was just curious if there was anything else i missed in class

Better solution:

I am sorry, but I will be unable to attend class today due to illness. I understand this absence will be unexcused, per the course syllabus, since I am not going to seek medical attention and will use this as one of my three allowed unexcused absences. I will check the course management system and my classmates to determine what I missed today. If I have any further questions, I will see you during your office hours tomorrow. I expect to return to the next class period.

Memo (One-Page Limit)

The following template (also provided online) is for a one-page memo. Your professor may ask you to adhere to this format or may suggest a different one.

NOTE

You should use a 10- or 12-point font, such as Times New Roman or Verdana, with 1 to 1.5 line spacing.

Margins should be set to 1 inch all around.

Be sure to use correct spelling and grammar.

Include the headings given here, in bold.

Be sure to keep this memo to a ONE-PAGE limit.

To:	Dr. Engineering
From:	Ima Tiger, Section 000 (IMT@school.edu)
Subject:	Memo Guidelines
Date:	May 21, 2018

Introduction: The first three or four sentences should explain the purpose behind the memo. You should attempt to explain what you were asked to do, what questions you are trying to solve, and what process you are attempting to determine.

Results: Place any experimental results, in tabular and/or graphical format, here. As space is limited, this normally only includes two items: two tables, two figures, or one table and one figure. Be sure that each is clear enough to stand alone, with one to two sentences of explanation. Be aware each table and figure should illustrate a *different* idea. Include a table caption at the top of each table, and a figure caption at the bottom of each figure. The caption should include a number and a word description. When a figure is used within a document, a title is not necessary on the graph and is replaced by the caption. The two items should be pasted side by side using the **Paste Special > JPEG** command or similar picture format command (PNG, Bitmap) and then sized appropriately.

Discussion: In this section, discuss how you obtained your data, the meaning of any trends observed, and the significance of your results. Refer to the tables or figures shown in results by name (Table 1 or Figure 1). Explain any errors in your data (if possible) and how your data differs from theory. If you are deciding among several alternatives, in addition to justifying your final selection be sure to explain why you did NOT choose the other options.

References: List any sources you use here. You may use a new page for references if necessary. Any reference format is acceptable; Modern Language Association (MLA) citation style is preferred.

Remember: This document has a ONE-PAGE limit!

EXAMPLE **4-5**

Sample of a poorly written student memo

To:	Dr. Engineering
From:	Ima Student
Subject:	Memo
Date:	April 1, 2018

Introduction: We are given the job to analyze the cost of upgrade a machine line, which produces widgets. We were given three companies to choose from, to figure out witch would be the best for the cost and its production. Just by graphing the variables would allow us to find our answer.

Results:

	Klein Teil	
Varible Cost	*0.95*	
Fixed Cost	*5.00E+06*	
Material Cost	*0.75*	
Energy Cost	*0.15*	
Labor Cost	*0.05*	
Selling Price	*3*	
apacity per da	*6500*	
antity Produc	*ein Teil Total Co*	*Revenue*
0.00E+00	*5.00E+06*	*0.00E+00*

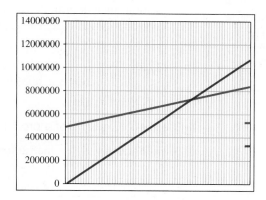

Discussion: We got the data by taking all variables from the information provided, then graphing the results together. This allowed us to see which machine line would provide the better outcome for the situation at hand. Considering the cost of the machine, material, labor, and energy into consideration with what would produce the quantity and quality product we're striving for. Figure 1 display all three solutions': total cost, revenue for us, and our breakeven point to ensure us of our choice. From observation of the graph we see that the Klein Teil machine is better. Its breakeven is at $2,400,000 and the profit is twice as much.

Summary: So to answer the question, Klein Teil would be our best option. The results yielded that the Klein Teil machine would give us the most quality for its price, a better production rate, and more money in return. From this research I hope you choose to take the Klein Teil machine.

Comments on this memo:
There are so many problems with this submission that we address only the major problems.

- *The subject line simply informs us that this is a memo.*
- *The introduction tells us very little about what the memo will address.*
- *The same data is presented in both the graph and the table. The formatting of the table is very poor; the formatting of the graph is worse.*
- *The discussion does not explain how the data was analyzed, and the justification of the final recommendation is essentially nonexistent.*
- *Similarly, the summary says almost nothing. Although not an appropriate summary, the three words "Buy Klein Teil" would probably have been more effective.*

How many more problems can you find in this sample memo?

Short Report (Two to Four Pages)

The following template (also provided online) is for a short report. Again, your professor may ask you to follow these guidelines or provide a somewhat modified version.

Introduction: Type the introduction here. This should be four or five sentences. What is the problem that will be addressed in this memo?

Procedure: Type the procedure here. This should be at most ¾ page. It may be in bulleted format. You should generalize the procedure used to include the basic steps, but you do not need to include every detail. The reader should gain an understanding of how you collected your data and performed your analysis.

Results Insert the results here, but do not discuss them or draw any conclusions. This may include a maximum of three illustrations in a combination of figures and tables. Be sure that each is clear enough to stand alone, with one or two sentences of explanation. Include a table caption at the top of the table and a figure caption at the bottom of the figure. The caption should include a number and a word description.

Be aware tables and figures should illustrate *different* ideas (Table 4-2 and Figure 4-2). Do *not* include large tables of raw data or *every* graph generated. This section should be a *sample* of those items, used to illustrate the points of your discussion.

Table 4-2 Example of a properly formatted table

Section	Instructor	E-mail	Time
−030	Dr. Stephan	e.stephan	M 8:00–9:55
−031	Dr. Park	w.park	M 12:20–2:15
−032	Dr. Sill	b.sill	W 8:00–9:55

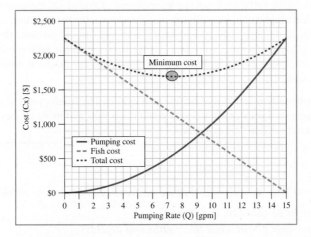

Figure 4-2

Example of a properly formatted figure.

Discussion Explain your results here. This can be up to a maximum of one page. Refer to the table and figure shown in results by name (Table 1 . . .). Be sure to include the items specifically requested in the original project description.

Summary What is the conclusion? This should be four to five sentences long, and answer the initial questions asked in the introduction and summarize any important findings.

References: List any sources you use here. Any reference format is acceptable; Modern Language Association (MLA) citation style is preferred.

Poster Presentation

The following template is for posters. Again, your professor may ask you to follow these guidelines or provide a somewhat modified version. The template provided online is a PowerPoint format but is meant to be printed. The default is set to $8\frac{1}{2} \times 11$ printing, which will allow you to submit this to your instructor without the need for a plotter. This could easily be changed, however, and this template be used to create a large poster.

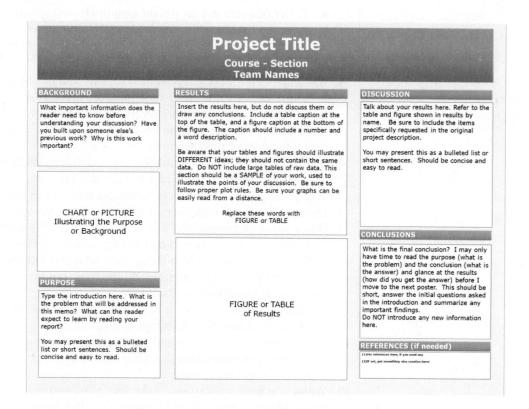

Dos and Don'ts of Poster Design (from http://www.ColinPurrington.com/tips/poster-design)

1. The number one mistake is to make your poster too long. Densely packed, high-word—count posters attract only those viewers who are excited by manuscripts pasted onto walls, and you typically don't want to talk to those types of people. They're weird. Aim for 800 words, and remember that fewer is fine, too.

2. Avoid titles with colons if you can: they are overused. Coloned titles are sometimes devised in order to inject humor into an otherwise mind-numbing poster topic (e.g., "Attack of the Crohn's: contribution of chromosome 16 allelic variants to inflammatory bowel disease progression"). The other motivation for using colons is to provide greater detail about the general topic introduced by the first clause, which is purposefully vague so as to interest a wider viewership (e.g., "Causes of obesity: additive effects of inactivity and ad libitum feeding on yearly weight gain in Homo sapiens"). Although humor and clarity are great,

it is better to achieve them without a grammatical crutch, especially if everyone else is using the same crutch. If you absolutely must have a colon in your title, just be sure it doesn't force you to spill onto a third line.

3. Format the title in sentence case (option A, below) so that trade names, Latin binomials, genes, alleles, etc. (that depend on formatting to convey meaning) are not obscured. Avoid italics (B), title case (C), and all caps (D). Be aware, however, that people of a particular age (really old) will Think My Advice Is Insane AND WILL DISAGREE (before the advent of computers, titles were typeset in title case and all caps, and opinions are slow to change).

   a. Effect of Lycra use on weight gain in Homo sapiens
   b. *Effect of Lycra use on weight gain in Homo sapiens*
   c. Effect of Lycra Use on Weight Gain in Homo Sapiens
   d. EFFECT OF LYCRA USE ON WEIGHT GAIN IN HOMO SAPIENS

4. Use a nonserif font (e.g., Helvetica) for title and headings and a serif font (e.g., Palatino) for body text. Serif-style fonts are much easier to read at smaller font sizes . . . that's why novels are rarely set with Helvetica and the like.

5. Do not add bullets to section headings. The use of a bolded, larger font is sufficient for demarcating sections.

6. The width of text boxes should be approximately 45–65 characters. Lines that are shorter or longer are harder to read quickly.

7. Whenever possible, use lists of sentences rather than blocks of text.

8. Use italics instead of underlining. Underlining draws too much attention to the word.

9. When using acronyms and numbers (e.g., ATP, 42) within the body of text, scale down the font size by a couple of points so that their sizes don't overpower the lowercase text, which they would do if you left them at the default size. Use of small caps will sometimes do the trick, but this effect varies with different fonts and with different software.

10. Set line spacing of all text to be exactly 1. Doing this protects the aesthetics if you have used super- or subscripted text.

11. Do not trust the tab button to insert the correct amount of space when you are indenting a paragraph (the default is usually too big). Set the tab amount manually, with the ruler feature.

12. When you have quotations, make sure your software hasn't used the double prime glyph instead. Double primes are the thingies used for inches (e.g., 5′11″) and mathematical formulas/formulae.

13. Correct any errors in spacing within and between words, especially before and after italicized text. Note that you can use a single space between sentences (the double-space convention was needed for typewriters, and we are slow to lose the habit). Use the Search/Replace feature to globally replace all double spaces with single spaces and to find locations where too many spaces occur between words.

14. Avoid dark backgrounds. They make your poster hard to read. Also, designing graphics is harder. It's better to just use a white background. You save on ink, too.

15. Avoid color combinations that render your poster unintelligible to those with color deficient alleles. Approximately 8% of males and 0.5% of females have some degree of color-vision deficiency. To test whether you've made a terrible mistake in color choice, you can run a version of your poster through the free Vischeck service (http://www.vischeck.com/vischeck/vischeckURL.php), or you can download their Photoshop plug-in that does the same thing. In general,

avoid using red and green together, and opt to use symbols and line patterns (e.g., dashed vs. solid) instead of colors for graph elements. See the Rigden article in the Useful Literature section of Collin Purrington's site for an excellent overview of color deficiency conditions and how to design for them.

16. Similarly, if you have a color sensitivity mutation and don't know it, you might inadvertently design posters that are difficult for wild types (the biology terms for those with typical alleles) to interpret. You can test your color perception online (http://www.maniacworld.com/color_blind_test.htm), by the way. White males of European descent are especially encouraged to test themselves.

17. Complete the entire poster on a single platform. Switching from PC to Mac or Mac to PC invites disaster, sometimes in the form of lost image files or garbled graph axes. Even if you are lucky enough to transfer content across platforms, switching in this way often creates printing problems in the future.

18. Give your graphs titles or informative phrases. You wouldn't do this in a manuscript for a journal, but for posters you want to guide the visitor.

19. If you can add miniature illustrations to any of your graphs, do it. Visual additions help attract and inform viewers much more effectively than text alone. Tables benefit from this trick as well.

20. Choose the right type of graph for the data you have and the point you are making.

21. Most graphing applications automatically give your graph an extremely annoying legend that you should immediately delete. Just directly label the different graph elements with the text tool.

22. Acronyms and other shorthands for genotypes, strains, and the like are great when talking to yourself but are terrible for communicating with others. Use general, descriptive terms that would make sense to somebody who is not familiar with your research area. You can always add the strain ID within parentheses: "Control genotype (Col-0)."

23. Y-axis labels aligned horizontally are much, much easier to read, and should be used whenever space allows. Football players and other viewers with fused neck vertebrae will be appreciative.

24. Format axis labels in sentence case (Not in Title Case and NOT IN ALL CAPS). People can read text formatted in sentence case faster.

25. Never give your graphs colored backgrounds, grid lines, or boxes. If your graphing program gives them to you automatically, get rid of them. If you are friends with any of the programmers who made software that has such settings as defaults, break their fingers so they can't code anymore.

26. Never display two-dimensional data in 3D. Three-dimensional graphs look adorable but obscure true differences among bar heights. 3D graphs belong in Time magazine and first grade. Again, if you know anyone on a software design group that makes 3D bar graphs the default output, make them suffer. They make the world suffer, so they deserve whatever you can dream up.

27. Make sure that details on graphs and photographs can be comfortably viewed from 6 feet away. A common mistake is to assume that axis labels, figure legends, and numbers on axes are somehow exempt from font-size guidelines. The truth is that the majority of viewers want to read only your figures.

28. If you include a photograph, add a thin gray or black border to make it more visually appealing and to help viewers quickly identify the extent of the image.

29. Give the source for any image that is not yours, and only use an image (illustration, photograph, etc.) that is fully public domain. When in doubt, ask the author/photographer/illustrator for permission, or buy it.

30. Use web graphics with caution. You need something high resolution so that it doesn't look pixelated (fuzzy) when printed.

31. If you can't find the perfect illustration or photograph, get one made. You can use it in multiple posters, future talks, and even in that great article you're writing for *National Geographic*. There are lots of illustrators and photographers out there for hiring.

32. Don't clutter the top of your poster with logos. If you are required by your mentor to include logos on your poster, put them on the bottom of the poster and make them small.

33. Format your literature-cited contents carefully. References that are only haphazardly formatted mark a poster, and thus you, as unprofessional and incapable of grasping the importance of details. When asking somebody to proof your poster, specifically ask them to be critical of your citation style. Keep your font size the same as the size of the normal body text; shrinking the font looks awful, even if everyone else is doing it.

34. Always write, "data are," not "data is." "Data" is a plural noun ("datum" is the singular). Many people roll their eyes at this advice and say that "data is" is acceptable because that's what people often say. Well, the data might support that fact, but the prevalence of bad grammar doesn't make bad grammar less badder [sic].

35. If you don't know the difference between "effect" and "affect," then don't use those words. The Oatmeal's "10 words you need to stop misspelling" http://www.theoatmeal.com/comics/misspelling) explains the difference nicely.

36. Resist the strange trend to use "woman" as an adjective. For example, don't write, "woman scientist" when you could just say, "female scientist." If you cannot resist the peer pressure, then at least be consistent and write, "man scientist," too.

37. This is probably obvious, but don't plagiarize.

Source: Courtesy of Colin Purrington. Used with permission.

CHAPTER 4

IN-CLASS ACTIVITIES

ICA 4-1

Critique the following student presentation, discussing improvement strategies.

Purpose

In this Business and Engineering Collaborative Project we were given to task to design a product that would make dorm life safer, more enjoyable, or more productive. Our design was driven by our desire to meet customers needs, which would correspond to the ability to make a profit. We will show you all the steps it took to make our product marketable. Finally, we would like to thank everyone for coming and we hope that you agree with us that our product is marketable and of course a masterpiece.

Idea Generation

Here are some of our best ideas....

- Insulated rooms (sound proof)
- Loft (size and cost efficient)
- Air Freshener (Make dorms more livable)
- Universal tack board (Tack board, phone holder, etc)
- Fan and light that doubles as one
- Improving air conditioning unit (Make rooms colder and warmer)
- Universal Bathroom carrier (Holds toothpaste, vitamins, floss, etc.)
- Fold down ironing board (Ironing board that folds down with cover behind it)
- Fold up coffee table (Coffee table that folds up and stores easily)
- Trash Compactor (Compacts trash so that students don't have to empty trash as often)
- Black-out shades (Shades that blocks out all light)
- Safe (Portable and light weight safe that makes storing valuables easy)
- Better Lighting (More lights and in better locations)
- Larger laundry bags (Some kind of bag that stores more clothes, but takes up less space)
- Door Bell (Some kind of portable and cheap door bell)

Screening

In our next step we took all the ideas that we had come up with and narrowed them down to a select three. The reason all the others were thrown out is because some of them already existed on the market, some of them would be too complex to the manufacturer, and some of them we didn't think that we had technical expertise needed to produce them. The three products we chose are....

- A cheap and size efficient loft.
- A Universal cabinet designed to combine all bathroom tasks to one
- A fold down ironing board with storage for cleaning supplies

---

Marketing Programs

- Each person worked 40 hours per week
- Insurance is $800 per quarter
- 30 people were employed for the quarter at $9.00 per hour
- Assembly average completing 90 lofts a day
- Raw materials for the loft cost $15.00 wholesale each
- Circular saw (8' ½") 168 per month = $2016.00 per year http://www.ecmantal.com/metal/products.html
- Drill ½" 144 per month = $1728 per year http://www.ecmantal.com/metal/products.html
- Palm Sander $75 per month = $900 per year http://www.esipowertool.com/equipment_dynaghtp
- Miscellaneous tools and material expenses $5000 per year
- Supervisor is paid $9000 per quarter
- The rent on our building is $2000 per quarter
- Miscellaneous overhead expenses are $200 per quarter
- 12 weeks a quarter (1 week off/quarter)

Production Processes

1st: The wood comes off of the trucks and goes strait to cutting

2nd: After the wood is cut, it is sent to be drilled.

3rd: Then all of the wood is sanded down, to make a nice surface.

4th: The loft is then put together by 5 employees.

5th: While the loft is still together it is tested for any flaws.

6th: The loft is then taken back apart by another 5 employees.

7th: The loft is then packaged and sent to be shipped.

8th: The last and final step, the loft is loaded into trucks.

Conclusion

Developing and introducing new products is frequently time-consuming, expensive, and risky. Thousands of new products are introduced annually but the failure rate is between 60 and 75 percent. If our product was put into production, we are confident that we would not be a part of that percentage. We think we have a feasible and well thought out marketing plan. We would like to thank you for coming and listening to our marketing pitch!

ICA 4-2

Critique the following student presentation, discussing improvement strategies.

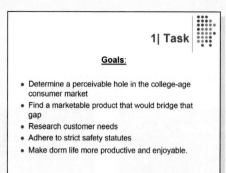

1| Task

Goals:

- Determine a perceivable hole in the college-age consumer market
- Find a marketable product that would bridge that gap
- Research customer needs
- Adhere to strict safety statutes
- Make dorm life more productive and enjoyable.

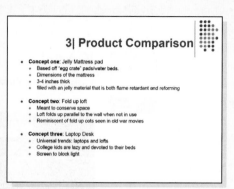

3| Product Comparison

- **Concept one**: Jelly Mattress pad
  - Based off "egg crate" pads/water beds.
  - Dimensions of the mattress
  - 3-4 inches thick
  - filled with an jelly material that is both flame retardant and reforming
- **Concept two**: Fold up loft
  - Meant to conserve space
  - Loft folds up parallel to the wall when not in use
  - Reminiscent of fold up cots seen in old war movies
- **Concept three**: Laptop Desk
  - Universal trends: laptops and lofts
  - College kids are lazy and devoted to their beds
  - Screen to block light

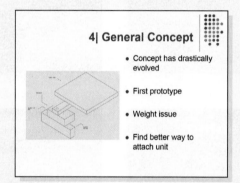

4| General Concept

- Concept has drastically evolved
- First prototype
- Weight issue
- Find better way to attach unit

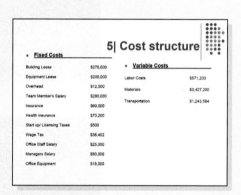

5| Cost structure

Fixed Costs		Variable Costs	
Building Lease	$270,000	Labor Costs	$571,200
Equipment Lease	$200,000	Materials	$3,427,200
Overhead	$12,000	Transportation	$1,243,584
Team Member's Salary	$280,000		
Insurance	$60,000		
Health Insurance	$70,200		
Start up/ Licensing Taxes	$500		
Wage Tax	$36,402		
Office Staff Salary	$25,000		
Managers Salary	$50,000		
Office Equipment	$18,000		

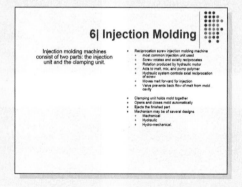

6| Injection Molding

Injection molding machines consist of two parts: the injection unit and the clamping unit.

- Reciprocation screw injection molding machine
  - most common injection unit used
  - Screw rotates and axially reciprocates
  - Rotation produced by hydraulic motor
  - Acts to melt, mix, and pump polymer
  - Hydraulic system controls axial reciprocation of screw
  - Moves melt forward for injection
  - Valve prevents back flow of melt from mold cavity
- Clamping unit holds mold together
- Opens and closes mold automatically
- Ejects the finished part
- Mechanism may be of several designs
  - Mechanical
  - Hydraulic
  - Hydro-mechanical.

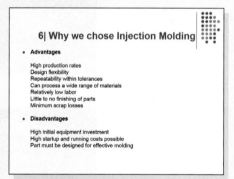

6| Why we chose Injection Molding

- **Advantages**

  High production rates
  Design flexibility
  Repeatability within tolerances
  Can process a wide range of materials
  Relatively low labor
  Little to no finishing of parts
  Minimum scrap losses

- **Disadvantages**

  High initial equipment investment
  High startup and running costs possible
  Part must be designed for effective molding

The following three ICAs are presentations you may be asked to make in this course. For your presentation, you may be required to choose one of the topics from the list provided, your instructor may specify different topics, or you may even be allowed to choose one of your own. Your instructor will make clear which options you have. Remember, you are giving a *technical* presentation. *Any discussion of the topic that is considered inappropriate by the instructor will be subject to a grade reduction.*

ICA 4-3

The purpose of this activity is to give you practice getting up in front of a group. The topics are informal, meant to simply to help you gain confidence in public speaking. The presentation topic will be assigned to you at random shortly before your scheduled presentation; your instructor will determine how shortly before. The presentation should last *one minute*. For your presentation, here are some possible approaches you may want to take. Suppose your topic is "WHALE."

Make it informative.
To give an informative talk, make a mental note of important things you know about whales. They are big. Some species are almost extinct. They are mammals. The blue whale is the largest animal that has ever lived on Earth. They were hunted for their blubber and oil that was used in lamps before electric lights. They hold their breath while under the water.

Tell a story.
If you know *Moby Dick*, you could give a brief outline of the most important points or of a particular scene in the book. You could tell of a visit to an aquarium where you saw how whales behave in that artificial habitat or of a whale watching trip where you saw how whales behave in the wild. You could make up a short bedtime story that you relate as one you would tell your child or that was told to you as a child.

Be innovative.
You could do something like this: Whale is spelled W-H-A-L-E. The W stands for "water" where the whales live (talk about the ocean for a while); H stands for "huge," which is what we think of when we say whale; A stands for "animal" and the whale is the largest; L stands for "large," which is a lot like huge; and E stands for "eating" since the whale has to eat a lot, or E could stand for "enormous" like huge and large.

Or

You could do something like this: I dreamed about a whale that was in the parking lot. It kept smiling, and let out an enormous burp every few minutes. Being an animal lover, I went up next to the whale and stroked its sides. I was amazed at how smooth it was.

Or

You could do something like this: I have always wanted to scuba dive and in particular to ride on the back of a killer whale. I know that it would be dangerous, but it would be something I could remember the rest of my life.

ICA 4-3: Presentation Topic Suggestions

- Person who impressed you
- A hobby
- Favorite team mascot
- Favorite course
- Importance of cars
- Importance of space exploration
- Why engineers are important
- Why calculus is important
- The worst insect
- How I would improve my school

- Coolest animal
- A childhood memory
- Someone I admire and why
- The best car
- Why I came to this college
- My dream job (after college)
- Why I want to be an engineer
- Place I would like to visit
- If I had a million dollars
- Favorite vacation

- Why I would like to be a professor
- The best thing about being in college
- What kind of boss I would make
- Something I learned this week
- The best thing about my school
- Advice for incoming first-year students
- A good book I have read
- My favorite year in grade school
- The farthest I have ever been from my hometown

ICA 4-4

The purpose of this activity is to discuss the graphical representation of various phenomena, such as "A total eclipse" or "Using a toaster." The presentations should last *one minute, 30 seconds.*

Elsewhere in this text, much time is spent on graphing mechanics and interpretation. You are to choose a topic from the following list and represent it graphically. A variety of presentations of the available data that tell different things are possible. Do not forget to use all your senses and imagination.

You must prepare a proper graph, incorporated in PowerPoint, and use it during your presentation. You are not allowed to copy a premade graph from the Internet! In your speech, you should discuss the process, explain how information is shown on the graph, and be prepared to answer questions about your process. *You are limited to three slides*: (1) a title slide, (2) an introductory slide, and (3) a graph. If needed, you will be allowed a single 3 × 5-inch index card with notes on one side only.

ICA 4-4: Presentation Topic Suggestions

- Moving a desk down a set of stairs
- Letting go of a helium balloon
- Feedback from an audio system
- A cow being picked up by a tornado
- A kangaroo hopping along
- A glass of water in a moving vehicle
- People in Florida
- People in Michigan
- The flight of a hot air balloon
- Using a toaster
- A train passing
- Dropping ice in a tub of warm water
- Slipping on a banana peel
- Sound echoing in a canyon
- Detecting a submarine using sonar
- Pouring water out of a bottle
- Boiling water in a whistling teapot
- Hammering nails
- Spinning a Hula-Hoop™
- Playing with a yo-yo
- Snow blowing over a roof
- Climbing a mountain
- Pumping air into a bicycle tire
- The plume from a smokestack
- Formation of an icicle
- A rabbit family
- Firing a bullet from a rifle
- Hiking the Appalachian Trail
- Eating a stack of pancakes
- A solar eclipse

- Taking a bath
- Throwing a ball
- Driving from Clemson to Greenville
- Diving into a swimming pool
- A person growing up
- Burning a pile of leaves
- Driving home from work
- An oak tree over the years
- Airplane from airport to airport
- Football game crowd
- Train passing through town
- A coastal river
- The moon
- Daily electric power consumption
- A typical day
- A thunderstorm
- A day in the life of a chicken
- Popping corn
- Feeding birds at a bird feeder
- Skipping a stone on water
- A burning candle
- Pony Express
- Talking on a cell phone
- Using Instant Messenger
- An engineer's salary
- Baseball in play
- Exercising
- Running of the bulls
- Drag racing

- D.O.R. possum ("dead on the road")
- Political affiliations
- Traffic at intersections
- River in a rainstorm
- A tiger hunting
- Hair
- Grades in calculus class
- Forest fire
- Cooking a Thanksgiving turkey
- Brushing your teeth
- Baking bread
- Eating at a fast-food restaurant
- Studying for an exam
- Power consumption of your laptop
- Power usage on campus
- Oil supply
- Air temperature
- Student attention span during class
- Baseball
- A diet
- Sleeping
- Bird migration
- Strength of concrete
- Human population
- A mosquito
- A video game
- Taking a test
- A trip to Mars
- The North Pole
- Space elevators

ICA 4-5

This presentation covers topics related to your future major. The presentations should last *three minutes*.

In 2000, the National Academy of Engineering compiled a list of "Greatest Engineering Achievements of the 20th Century." The following list contains some of the achievements listed, as well as some of the greatest failures of the past century. The full list can be found at http://www.greatachievements.org.

It is your responsibility to research the topic you choose so that you can intelligently discuss it in reference to the engineering involved and how it affects everyday life, both then and now.

You must prepare a PowerPoint presentation and use it in this speech. You should have between five and eight slides of which the first must be a title slide, and the last must be a slide listing references. Other visual aids may be helpful and are encouraged. If needed, you will be allowed a single 3 × 5-inch index card with notes on one side only.

ICA 4-5: Presentation Topic Suggestions

Electricity
- Grand Coulee Dam (1942)
- Electric light bulbs

Automobiles
- Mass production (1901: Oldsmobile; 1908: Ford)
- Octane ratings (1929)
- Goodrich "tubeless" tires (1947)

Airplanes
- Hindenburg (1937)
- Concord

Water Supply and Distribution
- Purification of water supply (1915: Wolman)
- Cuyahoga River pollution (1970s)
- Hinkley, California

Electronics
- Integrated circuits (1958: Kilby)
- Handheld calculator (1967: TI)

Radio and TV
- Color television (1928; 1954)
- Phonograph (Edison)

Agricultural Mechanization
- Mechanical cotton picker (1949: Rust)
- Irrigation equipment (1948: Zybach)

Computers
- Language compiler (1952: Hopper)
- Transistor (1947)

Telephones
- Touch-tone dialing (1961)

Air Conditioning and Refrigeration
- Air conditioning (1902: Carrier)
- Freezing cycle for food (1914: Birdseye)

Highways
- Pennsylvania turnpike
- Tacoma Narrows Bridge
- Golden Gate Bridge
- Chunnel

Spacecraft
- Space shuttle *Challenger* explosion
- Hubble telescope
- *Apollo 1* and *Apollo 13* failures
- *Gemini* spacecraft

Internet
- E-mail (1972: Tomlinson)
- WWW (1992)

Imaging
- Sonar (1915: Langevin)
- Radar (1940)
- Ultrasound (1958)
- MRI (1980)

Household Appliances
- Dishwasher (1932)
- Clothes dryer (1935: Moore)
- Microwave (1946)

Health
- Defibrillator (1932: Kauwenhoven)
- Artificial heart (1980: Jarvik)

Petroleum
- Distillation (1920s: Fischer and Tropsch)
- Ethyl gas (1921)
- Alaskan pipeline

Laser and Fiber Optics
- Chemical vapor deposition (1974)
- Optical fibers (1970: Maurer, Corning)

Nuclear Technologies
- Van de Graff generator (1937)
- Manhattan Project (1939–1945)
- Three Mile Island power plant (1979)

High-Performance Materials
- Vulcanization of rubber (1926)
- Polyvinyl chloride—PVC (1927)
- Teflon (1938: Plunkett)
- Kevlar (1971: Kwolek)

ICA 4-6 and 4-7 were written to address the question of what obligations humans may have to protect extraterrestrial life if it is discovered.

ICA 4-6

Critique the following writing assignment, discussing improvement strategies.

If life were to be discovered elsewhere in the universe, it would be our duty to protect that life only to the extent to which we do not compromise our own stability. The nation as a whole currently already has debt issues and is going through a depression. America could not afford to spend money on a life when they are struggling to take care of themselves. According to Hodges, "America has become more a debt 'junkie'—than ever before with total debt of $53 Trillion—and the highest debt ratio in history." (2) Globally, there are people who are dying from starvation each day. "On Tuesday September 11, 2001, at least 35,615 of our brother and sisters died from the worst possible death, starvation. Somewhere around 85% of these starvation deaths occur in children 5 years of age or younger." (1) What would make this new life form a higher priority than the neighbors right around the corner? If they have managed to survive this long we shouldn't worry about helping them until we have taken care of ourselves. Even if they are extremely advanced with many discoveries, we have not made, what good would it do to spend billions on reverse engineering but leave our own world in desperation. Now if we were to reach a point in the future in which we could help them; we would naturally have to approach with caution. Politically one of the worst first impressions is when new people meet and one group starts a biological plague. In 1493 right after Columbus came to America, he went back to Hispaniola, bringing livestock in order to start a colony there. "Influenza, probably from germs carried by the livestock, swept through the native people, killing many of them. Modern researchers believe that American Indian traders carried the disease to Florida and throughout the Caribbean." (3) We would not know what would hurt them as well as what could hurt us. In order to avoid this terrible introduction, we would need to do research with probes and satellites. These observations from afar would make a seamless introduction possible. After the introduction it would be top priority to keep diplomatic relations intact. Finding another life out in space would be great, but not if we made them our enemies.

ICA 4-7

Critique the following writing assignment, discussing improvement strategies.

For centuries, people have wondered if we were all alone in our universe. Still today, even with our advancements in science we still have not been able to find life anywhere else. Recently the Phoenix Lander *discovered what seems to be incontrovertible evidence that water exists on Mars. Since life is dependent on water eye brows have again been raised about the probability that life may exist, or has existed on Mars. So what if life is discovered somewhere other than on earth? Are we obligated to protect that life? Does that obligation depend on how advanced the life is? How do we stop ourselves from not destroying the non-terrestrial life? If life is discovered somewhere other than earth, some people will feel threaten, others will feel that maybe somehow human life could be supported by the non-terrestrial's planet and that life is elsewhere out there, maybe life like ours. As far as our obligation to protect that life, it is recommended that we do not interfere. We are having problems with our own planet, people pollute, litter, kill wildlife, start wildfires, cut down forests, and destroy whole ecosystems. Imagine what we will do to life if it doesn't matter to us if it lives or not. No matter how advance the life is we should leave the life to itself, unless the life was able to communicate with us somehow. The only way to keep ourselves from destroying the non-terrestrial life is to just to leave it alone. If we leave it, it wouldn't be our fault if it survives or eventually dies off. Yet still we haven't been able to find life anywhere besides earth so until we do we should try finding ways to turn our planet in the right direction before we lose what supports us.*

ICA 4-8 and 4-9 were written to address the question of whether we should pursue manned exploration of space or restrict such activities to unmanned robotic devices.

ICA 4-8

Critique the following writing assignment, discussing improvement strategies.

Manned, unmanned space travel has been a big controversy for many years; it's too risky or space bots gets just as much done as humans. I agree 100% that robots are far better off into space than humans; manned space travel is too dangerous. Almost all the money for such a mission would be spent simply to keep the people alive. (John Tierney) Space bots are more reliable, don't get tired, and do what it's programed to do, but more proficient. Things humans need for survival includes: air, food, water, things we must take with us for space travel, which adds to the cost for each individual we fly out into space. While space bots don't need any of these sources for survive, which makes space travel less expensive and more productive. Also the bots can retrieve more data because there aren't any stopping periods for breaks or any other reasons for that matter. It cost about 1.3 billion dollars per shuttle launched out into orbit, while it cost far more per individual we send out into space. Manned space travel is not all bad; things humans can do like, making quick and on the spot decision or human senses to evaluate our surroundings are things robots can't imitate. It's nothing like being somewhere physically and knowing what's there instead of watching it through another pair of eyes. By using unmanned space travel, in reality we can only see what the robot sees, but is a human life worth the risk just for data. A human life is priceless, it like time when it's gone; it's gone, so why not send robotic material into space to collect data. It's only scrap metal, but the life of any human is far more valuable just for the expense of useless data.

ICA 4-9

Critique the following writing assignment, discussing improvement strategies.

Since the early 1960s, humans have been venturing beyond the Earth's atmosphere, a few times as far as the moon, a quarter of a million miles away. NASA currently plans to return humans to the moon by 2020, and plans Mars missions after that. Much criticism has been leveled at the entire idea of manned spaceflight, claiming that unmanned craft can do essentially all of the jobs people can with less cost and less risk. There are of course two sides to the situation. The idea for space exploration and those against it.

Space flight is a very integral part of exploration and adventure. Discovering the worlds around us I a very key part to discovering space. NASA's mission is to pioneer the future in space exploration, scientific discovery and aeronautics research. To do that, thousands of people have been working around the world—and off of it—for almost 50 years, trying to answer some basic questions. What's out there in space? How do we get there? What will we find? What can we learn there, or learn just by trying to get there, that will make life better here on Earth? (NASA). The ideal of many working together to the common goal of space exploration makes it a very important part. Manned spaceflight can be very dangerous. Many deaths and tragedies have occurred because of the exploration of space.

I think that space should still be explored by mankind. With technology advances mentioned in the first 25 years of its existence, NASA conducted five manned spaceflight programs: Mercury, Gemini, Apollo, Skylab, and Shuttle.

The latter four programs produced spacecraft that had on-board digital computers. The Gemini computer was a single unit dedicated to guidance and navigation functions. Apollo used computers in the command module and lunar excursion module, again primarily for guidance and navigation. Skylab had a dual computer system for attitude control of the laboratory and pointing of the solar telescope. NASA's Space Shuttle is the most computerized spacecraft built to date, with five general-purpose computers as the heart of the avionics system and twin computers on each of the main engines. The Shuttle computers dominate all checkout, guidance, navigation, systems management, payload, and powered flight functions. The computers helped and as they advance manned spaceflight gets better. In the long run the risk is worth the gain for knowledge.

REVIEW QUESTIONS

1. When thinking about how you will phrase things for a talk, you should
   (A) Mirror your exact wording in your slides.
   (B) Be sure to use words that demonstrate your mastery of the topic.
   (C) Find a way to capture the attention of the audience.
   (D) Split your topic into as many points as you can.

2. Which statement is NOT good advice for preparing visual aids?
   (A) Choose a large font size.
   (B) Select high-contrast colors.
   (C) Fill empty space with images.
   (D) Use graphs rather than tables.

3. Which of the following sentences is the best example of good technical writing?
   (A) 43 of the bolts were .025 inches long.
   (B) Forty-three of the bolts were .025 inches long.
   (C) 43 of the bolts were 0.025 inches long.
   (D) Forty-three of the bolts were 0.025 inches long.

4. Which of the following sentences is the best example of good technical writing?
   (A) A water hammer happens when a surge in pressure reaches the end of a pipe; since water is incompressible below the saturated liquid/vapor line, the momentum is stopped over a short distance and the resulting forces are large.
   (B) A surge in water pressure can cause a large force. This force, called a water hammer, will cause the pipes to fail, if left unchecked.
   (C) A water hammer can easily break steel pipes.
   (D) A surge in water pressure can cause a large force. This force, called a water hammer, can result in amazing failures. It is one of my favorite modes of failure.

5. In researching a topic, which method is NOT a good way to validate the information on a webpage?
   (A) Search for other webpages on the same topic.
   (B) Follow the links on that website to see their supporting information.
   (C) Search for articles that have been peer reviewed.
   (D) Find official webpages on the topic, such as a .gov or company website.
   (E) Use the library to find books on the topic.

6. When writing a memo or short report, where should you place figures and tables?
   (A) Introduction
   (B) Results
   (C) Discussion
   (D) Appendix

7. When writing a memo or short report, where should you write about the importance of your graphs?
   (A) Introduction
   (B) Results
   (C) Discussion
   (D) Summary

Writing Assignments

For each question, write a response according to the directions given by your instructor. Each response should contain correct grammar, spelling, and punctuation. Be sure to answer the question completely, but choose your words carefully so as to not exceed the word limit if one is given. There is no right or wrong answer; your score will be based upon the strength of the argument you make to defend your position.

8. In August 2007, the space shuttle *Endeavor* suffered minor damage to the heat shield tiles during liftoff. After the *Columbia* burned up on reentry in 2003 due to damaged tiles, NASA developed a protocol for repairing damaged tiles while in orbit. After much consideration, NASA decided *not* to attempt to repair the tiles on *Endeavor* while in orbit on this mission.

Find a minimum of three different references discussing this incident and NASA's decision. Summarize the reasons both for and against repairing the tile while in orbit. If you had been on the team making this decision, would you have argued for or against the repair and why?

9. Research the failure of the Teton Dam. Describe what you think are the fundamental ethical issues involved in the failure. In terms of ethics, compare the Teton Dam failure in 1976 to the failure of levees in New Orleans in 2005.

10. In recent years, many studies have been conducted on the use of cell phones while driving. University of Utah psychologists have published a study showing that motorists who talk on handheld or hands-free cellular phones are as impaired as drunken drivers.

> *We found that people are as impaired when they drive and talk on a cell phone as they are when they drive intoxicated at the legal blood-alcohol limit of 0.08 percent, which is the minimum level that defines illegal drunken driving in most U.S. states," says study co-author Frank Drews, an assistant professor of psychology. "If legislators really want to address driver distraction, then they should consider outlawing cell phone use while driving.*

Strayer and Drews, "Human Factors." *Journal of the Human Factors and Ergonomics Society,* Summer 2006.

Another way to approach this problem is to force manufacturers of cell phones to create devices on the phones that would restrict usage if the user is driving, much like a safety on a gun trigger. Research and report current statistics on cell phone or other texting device usage during driving. Discuss the hazards (or lack of hazards) of driving while using a cell phone or other texting device. Discuss and justify whether cell phone manufacturers should be required to provide safety features on their devices to prevent usage while driving.

11. According to an article in the *Christian Science Monitor* (January 16, 2008), environmentalists claim that a development around a remote lake 140 miles north of Augusta, Maine, would emit 500,000 tons of carbon dioxide over 50 years, including estimated emissions from cars traveling to and from the development. An environmental group has presented this carbon footprint to the state and is requesting that the impact on the environment become part of the process for granting development permits.

As many as 35 states have adopted climate-action plans, but there are few cases like this in which environmental impact factors into government approval of land development. This could have a significant effect on engineers involved with land development, structures, landuse planning, or environmental impact assessments in the future. The original article can be found at http://www.csmonitor.com/2008/0116/p01s04-wogi.html.

Discuss both sides of this issue, and take a stance for or against mandating carbon footprint assessment for new developments. Justify your position, including information from at least three sources. Ideas to include in your essay are cases in other states, climate-action plans, calculation of a carbon footprint or the land-development approval process. Most importantly, consider how an engineer would view this issue.

Thank you to Dr. Lisa Benson for contributing this assignment.

12. From 2005 to 2009, much of the Southeast experienced a serious rainfall deficit. Lake levels were at record lows, water restrictions were debated and enacted, and the ecology of the area was changed. Amid all of this are nonessential services that are extremely heavy users of water. Perhaps the most egregious examples are sports fields, and in particular, golf courses. On average, an 18-hole golf course in the United States uses about 100 million gallons per year.

Summarize the arguments both in favor and against the heavy use of water for nonessential services, particularly in areas where water supply is limited. Consider the following:

- If you were in a position to recommend legislation for such water use, what policies would you recommend and why? Consider this from an engineer's point of view.
- Consider the impact on local environments for building dams for water supplies and hydropower. As an engineer, what would you consider to be the benefits and drawbacks of this type of project and why?

13. Since the early 1960s, humans have been venturing beyond Earth's atmosphere, a few times as far as the moon, a quarter of a million miles away. NASA currently plans to return humans to the moon by 2020, and plans Mars missions after that.

Much criticism has been leveled at the entire idea of manned spaceflight, claiming that unmanned craft can do essentially all the jobs people can with less cost and less risk.

Summarize the arguments on both sides of this issue.

14. In summer 2008, the *Phoenix Lander* discovered what seems to be incontrovertible evidence that water exists on Mars. Since life as we know depends on water (for a variety of reasons), this discovery raises the probability that life may exist, or did exist in the past, on Mars.

If life is discovered somewhere other than on Earth, what are our obligations to protect that life? Does that obligation depend on how "advanced" we perceive that life to be? What steps, if any, should be taken to ensure that we do not destroy the nonterrestrial life?

15. Advances in genetic engineering may make it possible to bring extinct species back to life.

(a) Should we attempt to restore populations of recently extinct animals, such as the passenger pigeon, Carolina parakeet, ivory-billed woodpecker, Tasmanian tiger, or Formosan clouded leopard, for which there are many preserved specimens from which DNA could be acquired? There is currently an effort to obtain enough DNA to reestablish a population of woolly mammoths. What are the ethical issues involved with bringing back a mammal extinct for 3500 years?

(b) It was announced in February 2009 that a first draft of the genome of *Homo sapiens neanderthalensis* (Neanderthal), comprising about two-thirds of the base pairs, has been completed. If future advances make it possible, should we attempt to bring Neanderthals back to life?

16. If you could go back in time and be part of any engineering achievement in the past, what would you choose?

Your choice must involve something that was accomplished before 1970, although it might still be in use today. You may go as far back in time as you wish, but not more recently than 1970.

Your essay must include the following:

- Why did you choose this specific thing?
- Discuss the actual development of the item in question. Include such issues as why the item was desired, what problems the designers confronted, and specific design decisions that were made.
- What effects has the development of this device (or process) had on society and the planet, including positive, negative, and neutral effects?
- A bibliography with at least three distinct entries. All may be online references, but no more than one may be Wikipedia or similar sites.

17. Choose a bridge innovation to research from the list of potential topics given in the alphabetic list that follows, and write a memo or presentation that includes the following:

- The name and location of the structure, when it was constructed, and other significant attributes
- If possible, a photo or sketch
- A summary table of the design features. Potential features to consider including, if appropriate:
  - Type of bridge design (arch, beam, suspension, etc.)
  - Dimensions (span, height)
  - Unique design features
  - Cost
  - Awards or superlatives (longest, highest, most cost-effective, costliest)
- A discussion of the aspects of the bridge design that contributed to its success (materials, structural design, geographical, topographical, or climatic challenges, aesthetic qualities, etc.)

  (a) Akashi Kaikyo Bridge, linking the islands of Honshu and Shikoku in Japan
  (b) Cooper River Bridge in Charleston, South Carolina
  (c) Forth Bridge, Scotland
  (d) Gateshead Millennium Bridge, spanning the River Tyne in England
  (e) Hanzhou Bay Bridge, China
  (f) Humber Bridge, England
  (g) Lake Pontchartrain Causeway, Louisiana
  (h) Mackinac Bridge, Michigan
  (i) Millau Viaduct, Millau, France
  (j) Natchez Trace Bridge in Franklin, Tennessee
  (k) Penobscot Narrows Bridge, Maine
  (l) Rio-Antirio Bridge, Greece
  (m) Sundial Bridge at Turtle Bay in Redding, California
  (n) Sunshine Skyway Bridge in Tampa, Florida
  (o) Sydney Harbor Bridge, Australia
  (p) Tower Bridge, London
  (q) Woodrow Wilson Bridge, Washington, D.C.
  (r) Zakim Bunker Hill Bridge in Boston in Massachusetts

18. Choose a major transportation structure failure to research from the list of potential topics in the alphabetic list that follows, and write a memo or presentation that includes the following:

- The name and location of the structure, when it was constructed, and other significant attributes
- If possible, a photo or sketch
- A summary of the event(s) surrounding the structure's failure, including dates. Discuss aspects of the design that contributed to its failure (materials, structural flaws, misuse, improper maintenance, climatic conditions, etc.). If it was rebuilt, repaired, or replaced, how was the design modified or improved?

  (a) Angers Bridge over the Maine River in Angers, France
  (b) Autoroute 19 Overpass, Quebec
  (c) Arroyo Pasajero Twin Bridges in Coalinga, California
  (d) Banqiao Dam, China
  (e) Charles de Gaulle Airport, France
  (f) Hartford Civic Center, Connecticut
  (g) Hyatt Regency hotel walkway in Kansas City, Missouri
  (h) Kemper Arena, Missouri
  (i) L'Ambiance Plaza in Bridgeport, Connecticut
  (j) Loncomilla Bridge, Chile
  (k) Mianus River Bridge in Greenwich, Connecticut

(l) Millennium Footbridge, London
(m) Sampoong Department Store, South Korea
(n) Seongsu Bridge, spanning the Han River in Seoul, South Korea
(o) Sgt. Aubrey Cosens VC Memorial Bridge in Latchford, Ontario
(p) Silver Bridge, between Point Pleasant, West Virginia, and Kanauga, Ohio
(q) The Big Dig in Boston, Massachusetts
(r) West Gate Bridge in Melbourne, Victoria, Australia

19. Choose an environmental issue to research from the following alphabetic list of potential topics, and write a memo or presentation that includes:

- A clear definition of the problem
- If possible, a photo or sketch
- Discussion of the issue; potential features to consider include, if appropriate:
  - Affected areas (local, region, continent, global)
  - Causes, sources
  - Effects (human, animal, vegetation, climate)
  - Prevention or reversal options (including feasibility and cost)
  - Important legislation (or lack thereof)

The topic should be discussed from an *engineering* viewpoint.

(a) Acid rain
(b) Air purification
(c) Bioremediation
(d) Clean Air Act
(e) Cuyahoga River pollution/Clean Water Act
(f) Deforestation
(g) Erosion control
(h) Groundwater pollution
(i) Kyoto Protocol
(j) Lake eutrophication/hypoxia
(k) Loss of wetlands/ecosystems/pollinators
(l) Ocean acidification/pollution
(m) Purification/safe drinking water
(n) RCRA (Resource Conservation and Recovery Act)
(o) Smog/particle reduction
(p) Soil contamination
(q) Solar energy
(r) Superfund/CERCLA
(s) Water Pollution Control Act
(t) Wind energy

20. Technology promises to revolutionize many aspects of life and society. In a memo or presentation, discuss the advantages and disadvantages of one or more of the following. You may need to research some of these topics first.

(a) Nano-robots designed to destroy cancer cells
(b) More powerful particle accelerators for probing the structure of matter
(c) The James Webb Space Telescope
(d) Cloaking technology at visible wavelengths
(e) Practical antigravity devices
(f) Drugs to inhibit soldiers' fears
(g) Nano-devices to rewrite an organism's DNA
(h) Practical (time and cost) interplanetary travel
(i) Artificial chlorophyll to harness energy from the sun
(j) Genetically engineered microbes to attack invasive species
(k) Direct neural interface to computers
(l) Replacement human limbs with full functionality

    (m) Terraforming Mars, Venus, or the moons of Jupiter or Saturn
    (n) Quantum computers
    (o) Room-temperature superconductors
    (p) Materials harder than diamonds
    (q) Artificial spider silk
    (r) Full-color night-vision goggles
    (s) Home-scale fusion reactors
    (t) Fully automated cars
    (u) Space elevators
    (v) Bionic implants for enhancing sight, sound, and/or smell

21. One critical writing skill is supporting your claims. This need not be long, written paragraphs. In some cases, a claim can be backed by a credible citation. In others cases, a graph is sufficient, and in still other cases, a longer logical argument is needed.

Find evidence for claims supporting or refuting the following statements assigned by your instructor. If you identify sources that have claims related to the statement that you disagree with, provide evidence to support your disagreement.

    (a) Compact fluorescent lightbulbs are much better for the planet.
    (b) Electric cars allow transportation without harmful carbon emissions.
    (c) Solar energy cannot ever be of any use, because solar cells produce less energy in their usable lifetime than it takes to manufacture them.
    (d) Motorcycles are better for the environment than cars.
    (e) Genetically modified foods are perfectly safe.
    (f) It is unsafe to live near power transmission lines.

CHAPTER 5

Estimation

Enrico Fermi was a Nobel laureate and one of many brilliant scientists and engineers involved in the Manhattan Project, which developed the first nuclear weapons during World War II. He wrote the following after witnessing the first test of an atomic bomb, called the Trinity Test (see Figure 5-1). Of particular note is the final paragraph.

On the morning of the 16th of July, I was stationed at the Base Camp at Trinity in a position about ten miles from the site of the explosion.

The explosion took place at about 5:30 A.M. I had my face protected by a large board in which a piece of dark welding glass had been inserted. My first impression of the explosion was the very intense flash of light, and a sensation of heat on the parts of my body that were exposed. Although I did not look directly towards the object, I had the impression that suddenly the countryside became brighter than in full daylight.

I subsequently looked in the direction of the explosion through the dark glass and could see something that looked like a conglomeration of flames that promptly started rising. After a few seconds the rising flames lost their brightness and appeared as a huge pillar of smoke with an expanded head like a gigantic mushroom that rose rapidly beyond the clouds probably to a height of the order of 30,000 feet. After reaching its full height, the smoke stayed stationary for a while before the wind started dispersing it.

About 40 seconds after the explosion the air blast reached me. I tried to estimate its strength by dropping from about six feet small pieces of paper before, during and after the passage of the blast wave. Since at the time there was no wind I could observe very distinctly and actually measure the displacement of the pieces of paper that were in the process of falling while the blast was passing. The shift was about 2½ meters, which, at the time, I estimated to correspond to the blast that would be produced by ten thousand tons of T.N.T.

*Citation: U.S. National Archives, Record Group 227, OSRD-S1 Committee, Box 82 folder 6, "Trinity."

Source: Everett Historical/Shutterstock

Figure 5-1 The first atomic bomb test, Alamogordo, New Mexico.

Before the test, no one knew what would happen. Speculation among the many people involved concerning the results of the test ranged from nothing (no explosion at all) to setting the planetary atmosphere on fire and destroying all life on Earth. When all the data from the blast were analyzed, the true strength of the blast was calculated to be 19 kilotons. By simply observing the behavior of falling bits of paper 10 miles from ground zero, Fermi's estimation of 10 kilotons was in error by less than a factor of 2.

After the war, Fermi taught at the University of Chicago, where he was noted for giving his students problems in which so much information was missing that a solution seemed impossible. Such problems have been named **Fermi problems**, and in general, they require the person considering them to determine an answer with far less information than would really be necessary to calculate an accurate value. Engineers are often faced with solving problems for which they do not have all the information. They must be adept at making initial estimates. This skill helps them identify critical information that is missing, develop their reasoning skills to solve problems, and recognize what a "reasonable" solution will look like.

Most practical engineering problems are better defined than Fermi problems and can be estimated more easily and accurately. The following are just a few examples of real engineering problems. See if you can estimate answers for these problems. These should be done without reference to the web or any other source of information.

- How many cubic yards of concrete are needed to pave 1 mile of interstate highway (two lanes each direction)?
- How many feet of wire are needed to connect the lighting systems in an automobile?
- What is the average flow rate in gallons per minute of gasoline moving from the fuel tank to the fuel injectors in an automobile cruising at highway speed?

An accomplished engineer knows the answer to most problems before doing any calculations. This does not mean an answer to three significant figures, but a general idea of the range of values that would be reasonable. For example, if you throw a baseball as high as possible, how long will it take for the ball to hit the ground? Obviously, an answer of a few milliseconds is unreasonable, as is several months. Several seconds seems more realistic. When you actually do a calculation to determine this time, you should ask yourself, "Is my answer reasonable?"

Sample Fermi Problems

- Estimate the total number of hairs on your head.
- Estimate the number of drops of water in all of the Great Lakes.
- Estimate the number of piano tuners in New York City.

EXAMPLE   5-1

Every year, numerous people run out of fuel while driving their vehicle on the road. Determine how many gallons of gasoline are carried to vehicles with empty fuel tanks each year in the United States so that the vehicles can be driven to the nearest gas station. Do this without reference to any other material, such as the Internet or reference books.

Estimations

In almost all cases, the first step is to estimate unknown pieces of information that are not available. In general, there are numerous paths to a solution, and different people may arrive at different answers. Often, the answer arrived at is only accurate to within an order of magnitude (a factor of 10) or less. Nonetheless, such problems can provide valuable insight not only into the problem itself but also into the nature of problem solving in general.

Remember, these are estimates, not accurate values. Someone else making these estimates might make different assumptions.

- *Number of people in the United States: 500,000,000 persons.*
- *Fraction of people in the United States that drive: seven drivers per 10 persons.*
- *Average times a person runs out of gas per year: one "out of gas" per 4 years per driver.*
- *Fraction of "out of gas" incidents in which gas is brought to the car (rather than pushing or towing the car to a station): 24 "bring gas to car" per 25 "out of gas."*
- *Average amount of gas carried to car: 1.5 gallons per "bring gas to car."*

Calculation

This is where you combine your estimates to arrive at a solution. In the process, you may realize that you need further information to complete the computation.

$$\text{Drivers in United States} = (5 \times 10^8 \text{ people})\left(\frac{7 \text{ drivers}}{10 \text{ people}}\right) = 3.5 \times 10^8 \text{ drivers}$$

$$\text{Number out of gas per year} = (3.5 \times 10^8 \text{ drivers})\left(\frac{1 \text{ out of gas}}{(4 \text{ years})(1 \text{ driver})}\right)$$

$$= 8.75 \times 10^7 \frac{\text{out of gas}}{\text{year}}$$

$$\text{Number bring gas to car per year} = \left(\frac{8.75 \times 10^7 \text{ out of gas}}{\text{year}}\right)\left(\frac{24 \text{ bring to car}}{25 \text{ out of gas}}\right)$$

$$= 8.4 \times 10^7 \frac{\text{bring gas to car}}{\text{year}}$$

$$\text{Amount of gas to cars} = \left(\frac{8.4 \times 10^7 \text{ bring gas to car}}{\text{year}}\right)\left(\frac{1.5 \text{ gallons of gas}}{\text{bring gas to car}}\right)$$

$$= 1.26 \times 10^8 \frac{\text{gallons}}{\text{year}}$$

Thus, we have estimated that about 125 million gallons of gas are taken to "out of fuel" vehicles each year in the United States. It would be perfectly valid to give the answer as "about 100 million" gallons, since we probably have only about one significant digit worth of confidence in our results, if that.
    A few things to note about this solution:

- *"Units" were used on all numerical values, although some of these units were somewhat contrived (e.g., "out of gas") to meet the needs of the problem. Keeping track of the units is critical to obtaining correct answers and will be highly emphasized, not only in this text but also throughout your engineering education and career.*
- *The units combine and cancel according to the regular algebraic rules. For example, in the first computation, the unit "persons" appeared in both the numerator and the denominator, and thus canceled, leaving "drivers."*
- *Rather than the computation being combined into one huge string of computations, it was broken into smaller pieces, with the results from one step used to compute the next step. This is certainly not an immutable rule, but for long computations it reduces careless errors and makes it easier to understand the overall flow of the problem.*
- *Fermi problems, like other engineering work, can benefit from the perspective of others. In this case, it is likely that if a team were working on this problem, someone on the team would know that the population of the United States is closer to 300,000,000.*

Plastic resin (shown in the picture, in pellet form) is used in many types of manufacturing methods including injection molding, extrusion, and blow molding. Injection molding describes the process by which resin is melted and "injected" into a closed mold, then cooled forming the final part. Extrusion is a continuous process by which resin is melted and pushed through an open mold to create shapes like a pipe or rod.

Most cars today use plastic fuel tanks made by blow molding. In blow molding, the resin is melted and pushed through an extrusion head that forms the plastic into a hollow shape. The shape is then pressurized with air and cooled in a mold to form the part. Most fuel tanks are formed with six layers of various plastic to increase toughness and eliminate permeation of the fuel through the tank.

Photo courtesy of E. Stephan

5.1 General Hints for Estimation

LEARN TO: Determine how much accuracy is needed in a particular situation
Identify the important variables affecting an estimate

As you gain more knowledge and experience, the types of problems you can estimate will become more complicated. Here we give you a few hints about making estimates.

- **Try to determine the accuracy required**.
  - Is order of magnitude enough? Is $\pm 25\%$?
  - What level of accuracy is needed to calculate a satellite trajectory?
  - What level of accuracy is needed to determine the amount of paint needed to paint a specified classroom?

The term "**orders of magnitude**" is often used when comparing things of very different scales, such as a small rock and a planet. By far the most common usage refers to factors of 10; for example, three orders of magnitude refer to a difference in scale of $10^3 = 1000$. If we wanted to consider the order of magnitude between 10,000,000 and 1000, we would calculate the logarithm of each value ($\log(10,000,000) = 7$ and $\log(1000) = 3$), thus there are $7 - 3 = 4$ orders of magnitude difference between 10,000,000 and 1000.

- Remember that a "**ballpark" value is often good enough** for an input parameter.
  - What is the square footage of a typical house?
  - What is the maximum high temperature to expect in Dallas, Texas, in July?
  - What is the typical velocity of a car on the highway?

- Always ask yourself **if it is better to err on the high side or the low side.**
  - **Safety and practical considerations**. Will a higher or lower estimate result in a safer or more reliable result?
    - If estimating the weight a bridge can support, it is better to err on the low side, so that the actual load it can safely carry is greater than the estimate.
    - For the bridge mentioned above, if estimating the load a single beam needs to support, it is better to err on the high side, thus giving a stronger beam than necessary. Be sure you understand the difference between these two points.
  - **Estimate improvement**. Can the errors cancel each other?
    - If estimating the product of two numbers, if one of the terms is rounded low, the other should be rounded high to counteract for the lower term.
    - If estimating a quotient, if you round the numerator term on the low side, should the denominator term be rounded low or high?
- Do not get bogged down with **second-order** or **minor effects**.
  - If estimating the mass of air in the classroom, do you need to correct for the presence of furniture?
  - In most instances, can the effect of temperature on the density of water be neglected?

The best way to develop your ability to estimate is through experience. An experienced painter can more easily estimate how much paint is needed to repaint a room because experience will have taught the painter such things as how many coats of one paint color it will take to paint over another, how different paint brands differ in their coverage, and how to estimate surface area quickly. In *Outliers: The Story of Success,* Malcolm Gladwell provides examples from diverse career pathways that demonstrate 10,000 hours of practice are required to develop world-class expertise in any area. Fortunately for aspiring engineers, much of this expertise can be developed starting at a young age and outside of formal schooling. For example, how many hours have you spent observing the effects of gravity? Of course, some important engineering concepts stem from phenomena that are not so easily observed, and some lend themselves to misinterpretation. As a result, it helps to have a systematic approach to developing estimates—particularly where we have less experience to guide us. Estimating an approximate answer of a calculation including known quantities is a valuable skill, such as approximating the square root of 50 as about 7, approximating the value of pi as 3 for quick estimates, etc. These mathematical approximations, however, assume that you have all the numbers to begin with and that you can use shortcuts to estimate the precisely calculated answer to save time or as a check against your more carefully calculated answer. Estimation is discussed here in a broader sense—estimating quantities that cannot be known, are complicated to measure, or are otherwise inconvenient to obtain. It is in these cases that the following strategies are recommended.

The Windows interface estimates the time needed to copy files. The estimate is dynamic and appears to be based on the total number of files and the assumption that each file will take the same amount of time to copy. As a result, when large files are copied, the estimate will increase—sometimes significantly. Similarly, as a large number of small files are copied, the estimate will decrease rapidly. A better estimation algorithm might be based on the percentage of the total file size.

5.2 Estimation by Analogy

LEARN TO: Recognize how to use analogy as a tool for estimation

One useful strategy for estimating a quantity is by comparison to something else we have measured previously or otherwise know the dimension of. The best way to prepare for this approach to estimation is to learn a large number of comparison measures for each type of quantity you might wish to estimate. Each of these comparison measures becomes an anchor point on that scale of measurement. This book provides some scale anchors for various measurable quantities—particularly in the case of power and energy, concepts with which many people struggle.

EXAMPLE 5-2

Estimate the size of a laptop computer using analogy.

Laptop computers come in different sizes, but it is not difficult to estimate the size of a particular laptop. Laptops were first called notebook computers—a good starting estimate would be to compare the particular laptop to notebook paper, which is 8.5 inches by 11 inches in the United States.

EXAMPLE 5-3

Estimate the size of an acre and a hectare of land using analogy.

American football field—playing area is 300 feet by 160 feet = approximately 50,000 square feet. An acre is 43,560 square feet. Using this data, we have a better sense of how much land an acre is—about 90% of the size of the playing area of an American football field. Soccer fields are larger, but vary in size. The largest soccer field that satisfies international guidelines would be about 2 acres.

A hectare, or 10,000 square meters, is equivalent to 108,000 square feet and is more than two acres—about the maximum size of the pitch in international rugby competition.

5.3 Estimation by Aggregation

LEARN TO: Recognize how to use aggregation as a tool for estimation

Another useful strategy is to estimate the quantity of something by adding up an estimate of its parts. This can involve multiplication in the case of a number of similarly sized parts, such as estimating the size of a tile by comparing it to your foot (estimation by analogy), counting the number of floor tiles across a room, and multiplying to estimate the total length of the room. In other cases, aggregation may involve adding together parts that are estimated by separate methods. For example, to estimate the volume of a two-scoop ice cream cone, you might estimate the volume of the cone and then separately estimate the volume of each scoop, assuming they are each spheres.

EXAMPLE   5-4

Estimate by aggregation the amount of money students at your school spend on pizza each year.

Ask students around you how often they purchase a pizza and how much it costs;

Convert this estimate into a cost per week;

Multiply your estimate by the number of weeks in an academic year;

Multiple that result by the number of students at your school.

5.4 Estimation by Upper and Lower Bounds

LEARN TO: Understand how upper and lower bounds can guide estimation

An important part of estimating is keeping track of whether your estimate is high or low. In the earlier example of estimating the volume of a two-scoop ice cream cone, we would have over-estimated, because one of the scoops of ice cream is pressed inside the cone. The effect of pressing the scoops together is not that important, because the same amount of ice cream is still there, but if the scoop is pressed into an ellipsoid, it may be difficult to estimate the original radius of the scoop.

Engineers frequently make conservative estimates, which consider the worst-case scenario. Depending on the situation, the worst case may be a lower limit (such as estimating the strength of a structure) or an upper limit (such as estimating how much material is needed for a project).

EXAMPLE   5-5

If you are to estimate how many gallons of paint are needed to paint the room you are in, what assumptions will you need to make? Where will you need to make assumptions to ensure that you have enough paint without running out?

In estimating the wall area, you should round up the length and height.

Noting that paint (for large jobs) is sold in five-gallon pails, you will want to round your final estimate to the next whole five-gallon pail.

Close estimates allow for subtracting 21 square feet per doorway (if the doors are not being painted the same color). In making a rough estimate, if there are not a lot of doorways, it would be conservative to leave in the door area.

5.5 Estimation Using Modeling

LEARN TO: Understand how models can guide estimation

In cases that are more complicated or where a more precise estimate is required, mathematical models and statistics might be used. Sometimes dimensionless quantities are useful for characterizing systems, sometimes modeling the relationship of a small number of variables is needed. At other times, extrapolating even a single variable from available data is all that is needed to make an estimate.

EXAMPLE 5-6

You would like to enjoy a bowl of peas, but they are too hot to eat. Spreading them out on a plate allows them to cool faster. Describe why this happens and devise a model of how much faster the peas on the plate will cool. Mice have a harder time keeping warm than elephants. Explain how this is related to the bowl of peas. How does this relate to the fact that smaller animals have higher heart rates? Canaries and hummingbirds can have heart rates of 1200 beats per minute, whereas human heart rates should not exceed 150 beats per minute even during exercise.

The peas cool faster when spread out because of the increase in surface area. The ratio of surface area (proportional to cooling) to volume (proportional to the heat capacity for a particular substance) is therefore important. Similarly, smaller animals have a harder time staying warm because they have a higher ratio of surface area to volume. The higher heart rate is needed to keep their bodies warm. This also relates to why smaller animals consume a much larger amount of food compared to their body mass.

EXAMPLE 5-7

A large sample of sunflower seeds is collected and their lengths are measured. Using that information, estimate the length of the longest sunflower seed you are likely to find if you measure one billion seeds.

Given a large sample, its average and standard deviation can be calculated. Assuming that the length of sunflower seeds is normally distributed, the one-in-a-billion largest sunflower seed would be expected to be six standard deviations greater than the sample average.

5.6 Significant Figures

LEARN TO: Define significant figures within a value
Understand how to determine the number of significant figures in calculations

Significant figures or sig figs are the digits considered reliable as a result of measurement or calculation. This is not to be confused with the number of digits or decimal places. The number of **decimal places** is simply the number of digits to the right of the decimal point. Example 5-8 illustrates these two concepts.

EXAMPLE   5-8

Number	Decimal Places	Significant Figures
376	0	3
376.0	1	4
376,908	0	6
3,760,000	0	3
3,760,000.	0	7
0.376	3	3
0.37600	5	5
0.0037600	7	5
376×10^{-6}	0*	3

*There is no universal agreement concerning whether numbers in scientific or engineering notation should be considered to have the number of decimal places indicated in the mantissa (as shown), or the number that would be present if the number were written out in standard decimal notation (6 in the last example).

For those who did not run across the term "mantissa" in high school (or have forgotten it)—the two parts of a number expressed in either scientific or engineering notation are the mantissa and the exponent. The mantissa is the part that gives the numerical values of the significant figures; the exponent specifies the location of the decimal point, thus the magnitude of the overall number. In the last row of the preceding table, the mantissa is 376 and the exponent is −6. The mantissa can also contain a decimal point, for example, 3.76×10^{-4}: in this case the mantissa is 3.76 and the exponent is −4.

The Meaning of "Significant"

All digits other than zero are automatically considered significant. Zero is significant when:

- It appears between two nonzero numbers
  - 306 has three significant figures
  - 5.006 has four significant figures
- It is a terminal zero in a number with a decimal point
  - 2.790 has four significant figures
  - 2000.0 has five significant figures

Zero is not significant when:

- It is used to fix a decimal place
  - 0.0456 has three significant figures
- It is used in integers without a decimal point that could be expressed in scientific notation without including that zero
  - 2000 has one significant figure (2×10^3)
  - 35,100 has three significant figures (3.51×10^4)

Comprehension Check   5-1

Determine the number of significant figures and decimal places for each value.

(a) 0.0050
(b) 3.00
(c) 447×10^9
(d) 75×10^{-3}
(e) 7,790,200
(f) 20.000

Calculation Rules

As an engineer, you will likely find that your job involves the design and creation of a product. It is imperative that your calculations lead to the design being reasonable. It is also important that you remember that others will use much of your work, including people with no technical training.

Engineers must not imply more accuracy in their calculations than is reasonable. To assist in this, there are many rules that pertain to using the proper number of significant figures in computations. These rules, however, are cumbersome and tedious. In your daily life as an engineer, you might use these rules only occasionally. The rules that follow provide a reference if you ever need them; in this text, however, you are simply expected to be *reasonable,* the concept of which is discussed in Section 5.8. In general, asking yourself if the number of significant figures in your answer is reasonable is usually sufficient. However, it is a good idea to be familiar with the rules, or at least know how to find them and use them if the need ever arises.

Multiplication and Division

A quotient or product should contain the same number of significant figures as the number with the fewest significant figures. Exact conversions do not affect this rule.

EXAMPLE   5-9

$(2.43)(17.675) = 42.95025 \cong 43.0$

- *2.43 has three significant figures.*
- *17.675 has five significant figures.*
- *The answer (43.0) has three significant figures.*

EXAMPLE   5-10

$(2.479 \text{ h})(60 \text{ min}/\text{h}) = 148.74 \cong 148.7 \text{ min}$

- *2.479 hours has four significant figures.*
- *60 minutes/hour is an exact conversion.*
- *The answer (148.7) has four significant figures.*

Addition and Subtraction

The answer resulting from an addition or subtraction should show significant figures only as far to the right as the least precise number in the calculation. For addition and subtraction operations, the least precise number should be that containing the lowest number of decimal places.

EXAMPLE 5-11

$$1725.463 + 489.2 + 16.73 = 1931.393$$

- *489.2 is the least precise.*
- *The answer should contain one decimal place: 1931.4.*

EXAMPLE 5-12

$$903{,}000 + 59{,}600 + 104{,}470 = 1{,}067{,}070$$

- *903,000 is the least precise.*
- *The answer should be carried to the thousands place: 1,067,000.*

Rounding

If the most significant figure dropped is 5 or greater, then increase the last digit retained by 1.

EXAMPLE 5-13

Quantity	Rounded to	Appears as
43.48	3 significant figures	43.5
43.48	2 significant figures	43
0.0143	2 significant figures	0.014
0.0143	1 significant figure	0.01
1.555	3 significant figures	1.56
1.555	2 significant figures	1.6
1.555	1 significant figures	2

At What Point In a Calculation Should I Round My Values?

Calculators are quite adept at keeping track of lots of digits—let them do what they are good at. In general, it is neither necessary nor desirable to round intermediate values in a calculation, and if you do, maintain at least two more significant figures for all intermediate values than the number you plan to use in the final result. The following example illustrates the risk of excessively rounding intermediate results:

$$\text{Evaluate } C = 10{,}000[\,0.6 - (5/9 + 0.044)\,]$$

- Using calculator with no intermediate rounding:

$$C = \mathbf{4.4}$$

- Rounding value in inner parenthesis to two significant figures:

$$C = 10{,}000(0.6 - (0.599555)) = 10{,}000(0.6 - (0.60)) = \mathbf{0}$$

- Rounding 9/5 to two significant figures:

$$C = 10{,}000(0.6 - (0.55555 + 0.044)) = 10{,}000(0.6 - (0.56 + 0.044))$$
$$= 10{,}000(0.6 - 0.604)) = \mathbf{-40}$$

As you can see, rounding to two significant figures at different points in the calculation gives dramatically different results: 4.4, 0, and −40. **Be very sure you know what effect rounding of intermediate values will have if you choose to do so!**

Some numbers, such as certain unit conversions, are considered exact by definition. Do not consider them in the determination of significant figures. In calculations with a *known constant,* such as pi (π), which is defined to an infinite number of significant figures, *include at least two more significant figures* in the constant than are contained in the other values in the calculation.

Comprehension Check   **5-2**

Express the answer to the following, using the correct number of significant digits.

(a) $102.345 + 7.8 - 169.05 =$
(b) $20. * 3.567 + 175.6 =$
(c) $(9.78 - 4.352)/2.20 =$
(d) $(783 + 8.98)/(2{,}980 - 1{,}387.2) =$

5.7 Reasonableness

LEARN TO: Describe the difference between accuracy, repeatability, and precision
Judge whether your answer is physically reasonable
Determine how many digits is reasonable in an answer

In the preceding discussion of estimation, the word "reasonable" was mentioned in several places. We consider two types of reasonableness in answers to problems in this section.

- **Physically reasonable.** Does the answer make sense in light of our understanding of the physical situation being explored or the estimates that we can make?
- **Reasonable precision.** Is the number of digits in the answer commensurate with the level of accuracy and precision available to us in the parameters of the problem?

When Is Something Physically Reasonable?

Here are a few hints to help you determine if a solution to a problem is physically reasonable.

First, ask yourself if the answer makes sense in the physical world.

- You determine that the wingspan of a new airplane to carry 200 passengers should be 4 feet. This is obvious rubbish.
- You determine that a sewage treatment plant for a community of 10,000 people must handle 100,000 pound-mass of sewage effluent per day. Since a gallon of water

weighs about 8 pound-mass, this is about 1.25 gallons per person per day, which is far too low to be reasonable.

- In an upper-level engineering course, you have to calculate the acceleration of a 1982 Volkswagen® Rabbit with a diesel engine. After performing your calculations, you find that the time required to accelerate from 0 to 60 miles per hour is 38 seconds. Although for a similarly sized gasoline engine, this is a rather low acceleration, for a small diesel engine, it is quite reasonable.

- You are designing playground equipment, including a swing set. The top support (pivot point) for the swings is 10 feet above the ground. You calculate that a child using one of the swings will make one full swing (forward, then backward) in 3.6 seconds. This seems reasonable.

If the final answer is in units for which you do not have an intuitive feel, convert to units for which you do have an intuitive feel.

- You calculate the speed of a pitched baseball to be 2×10^{13} millimeters per year. Converting this to miles per hour gives over 1400 miles per hour, obviously too fast for a pitched baseball.

- You are interested in what angle a smooth steel ramp must have before a wooden block will begin to slide down it. Your calculations show that the value is 0.55 radians. Is this reasonable? If you have a better feel for degrees, you should convert the value in radians to degrees, which gives 32 degrees; this value seems reasonable.

- You have measured the force of a hammer hitting a nail by using a brand-new sensor. The result is a value of 110 million dynes. Do you believe this value? Since few engineers work in dynes, it seems reasonable to convert this to units that are more familiar, such as pound-force. This conversion gives a value of 240 pound-force, which seems reasonable.

- You are told by a colleague that a ¾-inch pipe supplying water to a chemical process delivers 10 cubic meters of water per day. Converting to gallons per minute gives 1.8 gallons per minute, which seems completely reasonable.

If your solution is a mathematical model, consider the behavior of the model at very large and very small values.

- You have determined that the temperature (T in degrees Fahrenheit) of a freshly forged steel ingot can be described as a function of time (t in minutes) by this expression: $T = 2,500 - 10t$. Using this equation to calculate the temperature of the ingot, you discover that after less than 300 minutes (6 hours), the temperature of the ingot will be less than absolute zero!

- You have determined the temperature (T in degrees Celsius) of a small steel rod placed over a Bunsen burner with a flame temperature of 1000 degrees Celsius is given by $T = 960 - 939e^{-0.002t}$, where t is the time in seconds from the first application of the flame to the rod. At time $t = 0$, $T = 21$ degrees Celsius (since the exponential term will become $e = 1$). This seems reasonable, since it implies that the temperature of the rod at the beginning of the experiment is 21 degrees Celsius, which is about room temperature. As the value of time increases, the temperature approaches 960 degrees Celsius (since $e^{-\infty} = 0$). This also seems reasonable since the ultimate temperature of the rod will probably be a bit less than the temperature of the flame heating, because of inefficiencies in the heat transfer process.

- A large tank is filled with water to a depth of 10 meters, and a drain in the bottom is opened so that the water begins to flow out. You determine that the depth (D) of water in the tank is given by $D = 5t^{-0.1}$, where t is the time in minutes after the drain was opened at $t = 0$. As t increases, D approaches zero, as we would expect since all of the water will eventually drain from the tank. As t approaches zero, however, D approaches infinity, an obviously ridiculous situation; thus, the model is probably incorrect.

When Is an Answer Reasonably Precise?

First, we need to differentiate between the two terms: *accurate* and *precise*. Although laypersons tend to use these words interchangeably, an aspiring engineer should understand the difference in meaning as applied to measured or calculated values.

Accuracy is a measure of how close a calculation or measurement (or an average of a group of measurements) is to the actual value. For measured data, if the average of all measurements of a specific parameter is close to the actual value, then the measurement is accurate, whether or not the individual measurements are close to each other. The difference between the measured value and the actual value is the error in the measurement. Errors come about due to lack of accuracy of measuring equipment, poor measurement techniques, misuse of equipment, and factors in the environment (e.g., temperature or vibration).

Repeatability is a measure of how close together multiple measurements of the same parameter are, whether or not they are close to the actual value.

Precision is a combination of accuracy and repeatability and is reflected in the number of significant figures used to report a value. The more significant figures, the more precise the value is, assuming it is also accurate.

To illustrate these concepts, consider the distribution of hits on a standard bullseye target. The figure shows all four combinations of accuracy and repeatable.

- Neither repeatable nor accurate
- Repeatable, but not accurate
- Accurate, but not repeatable
- Both repeatable and accurate. This is called precise.

When considering if the precision of a numeric value is reasonable, always ask yourself the following questions:

- How many significant figures do I need in my design parameters? The more precision you specify in a design, the more it will cost and the less competitive it will be unless the extra precision is really needed.
  - You can buy a really nice 16-ounce hammer for about $20. If you wanted a 16 ± 0.0001-ounce hammer, it would probably cost well over a hundred dollars, possibly thousands.
- What are the inherent limitations of my measuring equipment? How much is the measurement affected by environmental factors, user error, etc.?
- The plastic ruler you buy at the discount store for considerably less than a dollar will measure lengths up to 12 inches with a precision of better than 0.1 inch, but not as good as 0.01 inch. On the other hand, you can spend a few hundred dollars for a high-quality micrometer and measure lengths up to perhaps 6 inches to a precision of 0.0001 inch.
- Pumps at gas stations all over the United States often display their gas price and the amount of gas pumped to three decimal places. When gas prices are high, it is extremely important to consumers that the pumps are correctly calibrated. The National Institute of Standards and Technology (NIST) requires that for every 5 gallons pumped, the amount must not be off by more than 6 cubic inches. To determine if a gas pump is calibrated correctly, you need to be able to see to three decimal places the amount pumped, since 6 cubic inches is approximately 0.026 gallons.

You should report values in engineering calculations in a way that does not imply a higher level of accuracy than is known. Use the fewest number of decimal places without reducing the usefulness of the answer. Several examples that illustrate this concept follow.

- We want to compute the area (A) of a circle. We measure the diameter (D) as 2.63 centimeters. We calculate

$$A = \frac{1}{4}\pi D^2 = \frac{1}{4}\pi(2.63)^2 = 5.432521 \text{ cm}^2$$

  The value of π is known to as many places as we desire, and ¼ is an exact number. It seems reasonable to give our answer as 5.4 or 5.43 square centimeters since the original data of diameter is given to two decimal places. Most of the time, reporting answers with two to four significant digits is acceptable and reasonable.

- We want to compute the area (A) of a circle. We measure the diameter (D) as 0.0024 centimeters. We calculate

$$A = \frac{1}{4}\pi D^2 = \frac{1}{4}\pi(0.0024)^2 = 0.0004446 \text{ cm}^2$$

  If we keep only two decimal places, our answer would be 0.00 square centimeters, which has no meaning. Consequently, when reporting numerical results, particularly those with a magnitude much smaller than 1, we use significant figures, not decimal places. It would be reasonable to report our answer as 0.00044 or 4.4×10^{-4} square centimeters.

- We want to determine a linear relationship for a set of data, using a standard software package such as Excel. The program will automatically generate a linear relationship based on the data set. Suppose that the result of this exercise is

$$y = 0.50236x + 2.0378$$

  While we do not necessarily have proof that the coefficients in this equation are nice simple numbers or even integers, a look at the equation above suggests that the linear relationship should probably be taken as

$$y = 0.5x + 2$$

- If calculations and design procedures require a high level of precision, pay close attention to the established rules regarding significant digits. If the values you generate are small, you may need more significant digits. For example, if all the values are between 0.02 and 0.04 and you select one significant figure, *all* your values will read 0.02, 0.03, or 0.04; going to two significant figures gives values such as 0.026 or 0.021 or 0.034.
- Calculators are often set to show eight or more decimal places.

  - If you measure the size of a rectangle as 2¹⁄₁₆ inches by 5⅛ inches, then the area is calculated to be 6.4453125 square inches since the calculator does not care about how many significant digits result. It is unreasonable that we can determine the area of a rectangle to seven decimal places when we made two measurements, the most accurate of which was 0.0625 inches, or four decimal places.
  - If a car has a mass of 1.5 tons, should we say it has a mass of 3010.29 pound-mass?

- Worksheets in Excel often have a default of six to eight decimal places. Two important reasons to use fewer are that (1) long decimal places are often unreasonable and

NOTE

In general, it is a good idea to set your calculator to show answers in engineering format (or generally less desirable, scientific format) with two to four decimal places.

(2) columns of numbers to this many decimal places make a worksheet difficult to read and unnecessarily cluttered.

Comprehension Check | **5-3**

In each of the following cases, a value of the desired quantity has been determined in some way, resulting in a number displayed on a calculator or computer screen. Your task is to round each number to a *reasonable* number of significant digits—*up* if a higher value is conservative, *down* if a lower value is conservative, and to the *nearest* value if it does not make a difference. Specify why your assumption is conservative.

(a) The mass of an adult human riding on an elevator   178.8 pounds
(b) The amount of milk needed to fill a cereal bowl   1.25 cups
(c) The time it takes to sing *Happy Birthday*   32.67 seconds

---

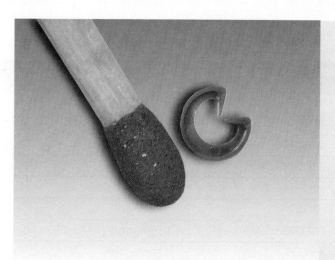

Increasingly, engineers are working at smaller and smaller scales. On the left, a vascular clamp is compared to the tip of a match. The clamp is made from a bio-absorbable plastic through the process of injection molding.

Photo courtesy of E. Stephan

5.8 Notation

LEARN TO: Report calculated numbers in standard, scientific, and engineering notation

When discussing numerical values, there are several different ways to represent the values. To read, interpret, and discuss values between scientists and engineers, it is important to understand the different styles of notation. For example, in the United States a period is used as the decimal separator and a comma is used as a digit group separator, indicating groups of a thousand (such as 5,245.25). In some countries, however, this notation is reversed (5.245,25), and in other countries, a space is used as the digit group separator (5 245.25). It is important to always consider the country of origin when interpreting written values. Several other types of notations are discussed next.

NOTE

Scientific Notation

$$\#.\#\#\# \times 10^N$$

N = integer

Engineering Notation

$$\#\#\#.\#\#\# \times 10^M$$

M = integer multiple of 3

Engineering Notation Versus Scientific Notation

In high school, you probably learned to represent numbers in scientific notation, particularly when the numbers were very large or very small. Although this is indeed a useful means of representing numeric values, in engineering, a slight modification of this notation, called engineering notation, is often more useful. This is particularly true when the value of a parameter can vary over many orders of magnitude. For example, electrical engineers routinely deal with currents ranging from 10^{-15} amperes or less to 10^2 amperes or more.

Scientific notation is typically expressed in the form $\#.\#\#\# \times 10^N$, where the digit to the left of the decimal point is the most significant nonzero digit of the value being represented. Sometimes, the digit to the right of the decimal point is the most significant digit instead. The number of decimal places can vary but is usually two to four. N is an integer, and multiplying by 10^N serves to locate the true position of the decimal point.

Engineering notation is expressed in the form $\#\#\#.\#\#\# \times 10^M$, where M is an integer multiple of 3, and the number of digits to the left of the decimal point is either 1, 2, or 3 as needed to yield a power of 10 that is indeed a multiple of 3. The number of digits to the right of the decimal point is typically between two and four.

EXAMPLE 5-14

Standard	Scientific	Engineering
43,480,000	4.348×10^7	43.48×10^6
0.0000003060	3.060×10^{-7}	306.0×10^{-9}
9,860,000,000	9.86×10^9	9.86×10^9
0.0351	3.51×10^{-2}	35.1×10^{-3}
0.0000000522	5.22×10^{-8}	52.2×10^{-9}
456200	4.562×10^5	456.2×10^3

Comprehension Check 5-4

Express each of the following values in scientific and engineering notation.

(a) 58,093,099
(b) 0.00458097
(c) 42,677,000.99

Calculator E Notation

Most scientific calculators use the uppercase "E" as shorthand for both scientific and engineering notation when representing numbers. To state the meaning of the letter "E" in English, it is read as "times 10 raised to the ___." For example, 3.707 E −5 would be read as "3.707 times 10 to the negative 5." When transcribing numbers from your calculator, in general it is best *not* to use the E notation, showing the actual power of 10 instead. Thus 3.707 E −5 should be written as 3.707×10^{-5}.

Never use a lowercase "e" for transcribing these values from the calculator, such as $3.707\ e^{-5}$, since this looks like you are multiplying by the number e ($\cong 2.717$) raised to the negative 5. If you do use a capital E (which is occasionally, though rarely, justifiable), *do not* superscript the number following the E (e.g., $3.707\ E^{-5}$) since this looks like you are raising some value E (whatever it may be) to a power.

Situations for Use of an Exponential Notation

In general, if the magnitude of a number is difficult to almost instantly determine when written in standard notation, use an exponential notation like scientific or engineering notation. Although there are no definite rules for this, if the magnitude is greater than 10,000 or less than 0.0001, you probably should consider using exponential notation. For larger numbers, using the comma notation can extend this range somewhat, for example, 85,048,900 is quickly seen to be 85 million plus a bit. However, there is no similar notation for very small numbers: 0.0000000483 is difficult to simply glance at and realize that it is about 48 billionths.

Note that it is never actually incorrect to use either exponential or standard notation; it is merely a matter of readability. To write 5×10^{20} as 500000000000000000000 is not wrong, but it may leave the readers' eyes vibrating trying to keep track of all the zeros. On the other hand, it is usually silly to write a number like 7 as 7×10^{0}.

Representation of Fractions and Use of Constants

Many of you have been previously taught that representing a numeric result exactly as a fraction is preferable to giving an approximate answer. This is seldom the preferred method of reporting values in engineering, for two reasons:

- Fractions are often difficult to glance at with instant comprehension of the actual value.
  - Quick! What does 727/41 equal? Did you immediately recognize that it is a little less than 18?
  - Is it easier to know the magnitude of 37/523 or 0.071?
- Seldom do engineers need a precision of more than three or four digits; thus, there is no need to try to represent values exactly by using the harder-to-read fractions.

Similarly, you may have learned earlier that when calculating with constants, such as pi, it is better to leave answers in terms of that constant. For the same reasons cited above, it is generally better to express such values as a decimal number, for example, 27 instead of 8.6π.

There are, of course, exceptions to these rules, but in general, a simple decimal number is more useful to engineers.

IN-CLASS ACTIVITIES

ICA 5-1

With your team, you are to determine common, readily available, or understood quantities to help you estimate a variety of parameters. For example, a two-liter bottle of soda weighs about four pounds (this is an understood quantity since almost everyone has picked up one of these many times), or the end joint of your middle finger is about an inch long (this is a readily available quantity since it goes everywhere you do). To determine the benchmarks or helpers, you may use whatever measuring tools are appropriate (rulers, scales, watches, etc.) to determine the values of the common objects or phenomena you choose. Try to determine at least two different estimation helpers for the following units:

- **Lengths:** millimeter, centimeter, inch (other than the example given above), foot, meter, kilometer
- **Areas:** square centimeter, square inch, square foot, square meter, acre
- **Volumes:** cubic centimeter, cubic inch, cubic foot, cubic meter, gallon
- **Weights and masses:** gram, newton, pound, kilogram, ton
- **Time:** second, minute, hour (the helpers you choose cannot be any form of device designed for measuring time).

ICA 5-2

Materials
    Ruler
    Tape measure
    Calipers

Procedure
The following measurements and estimations are to be completed individually, not using one set of measurements per team. This activity is designed to help you learn the size of common items to help you with future estimates. Be sure to give both the value and the unit for all measurements.

Measure the following:

- **(a)** Your height in meters
- **(b)** Your arm span (left fingertip to right fingertip) in meters
- **(c)** The length of your index finger, in centimeters
- **(d)** The width of your thumb, in centimeters
- **(e)** The width of the palm of your hand, in centimeters
- **(f)** The length of your shoe, in feet
- **(g)** The length of your pace, in feet. A pace is the distance between the toe of the rear shoe and the toe of the lead shoe during a normal step. Think about how to make this measurement before doing it.

Determine the following relationships:

- **(h)** How does your arm span measurement compare to your height?
- **(i)** How does your knee height compare to your overall height?
- **(j)** How does the length of your index finger compare to the width of your thumb?

Determine through estimation:

- **(k)** The height of a door in units of feet
- **(l)** The length of a car in units of yards

(m) The area of the floor of the classroom in units of square meters

(n) The volume of your body in units of gallons

ICA 5-3

(a) Estimate by aggregation how many gallons of gasoline are used by cars each year in the United States.

(b) Estimate by aggregation the volume of a person. A rough approximation of the volume of a person would be a cylinder approximately 1.75 meters tall with a radius of 0.25 meters. How different was your estimate from the cylindrical approximation?

(c) The website logging the progress of the Eagle Empowerment Youth Tour 2005 (http://www.eagle-empowerment.org/youthtour2005updates2.html) reported the height of the Empire State Building as 12,500 feet. Use estimation by analogy and estimation by aggregation to prove this is incorrect.

(d) Allow S to be the number of atoms along one edge of a cube of N total atoms. During X-ray diffraction, the scattering from each of those N atoms interacts with the scattering of every other atom. If a simulation of the diffraction from a cube with $S = 10$ takes 1 second to calculate, how long will a simulation of the diffraction from a cube with $S = 100$ take to calculate?

(e) If you were leaving on a trip of 1000 miles (1600 kilometers), and you could not stop for money along the way, how much cash would you need to carry to be able to buy gas during the trip? Estimate using upper and lower bounds.

ICA 5-4

(a) If estimating the amount of time to design a new product, should you err on the high side or the low side?

(b) If estimating the switching speed of the transistors for a faster computer, is it better to err on the low side (slower switching) or high side?

(c) How many square yards of fabric are needed to cover the seats in a typical minivan? Is it better to err on the high side or the low side?

(d) Estimate the dimensions of the classroom, in feet. Using these values and ignoring the fact that you would not paint over the windows and doors, estimate the gallons of paint required to paint the classroom. A gallon of paint covers 400 square feet. Would it be better to round the final answer up or down to the nearest gallon?

ICA 5-5

In each of the following cases, either display a value to the requested number of significant figures or display a calculated value to the appropriate number of significant figures.

(a) Round to three significant figures: 0.70973 kilograms.

(b) Area of a box measuring 1.15 centimeters long by 1.62 centimeters wide. Area is determined by length times width.

(c) You measure something using a ruler that has markings every 1/16 inches. You measure an object to 3 and 11/16 inches, commonly written as $3 - 11/16''$. Report its length in decimal format in inches.

(d) (536,000 meters) $\times$ (6576 meters)

ICA 5-6

In each of the following cases, either display a value to the requested number of significant figures or display a calculated value to the appropriate number of significant figures.

(a) 12.001 feet + 2.08 feet + 108.234 feet

(b) A piece of stone has a mass of 13.782 grams and a volume of 4.64 cubic centimeters. What is the density of the stone in units of grams per cubic centimeter? Density is determined by mass divided by volume.

(c) How many significant figures are there in the value 0.00470 centimeters?

(d) The mass of a rock sample is measured four times, yielding values of 24.996 grams, 25.008 grams, 25.011 grams, and 25.005 grams. What is the average mass of the sample in units of grams? The average value is determined by taking the sum of all samples and dividing by the total number of samples.

ICA 5-7

In each of the following cases, a value of some desired quantity has been determined in some way, resulting in a number displayed on a calculator or computer screen. Your task is to round each number to a *reasonable* number of significant digits—*up* if a higher value is conservative, *down* if a lower value is conservative, and to the *nearest* value if it does not make a difference.

(a) 4.36 gallons; amount of paint needed to cover a single room

(b) 1484.2 miles; the distance from Tampa to New York to estimate the amount of gas you need to purchase

(c) 1613 lumens; the brightness of a light bulb

(d) $20,144.52; cost of tuition, room and board at college this year

ICA 5-8

Complete the table to express each of the following values in standard, scientific, and engineering notation.

Standard	Scientific	Engineering
25,760,000		
	1.25×10^{-2}	
		349.6×10^{3}
0.0000005140		
	3.92×10^{9}	
		38.7×10^{-9}

REVIEW QUESTIONS

Use estimation to solve these Fermi-type problems.

1. How many cubic yards of concrete is required to construct one mile of interstate highway?

2. How many gallons of gasoline are burned per student when the students in this class leave for school break, assuming only one-way travel?

3. If all the land (both currently inhabited and all the inhabitable land) were divided equally among all the people now living, how many acres of land would each one have?

4. How many times do my rear tires rotate if I drive around the perimeter of campus?

5. How many gallons of water per day would be saved if everyone in the United States who does not turn off the faucet while they brush their teeth, did so?

6. In 1978, cars that got about 40 miles per gallon were readily available. If the average fuel economy of all cars sold since then was 40 miles per gallon (instead of the lower average mileage of the cars that were actually sold), how many billions of gallons of gas would have been saved in the United States since 1978?

7. How many toothpicks can be made from an 8-foot-long 2-inch by 4-inch board?

8. Noah's ark has been described as having the following dimensions: 300 cubits long × 50 cubits wide × 30 cubits high. If a cubit is 18 inches, how many people could fit into the ark?

9. Due to drought, the water level of a 10-acre pond is five feet below normal. If you wanted to fill the pond to normal capacity by using a hose connected to your kitchen faucet, how long would it take to fill the pond? Select an appropriate unit of time to report your answer.

10. A cubic meter of air has a mass of about 1.2 kilograms. What is the total mass of air in your home or in a designated building?

11. How many carrots are used to make all of the canned soup consumed in the United States in one year? How many acres are used to grow these carrots?

6

Solving Problems

All disciplines in engineering and science are chock full of problems to solve. These cover a myriad of topics; some are easy, some difficult, some tedious, and some straightforward. We often face new and previously unsolved problems. Some require highly accurate solutions, while others require just a "ballpark" value. Sometimes we face familiar problems, and sometimes the problem has absolutely no characteristics we have seen before.

What makes a good problem solver? The answer is simple: someone who can solve problems, often with an ease that belies the difficulty of the problem. Experience tells us that some people seem to have a problem-solving knack, while other, equally bright individuals have to work much harder at it. Determining what makes an effective problem solver is beyond the scope of this work. Suffice it to say that practice helps, and practicing certain techniques can help even more. It is the purpose of this short chapter to suggest some simple tools that facilitate problem solving. You provide the rest—practice, practice, practice!

6.1 Problem Types

LEARN TO: Categorize problems into one of several broad categories.

NOTE

Also mentioned below are some situations and conditions that inhibit your ability to solve problems—avoid these.

While not all problems are easily pigeonholed, there are several types that occur often in engineering situations. This is a good place to start. If the problem is written, read it over several times before deciding on the type. If it is given orally (as often happens in the work world) then just pause and repeat the basics to yourself before proceeding.

Some common problem types follow:

Ratio

We are interested in how a particular quantity varies when the conditions change.

EXAMPLE 6-1

If a 2-meter-long horizontal beam deflects 6 centimeters when a load of 2N is hung on the end, how much will a beam of the same material deflect under the same load if it is 4 meters long?

To quickly analyze this, we use the equation for beam deflection and write it twice, once for the initial situation and again for the new condition. Take the ratio of the two equations to find that many parameters will cancel, allowing an easy solution.

Equation of deflection:

$$\delta = \frac{P\,\ell^3}{3\,E\,I}$$

Here δ is the deflection, P is the load, ℓ is the length of the beam, E is Young's modulus, which depends on the material, and I is the moment of inertia. Now simply write this expression for the short beam (subscript s) and again for the longer beam (no subscript).

$$\delta = \frac{P\,\ell^3}{3\,E\,I}$$

Since the load, the shape (I), and the material (E) are all the same, much cancels, and we are left with

$$\frac{\delta}{\delta_s} = \left(\frac{\ell}{\ell_s}\right)^3 = \left(\frac{4\,m}{2\,m}\right)^3 = 8$$

This method is easy to use and shows clearly that there was no need to calculate the moment of inertia or to look up Young's modulus. The answer is 8.

Approximation

It is not always necessary to obtain an answer to four decimal places. In many situations in industry, an approximate answer (often called "quick and dirty" or "back of the envelope") is all that is needed. No need to spend the time or energy to provide more precision than necessary. When this is the case, it is possible to approximate in several ways: (1) use approximate input values, (2) use approximate equations, or (3) use an approximate geometry.

EXAMPLE 6-2

How far (d) will a falling body fall in 4.8 seconds if it is initially falling at 0.03 miles per second?

1. We may not need precise input values—(assume a fall time of 5 seconds).
2. We may be able to use an approximate governing equation. The exact equation follows:

$$d = v_o t + \frac{1}{2} g t^2$$

By approximating its initial fall speed (v_o) as zero, we can use only the second term. Knowing that $g = 9.8$ m/s^2 (or approximately 10 m/s^2) we quickly estimate that the distance fallen will be:

$$D = (1/2)(10\,m/s^2)(5\,s)^2 = 125\,m$$

A more detailed discussion of how to make and use estimations was given in Chapter 5 "Estimation."

Parametric

Often in design, it is required to study a wide range of input parameters (hence a **parametric** analysis), as well as the corresponding range of outputs. This can often conveniently be done using spreadsheets, graphical representations, and programming (as discussed in Part 3 and Part 4 of this text).

EXAMPLE 6-3

How does the range of a rocket vary with the initial speed (never to be greater than 600 miles per second) and launch angle?

Develop a spreadsheet (and corresponding graph) which shows the range for launch angles of 0, 5, 10, 15, 20, . . . 90 degrees for launch speeds of 100, 200, 300, . . . 600 m/s.

Ideal

Sometimes, particularly with straightforward problems, we are required to simply calculate a single, exact answer. We call these **ideal problems**.

EXAMPLE 6-4

How far does a vehicle travel if it moves at 36 miles per hour for 15 minutes?

distance = (speed)(time)
distance = (36 mph)(0.25 h) = 9 miles

Discussion

In conceptual studies, it is often important to simply develop a better understanding of how various parameters interact. In such situations, we may not need to obtain a numerical answer but rather to determine if certain variables are critical or if they may be neglected by discussing the overall problem statement. Such an analysis may then lead to a new design, quantification of material needed, or selection of parameters to be studied further.

EXAMPLE 6-5

Does the diameter of a bridge pier affect the amount of scour that occurs around its base during flooding conditions?

It is reasonable to assume that if larger piers are used, then fewer are needed to support the bridge? If more but smaller piers are used, how much of the cross section of the river is blocked compared to fewer but larger piers? We must find information relating the total bridge load on the piers and their strength (that would include the diameter, material, etc.). As more of the river cross section is blocked, the speed of the water between the piers increases, which would increase the scour (erosion). Once we have examined these variables, it may be possible to tell the bridge designer that it is better to use fewer large piers for support (or more smaller piers).

The problem solver first decides on problem type, picking from the five generic types:

Ratio	Compare a quantity with a known quantity
Approximation	Estimate values/use approximate equations
Parametric	Input parameters are studied over ranges
Ideal	Exact solution
Discussion	Talk about how parameters affect each other

This has the easily remembered acronym of RAPID and often provides a rapid means to get started solving a problem.

The RAPID process can help classify problems. Even when the type of problem type is known, different people are likely to approach solving problems in different ways. There are, however, some procedures and techniques that can help in developing a generally successful problem-solving approach. Although it is not possible to write down a specific recipe that will always work, some broad approaches will help. One approach to solving problems is presented here. Another approach is given in an Appendix, "Process."

6.2 SOLVEM — One Approach to Solving Problems

LEARN TO: Define the problem-solving methodology described by the acronym SOLVEM

One problem-solving approach has been given the acronym **SOLVEM**:

Sketch
Observations or **O**bjectives
List
Variables and
Equations
Manipulation

Note that this approach is equally useful for problems involving estimation and more precise calculations. Each step is described next.

Sketch

Figure 6-1 illustrates how a drawing can help you visualize a problem. In sketching a problem, you are subconsciously thinking about it. Be sure to draw the diagram large enough so that everything is clear, and label the things that you know about the problem in the diagram. For some problems, a before-and-after set of diagrams may be helpful. In very complex problems, you can use intermediate diagrams or subdiagrams as well.

Comprehension Check   **6-1**

We use SOLVEM to complete this problem in the Comprehension Checks in this chapter. Create a sketch for the following problem.

Calculate the mass in kilograms of gravel stored in a rectangular bin 18.5 feet by 25.0 feet. The depth of the gravel bin is 15 feet, and the density of the gravel is 97 pound-mass per cubic foot.

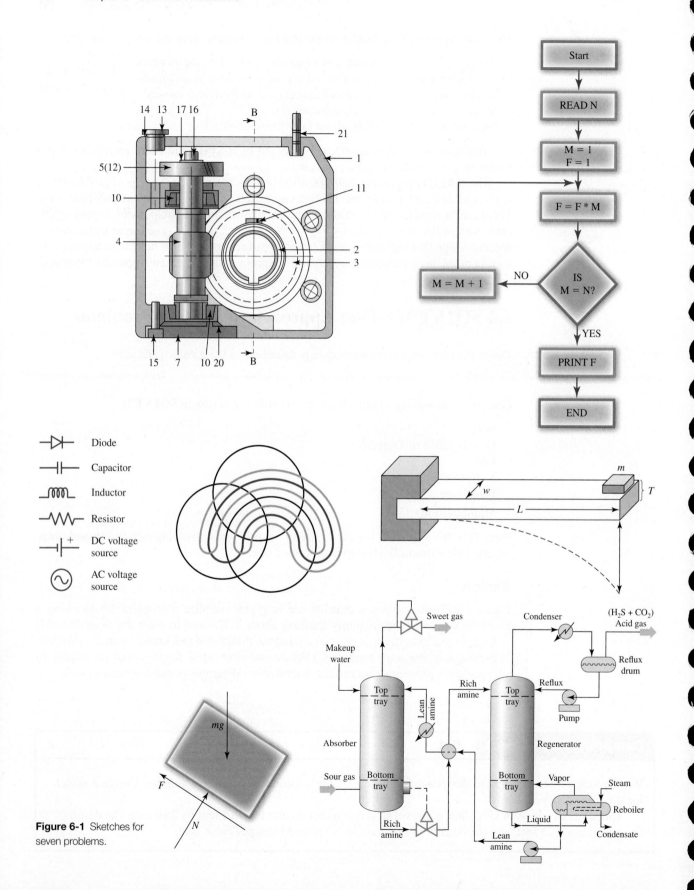

Figure 6-1 Sketches for seven problems.

Observations, Objectives

These can be in the form of simple statements, questions, or anything else that might acquaint you with the problem at hand. It often helps to divide your observations and objectives into several categories. Some of the easiest to remember follow:

NOTE

An essential part of stating observations is to state the objectives.

When making observations, do not forget that you have five senses.

- Objective to be achieved
- Observations about the problem geometry (size, shape, etc.)
- Observations about materials and material properties (density, hardness, etc.)
- Observations about parameters not easily sketched (temperature, velocity, etc.)
- Other miscellaneous observations that might be pertinent

You will almost always find after writing down some observations that you actually know more about the problem than you originally thought.

Here are some typical examples:

Objectives

- Find the velocity, force, flow rate, time, pressure, etc., for a given situation.
- Profitably market the device for less than $25.
- Fit the device into a 12-cubic-inch box.

Observations

PROBLEM GEOMETRY

- The liquid has a free surface.
- The submerged plate is rectangular.
- The support is vertical.
- The tank is cylindrical.
- The cross-sectional area is octagonal.
- The orbit is elliptical.

MATERIALS AND MATERIAL PROPERTIES

- The gate is steel.
- The coefficient of static friction is 0.6.
- The specific gravity is 0.65.
- Ice will float in water.
- The alloy superconducts at 97 kelvin.
- The alloy melts at 543 degrees Fahrenheit.

OTHER PARAMETERS

- If depth increases, pressure increases.
- If temperature increases, resistance increases.
- The flow is steady.
- The fluid is a gas and is compressible.
- The pulley is frictionless.
- The magnetic field is decreasing.
- Temperature may not fall below 34 degrees Fahrenheit.

MISCELLANEOUS

- The force will act to the right.
- Gravity causes the ball to accelerate.
- The sphere is buoyant.
- Drag increases as the speed increases.

Remember to include those quantities whose value is zero! Often such quantities are hidden with terms such as:

NOTE

Quantities whose value is zero contain valuable information about the problem.

- Constant (implies derivative = 0)
- Initially (at time = 0)
- At rest (no motion)
- Dropped (no initial velocity)
- At the origin (at zero position)
- Melts or evaporates (changes phase, temperature is constant)

Comprehension Check 6-2

State the objective and any relevant observations for the following problem.

Calculate the mass in kilograms of gravel stored in a rectangular bin 18.5 feet by 25.0 feet. The depth of the gravel bin is 15 feet, and the density of the gravel is 97 pound-mass per cubic foot.

Finally, one of the most important reasons to make *many* observations is that you often will observe the "wrong" thing. For example, write down things as you read this:

A bus contains 13 passengers.
At the first stop, four get off and two get on.
At the next stop, six get off and one gets on.
At the next stop, nobody gets off and five get on.
At the next stop, eight get off and three get on.
At the next stop, one gets off.
At the last stop, four get off and four get on.

After putting your pencil down and without looking again at the list, answer the question given below the textbox discussing "The Importance of Observations."

THE IMPORTANCE OF OBSERVATIONS

An excerpt adapted from The Crooked Man *by Sir Arthur Conan Doyle*

Dr. Watson writes: I looked at the clock. It was a quarter to twelve. This could not be a visitor at so late an hour. A patient, evidently, and possibly an all-night sitting. With a wry face I went out into the hall and opened the door. To my astonishment, it was Sherlock Holmes who stood upon my step.

"Ah, Watson, I hoped that I might not be too late to catch you."

"My dear fellow, pray come in."

"You look surprised, and no wonder! Relieved, too, I fancy! Hum! You still smoke the Arcadia mixture of your bachelor days, then! There's no mistaking that fluffy ash upon your coat. It's easy to tell that you've been accustomed to wear a uniform, Watson; you'll never pass as a pure-bred civilian as long as you keep that habit of carrying your handkerchief in your sleeve. Could you put me up for the night?"

"With pleasure."

"You told me that you had bachelor quarters for one, and I see that you have no gentleman visitor at present. Your hat-stand proclaims as much."

" MR. HENRY WOOD, I BELIEVE?"

Source: AF Fotografie/Alamy

"I shall be delighted if you will stay."

"Thank you. I'll find a vacant peg, then. Sorry to see that you've had the British workman in the house. He's a token of evil. Not the drains, I hope?"

"No, the gas."

"Ah! He has left two nail marks from his boot upon your linoleum just where the light strikes it. No, thank you, I had some supper at Waterloo, but I'll smoke a pipe with you with pleasure."

I handed him my pouch, and he seated himself opposite to me, and smoked for some time in silence. I was well aware that nothing but business of importance could have brought him to me at such an hour, so I waited patiently until he should come round to it.

"I see that you are professionally rather busy just now."

"Yes, I've had a busy day. It may seem very foolish in your eyes, but I really don't know how you deduced it."

"I have the advantage of knowing your habits, my dear Watson. When your round is a short one you walk, and when it is a long one you use a hansom (a carriage). As I perceive that your boots, although used, are by no means dirty, I cannot doubt that you are at present busy enough to justify the hansom."

"Excellent!"

"Elementary. It is one of those instances where the reason can produce an effect which seems remarkable to his neighbor, because the latter has missed the one little point which is the basis of the deduction."

While we cannot all be as observant as Sherlock Holmes, we can improve our powers of observation through practice. This will pay dividends as we seek to be engineers with high levels of analytical skills.

QUESTION: For the bus problem, how many stops did the bus make?
   The lesson here is that often we may be observing the wrong thing.

List of Variables and Constants

Go over the observations previously determined and list the variables that are important. It may help to divide the list into several broad categories—those related to the geometry of the problem, those related to the materials, and a properties category—although for some types of problems those categories may not be appropriate. Include in your list the written name of the variable, the symbol used to represent the quantity, and if the value of the variable is known, list the numeric value, including units. If a value is a constant you had to look up, record where you found the information.

Initial and Final Conditions

- Initial temperature (T_0) 60 [°F]
- Initial radius (r_i) 5 [cm]
- Mass of the object (m) 23 [kg]

Constants

- Acceleration of gravity (g) 32.2 [ft/s²]
- Ideal gas constant (R) 8314 [(Pa L)/(mol K)]

Geometry

- Length of beam (L) [m]
- Cross-sectional area of a pipe (A) [cm²]
- Volume of a reactor vessel (V) [gal]

MATERIALS

- Steel
- Polyvinyl chloride (PVC)
- Plasma
- Gallium arsenide
- Medium-density balsa wood

PROPERTIES

- Dynamic viscosity of honey (μ) 2500 [cP]
- Density of PVC (ρ) 1380 [kg/m³]
- Spring constant (k) 0.05 [N/m]
- Specific gravity   (SG)   1.34

Comprehension Check | **6-3**

Create a list of variables and constants for the following problem.

Calculate the mass in kilograms of gravel stored in a rectangular bin 18.5 feet by 25.0 feet. The depth of the gravel bin is 15 feet, and the density of the gravel is 97 pound-mass per cubic foot.

Equations

Only after completing the preceding steps (S-O-L-V) should you begin to think about the equations that might govern the problem. It is useful to make a list of the pertinent equations in a broad sense before listing specific expressions. For example:

- Conservation of energy
- Conservation of mass
- Conservation of momentum
- Frequency equations
- Ideal gas law
- Newton's laws of motion
- Stress–strain relations
- Surface areas of geometric solids
- Volumes of geometric solids
- Work, energy relations

You may need "subequations" such as:

- Distance = (velocity) (time)
- Energy = (power) (time)
- Force = (pressure) (area)
- Mass = (density) (volume)
- Voltage = (current) (resistance)
- Weight = (mass) (gravity)

For an equation, list the broad category of the equation (Hooke's law) and then the actual expression ($F = kx$) to help with problem recognition.

Do not substitute numerical values of the parameters into the equation right away. Instead, manipulate the equation algebraically to the desired form.

Comprehension Check | **6-4**

Create a list of equations for the following problem.

Calculate the mass in kilograms of gravel stored in a rectangular bin 18.5 feet by 25.0 feet. The depth of the gravel bin is 15 feet, and the density of the gravel is 97 pound-mass per cubic foot.

Manipulation

Most of the time, you need to manipulate pertinent equations before you can obtain a final solution. ***Do not substitute numerical values of the parameters into the equation right away***. Instead, manipulate the equation algebraically to the desired form. Often you will discover terms that will cancel, giving you a simpler expression to deal with. By doing this, you will:

- Obtain general expressions useful for solving other problems of this type
- Be less likely to make math errors
- Be able to judge whether your final equation is dimensionally consistent
- Better understand the final result

The SOLVEM acronym does not contain a word or step for numerical solution. In fact, this process helps you analyze the problem and obtain an expression or procedure so that you can find a numerical answer. The thought here is that engineers need training to be able to analyze and solve problems. If you can do everything except substitute numbers, you are essentially finished. As an engineer, you will "be paid the big bucks" for analysis, not for punching a calculator.

6.3 Representing Final Results

LEARN TO: Recognize where and when to apply reasonableness within SOLVEM

NOTE

Don't plug values into the equation until the final step.

When you have completed all the steps to SOLVEM, plug in values for the variables and constants and solve for a final answer. Be sure to use reasonableness. The final answer should include both a numeric value and its unit. In addition, it is often useful to write a sentence describing how the answer meets the objectives. Box your final answer for easy identification.

Repeated use of SOLVEM can help you develop a better gut-level understanding about the analysis of problems by forcing you to talk and think about the generalities of the problem before jumping in and searching for an equation into which you can immediately substitute numbers.

Comprehension Check | **6-5**

Manipulate and solve for the following problem, using the information from Comprehension Checks 1–4.

Calculate the mass in kilograms of gravel stored in a rectangular bin 18.5 feet by 25.0 feet. The depth of the gravel bin is 15 feet, and the density of the gravel is 97 pound-mass per cubic foot.

6.4 Avoiding Common Mistakes

LEARN TO: Adopt strategies that will assist in reducing errors in problem solving.

Erroneous or argumentative thinking can lead to problem-solving errors. For example,

- **I can probably find a good equation in the next few pages.** Perhaps you read a problem and rifle through the chapter to find the proper equation so that you can start substituting numbers. You find one that looks good. You do not worry about whether the equation is the right one or whether the assumptions you made in committing to the equation apply to the present problem. You whip out your calculator and produce an answer. *Don't do this!*

- **I hate algebra, or I cannot do algebra, or I have got the numbers, so let us substitute the values right in.** Many problems become much simpler if you are willing to do a little algebra before trying to find a numerical solution. Also, by doing some manipulation first, you often obtain a general expression that is easy to apply to another problem when a variable is given a new value. By doing a little algebra, you can also often circumvent problems with different sets of units. *Do some algebra!*

- **It is a simple problem, so why do I need a sketch?** Even if you have a photographic memory, you will need to communicate with people who do not. It is usually much simpler to sort out the various parts of a problem if a picture is staring you right in the face. *Draw pictures!*

- **I do not have time to think about the problem, I need to get this stuff finished.** Well, most often, if you take a deep breath and jot down several important aspects of the problem, you will find the problem much easier to solve and will solve it correctly. *Take your time!*

6.5 Examples of SOLVEM

EXAMPLE **6-6**

Estimate how many miles of wire stock are needed to make 1 million standard paper clips.

W

L

Sketch:

See the adjacent diagram.

Objective:

Determine the amount of wire needed to manufacture a million paper clips.

Observations:

- *Paper clips come in a variety of sizes.*
- *There are four straight segments and three semicircular sections in one clip.*
- *The three semicircular sections have slightly different diameters.*
- *The four straight sections have slightly different lengths.*

List of Variables and Constants:

- *L*   *Overall length of clip*
- *W*   *Overall width of clip*
- L_1, L_2, L_3, L_4 *Lengths of four straight sections*
- D_1, D_2, D_3 *Diameters of three semicircular sections*
- P_1, P_2, P_3 *Lengths of three semicircular sections*
- *A*   *Total amount (length) of wire per clip*

Estimations and Assumptions:

- *Length of clip:* $L = 1.5$ in
- *Width of clip:* $W = 3/8$
- *Diameters from largest to smallest*
  - $D_1 = W = 3/8$ in   • $D_3 = 1/4$ in
  - $D_2 = 5/6$ in

- *Lengths from left to right in sketch*
  - L_1 = To be calculated
  - L_2 = 0.8 in
  - $L_3 \approx L_4$ = 1 in

Equations:

- *Perimeter of semicircle:* $P = \pi D/2$ *(half of circumference of circle)*
- $L_1 = L - D_1/2 - D_2/2$
- *Total length of wire in clip:* $A = L_1 + L_2 + L_3 + L_4 + P_1 + P_2 + P_3$

Manipulation:

In this case, none of the equations need to be manipulated into another form.

Length of longest straight side:	$L_1 = 1.5 - 3/16 - 5/32 \approx 1.2$ in
Lengths of semicircular sections:	$P_1 = \pi 3/16 \approx 0.6$ in $P_2 = \pi 5/32 \approx 0.5$ in $P_3 = \pi 1/8 \approx 0.4$ in
Overall length for one clip:	$A = 1.2 + 0.8 + 1 + 1 + 0.6$ $\quad + 0.5 + 0.4 = 5.5$ in/clip
Length of wire for 1 million clips:	$(5.5 \times$ in/clip$)\,(1 \times 10^6$ clips$) = 5.5 \times 10^6$ in
Convert from inches to miles:	$(5.5 \times 10^6$ in$)\,(1$ ft$/12$ in$)$ $(1$ mile$/5{,}280$ ft$) \approx 86.8$ miles

One million, 1.5-inch paper clips require about 87 miles of wire stock.

EXAMPLE 6-7

A spherical balloon has an initial radius of 5 inches. Air is pumped in at a rate of 10 cubic inches per second, and the balloon expands. Assuming that the pressure and temperature of the air in the balloon remain constant, how long will it take for the surface area to reach 1000 square inches?

Sketch:

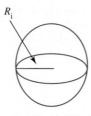
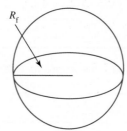

Objective:

Determine how long it will take for the surface area of the balloon to reach 1000 in²

Observations:

- *The balloon is spherical*
- *The balloon, thus its volume and surface area, gets larger as more air is pumped in*
- *The faster air is pumped in, the more rapidly the balloon expands*

List of Variables and Constants:

- *Initial radius:* $R_i = 5$ [in]
- *Final radius:* R_f [in]
- *Initial surface area:* A_i [in^2]
- *Final surface area:* A_f [in^2]
- *Change in volume:* ΔV [in^3]
- *Initial volume:* V_i [in^3]
- *Final volume:* V_f [in^3]
- *Fill rate:* $Q = 10$ [in^3/s]
- *Time since initial size:* t [s]

Equations:

- *Surface area of sphere:* $A = 4\pi R^2$
- *Volume of sphere:* $V = 4/3\ \pi R^3$
- *Change in volume:* $\Delta V = Qt$

Manipulation:

There are a few different ways to proceed. The plan used here is to determine how much the balloon volume changes as air is blown into the balloon and to equate this to an expression for the volume change in terms of the balloon geometry (actually the radius of the balloon).

Radius of balloon in terms of surface area:	$R = \left(\dfrac{A}{4\pi}\right)^{1/2}$
Final balloon radius in terms of surface area:	$R_f = \left(\dfrac{A_f}{4\pi}\right)^{1/2}$
Final volume of balloon in terms of surface area:	$V_f = \left(\dfrac{4\pi}{3}\right)\left(\dfrac{A_f}{4\pi}\right)^{3/2}$
Volume change in terms of air blown in:	$\Delta V = V_f - V_i = Qt$
Volume change in terms of geometry:	$V_f - V_i = \left(\dfrac{4\pi}{3}\right)\left(\dfrac{A_f}{4\pi}\right)^{3/2} - \left(\dfrac{4\pi}{3}\right)R_i^3$
Solve for time to blow up balloon:	$t = \left(\dfrac{4\pi}{3Q}\right)\left(\dfrac{A_f}{4\pi}\right)^{3/2} - \left(\dfrac{4\pi}{3Q}\right)R_i^3$
And simplifying:	$t = \left(\dfrac{4\pi}{3Q}\right)\left\{\left(\dfrac{A_f}{4\pi}\right)^{3/2} - R_i^3\right\}$

It takes just over 4 minutes to increase the volume to 1000 cubic inches.

NOTE

Whenever you obtain a result in equation form, you should check to see if the dimensions match in each term.

IMPORTANT

Recall that you should manipulate the equations *before* inserting known values. Note that the final expression for elapsed time is given in terms of initial radius (R_i), flow rate (Q), and final surface area (A_f). So you now have a general equation that can be solved for any values of these three parameters. If you had begun substituting numbers into equations at the beginning and then wanted to obtain the same result for different starting values, you would have to resolve the entire problem.

CHAPTER 6

IN-CLASS ACTIVITIES

ICA 6-1

Decide on the type of problem for each question.

(a) For the meter stick example, tests have been conducted for extensions of 0.2, 0.4, 0.6, and 0.8 meters with loads of 2, 4, 6, and 10 coins and 5 large books in each test. What will the deflection be for 0.5 meters and 8 coins?

(b) If no books are used to hold one end of the beam, what will be the maximum number of coins that can be placed on the tip?

(c) For the meter stick example, we have deflection data for a 0.7 meter extension with 6 coins. What will the deflection be for 3 coins?

(d) For the meter stick example above, will 5 stacked coins placed on the tip yield a greater or smaller deflection than 5 coins placed along the beam only one high?

ICA 6-2

Decide on the type of problem for each question.

(a) Given human reaction times as a function of age and information regarding the braking distances of cars as a function of speed, what are appropriate following distances for driving?

(b) Parachute manufacturers provide the fall speed of an open chute as a function of person weight. If we open our chute at 3000 ft, how long will it take to land?

(c) Which will cool more quickly, one cup of coffee or two cups half full?

(d) Estimate the force required to accelerate a mass of 61.4 kilograms at a rate of 2.9 m/s².

(e) At a party, we blow up a balloon. After three quick breaths, we pause and then wonder how much greater the surface area will be if we add another three breaths.

(f) Tom drives from his home to a nearby town that is 20 miles away at 30 miles per hour. He wishes to drive back home at a speed that will allow him to average 60 miles per hour for the round trip. How fast should he drive on the return trip?

(g) To construct a concrete patio, the soil is leveled and gravel is laid on the area. Concrete is poured on top of this. The thickness of the gravel layer needs to be determined based on the gravel size. For small pebble size (average size 4 mm), the layer must be the same thickness as the concrete. For medium size (average 16 mm), it must be two times thicker, and for large size (64 mm), it must be three times thicker. We need to prepare a design chart that will allow us to select the proper thickness of concrete immediately as the gravel is delivered. (We don't know what the gravel size will be until the shipment arrives.) Once on site, a quick test will provide the gravel size; we do know, however, it will be somewhere between 4 millimeters and 50 millimeters.

(h) A firefighter aims a stream of water just over the top of a 50-foot flagpole. If the firefighter stands 60 feet away from the pole, about how far does the water stream travel before hitting the ground?

(i) The drag generated by a parachute (which keeps us from falling faster) depends on the falling speed squared. If a 100-pound person falls at 10 feet per second, how fast will a 200-pound person fall?

ICA 6-3

Each of these items should be addressed by a team. List as many things about the specified items as you can determine by observation. One way to do this is to let each team member make one observation, write it down, and iteratively canvass the team until nobody can think of any more additions (it is fine to pass). Remember that you have five senses. Also note that observations are things you can actually detect during the experiment, *not* things you already know or deduce that cannot be observed.

To help, Examples A and B are given here before you do one activity (or more) on your own. Remember that not all observations will be important for a particular problem. Write them down anyway—they may trigger an observation that is important.

Example A: A Loudspeaker Reproducing Music

- Electrical signal is sent to speaker.
- Speaker vibrates from electrical signal.
- Speaker is gray.
- A magnet is involved.
- Speaker is circular.
- Speaker is about 10 inches in diameter.
- Speaker diaphragm is made of paper; vibrating air creates sound.
- Gravity acts on speaker.
- Speaker is in a "box."

Example B: Drinking a Soft Drink in a Can through a Straw

- Liquid is cold.
- Straw is cylindrical.
- Can is cylindrical.
- Liquid assumes the shape of the container.
- Gravity opposes the rise of the liquid.
- Moving liquid has kinetic energy.
- As the liquid rises, it gains potential energy.
- Table supports the weight of the can, liquid, and straw.
- Can is opaque.
- Liquid is brown.
- Liquid is carbonated (carbonic acid).
- Liquid contains caffeine.
- Friction between straw and lips allows you to hold it.
- Plan view of the can is a circle.
- Silhouette of the can is the same from any direction.
- Silhouette of the can is a rectangle.
- Can is painted.
- Can is metal.
- Liquid surface is horizontal.
- Liquid surface is circular.

From the following list or others as selected by the instructor, list as many observations as you can about each topic:

- **(a)** An object provided by the professor (placed on the desk)
- **(b)** Candle placed on the desk and lit
- **(c)** Ball dropped from several feet, allowed to bounce and come to rest
- **(d)** A weight on a string pulled to one side and released; watched until it comes to rest
- **(e)** Coin spinning on the desk
- **(f)** Cup of hot coffee placed on desk
- **(g)** A glass of cold water placed on desk
- **(h)** Book pushed across the desk
- **(i)** Large container guided smoothly up a ramp
- **(j)** Ruler hanging over the desk
- **(k)** Your computer (when turned off)
- **(l)** Your chair
- **(m)** Your classroom
- **(n)** A weight tied on a string and twirled

Final Assignment of this ICA: You have done several observation exercises. In these, you thought of observation as just a "stream of consciousness" with no regard to organization of your efforts. With your previous observation as a basis, generalize the search for observations into several (three to six or so) categories. The use of these categories should make the construction of a list of observations easier in the future.

Analyze the following problems using the SOLVEM method.

ICA 6-4

What diameter will produce a maximum discharge velocity of a liquid through an orifice on the side at the bottom of the cylindrical container? Consider diameters ranging from 0.2 to 2 meters.

ICA 6-5

A hungry bookworm bores through a complete set of encyclopedias consisting of n volumes stacked in numerical order on a library shelf. The bookworm starts inside the front cover of volume 1, bores from page 1 of volume 1 to the last page of the last volume, and stops inside the back cover of the last volume. Note that the bookworm starts inside the front cover of volume 1 and ends inside the back cover of volume n.

Assume that each volume has the same number of pages. For each book, assume that you know how thick the cover is and that the thickness of a front cover is equal to the thickness of a back cover; assume also that you know the total thickness of all the pages in the book. How far does the bookworm travel? How far will it travel if there are 13 volumes in the set and each book has 2 inches of pages and a $\frac{1}{8}$-inch thick cover?

ICA 6-6

Two cargo trains each leave their respective stations at 1:00 p.m. and approach each other, one traveling west at 10 miles per hour and the other on separate tracks traveling east at 15 miles per hour. The stations are 100 miles apart. Find the time when the trains meet, and determine how far the eastbound train has traveled.

ICA 6-7

Water drips from a faucet at the rate of 3 drops per second. What distance separates one drop from the following drop 0.65 seconds after the leading drop leaves the faucet? How much time elapses between impacts of the two drops if they fall onto a surface that is 6 feet below the lip of the faucet?

Your sketch should include the faucet, the two water drops of interest, and the impact surface. Annotate the sketch, labeling each item shown, and denote the relevant distances in symbolic form—for example, you might use d_1 to represent the distance from the faucet to the first drop.

ICA 6-8

During rush hour, cars back up when the traffic signal turns red. When cars line up at a traffic signal, assume that they are equally spaced (Δx) and that all the cars are the same length (L). You do not begin to move until the car in front of you begins to move, creating a reaction time (Δt) between the time the car in front begins to move and the time you start moving. To keep things simple, assume that when you start to move, you immediately move at a constant speed (v).

(a) If the traffic signal stays green for some time (t_g), how many cars (N) will make it through the light?

(b) If the light remains green for twice the time, how many more cars will get through the light?

(c) If the speed of each car is doubled when it begins to move, will twice as many cars get through the light? If not, what variable would have to go to zero for this to be true?

(d) For a reaction time of zero and no space between cars, find an expression for the number of cars that will pass through the light. Does this make sense?

ICA 6-9

Suppose that the earth were a smooth sphere and you could wrap a 25,000-mile-long band snugly around it. Now let us say that you lengthen the band by 10 feet, loosening it just a little. What would be the largest thing that could slither under the new band (assume that it is now raised above the earth's surface equally all the way around so that it doesn't touch anywhere): an amoeba, a snake, or an alligator?

ICA 6-10

Try to make at least five observations for the problems given in ICA 6-2. Sketch and label if possible. Note that for some problems, it may not be necessary to make any observations, and in others, five may not be required. Make a note of the two or three most important observations in each case.

ICA 6-11

Construct and label a simple sketch of the problems in ICA 6-2 if possible.

REVIEW QUESTIONS

Analyze the following problems using the SOLVEM method.

1. A motorcycle weighing 500 pounds-mass plus a rider weighing 300 pounds-mass produces the following chart. Predict a similar table if a 50-pound-mass dog is added as a passenger.

Velocity (v) [mi/h]	Time (t) [s]
0	0.0
10	2.3
20	4.6
30	6.9
40	9.2

2. A circus performer jumps from a platform onto one end of a seesaw, while his or her partner, a child of age 12, stands on the other end. How high will the child "fly"?

3. Your college quadrangle is 85 meters long and 66 meters wide. When you are late for class, you can walk (well, run) at 7 miles per hour. You are at one corner of the quad, and your class is at the directly opposite corner. How much time can you save by cutting across the quad rather than walking around the edge?

4. I am standing on the upper deck of the football stadium. I have an egg in my hand. I am going to drop it, and you are going to try to catch it. You are standing on the ground. Apparently, you do not want to stand directly under me; in fact, you would like to stand as far to one side as you can so that if I accidentally release it, it won't hit you on the head. If you can run at 20 feet per second and I am at a height of 100 feet, how far away can you stand and still catch the egg if you start running when I let go?

5. A 1-kilogram mass has just been dropped from the roof of a building. I need to catch it after it has fallen exactly 100 meters. If I weigh 80 kilograms and start running at 7 meters per second as soon as the object is released, how far away can I stand and still catch the object?

6. Neglect the weight of the drum in the following problem. A sealed cylindrical drum has a diameter of 6 feet and a length of 12 feet. The drum is filled exactly half-full of a liquid having a density of 90 pounds-mass per cubic foot. It is resting on its side at the bottom of a 10-foot-deep drainage channel that is empty. Suppose a flash flood suddenly raises the water level in the channel to a depth of 10 feet. Determine if the drum will float. The density of water is 62 pounds-mass per cubic foot.

Chapter 7

Fundamental Dimensions and Base Units

7.1 The Metric System

7.2 Other Unit Systems

7.3 Conversion Procedure for Units

7.4 Conversions Involving Multiple Steps

7.5 Conversions Involving "New" Units

7.6 Derived Dimensions and Units

7.7 Equation Laws

7.8 Conversion Involving Equations

Chapter 8

Universal Units

8.1 Force

8.2 Weight

8.3 Density

8.4 Amount

8.5 Temperature

8.6 Pressure

8.7 Gas Pressure

8.8 Energy

8.9 Power

8.10 Efficiency

8.11 Electrical Concepts

Chapter 9

Dimensionless Numbers

9.1 Constants With Units

9.2 Common Dimensionless Numbers

9.3 Dimensional Analysis

9.4 Rayleigh's Method

Source: World History Archive/Alamy

Engraving from Mechanics Magazine published in London in 1824.

Imagine you are in a small boat with a large stone in the bottom of the boat. The boat is floating in the swimming pool in the campus recreation center. What happens to the level of water in the pool if you throw the stone overboard? Assume no water splashes out of the pool or into the boat.

Archimedes was a Greek scientist and mathematician. Most people know Archimedes for his discovery of buoyancy. According to legend, the king asked Archimedes to determine if his new crown was made of pure gold. Before this, no method had been developed for measuring the density of irregularly shaped objects. While taking a bath, Archimedes noted that the water rose in proportion to the amount of his body in the tub. He shouted "Eureka (I have found it)!" and ran though the streets naked because he was so excited he forgot to get dressed. While Archimedes never recounts this tale himself, he does outline Archimedes' principle

Learning Objectives

The overall learning objectives for this unit include the following:

Chapter 7:

- Identify basic and derived dimensions and units.
- Express observations in appropriate units and perform conversions when necessary. Apply the laws governing equation development to aid in problem solutions.

Chapter 8:

- Apply basic principles from mathematical and physical sciences, such as the conservation of energy and the ideal gas law, to analyze engineering problems.
- Convert units for physical and chemical parameters such as density, energy, pressure,

and power as required for different systems of units.

- Use dimensions and units to aid in the solution of complex problems.

Chapter 9:

- Identify when a quantity is dimensionless.
- Using a graph of dimensionless groups, extract information from the plot about the physical system.
- Given a set of parameters, determine appropriate dimensionless groups using Rayleigh's Method.
- Determine the Reynolds Number; interpret the Reynolds Number for fluid flow in a pipe.

in his treatise *On Floating Bodies:* **A body immersed in a fluid is buoyed up by a force equal to the weight of the displaced fluid**. Before we can begin to answer the question of the boat and the stone (the answer is found on the final page of Chapter 8), we need to understand the principles of dimensions and units

"Give me a place to stand on, and I will move the Earth."

— Archimedes

In addition to buoyancy, Archimedes made many contributions to science, including the explanation of the lever, and is considered one of the greatest mathematicians.

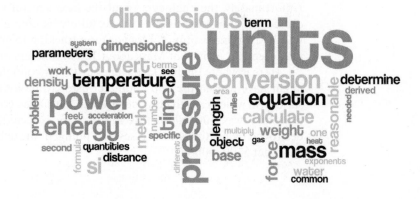

Lessons of the Mars Climate Orbiter

Some of you may have heard that the loss of the Mars Climate Orbiter (MCO) spacecraft in 1999 was due to a unit conversion error. The complete story is rather more complicated and illustrates a valuable lesson in engineering design. Most engineering failures are not due to a single mistake, since built-in redundancies and anticipation of failure modes make this unlikely. Three primary factors (plus bad luck) conspired to send the MCO off course.

Source: NASA

First, the spacecraft was asymmetrical, with the body of the spacecraft on one side and a large solar panel on the other. You might think shape is not an issue in the vacuum of space, but in fact it is, and the NASA engineers were well aware of the potential problems. The panel acted like a sail, causing the craft to slowly change its orientation and requiring the MCO to make occasional small corrections by firing thrusters onboard the craft. This was a perfectly manageable "problem."

Second, the software on the spacecraft expected thruster data in SI units, requiring the force expressed in newtons. On the Earth, a separate system calculated and sent instructions to the MCO concerning when and how long to fire its thrusters. The Earth-based system relied on software from an earlier Mars mission, and the thruster equations had to be modified to correct for the thrusters used on the new spacecraft. The original software had been written correctly, with the conversion factor from pound-force to newtons included. However, this conversion was neither documented nor obvious from the code, being buried in the equations. When the equations were rewritten, the programmers were unaware of the conversion factor and it was left out of the new code. This sent incorrect thruster-firing data to the MCO, specifically being too small by a factor of 4.45. This problem alone was manageable by comparing the calculated trajectory with tracking data.

Finally, after the third trajectory correction, the MCO entered "safe mode" while adjusting the solar panel, indicating a fault on the craft. At about the same time, the preliminary indications that the spacecraft trajectory was flawed began to come in. Unfortunately, the engineers spent the next several weeks trying to determine what caused the craft to enter safe mode, falsely assuming the preliminary trajectory data was in error and waiting for longer-term tracking to give a better estimate. In the end, the spacecraft arrived at Mars about 100 kilometers off course.

Here is where the bad luck comes in. Other configurations of the craft or trajectory might have caused the 100 kilometer error to be away from Mars or parallel to the surface, in which case the trajectory could have been corrected later. Unfortunately, the trajectory was 100 kilometers lower than expected, and the MCO was probably destroyed by heating and stresses as it plunged through the Martian atmosphere. Cost: well over $100 million.

Fundamental Dimensions and Base Units

As aspiring engineers you must learn to distinguish among many terms that laymen tend to use interchangeably. You must also understand the technical meaning of terms that are misunderstood by those untrained in science or engineering. One term that is often misunderstood is "dimension." To most people, a dimension refers to a straight line (length, one dimension), a flat surface (area, two dimensions), or a solid object (volume, three dimensions). Some folks might include time as a fourth dimension. The number of concepts classified as dimensions is far broader.

A **dimension** is a measurable physical idea; it generally consists solely of a word description with no numbers. A **unit** allows us to quantify a dimension—to state a number describing how much of that dimension exists in a specific situation. Units are defined by convention and related to an accepted standard.

- Length is a dimension. There are many units for length, such as mile, foot, meter, light-year, and fathom.

- Time is a dimension. There are many units for time, such as second, minute, hour, day, fortnight, year, and century.

- Temperature is a dimension. There are many units for temperature, such as Celsius, Fahrenheit, and kelvin.

The dimensions of length, time, and temperature are familiar to us, but in reality, we do not often use these words since they are fairly vague.

We do not say . . .	We do say . . .
It is really a long **length** to Lumberton.	Lumberton is about 175 **miles** away.
Bake the cake for a **time**.	Bake the cake for 35 **minutes**.
Set the oven to a high **temperature**.	Set the oven to 450 **degrees Fahrenheit**.

The difference between the left and the right columns is that the statements on the left refer to dimensions and those on the right refer to established standards or units.

7.1 The Metric System

LEARN TO: List the seven fundamental dimensions and their symbols
List the seven base SI units, their symbols, and the matching fundamental dimensions
Express units using the official SI rules

NOTE

Within this text, dimensions are shown in braces { } and units in brackets [ ].

The SI system (Le Système International d'Unités), commonly known as the metric system, is the standard set of units for most of the world. Originally developed by French scientists under King Louis XVI, the SI system was finalized by the international scientific community as the standard unit system in 1971. This system defines seven base units, from which all others are derived. Table 7-1 shows the seven base units and their corresponding fundamental dimensions.

Table 7-1 Fundamental dimensions and base units

Dimension	Symbol	Unit	Symbol
Mass	M	kilogram	kg
Length	L	meter	m
Time	T	second	s
Temperature	Θ	kelvin	K
Amount of substance	N	mole	mol
Light intensity	J	candela	cd
Electric current	I	ampere	A

SI Prefixes

The SI system is based upon multiples of 10. By using an **SI prefix** when reporting numbers, we avoid long strings of zeros. For example, instead of saying, "The distance to Atlanta is 198,000 meters," we would say, "The distance to Atlanta is 198 kilometers."

For a list of SI prefixes, refer to the inside cover of this book or to Table 7-2. Note that the abbreviations for all SI prefixes from kilo- down to yocto- are lowercase, whereas from Mega- up to Yotta- are uppercase.

Determining the appropriate SI prefix to use becomes simple when the number is placed in engineering notation: just examine the exponent. As a reminder, scientific and engineering notation are defined as follows:

Scientific notation is typically expressed in the form $\#.\#\#\# \times 10^N$, where the single digit to the left of the decimal point is the most significant nonzero digit of the value being represented. N is an integer, and multiplying by 10^N serves to locate the true position of the decimal point.

Engineering notation is expressed in the form $\#\#\#.\#\#\# \times 10^M$, where M is an integer multiple of 3, and the number of digits to the left of the decimal point is one, two, or three as needed to yield a power of 10 that is indeed a multiple of 3.

Table 7-2 SI prefixes (example: 1 millimeter [mm] = 1 × 10^{-3} meters [m])

Numbers Less than One			Numbers Greater than One		
Power of 10	Prefix	Abbreviation	Power of 10	Prefix	Abbreviation
10^{-1}	deci-	d	10^{1}	deca-	da
10^{-2}	centi-	c	10^{2}	hecto-	h
10^{-3}	milli-	m	10^{3}	kilo-	k
10^{-6}	micro-	μ	10^{6}	Mega-	M
10^{-9}	nano-	n	10^{9}	Giga-	G
10^{-12}	pico-	p	10^{12}	Tera-	T
10^{-15}	femto-	f	10^{15}	Peta-	P
10^{-18}	atto-	a	10^{18}	Exa-	E
10^{-21}	zepto-	z	10^{21}	Zetta-	Z
10^{-24}	yocto-	y	10^{24}	Yotta-	Y

EXAMPLE   7-1

Express the following values using scientific notation, engineering notation, and using the correct SI prefix.

Standard	Scientific	Engineering	With Prefix
(a) 43,480,000 m	4.348×10^{7} m	43.48×10^{6} m	43.48 Mm
(b) 0.0000003060 V	3.060×10^{-7} V	306.0×10^{-9} V	306.0 nV
(c) 9,860,000,000 J	9.86×10^{9} J	9.86×10^{9} J	9.86 GJ
(d) 0.0351 s	3.51×10^{-2} s	35.1×10^{-3} s	35.1 ms

Note that the numeric values of the mantissa are the same in the last two columns, and the exponent in engineering notation specifies the metric prefix.

Comprehension Check   7-1

Express the following values using an appropriate SI prefix such that there are only one, two, or three digits shown to the left of the decimal.

Standard	With Prefix
(a) 3,100 joules [J]	
(b) 26,510,000 watts [W]	
(c) 459,000 seconds [s]	
(d) 0.00000032 grams [g]	

Official SI Rules

When reporting units using the SI system, follow these official rules.

- **If a unit abbreviation appears as a capital letter, it has been named after a person; all other abbreviations appear as lowercase letters.** For example, the abbreviation "N" stands for "newton," the SI unit of force named after Isaac Newton.

  Correct: The book weighs 5 N.   Incorrect: The book weighs 5 n.
  Correct: The rod is 5 m long.   Incorrect: The rod is 5 M long.

  The one exception to this rule is the volumetric unit of liter. The abbreviation is shown as L, since a lowercase l can be confused with both the number 1 and the uppercase letter I.

- **Abbreviations of units are not shown as plural.**

  Correct: 10 centimeters = 10 cm   Incorrect: 10 centimeters ≠ 10 cms

- **Abbreviations are not shown with a period unless they appear at the end of a sentence.**

  Correct: The rod is 5 mm long.   Incorrect: The rod is 5 mm. long.

- **Abbreviations are written in upright Roman type (m, k, L) to distinguish them from mathematical variables (m, k, l), which are indicated by italics.**

- **One space separates the number and abbreviation, except with the degree symbol referring to an angle.**

  Correct: 5 mm or 5°   Incorrect: 5 mm or 5 °

- **Spaces or commas may be used to group digits by threes.**

  Correct: 1 000 000 or 1,000,000   Incorrect: 100,0000

- **Abbreviations for derived units formed by multiple units are joined by a space or interpunct (the center dot).** Care must be taken to avoid confusing SI prefixes with units.

  Correct: kg m or kg · m   Incorrect: kgm or kg * m

  This is particularly important when confusion might arise. For example, "ms" stands for millisecond, but "m s" stands for meter second. In cases like this, using a center dot (m·s) is preferable since it is less likely to be misunderstood.

- **Abbreviations for derived units formed by dividing units are joined by a virgule (the "slash" /) or shown with a negative exponent.** Care must be taken to appropriately display the entire denominator.

  Correct: $N/(m\ s^2)$ or $N\ m^{-1}\ s^{-2}$ Incorrect: $N/m\ s^2$ or $N/m/s^2$

- **Do not combine prefixes to form a compound prefix. Use a single prefix.**

  Correct: picojoules (pJ)   Incorrect: millinanojoules (mnJ)
  Correct: gigaseconds (Gs)   Incorrect: kilomegaseconds (kMs)

Comprehension Check | **7-2**

Indicate if the following units are correctly expressed according to the official SI rules. If the unit is incorrectly displayed, show the correction.

(a) Reading this sentence took 5 Secs.

(b) The average person's pupils are 60mms. apart.

(c) One gallon is the same as 380 microkiloliters.

7.2 Other Unit Systems

Prior to the adoption of the SI unit system by the scientific community, several other systems of units were used and are still used today, particularly in the United States. Great Britain officially converted to metric in 1965, but it is still common there to see nonmetric units used in communications for the general public.

It is important to know how to convert between all unit systems. Table 7-3 compares three systems. The system listed as AES (American Engineering System) is in common use by the general public in the United States. The USCS (United States Customary System) is commonly called "English" units.

Table 7-3 Comparison of unit systems, with corresponding abbreviations

Dimension	SI (MKS)	AES	USCS
Length	meter [m]	foot [ft]	foot [ft]
Mass	kilogram [kg]	pound-mass [lb$_m$]	slug
Time	second [s]	second [s]	second [s]
Temperature	Celsius [°C]	Fahrenheit [°F]	Fahrenheit [°F]
	kelvin [K]	Rankine [°R]	Rankine [°R]

Accepted Non-SI Units

The units in Table 7-4 are not technically in the SI system, but due to their common usage, are acceptable for use in combination with the base SI units.

Table 7-4 Acceptable non-SI units

Unit	Equivalent SI	Unit	Equivalent SI
Astronomical unit [AU]	1 AU = 1.4959787 × 10^{11} m	day [d]	1 d = 86,400 s
Atomic mass unit [amu]	1 amu = 1.6605402 × 10^{-24} g	hour [h]	1 h = 3600 s
Electronvolt [eV]	1 eV = 1.6021773 × 10^{-19} J	minute [min]	1 min = 60 s
Liter [L]	1 L = 0.001 m^3	year [yr]	1 yr = 3.16 × 10^7 s
		degree [°]	1° = 0.0175 rad or 1 rad = 57.3°

NOTE

1 liter does not equal 1 cubic meter!

7.3 Conversion Procedure for Units

We use conversion factors to translate from one set of units to another. Some common conversion factors can be found inside the cover of this book, categorized by dimension. Although many more conversions are available, all the work for a typical engineering class can be accomplished using the conversions found in this table.

LENGTH

1 m = 3.28 ft

1 km = 0.621 mi

1 ft = 12 in

1 in = 2.54 cm

1 mi = 5280 ft

1 yd = 3 ft

Let us examine the conversions found for the dimension of length, as shown in the box, beginning with the conversion: 1 meter [m] = 3.28 feet [ft]. By dividing both sides of this equation by 3.28 feet, we obtain

$$\frac{1 \text{ m}}{3.28 \text{ ft}} = 1$$

or in other words, "There is 1 meter per 3.28 feet." If we divide both sides of the original expression by 1 meter, we obtain

$$1 = \frac{3.28 \text{ ft}}{1 \text{ m}}$$

or in other words, "In every 3.28 feet there is 1 meter."

The number 1 is dimensionless, a pure number. *We can multiply any expression by 1 without changing the expression.* We use this rule to convert between unit systems.

For example, on a trip we note that the distance to Atlanta is 123 miles [mi]. How many kilometers [km] is it to Atlanta? From the conversion table, we can find that 1 kilometer [km] = 0.621 miles [mi], or

$$1 = \frac{1 \text{ km}}{0.621 \text{ mi}}$$

By multiplying the original quantity of 123 miles by 1, we can say

$$(123 \text{ mi}) (1) = (123 \text{ mi})\left(\frac{1 \text{ km}}{0.621 \text{ mi}}\right) = 198 \text{ km}$$

Note that we could have multiplied by the following relationship:

$$1 = \frac{0.621 \text{ mi}}{1 \text{ km}}$$

We would still have multiplied the original answer by 1, but the units would not cancel and we would be left with an awkward, meaningless answer.

$$(123 \text{ mi}) (1) = (123 \text{ mi})\left(\frac{0.621 \text{ mi}}{1 \text{ km}}\right) = 76 \frac{\text{mi}^2}{\text{km}}$$

We want to use the ratio that will make the unit we want to eliminate cancel from the original expression.

As a second example, we are designing a reactor system using 2-inch [in] diameter plastic pipe. The design office in Germany would like the pipe specifications in units of centimeters [cm]. From the conversion table, we find that 2.54 centimeter [cm] = 1 inch [in], or

$$1 = \frac{2.54 \text{ cm}}{1 \text{ in}}$$

By multiplying the original quantity of 2 inches by 1, we can say

$$(2 \text{ in}) (1) = (2 \text{ in})\left(\frac{2.54 \text{ cm}}{1 \text{ in}}\right) = 5 \text{ cm}$$

In a final example, suppose a car travels at 40 miles per hour [mi/h or mph]. Stated in words, "a car traveling at a rate of 40 mph will take 1 hour to travel 40 miles if the velocity remains constant." By simple arithmetic this means that the car will travel 80 miles in 2 hours or 120 miles in 3 hours. In general,

$$\text{Distance} = (\text{velocity})(\text{time elapsed at that velocity})$$

Suppose the car is traveling at 40 mph for 6 minutes. How far does the car travel? Simple calculation shows

$$\text{Distance} = (40)(6) = 240\ \text{miles}$$

TIME

1 d = 24 h

1 h = 60 min

1 min = 60 s

1 yr = 365 d

Without considering units, the preceding example implies that if we drive our car at 40 mph, we can cover the distance from Charlotte, North Carolina, to Atlanta, Georgia, in 6 minutes! What is wrong? Note that the velocity is given in miles per hour, and the time is given in minutes. We need to apply the conversion factor that 1 hour [h] = 60 minutes [min]. If the equation is written with consistent units attached, we get

$$\text{Distance} = \left(\frac{40\ \text{mi}}{\text{h}}\right)\left(\frac{6\ \text{min}}{}\left|\frac{1\ \text{h}}{60\ \text{min}}\right.\right) = 4\ \text{mi}$$

IMPORTANT CONCEPT

Be sure to *always* include units in your calculations and your final answer!

It seems more reasonable to say "traveling at a rate of 40 miles per hour for a time period of 6 minutes will allow us to go 4 miles."

To convert between any set of units, the following method demonstrated in Examples 7-2 to 7-8 is very helpful. This procedure is easy to use, but take care to avoid mistakes. If you use one of the conversion factors incorrectly, say, with 3 in the numerator instead of the denominator, your answer will be in error by a factor of 9.

Unit Conversion Procedure

1. Write the value and unit to be converted.
2. Write the conversion formula between the given unit and the desired unit.
3. Make a fraction, equal to 1, of the conversion formula in Step 2, such that the original unit in Step 1 is located so the original unit will cancel.
4. Multiply the term from Step 1 by the fraction developed in Step 3.
5. Cancel units, perform mathematical calculations, and express the answer in "reasonable" terms.

EXAMPLE   7-2

Convert the length 40 yards [yd] into units of feet [ft].

Method	Steps
(1) Term to be converted	40 yd
(2) Conversion formula	1 yd = 3 ft
(3) Make a fraction (equal to one)	$\dfrac{3\ \text{ft}}{1\ \text{yd}}$
(4) Multiply	$\dfrac{40\ \text{yd}}{}\left|\dfrac{3\ \text{ft}}{1\ \text{yd}}\right.$
(5) Cancel, calculate, be reasonable	120 ft

EXAMPLE   7-3

Convert the time 456,000 seconds [s] into units of minutes [min].

Method	Steps
(1) Term to be converted	456,000 s
(2) Conversion formula	1 min = 60 s
(3) Make a fraction (equal to one)	$\dfrac{1\text{ min}}{60\text{ s}}$
(4) Multiply	$\dfrac{456{,}000\text{ s}}{}\left\|\dfrac{1\text{ min}}{60\text{ s}}\right.$
(5) Cancel, calculate, be reasonable	7600 min

Comprehension Check   7-3

The highest mountain in the world is Mount Everest in Nepal. The peak of Mount Everest is 29,029 feet [ft] above sea level. Convert the height into units of miles [mi].

Comprehension Check   7-4

To be considered a full-time employee, companies in the United States required you work more than 30 hours [h] in a week. Convert the time into units of minutes [min].

7.4 Conversions Involving Multiple Steps

LEARN TO: Follow the 5-step conversion procedure to convert units when multiple steps are required

LENGTH

1 m = 3.28 ft

1 km = 0.621 mi

1 ft = 12 in

1 in = 2.54 cm

1 mi = 5280 ft

1 yd = 3 ft

Sometimes, more than one conversion factor is needed in a single quantity. We can multiply by several conversion factors, each one of which is the same as multiplying by 1. For example, suppose we determined that the distance to Atlanta is 123 miles [mi]. How many yards [yd] is it to Atlanta? From the conversion table, we do not have a direct conversion between miles and yards, but we see that both can be related to feet. We can find that 5280 feet [ft] = 1 mile [mi], or

$$1 = \frac{5280\text{ ft}}{1\text{ mi}}$$

We can also find that 1 yard [yd] = 3 feet [ft], or

$$1 = \frac{1\text{ yd}}{3\text{ ft}}$$

By multiplying the original quantity of 123 miles by 1 using the first set of conversion factors, we can say:

$$(123 \text{ mi})(1) = (123 \text{ mi})\left(\frac{5280 \text{ ft}}{1 \text{ mi}}\right) = 649{,}440 \text{ ft}$$

If we multiply by 1 again, using the second set of conversion factors and applying reasonableness:

$$(649{,}440 \text{ ft})(1) = (649{,}440 \text{ ft})\left(\frac{1 \text{ yd}}{3 \text{ ft}}\right) = 216{,}000 \text{ yd}$$

This is usually shown as a single step:

$$(123 \text{ mi})\left(\frac{5280 \text{ ft}}{1 \text{ mi}}\right)\left(\frac{1 \text{ yd}}{3 \text{ ft}}\right) = 216{,}000 \text{ yd}$$

EXAMPLE 7-4

Convert the length 40 yards [yd] into units of millimeters [mm].

Method	Steps
(1) Term to be converted	40 yd
(2) Conversion formula	1 yd = 3 ft 1 ft = 12 in 1 in = 2.54 cm 1 cm = 10 mm
(3) Make fractions (equal to one)	$\dfrac{3 \text{ ft}}{1 \text{ yd}} \quad \dfrac{12 \text{ in}}{1 \text{ ft}} \quad \dfrac{2.54 \text{ cm}}{1 \text{ in}} \quad \dfrac{10 \text{ mm}}{1 \text{ cm}}$
(4) Multiply	$\dfrac{40 \text{ yd}}{} \left\lvert \dfrac{3 \text{ ft}}{1 \text{ yd}} \right\rvert \dfrac{12 \text{ in}}{1 \text{ ft}} \left\lvert \dfrac{2.54 \text{ cm}}{1 \text{ in}} \right\rvert \dfrac{10 \text{ mm}}{1 \text{ cm}}$
(5) Cancel, calculate, be reasonable	37,000 mm

Here, there are several other viable solution paths. For example, we could have converted yards to feet, then feet to meters, then meters to millimeters.

EXAMPLE 7-5

Convert 55 miles per hour [mi/h or mph] to units of meters per second [m/s].

NOTE

We have two unit groups to convert here, miles to meters, and hours to seconds.

Method	Steps
(1) Term to be converted	55 mph
(2) Conversion formula	1 km = 0.621 mi 1 km = 1000 m 1 h = 60 min 1 min = 60 s
(3) Make fractions (equal to one) (4) Multiply	$\dfrac{55 \text{ mi}}{\text{h}} \left\lvert \dfrac{1 \text{ km}}{0.621 \text{ mi}} \right\rvert \dfrac{1{,}000 \text{ m}}{1 \text{ km}} \left\lvert \dfrac{1 \text{ h}}{60 \text{ min}} \right\rvert \dfrac{1 \text{ min}}{60 \text{ s}}$
(5) Cancel, calculate, be reasonable	24.6 m/s

EXAMPLE 7-6

Convert the power of 3,780,000 kilowatts [kW] into units of gigawatts [GW].

NOTE

Treat SI prefixes as units

Method	Steps
(1) Term to be converted	3,780,000 kW
(2) Conversion formula	$1\,kW = 1 \times 10^3\ W$ $1\,GW = 1 \times 10^9\ W$
(3) Make a fraction (equal to one)	$\dfrac{1 \times 10^3\ W}{1\ kW} \qquad \dfrac{1\ GW}{1 \times 10^9\ W}$
(4) Multiply	$\dfrac{3{,}780{,}000\ kW}{} \left\| \dfrac{1 \times 10^3\ W}{1\ kW} \right\| \dfrac{1\ GW}{1 \times 10^9\ W}$
(5) Cancel, calculate, be reasonable	3.78 GW

EXAMPLE 7-7

Convert the volume of 40 gallons [gal] into units of cubic feet [ft³].

By examining the "Volume" box in the conversion table, we see that the following facts are available for use:

$$1\ L = 0.264\ gal \qquad and \qquad 1\ L = 0.0353\ ft^3$$

By the transitive property, if a = b and a = c, then b = c. Therefore, we can directly write

$$0.264\ gal = 0.0353\ ft^3$$

VOLUME

$1\ L = 0.264\ gal$
$1\ L = 0.0353\ ft^3$
$1\ L = 33.8\ fl\ oz$
$1\ m^3 = 1000\ L$
$1\ mL = 1\ cm^3$

Method	Steps
(1) Term to be converted	40 gal
(2) Conversion formula	$0.264\ gal = 0.0353\ ft^3$
(3) Make fractions (equal to one)	$\dfrac{0.0353\ ft^3}{0.264\ gal}$
(4) Multiply	$\dfrac{40\ gal}{}\left\| \dfrac{0.0353\ ft^3}{0.264\ gal}\right.$
(5) Cancel, calculate, be reasonable	$5.3\ ft^3$

This picture shows 5-gallon water bottles made from polycarbonate. Millions of these bottles are made each year around the world to transport clean water to remote locations.

The use of polycarbonate to contain products for consumption has raised safety concerns because bisphenol A is leached from the plastic into the stored liquid. In July 2012, the US Food and Drug Administration banned the use of BPA in bottles and cups used by infants and small children.

Photo courtesy of E. Stephan

RULES OF THUMB

1 quart ≈ 1 liter	1 cubic foot ≈ 7.5 gallons
1 cubic meter ≈ 250 gallons	1 cubic meter ≈ 5, 55-gallon drums
1 cup ≈ 250 milliliters	1 golf ball ≈ 1 cubic inch

One frequently needs to convert a value that has some unit or units raised to a power, for example, converting a volume given in cubic feet to cubic meters. It is critical in this case that the power involved be applied to the *entire* conversion factor, both the numerical values and the units.

EXAMPLE   7-8

Convert 35 cubic inches [in^3] to cubic centimeters [cm^3].

NOTE

When raising a quantity to a power, be sure to apply the power to both the value and the units.

Method	Steps
(1) Term to be converted	35 in^3
(2) Conversion formula	1 in = 2.54 cm
(3) Make fractions (equal to one)	$\dfrac{(2.54 \text{ cm})^3}{(1 \text{ in})^3}$
(4) Multiply	$\dfrac{35 \text{ in}^3}{} \left| \dfrac{(2.54)^3 \text{ cm}^3}{1 \text{ in}^3}\right.$
(5) Cancel, calculate, be reasonable	574 cm^3

Note that in some cases, a unit that is raised to a power is being converted to another unit that has been defined to have the same dimension as the one raised to a power. This is difficult to say in words, but a couple of examples should clarify it.

If one is converting square meters [m^2] to acres, the conversion factor is *not* squared, since the conversion provided is already in terms of length squared: 1 acre = 4,047 m^2.

If one is converting cubic feet [ft^3] to liters [L], the conversion factor is *not* cubed, since the conversion provided is already in terms of length cubed: 1 L = 0.0353 ft^3.

Comprehension Check   7-5

In January 2008, *Scientific American* reported that physicists Peter Sutter and Eli Sutter of Brookhaven National Laboratories made a pipette to measure droplets in units of a zeptoliter [zL]. Previously, the smallest unit of measure in a pipette was an attoliter [aL]. Convert the measurement of 5 zeptoliters [zL] into units of picoliters [pL].

Comprehension Check	7-6

Officially, a hurricane is a tropical storm with sustained winds of at least 74 miles per hour [mi/h or mph]. Convert this speed into units of kilometers per minute [km/min].

Comprehension Check	7-7

Many toilets in commercial establishments have a value printed on them stating the amount of water consumed per flush. For example, a label of 2 Lpf indicates the consumption of 2 liters per flush. If a toilet is rated at 3 Lpf, how many flushes are required to consume 20 gallons [gal] of water?

7.5 Conversions Involving "New" Units

LEARN TO: Apply the 5-step conversion procedure to any units

In the past, many units were derived from common physical objects. The "inch" was the width of man's thumb, and the "foot" was the heel-to-toe length of a king's shoe. Obviously, when one king died or was deposed and another took over, the unit of "foot" changed, too. Over time, these units were standardized and have become common terminology.

New units are added as technology evolves; for example, in 1999 the unit of katal was added as an SI derived unit of catalytic activity used in biochemistry. As you proceed in your engineering field, you will be introduced to many "new" units. The conversion procedures discussed here apply to *any* unit in *any* engineering field.

EXAMPLE	7-9

According to the U.S. Food and Drug Administration (21CFR101.9), the following definition applies for nutritional labeling:

1 fluid ounce means 30 milliliters

Using this definition, how many fluid ounces [fl oz] are in a "U.S. standard" beverage can of 355 milliliters [mL]?

Method	Steps
(1) Term to be converted	355 mL
(2) Conversion formula	1 fl oz = 30 mL
(3) Make a fraction (equal to one)	$\dfrac{355 \text{ mL}}{} \Big| \dfrac{1 \text{ fl oz}}{30 \text{ mL}}$
(4) Multiply	
(5) Cancel, calculate, be reasonable	11.8 fl oz

EXAMPLE	7-10

The volume of water in a reservoir or aquifer is often expressed using the unit of acre-foot. A volume of 1 acre-foot is the amount of water covering an area of 1 acre to a depth of 1 foot [ft].

Lake Mead, located 30 miles southeast of Las Vegas, Nevada, is the largest man-made lake in the United States. It holds approximately 28.5 million acre-feet of water behind the Hoover Dam. Convert this volume to units of gallons [gal].

Method	Steps
(1) Term to be converted	28.5×10^6 acre feet
(2) Conversion formula	1 acre $= 4047 \text{ m}^2$ 1 m $= 3.28$ ft $1 \text{ m}^3 = 1000$ L 1 L $= 0.264$ gal
(3) Make a fraction (4) Multiply	$\dfrac{28.5 \times 10^6 \text{ acre ft}}{}$ $\left\vert \dfrac{4047 \text{ m}^2}{1 \text{ acre}} \right\vert \dfrac{1 \text{ m}}{3.28 \text{ ft}} \left\vert \dfrac{1000 \text{ L}}{1 \text{ m}^3} \right\vert \dfrac{0.264 \text{ gal}}{1 \text{ L}}$
(5) Cancel, calculate, be reasonable	9.3×10^{12} gal

Comprehension Check	7-8

A hogshead is a unit of volume describing a large barrel of liquid. Convert 10 hogsheads into units of cubic feet [ft³]. Conversion factor: 1 hogshead = 63 gallons [gal].

Comprehension Check	7-9

In NCAA basketball, a "three-point shot" is defined by an arc radius of 20 feet [ft], 9 inches [in]. Convert this length to units of cubits. Conversion factor: 1 cubit = 0.45 meters [m].

Comprehension Check	7-10

A boat is traveling at 20 knots. Convert this speed to units of meters per second [m/s]. Conversion factor: 1 knot = 1 nautical mile per hour; 1 nautical mile = 6076 feet [ft].

7.6 Derived Dimensions and Units

LEARN TO:	Identify a quantity as a fundamental or derived dimension and express the fundamental dimensions of the quantity using fractional or exponential notation
	Given the units of a quantity, determine the fundamental dimensions
	Given the fundamental dimensions of a quantity, determine the base SI units

With only the seven base dimensions in the metric system, all measurable things in the known universe can be expressed by various combinations of these concepts, called **derived dimensions**. As simple examples, area is length squared, volume is length cubed, and velocity is length divided by time.

As we explore more complex parameters, the dimensions become more complex. For example, the concept of force is derived from Newton's second law, which states that force is equal to mass times acceleration. Force is then used to define more complex dimensions such as pressure, which is force acting over an area, or work, which is force acting over a distance. As we introduce new concepts in later chapters, we introduce the dimensions and units for each parameter.

Sometimes, the derived dimensions become quite complicated. For example, electrical resistance is mass times length squared divided by both time cubed and current squared. Particularly in the more complicated cases like this, a **derived unit** is defined to avoid having to say things like "The resistance is 15 kilogram-meter squared divided by second cubed ampere squared." It is much easier to say "The resistance is 15 ohms," where the derived unit "ohm" equals one $(kg\ m^2)/(s^3\ A^2)$.

Within this text, dimensions are presented in exponential notation rather than fractional notation for written clarity.

Quantity	Fractional Notation	Exponential Notation
Velocity	$\dfrac{L}{T}$	$L^1\ T^{-1}$
Acceleration	$\dfrac{L}{T^2}$	$L^1\ T^{-2}$

VOLUME

1 L = 0.264 gal

1 L = 0.0353 ft³

1 L = 33.8 fl oz

1 m³ = 1000 L

1 mL = 1 cm³

One way to determine the dimensions of a quantity, such as volume, is to examine the common units used to express the quantity. While volume can be expressed in gallons, it can also be expressed as cubic feet or cubic meters. The units of cubic meters express volume in a manner easily transferred to dimensions. Remember, the boxes on the inside front cover of the textbook show units that have equivalent dimensions. The units of gallons and of cubic feet and of cubic meters are dimensionally equal to length cubed.

EXAMPLE	7-11

Determine the fundamental dimensions of the following quantities.

Quantity	Units	Equivalent Units	M	L	T	Θ	N	J	I
area	acres	m²	0	2	0	0	0	0	0
distance	yd	m	0	1	0	0	0	0	0
mass	slug	kg	1	0	0	0	0	0	0
temperature	°C	K	0	0	0	1	0	0	0
time	h	s	0	0	1	0	0	0	0

Currently, there are officially 22 named derived units in the SI system. All are named after famous scientists or engineers who are deceased. Five of the most common derived units can be found in Table 7-5 and on the back cover of the textbook. It is worth noting that numerous common derived dimensions do not have a corresponding named derived SI unit. For example, there is no named derived SI unit for the derived dimension velocity as there is for force (newton) or electrical resistance (ohm).

NOTE

One letter can represent several quantities in various engineering disciplines. For example, the letter "*P*" can indicate pressure, power, or vertical load on a beam. It is important to examine and determine the nomenclature in terms of the context of the problem presented.

Table 7-5 Common derived units in the SI system

Dimension	SI Unit	Base SI Units	Derived from
Force (F)	newton [N]	$1\,\text{N} = 1\dfrac{\text{kg m}}{\text{s}^2}$	$F = ma$ Force = mass times acceleration
Energy (E)	joule [J]	$1\,\text{J} = 1\,\text{N m} = 1\dfrac{\text{kg m}^2}{\text{s}^2}$	$E = Fd$ Energy = force times distance
Power (P)	watt [W]	$1\,\text{W} = 1\dfrac{\text{J}}{\text{s}} = 1\dfrac{\text{kg m}^2}{\text{s}^3}$	$P = E/t$ Power = energy per time
Pressure (P)	pascal [Pa]	$1\,\text{Pa} = 1\dfrac{\text{N}}{\text{m}^2} = 1\dfrac{\text{kg}}{\text{m s}^2}$	$P = F/A$ Pressure = force per area
Voltage (V)	volt [V]	$1\,\text{V} = 1\dfrac{\text{W}}{\text{A}} = 1\dfrac{\text{kg m}^2}{\text{s}^3\text{A}}$	$V = P/I$ Voltage = power per current

Similar to breaking down volume to be expressed as cubic meters, the named SI derived dimensions can be expressed in base SI units. Using the base SI units allows for the dimensions to be easily determined.

EXAMPLE 7-12

Determine the fundamental dimensions of the following quantity.

$F = ma$
$m\ [=]\ \text{kg}$
$a\ [=]\ \text{m/s}^2$

			Dimensions						
Quantity	Units	Equivalent Units	M	L	T	Θ	N	J	I
Force	newton	$\dfrac{\text{kg m}}{\text{s}^2}$	1	1	−2	0	0	0	0

By understanding that a newton is derived from Newton's second law, the fundamental dimensions become simple to determine. **The five common derived units in Table 7-5 occur so frequently in engineering calculations you will want to memorize each dimension and the equivalent base SI units.**

Comprehension Check 7-11

Determine the fundamental dimensions of the following quantities.

		Dimensions						
Quantity	Units	M	L	T	Θ	N	J	I
Density	$\text{lb}_\text{m}/\text{ft}^3$							
Evaporation	slug/h							
Flowrate	L/min							

Comprehension Check   **7-12**

Determine the fundamental dimensions of the following quantities.

		Dimensions						
Quantity	**Units**	**M**	**L**	**T**	**Θ**	**N**	**J**	**I**
Energy	calories [cal]							
Power	horsepower [hp]							
Pressure	atmospheres [atm]							

Dimensions can help us identify combinations of variables as a familiar quantity by examining their base SI units and fundamental dimensions.

EXAMPLE   **7-13**

Identify the quantity through the use of fundamental dimensions. Choose from the following quantities:

(A) Acceleration   **(B)** Energy   **(C)** Force

(D) Power   **(E)** Pressure   **(F)** Velocity

$$A\rho v^3$$

where:

A = area [acres]
ρ = density [kg/m³]
v = velocity [mi/h]

First, we can express each quantity individually in terms of fundamental dimensions:

A [=] acres *is dimensionally equivalent to* m² {=} L^2
ρ {=} M/L^3
v [mi/h] *is dimensionally equivalent to* m/s {=} L/T

Combining these quantities together in the given expression:

$$A\rho v^3 \; \{=\} \; \frac{L^2}{} \left| \frac{M}{L^3} \right| \left(\frac{L}{T}\right)^3 = \frac{L^2}{} \left| \frac{M}{L^3} \right| \frac{L^3}{T^3}$$

Note that since velocity is cubed in the original expression, the dimensions of velocity must be cubed.

This will simplify to:

$$\frac{M\,L^2}{T^3}$$

Using Table 7-5, this is equivalent to the dimensions of choice D, Power.

Comprehension Check	**7-13**

Identify the quantity through the use of fundamental dimensions. Choose from the following:

(A) Acceleration   **(B)** Energy   **(C)** Force

(D) Power   **(E)** Pressure   **(F)** Velocity

$$nRT$$

where:

n = amount [mol]
R = ideal gas constant [atm L/(mol K)]
T = temperature [K]

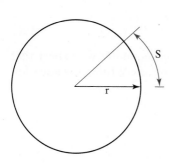

Special Unit: Radian

The derived unit of **radian** is defined as the angle at the center of a circle formed by an arc (S) equal in length to the radius (r) of that circle. In a complete circle there are 2π radians. Since by definition a radian is a length (S) divided by a length (r), it is a dimensionless ratio.

$$1 \text{ radian } [\text{rad}] = \frac{S}{r} \{=\} \frac{L}{L}$$

Thus, an angle has units, but is dimensionless! In addition to radians, another common unit used for angle is the degree [°]. There are 360° in a complete circle.

7.7 Equation Laws

LEARN TO: Determine if an expression is a valid using Plus law, Unit law, and Per law
Use the Plus law, Unit law, or Per law to determine dimensions of a quantity
Recognize that in equations, units must be consistent in order for resulting calculations to be valid

Equations are mathematical "sentences" composed of "words" (terms) that are strung together with "punctuation marks" (mathematical symbols, such as $+, -, \times, \div,$ and $=$). Just as there are rules in the English language that govern how a sentence is structured, there exists a set of "rules" for equations.

Addition and Subtraction

Suppose we are interested in the manufacture and use of sandpaper for furniture construction. We think for a while and then develop a list of the important quantities that affect the final product, along with their respective units and dimensions:

W	Wood removed	[in]	L
R	Roughness diameter	[mm]	L
D	Density of grains	[kg/m³]	$\dfrac{M}{L^3}$
A	Adhesive thickness	[mm]	L
H	How heavy the paper is	[N]	$\dfrac{M\,L}{T^2}$
O	Operation stroke length	[cm]	L
K	Kernel (grain) spacing	[mm]	L

Let us propose a simple equation with only plus and minus signs that could possibly relate several of these parameters. If we are interested in how heavy the product would be, we might assume this would depend on the thickness of the adhesive, the diameter of the roughness, and the grain density. We will try

$$H = A + R + D$$

Each of these terms represents something "real," and consequently we expect that each term can be expressed in terms of fundamental dimensions. Writing the equation in terms of dimensions given:

$$\frac{M\,L}{T^2} = L + L + \frac{M}{L^3}$$

IMPORTANT CONCEPT: PLUS LAW

Every term being added or subtracted in an equation must have the same dimension.

It is obvious that this is just terrible! We cannot add length and mass or time; as the adage goes, "You can't add apples and oranges!" The same holds true for dimensions. As a result of this observation, we see that this cannot possibly be a valid equation. This gives one important "law" governing equations, the **Plus law**.

Let us try this again with another equation to see if we can determine how effective the sandpaper will be, or how much wood will be removed after each stroke. We might assume this depends on the operation stroke length, the roughness diameter, and the spacing of the grains.

$$W = O + R + K$$

Substituting dimensions,

$$L = L + L + L$$

IMPORTANT CONCEPT: UNIT LAW

Every term in an equation must have the same units so that the arithmetic operations of addition and subtraction can be carried out.

We see that at least dimensionally, this can be a valid equation, based on the Plus law. Next, units can be inserted to give

$$\text{inches} = \text{centimeters} + \text{millimeters} + \text{millimeters}$$

Dimensionally, this equation is fine, but from the perspective of units, we cannot carry out the arithmetic above without first converting all the length dimensions into the same units, such as millimeters. We can state an important result from this observation forming the **Unit law**.

It is important to state a corollary to this observation. If two parameters have the same dimensions and units, it is not always meaningful to add or subtract them.

Two examples show this.

1. If Student A has a mass m_A [kilograms] and Student B a mass m_B [kg], then the total mass of both students [kg] is the sum of the two masses. This is correct and meaningful in both dimensions and units.

2. Suppose we assume that an equation to predict the mass of a car is this: mass of the car [kg] = mass of an oak tree [kg] + mass of an opossum [kg]. This equation has three terms; all with the dimension of mass and units of kilograms; thus, the terms can be added, although the equation itself is nonsense.

Consequently, the requirement that each term must have the same dimensions and units is a necessary, but not a sufficient, condition for a satisfactory equation.

Multiplication and Division

There are many ways to express the rate at which things are done. Much of our daily life is conducted on a "*per*" or rate basis. We eat 3 meals *per* day, have 5 fingers *per* hand, there are 11 players *per* team, 3 feet *per* yard, 4 tires *per* car, 12 fluid ounces *per* canned drink, and 4 people *per* quartet.

Although it is incorrect to add or subtract parameters with different dimensions, it is perfectly permissible to divide or multiply two or more parameters with different dimensions. This is another law of dimensions, the **Per law**.

When we say 65 miles *per* hour, we mean that we travel 65 miles in 1 hour. We could say we travel at 130 miles per 2 hours, and it would mean the same thing. Either way, this rate is expressed by the "*per*" ratio, distance per time.

One of the most useful applications of your knowledge of dimensions is in helping to determine if an equation is dimensionally correct. This is easy to do and only involves the substitution of the dimensions of every parameter into the equation and simplifying the resulting expressions. A simple application will demonstrate this process.

EXAMPLE   7-14

Is the following equation dimensionally correct?

$$t = \sqrt{\frac{d_{\text{final}} - d_{\text{initial}}}{0.5a}}$$

where t is time
d is distance
a is acceleration
0.5 is unitless

Determine the dimensions of each parameter:

Acceleration	*(a)*	$\{=\}\ L^1\ T^{-2}$
Distance	*(d)*	$\{=\}\ L$
Time	*(t)*	$\{=\}\ T$

NOTE

$L - L = L$

5 ft − 3 ft = 2 ft

Substitute into the equation: $T = \sqrt{(L - L)\left|\dfrac{T^2}{L}\right.}$

Simplifying $T = \sqrt{L\left|\dfrac{T^2}{L}\right.} = \sqrt{T^2} = T$

Yes, the equation is dimensionally correct. Both sides of the equation have the same dimensions.

EXAMPLE   7-15

Determine the fundamental dimensions of the following quantities:

j and k, in the expression: $t = j\,P + k$

where:

t = time [hours, h]
P = pressure [millimeters of mercury, mm Hg]

At first glance, it appears to many students that this problem is underdefined. There is a single equation but two unknown quantities. The equation laws are needed to provide a solution.

The Plus law tells us that every term in an equation must have the same dimensions.

The Unit law tells us that every term in an equation must have the same units to add or subtract terms.

Examining the equation provided and applying the Unit law, all terms must have units of hours, since the term on the left-hand side of the equation (t) has units of hours.

$$t = j\,P + k \qquad\qquad [h] = [h] + [h]$$

Knowing that the last term (k) has units of hours, we can determine the dimensions of k as time (T).

The second term (j P) must also have units of hours: $j\,P = [h]$

Substituting the known units of P [mm Hg]: $j\,[mm\ Hg] = [h]$

Solving for j: $j = [h]/[mm\ Hg]$

Substituting fundamental dimensions, noting that mm Hg is a unit of pressure and since the quantity is in the denominator, we can write the dimensions inverted:

$$j = \frac{T}{1}\left|\frac{LT^2}{M}\right.$$

Simplifying:
$$j = \frac{LT^3}{M}$$

Comprehension Check   7-14

The power absorbed by a resistor can be given by $P = I^2R$, where P is power in units of watts [W], I is electric current in amperes [A], and R is resistance in ohms [Ω]. Express the unit of ohms in terms of fundamental dimensions.

Comprehension Check   7-15

Indicate whether the following equation is dimensionally consistent (yes or no):

$$v = \sqrt{\frac{PE}{H\rho}}$$

where:

v = velocity [feet per second, ft/s]
PE = potential energy [joules, J]
H = height [feet, ft]
ρ = density [grams per cubic centimeter, g/cm^3]

7.8 Conversion Involving Equations

LEARN TO:	Solve an equation for a desired quantity using the 3-step procedure
	Recognize the importance of converting to base SI units in equation solutions

Engineering problems are rarely as simple as converting from one set of units to another. Normally, an equation is involved in the problem solution. To minimize the likelihood of mistakes, we adopt the following procedure for all problems. While this procedure may seem to overanalyze simple problems, it is relatively foolproof and will become more and more useful as the material progresses in difficulty.

Equation Procedure

1. Given a problem, first convert all parameters into base SI units, combinations of these units, or accepted non-SI units. Use the five-step conversion procedure previously described.
2. Perform all necessary calculations, as follows:
   (a) Determine the appropriate equation.
   (b) Insert the known quantities and units. Be sure to carry the units through until the end!
   (c) Calculate the desired quantity.
   This gives the answer in SI units.
3. Convert the final answer to the required units and express the answer in "reasonable" terms.

EXAMPLE	7-16

On a trip from Alphaville to Betaville, you can take two main routes. Route 1, which goes through Gammatown, is 50 kilometers [km]; however, you can only drive an average speed of 36 miles per hour [mph]. Route 2 travels along the freeway, at an average speed of 50 mph, but it is 65 kilometers [km]. How long does it take to complete each route? State the time for each route in minutes [min].

Step One: Convert to Base SI Units		
Method	**Route 1**	**Route 2**
(1) Term to be converted	36 mph	50 mph
(2) Conversion formula (3) Make a fraction (equal to one) (4) Multiply	$\dfrac{36\text{ mi}}{\text{h}}\left\|\dfrac{1\text{ km}}{0.621\text{ mi}}\right.$	$\dfrac{50\text{ mi}}{\text{h}}\left\|\dfrac{1\text{ km}}{0.621\text{ mi}}\right.$
(5) Cancel, calculate	58 km/h	81 km/h

Step Two: Calculate		
Method	**Route 1**	**Route 2**
(1) Determine appropriate equation	Distance = (velocity) (time) which can be rewritten as . . . Time = distance/velocity	
(2) Insert known quantities	Time $= \dfrac{50\text{ km}}{}\left\|\dfrac{\text{h}}{58\text{ km}}\right.$	Time $= \dfrac{65\text{ km}}{}\left\|\dfrac{\text{h}}{81\text{ km}}\right.$
(3) Calculate	Time = 0.86 h	Time = 0.8 h

NOTE

$$\frac{\dfrac{\text{km}}{\text{km}}}{\text{h}} = \frac{\text{km h}}{\text{km}} = \text{h}$$

Step Three: Convert from Base SI Units to Desired Units		
Method	**Route 1**	**Route 2**
(1) Term to be converted	0.86 h	0.8 h
(2) Conversion formula (3) Make a fraction (equal to one) (4) Multiply	$\dfrac{0.86\ \text{h}}{}\Bigg\vert\dfrac{60\ \text{min}}{1\ \text{h}}$	$\dfrac{0.8\ \text{h}}{}\Bigg\vert\dfrac{60\ \text{min}}{1\ \text{h}}$
(5) Cancel, calculate, be reasonable	52 min	48 min

EXAMPLE   7-17

You are designing a bottle to store juice for a large food manufacturing plant. The bottle is cylindrical in shape, with a 3-inch [in] diameter and a height of 0.45 meters [m]. What is the volume of the bottle in units of cubic centimeters [cm³]?

To solve this problem so the end result is in cubic centimeters, we must convert both the radius and height into units of centimeters before plugging the values into the equation.

Step One: Convert to Base SI Units		
Method	**Diameter**	**Height**
(1) Term to be converted	3 inches	0.45 meters
(2) Conversion formula (3) Make a fraction (equal to one) (4) Multiply	$\dfrac{3\ \text{in}}{}\Bigg\vert\dfrac{2.54\ \text{cm}}{1\ \text{in}}$	$\dfrac{0.45\ \text{m}}{}\Bigg\vert\dfrac{100\ \text{cm}}{1\ \text{m}}$
(5) Cancel, calculate	7.62 cm	45 cm

NOTE

The equation for the volume of a cylinder

$V_{\text{cylinder}} = \pi r^2 H$

Several common geometric formulas are provided in the end pages of this text.

Step Two: Calculate	
(1) Determine appropriate equation	$V_{\text{cylinder}} = \pi r^2 H$ $D = 2r$ so... $r = 1/2\ D$
(2) Insert known quantities	$r = 1/2\ (7.62\ \text{cm}) = 3.81\ \text{cm}$ $V = \pi(3.81\ \text{cm})^2\ (45\ \text{cm})$
(3) Calculate, be reasonable	$V = 2052\ \text{cm}^3$

Comprehension Check   7-16

A basketball has a diameter of approximately 27 centimeters [cm]. Find the spherical volume of the basketball in units of gallons [gal].

Comprehension Check   **7-17**

Eclipses, both solar and lunar, follow a cycle of just over 18 years [yr], specifically 6,585.32 days [d]. This is called the Saros Cycle. One Saros Cycle after any given eclipse, an almost identical eclipse will occur due to fact that the Earth, the Moon, and the Sun are in essentially the same positions relative to each other. The Sun, and the entire solar system, is moving relative to the Cosmic Microwave Background Radiation (the largest detectable frame of reference) at roughly 370 kilometers per second [km/s]. How far does our solar system travel through the universe in one Saros Cycle? Express your answer in the following units:

(a) meters [m], with an appropriate SI prefix such that there are only one, two, or three digits shown to the left of the decimal.
(b) light-years: one light year = 9.46×10^{15} meters [m].

CHAPTER 7

IN-CLASS ACTIVITIES

ICA 7-1

Express the following values using scientific notation, engineering notation, and using an appropriate SI prefix such that there are only one, two, or three digits shown to the left of the decimal.

Standard	Scientific	Engineering	With Prefix
(a) 389,589,000 joules [J]			
(b) 0.0000000008 pascals [Pa]			

ICA 7-2

Express the following values using scientific notation, engineering notation, and using an appropriate SI prefix such that there are only one, two, or three digits shown to the left of the decimal.

Standard	Scientific	Engineering	With Prefix
(a) 0.0698 meters [m]			
(b) 501,000,000,000 grams [g]			

ICA 7-3

Express the following values using scientific notation and engineering notation.

Standard	Scientific	Engineering
(a) 35.84 terameters [Tm]		
(b) 602 femtowatts [fW]		

ICA 7-4

Complete the following table.

	meters	centimeters	millimeters	micrometers	nanometers
Abbreviation	[m]				
Example	9×10^{-8}	9×10^{-6}	9×10^{-5}	0.09	90
(a)		50			
(b)				5	

ICA 7-5

Complete the following table.

	inches	feet	yards	meters	miles
Abbreviation	[in]				
(a)		90			
(b)					2

ICA 7-6

Which of the following is the longest distance?

(A) 26.4 miles [mi]
(B) 40 kilometers [km]
(C) 2,500 yards [yd]
(D) 100,000 feet [ft]

ICA 7-7

Complete the following table.

	cubic inch	fluid ounces	gallon	liter	cubic foot
Abbreviation	[in³]				
Example	716	400	3.12	11.8	0.414
(a)					3
(b)				5	

ICA 7-8

Which of the following is the largest volume?

(A) 50 gallons [gal]
(B) 100 liters [L]
(C) 1.5 cubic meters [m³]
(D) 2.5 cubic feet [ft³]

ICA 7-9

Complete the following table.

	miles per hour	kilometers per hour	yards per minute	feet per second
Abbreviation	[mph] or [mi/h]			
(a)		100		
(b)	55			

ICA 7-10

Which of the following is the fastest speed?

(A) 50 centimeters per second [cm/s]
(B) 2.5 kilometers per hour [km/h]
(C) 1 mile per hour [mi/h or mph]
(D) 125 feet per minute [ft/min]

ICA 7-11

Complete the following table.

	gallons per minute	cubic feet per hour	liters per second	fluid ounces per day
Abbreviation	[gpm] or [gal/min]			
(a)		15		
(b)	20			

ICA 7-12

Which of the following is the largest mass flowrate?

(A) 500 centigrams per hour [cg/h]
(B) 5 grams per minute [g/min]
(C) 80 milligrams per second [mg/s]
(D) 10 pounds-mass per day [lb$_m$/d]

ICA 7-13

Which of the following is the largest volumetric flowrate?

(A) 10 centiliters per minute [cL/min]
(B) 1 cubic inch per second [in^3/s]
(C) 10 gallons per hour [gal/h]
(D) 0.01 cubic foot per minute [ft^3/min]

ICA 7-14

In China, one "bu" is 1.66 meters [m]. The average height of a human is 5 feet [ft], 7 inches [in]. Convert this height to units of bu.

ICA 7-15

On April 20, 2010, the *DeepWater Horizon* oil drilling operation exploded, causing the largest marine oil spill ever recorded. When estimating the volume of an oil spill, if there are traces of color in the oil slick, it is estimated at 150 nanometers [nm] thick and a ratio of 1.5 liters per hectare [L/ha]. One hectare [ha] is equal to 2.471 acres. What is this ratio in gallons per square mile [gal/mi^2]?

ICA 7-16

In China, one "fen" is defined as 3.3 millimeters [mm]. Ten nanometers [nm] is the thickness of a cell membrane. Convert 10 nanometers [nm] to units of fen.

ICA 7-17

A blink of a human eye takes approximately 300–400 milliseconds [ms]. Convert 350 milliseconds [ms] to units of shake. One "shake" is equal to 10 nanoseconds [ns].

ICA 7-18

If the SI prefix system was expanded to other units, there would be such definitions as a "millihour," meaning 1/1000 of an hour. Convert 1 millihour [mh] to units of shake. One "shake" is equal to 10 nanoseconds [ns].

ICA 7-19

A "jiffy" is defined as 1/60 of a second [s]. Convert 20 jiffys to units of shake. One "shake" is equal to 10 nanoseconds [ns].

ICA 7-20

A "knot" is a unit of speed in marine travel. One knot is 1.852 kilometers per hour [km/h]. Rather than using a traditional unit system, we wish to use an unusual unit system called the FFF system: furlong–firkin–fortnight. One furlong is equal to 201 meters [m] and one fortnight is 14 days [d]. Convert the speed of 20 knots to units of furlong per fortnight.

ICA 7-21

The Earth's escape velocity is 7 miles per second [mi/s]. Rather than using a traditional unit system, we wish to use an unusual unit system called the FFF system: furlong–firkin–fortnight. One furlong is equal to 201 meters [m] and one fortnight is 14 days [d]. Convert the escape velocity to units of furlong per fortnight.

ICA 7-22

A manufacturing process uses 10 pound-mass of plastic resin per hour [lb_m/h]. Rather than using a traditional unit system, we wish to use an unusual unit system called the FFF system: furlong–firkin–fortnight. One firkin is equal to 40 kilograms [kg] and one fortnight is 14 days [d]. Convert this processing rate to units of firkin per fortnight.

ICA 7-23

Determine the fundamental dimensions of the following quantities.

	Quantity	Units	Dimensions						
			M	L	T	Θ	N	J	I
(a)	British thermal units per pounds-mass degree Fahrenheit	$\dfrac{BTU}{lb_m \ °F}$							
(b)	joule per gram	$\dfrac{J}{g}$							
(c)	watts per square meter degree Celsius	$\dfrac{W}{m^2 \ °C}$							

ICA 7-24

Determine the fundamental dimensions of the following quantities.

	Quantity	Common Units	Dimensions						
			M	L	T	Θ	N	J	I
(a)	Fuel consumption	$\dfrac{kg}{kW \ h}$							
(b)	Rate of drying	$\dfrac{lb_m}{ft^2 \ h}$							
(c)	Molar heat capacity	$\dfrac{cal}{mol \ °C}$							

ICA 7-25

Identify the following quantities through the use of fundamental dimensions. Choose from the following list:

(A) Acceleration   (C) Force   (E) Pressure
(B) Energy   (D) Power   (F) Velocity

(a) $P/(mg)$ where: m = mass [kg] g = gravity [m/s^2] P = power [W]
(b) mgv where: m = mass [kg] g = gravity [m/s^2] v = velocity [in/h]
(c) PV where: P = pressure [Pa] V = volume [m^3]

ICA 7-26

Football helmet safety is a current focus of research by engineers. To measure concussions, engineers are manufacturing helmets with force sensors to signal possible concussions. It is known that some players will experience up to 2000 of these potential concussion blows each season. To model the concussion effect on the brain, the following equations are considered. Indicate if each equation form is dimensionally consistent. More than one equation listed may be dimensionally consistent.

(a) $P = \dfrac{F}{V}$ P = pressure [=] pascal [Pa] F = force [=] newton [N]

V = volume [=] gallons [gal]

(b) $\dfrac{P}{v} = F_f - F_i$ P = power [=] watt [W] F = force [=] newton [N]; f = final, i = initial

v = velocity [=] meters per second [m/s]

(c) $P = \sqrt{\dfrac{F}{V}}$ P = dynamic viscosity [=] poise [g/(cm s)] F = force [=] newton [N]

V = voltage [=] volts [V]

ICA 7-27

For each equation listed, indicate if the equation is a correct mathematical expression based on dimensional considerations.

The following equations use these variables:

Acceleration (a) [=] m/s^2 Area (A) [=] acres   Energy (E) [=] cal

Mass (m) [=] kg   Power (P) [=] hp   Velocity (v) [=] mi/h or mph

(a) Acceleration = (velocity)^2/(area)$^{1/2}$ $a = \dfrac{v^2}{\sqrt{A}}$

(b) Energy = (mass) (velocity) (area)$^{1/2}$ $E = m\, v\, A^{0.5}$

(c) Power = (mass) (velocity)/(time) $P = \dfrac{m\, v}{t}$

ICA 7-28

What are the dimensions of the coefficient (k) in the following fictitious equations?

(a) Energy = $(k$ * mass)/(temperature * time)
(b) Force = $(k$ * area)/pressure
(c) Power = (amount * electric current)/$(k$ * light intensity)

ICA 7-29

When shipping freight around the world, most companies use a standardized set of containers to make transportation and handling easier. The 40-foot [ft] container is the most popular container worldwide. If the container is 2.4 meters [m] wide and has an enclosed volume of 2385 cubic feet [ft³], what is the height of the container in units of inches [in]?

ICA 7-30

A circular window has a 10-inch [in] radius. What is the surface area of one side of the window in units of square centimeters [cm²]?

ICA 7-31

A body traveling in a circle experiences an acceleration (a) of $a = v^2/r$, where v is the velocity of the body and r is the radius of the circle. We are tasked with designing a large centrifuge to allow astronauts to experience a high "g" forces similar to those encountered on takeoff. One "g" is defined as 9.8 meters per second squared [m/s²]. Design specifications indicate that our design must create at least 5 "g"s. If we use a radius of 30 feet [ft], what is the required velocity of the rotating capsule at the end of the arm, in units of meters per second [m/s]?

ICA 7-32

In NCAA basketball, the center circle diameter which encompasses the free throw line is 3.66 meters [m]. What is the area of the center circle, in units of square feet [ft²]?

ICA 7-33

Continental drift has an average velocity of 2 inches per year [in/yr]. At this rate, how far would a continental plate move in 12 hours [h]? Give your answer in units of meters [m], using an appropriate SI prefix such that there are only one, two, or three digits shown to the left of the decimal. Assume 1 year [yr] = 365 days [d].

ICA 7-34

In many engineering uses, the value of "g," the acceleration due to gravity, is taken as a constant. However, g is actually dependent upon the distance from the center of the Earth. A more accurate expression for g is:

$$g = g_0 \left(\frac{R_e}{R_e + A} \right)^2$$

Here, g_0 is the acceleration of gravity at the surface of the Earth, A is the altitude above the Earth's surface, and R_e is the radius of the Earth, approximately 6,380 kilometers [km]. Assume $g_0 = 9.8$ meters per second squared [m/s²]. What is the value of g at an altitude of 20 miles [mi] in units of meters per second squared [m/s²]?

REVIEW QUESTIONS

1. In 2001, the first iPod™ by Apple had a rated battery life of 10 hours [h] to run audio files. The 6th model, introduced in 2009, had rated battery life of 36 hours [h] to run audio files. If the average song is 3.5 minutes [min], how many more songs can you listen to using the 6th model iPod rather than the original iPod on a single battery charge?

2. New plastic fuel tanks for cars can be molded to many shapes, an advantage over the current metal tanks, allowing manufacturers to increase the tank capacity from 77 liters [L] to 82 liters [L]. What is this capacity increase in units of gallons [gal]?

3. The longest sea bridge, the Jiaozhou Bay Bridge in China, spans 26.4 miles [mi]. The longest sea bridge in the United States is the Lake Pontchartrian Causeway in Louisiana, which spans 41,940 yards [yd]. How much longer is the Jiaozhou Bridge, in units of feet [ft], than the Lake Pontchartrain Causeway?

4. The term "deep sea" refers to everything in the ocean below a depth of 200 meters [m]. It is estimated more than 90% of the living space on the planet exists at this depth. The deep sea is an area of great interest for explorers. If a submarine dives to a depth of 400 meters [m], how deep is this in units of miles [mi]?

5. A category F5 tornado can have wind speeds of 300 miles per hour [mph]. What is this velocity in units of meters per second [m/s]?

6. Some species of bamboo have been measured to grow at a rate of up to 4 inches per hour [in/h]. Express this growth rate in units of micrometers per second [μm/s].

7. A new hybrid automobile with regenerative braking has a fuel economy of 55 miles per gallon [mi/gal or mpg] in city driving. What is this fuel economy expressed in units of feet per milliliter [ft/mL]?

8. If a liquid evaporates at a rate of 50 kilograms per minute [kg/min], what is this evaporation rate in units of pounds-mass per second [lb_m/s]?

9. Acetone must be kept in a closed pressure tank due to evaporation. Acetone evaporates at a rate of 3 grams per minute [g/min]. If the tank holds 50 pound-mass [lb_m] of acetone and is allowed to remain open to the atmosphere, how many days [d] will it take for half of the acetone to evaporate?

10. The AbioCor™ artificial heart pumps at a rate of 10 liters per minute [L/min]. Express this rate in units of gallons per second [gal/s].

11. If a pump moves water at 70 gallons per minute [gal/min or gpm], what is the volumetric flow rate in units of cubic inches per second [in³/s]?

12. On April 20, 2010, the *DeepWater Horizon* oil drilling operation exploded, causing the largest marine oil spill ever recorded. Many estimates have been given to describe the volume of oil lost per day. One estimate, just before the well was capped, places the escaping oil at 8400 cubic meters per day [m³/d]. One barrel [bbl] of oil is 42 gallons [gal]. Convert this rate of escaping oil into barrels per hour [bbl/h].

13. One of the National Academy of Engineering Grand Challenges for Engineering is **Provide Access to Clean Water**. Only 5% of water is used for households—the majority is used for agriculture and industry. It is estimated that 528 gallons [gal] of water are required to produce food for one person for one day [d]. How many liters per year [L/yr] are required to feed one person?

14. The board foot is often used as a measure of volume for lumber, and is defined as the volume of a one-foot long board that is one-foot wide and one-inch thick [L × W × H = 1 foot × 1 foot × 1 inch]. What is the volume in cubic meters [m³] of a piling that is 135 board feet?

15. The oxgang is unit of area equal to 20 acres. Express an area of 12 oxgangs in units of square meters [m²].

16. In an effort to modernize the United States interstate system, the Department of Transportation proposes to change speed limits from miles per hour to "flashes." A flash is equal to 10 feet per second [ft/s]. On a car speedometer, what will the new range be in units of "flashes" if the old scale was set to a maximum of 120 miles per hour [mi/h or mph]?

17. Old Mississippi River paddle wheelers routinely measured the river depths to avoid running aground. They used the unit "fathoms," where 1 fathom = 6 feet [ft]. The pilot would sing out "mark three" when the river was 3 fathoms deep and "mark twain" at 2 fathoms. The American writer Samuel Clemens took this as his pen name, Mark Twain. If we take 2 fathoms as a new unit, "twain," express 60 miles per hour [mi/h or mph] in units of twains per second [twain/s].

18. The grain is a unit of mass equal to 64.8 milligrams [mg]. The dram is a unit of mass equal to 875/32 grains. What is 0.5 kilograms [kg] expressed in units of drams?

19. The hand [hh] is a traditional unit for measuring the height of horses and has been standardized to equal 4 inches [in]. The largest horse in recorded history was a Shire horse named Mammoth measuring 21.5 hands [hh]. Convert this height to feet [ft] and inches [in] (for example, 5.5 feet = 5 feet, 6 inches).

20. The Units Society Empire (USE) had defined the following set of "new" units: 1 foot [ft] = 10 toes. Convert 45 toes to units of meters [m].

21. The Units Society Empire (USE) had defined the following set of "new" units: 1 leap = 4 years [yr]. Convert 64 leaps to units of months.

22. The Units Society Empire (USE) had defined the following set of "new" units:

Length   1 car = 20 feet [ft]
Time   1 class = 50 minutes [min]

Determine X in the following expression: speed limit, 60 miles per hour [mi/h or mph] = X cars per class.

23. The Units Society Empire (USE) had defined the following set of "new" units:

Length   1 stride = 1.5 meters [m]
Time   1 blink = 0.3 seconds [s]

Determine X in the following expression: Boeing 747 cruising speed, 550 miles per hour [mi/h or mph] = X strides per blink.

Many common quantities with derived units do not have a unit name, such as velocity. For questions 7-24 to 7-29, express the variable shown in terms of fundamental dimensions.

24. Absorbed radiation dose (D)   [J/kg]

25. Electrical field (E)   [V/m]

26. Acoustic impedance (Z)   [Pa s/m]

27. Magnetic permeability (μ)   [V s/A]

28. Ideal gas constant (R)   [atm L/(mol K)]

29. Stefan-Boltzmann constant (σ)   [W/(m² K⁴)]

30. Analyze the following terms using fundamental dimensions, and choose the appropriate physical quantity from the list below.

(A) Acceleration   **(B)** Area   **(C)** Energy   **(D)** Force   **(E)** Length

(F) Power   **(G)** Pressure   **(H)** Time   **(I)** Velocity   **(J)** Other

(a) $\dfrac{\alpha\,\varphi}{\tau}$ $\alpha\ [=]$ volt meter minute [V m min] $\varphi\ [=]$ amperes per hour [A/h]

$\tau\ [=]$ atmosphere liter [atm L]

(b) $\dfrac{\alpha^2}{\lambda\,\beta\,\mu}$ $\alpha\ [=]$ newtons [N] $\lambda\ [=]$ millimeters of mercury [mm Hg]

$\beta =$ feet per second squared [ft/s^2] $\mu\ [=]$ slugs [slug]

31. We wish to analyze the velocity of a fluid exiting from an orifice in the side of a pressurized tank. Examine the equations below using fundamental dimensions and indicate for each if the equation is a valid or invalid equation; justify your answer for each case.

(a) $v = \dfrac{P}{\rho} + \sqrt{2gH}$

(b) $v = \sqrt{\dfrac{P}{\rho} + 2H}$

(c) $v = \sqrt{\dfrac{2P}{\rho} + 2gH}$

(d) $v = \sqrt{2P + 2gH}$

Here, the variables are as follows:

$g =$ gravity $[=]$ m/s^2
$H =$ height or depth of the fluid $[=]$ ft
$P =$ pressure $[=]$ atm
$v =$ velocity $[=]$ ft/min
$\rho =$ density $[=]$ kg/m^3

32. Wind energy uses large fans to extract energy from the wind and turn it into electric power. Examine the equations below using fundamental dimensions and indicate for each if the equation is a valid or invalid equation; justify your answer for each case.

(a) $P = \eta\,\rho\,A^2 v^2$

(b) $P = \eta\,\rho\,A\,v^2$

(c) $P = \eta\,\rho\,A\,v^3$

(d) $P = \eta\,\rho^2 A\,v$

(e) $P = \eta\sqrt{\rho\,A\,v^3}$

Here, the variables are as follows:

$A =$ area $[=]$ m^2
$P =$ power $[=]$ W
$v =$ velocity $[=]$ mi/h or mph
$\eta =$ efficiency $[=]$ unitless
$\rho =$ density $[=]$ kg/m^3

33. Based on the following equation, express the unknown variable in terms of fundamental dimensions.

α, in the expression: $\alpha \, \sigma \, C_p = k$

where σ [=] moles per ampere squared [mol/A²]

   C_p [=] calories per kilogram degrees Celsius [cal/(kg°C)]

   k [=] watts per meter [W/m]

34. Based on the following equation, express the unknown variable in terms of fundamental dimensions.

ε, in the expression: $k = \varepsilon \, \beta + \gamma$

where β [=] watts per square meter kelvin⁴ [W/(m²K⁴)]

   γ [=] moles per kilogram [mol/kg]

35. Based on the following equation, express the unknown variable in terms of fundamental dimensions.

q, in the expression: $L = \varepsilon \, P^2 + q$

where L [=] candela per cubic foot minute [cd/(ft³ min)]

   P [=] pascals [Pa]

36. Based on the following equation, answer the questions below.

$$P = \frac{8\xi L\phi}{\pi r^4} + \psi$$

where ξ = moles per centimeter second [mol/(cm s)]
   L [=] yards [yd]
   r [=] inches [in]
   P [=] pascals [Pa]
   The number 8 and the variable of π are unitless and dimensionless.

(a) What are the fundamental dimensions of ψ?
(b) What are the fundamental dimensions of ϕ?

37. Based on the following equation, answer the questions below.

$$\dot{\rho} = \frac{2\gamma\phi + \psi}{rg}$$

where $\dot{\rho}$ [=] moles per cubic foot [mol/ft³]
   γ [=] joules per kilogram [J/kg]
   r [=] inches [in]
   g [=] meters per second squared [m/s²]
   The number 2 is unitless and dimensionless.

(a) What are the fundamental dimensions of ψ?
(b) What are the fundamental dimensions of ϕ?

38. Based on the following equation, answer the questions below.

$$V = \frac{\gamma\, PT - \delta}{I}$$

where: V [=] volt [V]
$\quad\quad\ P$ [=] watt [W]
$\quad\quad\ I$ [=] amperes [A]
$\quad\quad\ T$ [=] kelvin per mole [K/mol]

(a) What are the fundamental dimensions of δ?
(b) What are the fundamental dimensions of γ?

39. A dust collection system operates using the following equation:

$$\varphi = \frac{v\, W\, L}{Q}$$

where: v = fluid velocity [=] meters per second [m/s]
$\quad\quad\ W$ = duct width [=] inches [in]
$\quad\quad\ L$ = duct length [=] feet [ft]
$\quad\quad\ Q$ = volumetric flowrate [=] gallons per minute [gal/min or gpm]

Use the equation and fundamental dimensions to determine the unknown quantity, ϕ. Select from the following choices:

(A) Area {=} L^2
(B) Efficiency {=} dimensionless
(C) Gravity {=} $L\, T^{-2}$
(D) Inverse time {=} T^{-1}
(E) Length {=} L
(F) Mass {=} M
(G) Temperature {=} Θ
(H) Viscosity {=} $M\, L^{-1}\, T^{-1}$

40. Determine if the following expression is a valid equation.

$$\kappa(T_w - T_f) = \mu v^2$$

where: κ = thermal conductivity [=] watts per meter kelvin [W/(m K)]
$\quad\quad\ T_w$ = wall temperature [=] degrees Fahrenheit [°F]
$\quad\quad\ T_f$ = fluid temperature [=] degrees Fahrenheit [°F]
$\quad\quad\ \mu$ = dynamic viscosity [=] grams per centimeter second [g/(cm s)]
$\quad\quad\ v$ = velocity [=] feet per minute [ft/min]

41. The largest hailstone is the United States was 44.5 centimeters [cm] in circumference in Coffeyville, Kansas. What is the diameter of the spherical hailstone in units of inches [in]?

42. The largest hailstone is the United States was 44.5 centimeters [cm] in circumference in Coffeyville, Kansas. What is the volume of the spherical hailstone in units of liters [L]?

43. How large a surface area in units of square feet [ft²] will 1 gallon [gal] of paint cover if we apply a coat of paint that is 0.1 centimeter [cm] thick?

44. How large a surface area in units of square feet [ft²] will 1 gallon [gal] of paint cover if we apply a coat of paint that is 0.1 inches [in] thick?

45. A box has a volume of 10 gallons [gal]. If two sides of the box measure 2.4 meters [m] by 2.4 feet [ft], what is the length of the third side of the box in units of inches [in]?

46. A 1.25-gallon [gal] quantity of molten (liquefied) plastic is going to be used to make a 1-foot [ft] by 8-inch [in] plastic block. If 0.75 liters [L] of the plastic is lost in the manufacturing process, how thick is the final plastic block, in units of inches [in]?

47. In American football, the playing field is 53.33 yards [yd] wide by 120 yards [yd] long. For a special game, the field staff want to paint the playing field orange. Of course, they will use biodegradable paint available for purchase in 25-gallon [gal] containers. If the paint is applied in a thickness of 1.2 millimeters [mm] in a uniform layer, how many containers of paint will they need to purchase?

48. For almost four millennia, the Pyramid of Khufu (aka The Great Pyramid of Giza) was the tallest man-made structure in the world, and it remains the only one of the Seven Wonders of the Ancient World that is still essentially intact. This is a square pyramid, with a base side length (L) of 813 feet [ft]. The volume (V) of the pyramid is 2.946×10^9 liters [L]. The volume (V) of a pyramid with a square base is given by

$$V = \frac{L^2 H}{3}$$

What is the height (H) of the pyramid in units of cubits? In Ancient Egypt, 1 cubit = 19.9 inches [in].

49. A torus is essentially a tube wrapped into a circle (a "doughnut" shape). The volume (V) of a torus can be calculated by

$$V = 2 \pi^2 R r^2$$

where R is the radius of the torus (from the center of the "tube" to the center of the hole in the middle) and r is the radius of the tube itself. If a torus has a volume of 2537 cubic inches [in³] with a torus radius (R) of 0.26 meters [m], what is the radius of the tube (r) in units of centimeters [cm]?

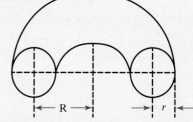

50. In many engineering uses, the value of "g," the acceleration due to gravity, is taken as a constant. However, g is actually dependent upon the distance from the center of the Earth. A more accurate expression for g is:

$$g = g_0 \left(\frac{R_e}{R_e + A} \right)^2$$

Here, g_0 is the acceleration of gravity at the surface of the Earth, A is the altitude above the Earth's surface, and R_e is the radius of the Earth, approximately 6,380 kilometers [km]. Assume $g_0 = 9.8$ meters per second squared [m/s²]. If the value of g is 9 meters per second squared [m/s²], what is the altitude in units of miles [mi]?

8

Universal Units

In the chapter on fundamental dimensions, the concepts of derived dimensions and units were introduced. Five of the most common named units were introduced in that chapter, and are so critical they are repeated here as Table 8-1. Recall that numerous common derived dimensions do not have a corresponding derived SI unit. For example, there is no named SI unit for the derived dimension velocity as there is for force (newton) or electrical resistance (ohm).

Table 8-1 Common derived units in the SI system

Dimension	SI Unit	Base SI Units	Derived from
Force (F)	newton [N]	$1\ \text{N} = 1\dfrac{\text{kg m}}{\text{s}^2}$	$F = ma$ Force = mass times acceleration
Energy (E)	joule [J]	$1\ \text{J} = 1\ \text{N m} = 1\dfrac{\text{kg m}^2}{\text{s}^2}$	$E = Fd$ Energy = force times distance
Power (P)	watt [W]	$1\ \text{W} = 1\dfrac{\text{J}}{\text{S}} = 1\dfrac{\text{kg m}^2}{\text{s}^3}$	$P = E/t$ Power = energy per time
Pressure (P)	pascal [Pa]	$1\ \text{Pa} = 1\dfrac{\text{N}}{\text{m}^2} = 1\dfrac{\text{kg}}{\text{m s}^2}$	$P = F/A$ Pressure = force per area
Voltage (V)	volt [V]	$1\ \text{V} = 1\dfrac{\text{W}}{\text{A}} = 1\dfrac{\text{kg m}^2}{\text{s}^3\text{A}}$	$V = P/I$ Voltage = power per current

8.1 Force

LEARN TO: Identify a force quantity when it is expressed in base SI units
Convert from one unit of force to another
Determine the final quantity if given two quantities: force, acceleration, weight

When you push a grocery cart, it moves. If you keep pushing, it keeps moving. The longer you push, the faster it goes; the velocity increases over time, meaning that it accelerates. If you push a full grocery cart that has a high mass, it does not speed up as much, meaning it accelerates less than a cart with low mass. Simply put, the acceleration (a) of a body depends on the force (F) exerted on it and its mass (m). This is a simple form of Newton's second law of motion and is usually written as $F = ma$ (see Table 8-2.)

IMPORTANT CONCEPT

$F = ma$

SI unit of force =
newton [N]

Table 8-2 Dimensions of force

Quantity	Common Units	Exponents						
		M	**L**	**T**	**Θ**	**N**	**J**	**I**
Force	N	1	1	−2	0	0	0	0

The SI unit of force, the **newton** [N], is defined as the force required to accelerate a mass of one kilogram at a rate of one meter per second squared (see Table 8-3). It is named for Sir Isaac Newton (1643–1727). Newton's *Principia* is considered one of the world's greatest scientific writings, explaining the law of universal gravitation and the three laws of motion. Newton also developed the law of conservation of momentum, the law of cooling, and the reflecting telescope. He shares credit for the development of calculus with Gottfried Leibniz.

In the SI system, mass, length, and time are base units and force is a derived unit; force is found from combining mass, length, and time using Newton's second law. The SI system is called *coherent*, because the derived unit is set at one by combining base units. The AES system is considered non-coherent as it uses units that do not work together in the same fashion as the SI units do. There are two uses of the term "pound" in the AES system, which occurred in common usage long before Newton discovered gravity. To distinguish mass in pounds and force in pounds, the unit of mass is given as pound-mass (lb_m) and the unit of force is given as pound-force (lb_f). One pound-force is the amount of force needed to accelerate one pound-mass at a rate of 32.2 feet per second squared. Since this relationship is not easy to remember or use in conversions, we will stick with SI units for problem solving, following the procedure discussed in the chapter on fundamental dimensions.

NOTE

In general, a "pound" can be used as a unit of mass or force. For distinction, the following convention is used:

– pound-mass [lb_m]

– pound-force [lb_f]

Table 8-3 Definitions of force units in various unit systems

Unit System	Mass	Acceleration	Force	
SI	1 kg	1 m/s²	1 N = 1 kg m/s²	coherent
AES	1 lb_m	32.2 ft/s²	1 lb_f = 32.2 lb_m ft/s²	non-coherent

EXAMPLE   8-1

A professional archer is designing a new longbow with a full draw weight of 63 pounds-force [lb_f]. The draw weight is the amount of force needed to hold the bowstring at a given amount of draw, or the distance the string has been pulled back from the rest position. What is the full draw weight of this bow in units of newtons [N]?

Convert to Base SI Units	Steps
(1) Convert term	63 lb_f
(2) Apply conversion formula	1 N = 0.225 lb_f
(3) Make a fraction (4) Multiply	$\dfrac{63\ lb_f}{} \left\| \dfrac{1\ N}{0.225\ lb_f}\right.$
(5) Cancel, calculate, be reasonable	280 N

EXAMPLE 8-2

NOTE

The term **thrust** is used to express a force used to push an object very quickly, such as a rocket or an airplane.

A ship is being designed to use an engine that runs continuously, providing a small but constant acceleration of 0.06 meters per second squared [m/s²]. If the ship has a mass of 30,000 kilograms [kg], what is the thrust (force) provided by the engine in units of pounds-force [lb$_f$]?

Calculate Required Parameters	Steps
(1) Determine appropriate equation	$F = m\,a$
(2) Insert known quantities	$F = \dfrac{30{,}000 \text{ kg}}{}\left\lvert \dfrac{0.06 \text{ m}}{s^2}\right.$
(3) Calculate	$F = 1800 \dfrac{\text{kg m}}{s^2}$

This is apparently our final answer, but the units are puzzling. If the unit of force is the newton, and if this is a valid equation, then our final result for force should be newtons. If we consider the dimensions of force

		Exponents		
Quantity	Common Units	M	L	T
Force	N	1	1	−2

A unit of force has dimensions F {=} ML/T², which in terms of base SI units would be F [=] kg m/s². As this term occurs frequently, it is given the special name "newton" (see Table 8-1). Anytime we see the term [kg m/s²], we know we are dealing with a force equal to a newton.

IMPORTANT CONCEPT

$1 \text{ N} = 1 \text{ (kg m)} / s^2$

(3) Calculate	$F = \dfrac{1800 \text{ kg m}}{s^2}\left\lvert \dfrac{1 \text{ N}}{1 \frac{\text{kg m}}{s^2}}\right. = 1800 \text{ N}$

Convert to Desired Units	Steps
(1) Convert term	1800 N
(2) Apply conversion formula	$1 \text{ N} = 0.225 \text{ lb}_f$
(3) Make a fraction	$\dfrac{1800 \text{ N}}{}\left\lvert \dfrac{0.225 \text{ lb}_f}{1 \text{ N}}\right.$
(4) Multiply	
(5) Cancel, calculate, be reasonable	405 lb_f

Comprehension Check 8-1

The engine on a spacecraft nearing Mars can provide a thrust of 15,000 newtons [N]. If the spacecraft has a mass of 750 kilograms [kg], what is the acceleration of the spacecraft in miles per hour squared [mi/h²]?

8.2 Weight

The **mass** of an object is a fundamental dimension. Mass is a quantitative measure of how much of an object there is, or in other words, how much matter it contains. The **weight** (w) of an object is a force equal to the mass of the object (m) times the acceleration of **gravity** (g). This is a special application of Newton's second law, where acceleration is equal to gravity.

While mass is independent of location in the universe, weight is dependent upon both mass and gravity (Table 8-4).

On the Earth, gravity is approximately 9.8 meters per second squared [m/s²]. On the moon, gravity is approximately 1.6 meters per second squared [m/s²]. A one kilogram [kg] object acted on by Earth's gravity would have a weight of 9.8 newtons [N], but on the moon it would have a weight of 1.6 newtons [N]. Unless otherwise stated, assume all examples take place on the Earth.

IMPORTANT CONCEPT

Weight is a FORCE

$w = mg$

SI unit of weight = newton [N]

NOTE

Objects in space are weightless, not massless.

IMPORTANT CONCEPT

Mass of an object: A quantitative measure of how much of an object there is.

Weight of an object: A quantitative measure of the force exerted on the object due to gravity.

Table 8-4 Dimensions of weight

| Quantity | Common Units | \multicolumn{7}{c}{Exponents} |
		M	L	T	Θ	N	J	I
Weight	N	1	1	−2	0	0	0	0

DEVILISH DERIVATION

Newton's law of universal gravitation states

$$F = G\frac{m_1 m_2}{r^2}$$

where:
  G is universal gravitational constant $G = 6.673 \times 10^{-11}\,(\text{N m}^2)/\text{kg}^2$
  m is the mass
  r is the distance between the centers of mass of two bodies

On the Earth, the distance between the center of a body and the center of the Earth is approximately the radius of the Earth, r_e.

The mass of one of the bodies can be defined as is the mass of the Earth. Rewrite the equation:

$$F = m\left[G\frac{m_e}{r_e^2}\right]$$

The quantity in square brackets is a constant (call it "g"). We call the force "the weight (w) of the body." So,

$$w = mg$$

This is the common equation that relates weight and mass. The value for g on Earth is calculated to be 9.8 meters per second squared [m/s²], or 32.2 feet per second squared [ft/s²].

EXAMPLE   8-3

What is the weight of a 225-kilogram [kg] bag of birdseed in units of newtons [N]?

Calculate Required Parameters	Steps
(1) Determine appropriate equation	$w = mg$
(2) Insert known quantities	$w = \dfrac{225 \text{ kg}}{} \bigg\| \dfrac{9.8 \text{ m}}{s^2}$
(3) Calculate, be reasonable	$w = 2205 \dfrac{\text{kg m}}{s^2} \bigg\| \dfrac{1 \text{ N}}{\frac{1 \text{ kg m}}{s^2}} = 2205 \text{ N}$

Comprehension Check   8-2

The mass of the human brain is 1360 grams [g]. State the weight of the human brain in units of newtons [N] on the Earth.

Comprehension Check   8-3

The mass of the human brain is 1360 grams [g]. State the weight of the human brain in units of newtons [N] on the moon. The gravity on the moon is 1.6 meters per second squared [m/s²].

8.3 Density

LEARN TO: Determine the density in any required units if given specific gravity
Recall values for density of water and limits of density for solids, liquids, and gases
Determine the final quantity if given two quantities: density, mass, volume

IMPORTANT CONCEPT

$\rho = m/V$

$\gamma = w/V$

Density (ρ, Greek letter rho) is the mass of an object (m) divided by the volume (V) the object occupies (V). Density should not be confused with weight—think of the old riddle: which weighs more, a pound of feathers or a pound of bricks? The answer is they both weigh the same amount, one pound, but the density of each is different. The bricks have a higher density than the feathers, since the same mass takes up less space.

Specific weight (γ, Greek letter gamma) is the weight of an object (w) divided by the volume the object occupies (V). Table 8-5 compares density and specific weight in terms of fundamental dimensions.

Table 8-5 Dimensions of density and specific weight

Quantity	Common Units	Exponents						
		M	L	T	Θ	N	J	I
Density	kg/m³	1	−3	0	0	0	0	0
Specific weight	N/m³	1	−2	−2	0	0	0	0

EXAMPLE   8-4

NOTE

Upon conversion from units of grams per cubic centimeter to pound-mass per cubic foot, the answer should be ≈ 60 times larger.

The density of sugar is 1.61 grams per cubic centimeter [g/cm³]. What is the density of sugar in units of pound-mass per cubic foot [lb_m/ft^3]?

Convert to Desired Units	Steps
(1) Term to be converted	1.61 g/cm³
(2) Conversion formula	
(3) Make fractions	$1.61\dfrac{g}{cm^3}\left|\dfrac{2.205\ lb_m}{1000\ g}\right|\dfrac{1000\ cm^3}{0.0353\ ft^3}$
(4) Multiply	
(5) Cancel, calculate, be reasonable	101 lb_m/ft^3

EXAMPLE   8-5

NOTE

Upon conversion from units of grams per cubic centimeter to kilograms per cubic meter, the answer should be 1000 times larger.

The density of a biofuel blend is 0.72 grams per cubic centimeter [g/cm³]. What is the density of the biofuel in units of kilograms per cubic meter [kg/m³]?

Convert to Desired Units	Steps
(1) Term to be converted	0.72 g/cm³
(2) Conversion formula	
(3) Make fractions	$0.72\dfrac{g}{cm^3}\left|\dfrac{1\ kg}{1000\ g}\right|\dfrac{100^3\ cm^3}{1\ m^3}$
(4) Multiply	
(5) Cancel, calculate, be reasonable	720 kg/m³

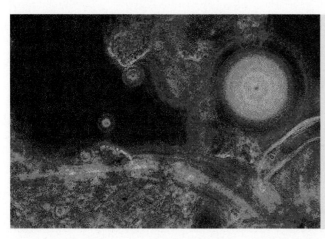

A vast array of valuable compounds can be formed by microbial cultures. Oil produced by the fungi Pythium irregulare *can be extracted and used for biofuels or pharmaceutical compounds. Biosystems engineers culture the microorganism, design the bioreactor, and extract the valuable compounds using sustainable, ecoprocessing techniques.*

Photo courtesy of C. Drapcho

EXAMPLE   8-6

What is the weight of water, in units of pounds-force [lb$_f$], in a 55-gallon [gal] drum completely full? Assume the density of water to be 1 gram per cubic centimeter [g/cm³]. Ignore the weight of the drum.

Convert to Base SI Units	Steps	
(1) Term to be converted	55 gal	1 g/cm³
(2) Conversion formula		
(3) Make fractions	$\dfrac{55\ \text{gal}}{} \Big\| \dfrac{1\ \text{L}}{0.264\ \text{gal}} \Big\| \dfrac{1\ \text{m}^3}{1000\ \text{L}}$	$\dfrac{1\ \text{g}}{\text{cm}^3} \Big\| \dfrac{1\ \text{kg}}{1000\ \text{g}} \Big\| \dfrac{100^3\ \text{cm}^3}{1\ \text{m}^3}$
(4) Multiply		
(5) Cancel, calculate	0.208 m³	1000 kg/m³

Calculate Required Parameters	Steps
(1) Determine appropriate equation	$w = m\,g$
(2) Insert known quantities	$w = \dfrac{m}{}\Big\|\dfrac{9.8\ \text{m}}{\text{s}^2}$

NOTE

Here, we must first determine mass, then use this to determine weight. We repeat the process to *Calculate Required Parameters* twice.

	Steps
(1) Determine appropriate equation	$m = \rho V$
(2) Insert known quantities	$m = \dfrac{1000\ \text{kg}}{\text{m}^3}\Big\|\dfrac{0.208\ \text{m}^3}{}$
(3) Calculate, be reasonable	$m = 208\ \text{kg}$
(2) Insert known quantities	$w = \dfrac{208\ \text{kg}}{}\Big\|\dfrac{9.8\ \text{m}}{\text{s}^2}$
(3) Calculate	$w = 2038\dfrac{\text{kg m}}{\text{s}^2}\Big\|\dfrac{1\ \text{N}}{\frac{1\ \text{kg m}}{\text{s}^2}} = 2038\ \text{N}$

Convert to Desired Units	Steps
(1) Term to be converted	2038 N
(2) Conversion formula	
(3) Make a fraction	$\dfrac{2038\ \text{N}}{}\Big\|\dfrac{0.225\ \text{lb}_f}{1\ \text{N}}$
(4) Multiply	
(5) Cancel, calculate, be reasonable	460 lb$_f$

IMPORTANT CONCEPT

$$SG = \frac{\rho_{\text{object}}}{\rho_{\text{water}}}$$

Specific Gravity

In technical literature, density is rarely given; instead, the **specific gravity** is reported. The specific gravity (SG) of an object is a dimensionless ratio of the density of the object to the density of water, shown in Table 8-6 in terms of fundamental dimensions. It is convenient to list density in this fashion so *any* unit system may be applied by our choice of the units of the density of water. The specific gravities of several common substances are listed in Table 8-7.

Table 8-6 Dimensions of specific gravity

Quantity	Common Units	Exponents						
		M	**L**	**T**	**Θ**	**N**	**J**	**I**
Specific gravity	—	0	0	0	0	0	0	0

Density of water

= 1 g/cm^3

= 1 kg/L

= 1000 kg/m^3

= 62.4 lb$_m$/ft^3

= 1.94 slug/ft^3

Table 8-7 Specific gravity values for common substances

Liquids	SG	Solids	SG
Acetone	0.785	Aluminum	2.70
Citric acid	1.67	Baking soda	0.689
Gasoline	0.739	Concrete	2.30
Glycerin	1.26	Copper	8.96
Iodine	4.93	Gold	19.3
Mercury	13.6	Graphite	2.20
Olive oil	0.703	Iron	7.87
Propane	0.806	Lead	11.4
Sea water	1.03	Polyvinyl chloride (PVC)	1.38
Water	1.00	Silicon	2.33

IMPORTANT CONCEPT

Specific Gravity Limits

0.5 < Solids < 23

Liquids ~1
*exceptions:

   iodine, 4.93

   mercury, 13.6

0.0001 < Gases < 0.001

When calculating or considering specific gravities, it is helpful to keep in mind the range of values that you are likely to have.

The densest naturally occurring elements at normal temperature and pressure are osmium and iridium, both with a specific gravity close to 22.6. The *densest substances that a normal person is likely to encounter are platinum (SG = 21.5) and gold (SG = 19.3)*. Thus, if you calculate a specific gravity to be higher than 23, you have almost certainly made an error.

Most liquids are similar to water, with a specific gravity around 1. One notable exception is mercury, with a specific gravity of 13.

On the lower end of the scale, the *specific gravity of air is about 0.001*, whereas hydrogen has a specific gravity of slightly less than 0.0001. Therefore, if you get a specific gravity value less than about 10^{-4}, you need to check your work very carefully.

EXAMPLE   8-7

NOTE

Since we want the final units in kg/m^3, we choose a density of water value = 1000 kg/m^3

The specific gravity of butane is 0.599. What is the density of butane in units of kilograms per cubic meter [kg/m^3]?

Calculate Required Parameters	Steps
(1) Determine appropriate equation	$\rho_{object} = (SG)(\rho_{water})$
(2) Insert known quantities	$\rho_{object} = (0.599)\left(1000\dfrac{kg}{m^3}\right)$
(3) Calculate, be reasonable	$\rho_{object} = 599\dfrac{kg}{m^3}$

EXAMPLE 8-8

Mercury has a specific gravity of 13.6. What is the density of mercury in units of slugs per liter [slug/L]?

NOTE

Since we want the final units in slug/L, we choose a density of water value = 1.94 slug/ft³, then convert ft³ to L.

Calculate Required Parameters	Steps
(1) Determine appropriate equation	$\rho_{object} = (SG)(\rho_{water})$
(2) Insert known quantities	$\rho_{object} = (13.6)\left(1.94\dfrac{slug}{ft^3}\right)$
(3) Calculate	$\rho_{object} = 26.384\dfrac{slug}{ft^3}$

Convert to Desired Units	Steps
(1) Term to be converted	26.384 slug / ft³
(2) Conversion formula	
(3) Make a fraction	$\dfrac{26.384\ slug}{ft^3}\left\|\dfrac{0.0353\ ft^3}{1\ L}\right.$
(4) Multiply	
(5) Cancel, calculate, be reasonable	0.931 slug/L

Comprehension Check 8-4

Convert 50 grams per cubic centimeter [g/cm³] into units of pounds-mass per cubic foot [lb$_m$/ft³].

Comprehension Check 8-5

A 75-gram [g] cylindrical rod is measured to be 10 centimeters [cm] long and 2.5 centimeters [cm] in diameter. What is the specific gravity of the material? Recall that the volume of a cylinder can be determined by: $V = \pi r^2 H$, where r is the radius and H is the height.

8.4 Amount

IMPORTANT CONCEPT

A mole is made of 6.022×10^{23} fundamental units. It is a term defining a quantity, similar to the term "dozen."

NOTE

1 atomic mass unit [amu] = 1 Dalton [Da] = 1.66×10^{-24} g

IMPORTANT CONCEPT

$MW = m/n$

NOTE

If Element Z has an atomic mass of X amu, there are X grams per mole of Element Z.

Some things are very large and some are very small. Stellar distances are so large that it becomes inconvenient to report values such as 235 trillion miles, or 6.4×10^{21} feet when we are interested in the distance between two stars or two galaxies. To make things better, we use a new unit of length that itself is large—the distance that light goes in a year; this is a very long way, 3.1×10^{16} feet. As a result, we do not have to say that the distance between two stars is 620,000,000,000,000,000 feet, we can just say that they are 2 light-years apart.

This same logic holds when we want to discuss very small things such as molecules or atoms. Most often we use a constant that has been named after Amedeo Avogadro, an Italian scientist (1777–1856) who first proposed the idea of a fixed ratio between the amount of substance and the number of elementary particles. If we have 12 of something, we call it a dozen. If we have 20, it is a score. If we have 6.022×10^{23} of anything, we have a mole. If we have 6.022×10^{23} baseballs, we have a mole of baseballs. If we have 6.022×10^{23} elephants, we have a mole of elephants, and if we have 6.022×10^{23} molecules, we have a mole of molecules. Of course, the mole is never used to define amounts of macroscopic things like elephants or baseballs, being relegated to the realm of the extremely tiny.

The mass of a nucleon (neutron or proton) is about 1.66×10^{-24} grams. To avoid having to use such tiny numeric values when dealing with nucleons, physicists defined the **atomic mass unit** [amu] to be approximately the mass of one nucleon. Technically, it is defined as one-twelfth of the mass of a carbon twelve atom. In other words, 1 amu = 1.66×10^{-24} g, which is also known as a **Dalton** [Da].

If there is $(1.66 \times 10^{-24}$ g)/(1 amu), then there is (1 amu)/(1.66×10^{-24} g). Dividing this out gives 6.022×10^{23} amu/g. This numeric value is used to define the **mole** [mol]. One mole of a substance (usually an element or compound) contains exactly 6.022×10^{23} fundamental units (atoms or molecules) of that substance. In other words, there are 6.022×10^{23} fundamental units per mole. This is often written as

$$N_A = 6.022 \times 10^{23} \text{ mol}^{-1}$$

This is called Avogadro's constant or **Avogadro's number**, symbolized by N_A. So why is this important? Consider combining hydrogen and oxygen to get water (H_2O). We need twice as many atoms of hydrogen as atoms of oxygen for this reaction; thus, for every mole of oxygen, we need two moles of hydrogen.

The problem is that it is difficult to measure a substance directly in moles, but it is easy to measure its mass. *Avogadro's number affords a conversion path between moles and mass.* Consider hydrogen and oxygen in the above. The atomic mass of an atom in atomic mass units is approximately equal to the number of nucleons it contains. Hydrogen contains one proton, and thus has an atomic mass of 1 amu. We can also say that there is 1 amu per hydrogen atom. Oxygen has an atomic mass of 16; thus, there are 16 amu per oxygen atom. Since atomic mass refers to an individual specific atom, the term **atomic weight** is used interchangably with it, representing the average value of all isotopes of the element. This is the value commonly listed on periodic tables.

$$\text{Hydrogen:} \quad \frac{1 \text{ amu}}{\text{H atom}} \left| \frac{1 \text{ g}}{6.022 \times 10^{23} \text{ amu}} \right| \frac{6.022 \times 10^{23} \text{atom}}{1 \text{ mol}} = \frac{1 \text{ g}}{1 \text{ mol H}}$$

$$\text{Oxygen:} \quad \frac{16 \text{ amu}}{\text{O atom}} \left| \frac{1 \text{ g}}{6.022 \times 10^{23} \text{ amu}} \right| \frac{6.022 \times 10^{23} \text{ atom}}{1 \text{ mol}} = \frac{16 \text{ g}}{1 \text{ mol O}}$$

The numerical value for the atomic mass or atomic weight of a substance is the same as the number of grams in one mole of that substance, often called the **molar mass**. Avogadro's number is the link between the two. Hydrogen has a molar mass of 1 gram per mole; oxygen has a molar mass of 16 grams per mole.

When groups of atoms react together, they form molecules. Consider combining hydrogen and oxygen to get water (H_2O). Two atoms of hydrogen combine with one atom of oxygen, so 2(1 amu H) + 16 amu O = 18 amu H_2O. The **molecular mass** of water is 18 amu. By an extension of the example above, we can also state that one mole of water has a mass of 18 grams, called the **formula weight**.

The difference between these ideas is summarized in Table 8-8.

Table 8-8 Definitions of "amount" of substance

The quantity . . .	measures the . . .	in units of . . .	and is found by ...
Atomic mass	Mass of one atom of an individual isotope of an element	[amu]	Direct laboratory measurement
Atomic weight	Average mass of all isotopes of an element	[amu]	Listed on Periodic Table
Molar mass	Mass of one mole of the atom	[g/mol]	Listed on Periodic Table
Molecular mass or molecular weight	Sum of average weight of isotopes in molecule	[amu]	Combining atomic weights of individual atoms represented in the molecule
Formula weight	Mass of one mole of the molecule	[g/mol]	Combining molar mass of individual atoms represented in the molecule

This text assumes that you have been exposed to these ideas in an introductory chemistry class and so does not cover them in any detail. In all problems presented, you will be given the atomic weight of the elements or the formula weight of the molecule, depending on the question asked. This topic is briefly introduced because Avogadro's number (N_A) is important in the relationship between several constants, including the following:

- The gas constant (R [=] J/(mol K)) and the Boltzmann constant (k [=] J/K), which relates energy to temperature: $R = kN_A$.
- The elementary charge (e [=] C) and the Faraday constant (F [=] C/mol), which is the electric charge contained in one mole of electrons: $F = eN_A$.

EXAMPLE | **8-9**

Let us return to the problem of combining hydrogen and oxygen to get water. Assume you have 50 grams [g] of oxygen with which you want to combine the proper mass of hydrogen to convert the oxygen completely to water. The atomic weight of hydrogen is 1 and the atomic weight of oxygen is 16.

Calculate Desired Parameters	Steps
(1) Determine appropriate equation	$m = n\,MW$ so $n = \dfrac{m}{MW}$
(2) Insert known quantities	$n_O = \dfrac{50\text{ g O}}{} \bigg| \dfrac{1\text{ mol O}}{16\text{ g O}}$
(3) Calculate, reasonable	$n_O = 3.125$ mol O

By stoichiometry, oxygen (O) and hydrogen (H) combine in a one:two ratio to form water (H_2O). We need twice as many moles of hydrogen as oxygen.

$$n_H = 2n_O = 6.25 \text{ mol H}$$

Calculate Desired Parameters	Steps
(1) Determine appropriate equation	$m = n\,MW$
(2) Insert known quantities	$m_H = \dfrac{6.25 \text{ mol H}}{} \left\| \dfrac{1 \text{ g H}}{1 \text{ mol H}} \right.$
(3) Calculate, reasonable	$m_H = 6.25 \text{ g H}$

EXAMPLE 8-10

Acetylsalicylic acid (aspirin) has the chemical formula $C_9H_8O_4$. How many moles [mol] of aspirin are in a 1-gram [g] dose? Use the following facts:

- Atomic weight of carbon = 12
- Atomic weight of hydrogen = 1
- Atomic weight of oxygen = 16

Calculate Desired Parameters	Steps
(1) Determine appropriate equation	$FW_{aspirin} = 9(MW_C) + 8(MW_H) + 4(MW_O)$
(2) Insert known quantities	$FW_{aspirin} = 9 \text{ molecules} \left\| \dfrac{12\dfrac{\text{g C}}{\text{mol}}}{\text{molecule}} \right. + 8 \text{ molecules} \left\| \dfrac{1\dfrac{\text{g H}}{\text{mol}}}{\text{molecule}} \right.$ $+ 4 \text{ molecules} \left\| \dfrac{16\dfrac{\text{g O}}{\text{mol}}}{\text{molecule}} \right.$
(3) Calculate, reasonable	$FW_{aspirin} = 180\dfrac{\text{g}}{\text{mol}}$

Calculate Desired Parameters	Steps
(1) Determine appropriate equation	$m = n\,FW$ so $n = \dfrac{m}{FW}$
(2) Insert known quantities	$n_{aspirin} = \dfrac{1\text{g}}{\text{dose}} \left\| \dfrac{\text{mol}}{180 \text{ g}} \right.$
(3) Calculate, reasonable	$n_{aspirin} = 5.56 \times 10^{-3}\dfrac{\text{mol}}{\text{dose}} = 5.56\dfrac{\text{mmol}}{\text{dose}}$

EXAMPLE | **8-11**

Many gases exist as diatomic compounds in nature, meaning two of the atoms are attached to form a molecule. Hydrogen, oxygen, and nitrogen all exist in a gaseous diatomic state under standard conditions.

Assume there are 100 grams [g] of nitrogen gas in a container. How many moles [mol] of nitrogen (N_2) are in the container? Atomic weight of nitrogen (N) = 14.

Calculate Desired Parameters	Steps
(1) Determine appropriate equation	$FW_{N2} = 2(MW_N)$
(2) Insert known quantities	$FW_{N2} = 2 \text{ molecules} \left\| \dfrac{14\frac{g\,N}{mol}}{molecule} \right.$
(3) Calculate, reasonable	$FW_{N2} = 28\frac{g}{mol}$

Calculate Desired Parameters	Steps
(1) Determine appropriate equation	$m = n\,FW$ so $n = \dfrac{m}{FW}$
(2) Insert known quantities	$n_{N2} = \dfrac{100\,g}{} \left\| \dfrac{mol}{28\,g} \right.$
(3) Calculate, reasonable	$n_{N2} = 3.57\,mol$

Comprehension Check | **8-6**

Determine the mass in units of grams of 0.025 moles [mol] of caffeine (formula: $C_8H_{10}N_4O_2$). The components are hydrogen (H, amu = 1); carbon (C, amu = 12); nitrogen (N, amu = 14); and oxygen (O, amu = 16).

Comprehension Check | **8-7**

Determine the amount in units of moles [mol] of 5 grams [g] of a common analgesic acetaminophen (formula: $C_8H_9NO_2$). The components are hydrogen (H, amu = 1); carbon (C, amu = 12); nitrogen (N, amu = 14); and oxygen (O, amu = 16).

8.5 Temperature

LEARN TO: Convert a specific temperature value from one unit of temperature to another
Convert a material property value from one unit of temperature to another
Recall the temperature properties of water and the limits of the four common temperature scales (°C, °F, K, °R)

Temperature was originally conceived as a description of energy: heat (thermal energy) flows spontaneously from "hot" to "cold." But how hot is "hot"? The thermometer was devised as a way to measure the "hotness" of an object. As an object gets warmer, it usually expands. In a thermometer, a temperature is a level of hotness that corresponds to the length of the liquid in the tube. As the liquid gets warmer, it expands and moves up the tube. To give temperature a quantitative meaning, numerous temperature scales have been developed.

Many scientists, including Isaac Newton, have proposed temperature scales. Two scales were originally developed about the same time—**Fahrenheit** [°F] and **Celsius** [°C]—and have become widely accepted by the general public. Gabriel Fahrenheit (1686–1736), a German physicist and engineer, developed the Fahrenheit scale in 1708. Anders Celsius (1701–1744), a Swedish astronomer, developed the Celsius scale in 1742. The properties of each scale are in Table 8-9.

Table 8-9 Properties of water

Scale	Freezing Point	Boiling Point	Divisions Between Freezing and Boiling
Fahrenheit [°F]	32	212	180
Celsius [°C]	0	100	100
Kelvin [K]	273	373	100
Rankine [°R]	492	672	180

You may wonder why the Celsius scale seems so reasonable, and the Fahrenheit scale so random. Actually, Mr. Fahrenheit was just as reasonable as Mr. Celsius. Mr. Celsius set the freezing point of water to be 0 and the boiling point to be 100. Mr. Fahrenheit took as 0 a freezing mixture of salt and ice, and as 100 body temperature. With this scale, it just so happens that the freezing and boiling points of water work out to be strange numbers.

Some units can cause confusion depending on how they are used in calculation. One of those is temperature. One reason for this is that we use temperature in two different ways: **(1) reporting an actual temperature value and (2) discussing the way a change in temperature affects a material property.** To clarify, we resort to examples.

Calculating Temperature Values

When an actual temperature reading is reported, such as "the temperature in this room is 70°F," how do we determine this reading in another temperature scale? The scales have different zero points, so they cannot be determined using a single conversion factor, but require a formula. Most of you are familiar with the formula to calculate between Fahrenheit and Celsius, but this equation is cumbersome to remember.

$$T[°F] = \frac{9}{5} T[°C] + 32$$

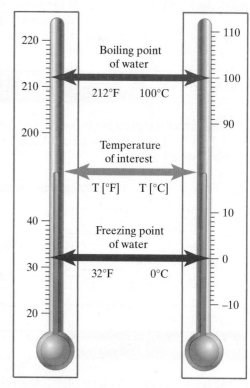

Fahrenheit   **Celsius**

Let us imagine we have two thermometers, one with the Fahrenheit scale and the other with the Celsius scale. We set two thermometers side by side so that the freezing point and the boiling point of water are at the same location on both thermometers. We are interested in the relationship between these two scales. From this figure we see that the fraction of the distance from the freezing point to the boiling point in both scales is the same. This means that we can write

$$\frac{T[°F] - 32}{212 - 32} = \frac{T[°C] - 0}{100 - 0}$$

This relationship is really all we need to know to relate a temperature in degrees Fahrenheit to degrees Celsius. You can easily do the algebra to calculate from Fahrenheit to Celsius, or vice versa. By remembering this form, you do not have to remember if the value is 9/5 or 5/9, or to add or subtract 32. This formula is determined by the method of interpolation.

There are numerous other temperature scales, but two are worth mentioning: **kelvin** [K] and **degrees Rankine** [°R]. The kelvin scale is named for First Baron William Thomson Kelvin (1824–1907), an English mathematician and physicist. Kelvin first proposed the idea of "infinite cold," or absolute zero, in 1848, using the Celsius scale for comparison. The Rankine scale is named for William J. M. Rankine (1820–1872), a Scottish engineer and physicist, who proposed an analogy to the kelvin scale, using the Fahrenheit scale for comparison. Both men made significant contributions to the field of thermodynamics.

The kelvin and Rankine scales are "absolute," which means that at absolute zero, the temperature at which molecules have minimum possible motion, the temperature is zero. *Absolute temperature scales therefore have no negative values.*

In the kelvin scale, the degree sign is not used; it is simply referred to as "kelvin," not "degrees kelvin." It is the base SI unit for temperature and the most frequently used temperature unit in the scientific community.

EXAMPLE | **8-12**

The hottest temperature in the United States ever recorded by the National Weather Service, 56.7 degrees Celsius [°C], occurred in Death Valley, California, on July 10, 1913. State this value in units of degrees Fahrenheit [°F].

Calculate Required Parameters	Steps
(1) Determine appropriate equation	$\dfrac{T[°F] - 32}{212 - 32} = \dfrac{T[°C] - 0}{100 - 0}$
(2) Insert known quantities	$\dfrac{T[°F] - 32}{180} = \dfrac{56.7 - 0}{100}$
(3) Calculate, be reasonable	$T = 134°F$

Comprehension Check | **8-8**

The temperature of dry ice is –109.3 degrees Fahrenheit [°F]. Convert this temperature into units of kelvins [K].

When properties that contain temperature are converted:

$$\frac{1°C}{1.8°F} \quad \frac{1K}{1°C} \quad \frac{1°R}{1°F}$$

For this type of conversion, we read the units under consideration as "*per* degree Fahrenheit," with the clue being the word "*per.*"

Conversions Involving Temperature Within a Material Property

When considering how a change in temperature affects a material property, we use a scalar conversion factor. In general, we encounter this in sets of units relating to the property of the material; for example, the units of the thermal conductivity are given by W/(m K), which is read as "watts per meter kelvin." When this is the case, we are referring to the size of the degree, not the actual temperature.

To find this relationship, remember that between the freezing point and the boiling point of pure water, the Celsius scale contains 100 divisions, whereas the Fahrenheit scale contains 180 divisions. The conversion factor between Celsius and Fahrenheit is $100°C \equiv 180°F$, or $1°C \equiv 1.8°F$.

EXAMPLE 8-13

The specific heat (C_p) is the ability of an object to store heat. Specific heat is a material property, and values are available in technical literature. The specific heat of copper is 0.385 J/(g °C), which is read as "joules per gram degree Celsius." Convert this to units of J/(lb$_m$°F), which reads "joules per pound-mass degree Fahrenheit."

NOTE

Specific heat [J/(g K)]

Air = 1.012

Aluminum = 0.897

Copper = 0.385

Helium = 5.1932

Lead = 0.127

Water = 4.184

Convert to Desired Units	Steps
(1) Term to be converted	$0.385\dfrac{J}{g\,°C}$
(2) Conversion formula	
(3) Make a fraction	$\dfrac{0.385\,J}{g\,°C}\left\|\dfrac{1000\,g}{2.205\,lb_m}\right\|\dfrac{1\,°C}{1.8\,°F}$
(4) Multiply	
(5) Cancel, calculate, be reasonable	$97\dfrac{J}{lb_m\,°F}$

A note of clarification about the term "per"—when reading the sentence: "Gravity on earth is commonly assumed to be 9.8 meters per second squared," there is often little confusion in translating the words to symbols: $g = 9.8$ m/s^2. For a more complex unit, however, this can present a challenge. For example, the sentence "The thermal conductivity of aluminum is 237 calories per hour meter degree Celsius," can be confusing because it can be interpreted as:

$$k = 237\,\frac{cal}{h\,m\,°C} \quad \text{or} \quad k = 237\,\frac{cal}{h}(m\,°C) \quad \text{or} \quad k = 237\,\frac{cal}{h\,m}°C$$

Officially, according to SI rules, *when writing out unit names anything following the word "per" appears in the denominator of the expression.* This implies the first example listed is correct.

Comprehension Check 8-9

The specific heat capacity of copper is 0.09 British thermal units per pound-mass degree Fahrenheit [BTU/(lb$_m$ °F)]. Convert into units of British thermal units per gram kelvin [BTU/(g K)].

8.6 Pressure

LEARN TO:	Determine final quantity if given four of the following: total pressure, hydrostatic pressure, density, gravity, height
	Describe Pascal's law
	Recall the common values for atmospheric pressure

IMPORTANT CONCEPT

$P = F/A$

SI unit of pressure = pascal [Pa]

Pressure is defined as force acting over an area, where the force is perpendicular to the area. In SI units, a **pascal** [Pa] is the unit of pressure, defined as one newton of force acting on an area of one square meter, expressed in terms of fundamental dimensions in Table 8-10. The unit pascal is named after Blaise Pascal (1623–1662), a French mathematician and physicist who made great contributions to the study of fluids, pressure, and vacuums. His contributions with Pierre de Fermat on the theory of probability were the groundwork for calculus.

Table 8-10 Dimensions of pressure

Quantity	Common Units	Exponents						
		M	**L**	**T**	**Θ**	**N**	**J**	**I**
Pressure	Pa	1	−1	−2	0	0	0	0

EXAMPLE   8-14

An automobile tire is pressurized to 40 pound-force per square inch [psi or lb_f/in^2]. State this pressure in units of atmospheres [atm].

PRESSURE

1 atm = 1.01325 bar
= 33.9 ft H_2O
= 29.92 in Hg
= 760 mm Hg
= 101,325 Pa
= 14.7 psi

Convert to Desired Units	Steps
(1) Term to be converted	40 psi
(2) Conversion formula	1 atm = 14.7 psi
(3) Make a fraction	$\dfrac{40\ psi}{}\ \bigg|\ \dfrac{1\ atm}{14.7\ psi}$
(4) Multiply	
(5) Cancel, calculate, be reasonable	2.7 atm

Comprehension Check   8-10

If the pressure is 250 feet of water [ft H_2O], what is the pressure in units of inches of mercury [in Hg]?

The general term **fluid** applies to a gas, such as helium or air, or a liquid, such as water or honey. In this chapter, we consider four forms of pressure, all involving fluids.

- **Atmospheric pressure**—the pressure created by the weight of air above an object.
- **Hydrostatic pressure**—the pressure exerted on a submerged object by the weight of a liquid above it.
- **Total pressure**—the summation of atmospheric and hydrostatic pressure.
- **Gas pressure**—the pressure created by the movement of gas molecules inside a closed container.

Atmospheric Pressure

Atmospheric pressure results from the weight of the air above an object which varies with both altitude and weather patterns. Standard atmospheric pressure is an average air pressure at sea level, defined as one atmosphere [atm], approximately equal to 14.7 pound-force per square inch [psi].

Pressure Measurement

When measuring pressure, two types of reference points are commonly used.

Absolute pressure uses a perfect vacuum as a reference point. Most meteorological readings are given as absolute pressure, using units of atmospheres or bars. When solving problems dealing with hydrostatic or gas pressure, the pressure given should be considered as absolute.

Gauge pressure uses the local atmospheric pressure as a reference point. Measurements such as tire pressure and blood pressure are given as gauge pressure.

Absolute pressures are distinguished by an "a" after the pressure unit, such as "psia" to signify "pound-force per square inch absolute." Gauge pressure readings are distinguished by a "g" after the pressure unit, such as "psig" to signify "pound-force per square inch gauge." When using instrumentation to determine the pressure, be sure to note whether the device reads absolute or gauge pressure.

Gauge pressure, absolute pressure, and atmospheric pressure are related by

$$P_{absolute} = P_{gauge} + P_{atmospheric}$$

For example, if we have a reading of 35 psig, this would be 49.7 psia assuming an atmospheric pressure of 14.7 psi.

$$13.5 \text{ psig} + 14.7 \text{ psi} = 49.7 \text{ psia}$$

If a gauge pressure being measured is less than the local atmospheric pressure, this is usually referred to as **vacuum pressure**. A perfect vacuum is defined as 0 psia. Thus, a perfect vacuum created at sea level on the Earth would read −14.7 psig, or 14.7 psig vacuum pressure.

As another example, if we have a reading of 10 psig vacuum pressure, this would be 4.7 psia assuming an atmospheric pressure of 14.7 psi.

$$-10 \text{ psig} + 14.7 \text{ psi} = 4.7 \text{ psia}$$

To illustrate the effect of local atmospheric pressure, consider the following scenario. You fill your automobile's tires to 35 psig on the shore of the Pacific Ocean in Peru, and then drive to Lake Titicaca on the Bolivian border at about 12,500 feet above sea level. The absolute pressure in the tires must remain the same in both locations, so your tire pressure now reads about 40 psig due to the decreased atmospheric pressure.

NOTE

Car tires are inflated with between 30 and 40 psi (gauge pressure).

"Normal" blood pressure is 120 mm Hg/80 mm Hg (gauge pressure).

At the shore: 35 psig + 14.7 psi = 49.7 psia

At the lake: 49.7 psia − 9.5 psi = 40.2 psig

Occasionally in industry, it may be helpful to use a point of reference other than atmospheric pressure. For these specific applications, pressure may be discussed in terms of **differential pressure**, distinguished by a "d" after the pressure unit, such as "psid."

Comprehension Check	8-11

A car tire is inflated to a gauge pressure of 35 pound-force per square inch [psi]. What is the absolute pressure rating of the tire in units of atmospheres [atm]?

Hydrostatic Pressure

Hydrostatic pressure (P_{hydro}) results from the weight of a liquid pushing on an object. *Remember, weight is a force!* A simple way to determine this is to consider a cylinder with a cross-sectional area (A) filled with a liquid of density ρ.

IMPORTANT CONCEPT

Pascal's Law

$P_{\text{hydro}} = \rho g H$

The pressure (P) at the bottom of the container can be found by **Pascal's law**, named after (once again) Blaise Pascal. Pascal's law states the hydrostatic pressure of a fluid is equal to the force of the fluid acting over an area.

IMPORTANT CONCEPT

How does the term $\rho\, g\, H$ equal a pressure?

$$P_{\text{hydro}} = \frac{F}{A}$$

Weight is a force

$$= \frac{w}{A}$$

$w = m\, g$

$$= \frac{m\, g}{A}$$

$\rho = m/V$

$$= \frac{\rho V g}{A}$$

$V = A H$

$$= \frac{\rho\, (A\, H)\, g}{A}$$

$$P_{\text{hydro}} = \rho\, g\, H$$

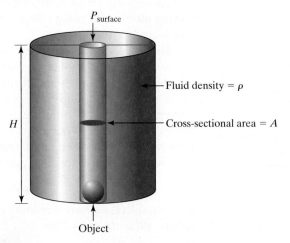

Recreational scuba diving takes place at depths between 0 and 20 meters. At deeper depths, additional training is usually required because of the increased risk of narcosis, a state similar to alcohol intoxication. The relationship between the depth and the level of narcosis is called the "Martini Effect," as it is said divers feel like they have drunk one martini for every 20 meters they descend.

Photo courtesy of E. Stephan

EXAMPLE   8-15

We want to know the hydrostatic pressure in a lake at a depth of 20 feet [ft] in units of pascals [Pa].

NOTE

Density of water = 1000 kg/m³

For hydrostatic pressure, we need to know the density of the fluid in the lake. Since a density is not specified, we assume the density to be 1000 kilograms per cubic meter [kg/m³] for water.

Convert to Base SI Units	Steps
(1) Term to be converted	20 ft
(2) Conversion formula	
(3) Make a fraction	$\dfrac{20\,\text{ft}}{}\left|\dfrac{1\,\text{m}}{3.28\,\text{ft}}\right.$
(4) Multiply	
(5) Cancel, calculate	6.1 m

Calculate Required Parameters	Steps
(1) Determine appropriate equation	$P_{\text{hydro}} = \rho\, g\, H$
(2) Insert known quantities	$P_{\text{hydro}} = \dfrac{1000\,\text{kg}}{\text{m}^3}\left|\dfrac{9.8\,\text{m}}{\text{s}^2}\right|\dfrac{6.1\,\text{m}}{}$
(3) Calculate, be reasonable	$P_{\text{hydro}} = 59{,}760\dfrac{\text{kg}}{\text{m s}^2}$

This is apparently our final answer, but the units are puzzling. If the units of pressure are pascals and if this is a valid equation, then our final result for pressure should be pascals. If we consider the dimensions of pressure:

Quantity	Common Units	Exponents		
		M	**L**	**T**
Pressure	Pa	1	−1	−2

A unit of pressure has dimensions, $P\{=\}\, M/(LT^2)$, which in terms of base SI units would be $P[=]\text{kg}/(\text{m s}^2)$. As this term occurs frequently, it is given the special name "pascal." When we see this term, we know we are dealing with a pressure equal to a pascal.

(3) Calculate, be reasonable	$P_{\text{hydro}} = 59{,}760\dfrac{\text{kg}}{\text{m s}^2}\left|\dfrac{1\,\text{Pa}}{1\frac{\text{kg}}{\text{m s}^2}}\right. = 59{,}760\,\text{Pa}$

Total Pressure

We need to realize that Pascal's law is only a part of the story. Suppose we dive to a depth of 5 feet in a swimming pool and measure the pressure. Now we construct an enclosure over the pool and increase the pressure of the air above the water surface from 1 atmosphere [atm] to 3 atmospheres [atm]. When we dive back to the 5-foot [ft] depth, the pressure will have increased by 2 atmospheres [atm].

Consequently, we conclude that total pressure at any depth in a fluid is the sum of hydrostatic pressure and surface pressure.

IMPORTANT CONCEPT

$P_{\text{total}} = P_{\text{surface}} + P_{\text{hydro}}$

$P_{\text{total}} = P_{\text{surface}} + \rho\, g\, H$

EXAMPLE   8-16

When you dive to the bottom of a pool, at 12 feet [ft] under water, how much total pressure do you feel in units of atmospheres [atm]?

Convert to Base SI Units	Steps
(1) Term to be converted	12 ft
(2) Conversion formula	
(3) Make a fraction	$\dfrac{12 \text{ ft}}{} \bigg| \dfrac{1 \text{ m}}{3.28 \text{ ft}}$
(4) Multiply	
(5) Cancel, calculate	3.66 m

NOTE

Density of water = 1000 kg/m³

Surface pressure = 1 atm

For hydrostatic pressure, we need to know the density of the fluid in the pool. Since a density is not specified, we assume the density to be 1000 kilograms per cubic meter [kg/m³] for water.

For total pressure, we need to know the surface pressure on top of the pool. Since a surface pressure is not specified, we assume the pressure to be 1 atmosphere [atm] or 101,325 pascals.

Calculate Required Parameters	Steps
(1) Determine appropriate equation	$P_{total} = P_{surface} + \rho\, g\, H$
(2) Insert known quantities	$P_{total} = \dfrac{101{,}325 \text{ kg}}{\text{m s}^2} + \dfrac{1000 \text{ kg}}{\text{m}^3} \bigg| \dfrac{9.8 \text{ m}}{\text{s}^2} \bigg| \dfrac{3.66 \text{ m}}{}$
(3) Calculate	$P_{total} = 137{,}193 \dfrac{\text{kg}}{\text{m s}^2} \bigg| \dfrac{1 \text{ Pa}}{\frac{1 \text{ kg}}{\text{m s}^2}} = 137{,}193 \text{ Pa}$

Convert to Desired Units	Steps
(1) Term to be converted	137,193 Pa
(2) Conversion formula	
(3) Make a fraction	$\dfrac{137{,}193 \text{ Pa}}{} \bigg| \dfrac{1 \text{ atm}}{101{,}325 \text{ Pa}}$
(4) Multiply	
(5) Cancel, calculate, be reasonable	1.35 atm

Comprehension Check   8-12

An object is completely submerged in a liquid with a specific gravity of 0.75 at a depth of 3 meters [m]. What is the total pressure on the object? State your answer in atmospheres [atm].

8.7 Gas Pressure

LEARN TO:	Describe the ideal gas law
	Determine final quantity if given three of the following: amount, pressure, temperature, volume
	Recall the common values for ideal gas constant

IMPORTANT CONCEPT

Ideal Gas Law

$$P V = n R T$$

Only absolute temperature units (K or °R) can be used in the ideal gas equation.

NOTE

$$R = 8314 \frac{Pa\,L}{mol\,K}$$

$$= 0.08206 \frac{atm\,L}{mol\,K}$$

Gas pressure results when gas molecules impact the inner walls of a sealed container. The **ideal gas law** relates the quantities of pressure (P), volume (V), temperature (T), and amount (n) of gas in a closed container:

$$P V = n R T$$

In this equation, R is a fundamental constant called the **gas constant**. It can have many different numerical values, depending on the units chosen for pressure, volume, temperature, and amount, just as a length has different numerical values, depending on whether feet or meters or miles is the unit being used. Scientists have defined an "ideal" gas as one where one mole [mol] of gas at a temperature of 273 kelvins [K] and a pressure of one atmosphere [atm] will occupy a volume of 22.4 liters [L]. Using these values to solve for the constant R yields

$$R = \frac{P V}{n T} = \frac{1\,[\,atm\,]\;22.4\,[\,L\,]}{1\,[\,mol\,]\;273\,[\,K\,]} = 0.08206 \frac{atm\,L}{mol\,K}$$

Note that we must *use* absolute *temperature units in the ideal gas equation.* We cannot begin with relative temperature units and then convert the final answer. Also, all pressure readings must be in absolute, not gauge or differential, units.

In previous chapters, we have suggested a procedure for solving problems involving equations and unit conversions. For ideal gas law problems, we suggest a slightly different procedure.

Ideal Gas Law Procedure

1. Examine the units given in the problem statement. Choose a gas constant (R) that contains as many of the units given in the problem as possible.
2. If necessary, convert all parameters into units found in the gas constant (R) that you choose in Step 1.
3. Solve the ideal gas law for the variable of interest.
4. Substitute values and perform all necessary calculations.
5. If necessary, convert your final answer to the required units and apply reasonableness.

EXAMPLE 8-17

A container holds 1.43 moles [mol] of nitrogen (formula: N_2) at a pressure of 3.4 atmospheres [atm] and a temperature of 500 degrees Fahrenheit [°F]. What is the volume of the container in liters [L]?

Ideal Gas Law Procedure	Steps
(1) Choose ideal gas constant	Given units: mol, atm, °F, L Select R: 0.08206 $\dfrac{\text{atm L}}{\text{mol K}}$
(2) Convert to units of chosen R	$T = 500°F = 533\ K$
(3) Solve for variable of interest	$V = \dfrac{n\,R\,T}{P}$
(4) Calculate	$V = \dfrac{1.43\ \text{mol}}{} \left\| \dfrac{0.08206\ \text{atm L}}{\text{mol K}} \right\| \dfrac{533\ K}{} \left\| \dfrac{}{3.4\ \text{atm}} \right.$
(5) Convert, be reasonable	$V = 18.4\ L$

EXAMPLE 8-18

A 12-liter [L] container holds nitrogen (formula: N_2) at a temperature of 160 degrees Celsius [°C]. If the gauge pressure reads 350 kilopascals [kPa], what is the amount, in units of moles [mol], of gas in the container?

Ideal Gas Law Procedure	Steps
(1) Choose ideal gas constant	Given units: L, °C, kPa Select $R = 8314\dfrac{\text{Pa L}}{\text{mol K}}$
(2) Convert into units of chosen R	$P = \dfrac{350\text{k Pa}}{} \left\| \dfrac{1000\ \text{Pa}}{\text{k Pa}} \right. = 350{,}000\ \text{Pa}$ $T = 160\ °C = 433\ K$
(3) Solve for variable of interest	$n = \left(\dfrac{P\,V}{R\,T}\right)$
(4) Calculate	$n = \dfrac{350000\ \text{Pa}}{} \left\| \dfrac{12\ L}{} \right\| \dfrac{\text{mol K}}{8314\ \text{Pa L}} \left\| \dfrac{}{433\ K} \right.$
(5) Convert, be reasonable	$n = 1.17\ \text{mol}$

Comprehension Check 8-13

A 5-gallon [gal] container holds 35 grams [g] of nitrogen (formula: N_2, molecular weight = 28 grams per mole [g/mol]) at a temperature of 400 kelvins [K]. What is the container pressure in units of kilopascals [kPa]?

EXAMPLE   8-19

A gas originally at a temperature of 300 kelvins [K] and 3 atmospheres [atm] pressure in a 3.9-liter [L] flask is cooled until the temperature reaches 284 kelvins [K]. What is the new pressure of gas in atmospheres [atm]?

Since the volume and the mass of the gas remain constant, we can examine the ratio between the initial condition (1) and the final condition (2) for pressure and temperature. The volume of the container (V) and the amount of gas (n) are constant, so $V_1 = V_2$ and $n_1 = n_2$.

Ideal Gas Law Procedure	Steps
(1) Choose ideal gas constant	Given units: mol, K, L Select R: $0.08206 \dfrac{\text{atm L}}{\text{mol K}}$
(2) Convert to units of chosen R	None needed
(3) Solve for variable of interest, eliminated any variables that remain constant between the initial and final state	$\dfrac{P_1 V_1}{P_2 V_2} = \dfrac{n_1 R T_1}{n_2 R T_2}$ $\dfrac{P_1}{P_2} = \dfrac{T_1}{T_2}$
(4) Calculate	$\dfrac{3 \text{ atm}}{P_2} = \dfrac{300 \text{ K}}{284 \text{ K}}$
(5) Convert, be reasonable	$P_2 = 2.8 \text{ atm}$

Comprehension Check   8-14

An 8-liter [L] container holds nitrogen (formula: N_2, molecular weight = 28 grams per mole [g/mol]) at a pressure of 1.5 atmospheres [atm] and a temperature of 310 kelvins [K]. If the gas is compressed by reduction of the volume of the container until the gas pressure increases to 5 atmospheres [atm] while the temperature and amount are held constant, what is the new volume of the container in units of liters [L]?

8.8 Energy

LEARN TO:	Determine final quantity if given the other terms in an energy expression
	Convert from one unit of energy to another
	Select the energy units appropriate to a context

Energy is an abstract quantity with several definitions, depending on the form of energy being discussed. You may be familiar with some of the following types of energy.

IMPORTANT CONCEPT

Work
$$W = F\Delta x$$

Potential Energy
$$PE = mg\Delta H$$

Kinetic Energy
$$KE_T = \tfrac{1}{2}m(v_f^2 - v_i^2)$$

Thermal Energy
$$Q = mC_p\Delta T$$

Types of Energy

- **Work** (W) is energy expended by exertion of a force (F) over a change in distance (Δx). As an example, if you exert a force on (push) a heavy desk so that it slides across the floor.
- **Potential energy** (PE) is a form of work done by moving a weight (w) over a change in a vertical distance (ΔH). Recall that weight is mass (m) times gravity (g). Note that this is a special case of the work equation, where force is weight and distance is height.
- **Kinetic energy** (KE) is a form of energy possessed by an object in motion. If a constant force is exerted on a body, then by $F = ma$, we see that the body experiences a constant acceleration, meaning the velocity increases linearly with time. Since the velocity increases as long as the force is maintained, work is being done on the object. Another way of saying this is that the object upon which the force is applied acquires kinetic energy by a change in velocity, also called **energy of translational motion**. For a nonrotating body moving with some velocity (v) the kinetic energy can be calculated by $KE_T = (\tfrac{1}{2})mv^2$, assuming the body was initially at rest ($v = 0$).
- **Thermal Energy** or heat (Q) is energy associated with a change in temperature (ΔT). It is a function of the mass of the object (m) and the specific heat (C_p), which is a property of the material being heated.

Where Does $KE = \tfrac{1}{2}mv^2$ Come From?

If a constant force is applied to a body,

- That body will have a constant acceleration (remember $F = ma$).
- Its average velocity is the average of its initial and final values.

Average velocity	$v = (v_f + v_i)/2$
Distance traveled	$d = vt$
Work done	$W = Fd$
	$= (ma)d$
	$= mavt$

Acceleration is the change in velocity over time, or $a = (v_f - v_i)/t$
Substituting for a and v in the work equation

$$W = mavt = m[(v_f - v_i)/t][(v_f + v_i)/2][t]$$
$$W = (\tfrac{1}{2})m(v_f^2 - v_i^2)$$

This is given the name *kinetic energy*, or the energy of motion. Remember, work and energy are equivalent. This expression is for the *translation* of a body only.

Calories and BTUs and Joules—Oh My!

The SI unit of energy is **joule**, defined as one newton of force acting over a distance of one meter (Table 8-11). The unit is named after James Joule (1818–1889), an English physicist responsible for several theories involving energy, including the definition of the mechanical equivalent of heat and Joule's law, which describes the amount of electrical energy converted to heat by a resistor (an electrical component) when an electric current flows through it. In some mechanical systems, work is described in units of foot pound-force [ft lb$_f$].

For energy in the form of heat, units are typically reported as British thermal units and calories instead of joules. A **British thermal unit** [BTU] is the amount of heat required to raise the temperature of one pound-mass of water by one degree Fahrenheit. A **calorie** [cal] is amount of heat required to raise the temperature of one gram of water by one degree Celsius.

Table 8-11 Dimensions of energy

Quantity	Common Units	Exponents						
		M	**L**	**T**	**Θ**	**N**	**J**	**I**
Energy	J	1	2	−2	0	0	0	0

EXAMPLE   8-20

A 50-kilogram [kg] load is raised vertically a distance of 5 meters [m] by an electric motor. How much energy in units of joules [J] was required to raise the load?

First, we must determine the type of energy. The parameters we are discussing include mass (kilograms) and height (meters). Examining the energy formulas given above, the equation for potential energy fits. Also, the words "load is raised vertically a distance" fits with our understanding of potential energy.

Calculate Required Parameters	Steps
(1) Determine appropriate equation	$PE = m\,g\,\Delta H$
(2) Insert known quantities	$PE = \dfrac{50\ \text{kg}}{}\left\|\dfrac{9.8\ \text{m}}{\text{s}^2}\right\|\dfrac{5\ \text{m}}{}$
(3) Calculate, be reasonable	$PE = 2450\ \dfrac{\text{kg m}^2}{\text{s}^2}$

This is apparently our final answer, but the units are puzzling. If the units of energy are joules and if this is a valid equation, then our final result for energy should be joules. If we consider the dimensions of energy:

Quantity	Common Units	Exponents		
		M	**L**	**T**
Energy	J	1	2	−2

A unit of energy has dimensions $E\{=\} M L^2/T^2$, which in terms of base SI units would be $E[=]kg\ m^2/s^2$. As this term occurs frequently, it is given the special name "joule." Anytime we see this term ($kg\ m^2/s^2$), we know we are dealing with an energy, equal to a joule.

(3) Calculate, be reasonable	$PE = 2450\dfrac{kg\ m^2}{s^2}\left\vert\dfrac{1\ J}{1\frac{kg\ m^2}{s^2}}\right. = 2450\ J$

EXAMPLE   8-21

In the morning, you like to drink your coffee at a temperature of exactly 70 degrees Celsius [°C]. The mass of the coffee in your mug is 470 grams [g]. To make your coffee, you had to raise the temperature of the water by 30 degrees Celsius [°C]. How much energy in units of British thermal units [BTU] did it take to heat your coffee? The specific heat of water is 4.18 joules per gram degree Celsius [J/(g °C)].

First, you must determine the type of energy we are using. The parameters discussed include mass, temperature, and specific heat. Examining the energy formulas given above, the equation for thermal energy fits. Also, the words "How much energy . . . did it take to heat your coffee" fits with an understanding of thermal energy.

Calculate Required Parameters	Steps
(1) Determine appropriate equation	$Q = m\ C_p\ \Delta T$
(2) Insert known quantities	$Q = \dfrac{470\ g}{}\left\vert\dfrac{4.18\ J}{g\ °C}\right\vert\dfrac{30\ °C}{}$
(3) Calculate	$Q = 59{,}370\ J$

Convert to Desired Units	Steps
(1) Term to be converted	$59{,}370\ J$
(2) Conversion formula	
(3) Make a fraction	$\dfrac{59{,}370\ J}{}\left\vert\dfrac{9.48 \times 10^{-4}\ BTU}{1\ J}\right.$
(4) Multiply	
(5) Cancel, calculate, be reasonable	$56\ BTU$

Comprehension Check	**8-15**

You push an automobile with a constant force of 20 pound-force [lb$_f$] until 1,500 joules [J] of energy has been added to the car. How far did the car travel in units of meters [m] during this time? You may assume that frictional losses are negligible.

Comprehension Check	**8-16**

One gram of material A is heated until the temperature rises by 10 kelvins [K]. If the same amount of heat is applied to one gram [g] of material B, what is the temperature rise of material B in units of kelvins [K]?
  The specific heat (C_p) of material A = 4 joules per gram kelvin [J/(g K)]
  The specific heat (C_p) of material B = 2 joules per gram kelvin [J/(g K)]

8.9 Power

LEARN TO:	Convert from one unit of power to another
	Determine the final quantity if given two of the following: energy, power, time

IMPORTANT CONCEPT

$P = E / t$

SI unit of power = watt [W]

NOTE

Power is the RATE at which energy is delivered over time.

Power is defined as energy per time (Table 8-12). The SI unit of power is **watt**, named after James Watt (1736–1819), a Scottish mathematician and engineer whose improvements to the steam engine were important to the Industrial Revolution. He is responsible for the definition of **horsepower** [hp], a unit of power originally used to quantify how the steam engine could replace the work done by a horse.

Table 8-12 Dimensions of power

Quantity	Common Units	Exponents						
		M	L	T	Θ	N	J	I
Power	W	1	2	−3	0	0	0	0

To help understand the relationship between energy and power, imagine the following. Your 1000-kilogram car has run out of gas on a level road. There is a gas station not far ahead, so you decide to push the car to the gas station. Assume that you intend to accelerate the car up to a speed of one meter per second (about 2.2 miles per hour), and then continue pushing at that speed until you reach the station. Ask yourself the following questions:

- Can I accelerate the car to one meter per second in one minute?
- On the other hand, can I accelerate it to one meter per second in one second?

Most of you would probably answer "yes" to the first and "no" to the second, but why? Well, personal experience! But that is not really an explanation. Since the change in kinetic energy is the same in each case, to accelerate the car in one second, your body would have to generate energy at a rate 60 times greater than the rate required if you accelerated it in one minute. The key word is rate, or how much energy your body can produce per second. If you do the calculations, you will find that for the one-minute scenario, your body would have to produce about $1/90$ horsepower, which seems quite reasonable. On the other hand, if you try to accomplish the same acceleration in one second, you would need to generate $2/3$ horsepower. Are you two-thirds as powerful as a horse?

As another example, assume that you attend a class on the third floor of the engineering building. When you are on time, you take 2 minutes to climb to the third floor. On the other hand, when you are late for class, you run up the three flights in 30 seconds. In which case do you do the most work (expend the most energy)? In which case do you generate the most power?

EXAMPLE 8-22

A 50-kilogram [kg] load is raised vertically a distance of 5 meters [m] by an electric motor in 60 seconds [s]. How much power in units of watts [W] does the motor use, assuming no energy is lost in the process?

This problem was started in Example 8-20. The energy used by the system was found to be 2450 joules [J], the analysis of which is not repeated here.

Calculate Required Parameters	Steps
(1) Determine appropriate equation	$P = \dfrac{E}{t}$
(2) Insert known quantities	$P = \dfrac{2450 \text{ J}}{60 \text{ s}} \left\| \dfrac{1 \text{ W}}{1 \frac{\text{J}}{\text{s}}} \right.$
(3) Calculate, be reasonable	$P = 41 \text{ W}$

Note that since power = energy/time, energy = power * time. We pay the electric company for energy calculated this way as kilowatt-hours. If power is constant, we can obtain the total energy involved simply by multiplying the power by the duration of time that power is applied. If power is *not* constant, we would usually use calculus to determine the total energy, but that solution is beyond the scope of this book.

Comprehension Check 8-17

A motor with a power of 100 watts [W] is connected to a flywheel. How long, in units of hours [h], must the motor operate to transfer 300,000 joules [J] to the flywheel?

8.10 Efficiency

LEARN TO: Recall the limits of efficiency

Determine the final quantity if given three of the following: efficiency, energy, power, time

IMPORTANT CONCEPT

Efficiency is always less than 100%.

IMPORTANT CONCEPT

η = Output/Input

Input = Output + Loss

Input = quantity required by mechanism to operate

Output = quantity actually applied to task

Loss = quantity wasted during the application

Efficiency (η, Greek letter eta) is a measure of how much energy or power is lost in a process. In a perfect world, efficiency would always be 100%. All energy put into a process would be recovered and used to accomplish the desired task. We know that this can never happen, so *efficiency is always less than 100%*. If a machine operates at 75% efficiency, 25% of the energy put into the process is lost.

The use of the terms "input" and "output" require some explanation. The **input** is the quantity of energy or power supplied to the mechanism from some source that the mechanism requires to operate. The **output** is the amount of energy or power that is actually applied to a task by the mechanism. Note that the rated power or power listed on the device label refers to the input power—the power needed to operate the device. In an ideal, 100% efficient system, the input and output would be equivalent. In an inefficient system (the real world), the input is equivalent to the sum of the output and the power or energy lost.

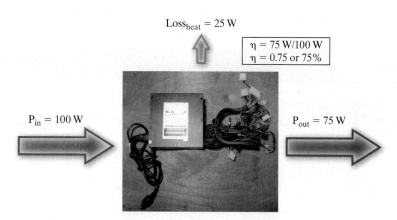

$\text{Loss}_{\text{heat}}$ = 25 W

η = 75 W/100 W
η = 0.75 or 75%

P_{in} = 100 W

P_{out} = 75 W

Photo courtesy of E Stephan

Orders of Magnitude

Table 8-13 gives you an idea of orders of magnitude of power and energy as related to real-world objects and phenomena. All values are approximate and, in most cases, have been rounded to only one significant figure. Thus, if you actually do the calculations from power to energy, you will find discrepancies. A few things of possible interest:

- U.S. power consumption (all types) is one-fifth of the total world power consumption.
- The Tsar Bomba generated 1.5% of the power of the sun, but only lasted 40 nanoseconds.
- Total human power consumption on the planet is about 0.01% (1/10,000) of the total power received from the sun.

Table 8-13 Order of magnitude for power and energy comparison

Power	"Device"	Energy per Hour	Energy per Year
10 fW	Minimum reception power for cell phone	40 pJ	300 nJ
	Single human cell	4 nJ	30 mJ
10 mW	DVD laser	40 J	300 kJ
500 mW	Cell phone microprocessor	2 kJ	15 MJ
750 W	Power per square meter bright sunshine	3 MJ	25 GJ
20 kW	Average U.S. home	80 MJ	600 GJ
100 kW	Typical automobile	400 MJ	3 TJ
150 MW	Boeing 747 jet	500 GJ	5 PJ
1 GW	Large commercial nuclear reactor	4 TJ	30 PJ
20 GW	Three Gorges hydroelectric dam (China)	80 TJ	600 PJ
4 TW	U.S. total power consumption	15 PJ	100 EJ
20 TW	Total human power consumption	80 PJ	600 EJ
100 TW	Average hurricane	400 PJ	(NA)
200 PW	Total power received on the Earth from sun	1 ZJ	6 YJ
400 YW	Total power of the sun	10^6 YJ	10^9 YJ

EXAMPLE   8-23

A standard incandescent light bulb has an efficiency of about 5%; thus, $\eta = 0.05$. An incandescent bulb works by heating a wire (the filament) inside the bulb to such a high temperature that it glows white.

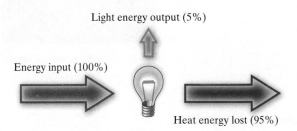

Light energy output (5%)

Energy input (100%)

Heat energy lost (95%)

If a 100-watt [W] bulb is turned on for 15 minutes [min], how much energy, in units of joules [J], is "lost" as heat during the 15-minute [min] period?

Calculate Required Parameters	Steps
(1) Determine appropriate equation	$E = P\,t$
(2) Insert known quantities	$E = (P)(15 \text{ min})$

(1) Determine appropriate equation	$P_{in} = P_{out}/\eta$ $P_{loss} = P_{in} - P_{out}$
(2) Insert known quantities	$100\ W = P_{out}/0.05$
(3) Calculate	$P_{out} = 5\ W$ $P_{loss} = 100\ W - 5\ W = 95\ W$
(2) Insert known quantities	$E = (95\ W)(15\ min)$
(3) Calculate, be reasonable	$E = 95\dfrac{J}{s}\left|\dfrac{15\ min}{}\right.\left|\dfrac{60\ s}{1\ min}\right. = 85{,}500\ J$

EXAMPLE   8-24

Over the past few decades, the efficiency of solar cells has risen from about 10% to the most recent technologies achieving about 40% conversion of solar energy to electricity. The losses are due to several factors, including reflectance and resistive losses, among others.

Assume you have an array of solar cells mounted on your roof with an efficiency of 28%. If the array is delivering 750 watts [W] of electricity to your home, how much solar power, in units of watts [W], is falling on the photoelectric cells?

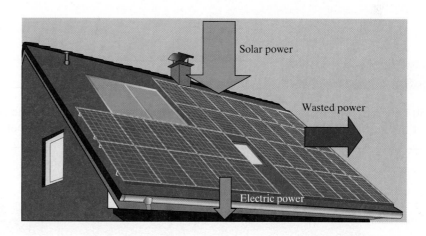

NOTE

How much power is wasted by the array of solar cells? If the array received 2680 W and delivered 750 W, then the difference is 2680 − 750 = 1930 W; thus, 1930 W are wasted.

Calculate Required Parameters	Steps
(1) Determine appropriate equation	$P_{in} = P_{out}/\eta$
(2) Insert known quantities	$P_{in} = 750\ W/0.28$
(3) Calculate, be reasonable	$P_{in} = 2680\ W$

EXAMPLE   8-25

If your microwave takes 2 minutes [min] to heat your coffee in Example 8-21, how many watts [W] of power does your microwave require, assuming that it is 80% efficient? Remember that our answer was 59,370 joules [J], before we converted the final answer to units of British Thermal Units [BTU].

Calculate Required Parameters	Steps
(1) Determine appropriate equation	$P_{in} = P_{out}/\eta$
(2) Insert known quantities	$P_{in} = P_{out}/0.8$
(1) Determine appropriate equation	$P_{out} = E/t$
(2) Insert known quantities	$P_{out} = \dfrac{59,370 \text{ J}}{}\left\|\dfrac{}{2 \text{ min}}\right\|\dfrac{1 \text{ min}}{60 \text{ s}}\left\|\dfrac{1 \text{ W s}}{1 \text{ J}}\right.$
(3) Calculate	$P_{out} = 493 \text{ W}$
(2) Insert known quantities	$P_{in} = 493 \text{ W}/0.8$
(3) Calculate, be reasonable	$P_{in} = 615 \text{ W}$

Comprehension Check   8-18

A motor with an input power of 100 watts [W] is connected to a flywheel. How long, in units of hours [h], must the motor operate to transfer 300,000 joules [J] to the flywheel, assuming the process is 80% efficient?

Comprehension Check   8-19

If a 50-kilogram [kg] load was raised 5 meters [m] in 50 seconds [s], determine the minimum rated wattage [W] of the motor needed to accomplish this task, assuming the motor is 80% efficient.

EXAMPLE   8-26

A simple two-stage machine is shown in the diagram below. Initially, an electric motor receives power from the power grid, accessed by being plugged into a standard electrical wall socket. The power received by the motor from the wall socket is the "input" power or the power the motor uses or requires (Point A).

    The spinning drive shaft on the motor can then be used to power other devices, such as a hoist or a vacuum cleaner or a DVD drive; the power available from the spinning shaft is the "output" power of the motor (Point C). In the process of making the drive shaft spin, however, some of the input power is lost (Point B) because of both frictional and ohmic heating as well as other wasted forms such as sound.

The power available from the spinning shaft of the motor (Point C) is then used to operate some device: the hoist or vacuum cleaner or DVD drive (Point E). In other words, the output power of the motor is the input power to the device it drives. This device will have its own efficiency, thus wasting some of the power supplied to it by the motor (Point D).

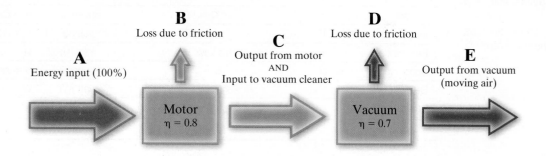

NOTE

The overall efficiency of two linked devices is the product of the two efficiencies.

Assume that a motor with an efficiency of 80% is used to power a vacuum cleaner that has an efficiency of 70%. If the input power to the motor (Point A) is one-half horsepower, what is the output power of the vacuum (Point E)?

*The output from the motor (Point C) is $P_C = \eta\, P_A = 0.8\,(0.5\ hp) = 0.4\ hp$.
This is the input to the vacuum, so the output from the vacuum (Point E) is*

$$P_E = \eta\, P_C = 0.7\,(0.4\ hp) = 0.28\ hp$$

What is the overall efficiency of this machine?

The input power (Point A) is 0.5 hp, and the output power (Point E) is 0.28 hp; thus, the efficiency of this linked system is:

$$\eta = P_A/P_E = 0.28\ hp/0.5\ hp = 0.56,\ \textit{or an efficiency of 56\%.}$$

8.11 Electrical Concepts

LEARN TO: Describe the relationship between electric charge and electric current
Use relationships between common electrical properties and their units to facilitate problem solution

The basic concepts of electricity and electrical devices are perhaps less familiar to most students than are many of the other physical phenomena covered previously in this text. This is due partly to lack of practical experience, and partly to the fact that in general these phenomena are themselves invisible, only their effects being perceptible. These effects range from receiving an electric shock to the almost magical performance of touch-screen devices. Table 8-14 summarizes the concepts discussed in this section, and Table 8-15 shows the representations of electrical devices in schematic drawings.

Table 8-14 Summary of electrical properties

Property	Symbol	Related Equations	Typical Units	Equivalent Units
Charge	Q	Coulomb's Law: $\|F\| = k_e \dfrac{\|Q_1 Q_2\|}{r^2}$	coulomb [C]	C = A s
Current	I	$Q = I\,t$	ampere [A]	A = C/s
Voltage	V		volt [V]	V = J/C
Resistance	R	Ohm's Law: $V = I\,R$	ohm [Ω]	Ω = V/A
Conductance	G	$G = 1/R$	siemens [S]	S = A/V
Electric power	P	$P = V\,I = I^2\,R = V^2/R$	watt [W]	W = V A
Capacitance	C	$Q = C\,V \qquad E_C = \frac{1}{2}C\,V^2$	farad [F]	F = C/V
Inductance	L	$E_L = \frac{1}{2}L\,I^2$	henry [H]	H = V s/A

Table 8-15 Schematic symbols

Device	Schematic Symbol	Notes
Battery (Single Cell & Multi-cell)		The battery terminal with the longer line next to it is always the positive.
Resistor		
Capacitor		For polarized capacitors, the positive terminal is the one with the straight line, not the curved line.
Inductor		The inductor symbol can also be drawn with curved bumps rather than loops.

Electric Charge

Electrons and protons, as well as some subatomic particles, have a property known as **electric charge**. On a small scale, charge can be measured in terms of the elementary charge (e). The magnitude of the elementary charge on either an electron or a proton is 1, and by convention, the charge of a proton is called positive ($e = +1$) and that of an electron negative ($e = -1$).

A force acts on a charged particle when in the vicinity of another charged particle. If the charges are alike, both positive or both negative, the force is a repulsive force, and the charges tend to accelerate away from each other. If the charges are unlike, one positive and one negative, the force is attractive with the particles tending to accelerate toward each other.

The value of the elementary charge (e) is inconveniently small, so **charge** (Q) is generally quantified using the derived unit **coulomb** [C]. A charge of one coulomb represents the total charge on approximately 6.24×10^{18} protons. Another way to say this: the elementary charge of a single electron is $1/6.24 \times 10^{18} = 1.6 \times 10^{-19}$ C.

The actual force exerted on a charged object varies with both the amount of charge on each object (Q_1 and Q_2) and the distance (r) between the charges. This relationship is defined by **Coulomb's law**, named after the French physicist Charles-Augustin de Coulomb (1736 – 1806) who first described and quantified the attractive and repulsive electrostatic force.

$$|F| = k_e \frac{|Q_1 Q_2|}{r^2}$$

In this equation, k_e is Coulomb's constant, and is approximately equal to 9×10^9 N m^2/C^2. This version of Coulomb's law is actually a specific case for two charges only. In general, there are more than two charges in three-dimensional space and they require three-dimensional vector representation. Such mathematics, however, are beyond the scope of this course.

Electric Current

Electric current is superficially analogous to a current of water. Just as a current of water is a movement of water molecules in a pipe or channel, electric current is a movement of electric charge in a wire or other solid material.

Electric current (I) is measured in **amperes** [A], named for Andre-Marie Ampere, (1775 – 1836), the French physicist who is credited with discovering electromagnetism.

The derived unit coulomb is defined in terms of the ampere as one ampere second. This may be easier to understand by rephrasing this as a current of one ampere represents a movement of one coulomb of charge past any given point in the wire every second.

To put the magnitude of the ampere in context: for those who have received an electric shock by sticking your finger in a light socket, for instance, you realize the sensation is rather unpleasant. In general, in domestic appliances, such a shock is typically about 0.005 amperes. This level of current is small enough that although unpleasant, your muscles will still respond to the commands from your brain, and you can release the live wire or pull your hand away. However, in circuits powering large appliances such as stoves or air conditioners the current from a typical shock is roughly twice that value, or 0.01 amperes. This is very close to the current level that will overload your nervous system so that your muscles will no longer obey and you become unable to let go. This is FAR more dangerous.

Voltage

To really understand voltage requires knowledge of the concept of the electric field. This is unfortunately a bit too complicated for the limited time and space we have here, so we will merely attempt to help you develop a feel for how voltage affects other electrical parameters. You will study electric fields in some depth in physics, typically the second physics course, and may learn even more in other courses, particularly if you choose to study electrical or computer engineering.

Voltage (V) is quantified using units of volts [V], and is a measure of how much work is required to move an electric charge in the vicinity of other electric charges. The unit of volt is named for Italian physicist Alessandro Volta, (1745 – 1827), who possibly invented the first chemical battery, called a voltaic pile. A somewhat inaccurate explanation of voltage, although useful in understanding it, is that voltage is what pushes the charges around to create current. Some decades ago, voltage was commonly called electromotive force (EMF), but this has fallen out of favor for a variety of reasons, not least of which is that voltage is not a force, being dimensionally quite different.

One volt is defined as one joule per coulomb. In other words, if one joule of energy is required to move one coulomb of charge from one place to another, the voltage between those two points is one volt.

To move a solid object, work equals force times the distance through which that force moves an object: $W = F\Delta x$. Similarly, to move electric charge, work equals electric charge times the difference in voltage through which that charge moves: $W = Q\Delta V$.

IMPORTANT CONCEPT

SI unit of voltage = volt [V]

$1\ V = 1\ J\ /\ C$

IMPORTANT CONCEPT

$W = Q\ \Delta V$

Electrical Resistance

Resistance is a measure of how difficult it is to move charges through a material. In some substances, such as many metals, electrons can move quite easily. In other materials such as glass or air, considerable force, thus considerable voltage, is required to make electrons move therein.

Resistance (R) is quantified using units of ohms [Ω], where one ohm is defined as one volt per ampere. For example, if a 1-volt battery were connected to a device having a resistance of one ohm, one ampere of current would flow through it. The ohm is named for Georg Simon Ohm, the German physicist who first described the relationship linking voltage, current, and resistance.

Resistance relates the voltage across a device to the current through the device. **Take particular note of the choice of prepositions – across and through**. Understanding this choice will help you understand voltage and current.

IMPORTANT CONCEPT

SI unit of resistance = ohm [Ω]

$1\ \Omega = 1\ V\ /\ A$

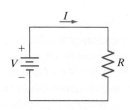

NOTE

When denoting a current (I) on a circuit diagram, an arrow is used to indicate the direction of the current.

IMPORTANT CONCEPT

Ohm's Law
$V = IR$

IMPORTANT CONCEPT

$G = 1/R$
SI unit of conductance = siemens or mho [S or ℧]
$1\,S = 1/\Omega = 1\,A/V$

Electric current is the movement of charge, typically electrons moving **THROUGH** a substance. Voltage is to some extent a measure of the force being exerted on the moving charges by forces at either end of the device. *This is where it can be a little confusing, particularly without using electric fields in the discussion.* However, imagine that on each side of a device is an accumulation of charge, each exerting a force on the electrons inside the device. Each of those forces might be a "push" or a "pull," and the total force on the electrons in the device is the difference in these forces. The difference in the forces from one side of the device to the other is referred to as the voltage **ACROSS** the device.

As an analogy, if you are trying to push a sofa across the room, but someone else is trying to push the sofa in the opposite direction with the same force, the net force is zero and the sofa does not move. If one person pulls and the other pushes, however, the sofa will move quicker than with either person alone.

Resistance is related to current and voltage by **Ohm's law**. Note the following implications of Ohm's law:

- To maintain a specific current through a resistance requires a voltage proportional to the resistance. A larger resistance makes it harder to "push" the electrons through the device, thus a larger voltage is required.
- For a given voltage, current is inversely proportional to resistance. If the resistance increases, the voltage cannot "push" as many electrons through the device per second, so the current must decrease.

In some contexts, it is computationally simpler to use conductance instead of resistance. **Conductance** (G) is measured in siemens [S] and is simply the reciprocal of resistance. An older unit for conductance that you might find, particularly in older references, is the mho (ohm spelled backwards) and is represented by an upside-down omega [℧]. The unit siemens is named for the German inventor Ernst Werner von Siemens who, among other things, built the first electric elevator and founded the company known today as Siemens AG.

EXAMPLE   8-27

The voltage across a resistor is 15 volts [V], and the current through it is 6 milliamps [mA].

What is the value of the resistance in units of ohms? Express the answer using an appropriate SI prefix such that there are only one, two, or three digits shown to the left of the decimal.

Calculate Desired Parameters	Steps	
(1) Determine appropriate equation	$V = IR$ so $R = \dfrac{V}{I}$	
(2) Insert known quantities	$R = \dfrac{15\,V}{6\,mA}\left\|\dfrac{1000\,mA}{1\,A}\right\|\dfrac{1\,\Omega A}{V}$	
(2) Calculate, reasonable	$R = 2500\,\Omega = 2.5\,k\Omega$	

Comprehension Check   8-20

The current through a 12 kilo-ohms [kΩ] resistor is 25 microamps [μA]. What is the voltage [V] across the resistor?

Electric Power

Conceptually, **electric power** is perhaps easiest to understand by examining the formula for gravitational potential energy. A mass has its potential energy increased by expending energy to lift it higher above the surface of the planet, since the mass of the object and the mass of the planet are mutually attracting each other. Similarly, forcing electrons closer to other electrons stores potential energy since they are mutually trying to repel each other. Recall our discussion of batteries. For each electron that is transferred to the negative terminal of the battery by the chemical reaction, a little bit of energy is "stored" in the battery. This is effectively electrical potential energy. The more electrons per second that are jammed together, the more energy per second is stored. Current is measured in charge (electrons) per second, power is proportional to current: $P \propto I$ or $P = XI$ where X is the proportionality constant.

Now think back to voltage. Voltage is a measure of how much energy is used to move a given amount of charge: one volt is one joule per coulomb. Therefore, voltage is the proportionality constant and $P = VI$

EXAMPLE   8-28

A semiconductor diode has 500 millivolts [mV] across it and 700 microamps [μA] of current through it. How much power is the diode absorbing, in units of watts [W]? Express the answer using an appropriate SI prefix such that there are only one, two, or three digits shown to the left of the decimal.

Calculate Desired Parameters	Steps
(1) Determine appropriate equation	$P = VI$
(2) Insert known quantities	$P = \left(\dfrac{500 \text{ mV}}{}\middle\vert \dfrac{1 \text{ V}}{1000 \text{ mV}}\right)\left(\dfrac{700 \text{ μA}}{}\middle\vert \dfrac{1 \text{ A}}{1 \times 10^6 \text{ μA}}\right)\middle\vert \dfrac{1 \text{ W}}{\text{V A}}$
(3) Calculate, reasonable	$P = 3.5 \times 10^{-4} \text{ W} = 350 \text{ μW}$

For resistors, the electrical power absorbed is usually converted to heat, and we can use Ohm's law to replace either the voltage or the current in this power relationship:

$$P = VI = (IR)I = I^2R \quad \text{or} \quad P = VI = V(V/R) = V^2/R$$

Note that these two relationships expressed in terms of resistance are only valid for resistors, not for other electrical components. However, it gives us a means to quickly calculate the power absorbed by a resistor when we know only the voltage or current, but not both.

Comprehension Check   8-21

A device contains a 1000-ohm [Ω] resistor with 120 volts [V] across it. What is the minimum power rating, in units of watts [W], of the device?

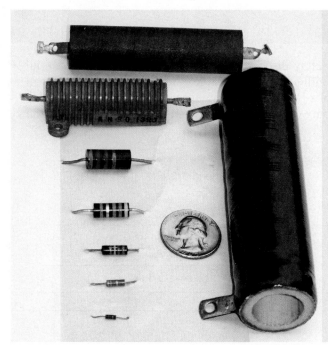

When specifying resistors, both the resistance and the wattage must be given. If you connect a resistor rated at 1 W in a circuit in which it will have to dissipate 100 W, it will literally burst into flames, or at least perform an imitation of popcorn by exploding!

From bottom to top, the power rating of these resistors is ⅛ W, ¼ W, ½ W, 1 W, 2 W, 15 W, 25 W, and on the right, 50 W.

Photo courtesy of W. Park

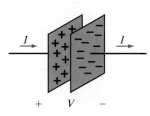

IMPORTANT CONCEPT

$Q = C V$

$E = 1/2\ C V^2$

SI units of capacitance = farad [F]

$1\ F = 1\ C / V$

NOTE

One farad is a LARGE amount of capacitance. Most capacitors are much smaller than this, and are usually measured in microfarads [µF], nanofarads [nF], or even picofarads [pF].

Capacitance

Another simple electrical device is the capacitor. The capacitor is formed by arranging two conducting, low resistance plates very close together, but separated by an insulator with extremely high resistance. Each plate has a wire connected to it. If a current is run into one of the plates of the capacitor, the charges accumulate on that plate since they cannot cross the insulating barrier to the other plate.

While accumulating negative charge on one plate, the negative electrons in the other plate are repelled, leaving behind an overall positive charge. This superficially gives the appearance that the current is going through the capacitor. However, the electrons entering one plate and leaving the other are different electrons, and a charge separation is accumulated on the plates of the capacitor.

The voltage across the capacitor depends not only on the total charge stored, but also on the physical construction of the device, particularly the surface area of the plates. The charge (Q) stored in a capacitor is proportional to the voltage (V) across it. C is the proportionality constant relating charge and voltage, called capacitance.

Capacitance is measured in units of **farads** [F], where one farad equals one coulomb per volt. In other words, if a capacitor was storing a charge of one coulomb, and the resulting voltage across its plates was one volt, the capacitance of the device would be one farad.

Since capacitors can contain a separation of charge, they can store energy. The energy stored in a capacitor can be calculated by $E_C = \frac{1}{2}CV^2$. Note that the energy is proportional to the square of the voltage.

EXAMPLE   8-29

A 0.01-microfarad [μF] capacitor is initially completely discharged (no stored charge, thus an initial voltage across it of zero). If a constant current of 4 milliamps [mA] charges the capacitor, how long will be required to change the voltage across it to 5 volts [V]? Express the answer in units of seconds [s], using an appropriate SI prefix such that there are only one, two, or three digits shown to the left of the decimal.

Calculate Desired Parameters	Steps
	$Q = CV$ and $Q = It$
(1) Determine appropriate equation	$t = \dfrac{CV}{I}$
(2) Insert known quantities	$t = \left(\dfrac{0.01\,\mu F}{1 \times 10^6\,\mu F}\Bigg\vert \dfrac{1\,F}{1 \times 10^6\,\mu F}\right)\left(\dfrac{5\,V}{}\right)\left(\dfrac{}{4\,mA}\Bigg\vert\dfrac{1000\,mA}{1\,A}\right)\Bigg\vert\dfrac{1\,C}{F\,V}\Bigg\vert\dfrac{1\,As}{1\,C}$
(3) Calculate, reasonable	$t = 1.25 \times 10^{-5}\,s = 12.5\,\mu s$

Comprehension Check   8-22

A constant current transfers 12×10^{14} electrons onto one plate of a capacitor in 3 minutes [min].

(a) What is the current, in units of amperes [A]? Express the answer using an appropriate SI prefix such that there are only 1, 2, or 3 digits shown to the left of the decimal.

(b) If the capacitor has a capacitance of 6.8 microfarads [μF], what is the voltage in units of volts [V] across the capacitor?

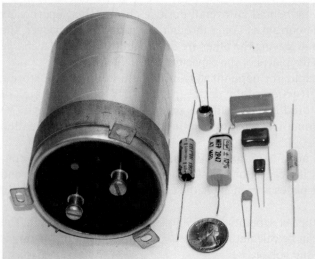

When specifying capacitors, both the capacitance and the maximum voltage must be given. If you connect a capacitor rated at 25 V in a circuit in which it will have 200 V across it, the insulator between the plates will probably fail and the capacitor will be destroyed; it might even explode! The voltage ratings of the capacitors shown range from 25 V to 500 V. The big 10,000 μF on the left is rated at 50 V.

Photo courtesy of W. Park

IMPORTANT CONCEPT

$V = L \, dI/dt$

$E = 1/2 \, L \, V^2$

SI units of inductance =
henry [H]

$1 \, H = 1 \, (V \, s) / A$

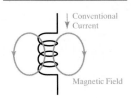

Conventional
Current

Magnetic Field

NOTE

Just as a pendulum trans-
fers energy back and
forth between kinetic and
potential forms thus cre-
ating a physical oscilla-
tion, an inductor and a
capacitor connected
together can swap energy
back and forth from a
magnetic field to an electric field forming an electrical oscillation. For
example, the Theremin,
an electronic musical
instrument that is played
without touching it, relies
on such oscillations.

Inductance

If a current is moving through a wire, a magnetic field is generated surrounding that wire. If a wire is placed in a *changing* magnetic field, a current is induced in the wire. This is the physical basis for the final electrical device we will discuss, the inductor.

In its simplest form, an inductor is just a coil of wire. If a current flows through the coil, the magnetic field generated in each loop adds to the magnetic field generated in every other loop, generating a stronger magnetic field. This is the basis of the electromagnet. Industrial electromagnets can concentrate the magnetic field to a large enough value to pick up huge objects, like cars and trucks. However, in electrical engineering, inductors are more often used for their ability to store energy in the form of a magnetic field.

Note that once a magnetic field has been created in an inductor, if the source of the current that created the magnetic field is removed, the magnetic field will begin to collapse. However, a wire in a changing magnetic field will have a current induced in it, thus allowing the stored energy to be transferred to another place and form.

Inductance (L) is measured in units of **henrys** [H]. The voltage across an inductor is equal to the inductance of the device times the rate of change of current through the inductor. If the current in an inductor changes by one ampere per second, and a voltage is developed across the inductor of one volt, then the inductance is one henry.

The total energy stored in an inductor can be calculated by $E_L = \frac{1}{2} L I^2$. Note that the energy is proportional to the square of the current. You might note the similarity of this to both the formulae for energy stored in a capacitor $E_C = \frac{1}{2} C V^2$ and kinetic energy $E = \frac{1}{2} m v^2$.

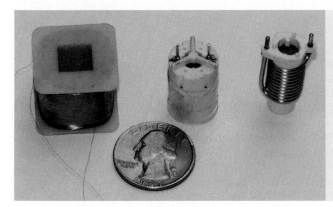

When specifying inductors, both the inductance and the maximum current must be given. If you connect an inductor rated at 5 mA in a circuit which will have 2 A through it, the wire will probably melt – a BAD idea!

The inductor shown on the left is wound with many thousands of turns of extremely small wire and its current capacity is rated in microamperes. The one on the right has only ten turns of rather thick wire and can handle about an ampere.

Photo courtesy of W. Park

EXAMPLE   8-30

The energy stored in a 50-millihenry [mH] inductor is 75 millijoules [mJ]. If all of this energy is transferred to a 15-microfarad [μF] capacitor, what is the voltage [V] across the capacitor?

Calculate Desired Parameters	Steps
(1) Determine appropriate equation	$E_L = E_C$ $E_L = \frac{1}{2} C V^2$ $V = \sqrt{\dfrac{2\,E_L}{C}}$
(2) Insert known quantities	$V = \sqrt{2\dfrac{75\,\text{mJ}}{15\,\mu\text{F}}\left\|\dfrac{1 \times 10^6\,\mu\text{F}}{1\,\text{F}}\right\|\dfrac{1\,\text{J}}{1000\,\text{mJ}}\left\|\dfrac{1\,\text{VC}}{\text{J}}\right\|\dfrac{1\,\text{FV}}{\text{C}}}$
(3) Calculate, reasonable	$V = 100\,\text{V}$

Comprehension Check   8-23

A 100-millihenry [mH] inductor has a current through it of 750 milliamperes [mA]. If the inductor discharges all of its energy into an initially discharged 5-microfarad [μF] capacitor, what is the voltage in units of volts [V] across the capacitor?

Comprehension Check   8-24

Express $\dfrac{1}{\sqrt{L\,C}}$ in terms of base SI units, where L is in henrys [H], and C is in farads [F].

SO WHAT WAS THE ANSWER TO THE BOAT AND THE STONE QUESTION FROM THE INTRODUCTION?

This is a problem that can be analyzed without the need for any equations or formal mathematics. We will start with Archimedes' principle that states simply: "When a body is submerged in a fluid, the buoyancy force (which pushes the body upward) is equal to the weight of the fluid which the body displaces."

If an object is floating at equilibrium in a fluid, the buoyancy force is equal to the weight of the object. The upward buoyancy force and downward weight of the object are the same, so the net force on the object is zero. For example, when a 500 pound-mass boat is placed in a pool, its weight acts downward (with a value of 500 pound-force if at sea level on the Earth), and it sinks deeper and deeper, displacing more and more fluid, until it has displaced 500 pound-mass of fluid. At that point, the buoyancy force is 500 pound-force, and this opposes the weight of the boat so that it rests at equilibrium.

If the body we place in a fluid weighs more than the fluid it displaces when completely submerged, then the body sinks, but it still experiences a buoyancy force equal to the weight of the fluid displaced. This is one reason astronauts train under water—the net downward force of themselves and of objects they are manipulating is greatly reduced although their mass is the same.

If we place the same 500 pound-mass boat in a pool, and put a 200 pound-mass person in it, the boat will sink in the fluid more deeply than before, until it displaces 700 pound-mass of fluid. At this new level of submergence, the upward and downward forces are equal and the new "system" of boat and person will rest at equilibrium. If we add a stone in the boat, the new system of the boat, the person, and the stone will sink even deeper until it displaces a mass of fluid equal to the combined masses of the boat, the person, and the stone. In each of these situations the displaced fluid has to go somewhere, so the level of the liquid in the pool rises.

If we assume that the stone will sink when tossed overboard, it will displace a volume of liquid equal to the volume of the stone. Since the stone is denser than the liquid (has more mass per volume), it now displaces less liquid than it did when it was in the boat. Thus, when the stone is placed in the pool, the water level will drop.

IN-CLASS ACTIVITIES

ICA 8-1

The space shuttle fleet was designed with two booster stages.

(a) If the first stage provides a thrust of 5.25 mega-newtons [MN] and the space shuttle has a mass of 4,470,000 pound-mass [lb_m], what is the acceleration of the spacecraft in miles per hour squared [mi/h^2]?

(b) If the second stage provides a thrust of 75 kilo-newtons [kN] and the space shuttle has an acceleration of 15,000 miles per hour squared [mi/h^2], what is the mass of the spacecraft in units of pounds-mass [lb_m]?

ICA 8-2

Complete the following table assuming the object is on Earth.

	pound-mass	kilogram	newton	pound-force
Abbreviation		[kg]		
(a)	10			
(b)			700	

ICA 8-3

Rank the following from largest mass to smallest mass.
Gravity on the moon = 1.6 meters per second squared [m/s^2]
Gravity on Mars = 3.7 meters per second squared [m/s^2]

(A) 1-kilogram [kg] object on Earth
(B) 2.75-pound-mass [lb_m] object on the moon
(C) 1.25-pounds-force [lb_f] object on the moon
(D) 15-newton [N] object on Earth
(E) 15-newton [N] object on Mars

ICA 8-4

A typical American football weighs 14.5 ounces [oz]. If the ball is taken to the moon, what is the mass in units of grams [g] of the ball on the moon?
Gravity on the moon = 1.6 meters per second squared [m/s^2].
An additional definition you may need: *16 ounces [oz] = 1 pound-mass [lb_m]*

ICA 8-5

The weight of the Hubble Space Telescope (HST) on Earth was 24,500 pounds-force [lb_f]. What is the mass of the telescope in grams [g], using an appropriate SI prefix such that there are only 1, 2, or 3 digits shown to the left of the decimal? At the HST current orbital location of 347 miles [mi] above the Earth, it experiences an acceleration due to gravity of 8.169 meters per second squared [m/s^2].

ICA 8-6

The value of gravitational acceleration on Earth, 9.8 meters per second squared [m/s²] is actually an average value since it varies from place to place on the planet. In Mexico City, gravity is 9.766 meters per second squared [m/s²], whereas in Helsinki gravity is 9.825 meters per second squared [m/s²].

(a) How many pounds does a 75-kilogram [kg] person "gain" or "lose" if they travel from Mexico City to Helsinki? Correctly identify if the change is a gain (positive), loss (negative), or no change is observed for both the units of pounds referring to pound-force [lb$_f$] or pound-mass [lb$_m$].

(b) If this person was transported to Titan, how much weight does a 75-kilogram [kg] person "gain" or "lose" if they travel from Helsinki to Titan? The gravity on Titan is 1.352 meters per second squared [m/s²]. Choose the statement below that describes the effect on the weight of the person.

   (A) The weight will increase.
   (B) The weight will decrease.
   (C) The weight will remain the same.
   (D) There is not enough information provided.

ICA 8-7

Complete the following table:

	Compound	SG	Density [lb$_m$/ft³]	Density [g/cm³]	Density [kg/m³]
(a)	Tetrachloroethane		100		
(b)	Chloroform	1.489			

ICA 8-8

If the density of sodium is 98 kilograms per cubic meter [kg/m³], what is this in slugs per gallon [slug/gal]?

ICA 8-9

The specific gravity of iodine is 4.927. State the density in units of slugs per liter [slug/L], assuming the object is on the moon.

   Gravity on the moon = 1.6 meters per second squared [m/s²].

ICA 8-10

Complete the following table. Assume you have a cube composed of each material, with "length" indicating the length of one side of the cube.

	Material	Mass [g]	Length [in]	Volume [cm³]	Specific Gravity
(a)	Zinc			25	7.14
(b)	Copper	107			8.92

ICA 8-11

Complete the following table. Assume you have a cylinder, composed of each material, with "radius" and "height" indicating the dimensions of the cylinder.

	Material	Mass [kg]	Radius [cm]	Height [mm]	Specific Gravity
(a)	Aluminum		125	750	2.7
(b)	Titanium	8,000		1,000	4.54

ICA 8-12

Fluid A fills a one-gallon [gal] container and weighs 50 newtons [N]. The density of Fluid B is 1125 kilograms per cubic meter [kg/m³] and fills a one-gallon [gal] container. Fluid C fills a two-liter [L] container and has a specific gravity of 2. Which container has a higher mass? Assume the weight of the container is negligible.

(A) The container holding Fluid A
(B) The container holding Fluid B
(C) The container holding Fluid C
(D) The containers are of equal mass
(E) The answer cannot be determined from the information given.

ICA 8-13

A golden bar of metal (5 centimeters [cm] by 18 centimeters [cm] by 4 centimeters [cm]) being transported by armored car is suspected of being fake, made from a less valuable metal with a thin coating of pure gold. The bar is found to have a mass of 2.7 kilograms [kg]. If the specific gravity of gold is 19.3, is the bar fake? Justify your answer.

ICA 8-14

A lab reports the density of a new element is X kilograms per cubic foot [kg/ft³] and Y grams per cubic meter [g/m³]. Which of the following statements is true?

(A) $X > Y$
(B) $X < Y$
(C) $X = Y$
(D) The answer cannot be determined

ICA 8-15

You have been hired as an intern for a plastics manufacturing company. When researching several types of plastics (Acrylic, HDPE, PET) to use for a new bottle design, the following facts were determined by a previous intern. Your boss has asked you to check the work of the previous intern.

The previous intern listed several values of the specific gravity of a plastic material. Which of the following is not a physically possible specific gravity of the plastic? Explain your selection.

(A) 0.00085
(B) 0.00125
(C) 0.85
(D) 1.25
(E) 85
(F) 125

ICA 8-16

A submersible robot is exploring one of the methane seas on Titan, Saturn's largest moon. It discovers a number of small spherical structures on the bottom of the sea at depth of 7 meters [m], and selects one for analysis. The sphere selected has a volume of 1.4 cubic centimeters [cm³] and a density of 2.05 grams per cubic centimeter [g/cm³]. When the rock is returned to Earth for analysis, what is the weight of one sphere in newtons [N]? Gravity on Titan is 1.467 meters per second squared [m/s²]. The density of methane is 0.712 grams per liter [g/L].

ICA 8-17

Several students calculate the density of a gas, obtaining the following answers. Some of the answers may be incorrect, indicating the students made an error. Determine whether the density for each answer is too low, reasonable, or too high for a gas.

(a) 0.4 micrograms per cubic centimeter [μg/cm³]
(b) 0.335 pound-mass per cubic foot [lb$_m$ / ft³]
(c) 700 kilogram per liter [kg / L]
(d) 3.05 kilogram per cubic meter [kg / m³]

ICA 8-18

You have been working to develop a new fictitious compound in the lab. Determine the amount in units of moles [mol] of 20 grams [g] of this compound. The compound has the formula: $X_2Y_2Z_7$, where the components are X, amu = 47; Y, amu = 42; Z, amu = 16.

ICA 8-19

Determine the mass in units of grams [g] of 0.35 moles [mol] of a new fictitious compound you have developed in the lab. The formula is $A_5B_8C_2D_3$. The components are A, amu = 3; B, amu = 22; C, amu = 36; and D, amu = 54.

ICA 8-20

A **eutectic alloy** of two metals contains the specific percentage of each metal that gives the lowest possible melting temperature for any combination of those two metals. Eutectic alloys are often used for soldering electronic components to minimize the possibility of thermal damage. In the past, the most common eutectic alloy used in this application has been 63% Sn, 37% Pb, with a melting temperature of about 361 degrees Fahrenheit [°F]. To reduce lead pollution in the environment, many other alloys have been tried, including those in the table below. Complete the following table:

	Compound	Eutectic Temperature			
		[°F]	[°C]	[K]	[°R]
(a)	91% Sn, 9% Zn			472	
(b)	96.5% Sn, 3.5% Ag				890

ICA 8-21

Which of the following plastic has the highest melting temperature? You must prove your answer for credit!

(A) Plastic #1 at 150 degrees Fahrenheit [°F].
(B) Plastic #2 at 423 kelvins [K].
(C) Plastic #3 at 710 degrees Rankine [°R].

ICA 8-22

The outside air temperature of an airplane typically reaches a low of −60 degrees Celsius [°C] during the cruising phase of a flight. What is the outside air temperature in units of degrees Rankine [°R]?

ICA 8-23

The inside of an airplane is kept at 75 degrees Fahrenheit [°F], or what most people would consider room temperature. What is the difference in temperature between the outside of the plane at cruising altitude (−60 degrees Celsius [°C]) and the inside of the plane at room temperature, in units of kelvins [K]?

ICA 8-24

The Newton [°N] temperature scale was devised by Isaac Newton, using the principles of thermodynamics. He defined zero degrees as melting snow and boiling water as 33 degrees. By comparison to the Fahrenheit [°F] scale:

$$T[°N] = (T[°F] - 32)*11/60$$

Circle all the physically possible temperature readings from the choices listed.

(A) −150 °N
(B) −75 °N
(C) 30 °N
(D) 800 °N
(E) 1500 °N

ICA 8-25

In the Spring of 2004, NASA discovered a new planet beyond Pluto named Sedna. In the news report about the discovery, the temperature of Sedna was reported as "never rising above −400 degrees." What are the units of the reported temperature? Justify your choice.

ICA 8-26

Is there a physical condition at which a Fahrenheit thermometer and Celsius thermometer will read the same numerical value? If so, what is this value?

ICA 8-27

Complete the table:

	atmosphere	pascal	inches of mercury	pound-force per square inch
Abbreviation	[atm]			
(a)			30	
(b)				50

ICA 8-28

If the pressure is 250 feet of water [ft H_2O], what is the pressure in inches of mercury [in Hg]?

ICA 8-29

A "normal" blood pressure has a gauge pressure of 120 millimeters of mercury [mm Hg] (systolic reading) over 80 millimeters of mercury [mm Hg] (diastolic reading). Determine the differential pressure, or the difference between the systolic and diastolic pressures, in units of atmospheres [atm].

ICA 8-30

In all football sports, the ball is inflated with air to a specified pressure. Rank the following pressures in order from lowest to highest.

(A) American football   13 pound-force per square inch [psi]

(B) Association football (soccer)   0.0855 megapascal [MPa]

(C) Gaelic football   0.825 kilogram-force per square centimeter [kg_f/cm^2]

> *An additional definition you may need:*
> *1 technical atmosphere [at] = 1 kilogram-force per square centimeter [kg_f/cm^2] = 98,066.5 pascals [Pa]*

ICA 8-31

If a force of 15 newtons [N] is applied to a surface and the pressure is measured as 4,000 pascals [Pa], what is the area of the surface in units of square meters [m^2]?

ICA 8-32

You wish to submerge an object in a liquid at sea level on Earth at a depth of 75 feet [ft] without exceeding a total pressure of 3.1 atmospheres [atm]. You are able to choose from eight liquids, listed below with their specific gravities. Circle the letter for all liquids that could be used to meet these requirements.

(A) Nitric Acid	1.560		**(E)** Turpentine	0.868
(B) Trichlor ethylene	1.470		**(F)** Propanol	0.804
(C) Glycerol	1.126		**(G)** Methanol	0.791
(D) Styrene	0.903		**(H)** Petroleum Ether	0.640

ICA 8-33

(a) Determine the maximum volume in gallons [gal] of olive oil that can be stored in a closed cylindrical silo with a diameter of 3 feet [ft] so the total pressure at the bottom of the container will not exceed 20 pound-force per square inch [psi]. Assume the height of the tank is sufficient to store the amount of olive oil required, and the surface pressure is 1 atmosphere [atm]. The specific gravity of olive oil is 0.86.

(b) Assume the tank described above is now pressurized using 1.3 atmospheres [atm] of surface pressure to prevent evaporation. Choose the statement below that described the effect to the maximum volume of olive oil able to be stored in the tank so the total pressure at the bottom of the container will not exceed 20 pound-force per square inch [psi]. Assume the height of the tank is sufficient to store the amount of olive oil required.

(A) The volume held in the tank will increase.
(B) The volume held in the tank will decrease.
(C) The volume held in the tank will remain the same.
(D) There is not enough information provided.

ICA 8-34

A sensor is submerged in a silo to detect any bacterial growth in the stored fluid. The stored fluid has a density of 2.2 grams per cubic centimeters [g/cm³]. What is the hydrostatic pressure felt by the sensor at a depth of 30 meters [m] in units of atmospheres [atm]?

ICA 8-35

Complete the table, using the ideal gas law:

	Compound	Mass [lb$_m$]	MW [g/mol]	Amount [mol]	Pressure [Pa]	Volume [gal]	Temperature [°C]
(a)	Acetylene	0.1	26		303,975		−23
(b)	Naphthalene	0.07		0.25	131,723	1.32	

ICA 8-36

A 10-liter [L] flask contains 1.3 moles [mol] of an ideal gas at a temperature of 20 degrees Celsius [°C]. What is the pressure in the flask in units of atmospheres [atm]?

ICA 8-37

An ideal gas in a 1.25-gallon [gal] container is at a temperature of 125 degrees Celsius [°C] and pressure of 2.5 atmospheres [atm]. If the gas is oxygen (formula: O_2, molecular weight = 32 grams per mole [g/mol]), what is the mass of gas in the container in units of grams [g]?

ICA 8-38

An ideal gas is kept in a 10-liter [L] container at a pressure of 1.5 atmospheres [atm] and a temperature of 310 kelvins [K]. If the gas is compressed until its pressure is raised to 3 atmospheres [atm] while holding the temperature constant, what is the new volume in units of liters [L]?

ICA 8-39

Football A is filled with 50 grams [g] of Gas X, resulting in a pressure of 48,263 pascals [Pa]. The molecular weight of Gas X is 25 grams per mole [g/mol]. Football B is filled with 32 grams [g] of Gas Y, resulting in a pressure of 55,142 pascals [Pa]. Assuming that footballs A and B are identical, with the same volume and at the same temperature, what is the molecular weight of Gas Y in units of grams per mole [g/mol]?

ICA 8-40

Determine the specific gravity of chlorine (Cl_2) gas at 25 degrees Celsius [°C] and a pressure of 0.6 atmospheres [atm]. Assume chlorine is an ideal gas and obeys the ideal gas law. The molecular weight of chlorine is 70 grams per mole [g/mol].

ICA 8-41

Complete the table for specific heat conversions:

	Compound	[cal/(g °C)]	[BTU/(lb$_m$ °F)]	[J/(kg K)]
(a)	Benzene		0.0406	
(b)	Mercury	0.03325		

ICA 8-42

Which of the following requires the expenditure of more work? You must show your work to receive credit.

(A) Lifting a 100-newton [N] weight a height of 4 meters [m].
(B) Exerting a force of 50 pounds-force [lb$_f$] on a sofa to slide it 30 feet [ft] across a room.

ICA 8-43

Which object—A, B, or C—has the most potential energy when held a distance H above the surface of the ground? You must show your work to receive credit.

(A) Object A: mass = 1 kilogram [kg]   height = 3 meters [m]
(B) Object B: mass = 1 slug   height = 3 feet [ft]
(C) Object C: mass = 1 gram [g]   height = 1 centimeter [cm]

ICA 8-44

Measurements indicate that boat A has twice the kinetic energy of boat B of the same mass. How fast is boat A traveling if boat B is moving at 30 knots? 1 knot = 1 nautical mile per hour [nmi/h]; 1 nautical mile [nmi] = 6,076 feet [ft].

ICA 8-45

A 10-gram [g] rubber ball is released from a height of 6 meters [m] above a flat surface on the moon. Gravitational acceleration on the moon is 1.62 meters per second squared [m/s^2]. Assume that no energy is lost from frictional drag. What is the velocity, in units of meters per second [m/s], of the rubber ball the instant before it strikes the flat surface?

ICA 8-46

If a ball is dropped from a height (H) its velocity will increase until it hits the ground (assuming that aerodynamic drag due to the air is negligible). During its fall, its initial potential energy is converted into kinetic energy. If the mass of the ball is doubled, how will the impact velocity change?

ICA 8-47

Eight hundred kilograms [kg] of paint solvent will be stored to remove a paint display on a field. If the specific heat capacity of the solvent is 2200 joules per kilogram kelvin [J/(kg K)], how much energy in megajoules [MJ] is needed to increase the temperature of the solvent from −15 degrees Celsius [°C] to 10 degrees Celsius [°C]?

ICA 8-48

We go out to sunbathe on a warm summer day. If we soak up 100 British thermal units per hour [BTU/h] of energy, how much will the temperature of 132 pound-mass [lb_m] person increase in 2 hours [h] in units of degrees Celsius [°C]? We assume that since our bodies are mostly water they have the same specific heat as water, 4.18 joules per gram degree Celsius [J/(g °C)].

ICA 8-49

The specific heats of aluminum and iron are 0.214 and 0.107 calories per gram degree Celsius [cal/(g °C)], respectively. If we add the same amount of energy to a cube of each material of the same mass and find that the temperature of the aluminum increases by 30 degrees Fahrenheit [°F], how much will the iron temperature increase in degrees Fahrenheit [°F]?

ICA 8-50

Complete the table for thermal conductivity conversions:

	Compound	[W/(m °C)]	[BTU/(ft h °F)]	[cal/(cm min K)]
(a)	Zinc		122	
(b)	Silver	420		

ICA 8-51

The thermal conductivity of a plastic is 0.325 British thermal units per foot hour degree Fahrenheit [BTU/(ft h °F)]. Convert this value in units of watts per meter kelvin [W/(m K)].

ICA 8-52

The heat transfer coefficient of steel is 25 watts per square meter degree Celsius [W/(m² °C)]. Convert this value into units of calories per square centimeter second kelvin [cal/(cm² s K)].

ICA 8-53

Complete the table. This problem involves the power required to raise a mass a given distance in a given amount of time assuming 100% efficiency.

	Mass [lb_m]	Distance [ft]	Energy [J]	Time [min]	Power [hp]
(a)	220	15		0.5	
(b)		35		2	0.134

ICA 8-54

When we drive our car at 100 feet per second [ft/s], we measure an aerodynamic force (called drag) of 66 pound-force [lb_f] that opposes the motion of the car. How much horsepower [hp] is required to overcome this drag?

ICA 8-55

The power required by an airplane is given by $P = Fv$, where P is the engine power, F is the thrust, and v is the plane speed. Which of the following planes has the most power? You must show your work to receive credit.

(A) Plane A: Thrust = 2,000 pound-force [lb_f]   Speed = 200 meters per second [m/s]

(B) Plane B: Thrust = 13,000 newtons [N]   Speed = 500 feet per second [ft/s]

ICA 8-56

Complete the table. This problem involves the power required to raise a mass a given distance in a given amount of time.

	Mass [lb_m]	Distance [ft]	Energy [J]	Time [min]	Power [hp]	Efficiency [%]
(a)	1,875	145			0.268	85
(b)	200		4,000	1		62

ICA 8-57

A 100-watt [W] motor (60% efficient) is available to raise a load 5 meters [m] into the air. If the task takes 65 seconds [s] to complete, how heavy was the load in units of kilograms [kg]?

ICA 8-58

When boiling water, a hot plate takes an average of 8 minutes [min] and 55 seconds [s] to boil 100 milliliters [mL] of water. Assume the temperature in the lab is 75 degrees Fahrenheit [°F]. The hot plate is rated to provide 283 watts [W]. The specific heat capacity of water is 4.18 joules per gram degree Celsius [J/(g °C)]. How efficient is the hot plate?

ICA 8-59

One problem with solar energy is that any given point on the planet is illuminated by the sun for only half of the time at best. It would be helpful, if there were a simple, affordable, and efficient means for storing any excess energy generated on sunny days for use during the night or on cloudy days.

   You are investigating the electrodes used in electrolysis cells as part of a three-stage process for solar energy collection and storage.

1. Convert sunlight to electricity with photovoltaic cells.
2. Use the electricity generated in an electrolysis cell to split water into its component elements, hydrogen and oxygen. The hydrogen can be stored indefinitely. The oxygen can simply be released into the atmosphere.
3. Use a fuel cell to recombine the stored hydrogen with oxygen from the atmosphere to generate electricity.

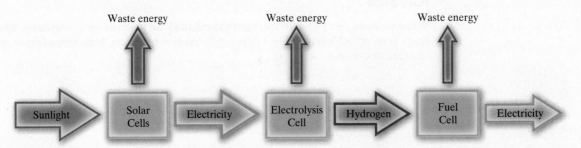

You have obtained an array of new high-efficiency, thin-film photovoltaic cells with an efficiency of 41%. The efficiency of fuel cells varies with the current demands placed on them, but the cells you have obtained yield an overall efficiency of 37% at the anticipated load.

Assume the total solar power on the solar cells is 2000 watts [W]. You conduct four experiments, each with a different alloy of palladium, platinum, gold, copper, and/or silver for the electrodes in the electrolysis cell. The final output power from the fuel cell is measured for each case, and the results are tabulated below. Determine the efficiency of each electrolysis cell and complete the table.

Alloy	Output Power (P_o) [W]	Electrolysis Cell Efficiency (η)
(a) Alloy A	137	
(b) Alloy B	201	
(c) Alloy C	67	
(d) Alloy D	177	

ICA 8-60

For each row below, the resistor absorbs one joule [J] of energy over some amount of time. Fill in all of the blanks.

	Voltage	Current	Resistance	Power	Time to absorb 1 J
(a)		20 μA		600 μW	
(b)			12 kΩ	3 mW	
(c)	5 V				10,000 s

ICA 8-61

A resistor is dissipating 125 milliwatts [mW] of power. The voltage across the resistor is 12 volts [V].

(a) What is the value of the resistance? Express the answer in units of ohms [Ω] using an appropriate SI prefix such that there are only one, two, or three digits shown to the left of the decimal.

(b) What is the current through the resistor? Express the answer in units of amperes [A] using an appropriate SI prefix such that there are only one, two, or three digits shown to the left of the decimal.

ICA 8-62

An average cloud-to-ground lightning strike delivers about 0.5 (half) a gigajoule [GJ] of energy to the Earth. Assuming that 25 coulombs [C] of charge was delivered with a current of 40 kilo-amperes [kA], answer the following questions. Express the answers using an appropriate SI prefix such that there are only 1, 2, or 3 digits shown to the left of the decimal.

(a) How long did the lightning strike last, in units of seconds [s]?
(b) What was the voltage between the cloud and the ground in units of volts [V]?
(c) If an experimental lightning capturing system was able to store this energy with an efficiency of 15%, how many such strikes would be necessary to power an average American home for one year, assuming the average yearly energy usage is 600 gigajoules [GJ]?

ICA 8-63

Most electric stoves in the United States operate on 240 volts [V]. If such a stove took 275 seconds [s] to heat one quart [qt] of water from an initial temperature of 70 degrees Fahrenheit [°F] to boiling, what was the current in units of amperes [A] if the efficiency was 37%?

ICA 8-64

A 3-volt [V] battery delivers a constant current of 100 milliamperes [mA] to the bulb in a flashlight for 20 minutes [min].

(a) What is the total charge in units of coulombs [C] that passes through the flashlight bulb?
(b) What is the total energy delivered to the flashlight bulb in units of joules [J]?

ICA 8-65

For each row below, a constant current of 20 microamperes [μA] is used to charge the capacitor from an initially discharged state in some length of time. Complete the table.

	Capacitance	Voltage	Energy	Charge	Time to charge
(a)		7.5 V		375 nC	
(b)	200,000 μF				1.75 ms
(c)			6.97 pJ	215 pC	

ICA 8-66

A constant current charges a 20-nanofarad [nF] capacitor to 5 volts [V] in 10 microseconds [μs].

(a) Determine the current. Express the answer in units of amperes [A] using an appropriate SI prefix such that only one, two, or three digits shown to the left of the decimal.
(b) What is the total energy stored in the capacitor? Express units of joules [J] using an appropriate SI prefix such that only one, two, or three digits shown to the left of the decimal.

ICA 8-67

When a constant voltage is used to charge an inductor from an initially discharged state, the current through the inductor at time t can be determined by $I = V\,t/L$. For each of the following inductors, a constant voltage of 10 volts [V] is used to charge the inductor from an initially discharged state in some amount of time. Complete the table.

	Inductance	Current	Energy	Time to charge
(a)		5 A	25 J	
(b)		2 mA		500 ps
(c)	25 μH		5 pJ	

ICA 8-68

(a) What is the average power, in units of watts [W], needed to increase the voltage across a 2-nanofarad [nF] capacitor from 5 volts [V] to 15 volts [V] in 0.5 microseconds [μs]?
(b) What is the average power, in units of watts [W], needed to increase the current through a 25-millihenry [mH] inductor from 30 milliamperes [mA] to 100 milliamperes [mA] in 400 microseconds [μs]?

ICA 8-69

Materials

Bag of cylinders   Scale   Calipers   Ruler

Procedure

For each cylinder, record the mass, length, and diameter in the table provided.

Analysis

- Calculate the volume and density for each cylinder, recording the results in the table provided.
- Rank the rods in order of increasing density, with the least dense rod first on the list.
- Using the density, determine the material of each rod. Your professor will provide you more information on the possible materials in your cylinder bag.

Data Worksheet

	Measured Values			Calculated Values	
Description	Mass	Length	Diameter	Volume	Density
Units					
Rod 1					
Rod 2					
Rod 3					
Rod 4					
Rod 5					

Rank Rods Increasing Density	Density	Material

ICA 8-70

Materials

25 mL graduated cylinder   Scale   Paper towels

Unknown liquids   Basket labeled "Wash"   Water bottle

Wastewater bucket

Procedure

Record the following data in the table provided:

1. Weigh the empty graduated cylinder and record the value.
2. Pour 15 milliliters [mL] of water into the cylinder.
3. Weigh the cylinder with water and record the value.
4. Pour the water into the wastewater bucket.
5. Pour 15 milliliters [mL] of unknown liquid (UL) 1 into a cylinder.
6. Weigh the cylinder with UL 1 and record the value.

7. Pour the UL back into the original container.
8. Place the graduated cylinder in the "Wash" basket if more than one cylinder is available. If only one cylinder is being used, wash the cylinder with water so no trace of the UL is left in the cylinder.
9. Repeat Steps 5–9 with each unknown liquid provided.

Analysis

- Calculate the density and specific gravity for each fluid, recording the results in the table provided.
- Rank the fluids in order of increasing density, with the least dense fluid first on the list.
- Using the specific gravity, determine the type of liquid in each container. Your professor will provide you more information on the possible fluids.

Data Worksheet

| | Measured | | | Calculated | |
Description	Total Mass	Volume	Liquid Mass	Density	Specific Gravity
Units					
Empty cylinder					
Water					
UL 1					
UL 2					
UL 3					
UL 4					
UL 5					

Rank Liquids Increasing Density	Density	Liquid

REVIEW QUESTIONS

1. The National Football League specifies that all footballs must be Wilson brand. Wilson footballs are manufactured in Ada, Ohio, to very strict criteria. The tear strength of the interior bladder, made of urethane, is 192.2 kilonewtons per meter [kN/m]. Convert this tear strength measurement into units of pound-force per inch [lb_f/in].

2. A space probe is built with a mass of 1700 pound-mass [lb_m] before launch on Earth. The probe is powered by four ion thrusters, each capable of generating 225 millinewtons [mN] of thrust. Using Newton's second law, the acceleration (a) of the craft is equal to the force (F) divided by the mass (m):

$$a = \frac{F}{m}$$

   The velocity (v) of an object increases as an object accelerates. If the object starts at rest, the final velocity is given by the acceleration multiplied by the length of time the acceleration is applied to the object (t):

$$v = a\,t = \frac{F\,t}{m}$$

   Using this equation, how many weeks will the thrusters have to operate for the probe, initially at rest, to reach a velocity of 420 miles per minute [mi/min]? You may assume the initial velocity of the probe is zero miles per minute [0 mi/min].

3. A person who has a weight of 165 pound-force [lb_f] on Earth is travelling to Mars in a spacecraft. As long as the engines on the spacecraft are not operating, the person is essentially weightless during the voyage. During a course correction, the spacecraft undergoes an acceleration of 0.72 g. The term "g" is the Earth-normal gravity, so 1 g is 9.8 meters per second squared [m/s^2]. This acceleration makes it feel like there is gravity in the spacecraft, and the person will have a perceived weight during the acceleration period, rather than feeling weightless. What is the person's mass, in units of kilograms [kg], in the spacecraft while travelling to Mars when the engines are not running?

4. A person who has a weight of 165 pound-force [lb_f] on Earth is travelling to Mars in a spacecraft. As long as the engines on the spacecraft are not operating, the person is essentially weightless during the voyage. During a course correction, the spacecraft undergoes an acceleration of 0.72 g. The term "g" is the Earth-normal gravity, so 1 g is 9.8 meters per second squared [m/s^2]. This acceleration makes it feel like there is gravity in the spacecraft, and the person will have a perceived weight during the acceleration period, rather than feeling weightless. What is the person's perceived weight in units of newtons [N] during the course correction while the engines are running at an acceleration of 0.72 g?

5. A person who has a weight of 165 pound-force [lb_f] on Earth is travelling to Mars in a spacecraft. As long as the engines on the spacecraft are not operating, the person is essentially weightless during the voyage. During a course correction, the spacecraft undergoes an acceleration of 0.72 "g". The term "g" is the Earth-normal gravity, so 1 "g" is 9.8 meters per second squared [m/s^2]. This acceleration makes it "feel" like there is gravity in the spacecraft, and the person will have a perceived weight during the acceleration period, rather than feeling weightless.

   Once the person reaches Mars, what is the mass, in units of kilograms [kg], and weight in units of pound-force [lb_f], of the person on Mars? The gravity of Mars is 3.71 meters per second squared [m/s^2].

6. *The European Space Agency launched a probe called* Rosetta *in March 2004. In August 2014,* Rosetta *reached its destination: a comet called 67P/Churyumov-Gerasimenko.* Rosetta *is the first spacecraft to rendezvous with a comet.*

   The *Rosetta* has a launch mass on Earth of 3000 kilograms [kg]. In order to escape Earth's gravity, the spacecraft must accelerate at a rate of 25,000 miles per hour squared [mi/h^2]. How much thrust, in units of newtons [N], must the engines provide to reach this acceleration.

7. *The European Space Agency launched a probe called* Rosetta *in March 2004. In August 2014,* Rosetta *reached its destination: a comet called 67P/Churyumov-Gerasimenko.* Rosetta *is the first spacecraft to rendezvous with a comet. It contained a lander craft, called* Philae.

   The *Rosetta* has a launch mass on Earth of 3000 kilograms [kg] and the *Philae* has a launch mass of 220 pound-mass [lb$_m$]. In order to escape Earth's gravity, the combined space-craft must accelerate at a rate of 25,000 miles per hour squared [mi/h^2]. How much thrust, in units of newtons [N], must the engines provide to reach this acceleration?

8. *The European Space Agency launched a probe called* Rosetta *in March 2004. In August 2014,* Rosetta *reached its destination: a comet called 67P/Churyumov-Gerasimenko.* Rosetta *is the first spacecraft to rendezvous with a comet. It contained a lander craft, called* Philae.

   The *Philae* has a launch mass of 220 pound-mass [lb$_m$]. If the *Philae* was deployed to Mars instead of Comet 67P, what would the weight of the lander be after landing on the Martian surface, in units of pound-force [lb$_f$]? The gravity of Mars is 3.71 meters per second squared [mi/s^2].

9. *The European Space Agency launched a probe called* Rosetta *in March 2004. In August 2014,* Rosetta *reached its destination: a comet called 67P/Churyumov-Gerasimenko.* Rosetta *is the first spacecraft to rendezvous with a comet. It contained a lander craft, called* Philae.

   The Philae *has several scientific instruments onboard, used to study the comet. ALICE is an ultraviolet imaging spectrograph, used to search for noble gases in the comet core. The MIRO measures microwave emissions in an attempt to measure substances like water, ammonia, or carbon dioxide. MIDAS is a dust analysis instrument, which will collect dust particles from the comet.*

   Complete the table by determining the required information about these instruments. Note that you do not need to complete the entire table, just the values indicated. The gravity of Comet 67P is estimated at 1 millimeter per second squared [mm/s^2]. The gravity of Mars is 3.71 meters per second squared [m/s^2].

Instrument	Mass on . . . . [kg]			Weight on . . . . [lb$_f$]		
	Earth	Mars	Comet 67P	Earth	Mars	Comet 67P
ALICE	(a)		(b)	6.8		(c)
MIRO		(d)		(e)	15.4	
MIDAS	8.3	(f)			(g)	(h)

10. A basketball has a mass of approximately 624 grams [g] and a volume of 0.25 cubic feet [ft^3]. Determine the density of the basketball in units of slug per gallon [slug/gal].

11. A website you discover lists the density of a plastic as 7.93 pound-mass per gallon [lb$_m$/gal]. which is a very strange way to list the density of a solid. State the specific gravity of the plastic.

12. Consider the following strange, but true, units:

1 arroba = 11.5 kilograms [kg]   1 peck = 9 liters [L]

A basketball has a mass of approximately 624 grams [g] and a volume of 0.25 cubic feet [ft³]. Determine the density of the basketball in units of arroba per peck.

13. Consider the following strange, but true, units:

1 batman = 3 kilograms [kg]   1 hogshead = 63 gallons [gal]

A basketball has a mass of approximately 624 grams [g] and a volume of 0.25 cubic feet [ft³]. Determine the density of the basketball in units of batman per hogshead.

14. A cube of material X, 1 inch [in] on all sides, has a mass of 0.05 kilograms [kg]. Determine the specific gravity of material X.

15. The density of gasoline is 0.72 grams per cubic centimeter [g/cm³]. What is the mass in units of kilograms [kg] of a 5-gallon [gal] container filled completely with gasoline? Ignore the mass of the container.

16. The Eco-Marathon is an annual competition sponsored by Shell Oil, in which participants build special vehicles to achieve the highest possible fuel efficiency. The Eco-Marathon is held around the world with events in the United Kingdom, Finland, France, Holland, Japan, and the United States.

A world record was set in Eco-Marathon by a French team in 2003 called Microjoule with a performance of 10,705 miles per gallon [mi/gal]. The Microjoule runs on ethanol. If the cars are given 100 grams [g] of ethanol (specific gravity = 0.789) and driven until the fuel runs out, how far did the Microjoule drive in kilometers [km]?

17. A spaceship has a mass of 3,250,000 metric tons. One metric ton is 1,000 kilograms [kg]. If this ship landed on Planet Z with a gravity of 7.4 meters per second squared [m/s²], what is the weight of the ship in units of newtons [N] on Planet Z? Express the answers using an appropriate SI prefix such that there are only one, two, or three digits shown to the left of the decimal.

18. Soccer is also called "association football." A soccer ball is a sphere, with a circumference of 70 centimeters [cm]. In developing a new material to use for a soccer ball when NASA attempts to colonize Mars, you test a material with a specific gravity of 1.21. Assume the ball will be a solid sphere made from this material. What is the weight in units of newtons [N] of the ball on Mars? Recall gravity on Mars is 3.7 meters per second squared [m/s²].
*Recall the circumference of a circle is equal to the diameter times pi: $C = \pi D$

19. A rod on the surface of Jupiter's moon Callisto has a volume of 0.3 cubic meters [m³]. Determine the weight of the rod in units of pound-force [lb$_f$]. The specific gravity of the material is 4.7. Gravitational acceleration on Callisto is 1.25 meters per second squared [m/s²].

20. *The European Space Agency launched a probe called* Rosetta *in March 2004. In August 2014,* Rosetta *reached its destination: a comet called 67P/Churyumov-Gerasimenko.* Rosetta *is the first spacecraft to rendezvous with a comet. It contained a lander craft, called* Philae.

The *Philae* lander has several scientific instruments onboard, used to study the comet. ALICE is an ultraviolet imaging spectrograph, used to search for noble gases in the comet core. The ALICE instrument uses potassium bromide as one of the compounds in the analysis of noble gases. Potassium bromide has a specific gravity of 2.74. Assume the lander has a liquid sample of potassium bromide onboard that fills a 1.25-gallon [gal] container. What is the mass of the potassium bromide sample in the container, in units of kilograms [kg]?

21. One area of concern in athletics is the use of steroids. One anabolic steroid called *tetrahydrogestrinone* or *THG* has been in the news for several years due to the popularity of the drug among athletes. THG has a molecular mass of 312.5 grams per mole [g/mol]. A test is being developed to determine if an athlete has used THG. The detection limit is 2 nanograms per milliliter [ng/mL]. In a sample of 0.05 liters [L], how many moles [mol] of THG must be present to surpass the detection limit? Place your answer in scientific notation, with the form $Z.ZZ \times 10^X$.

22. When an object is made of the plastic polyester (PET), it uses antimony as a catalyst. After production, antimony can leach into food and drink stored in PET containers. Usually, if stored correctly for a short amount of time, the amount of antimony found in liquids is well below the safety limits. However, there have been reports of as much as 44.7 micrograms per liter [μg/L] of antimony found in fruit juice concentrates. How much antimony, in units of moles [mol], could be found in one gallon [gal] of fruit juice concentrate at this level of contamination? The molecular mass of antimony is 121.76 grams per mole [g/mol].

23. The largest temperature decline during a 1 day (24 hour) time period was 56 degrees Celsius [°C] in Browning, Montana. Express this as degrees Fahrenheit per minute [°F/min].

24. If we increase the temperature in a reactor by 90 degrees Fahrenheit [°F], how many degrees Celsius [°C] will the temperature increase?

25. We are making a cup of coffee and want the temperature to be just right, so we measure the temperature with both Fahrenheit and Celsius thermometers. The Fahrenheit meter registers 110 degrees Fahrenheit [°F], but you prefer to it to be slightly hotter at 119 degrees Fahrenheit [°F], so we heat it up a little. How much will the Celsius thermometer increase when we make this change?

26. We want to construct a thermometer using mercury. As the mercury in the bulb is heated, it expands and moves up the thin capillary tube connected to the bulb. The symbol used for the coefficient of volume expansion of a substance due to a temperature increase is β. It is used in the following equation:

$$\Delta V = \beta V (\Delta T)$$

Here, ΔV is the increase in volume, V is the original volume, and ΔT is the temperature increase. The value of β for mercury is 1.8×10^{-4} [1/degree Celsius]. If the bulb contains 0.2 milliliters [mL] and the tube has a diameter of 0.2 millimeters [mm], how much will the mercury rise in the tube in units of centimeters [cm] if we increase the temperature from 30 degrees Fahrenheit [°F] to 70 degrees Fahrenheit [°F]?

27. A "normal" blood pressure has a gauge pressure of 120 millimeters of mercury [mm Hg] (systolic reading) over 80 millimeters of mercury [mm Hg] (diastolic reading). Convert 120 millimeters of mercury into units of pound-force per square inch [lb_f/in^2 or psi].

28. The force on the inside of a cork in a champagne bottle is 10 pound-force [lb_f]. If the cylindrical cork has a diameter of 0.5 inches [in], what is the pressure inside the bottle acting on the bottom surface of the cork in units of feet of water [ft H_2O]?

29. One of the National Academy of Engineering Grand Challenges for Engineering is **Develop Carbon Sequestration Methods**. The NAE defines carbon sequestration as "capturing the carbon dioxide produced by burning fossil fuels and storing it safely away from the atmosphere." The most promising storage location is underground, possibly in sedimentary brine formations. You are assigned to develop instrumentation to measure the properties of a brine formation, located 800 meters [m] deep. Assume the instruments will feel an equivalent amount of pressure to the amount of hydrostatic pressure felt at the bottom of an 800-meter [m] high column of brine, with a specific gravity of 1.35. To what hydrostatic pressure, in units of atmospheres [atm], must the instrumentation be built to withstand?

30. A new gel is being developed to use inside padding and helmets to cushion the body from impacts. The gel is stored in a 4.1 cubic meters [m³] cylindrical tank with a diameter of 2 meters [m]. The tank is pressurized to 1.3 atmosphere [atm] of surface pressure to prevent evaporation. A total pressure probe located at the bottom of the tank reads 60 feet of water [ft H₂O]. What is specific gravity of the gel contained in the tank?

31. A 2-liter [L] soda bottle, made of PET, will fail at approximately 150 pound-force per square inch [psi] of pressure. If you were to dive straight down into the ocean with a 2-liter [L] bottle, at what depth in units of feet [ft] would the bottle fail? Assume the specific gravity of ocean water is 1.025.

32. A storage tank is being designed for iodine. The storage tank is to be spherical and will withstand a maximum hydrostatic pressure of 1375 millimeters of mercury [mm Hg]. What is the diameter of the tank in meters [m]? The density of iodine is 5010 kilograms per cubic meter [kg/m³].

33. The specific gravity of honey is 1.43. A cylindrical container of honey 10 feet [ft] deep is inside a dome pressurized to a surface pressure of 0.06 atmospheres [atm] on the surface of Pluto. The total pressure at the bottom of the container is 67 millimeters of mercury [mm Hg]. What is the gravitational acceleration of Pluto, in units of meters per second squared [m/s²]?

34. NASA is designing a mission to explore Titan, the largest moon of Saturn. Titan has numerous hydrocarbon lakes containing a mix of liquid methane and liquid ethane in unknown proportions. As part of the mission, a small submersible vehicle will explore Kraken Mare, the largest of these lakes. Assuming the maximum depth of Kraken Mare is less than 400 meters [m], how much total pressure in units of atmospheres [atm] must the submersible be designed to withstand? Assume the surface pressure on Titan is 147 kilopascals [kPa], the surface temperature is 94 kelvins [K], and the gravity is 1.35 meters per second squared [m/s²]. The specific gravity of liquid methane is 0.415 and the specific gravity of liquid ethane is 0.546.

35. Airspeed (v) is related to dynamic pressure using the following formula: $P_{dynamic} = \frac{1}{2} \rho v^2$. Determine the dynamic pressure, in units of pascals [Pa], for an aircraft moving at an airspeed of 600 miles per hour [mi/h or mph]. Air has a specific gravity of 0.0012.

36. When a flowing fluid is stopped, its pressure increases. This is called stagnation pressure. The stagnation pressure is determined by: $P_{stagnation} = \frac{1}{2} \rho v^2 + P_{surface}$, where ρ is the fluid density, v the fluid velocity, and $P_{surface}$ the atmospheric pressure. Calculate the stagnation pressure in units of atmospheres [atm] for acetone flowing at a velocity of 15 feet per second [ft/s]. Assume the specific gravity of acetone is 0.785.

37. A 10-liter [L] flask contains 5 moles [mol] of gas at a pressure of 15 atmospheres [atm]. What is the temperature in the flask in units of kelvins [K]?

38. A company that produces footballs uses a proprietary mixture of ideal gases to inflate their footballs. If the temperature of 230 grams [g] of gas mixture in a 15-liter [L] tank is maintained at 465 degrees Rankine [°R] and the tank is pressurized to 135 pound-force per square inch [psi], what is the molecular weight of the gas mixture in units of grams per mole [g/mol]?

39. A 5-liter [L] container holds nitrogen (formula: N_2, molecular weight = 28 grams per mole [g/mol]) at a pressure of 1.1 atmospheres [atm] and a temperature of 400 kelvins [K]. What is the mass of nitrogen in the container, in units of grams [g]?

40. A typical hot air balloon has a volume of 100,000 cubic feet [ft³]. Air has a specific gravity of 0.00129 and a molecular mass of 28.97 grams per mole [g/mol] at one atmosphere [atm] and 533 degrees Rankine [°R]. What is the weight in units of pound-force [lb$_f$] of the air in such a balloon?

41. An ideal gas, kept in a 5-liter [L] container at 300 kelvins [K], exhibits a pressure of 2 atmospheres [atm]. If the volume of the container is decreased to 2.9 liters [L], but the amount and temperature remains the same, what is pressure in the new container in units of atmospheres [atm]?

42. A container holding 1.5 moles [mol] of oxygen (formula: O_2, molecular weight = 32 grams per mole [g/mol]) at a pressure of 1.5 atmospheres [atm] and a temperature of 310 kelvins [K] is heated to 420 kelvins [K], while maintaining constant amount and volume. What is the new pressure inside the container in units of pascals [Pa]?

43. A spacecraft is equipped with nine sensors, designed to gather data to attempt to reconstruct how the Martian planet may have lost its atmosphere. A reaction vessel is used as part of a chemical analysis process. The chemical reaction proceeds most efficiently at a temperature of 27 degrees Celsius [°C]. The 1.2-gallon [gal] reactor initially contains 0.3 moles [mol] of an ideal gas at a pressure of 1.95 atmospheres [atm]. To reach the ideal temperature for the reaction, the pressure in the reactor will be changed from the current conditions, while the volume and amount will remain constant. Determine the change (increase or decrease) in the pressure of the reactor in units of pascals [Pa] necessary to allow the reaction to proceed at the most efficient temperature.

44. A robotic rover on Mars finds a spherical rock with a diameter of 10 centimeters [cm]. The rover picks up the rock and lifts it 20 centimeters [cm] straight up. The resulting potential energy of the rock relative to the surface is 2 joules [J]. Gravitational acceleration on Mars is 3.7 meters per second squared [m/s²]. What is the specific gravity of the rock?

45. On February 15, 2013, Asteroid 2012 DA_{14} passed within 17,200 miles [mi] of the surface of the Earth at a relative speed of 7.8 kilometers per second [km/s]. This is considerably closer than the orbit of geosynchronous satellites (26,200 miles). This is the closest recorded approach of an object this large.

   The asteroid 2012 DA_{14} was estimated to have a diameter of 30 meters [m] and a specific gravity of 3. If 2012 DA_{14} had hit the Earth, what is the total amount of energy that would have been released (i.e., what was the kinetic energy of the asteroid)? Express your answer in megatons [Mton]. One megaton is the energy released by one million metric tons of TNT explosive. A metric ton equals 1000 kilograms [kg], and the explosive energy of TNT is 4184 joules per gram [J/g].

46. If a 10-kilogram [kg] rotating solid cylinder moves at a velocity (v), it has a kinetic energy of 36 joules [J]. Determine the velocity the object is moving in units of meters per second [m/s] if the kinetic energy is given by $KE = \frac{1}{2} mv^2 + \frac{1}{4} mv^2$.

47. If a ball is dropped from a height its velocity will increase until it hits the ground, assuming that aerodynamic drag due to the air is negligible. During its fall, its initial potential energy is converted into kinetic energy. If the ball is dropped from a height of 800 centimeters [cm], and the impact velocity is 41 feet per second [ft/s], determine the value of gravity in units of meters per second squared [m/s²].

48. A ball is thrown vertically into the air with an initial kinetic energy of 2,500 joules [J]. As the ball rises, it gradually loses kinetic energy as its potential energy increases. At the top of its flight, when its vertical speed goes to zero, all of the kinetic energy has been converted into potential energy. Assume that no energy is lost to frictional drag, etc. How high does the ball rise in units of meters [m] if it has a mass of 5 kilograms [kg]?

49. The maximum radius (R) a falling liquid drop can have without breaking apart is given by the equation $R = \sqrt{\sigma/(g\rho)}$, where σ is the liquid surface tension, g is the acceleration due to gravity, and ρ is the density of the liquid. For an unknown liquid, determine the surface tension in units of joules per meter squared [J/m²] if the maximum radius of a drop is 0.8 centimeter [cm] and the specific gravity of the liquid is 2.9.

50. A heat recovery system (HRS) is used to conserve heat from the surroundings and supply it to the Mars Rover. The HRS fluid loops use Freon as the working fluid.

Some facts about Freon:

- The specific heat of Freon is 74 joules per mole kelvin [J/(mol K)]
- The specific gravity of Freon is 1.49
- The molecular mass of Freon is 120 grams per mole [g/mol]

The instrumentation must be kept as a temperature greater than −67 degrees Fahrenheit [°F] to avoid damage. The temperature in the area of Mars where the rover is exploring is 189 kelvins [K]. If the system must remove 47.4 British Thermal Units [BTU] of energy, what volume of Freon is needed in units of liters [L]?

51. At noon on a clear day in midsummer, a cylindrical titanium plate is placed in the sun. The plate is painted flat black so that it will absorb most of the energy from the sunlight rather than reflecting it. The plate is 5 centimeters [cm] in diameter, one centimeter [cm] thick, has a specific gravity of 4.5, a molecular weight of 47.9 grams per mole [g/mol] and has a specific heat capacity of 25 joules per mole kelvin [J/(mol K)]. If the temperature of the plate increased by 17 degrees Fahrenheit [°F], how much energy in units of calories [cal] did it absorb from the sun, assuming no heat is reradiated to the surroundings?

52. In a factory, various metal cylinders are forged and then plunged into a liquid to quickly cool the metal. The metal pieces vary in mass, and are produced at a temperature of 572 degrees Fahrenheit [°F]. The liquid used to cool the metal is glycerol. The initial temperature of glycerol is 20 degrees Celsius [°C]. At the end, glycerol and the metal will be at the same temperature. Determine the final temperature of the metal and the glycerol in units of degrees Fahrenheit [°F].

Material Property	Glycerol	Cadmium
Specific heat	221.9 J/(mol K)	26 J/(mol K)
Specific gravity	1.261	8.65
Molecular weight	92 g/mol	112.4 g/mol
Volume	1.25 gal	
Dimensions		Radius = 4 cm Height = 15 cm

53. A 3-kilogram [kg] projectile traveling at 100 meters per second [m/s] is stopped by being shot into an insulated tank containing 100 kilograms [kg] of water. If the kinetic energy of the projectile is completely converted into thermal energy with no energy lost, how much will the water increase in temperature in units of degrees Celsius [°C]? The specific heat of water is 1 calorie per gram degree Celsius [cal/(g °C)].

54. Plutonium isotopes undergo decay, producing heat. Plutonium isotope 239 (Pu-239) has 1.9 watts per kilogram [W/kg] of decay heat. How much heat, in units of calories [cal], will 1.25 moles [mol] of Pu-239 release after decaying for three hours [h]? Assume plutonium has a molecular weight of 244 grams per mole [g/mol].

55. The power required by an airplane is given by $P = Fv$, where P is the engine power, F is the thrust, and v is the plane speed. At what speed, in units of miles per hour [mi/h or mph], will a 500-horsepower [hp] engine with 1000 pound-force [lb$_f$] of thrust propel the plane?

56. When gasoline is burned in the cylinder of an engine, it creates a high pressure that pushes on the piston. If the pressure is 100 pound-force per square inch [psi], and it moves the 3-inch [in] diameter piston a distance of 5 centimeters [cm] in 0.1 seconds [s], how much horsepower [hp] does this action produce?

57. *A spacecraft utilized Earth's gravity and increasing large orbital paths around Earth to minimize the use of propellant to escape Earth's orbit. This method involves the use of several short firings of the engines to push the rocket a smaller distance rather than one long firing with a large consumption of fuel to push the rocket the entire distance.* One of the maneuvers used 440 newtons [N] of thrust for 570.6 seconds [s], resulting in a rise in orbit of 7055 miles [mi]. How much power, in units of watts [W], did this maneuver require? Express the answers using an appropriate SI prefix such that there are only one, two, or three digits shown to the left of the decimal. Assume the process is 100% efficient.

58. A spacecraft is fueled using hydrazine (N_2H_4; molecular weight of 32 grams per mole [g/mol]) and carries 1640 kilograms [kg] of fuel. On a mission to orbit a planet, the fuel will first be warmed from −186 degrees Fahrenheit [°F] to 78 degrees Fahrenheit [°F] before being used after the long space flight to reach the planet. The specific heat capacity of hydrazine is 0.099 kilojoules per mole kelvin [kJ/(mol K)]. If there is 300 watts [W] of power available to heat the fuel, how long will the heating process take in units of hours [h]?

59. A 100-watt [W] motor (60% efficient) is used to raise a 100-kilogram [kg] load 5 meters [m] into the air. How long, in units of seconds [s], will it take the motor to accomplish this task?

60. You need to purchase a motor to supply 400 joules [J] in 10 seconds [s]. All of the motors you can choose from are 80% efficient. What is the minimum wattage [W] on the motor you need to choose?

61. *The* Curiosity *rover landed on Mars on August 6, 2012, and is currently exploring the surface.* Curiosity *uses a radioisotope thermoelectric generator (RTG) to provide power for all operations.* The *Curiosity* RTG has a maximum output of 110 watts [W], and loses 2000 watts [W] as heat. By comparison, *Viking 1* spacecraft launched in 1975. The *Viking 1* has an input power of 0.75 horsepower [hp] and loses 525 watts [W] to heat. Which device is more efficient? You must determine the efficiency of each device to prove your answer.

62. A robotic rover on Mars finds a spherical rock with a diameter of 10 centimeters [cm]. The rover picks up the rock and lifts it 20 centimeters [cm] straight up. The rock has a specific gravity of 4.75. The gravitational acceleration on Mars is 3.7 meters per second squared [m/s^2]. If the robot's lifting arm has an efficiency of 40% and required 10 seconds [s] to raise the rock 20 centimeters [cm], how much power in units of watts [W] did the arm use?

63. Consider the following strange, but true, unit:

1 donkeypower = 0.33 horsepower [hp]

A certain motor is rated to supply an input power of 2500 calories per minute [cal/min] at an efficiency of 90%. Determine the amount of output power available in units of donkeypower.

64. When boiling water, a hot plate takes an average of 8 minutes [min] and 55 seconds [s] to boil 100 milliliters [mL] of water. Assume the temperature in the lab is 75 degrees Fahrenheit [°F]. The hot plate is rated to provide 283 watts [W]. If we wish to boil 100 milliliters [mL] of acetone using this same hot plate, how long do we expect the process to take? Acetone has a boiling point of 56 degrees Celsius [°C]. The specific heat capacity of water is 4.18 joules per gram degree Celsius [J/(g °C)]. Acetone has a specific gravity of 0.785 and a specific heat capacity of 2.15 joules per gram degree Celsius [J/(g °C)]. [*Hint:* You must determine the efficiency of the hotplate.]

65. You are part of an engineering firm on contract by the U.S. Department of Energy's Energy Efficiency and Renewable Energy task force to measure the power efficiency of home appliances. Your job is to measure the efficiency of stove-top burners. In order to report the efficiency, you will place a pan containing one gallon [gal] of room temperature water on their stove, record the initial room temperature, turn on the burner, and wait for it to boil. When the water begins to boil, you will record the time it takes the water to boil and look up the power for the burner provided by the manufacturer. The specific heat capacity of water is 4.18 joules per gram degree Celsius [J/(g °C)]. After measuring the following stove-top burners, what is the efficiency of each burner?

	Room Temp [°F]	Time to Boil [min]	Rated Burner Power [W]
(a)	72	21	1500
(b)	69	18	1350

66. The power available from a wind turbine is calculated by the following equation:

$$P = \frac{1}{2} A \rho v^3$$

where P = power [watts], A = sweep area (circular) of the blades [square meters], ρ = air density [kilograms per cubic meter], and v = velocity [meters per second]. The world's largest sweep area wind turbine generator in Spain has a blade diameter of 420 feet [ft]. The specific gravity of air is 0.00123. Assuming a velocity of 30 miles per hour [mi/h] and the power produced is 5 megawatts [MW], determine the efficiency of this turbine.

67. The computer system on the Indian Mars Orbiter Mission (MOM) requires a 5-volt [V] power supply capable of delivering a peak power of 205 watts [W]. What is the maximum current, in units of amperes [A], the power supply needs to be able to deliver to the onboard computer?

68. An electric winch operates on 120 volts [V] and draws 4 amperes [A] of current. The winch has an efficiency of 68%. The winch is used to lift an object that weighs 440 pound-force [lb$_f$] for 20 seconds [s]. Determine the height the object is lifted to, in units of meters [m].

69. A robot is exploring Charon, the dwarf planet Pluto's largest moon. Gravity on Charon is 0.278 meters per second [m/s^2]. During its investigations, the robot picks up a small spherical rock for inspection. The rock has a diameter of 4 centimeters [cm] and is lifted 15 centimeters [cm] above the surface. The specific gravity of the rock is 10.3. The mechanism lifting the rock is powered by a 10-volt [V] power supply and draws 1.75 milliamperes [mA] of current. It requires 13 seconds [s] to perform this lifting task. What is the efficiency of the robot?

70. A 10,000-microfarad [μF] capacitor is charged to 25 volts [V]. If the capacitor is completely discharged through an iron rod 0.2 meters [m] long and 0.25 centimeter [cm] in diameter, resulting in 90% of the stored energy being transferred to the rod as heat, how much does the temperature of the rod increase? Give your answer in units of kelvins [K].
  Data you may need:
  - Specific gravity of iron: SG = 7.874
  - Specific heat of iron: C_P = 0.450 J/(g K)

71. A 250,000-microfarad [μF] capacitor is used to keep the memory circuits of a hand-held device active in stand-by mode during power failures or when the battery is being changed. If the memory circuits require a constant current of 40 micro-amperes [μA] at 1.5 volts [V] in standby mode and the capacitor must maintain a stored voltage of at least 2.3 volts [V] to maintain the required current to the memory circuits, how many hours [h] can the capacitor keep the memory active if it was initially charged to 5 volts [V]? Assume the standby circuit is 75% efficient.

72. A 250-kilo-ohm [kΩ] resistor and a 0.2-microfarad [μF] capacitor rated at 250 volts [V] are connected in series, meaning that all of the current through the resistor enters the capacitor.

  (a) Assuming that the capacitor is initially fully discharged and that a constant voltage of 5 volts [V] is maintained across the resistor, how long, in units of seconds [s], can this circuit operate before the voltage rating of the capacitor is exceeded?

  (b) If the capacitor remains fixed at 0.2 microfarads [μF], to what value must the resistor be changed so that the rated capacitor voltage will not be exceeded for one minute

[min]? Express your answer in ohms [Ω] using an appropriate SI prefix such that there are only one, two, or three digits shown to the left of the decimal.

(c) If the resistor remains fixed at 250 kilo-ohms [kΩ], to what value must the capacitor be changed so that the rated capacitor voltage will not be exceeded for one minute [min]? Express your answer in farad [F] using an appropriate SI prefix such that there are only one, two, or three digits shown to the left of the decimal.

73. A 0.05-microfarad [μF] capacitor discharges through a 47-kilo-ohm [kΩ] resistor.

(a) If the initial capacitor voltage was 15 volts [V], what will be the capacitor voltage after the resistor has absorbed 4 micojoules [μJ] of energy, and by how much did the capacitor voltage change? Express both quantities in units of volts [V].

(b) Repeat part (a) if the initial capacitor voltage was doubled to 30 volts [V].

74. A variable capacitor has a range from 100 picofarads [pF] to 250 picofarads [pF]. If the capacitor is set to its maximum capacitance value and has a voltage across it of 3 volts [V], what will be the voltage across it if the capacitance is adjusted to the minimum value, assuming no current enters or leaves the capacitor in the process? Express your answer in units of volts [V].

75. A constant voltage of 5 volts [V] is applied across a 250-millihenry [mH] inductor until the current through the inductor is 200 microamperes [μA].

(a) For how much time was the voltage applied to the inductor? Express the answer in units of seconds [s] using an appropriate SI prefix such that there are only one, two, or three digits shown to the left of the decimal.

(b) What is the total energy stored in the inductor? Express the answer in units of joules [J] using an appropriate SI prefix such that there are only one, two, or three digits shown to the left of the decimal.

76. A 20-millihenry [mH] inductor, a 0.5-microfarad [μF] capacitor, and a 25-kilo-ohm [kΩ] resistor are connected in a loop, so that all current through any component goes through both of the others. If the inductor had an initial current of 40 milliamperes [mA] and the capacitor had an initial voltage of 9 volts [V], how much energy will the resistor have absorbed after both the inductor and the capacitor have fully discharged? Express your answer in joules [J] using an appropriate SI prefix such that there are only one, two, or three digits shown to the left of the decimal.

Dimensionless Numbers

NOTE

Within this text, dimensions are shown in braces { } and units in brackets [ ].

Recall that in the chapter on Fundamental Dimensions and Base Units, we discussed the concept of dimensions. A **dimension** is a measurable physical idea; it generally consists solely of a word description with no numbers. A **unit** allows us to quantify a dimension, to state a number describing how much of that dimension exists in a specific situation. Units are defined by convention and related to an accepted standard. Table 9-1 shows the seven base units and their corresponding fundamental dimensions.

Table 9-1 Fundamental dimensions and base units

Dimension	Symbol	Unit	Symbol
Mass	M	kilogram	kg
Length	L	meter	m
Time	T	second	s
Temperature	Θ	kelvin	K
Amount of substance	N	mole	mol
Light intensity	J	candela	cd
Electric current	I	ampere	A

9.1 Constants with Units

LEARN TO: Understand the concept of physical constants
Recognize the difference between fundamental constants and material constants

For some constants, their values are always the same regardless of the situation. Several **fundamental constants** used in various engineering applications are found in Table 9-2.

You will encounter many of these constants in your later studies of engineering. Two that you may already be familiar with are described below.

Table 9-2 Selected fundamental constants

Property	Symbol	Value
Avogadro constant	N_A	6.022×10^{23} mol^{-1}
Boltzmann constant	k	1.38065×10^{-23} J/K
Faraday constant	F	9.65×10^4 C/mol
Ideal gas law constant	R	8314 (Pa L)/(mol K)
		0.08206 (atm L)/(mol K)
Planck constant	h	6.62×10^{-34} J s
Speed of light in a vacuum	c	3×10^8 m/s
Stefan-Boltzmann constant	σ	5.67×10^{-8} W/(m^2 K^4)
Universal gravitational constant	G	6.67×10^{-11} (N m^2)/kg^2

Universal Gravitation Constant

When the centers of two bodies of mass (m_1 and m_2) are separated by some radius (r), then the force (F) tending to pull them toward each other is given by Newton's **Law of Universal Gravitation**, named after Isaac Newton, the famous English scientist who is responsible for the concepts of gravitation, laws of motion, and along with Gottfried Leibniz, differential calculus. The **universal gravitational constant** (G) is a proportionality constant whose value does not change, regardless of the bodies being compared.

$$F = G\frac{m_1 m_2}{r^2}$$

Ideal Gas Law Constant

The **ideal gas law** relates the quantities of pressure (P), volume (V), temperature (T) and amount (n) of gas in a container. This law was first proposed by Benoît Clapeyron, a French engineer who made great contributions to the field of thermodynamics. The **ideal gas law constant** (R) is the relationship found by "ideal" gas behavior, where 1 mole [mol] of gas occupies a volume of 22.4 liters [L] at a temperature of zero degrees Celsius [°C] and a pressure of 1 atmosphere [atm].

$$PV = nRT$$

Another type of "constant" maintains the same value as long as the physical situation remains the same. These **material constants** are found in equations that describe how matter and/or energy behave and are a property of the material involved. Several simple examples of how varying conditions cause changes in such "constants" are given below and summarized in Table 9-3. The values of many constants are well documented and are readily available in the literature. Several of these have been discussed in the chapter on Universal Units; a short reminder is provided here.

Table 9-3 Selected material constants

Property	Symbol	Typical Units	Material	Value
Gravitational acceleration	g	$\dfrac{m}{s^2}$	Earth	9.8
			moon	1.6
Density	ρ	$\dfrac{g}{cm^3}$	air	0.00129
			mercury	13.6
			silicon carbide	3.1
			water	1.0
Specific heat	C_p	$\dfrac{J}{g\,K}$	air	1.005
			mercury	0.14
			silicon carbide	0.75
			water	4.18
Thermal conductivity	k	$\dfrac{W}{m\,K}$	air	0.0243
			mercury	8.34
			silicon carbide	120
			water	0.607

Acceleration of Gravity

If the Law of Universal Gravitation is written for a small body (subscript b) and the earth (subscript e) as we hold the body close to the earth, we obtain the following equation. Since the term in parentheses is a constant specific to Earth, we can replace the three parameters in parentheses by a single constant, called **gravity** (g).

$$F = m_b\left(G\frac{m_e}{r_e^2}\right) = m_b g$$

If the values for universal gravitational constant, the mass and the radius of the Earth are substituted into the expression in parenthesis, the resulting value for g will be 9.8 meters per second squared [m/s²]. If you go to Earth's moon, the terms represented by m_e and r_e are much smaller, and gravity in this case is about one-sixth of the value on Earth, or about 1.6 meters per second squared [m/s²].

Density

The relationship between the mass (m) of an object and the volume (V) the object occupies is called **density** (ρ, Greek letter rho) and has a dimension of mass per volume. For example, the density of potassium is 0.86 grams per cubic centimeter [g/cm³], whereas the density of gold is 19.3 grams per cubic centimeter [g/cm³].

$$\rho = \frac{m}{V}$$

Specific Heat

The **specific heat** of a material indicates how much energy must be added to a given mass of material in order to cause the temperature to increase by a specified amount. To be a bit more precise, the thermal energy (Q) associated with a change in temperature (ΔT) is a function of the mass of the object (m) and the specific heat (C_p).

$$Q = mC_p\Delta T$$

For example, to raise the temperature of one gram of liquid mercury by 1 degree Celsius requires 0.14 joules of energy ($C_p = 0.14$ J/(g °C)). For comparison, liquid water has a specific heat of 4.18 J/(g °C). This means water requires 30 times as much energy to increase its temperature by one degrees Celsius compared with the same mass of mercury. This high value of specific heat is one of the reasons that liquid water is critically important to life as we know it.

Thermal Conductivity

When one side of an object is hotter than the other side, heat will flow spontaneously through the object from the high temperature to the low temperature in a phenomenon called **conduction**. The rate of heat transfer (Q/t) is a function of the cross-sectional area (A), the distance across which the heat travels (d), and the difference between the high temperature and the low temperature (ΔT). This model is called **Fourier's law**, named for Joseph Fourier, a French physicist who made many contributions to heat flow and mathematics. Thermal conductivity (κ) is a material property that denotes the ability of a material to conduct heat. A material with a high thermal conductivity readily transports heat, whereas a material with a low thermal conductivity retards heat flow.

$$\frac{Q}{t} = -\kappa A \frac{\Delta T}{d}$$

Comprehension Check | **9-1**

The heat loss (Q/t, in units of joules per second [J/s]) from the surface of a hot liquid is given by:

$$\frac{Q}{t} = hA(T - T_0)$$

Choose the correct set of fundamental dimensions for the parameter (h) from the choices below if the area (A) is given in units of square meters [m^2]. Both the temperature of the liquid (T) and the ambient temperature (T_0) are measured in degrees Celsius [°C].

(A) $M\,T^{-3}$

(B) $M\,T^{-3}\Theta^{-1}$

(C) $M\,L^2\,T^{-1}\Theta$

(D) $L^2\,T^{-1}$

9.2 Common Dimensionless Numbers

LEARN TO: Understand when a quantity is dimensionless

Sometimes, we form the ratio of two parameters, where each parameter has the same dimensions. Sometimes, we form a ratio with two groups of parameters, where each group has the same dimensions. The final result in both cases is dimensionless. Several common examples are given below, summarized in Table 9-4.

Pi (π): One example is the parameter π, used in the calculation of a circumference or area of a circle. The reason π is dimensionless is that it is actually defined as the ratio of the circumference (C) of a circle to its diameter (D):

$$\pi = \frac{C}{D} = \frac{\text{circumference}}{\text{diameter}} \{=\} \frac{\text{length}}{\text{length}} = \frac{L^1}{L^1} = L^0$$

Specific Gravity (SG): The specific gravity is the ratio of the density of an object to the density of water.

$$\text{Specific gravity} = \frac{\text{density of the object}}{\text{density of water}} = \frac{\text{mass/volume}}{\text{mass/volume}} \{=\} \frac{\{M/L^3\}}{\{M/L^3\}} = M^0 L^0$$

Mach Number (Ma): We often describe the speed at which an airplane or rocket travels in terms of the Mach number, named after Ernst Mach, an Austrian physicist. This number is the ratio of the speed of the object compared with the speed of sound in air.

$$\text{Mach number} = \frac{\text{speed of the object}}{\text{speed of sound in air}} \{=\} \frac{\{L/T\}}{\{L/T\}} = L^0 T^0$$

NOTE

Often, when a ratio with two groups of parameters with the same dimensions occurs frequently in calculations, the ratio is given a special name. A dimensionless number is often named for a famous scientist followed by the word "number," like the Mach number.

When describing the speed, "Mach 2" indicates the object is traveling twice the speed of sound.

Table 9-4 Some common dimensionless parameters

Name	Phenomena Ratio	Symbol	Expression
Coefficient of friction	Sideways force (F)/weight of object (w) [object static or kinetic (object sliding)]	μ_{st} and μ_k	F/w
Drag coefficient	Drag force (F_d)/inertia force (ρ, density; v, speed; A, body area)	C_d	$F_d/(\frac{1}{2}\rho v^2 A)$
Mach number	Object speed (v)/speed of sound (v_{sound})	Ma	v/v_{sound}
Pi	Circle circumference (C)/circle diameter (D)	π	C/D
Poisson's ratio	Transverse contraction (ε_{trans})/longitudinal extension (ε_{long})	v	$\varepsilon_{trans}/\varepsilon_{long}$
Specific gravity	Object density/density of water	SG	ρ/ρ_{H_2O}

Comprehension Check **9-2**

A simple expression for the velocity of molecules in a gas is:

$$v = K\sqrt{\frac{\phi}{\rho}}$$

Assume the constant K is dimensionless, the velocity (v) is given in feet per second [ft/s], and density (ρ) in grams per cubic centimeter [g/cm³]. What are the fundamental dimensions of the variable ϕ?

We must remind ourselves that it is essential to use the appropriate dimensions **and** units for every parameter. Suppose that we are interested in computing the sine of an angle. This can be expressed as a dimensionless number by forming the ratio of the length of the opposite side divided by the length of the hypotenuse of a right triangle.

$$\sin(x) = \frac{\text{length opposite side}}{\text{length hypotenuse}} \ \{=\} \ \frac{L}{L} = L^0$$

In addition to the ratio of two lengths, you may know from one of your math classes that the sine can be also be expressed as an infinite series given by:

$$\sin(x) = x - \frac{x^3}{3!} + \frac{x^5}{5!} - \frac{x^7}{7!} + \cdots$$

Let us suppose that the argument x had the units of length, say, feet. The units in this series would then read as:

$$\text{ft} - \frac{\text{ft}^3}{3!} + \frac{\text{ft}^5}{5!} - \frac{\text{ft}^7}{7!} + \cdots$$

LAW OF ARGUMENTS

Any function that can be computed using a series must employ a *dimensionless* argument.

This includes all the trigonometric functions, logarithms, and e^x, where e is the base of the natural logarithm.

We already know that we cannot add two terms unless they have the same units; recall the Plus law from the chapter on Fundamental Dimensions. The only way we can add these terms, all with different exponents, is if each term is dimensionless. Consequently, when we calculate $\sin(x)$, we see that the x must be dimensionless, which is why we use the unit of radians. This conclusion is true for any function that can be computed using a series form, leading to the **Law of Arguments**.

EXAMPLE   9-1

What are the dimensions of k in the following equation, where d is distance and t is time?

$$d = Be^{kt}$$

Since exponents must be dimensionless, the product of k and t must not contain any dimensions. The dimensions of time are $\{T\}$

$$kT^1 \{=\} T^0$$

Solving for k yields:

$$k \{=\} T^{-1}$$

k is expressed in dimensions of inverse time or "per time."

Comprehension Check	9-3

What are the dimensions of the value 6 in the following equation, assuming T is temperature [kelvin, K] and P is algae population [gram per milliliter, g/mL]?

$$T = 102\ e^{-6P}$$

9.3 Dimensional Analysis

LEARN TO: Understand the reasoning behind using dimensional analysis to simplify problem solutions

Dimensionless quantities are generated as a result of a process called **dimensional analysis**. As an example, suppose we want to study rectangles, assuming that we know nothing about rectangles. We are interested in the relationship between the area of a rectangle (A), the width of the rectangle (W), and the perimeter of the rectangle (P). We cut out a lot of paper rectangles and ask students in the class to measure the area, the perimeter, and the width (Table 9-5).

If we graph the area against the perimeter, we obtain Figure 9-1. From this, we see that the data are scattered. We would not have a great deal of confidence in drawing conclusions about how the area depended on the perimeter of the rectangle. The best we could do is to make a statement such as, "It seems that the larger the perimeter, the larger the area." However, close examination of the data table shows that as the perimeter increases from 8.75 to 10.25 centimeters and from 16.25 to 17 centimeters, the area

Table 9-5 Rectangle measurements

Perimeter (P) [cm]	Area (A) [cm²]	Width (W) [cm]
4.02	1.0	1.1
8.75	4.7	1.9
6	2.3	1.55
13.1	6.0	1.1
17.75	19	5.25
10.25	1.2	0.25
12.1	3.0	5.5
6	0.3	2.9
16.25	15.4	5.1
17	7.8	1.05

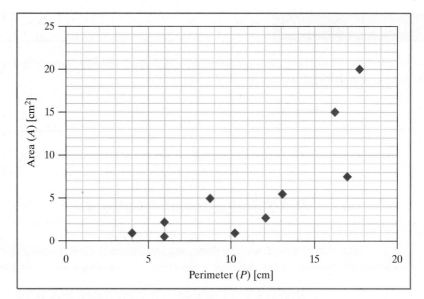

Figure 9-1 Graph of rectangle data.

actually decreases in each case. One reason for this problem is that our plot has omitted one important parameter: the width.

Analysis shows that one way in which to generalize plots of this type is to create dimensionless parameters from the problem variables. In this case, we have perimeter with dimension of length, width with the dimension of length, and area with the dimension of length squared. A little thought shows that we could use the ratio of P/W (or W/P) instead of just P on the abscissa. The ratio W/P has the dimensions of length/length, so it is dimensionless. It does not matter whether this is miles/miles, or centimeters/centimeters, the ratio is dimensionless. Similarly, we could write $A/(W^2)$, and this would also be dimensionless.

These ratios are plotted and shown in Figure 9-2. The scatter of Figure 9-1 disappears and all the data appear along a single line.

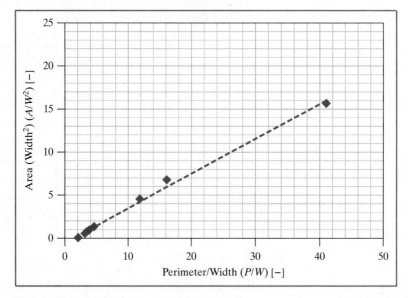

Figure 9-2 Dimensionless graph of rectangle data.

To understand how to read data from Figure 9–2, let us examine the following question. If a rectangle has a perimeter of 20 feet [ft] and a width of 2 feet [ft], what is the area in square feet [ft²]?

Step 1: $P/W = (20\ \text{ft})/(2\ \text{ft}) = 10$ (with no units).

Step 2: From the chart, at a P/W value of 10, we read a value from the line of $A/(W^2) = 3.5$.

Step 3: Calculate A from this as $A = 3.5 * (2\ \text{ft} * 2\ \text{ft}) = 14\ \text{ft}^2$.

Some of you may be thinking that we made this problem unnecessarily difficult. After all, anyone who manages to get to college knows that the "sensible" measurements to make are length, width, and area. However, many phenomena are far more complicated than simple rectangles, and it is often not at all obvious what parameters should be measured to characterize the behavior of the system we are studying. In situations of this type, dimensionless analysis can become a powerful tool to help us understand *which* parameters affect the behavior of the system and *how* they affect it. With this in mind, let us look at a slightly more complicated example.

EXAMPLE   9-2

A not-so-famous scientist, Dr. Triticale, decided to apply his scientific skills to cooking. He had always been fascinated with the process of cooking pancakes, so it seemed reasonable that he start there. He wanted to learn how to flip the flapjacks in a graceful arc in the air and then catch them.

He spent long summer days pondering this process until he finally was able to produce a list of the parameters that he felt were important. He kept asking himself, "If I change this parameter, will the trajectory of the pancake change?" If he could answer "Yes!" or even "Probably," he then considered the parameter as important enough to include on his list. As he saw it, these parameters were:

Speed of the frying pan, U	Mass of the flapjack, m
Height of the flip, H	Gravity (it pulls the flapjack back down), g

He then wrote this dependency in equation form as $H = f:(U, m, g)$.

Dr. Triticale realized that while he felt that gravity was important, it would not be easy to change the value of gravity in his tests (he could have gone to a high mountain or the moon, but this was too hard). His plan was to do many tests (and consequently eat many pancakes). He would make many measurements for many different flipping speeds and pancake masses, and try to fit a curve to the data.

Based on his work with conversion factors and his knowledge of the Per law, he reasoned that it is acceptable to multiply parameters with different dimensions. It is also fine to raise a parameter (and its associated units) to a power. Based on his understanding of the Plus law, he knew it is not acceptable to add parameters with different dimensions.

Using this information, he decided that it would be permissible to try to "fit" the dependence of pancake flipping to the important parameters raised to different powers and multiplied together. This would create a single term like $k_1 U^{a_1} m^{b_1} g^{c_1}$.

He also knew that if this term were made to have the same dimensions as H, it just might be a legitimate expression. In fact, if this were the case, he could use many terms, each of which had the dimensions of H and add them all together. While this might not be a valid equation, at least it would satisfy the Per and Plus laws, and with many terms he would have a good chance of his equation fitting the data. So, he boldly decided to try the following series:

$$H = k_1 U^{a_1} m^{b_1} g^{c_1} + k_2 U^{a_2} m^{b_2} g^{c_2} + k_3 U^{a_3} m^{b_3} g^{c_3} + \cdots$$

He needed to determine the values of the dimensionless k constants as well as all of the exponents. He knew that all the terms on the right-hand side must have the same dimensions, or they could not be added together. He also knew that the dimensions on the left and right sides must match.

With this, he then realized that he could examine the dimensions of any term on the right-hand side since each had to be the same dimensionally. He did this by comparing a typical right-hand term with the left-hand side of the equation, or

$$H = kU^a m^b g^c$$

The next step was to select the proper values of a, b, and c, so that the dimensions of the right-hand side would match those on the left-hand side. To do this, he substituted the dimensions of each parameter:

$$L^1 M^0 T^0 = \{LT^{-1}\}^a \{M\}^b \{LT^{-2}\}^c = \{L\}^{a+c} \{M\}^b \{T\}^{-a-2c}$$

For this to be dimensionally correct, the exponents for L, M, and T on the right and left would have to match, or

$$L: 1 = a + c \quad M: 0 = b \quad T: 0 = -a - 2c$$

This yields

$$a = 2 \quad b = 0 \quad c = -1$$

From this, Dr. Triticale settled on a typical term as

$$k U^2 m^0 g^{-1}$$

Finally, he wrote the curve fitting equation (with a whole series of terms) as

$$H = \sum_{i=1}^{\infty} k_1 U^2 g^{-1} + k_2 U^2 g^{-1} + k_3 U^2 g^{-1} + \cdots = \frac{U^2}{g} \sum_{i=1}^{\infty} k_i = (K)\left(\frac{U^2}{g}\right)$$

*Now, armed with this expression, he was sure that he could flip flapjacks with the best, although he knew that he would have to conduct many experiments to make sure the equation was valid (and to determine the value of K). What he did not realize was that he had just performed a procedure called **dimensional analysis**.*

9.4 Rayleigh's Method

LEARN TO: Determine appropriate dimensionless numbers using Rayleigh's method
Understand the physical significance of the Reynolds number as it applies to pipe flow
Determine final quantity if given four of following: density, diameter, Reynolds number, velocity, viscosity

In this section we formalize the discussion presented in Example 9-2 by introducing a method of dimensional analysis devised by Lord Rayleigh, John William Strutt, the third Baron Rayleigh. Three detailed examples illustrate his approach to dimensionless analysis:

- Example 9-3, in which we analyze factors affecting the distance traveled by an accelerating object
- Example 9-4, in which we determine the most famous named dimensionless number, the Reynolds number
- Example 9-5, in which we simplify one use of Rayleigh's method

No matter the problem, the way we solve it will use the same 8-step method:

Rayleigh's Method

Step 1: Write each variable and raise each to an unknown exponent (use *all* the variables, even the dependent variable). Order and choice of exponent do not matter.

Step 2: Substitute dimensions of the variables into Step 1. Be sure to raise each dimension to the proper exponent groups from Step 1.

Step 3: Group by dimension.

Step 4: Exponents on each dimension must equal zero for dimensionless numbers, so form a set of equations by setting the exponent groups from Step 3 for each dimension equal to zero.

Step 5: Solve the simultaneous equations (as best as you can).

   Hint: Number of unknowns − number of equations = number of groups

Step 6: Substitute results of Step 5 back into Step 1 exponents.

Step 7: Group variables by exponent. These resulting groups are your dimensionless numbers.

Step 8: Be sure to *check* it out!! Are *all* of the ratios really dimensionless?

   Hint: If the resulting groups are *not* dimensionless, you most likely goofed in either Step 2 or Step 5!

Rayleigh's analysis is quite similar to the Buckingham Pi method, another method to determine dimensionless groups. Rayleigh's method is, however, a bit more direct and often seems less "mysterious" to those who are new to dimensional analysis. Both methods use a general form with multiplied and exponentiated variables. Any inspection of physics, engineering, and mathematical texts reveal many examples of this form of equation governing a myriad of behaviors.

EXAMPLE 9-3

To develop an understanding of how initial velocity, acceleration, and time all affect the distance traveled by an accelerating object, we conduct some experiments and then analyze the resulting data. We asked a student to conduct a series of tests for us. She observed 25 different moving bodies with a wide range of initial velocities and different accelerations. For each, she measured the distance the bodies traveled for some prescribed time interval. Results are given in Table 9-6. A portion of this table is shown below.

Table 9-6 Position of a body as a function of initial velocity, acceleration, and time

Test	Initial Velocity (v_0) [m/s]	Acceleration (a) [m/s²]	Time (t) [s]	Distance (d) [m]
1	3	1	6	36
2	3	2	6	54
3	1.5	5	6	99
4	5	4	6	44
5	5	3	8	136

In addition, we would like to use this data set to help make predictions of the distance traveled by other bodies under different conditions. For example, we might want to answer the following question: What is the acceleration needed to travel 4800 meters [m] in 200 seconds [s], if the initial velocity is 8 meters per second [m/s]?

There are several independent variables (initial velocity, acceleration, and time), so it is not obvious what to plot. We can write the dependency as

$$d = f: (v_0, a, t)$$

We anticipate that it is difficult to draw conclusions regarding the interdependence of all of these variables. Realizing this, we plot distance against time without worrying about the initial velocity and the acceleration.

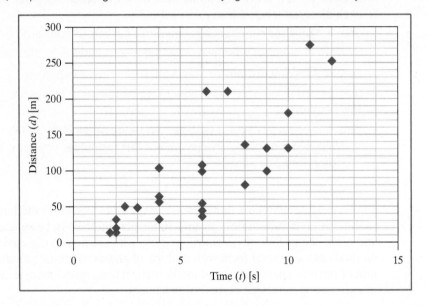

It seems that, in general, the longer one travels, the farther one goes. Upon closer inspection, however, it is obvious that this is not always the case. For example, for a travel time of 6 seconds, the distance traveled varies from about 25 to 210 meters. Since enough tests were not conducted with systematic variation of the initial velocity or acceleration, it is not possible to do much better than this. We certainly have no hope of answering the questions above with any confidence. In addition, since the values are so scattered, we realize that we have no good way to determine if any of our measurements were bad. With this disheartening conclusion, we perform a dimensional analysis in an attempt to place the parameters into fewer groups.

Step 1: *Raise each variable to a different unknown power, using symbols for the exponents that do not already appear in the problem, and then multiply all of these individual terms together. For example, since the current problem has both an a and d used as variables, we should not use a and d for the exponents; hence we choose the letters p through s. The order in which we list the variables and assign exponents is completely random.*

$$d^p v_0^q a^r t^s$$

Step 2: *Substitute the correct dimensions for each variable.*

$$\{L\}^p \left\{\frac{L}{T}\right\}^q \left\{\frac{L}{T^2}\right\}^r \{T\}^s$$

Step 3: *Expand the expression to have each dimension as a base raised to some power.*

$$\{L\}^{p+q+r}\{T\}^{-q-2r+s}$$

Step 4: *For the expression in Step 3 to be dimensionless, each exponent must equal zero, or*

$$p + q + r = 0 \ and \ -q - 2r + s = 0$$

NOTE

There are many ways to solve this and all will lead to two dimensionless ratios. If you do not like the plot you get from doing it one way, try solving for different exponents and see if that provides a better plot. All will be correct, but some are easier to use than others.

Step 5: *Solve for the exponents. In this case, we have two equations and four unknowns, so it is not possible to solve for all the unknowns in terms of an actual number. We must be satisfied with finding two of the exponents in terms of the other two. If a variable appears in all or most of the equations, that may be a good one to begin with.*

In our example, we solve the second equation for q.

$$q = -2r + s$$

Substituting for q into the other equation gives

$$p - 2r + s + r = 0$$

thus,

$$p = r - s$$

At this point, we have defined p and q in terms of the other two variables, r and s.

Step 6: *Substitute into the original expression.*

$$d^{\,r-s}v_0^{-2r+s}a^{\,r}t^{\,s}$$

Note that all of the exponents are now expressed in terms of only two variables, r and s.

Step 7: *Simplify by collecting all terms associated with the remaining exponential variables (r and s in this case).*

$$\left(\frac{d\,a}{v_0^2}\right)^r\left(\frac{t\,v_0}{d}\right)^s$$

Step 8: *The simplification in Step 7 gives the dimensionless ratios we are looking for. Dropping the exponents assumed in Step 1 gives the following groups:*

$$\left(\frac{d\,a}{v_0^2}\right) and \left(\frac{t\,v_0}{d}\right)$$

We need to double-check both groups are dimensionless. Before plotting we make two additional observations. (1) The variables of distance and initial velocity appear in both quantities. This may not always be desirable. (2) The initial velocity appears in the denominator of the first ratio. This may cause problems if we are examining data in which the initial velocity is very small, making the ratio very large. While dimensional analysis is much more involved than the examples given here, there are several important facts for you to remember.

First, since the results of the dimensional analysis produces dimensionless ratios, these ratios may be used as they appear above or they may be inverted. In other words, for this example, we can use $\dfrac{d\,a}{v_0^2}$ or $\dfrac{v_0^2}{d\,a}$ equally well.

To eliminate the problem of very small initial velocity values, the second form is preferable for our work here. As a side note, if we are interested in the behaviors at very small times, then we would prefer for time to appear in the numerator of the second ratio.

Second, it is permissible to alter the form of one of the ratios by multiplying it by the other one or by the inverse of the other one or by the other one squared, etc. This will change the form of the first ratio and may produce results that are easier to interpret. A simple example can be used to show this. For the two ratios here, multiply the first ratio by the second ratio squared. This yields a "new" first ratio as $[(a\,t^2)/d]$, and this could be used along with the second ratio $[(t\,v_0)/d]$. This result may have the advantage of initial velocity appearing in only one of the ratios.

To continue this example, we plot one variable against the other. This result is shown below:

NOTE

There seems to be one "bad" data point. Would you have been able to pick out this point from the original data or from the dimensional plot?

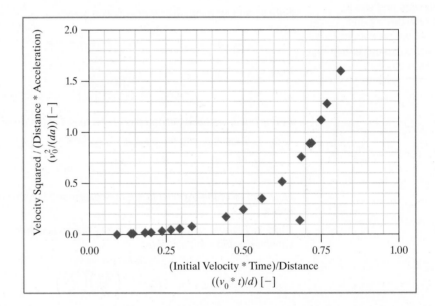

Now the scatter from the original dimensional plot is gone, and all but one data point seems to lie on a smooth curve. We can use this plot to determine the relationship between distance traveled, acceleration, time, and initial velocity. Let us see how to do this for the question we posed earlier.

What acceleration is required to go 4800 meters [m] in 200 seconds [s] if the initial car velocity is 8 meters per second [m/s]?

On the abscissa: $[(8\text{ m/s})\,(200\text{ s})]/(4800\text{ m}) = 0.33$

Reading from the graph at this point: $v_0^2/(d\,a) = 0.075$

Solving for acceleration: $a = (8\text{ m/s})^2/[(0.075)\,(4800\text{ m})] = 0.18\text{ m/s}^2$

Several important conclusions can be drawn from this exercise.

- Dimensionless parameters often allow us to present data in an easily interpretable fashion when the raw data have no recognizable pattern.
- When we use the exponent approach to find dimensionless parameters, we need to remember that the exponents could be either positive or negative. Thus, in the case of the rectangle, P/W is just as good as W/P. You can always try both to see which gives the best-looking results. The choice is yours, and sometimes depends on whether one of the variables goes to zero; since you cannot divide by zero, that variable should not be in the denominator, if possible.
- If we collapse the data by using dimensionless parameters so that a single curve can fit through the resulting points, *bad data points will usually become obvious.*
- Finally, this approach can reduce literally thousands of different measurements into one simple curve. In the case of the car, the single line we obtained will work for all possible combination of times, initial velocities, accelerations, and distance. If we used dimensional plots, we would need more plots, with many lines on each plot. This would require an entire book of plots rather than the single plot with a single line that we obtained above by dimensionless ratios.

EXAMPLE 9-4

To classify the smoothness of a flowing fluid, Osborne Reynolds developed the now famous dimensionless quantity of **Reynolds number**. His theory stated that the smoothness or roughness (a lot of eddies or swirling) of a fluid depended upon:

How fast the fluid was moving (velocity)	$v\ [=]\ \text{m/s}$
The density of the fluid	$\rho\ [=]\ \text{kg/m}^3$
The diameter of the pipe	$D\ [=]\ \text{m}$
How hard it was to move the fluid (viscosity)	$\mu\ [=]\ \text{g/(cm s)}$

Reynolds knew the smoothness depended upon these quantities:

$$\text{Smoothness of the flow} = f{:}\ (v, \rho, D, \mu)$$

But how did they depend on one another? We could write the four variables above as

$$v^a \rho^b D^c \mu^d$$

and if this was dimensionless, it would appear as $M^0 L^0 T^0$.
To make this grouping dimensionless, first we substitute in the dimensions of the four variables to obtain:

$$\left\{\frac{L}{T}\right\}^a \left\{\frac{M}{L^3}\right\}^b \{L^c\} \left\{\frac{M}{LT}\right\}^d = M^{b+d} L^{a-3b+c-d} T^{-a-d}$$

If this is to be dimensionless, then the exponents on all of the dimensions must equal zero, therefore:

M: $\quad b + d = 0$

L: $\quad a - 3b + c - d = 0$

T: $\quad -a - d = 0$

This gives three equations in four unknowns, so we will have to solve for three of the variables in terms of the fourth. In this example, we solve for the three unknowns a, b, c in terms of d:

M: $b = -d$

T: $a = -d$

L: $c = -a + 3b + d = d - 3d + d = -d$

Substituting these back into the original parameters gives:

$$v^{-d}\rho^{-d}D^{-d}\mu^{d}$$

We see that there is one dimensionless group, since all the parameters have an exponent of d.

$$\frac{\mu}{v\rho D}\{=\} \frac{\dfrac{M}{LT}}{\dfrac{L}{T}\dfrac{M}{L^3}L} = \frac{M}{LT}\left|\frac{T}{L}\right|\frac{L^3}{M}\left|\frac{}{L}\right| = \text{dimensionless}$$

Since the variables of diameter and velocity can approach zero, the Reynolds number is commonly written as follows:

$$Re = \frac{\rho D v}{\mu}$$

If the Reynolds number has a value less than 2000, the flow is described as **laminar**, meaning it moves slowly and gently with no mixing or churning. If the Reynolds number has a value greater than 10,000, the flow is described as **turbulent**, meaning it moves quickly with much mixing and churning (lots of eddies) occurring. The region in between 2000 and 10,000 is called the **transition region**.

NOTE

The Reynolds number is used to describe fluid flow.

$Re < 2000 = \text{laminar}$

$2000 < Re < 10,000$
$\quad = \text{transitional}$

$Re > 10,000 = \text{turbulent}$

EXAMPLE   9-5

Suppose we conduct an experiment with a ball that we throw from the top of a tall tower of height H. We throw it directly downward with some initial velocity v, and then measure the elapsed time t until it hits the ground. We vary the initial height and the initial velocity. The variables of interest in this problem are H, v, and t. A little thought leads us to include g, since it is the force of gravity that causes the ball to fall in the first place. Using Rayleigh's method, find a set of dimensionless ratios that can be used to correlate our data.

Step 1: *Write each variable and raise each to an unknown exponent (use all the variables, even the dependent variable).*

$$t^{a}H^{b}v^{c}g^{d}$$

Step 2: *Substitute dimensions of the variables into Step 1. Be sure to raise each dimension to the proper exponent from Step 1.*

$$t^{a}\{=\} T^{a} \quad H^{b}\{=\} L^{b} \quad v^{c}\{=\} L^{c}T^{-c} \quad g^{d}\{=\} L^{d}T^{-2d}$$

Step 3: *Group by dimension.*

$$L^{b+c+d} T^{a-c-2d}$$

Step 4: *Exponents on each dimension must equal zero for dimensionless numbers! Form a set of equations by setting the exponents for each dimension equal to zero.*

$$b + c + d = 0 \qquad a - c - 2d = 0$$

Step 5: *Solve the simultaneous equations (as best as you can).*

$$b = -c - d \qquad a = c + 2d$$

Step 6: *Substitute results of Step 5 back into Step 1 exponents.*

$$t^{\,c+2d}\, H^{-c-d}\, v^{\,c}\, g^{\,d}$$

Step 7: *Group variables by exponent. These resulting groups are your dimensionless numbers.*

$$\left[\frac{v\,t}{H} \right] \qquad \left[\frac{g\,t^2}{H} \right]$$

Step 8: *Be sure to check it out!! Are all of the ratios really dimensionless?*

Comprehension Check   9-4

The Euler number is a function of the pressure drop, velocity, and density. Determine the form of the Euler number with Rayleigh's method.

Pressure drop	ΔP	pascal [Pa]
Density	ρ	grams per cubic centimeter [g/cm³]
Velocity	v	meters per second [m/s]

NOTE

We are not sure that the results of this technique are *physically correct*, only that they are *dimensionally correct* until we conduct experiments.

At the beginning of the analysis, when in doubt about the importance of a parameter, put it in the list of important parameters.

Always remember that we initiate this procedure simply by providing a list of parameters we think are important to the situation at hand. If we omit an important parameter, our final result will not be physically correct, even if it is dimensionally correct. Consequently, if we select an improper parameter, then when tests are conducted, we will discover that it was not important to the problem and we can drop it from further consideration. *We cannot decide whether any variable is important until we conduct some experiments.*

Consequently, if we are *sure* that a parameter is important, then we know it should *not* drop from the analysis. The only way it can be retained is if at least one other parameter contains the missing dimension. In this case, we need to ask ourselves what other parameters might be important, add them to our list, and rework the analysis.

Dimensional analysis helps us organize data by allowing us to plot one-dimensionless parameter against another, resulting in one line on a single plot. This is a powerful result, and reduces a problem of multiple initial parameters to one containing only two.

CHAPTER 9

IN-CLASS ACTIVITIES

ICA 9-1

Complete the following table.

| | Quantity | SI Units | Dimensions | | | | | | |
			M	L	T	Θ	N	J	I
Example	Acoustic impedance	(Pa s)/m	1	−2	−1	0	0	0	0
(a)	Circuit resistance	V/A							
(b)	Luminous efficacy	cd/W							
(c)	Thermal conductivity	cal/(cm s °C)							

ICA 9-2

Complete the following table.

| | Quantity | SI Units | Dimensions | | | | | | |
			M	L	T	Θ	N	J	I
Example	Acoustic impedance	(Pa s)/m	1	−2	−1	0	0	0	0
(a)	Inductance	J/A^2							
(b)	Molarity	mol/kg							
(c)	Wire resistivity	V m/A							

ICA 9-3

Calculate the numerical value of each of the dimensionless parameters listed in the table. Be sure to check that the ratio is actually dimensionless after you insert the values.

	Situation	Name	Expression	Value
(a)	Hot water	Prandtl number, Pr	$\dfrac{\mu C_p}{k}$	
(b)	Water in a river	Froude number, Fr	$\dfrac{v_W}{\sqrt{gH}}$	

Property	Symbol	Units	Value
Dynamic viscosity	μ	kg/(m s)	4×10^{-4}
Thermal conductivity	k	W/(m K)	0.7
Specific heat	C_p	cal/(g °C)	1
Water depth	H	m	3
Water speed	v_w	cm/s	210

ICA 9-4

Calculate the numerical value of each of the dimensionless parameters listed in the table. Be sure to check that the ratio is actually dimensionless after you insert the values.

	Situation	Name	Expression	Value
(a)	Air over a flat plate	Nusselt number, Nu	$\dfrac{h\,L}{k}$	
(b)	Wind making a wire "sing"	Strouhal number, St	$\dfrac{\omega\,D_{wire}}{v_a}$	

Property	Symbol	Units	Value
Heat transfer coefficient	h	W/(m² °C)	20
Thermal conductivity	k	W/(m K)	0.025
Plate length	L	ft	2
Air speed	v_a	mph	60
Oscillation frequency	ω	Hz or cycles/s	140
Wire diameter	D_{wire}	mm	20

ICA 9-5

A fluid with a specific gravity of 0.91 and a viscosity of 0.38 pascal seconds [Pa s] is pumped through a 25-millimeter [mm] diameter smooth pipe at an average velocity of 2.6 meters per second [m/s]. Determine the Reynolds number in the pipe for the system and indicate if the flow is laminar, transitional, or turbulent.

ICA 9-6

Brine, with a density of 1.25 grams per cubic centimeter [g/cm³] and a viscosity of 0.015 grams per centimeter second [g/(cm s)] is pumped through a 5-centimeter [cm] radius steel pipe at an average velocity of 15 centimeters per second [cm/s]. Determine the Reynolds number in the pipe for the system and indicate if the flow is laminar, transitional, or turbulent.

ICA 9-7

When a simple turbine is used for mixing, the following variables are involved:

P = Power requirement	[=] watt [W]
N = Shaft speed	[=] hertz [Hz = $\frac{1}{s}$]
D = Blade diameter	[=] meters [m]
W = Blade width	[=] meters [m]
ρ = Liquid density	[=] kilograms per meter cubed [kg/m³]

Determine a set of dimensionless groups using Rayleigh's method.

ICA 9-8

History suggests that Newton developed his idea about gravity after watching an apple fall from a tree. Using the following variables and Rayleigh's method, determine a dimensionless number—the Newton number—to describe falling objects:

w = Weight of the apple	[=] newtons [N]
s = Speed of the apple at impact to the ground	[=] meters per second [m/s]
d = Apple diameter	[=] inches [in]
ρ = Apple density	[=] pound-mass per cubic foot [lb_m/ft^3]
T = Temperature of the air	[=] degrees Fahrenheit [°F]

ICA 9-9

It is proposed to create a set of dimensionless numbers used to describe the phenomena of reaching the escape velocity necessary to orbit the Earth. A set of names is proposed, based on famous astronauts:

- The Gagarin number, after Yuri Gagarin, a Russian astronaut and the first human to achieve escape velocity and orbit the Earth in outer space;
- The Valentina number, after Valentina Tereshkova, a Russian astronaut who was the first female to orbit the Earth;
- The Shepard number, after Alan Shepard, the first American to orbit the Earth; and
- The Ride number, after Sally Ride, the first American woman in space.

To begin the analysis, assume the following variables are important:

g = Gravitational pull between planet and rocket	[=] meters per second squared [m/s²]
n_r = Amount of rocket fuel	[=] moles [mol]
w_p = Weight of planet	[=] pounds-force [lb_f]
d = Diameter of planet	[=] miles [mi]
G = Newton's gravitational constant	[=] newton meters squared per kilogram squared [N m²/kg²]
v = Velocity of rocket	[=] miles per hour [mph or mi/h]
η = Efficiency of the rocket engine	[=] unitless

Determine a set of dimensionless groups using Rayleigh's method.

ICA 9-10

We assume that the total storm water runoff (R) from a plot of land depends on the length of time that it rains (t), the area of the land (A), and the rainfall rate. A portion of the data is shown below. Please download the Excel file to access the complete data set.

Rainfall Rate (r) [in/h]	Land Area (A) [acres]	Rainfall Duration (t) [h]	Measured Runoff (R) [ft³]
0.50	2	3.0	49
0.30	14	2.5	172
0.78	87	4.1	4563

(a) Using the data, construct a plot of the runoff versus the time that it rains. You should see that this plot is of little help in understanding the relationships between the various parameters.

(b) Using this plot, estimate the total runoff from 200 acres if rain falls for 3 hours [h] at a rate of 1.2 inches per hour [in/h]. Is this even possible using this plot?

(c) Using the variables in the table, complete a dimensional analysis to help you plot the data. Plot the resulting ratios, with the ratio containing the total runoff on the ordinate. This plot should collapse the values to a single line. Draw a smooth curve through the values.

(d) Use this line to answer question (b) again.

ICA 9-11

We are interested in analyzing the velocity of a wave in water. By drawing a sketch of the wave and labeling it, we decide that the velocity depends on the wavelength (λ), the depth of the water (H), the density of the water (ρ), and the effect of gravity (g). We have measured wave speeds in many situations. A portion of the data is shown below. Please download the Excel file to access the complete data set.

Water Depth (H) [m]	Wave Length (λ) [m]	Velocity (v) [m/s]
1.0	10.0	7.4
9.0	40.0	18.7
0.2	13.0	3.5
33.0	30.0	17.1

(a) Construct a plot of wave velocity versus either wave length or water depth. You will see substantial scatter.

(b) Using your plot, estimate the wave velocity for a wave length of 25 meters [m] in water that is 25 meters [m] deep. Is this even possible using this plot?

(c) Perform a dimensional analysis on the parameters ($v, \rho, H, g,$ and λ). After calculating the new dimensionless ratio values, make a dimensionless plot.

(d) Recalculate the answer to question (b).

REVIEW QUESTIONS

1. While researching fluid dynamics, you come across a reference to the dimensionless number called the *Grashof number,* given by the equation below.

$$Gr = \frac{g\,\beta\,(T_S - T_b)\,D^3}{\nu^2}$$

where:

$D =$ pipe diameter $[=]$ ft
$g \;=$ acceleration due to gravity $[=]$ m/s^2
$T =$ temperature of the surface (T_s) and bulk fluid (T_b) $[=]$ K
$\nu \;=$ kinematic viscosity $[=]$ cm^2/s

What are the dimensions of beta, β?

2. While researching fluid dynamics, you come across a reference to the dimensionless number called the *capillary number,* given by the equation below.

$$Ca \;=\; \frac{\mu\,v}{\gamma}$$

where:

$\mu \;=$ fluid viscosity $[=]$ g/(m s)
$v \;=$ velocity $[=]$ ft/s

What are the dimensions of gamma, γ?

3. While researching fluid dynamics, you come across a reference to the dimensionless number called the *Laplace number,* given by the equation below.

$$La \;=\; \frac{\delta\,\rho\,L}{\mu^2}$$

where:

$\rho =$ fluid density $[=]$ kg/m^3
$\mu =$ fluid viscosity $[=]$ g/(m s)
$L =$ length $[=]$ ft

What are the dimensions of delta, δ?

4. The Arrhenius number (Ar) is the dimensionless parameter describing the ratio of activation energy to thermal energy, often used in chemistry. It depends on the following quantities:

$$E_a = \text{activation energy } [=] \text{ J/mol}$$
$$R = \text{ideal gas constant } [=] \text{ (atm L)/(mol K)}$$
$$T = \text{temperature } [=] \text{ K}$$

Use your knowledge of dimensions to determine the proper form of the Arrhenius number, choosing from the options below.

(A) $Ar = \dfrac{E_a}{RT}$

(B) $Ar = E_a\,RT$

(C) $Ar = \dfrac{RT^2}{E_a}$

(D) $Ar = \dfrac{E_a\,T}{R}$

5. The Biot number (Bi) is the dimensionless parameter describing if the temperature of an object will vary significantly in space. It depends on the following quantities:

$$h = \text{heat transfer coefficient } [=] \text{ W/(m}^2\,{}^\circ\text{C)}$$
$$L_C = \text{volume of object/surface area of object } [=] \text{ m}^3/\text{m}^2 = \text{m}$$
$$k = \text{thermal conductivity } [=] \text{ W/(m K)}$$

Use your knowledge of dimensions to determine the proper form of the Biot number, choosing from the options below.

(A) $Bi = \dfrac{h\,L_c}{k}$

(B) $Bi = h\,L_c\,k$

(C) $Bi = \dfrac{k\,L_c}{h}$

(D) $Bi = \dfrac{h\,k}{L_c^2}$

6. A biodegradable fuel having a specific gravity of 0.95 and a viscosity of 0.04 grams per centimeter second [g/(cm s)] is draining by gravity from the bottom of a tank. The drain line is a plastic 3-inch [in] diameter pipe. The velocity is 5.02 meters per second [m/s]. Determine the Reynolds number in the pipe for the system and indicate if the flow is laminar, in transition, or turbulent.

7. A sludge mixture having a specific gravity of 2.93 and a viscosity of 0.09 grams per centimeter second [g/(cm s)] is pumped from a reactor to a holding tank. The pipe is a 2½-inch [in] diameter pipe. The velocity is 1.8 meters per second [m/s]. Determine the Reynolds number in the pipe for the system and indicate if the flow is laminar, in transition, or turbulent.

8. Water (specific gravity = 1.02; viscosity = 0.0102 grams per centimeter second [g/(cm s)]) is pumped through a 0.5-meter [m] diameter pipe. If the Reynolds number is 1800 for the system, determine the velocity of the water in units of meters per second [m/s].

9. We have been asked to determine a set of dimensionless numbers to evaluate the speed of a wave in a shallow liquid.

ρ = Density of the liquid	[=] grams per cubic centimeter [g/cm^3]
D = Depth of the liquid	[=] feet [ft]
v = Velocity of the wave	[=] miles per hour [mph or mi/h]
σ = Surface tension of the liquid	[=] joules per square meter [J/m^2]
μ = Viscosity of the liquid	[=] pascal seconds [Pa s]
T = Temperature of the liquid	[=] kelvin [K]

   (a) Analyze this system using Rayleigh's method.
   (b) How does this analysis change if gravity (g, in meters per second squared [m/s^2]) is included?

10. The Peclet number is used in heat transfer in general and forced convection calculations in particular. It is a function of the two other dimensionless groups, the Reynolds number and the Prandtl number. Determine the functional form of these dimensionless groups, using Rayleigh's method. The problem depends on the following variables:

ρ = Liquid density	[=] kg/m^3
C_p = Specific heat of liquid	[=] J/(g °C)
μ = Liquid viscosity	[=] kg/(m s)
α = Thermal diffusivity	[=] m^2/s
k = Thermal conductivity of the plate	[=] W/(m °C)
x = Distance from edge of the plate	[=] m
v = Liquid velocity	[=] m/s

11. When a fluid flows slowly across a flat plate and transfers heat to the plate, the following variables are important. Analyze this system using Rayleigh's method.

ρ = Liquid density	[=] kg/m^3
C_p = Specific heat of liquid	[=] J/(g °C)
μ = Liquid viscosity	[=] kg/(m s)
k = Thermal conductivity of the plate	[=] W/(m °C)
h = Heat transfer coefficient	[=] W/(m^2 °C)
x = Distance from edge of the plate	[=] m
v = Liquid velocity	[=] m/s

12. In modeling the flow of liquid in a piping system, you decide to try to develop some dimensionless groups to determine the interaction between variables. You decide the following variables are important:

Q = Volumetric flowrate	[=] gallons per minute [gal/min]
ν = Kinematic viscosity	[=] centimeters squared per second [cm^2/s]
μ = Dynamic viscosity	[=] pascal seconds [Pa s]
ρ = Density	[=] kilograms per cubic meter [kg/m^3]
v = Velocity	[=] feet per second [ft/s]
D = Diameter	[=] millimeters [mm]
m = Mass of fluid	[=] kilograms [kg]

Use Rayleigh's method to determine a set of dimensionless groups.

13. A projectile is fired with an initial velocity (v_0) at an angle (θ) with the horizontal plane. Find an expression for the range (R). A portion of the data is shown below. Please download the Excel file to access the complete data set.

Launch Angle (θ) [°]	Launch Speed (v_0) [m/s]	Measured Range (R) [m]
4	70	73
50	50	230
22	8	4.4
88	100	77

(a) Complete a dimensional analysis of this situation. In this case, you would assume that the important parameters are θ, v_0, and R. Upon closer examination, however, it would seem that the range on Earth and on the moon would be different. This suggests that gravity is important, and you should include g in the list of parameters. Finally, since it is not clear how to include θ, you could omit it and replace the velocity by v_x and v_z, where x is the distance downrange and z the height. When you complete the analysis, you should find that these four parameters will be grouped into a single dimensionless ratio.

(b) Use the data from the table to calculate the numerical value of the ratio for each test. Note that $v_x = v \cos(\theta)$ and that you can find a similar expression for v_z. Insert these expressions into your dimensionless ratio.

(c) Assuming that you performed the dimensional analysis correctly, you should find that the ratio you obtained will always give the same value (at least nearly, within test-to-test error). Calculate the average value of the tests, and if it is nearly an integer, use the integer value.

(d) Finally, set this ratio equal to this integer, and then solve for the range R. Write your final equation for the range (i.e., $R = $ xxxxx).

14. The drag on a body moving in a fluid depends on the properties of the fluid, the size and the shape of the body, and probably most importantly, the velocity of the body. We find that for high velocities, the fluid density is important but the "stickiness" (or viscosity) of the fluid is not. The frontal area of the object is important. You might expect that there will be more drag on a double-decker bus moving at 60 miles per hour than on a sports car.

The following table gives some data for tests of several spheres placed in air and in water. A portion of the data is shown below. Please download the Excel file to access the complete data set. The terminal velocity, the point at which the velocity becomes constant when the weight is balanced by the drag, is shown.

Object	Drag (F) [lb$_f$]	Velocity (v) [ft/s]	Diameter (D) [in]	Fluid
Table tennis ball	0.005	12	1.6	Air
Bowling ball	6	60	11	Air
Table tennis ball	0.0028	0.33	1.6	Water
Cannon ball	31	6.2	9	Water

(a) Complete a dimensional analysis of this situation and replot the data. First, recognize that the important parameters are the ball diameter (use the silhouette area of a circle), the density of the fluid, the drag, and the velocity. You will find a single dimensionless ratio that combines these parameters.

(b) Compute the value of this ratio for the eight tests. Be sure in your analysis that you use consistent units so that the final ratio is truly unitless.

(c) Use this result to help you answer the question: What is the drag on a baseball in gasoline (specific gravity = 0.72) at 30 feet per second [ft/s]?

Scrupulous Worksheets

Chapter 10

Excel Workbooks

10.1 Cell References

10.2 Functions in Excel

10.3 Logic and Conditionals

10.4 Lookup and Data Validation

10.5 Conditional Formatting

10.6 Sorting and Filters

Chapter 11

Graphical Solutions

11.1 Graphing Terminology

11.2 Proper Plots

11.3 Available Graph Types in Excel

11.4 Graph Interpretation

11.5 Meaning of Line Shapes

11.6 Graphical Solutions

Chapter 12

Models and Systems

12.1 Proper Plot Rules for Trendlines

12.2 Linear Functions

12.3 Linear Relationships

12.4 Combinations of Linear Relationships

12.5 Power Functions

12.6 Exponential Functions

Chapter 13

Mathematical Models

13.1 Selecting a Trendline Type

13.2 Interpreting Logarithmic Graphs

13.3 Proper Plot Rules for Log Plots

13.4 Converting Scales to Log in Excel

13.5 Dealing with Limitations of Excel

Chapter 14

Statistics

14.1 Histograms

14.2 Statistical Behavior

14.3 Distributions

14.4 Cumulative Distribution Functions

14.5 Statistical Process Control (SPC)

14.6 Statistics in Excel

14.7 Statistics in MATLAB

Microsoft Excel is a worksheet computer program used internationally for an incalculable number of different applications. A **worksheet** is a document that contains data separated by rows and columns. The idea of using a worksheet to solve different types of problems originated before the advent of computers in the form of bookkeeping ledgers. The first graphical worksheet computer program for personal computers, VisiCalc, was released in 1979 for the Apple II® computer.

Modern worksheet computer programs like Excel are significantly more powerful than earlier versions like VisiCalc; a comparison of the interface is shown in Figure P3-1. Excel contains text-formatting controls, built-in functions to perform common calculations, and a number of different plotting capabilities that make it an extremely powerful data analysis tool for engineers. Part 3 introduces the Microsoft Excel interface, the formatting controls used to create organized worksheets, and many built-in functions to assist in analyzing data or performing calculations on data contained in the worksheet.

Figure P3-1 Comparison of VisiCalc and Excel interfaces.

Learning Objectives

The overall learning objectives for this part include the following:

Chapter 10:
- Use Microsoft Excel to enhance problem solution techniques, including entering, sorting, and formatting data in a worksheet;
- Applying functions, including mathematical, statistical, and trigonometric;
- Read, write, and predict conditional statements, lookup functions, and data validation statements;
- Use conditional formatting, sorting, and filtering to aid in problem solutions.

Chapter 11:
- Use graphical techniques to create "proper" plots, sketch functions, and determine graphical solutions to problems.
- Create and format data into graphs using Microsoft Excel.

Chapter 12:
- Describe and interpret mathematical models in terms of physical phenomena.

- Given a graph, determine the type of trendline shown and interpret the physical parameters of the experimental system.

Chapter 13:
- Determine an appropriate mathematical model to describe experimental data using physical knowledge and logarithmic plots, then apply the model to form graphical solutions to engineering problems.
- Given a logarithmic plot, determine the equation of the trendline.
- Use Microsoft Excel to model experimental data by creating logarithmic plots.

Chapter 14:
- Apply basic concepts of statistics to experimental data.
- Use statistical and graphical functions and in Microsoft Excel and MATLAB to enhance solution techniques.

A successful engineer must rely on knowledge of the way things work in order to develop solutions to problems, whether ameliorating climate change or trapping cockroaches. In many cases, the behavior of systems or phenomena can be described mathematically. These mathematical descriptions are often called mathematical models. The variables in the model vary with respect to one another in the same way that the corresponding parameters of the real physical system change.

As a very simple example, imagine you are driving your car on a country road at a constant speed of 30 miles per hour. You know that at this speed, you travel one-half mile every minute. If you drive at this speed for 44 minutes, you cover a distance of 22 miles.

A mathematical model for this is $d = 0.5t$, where d is distance in miles, t is time in minutes, and the value 0.5 has units of miles per minute. If you substitute *any* number of minutes in this equation for time (including 44), the distance (in miles) will be exactly half of the time numerical value. This allows you to predict what would happen in the "real world" of cars and roads without having to actually go out and drive down the road to determine what would happen if you drove 30 miles per hour for 44 minutes.

Needless to say, the mathematical descriptions for some physical systems can be extremely complicated, such as models for the weather, global economic fluctuations, or the behavior of plasma in an experimental fusion reactor.

As it turns out, a significant number of phenomena important in engineering applications can be described mathematically with only three simple types of models. Also in Part 3, we introduce these three models and their characteristics, as well as discuss the use of Excel to determine a mathematical model from a set of data determined by experimentation.

A few notes about this part of the book:

- Within the examples given in this portion of the text, note that any information you are asked to type directly into Excel will be found in quotations. Do not type the quotation marks, type only the information found within the quotation marks.
- In hardcopy, the data needed to create a chart will be shown in columns or rows, depending on the size of the data, to efficiently use space and save a few trees by using less textbook paper. In the worksheets containing the starting data online, the data will be shown in columns.
- Files available online are indicated by the symbol ⊠ Excel.
- ✎ This symbol indicates directions for an important process to follow. Step-by-step instructions are given once for each procedure.
- ⌘ This symbol indicates special instructions for Mac OS users.

Excel Workbooks

The following is an example of the level of knowledge of Excel needed to proceed with this chapter. *If you are not able to quickly re-create an Excel worksheet similar to the one shown, including equations and formatting, please review worksheet basics in Appendix B online before proceeding.*

Begin with a new worksheet. Add correct header information (date, name, course, purpose/problem statement).

In row 5, add the following headers:

- Mass (m) [g]
- Height (H) [ft]
- Potential Energy (PE) [J]

- Time (t) [min]
- Power (P) [W]

Color the cells of row 5 the cell shade and font color of your choice.

Add the following data:

Mass [g]	Height [ft]	Time [min]
10	5	1
50	8	0.5
75	10	2.5

Calculate the corresponding potential energy and power terms in row 6. Be sure to watch your units!

Choose an appropriate (reasonable) way to display the data in terms of number format.

Copy the equations from row 6 down to row 8 using the fill handle.

Add a border to all cells in columns A–E, rows 5–8.

Center all the information within each column.

A sample worksheet follows.

	A	B	C	D	E
1	Date		Course - Section		Name
2	Purpose: This worksheet demonstrates the skills necessary to proceed with the Excel Workbooks chapter.				
3					
4					
5	Mass (m) [g]	Height (H) [ft]	Potential Energy (PE) [J]	Time (t) [min]	Power (P) [W]
6	10	5	0.15	1	0.002
7	50	8	1.20	0.5	0.040
8	75	10	2.24	2.5	0.015

10.1 Cell References

LEARN TO: Create an Excel worksheet that implements relative, absolute, and mixed cell addressing
Understand how formulas execute when written and copied using cell addressing

At some point prior to starting their first engineering class, many students have had some exposure to Microsoft Excel, whether used directly in a computer applications course in high school, or even just from an awareness perspective, seeing their parents, teachers, or other person in their life use Microsoft Excel to manage a budget, record grades in a classroom, or even to create an inventory of items for tax or insurance requirements. One of the basic building blocks of any worksheet application is the ability to build referential analytic tools so that calculations can be made quickly and dynamically across a large (or small) set of data. At the lowest level, the primary method for building out referential analysis in a worksheet application like Microsoft Excel is through **cell references**. Although being a self-defining term, cell references are essential for expressing how variables in an equation or formula (for example) can be expressed in a worksheet, as seen in Example 10-1.

EXAMPLE   10-1

Suppose we are given a list of *xy* coordinates in a worksheet. We want to calculate the distance between each point. We can find the distance between two *xy* coordinates by using Pythagoras's theorem:

$$d = \sqrt{(x_2 - x_1)^2 + (y_2 - y_1)^2}$$

	A	B	C	D	E	F	G	H	I
2	This example demonstrates how to handle Excel's order of operations and cell references.								
3									
4									
5		Point 1			Point 2				
6	X	Y		X	Y				
7	27	20		25	10				
8	25	4		7	8				
9	4	6		24	3				
10	25	26		13	24				
11	19	24		26	1				
12	29	10		0	5				
13	7	29		13	13				
14	3	20		19	16				
15	20	7		5	17				
16	20	26		19	3				
17	13	15		13	14				
18	23	22		17	25				
19	3	27		10	22				
20	30	16		30	17				
21									

To solve this problem, we must adhere to the default behavior of Excel to properly calculate the distance between the coordinates. First, we must observe the order of operations that Excel follows to determine how we need to write our equations. Second, we must determine how to use **cell references** to translate the x_2, x_1, y_2, and y_1 values in the equation shown above into locations in our worksheet.

Let us rewrite Pythagoras's theorem in the notation shown above using what we know about order of operations in Excel:

$$d = ((x_2 - x_1)\wedge2 + (y_2 - y_1)\wedge2)\wedge(1/2)$$

Let us calculate the distance between point 1 and point 2 in column G. In cell G7, we need to translate the equation into an equation that replaces the x_1, y_1, and x_2, y_2 variables with addresses to cells in the worksheet. Since each row represents a single calculation, we know that for the first data pair, x_1 is located in cell A7, y_1 is in B7, x_2 is in D7, and y_2 is in E7.

The equation we need to type into cell G7 becomes

$$= ((D7 - A7)\wedge2 + (E7 - B7)\wedge2)\wedge(1/2)$$

If we copy that equation down for the other pairs of xy coordinates, our sheet should now contain a column of all the distance calculations.

	A	B	C	D	E	F	G	H	I
1	**Distance Between XY Coordinates**								
2	This example demonstrates how to handle Excel's order of operations and cell references.								
3									
4									
5		**Point 1**			**Point 2**				
6	**X**	**Y**		**X**	**Y**		**Distance**		
7	27	20		25	10		10.20		
8	25	4		7	8		18.44		
9	4	6		24	3		20.22		
10	25	26		13	24		12.17		
11	19	24		26	1		24.04		
12	29	10		0	5		29.43		
13	7	29		13	13		17.09		
14	3	20		19	16		16.49		
15	20	7		5	17		18.03		
16	20	26		19	3		23.02		
17	13	15		13	14		1.00		
18	23	22		17	25		6.71		
19	3	27		10	22		8.60		
20	30	16		30	17		1.00		
21									

Suppose we start off with a slightly modified worksheet that requires us to calculate the distance between all the points in the first column of xy values to a single point in the second column.

*We can calculate the distance between all the points in the first column to the single point through the use of absolute addressing. An **absolute address** allows an equation to reference a single cell that will remain constant regardless of where the equation is copied in the worksheet. An absolute reference is indicated by a dollar sign ($) in front of the row and column designators. In this example, we want to use an absolute reference on cells D7 and E7 in all distance calculations. The equation we need to type in cell G7 becomes:*

$$= (($D$7 - A7)\wedge2 + ($E$7 - B7)\wedge2)\wedge(1/2)$$

	A	B	C	D	E	F	G	H	I
1	**Distance Between XY Coordinates**								
2	This example demonstrates how to handle Excel's order of operations and cell references.								
3									
4									
5	**Point 1**			**Point 2**					
6	**X**	**Y**		**X**	**Y**		**Distance**		
7	27	20		25	10		10.20		
8	25	4					6.00		
9	4	6					21.38		
10	25	26					16.00		
11	19	24					15.23		
12	29	10					4.00		
13	7	29					26.17		
14	3	20					24.17		
15	20	7					5.83		
16	20	26					16.76		
17	13	15					13.00		
18	23	22					12.17		
19	3	27					27.80		
20	30	16					7.81		
21									

Relative Addressing

- A **relative cell address** used in a formula will always *refer to the cell in the same relative position* to the cell containing the formula, no matter where the formula is copied in the worksheet. For example, if "=B2" is typed into cell C4 and then copied to cell C7, the formula in cell C7 would read "=B5". In this case, the cell reference is to call the cell two rows up and one cell to the left. Note that throughout this book and appendix materials, double quotes may surround expressions to be typed into cells (for example, "=B5"), but those double quotes should not actually be typed into the cell as they are only used to indicate the starting and stopping points of what should be typed into the cell.
- When we insert or change cells, the formulas automatically update. This is one of a worksheet's major advantages: easily applying the same calculation to many different sets of data.

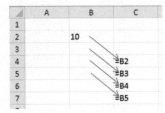

Relative Addressing

Relative Addressing

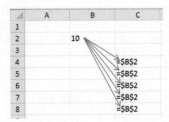

Absolute Addressing

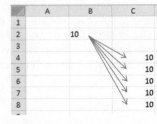

Absolute Addressing

Absolute Addressing

- Absolute addressing is indicated by the presence of a dollar sign ($) immediately before both the column and row designators in the formula (e.g., $C$5; $AB$10).
- An **absolute cell address** will *always refer to the same cell* if the formula is copied to another location. For example, if "=$B$2" is typed into cell C4 and then copied to cell C7, the formula in cell C7 would read "=$B$2".

Mixed Addressing

- In **mixed addressing**, *either the row or the column designator is fixed* (by the $), but the other is relative (e.g., $C5; AB$10; $AB10).
- It may not be immediately obvious why this capability is desirable, but many problems are dramatically simplified with this approach. We will study this in more detail later.

Comprehension Check   10-1

Type "5" in cell E22 and "13" in cell E23; type "=E22 + 4" in cell F22. Copy cell F22 to cell F23.

- Is this an example of absolute, mixed, or relative addressing?
- What is displayed in cell F23?

Comprehension Check   10-2

Type "45" into cell G22 and "=$G$22 + 10" in cell H22. Copy cell H22 down to row 26 using the fill handle.

- Is this an example of absolute, mixed, or relative addressing?
- What is displayed in cell H26?

Comprehension Check   10-3

Type "40" into cell A28 and "=A$28 + 10" in cell D28. Copy cell D28 down to row 30 using the fill handle. Copy cell D28 across to column F using the fill handle.

- Is this an example of absolute, mixed, or relative addressing?
- What is displayed in cell D30? What is displayed in cell F28?

Comprehension Check   10-4

Type "40" into cell A28 and "=$A28 + 5" in cell G28. Copy cell G28 down to row 30 using the fill handle. Copy cell G28 across to column J using the fill handle.

- Is this an example of absolute, mixed, or relative addressing?
- What is displayed in cell G30? What is displayed in cell J28?

10.2 Functions in Excel

LEARN TO: Properly use Excel functions, especially those listed in tables in this section
Understand limitations of certain functions, especially trig function arguments
Given an Excel equation with built-in functions, predict the output

Hundreds of functions are built into Excel. Tables 10-1 through Table 10-4 list a few functions commonly used in engineering applications. Table 10-5 contains common error messages you may encounter. There are several things you should note when using these functions.

- You must make certain to *use the correct name of the function.* For example, the average function is written as AVERAGE and cannot be abbreviated AVE or AVG.
- *All functions must be followed by parentheses.* For example, the value of π is given as PI(), with nothing inside the parentheses.
- The *argument* of the function (the stuff in the parentheses) can include numbers, text, expressions, or cell references, as long as they are appropriate for the function.
- Many functions can *accept a list or range of cells as the argument.* These can be expressed as a list separated by commas [e.g., A6, D7, R2, F9], as a rectangular block designated by the top-left cell and bottom-right cell separated by a colon [e.g., D3:F9], or as a mixed group [e.g., A6, R2, D3:F9]. To insert cells into a formula, type the formula up to the open parenthesis and select the desired cells. You can also type in the references directly into the formula.
- Most functions will also *accept another function as the argument.* These can be fairly simple [e.g., SIN(RADIANS(90))] or more complicated [e.g., AVERAGE(SQRT(R2), COS(S4 + C4), MIN(D3:F9) + 2)].
- Some functions, such as trigonometric functions, require specific arguments. *Trigonometric functions must have an argument in units of radians, not units of degrees.* Be sure you are aware of any limitations of the functions you are using. Look up an unfamiliar function in the **HELP** menu.
- Note that *some functions can be expressed in several different ways.* For example, raising the number 2 to the fifth power can be written as = 2 ^ 5 or as POWER(2,5).

Table 10-1 Trigonometric functions in Excel

Function as Written in Excel	Definition
ACOS (cell)	Calculates the inverse cosine of a number (also ASIN)
COS (angle in radians)	Calculates the cosine of an angle (also SIN)
DEGREES (angle in radians)	Converts radians to degrees
PI ()	Calculates pi (π) to about 15 significant figures
RADIANS (angle in degrees)	Converts degrees to radians

Table 10-2 Mathematical functions in Excel

Function as Written in Excel	Definition
EXP (cell)	Raises e (base of the natural log) to the power "cell" (inverse of LN)
POWER (cell, power)	Raises the cell to "power"
PRODUCT (cells)	Finds the product of a list of cells
SQRT (cell)	Finds the square root of cell
SUM (cells)	Finds the sum of a list of cells

Table 10-3 Statistical functions in Excel

Function as Written in Excel	Definition
AVERAGE (cells)	Finds the mean or average value of a list of cells
MAX (cells)	Finds the maximum value in a list of cells
MEDIAN (cells)	Finds the median value of a list of cells
MIN (cells)	Finds the minimum value in a list of cells
STDEV.P (cells)	Finds the standard deviation value of a list of cells
VAR.P (cells)	Finds the variance value of a list of cells

Table 10-4 Miscellaneous functions in Excel

Function as Written in Excel	Definition
COUNT (cells)	Counts number of cells that contain numeric or date values
COUNTA (cells)	Counts number of cells that are not blank and that do not contain an error, including text, date, numeric values
COUNTIF (cells, criteria)	Counts number of cells that meet the stated criteria, such as a numerical value, text, or a cell reference
COUNTIFS (cells1, criteria1, cells2, criteria2, . . .)	Counts number of cells that meet multiple stated criteria, such as a numerical value, text, or a cell reference
INTERCEPT (y values, x values)	Calculates linear line for range of (x, y) pairs and returns the intercept value of y (where $x = 0$)
ROUND (cell, number of decimal places)	Rounds a number to a specific number of decimal places
ROUNDUP (cell, number of decimal places)	Rounds a number up to a specific number of decimal places
ROUNDDOWN (cell, number of decimal places	Rounds a number down to a specific number of decimal places
SLOPE (y values, x values)	Calculates linear line for range of (x, y) pairs and returns the slope value
TRUNC (cell, number of digits)	Truncates a number to a specific number of digits

Table 10-5 Common error messages in Excel and possible solutions

Error	Explanation	Possible Fix	Example
#####	Column is not wide enough to display a number.	Make column wider.	−125,000,500 will not fit in a cell with a standard width.
#DIV/0!	Formula has resulted in division by zero.	Check values in denominator of formula contained in the cell.	If cell A1 contains 12 and cell A2 is empty, the formula = A1/A2 will return #DIV/0!

(continued)

Table 10-5 Common error messages in Excel and possible solutions (*continued*)

Error	Explanation	Possible Fix	Example
#NAME?	Excel does not recognize something you have typed.	Check spelling! Check operators for missing : Check for missing " " around text	Formula names: MXA should be MAX PI should be PI() Range of cells: A2B3 should be A2:B3
#NULL!	You specify a set of cells that do not intersect.	Check formulas for spaces, missing commas.	= SUM(A2:A5 B4:B6) will return this error; fix as = SUM(A2:A5,B4:B6)
#VALUE!	Formula contains invalid data types.	Arguments of functions must be numbers, not text Sometimes, part of a required function is missing; check for all required elements.	If cell A2 contains "2 grams" and cell A3 contains 3, the formula = A2 + A3 will result in this error since A2 is text (the word grams makes the cell text, not a number). = VLOOKUP(A2:B5,2,FALSE) will result in this error since a lookup function must contain four parts in the argument, not three.
#N/A	Formula has called a value that is not available.	Check for lookup value in data table (see Section 10.4).	If A2 contains 11, and the data table contains values 1 to 10 in the first column, this error will appear since the value 11 is not in the first column of the data table.
#REF!	Invalid cell reference.	Check operators for missing * or /. Check formula for data table size and number of column to return (see Section 10.4).	Operators: (A7)(B6) should be (A7)*(B6) = VLOOKUP(A2, A2:B5,3,FALSE) will return this error because there are not three columns available in the lookup table.
#NUM!	Formula results in invalid numeric values.	Check that the numerical result expected is between -1×10^{307} and 1×10^{307}.	If the calculation results in a value outside the range given, such as 2×10^{400}, this error will appear.

Handling Calculation Errors: IFERROR

Especially when dealing with worksheets that rely on user interaction to create meaningful information or analysis, there are often scenarios that will result in calculations that are not possible or might result in an error in a cell calculation. If you see cells in your worksheet that contain values like #DIV/0!, #N/A, or other messages that begin with the # symbol, that means Excel was not able to calculate or look up the expression typed into the cell. The IFERROR function will allow the programmer of an Excel worksheet to specify what value should appear in a cell if there is a calculation error in the worksheet. The IFERROR function is often used when dealing with lookup statements or iterative expressions where error messages in cells might throw off the intended result of the calculation. For example, if you type = A1/A2 into cell B1 and it results in #DIV/0! you could type the following instead:

$$= IFERROR(A1/A2,0)$$

This function will check to see if A1/A2 results in an error message. If it does not generate an error, the resulting value of A1/A2 will appear in the cell; otherwise the value 0 will appear in the cell. It is worth noting that "0" in the preceding formula can be replaced with any valid Excel commands, including function calls, conditional statements, lookup statements, or simply hardcoding a value like 0 as shown above. For example, all of the following are valid IFERROR expressions:

$$= IFERROR(A1/A2,MAX(A1,A2))$$
$$= IFERROR(IF(B2<3,A1/A2,B2),0)$$
$$= IFERROR(VLOOKUP(B2,A15:F20,3,FALSE),0)$$

In the final example, if the lookup value of B2 is not found in the table located in A15:F20, the formula will return the value 0 rather than the error message #N/A.

EXAMPLE   10-2

Assume we are studying the number of accidents that occur during different times of the day. Using the data given in the Excel workbook collected each week for two years, we want to use Excel to analyze our data to determine the average, minimum, or maximum number of accidents, as well as a few other items that might be of significance.

	A	B	C	D	E	F	G
1	Vehicular Accidents						
2	This worksheet demonstrates the proper use of Excel functions						
3							
4	Week	Number Accidents	Total Accidents		Samples Greater than Mean		
5	Y1 - 1	161	Total Samples				
6	Y1 - 2	209	Mean				
7	Y1 - 3	212	Median		Samples Between	180	200
8	Y1 - 4	62	Variance				
9	Y1 - 5	154	Standard Deviation				
10	Y1 - 6	68					
11	Y1 - 7	249					
12	Y1 - 8	33					
13	Y1 - 9	86					

NOTE

The ROUND function refers to number of decimal places, although the Excel help menu calls this "num_digits." Be sure to always read ALL the help menu file when using a new function.

Total accidents: = SUM(B5:B108)
Total samples: = COUNT(B5:B108)
Mean: = AVERAGE(B5:B108)
Median: = MEDIAN(B5:B108)
Variance: = VAR.P(B5:B108)
Standard deviation: = STDEV.P(B5:B108)

Note that decimal values appear when we calculate the mean, median, variance, and standard deviation of the accident data. Since it makes sense to round these values up to the nearest whole number, we need to type those functions as the argument to a rounding function. Start by modifying the equation for the mean by typing the **ROUND** function. Notice that as you start typing the ROUND function in the cell, a drop-down menu with a list of all of the functions that start with the letters ROUND *appears below the cell. Note that Excel contains a function called **ROUNDUP** that will round a number up to the nearest whole value away from zero.*

Total Accidents	15124	Samples Greater than Mean
Total Samples	104	
Mean	=round	
Median	(ROUND	Rounds a number to a specified number of digits	2(
Variance	(ROUNDDOWN	
Standard Deviation	(ROUNDUP	

After we select the ROUNDUP function, a new box below the cell documents the arguments the function requires. Note that we need to provide the value we want to round as the first argument and the number of decimal places to which we want to round the number (in this case, 0).

Total Accidents	15124	Samples Greater than Mean
Total Samples	104	
Mean	=roundup(	
Median	ROUNDUP(**number**, num_digits) Between	
Variance	4229.95562	
Standard Deviation	65.038109	

The new function we need to type ultimately becomes

= ROUNDUP(AVERAGE(B5:B108), 0)

Repeat this with the equations for calculating the median, variance, and the standard deviation.

Suppose we want to determine how many of the samples reported accidents greater than the calculated average number of accidents. Note that the **COUNTIF** function requires a "criteria" argument, which can take on a number of different values. For example, if we want to count the number of values greater than 200 in the range B5:B108, we need to type the criteria ">200" (in double quotes) as the second argument to the COUNTIF function.

= COUNTIF(B5:B108,">200")

*In this example, we want to compare our COUNTIF result to a value calculated in a different cell. Since we cannot type cell references inside of double quotes (">E21"), we need to use the **ampersand** operator (&) to **concatenate** the logical operator to the cell reference (">"&E21).*

Samples Greater than Mean: = COUNTIF(B5:B108,">"&D6)

Similarly, we could use the **COUNTIFS** function to calculate the number of samples that have a number of accidents between (and including) 180 and 200. COUNTIFS is a special function that contains a variable number of arguments, with a minimum of two arguments required (range1, criteria1) to use the function. Since we have two criteria that must be met (>180 and <200), we must pass in four arguments to the COUNTIFS function (range1, critera1, range2, criteria2). In this example, range1 and range2 must be the same range of cells since we are enforcing the criteria on the same set of data. We will place the bounds in the worksheet as follows:

Lower Bound in F7: 180   Upper Bound in G7: 200

Samples Between: = COUNTIFS(B5:B108,">="&F7, B5:B108, "<="&G7)

Your final worksheet should appear as shown.

	A	B	C	D	E	F	G
1	Vehicular Accidents						
2	This worksheet demonstrates the proper use of Excel functions						
3							
4	Week	Number Accidents	Total Accidents	15124	Samples Greater than Mean		
5	Y1 - 1	161	Total Samples	104	56		
6	Y1 - 2	209	Mean	146			
7	Y1 - 3	212	Median	152	Samples Between	180	200
8	Y1 - 4	62	Variance	4230	9		
9	Y1 - 5	154	Standard Deviation	66			
10	Y1 - 6	68					
11	Y1 - 7	249					
12	Y1 - 8	33					
13	Y1 - 9	86					

In Example 10-2, we touched on the concept of concatenating cell references to text to build out logic for use in any logical counting functions like COUNTIF or COUNTIFS. In-line concatenation using the **ampersand (&)** symbol is an important and recurring concept throughout the use of Microsoft Excel. In addition to using the ampersand for concatenation, Excel also provides a built-in function named **CONCAT** that accepts a range of cells or values as an argument and returns the result as text. For example, the expression = **CONCAT("Excel"," ","is"," ","Excellent")** would display the text **Excel is Excellent** in the cell, which would be equivalent to typing = **"Excel"&" "&"is"&" "&"Excellent"** in the same cell. The use of a single character to replace the use of an entire function (like the use of the ampersand in place of the function CONCAT) is referred to as a **short-cut operator**.

Comprehension Check   10-5

Launch a new worksheet. Type the following Excel expressions into the specified cells. Be certain you understand *why* each of the following yields the specific result. Note that not all functions shown in this table are valid Excel functions. If the formula returns an error, how can the formula be changed to correctly display the desired result?

In Cell . . .	Enter the Formula . . .	The Cell Will Display . . .
A1	= SQRT(169)	
A2	= MAX(5, 8, 20/2, 5 + 7)	
A3	= AVERAGE(15, SQRT(400), 25)	
A4	= POWER(2, 5)	
A5	= PI()	
A6	= PI	
A7	= PRODUCT(2, 5, A2)	
A8	= SUM(2 + 7, 3 * 2, A1:A3)	
A9	= RADIANS(90)	
A10	= SIN(RADIANS(90))	
A11	= SIN(90)	
A12	= ACOS(0.7071)	
A13	= DEGREES(ACOS(0.7071))	
A14	= CUBRT(27)	

EXAMPLE 10-3

Video Solutions   Excel

The maximum height (H) an object can achieve when thrown can be determined from the velocity (v) and the launch angle with respect to the horizontal (θ):

$$H = \frac{v^2 \sin (\theta)}{2g}$$

Note the use of a cell (E7) to hold the value of the acceleration due to gravity. This cell will be referenced in the formulae instead of our inserting the actual value into the formulae. This will allow us to easily work the problem in a different gravitational environment (e.g., Mars) simply by changing the one cell containing the gravitational constant.

	A	B	C	D	E	F	G
1	Basic Examples of Trig Functions and Cell Addressing						
2	The following data is used to illustrate built-in trig functions and mixed references						
3							
4							
5							
6							
7	Planet	Earth		Gravity (g)	9.8	[m/s²]	
8							
9	Velocity	Angle (θ) [degrees]					
10	(v)[m/s]	50	60	70	80		
11	10						
12	12						
13	14						
14	16						
15	18						
16	20						
17							

For the following, assume that the angle 50° is in cell B10. After setting up the column of velocities and the row of angles, we type the following into cell B11 (immediately below 50°)

=$A11^2 * SIN(RADIANS(B$10))/(2*$E$7)

Note the use of absolute addressing (for gravity) and mixed addressing (for angle and velocity). For the angle, we allow the column to change (since the angles are in different columns) but not the row (since all angles are in row 10). For the velocity, we allow the row to change (since the velocities are in different rows) but the column is fixed (since all velocities are in column A). This allows us to write a single formula and replicate it in both directions.

The sine function requires an argument in units of radians, and the angle is given in units of degrees in the problem statement. In this example, we used the RADIANS function to convert from degrees into radians. Another method is to use the relationship 2π radians is equal to 360 degrees, or

=$A11^2 * SIN((2 * PI()/360) * B$10)/(2*$E$7)

We replicate the formula in cell B11 across the row to cell E11, selecting all four formulae in row 11 and replicating to row 16. If done correctly, the values should appear as shown.

Velocity (v) [m/s]	Angle (θ) [°]			
	50	60	70	80
20	3.91	4.42	4.79	5.02
12	5.63	6.36	6.90	7.24
14	7.66	8.66	9.40	9.85
16	10.01	11.31	12.27	12.86
18	12.66	14.32	15.53	16.28
20	15.63	17.67	19.18	20.10

Here, we consider the planet to be Mars with a gravity of 3.7 meters per second squared in cell E7. The worksheet should automatically update, and the values should appear as shown.

Velocity (v) [m/s]	Angle (θ) [°]			
	50	60	70	80
10	10.35	11.70	12.70	13.31
12	14.91	16.85	18.29	19.16
14	20.29	22.94	24.89	26.08
16	26.50	29.96	32.51	34.07
18	33.54	37.92	41.14	43.12
20	41.41	46.81	50.79	53.23

Now, we consider the planet to be moon with a gravity of 1.6 meters per second squared in cell E7. The worksheet should automatically update, and the values should appear as shown.

Velocity (v) [m/s]	Angle (θ) [°]			
	50	60	70	80
10	23.94	27.06	29.37	30.78
12	34.47	38.97	42.29	44.32
14	46.92	53.04	57.56	60.32
16	61.28	69.28	75.18	78.78
18	77.56	87.69	95.14	99.71
20	95.76	108.25	117.46	123.10

Comprehension Check   10-6

As part of the design of a high-performance engine, you are analyzing properties of spherical ceramic ball bearings. Since many ceramic materials are considerably less dense than the metals typically used in such applications, the centrifugal load added by the bearings can be significantly reduced by the use of ceramics.

	A	B	C	D	E	F	G
1							
2		Mass of Ball Bearings (m) [g]					
3		Specific Gravity					
4	Radius (r) [cm]	3.12	3.18	3.22	3.31	3.37	
5	1.00	13.1	13.3	13.5	13.9	14.1	
6	1.05	15.1	15.4	15.6	16.1	16.3	
7	1.10	17.4	17.7	18.0	18.5	18.8	
8	1.15	19.9	20.3	20.5	21.1	21.5	
9	1.20	22.6	23.0	23.3	24.0	24.4	
10	1.25	25.5	26.0	26.3	27.1	27.6	
11	1.30	28.7	29.3	29.6	30.5	31.0	

Which of the following could be typed in cell B5 and copied across to cell F5, then down to cell F11 to calculate the masses of the various ball bearings shown in the table? If more than one answer is correct, indicate all that apply.

A. = 4/3 * PI * $A5^3 * $B4
B. = 4/3 * PI() * $A5^3 * B$4
C. = 4/3 * PI() * $A$5^3 * $B$4
D. = 4/3 * PI * A$5^3 * B$4
E. = 4/3 * PI() * A5^3 * B4

10.3 Logic and Conditionals

LEARN TO: Create IF statements in Excel to create conditional results
Generate compound logic to develop complex conditions
Predict the output of an IF statement

Outside of the realm of computing, logic exists as a driving force for decision making. Logic transforms a list of arguments into outcomes based on a decision.

Arguments ⟶ Decision ⟶ Outcomes

Some examples of everyday decision making:

- If the traffic light is red, stop. If the traffic light is yellow, slow down. If the traffic light is green, go.

Argument	Decision	Outcomes
three traffic bulbs	is bulb lit?	stop, slow, go

- If the milk has passed the expiration date, throw it out; otherwise, keep the milk

Argument	Decision	Outcomes
expiration date	before or after?	garbage, keep

To bring decision making into our perspective on problem solving, we need to first understand how computers make decisions. **Boolean logic** exists to assist in the decision-making process, where each argument has a binary result and our overall outcome exhibits binary behavior. **Binary behavior**, depending on the application, is any sort of behavior that results in two possible outcomes.

In computing, we often refer to the outcome of Boolean calculations as "yes" and "no." Alternatively, we may refer to the outcomes as "true" and "false," or "1" and "0."

To determine the relationship between two cells (containing numbers or text), we have a few operators, listed in Table 10-6, that allow us to compare two cells to determine whether or not the comparison is true or false.

Table 10-6 Relational operators in Excel

Operator	Meaning
>	Greater than
<	Less than
> =	Greater than or equal to
< =	Less than or equal to
=	Equal to
<>	Not equal to

These relational operators are usually placed between two different cells to determine the relationship between the two values. This expression of **cell–operator–cell** is typically called a **relational expression**. If more than two relational expressions are needed to form a decision, relational expressions can be combined by means of logical operators to create a **logical expression.** To connect the Boolean arguments to make a logical decision, we have a few logical operators that allow us to relate our arguments to determine a final outcome.

- **AND:** The AND logical operator enables us to connect two Boolean arguments and return the result as TRUE if and only if *both* Boolean arguments have the value of TRUE. In Excel, the AND function accepts more than two arguments and is TRUE if all the arguments are TRUE.
- **OR:** The OR logical operator enables us to connect two Boolean arguments and return the result as TRUE if *only one* of the Boolean arguments has the value of TRUE. In Excel, the OR function accepts two or more arguments and is TRUE if at least one of the arguments is TRUE.
- **NOT:** The NOT logical operator enables us to invert the result of a Boolean operation. In Excel, the NOT function accepts one argument. If the value of that argument is TRUE, the NOT function returns FALSE. Likewise, if the argument of the function is FALSE, the NOT function returns TRUE.

Conditional statements are commands that give some decision-making authority to the computer. Specifically, the user asks the computer a question using conditional statements, and then the computer selects a path forward based on the answer to the question. Some sample statements follow.

- If the water velocity is fast enough, switch to an equation for turbulent flow!
- If the temperature is high enough, reduce the allowable stress on this steel beam!

- If the RPM level is above red line, issue a warning!
- If your grade is high enough on the test, state: You Passed!

In these examples, the comma indicates the separation of the condition and the action that is to be taken if the condition is true. The exclamation point marks the end of the statement. Just as in language, more complex conditional statements can be crafted with the use of "else" and "otherwise" and similar words. In these statements, the use of a semicolon introduces a new conditional clause, known as a nested conditional statement. For example:

- If the collected data indicate the process is in control, continue taking data; otherwise, alert the operator.
- If the water temperature is at or less than 10 degrees Celsius, turn on the heater; or else if the water temperature is at or greater than 80 degrees Celsius, turn on the chiller; otherwise, take no action.

Single Conditional Statements

In Excel, conditional statements can be used to return a value within a cell based upon specified criteria. The IF conditional statement within Excel takes the form

$$= IF(\text{logical test, value if true, value if false})$$

Every statement must contain three and only three parts:

1. **A logical test, or the question to be answered**
   The answer to the logical test must be TRUE or FALSE.
   Is the flow rate in Reactor #1 higher than Reactor #5?

2. **A TRUE response,** if the answer to the question is yes
   Show the number 1 to indicate Reactor #1.

3. **A FALSE response,** if the answer to the question is no
   Show the number 5 to indicate Reactor #5.

The whole statement for the preceding example would read:

$$= IF(B3 > B4, 1, 5)$$

	A	B	C
1			
2			
3	Reactor #1 Flowrate	10	[gpm]
4	Reactor #5 Flowrate	25	[gpm]
5	Maximum Flowrate in Reactor #	5	

Special Things to Note

- **To leave a cell blank, type a set of quotations with nothing in between ("").** For example, the statement = IF(C3 > 10, 5, "") is blank if C3 is less than 10.
- **For display of a text statement, the text must be stated within quotes** (*"text goes in here"*). For example, the statement = IF(E5 > 10, 5, "WARNING") would display the word WARNING if E5 is less than 10.

EXAMPLE 10-4

For the following scenarios, write a conditional statement to be placed in cell B5 to satisfy the conditions given. Following each statement are sample outcomes of the worksheet in different scenarios.

(a) Display the pressure difference between upstream station 1 (displayed in cell B3) and downstream station 2 (displayed in cell B4) if the pressure difference is positive; otherwise, display the number 1.

	A	B	C
1			
2			
3	Station #1 Pressure	2.4	[atm]
4	Station #2 Pressure	2.8	[atm]
5	Pressure Difference	1	[atm]

	A	B	C
1			
2			
3	Station #1 Pressure	3.2	[atm]
4	Station #2 Pressure	2.8	[atm]
5	Pressure Difference	0.4	[atm]

Answer: = IF((B3 > B4) > 0, B3 − B4, 1)

(b) Display the value of the current tank pressure if the current pressure is less than the maximum tank pressure; otherwise, display the word "MAX".

	A	B	C
1			
2			
3	Maximum Tank Pressure	5	[atm]
4	Current Tank Pressure	2	[atm]
5	Pressure Status	2	[atm]

	A	B	C
1			
2			
3	Maximum Tank Pressure	5	[atm]
4	Current Tank Pressure	10	[atm]
5	Pressure Status	MAX	[atm]

Answer: = IF(B3 > B4, B4, "MAX")

(c) If the sum of the temperature values shown in cells B2, B3, and B4 is greater than or equal to 100, leave the cell blank; otherwise, display a warning to the operator that the temperature is too low.

	A	B	C
1			
2	Temperature Reading #1	25	[°C]
3	Temperature Reading #2	50	[°C]
4	Temperature Reading #3	45	[°C]
5	Cumulative Temperature		

	A	B	C
1			
2	Temperature Reading #1	25	[°C]
3	Temperature Reading #2	10	[°C]
4	Temperature Reading #3	45	[°C]
5	Cumulative Temperature	Too Low	

Answer: = IF(SUM(B2:B4) > = 100, "", "Too Low")

Comprehension Check | **10-7**

Evaluate the following expressions. What is the final results that would occur when the formula is evaluated using the worksheet shown?
Comparison A: = IF(B2 + B3 <= 2*B9, B3 + B4, MIN(B2:B9))
Comparison B: = IF(B5 > B6, B7, "")
Comparison C: = IF(B9 <> B8, "B9", B9/B8)

	A	B	C	D	E	F
1						
2	Value 1	4				
3	Value 2	13			Comparison A	
4	Value 3	19			Comparison B	
5	Value 4	18			Comparison C	
6	Value 5	21				
7	Value 6	10				
8	Value 7	6				
9	Value 8	17				

Nested Conditional Statements

If more than two outcomes exist, the conditional statements in Excel can be nested. The nested IF conditional statement within Excel can take the form

---

= IF(logical test #1, value if #1 true, IF(logical test #2, value if #2 true,
value if both false))

Note that the number of parentheses must match (open and closed) and must be placed in the proper location. Recall that every statement must contain three and only three parts. For the first IF statement, they are:

1. **The first logical test, or the first question to be answered**
   The answer to the logical test must be TRUE or FALSE.
   Is the score for Quiz #1 less than the score for Quiz #2?

2. **A true response**, or what to do if the answer to the first question is yes
   Show the score for Quiz #1.

3. **A false response**, or what to do if the answer to the first question is no
   Proceed to the logical question for the second IF statement.

For the second IF statement, the three parts are:

1. **The second logical test, or the second question to be answered**
   The answer to the logical test must be TRUE or FALSE.
   Is the score for Quiz #2 less than the score for Quiz #1?

2. **A true response**, or what to do if the answer to the second question is yes
   Show the score for Quiz #2.

3. **A false response**, or what to do if the answer to the second question, and by default both questions, is no
   Show the text "Equal".

The whole statement typed in cell B5 for the above example would read

$$= IF(B3 < B4, B3, IF(B3 > B4, B4, \text{"Equal"}))$$

	A	B	C
1			
2			
3	Quiz Grade #1	70	
4	Quiz Grade #2	70	
5	Lowest Quiz Score	Equal	

	A	B	C
1			
2			
3	Quiz Grade #1	90	
4	Quiz Grade #2	70	
5	Lowest Quiz Score	70	

	A	B	C
1			
2			
3	Quiz Grade #1	50	
4	Quiz Grade #2	70	
5	Lowest Quiz Score	50	

There can be a maximum of 64 nested IF statements within a single cell. The nested IF can appear as either the true or false response to the first IF logical test. In the preceding example, only the false response option is shown.

EXAMPLE 10-5

Write the conditional statement to display the state of water (ice, liquid, or steam) based upon temperature displayed in cell B4, given in degrees Celsius. Following are sample outcomes of the worksheet in different scenarios.

	A	B	C
1			
2			
3			
4	Temperature of Mixture	75	[°C]
5	State of Mixture	Liquid	

	A	B	C
1			
2			
3			
4	Temperature of Mixture	110	[°C]
5	State of Mixture	Steam	

	A	B	C
1			
2			
3			
4	Temperature of Mixture	10	[°C]
5	State of Mixture	Ice	

Here, there must be two conditional statements because there are three responses:

- *If the temperature is less than or equal to zero, display "Ice";*
- *If the temperature is greater than or equal to 100, display "Steam";*
- *Otherwise, display "Liquid".*

Answer: $= IF(B4 <= 0, \text{"Ice"}, IF(B4 >= 100, \text{"Steam"}, \text{"Liquid"}))$

Comprehension Check   10-8

Continue the example in Comprehension Check 10-6. The following is typed into cell G5, then copied down to cell G11:

$$= \text{IF}(\text{MAX}(B5:F5) > \text{AVERAGE}(\$D\$5:\$D\$11), \text{IF}$$
$$\text{MIN}(B5:F5) > \text{AVERAGE}(\$D\$5:\$D\$11),\text{"X"},\text{"Z"}),\text{"Y"})$$

(a) Which of the following will appear in cell G9?
(b) Which of the following will appear in cell G8?
(c) Which of the following will appear in cell G7?

Choose from:

A. X
B. Y
C. Z
D. An error message will appear
E. The cell will be blank

	A	B	C	D	E	F	G
1							
2		Mass of Ball Bearings (m) [g]					
3		Specific Gravity					
4	Radius (r) [cm]	3.12	3.18	3.22	3.31	3.37	
5	1.00	13.1	13.3	13.5	13.9	14.1	
6	1.05	15.1	15.4	15.6	16.1	16.3	
7	1.10	17.4	17.7	18.0	18.5	18.8	
8	1.15	19.9	20.3	20.5	21.1	21.5	
9	1.20	22.6	23.0	23.3	24.0	24.4	
10	1.25	25.5	26.0	26.3	27.1	27.6	
11	1.30	28.7	29.3	29.6	30.5	31.0	

Simplifying Nested Conditionals for Range Assessments

In situations where you want to design some sort of conditional expression that slices up a range of numbers into different categories of information, depending on the number of categories, it could lead to very large and tedious nested IF statements, as described in the previous section. For example, if we wanted to categorize the number of errors recorded by a computer application as "Urgent Fix Needed," "Critical," "Important," "Somewhat Important," "Less Important," and so on, it's common to cringe at the thought of building a large chained logical expression to categorize this information. This pain point is why Microsoft introduced the IFS function in Microsoft Excel 2016. To demonstrate the functionality of IFS, we will start by quantifying the computer application error count scenario referenced above, demonstrate the solution of the problem using the standard approach of nesting IF statements, and then finally show the simplification of using the IFS function for chained logic. Let's assume the following category mapping of computer application errors: "Urgent Fix Needed" is anything more than 100 errors, "Critical" is more than 75 but less than or equal to 100 errors, "Important" is more than 25 but less than or equal to 75 errors, "Somewhat Important" is more than 15 but less than or equal to 25 errors, "Less Important" is more than 5 but less than or equal to 15 errors; otherwise the categorization is "Acceptable". Assume that cell A5 contains the number of computer application errors. This logical expression with nested IF statements looks like the following:

=IF(A5>100,"Urgent Fix Needed", IF(A5>75,"Critical", IF(A5>25,"Important", IF(A5>15,"Somewhat Important", IF(A5>5,"Less Important", "Acceptable")))))

As promised, this expression is long, accident prone, and contains a disorienting number of trailing closing parenthesis symbols at the end of the expression. The IFS conditional statement within Excel 2016 can take the form

---

=IF(logical test #1, value if #1 true, logical test #2, value if #2 true, . . .)

Unlike standard nested IF statements, the IFS function can accept 127 logical tests, rather than the limitation of 64 nested IF statements. Unlike the IF statement, there is no "value if false," so the recommended approach for "else" scenarios is to include a logical test that is always true (TRUE, 1=1, etc.) at the end with "else" value. This is important because the IFS function will return the "#N/A" error if no logical test in the IFS expression evaluates to true. To demonstrate the syntax of the IFS function, the computer application error categorization expression rewritten using the IFS function is simplified to the following expression:

=IFS(A5>100,"Urgent Fix Needed", A5>75,"Critical", A5>25,"Important", A5>15,"Somewhat Important", A5>5,"Less Important", 1=1,"Acceptable")

Note that the preceding expression, compared to the logical equivalent written with nested IF statements, contains fewer characters, fewer parentheses, and is significantly easier to read and quickly troubleshoot.

Compound Conditional Statements

If more than two logic tests exist for a single condition, conditional statements can be linked together by AND, OR, and NOT functions. Up to 255 logical tests can be compared in a single IF statement (only two are shown in the following box). The compound IF conditional statement takes the form

---

= IF(AND(logical test #1, logical test #2), value if both tests are true, value if either test is false)

= IF(OR(logical test #1, logical test #2), value if either test is true, value if both tests are false)

EXAMPLE 10-6

Write the conditional statement that meets the following criteria:

(a) If the product has cleared all three quality checks (given in cells B2, B3, and B4) with a score of 80 or more on each check, mark the product as "OK" to ship; otherwise, mark the product as "Recycle."

	A	B	C
1			
2	Quality Check #1 Rating	90	
3	Quality Check #2 Rating	80	
4	Quality Check #3 Rating	85	
5	Mark Product	OK	

	A	B	C
1			
2	Quality Check #1 Rating	60	
3	Quality Check #2 Rating	80	
4	Quality Check #3 Rating	85	
5	Mark Product	Recycle	

Answer: = IF(AND(B2 >= 80, B3 >= 80, B4 >= 80), "OK", "Recycle")

(b) If the product has cleared all three quality checks (given in cells B2, B3, and B4) with a minimum score of 80 on each check, mark the product as "OK" to ship; otherwise, if the product scored a 50 or below on any check, mark the product as "Rejected"; otherwise, mark the product as "Rework."

	A	B	C
1			
2	Quality Check #1 Rating	90	
3	Quality Check #2 Rating	80	
4	Quality Check #3 Rating	85	
5	Mark Product	OK	

	A	B	C
1			
2	Quality Check #1 Rating	40	
3	Quality Check #2 Rating	80	
4	Quality Check #3 Rating	85	
5	Mark Product	Rejected	

	A	B	C
1			
2	Quality Check #1 Rating	60	
3	Quality Check #2 Rating	80	
4	Quality Check #3 Rating	85	
5	Mark Product	Rework	

Answer: = IF(AND(B2 >= 80, B3 >= 80, B4 >= 80), "OK", IF(OR(B2 <= 50, B3 <= 50, B4 <= 50), "Rejected", "Rework"))

Comprehension Check 10-9

Continue the example in Comprehension Check 10-6 above. Which of the following could be typed in cell H7 that will result in OK appearing in H7 if the mass in cell C7 is between 27.5 grams and 30 grams inclusive, but leave H7 blank otherwise? If more than one answer is correct, check all that apply.

A. =IF(C7<27.5,"", IF(C7>30,"","OK"))
B. =IF(C7<27.5,"OK", IF(C7>30,"OK",""))
C. =IF(C7<27.5 OR C7>30),"","OK")
D. =IF(OR(C7<27.5,C7>30),"","OK")
E. =(IF(C7<27.5) OR IF(C7>30),"","OK")

	A	B	C	D	E	F	G
1							
2		Mass of Ball Bearings (m) [g]					
3		Specific Gravity					
4	Radius (r) [cm]	3.12	3.18	3.22	3.31	3.37	
5	1.00	13.1	13.3	13.5	13.9	14.1	
6	1.05	15.1	15.4	15.6	16.1	16.3	
7	1.10	17.4	17.7	18.0	18.5	18.8	
8	1.15	19.9	20.3	20.5	21.1	21.5	
9	1.20	22.6	23.0	23.3	24.0	24.4	
10	1.25	25.5	26.0	26.3	27.1	27.6	
11	1.30	28.7	29.3	29.6	30.5	31.0	

10.4 Lookup and Data Validation

LEARN TO: Use a lookup function to merge data given two data tables with at least one common field
Predict the output if given a lookup statement
Create a validation protocol for data

The lookup function enables Excel to locate information from a table of data in a worksheet. There are two lookup functions: VLOOKUP, which searches vertically, and HLOOKUP, which searches horizontally. In the following example, we focus on VLOOKUP, but the same principles could easily be applied to HLOOKUP. To use the VLOOKUP function, we need to pass in four different arguments:

> VLOOKUP(lookup_value, table_array, col_index_num, [range_lookup])

- The *lookup_value* argument is the value we want to look up in the table. Typically, this value is a string, but it can be a numerical value. Note that whatever we use as the *lookup_value*, Excel will perform a case-insensitive search of the data for the value, which means that any special characters used in the string, like punctuation or spaces, must appear the same in the *lookup_value* and the table, and must be a unique identifier in the first column of the table.
- The *table_array* is the range of cells that encapsulates the entire data table we want to search. Since we are using VLOOKUP, it is important to realize that our *table_array* must have at least two columns of data. Note that the *lookup_value* we are passing in to the VLOOKUP function will only search the first column of the *table_array*, so it might be necessary to move the data around.
- The *col_index_num* argument is the column number that contains the data we want as a result of our search. By default, Excel will refer to the first column where the *lookup_value* is located as the number 1, so the *col_index_num* will typically be a number greater than 1.
- The last argument, [*range_lookup*], is an optional argument as indicated by the square brackets. This argument tells the function what type of search to perform and can only take on two values: TRUE or FALSE. In most cases, you will want to list this as FALSE.

- Passing in TRUE tells Excel to conduct an approximate search of the data. That is, Excel will search the data table for the largest value that is less than the *lookup_value* and use that result as the selected value. Note that for an approximate search, the first column of the *table_array* must be sorted in ascending order.
- Passing in FALSE tells Excel to conduct an exact search of the data. The data need not be sorted for this option. If an exact match is not found, the function returns an error.
- If we do not specify TRUE or FALSE, Excel attempts to match the data exactly, and if a match is not found, Excel returns an approximate value. This may give undesired results. It is good practice to tell Excel which searching algorithm to use to search the *table_array*.

Assume we are given the following table of data on students. To determine what Sally's eye color is from (column C) in cell A5, we could type

= VLOOKUP("Sally", A1:D4, 3, FALSE)

since the data are unsorted and we are looking for an exact match on Sally.

	A	B	C	D
1	Joe	18	Blue	EE
2	John	19	Brown	ME
3	Sally	18	Brown	IE
4	Julie	18	Blue	CE

One motivation for using lookup expressions with functions like VLOOKUP is to reduce the need for unnecessary retyping or copying data from the original source. Using data from a primary source to transform data in order to solve a problem helps improve data quality and can reduce errors in calculations or typos that occur during manual transposition of data. Another mechanism for improving data quality is the use of **data validation** tools in Excel worksheets. The data validation tools available in Microsoft Excel, as demonstrated in the second part of Example 10-7, allow worksheets to be created with requirements surrounding the values typed into specific cells. For example, cells can be created that restrict cell values to contain a range of values or restrict values to a selection of one item from a list of items. Similarly, the data validation features of Microsoft Excel can allow for custom input or error messages to display as pop-up messages to provide additional feedback when conditions are triggered based on cell values.

EXAMPLE   10-7

Digital audio is a relatively new medium for storing and reproducing music. Before albums were sold on CD and other digital media formats, analog recordings were commonly sold as vinyl records, eight-track tapes, and cassette tapes. We want to build a worksheet to help us compare these different media formats to observe how information storage has progressed over the past 50 years. Note the following media equivalencies:

- A 74-minute CD (44.1 kilohertz, 2 channel, 16-bit digital audio) can hold 650 MB of data.
- A single-sided, single-layer DVD can hold 4.7 GB of data (~4813 MB, 547 minutes of 44.1 kilohertz, two-channel, 16-bit digital audio).
- A single-sided, single-layer Blu-ray disc can hold 25 GB of data (~25,600 MB, 2914 minutes of 44.1 kilohertz, two-channel, 16-bit digital audio).
- A 7-inch vinyl record recorded at 45 rpm can hold 9 minutes of music.
- A 7-inch vinyl record recorded at $33\frac{1}{3}$ rpm can hold 12 minutes of music.
- A 12-inch vinyl record recorded at 45 rpm can hold 24 minutes of music.
- A 12-inch vinyl record recorded at $33\frac{1}{3}$ rpm can hold 36 minutes of music.
- An eight-track tape can hold 46 minutes of music.
- A typical cassette tape can hold 60 minutes of music.

To determine audio equivalencies between these different storage formats, we first create a worksheet. We want to allow the user to input the media type and quantity of the desired format to be converted. To complete the comparison, it would seem like each calculation requires a statement with nine questions to ask (Is it a CD? Is it a DVD? Is it a Blu-ray? . . .).

	A	B	C	D	E
1	**Digital Audio Media**				
2	This worksheet demonstrates the use of VLOOKUP and data validation				
3					
4					
5	Quantity	Format	*is equivalent to*	Quantity	Format
6					CD
7					DVD
8					Blu-ray Disc
9	**Storage Information**				7" @ 45 rpm
10	Format	Length [min]			7" @ 33 1/3 rpm
11	CD	74			12" @ 45 rpm
12	DVD	547			12" @ 33 1/3 rpm
13	Blu-ray Disc	2914			8-track tape
14	7" @ 45 rpm	9			Cassette tape
15	7" @ 33 1/3 rpm	12			
16	12" @ 45 rpm	24			
17	12" @ 33 1/3 rpm	36			
18	8-track tape	46			
19	Cassette tape	60			

Rather than requiring the user to type the name of the media each time (CD, DVD, Blu-ray, etc.), Excel can do **data validation**, *so we can give the user of our worksheet a drop-down menu from which to select the media. We need to add a table that contains the name of each media type along with the length of the audio we can fit on each media. We will place this table below our initial data, in cells A10:B19.*

Next, we need to calculate the quantity of each item. Since the name of the media will appear in cell B6, we use that as the lookup value in our VLOOKUP statement. To calculate the quantity for each equivalent media, we look up the length of the format specified in B6, divide that by the length of each media given in column E, and multiply that by the number of the original media provided in A6. Note that we need to round this number up since it does not make sense to have a noninteger value in our count.

For CDs, the calculation in cell D6 should be

$$= \text{ROUNDUP(VLOOKUP(\$B\$6, \$A\$11:\$B\$19, 2, FALSE)/}$$
$$\text{VLOOKUP(E6, \$A\$11:\$B\$19, 2, FALSE) *\$A\$6, 0)}$$

The next step to finish our worksheet is to include a drop-down menu of the different media formats. To insert data validation on the media format, we click cell B6 and go to **Data > Data Tools > Data Validation**.

The **Data Validation** *window is displayed. Under the* **Settings** *tab, the* **Allow:** *menu lets us specify the type of data that can be provided in the cell we selected. Since we want to restrict the data to a list of values, we select* **List**.

Under the **Source:** option, we select the range of all of the media types, $A$11:$A$19, and click **OK** to close the Data Validation window.

Notice the drop-down handle next to cell B6. When the user of the worksheet clicks B6, a drop-down menu appears that lists all of the possible media types so that the user can quickly select an item from the list. Furthermore, this feature prevents the user from typing items that are not on the list, making a typo, or entering any other information that will cause an error in calculations that rely on the value in B6.

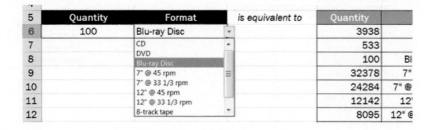

In addition to controlling the input type to a cell, it is also possible to give feedback to the person using the worksheet using pop-up messages. In this example, the quantity cannot be a negative number, so we need to bring up the Data Validation window again and restrict the input to only allow whole numbers that are greater than or equal to zero.

Next, we need to click the Input Message tab to type in a message that will appear below the cell when the person using our worksheet clicks on the cell to type in a quantity.

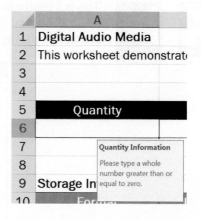

Finally, we need to click the Error Alert tab to provide the message that should pop up when an invalid number is typed into the cell.

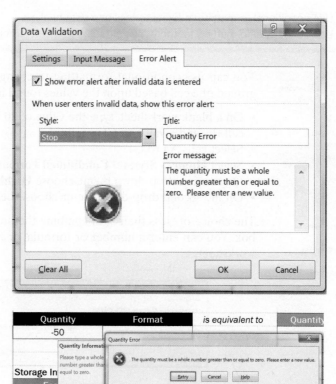

Comprehension Check | 10-10

This is a continuation of the worksheet you created in Example 10-3. Modify it to use VLOOKUP and data validation to allow the user of the worksheet to select the planet and automatically fill in the gravity for each planet.

Planet	Gravity (g) [m/s^2]
Earth	9.8
Jupiter	24.8
Mars	3.7
Mercury	3.7
Moon	1.6
Neptune	11.2
Pluto	0.7
Saturn	10.4
Uranus	8.9
Venus	8.9

10.5 Conditional Formatting

LEARN TO: Use conditional formatting in Excel to facilitate data analysis
Use conditional formatting to apply multiple rules to create compound logic analysis

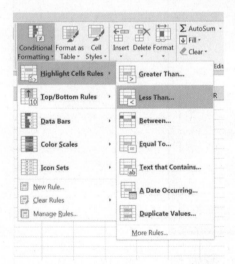

You can use conditional formatting to change the font color or background of a cell based upon the values found in that cell. As an example:

- On a blank worksheet, type the value of 20 in cell A4, a value of 30 in cell B4, and a value of 50 in cell C4.
- Select cells A4 to C4.
- Select **Home > Styles > Conditional Formatting**.
- On the first drop-down menu, choose **Highlight Cells Rules**.
- On the second drop-down menu, choose **Less Than**.

The choice of "less than" will combine the next two boxes into a single box. You can enter a number or formula, or reference a cell within the worksheet.

For this example, enter the value "25." Note: If you enter a formula, the same rules apply for absolute and relative referencing. In addition, if you select a cell within the worksheet, the program automatically defaults to an absolute reference.

- Select the formatting you want to apply when the cell value meets the condition or the formula returns the value TRUE using the drop-down menu shown after the word "with." The default is set to "Light Red Fill with Dark Red Text." You can change the font, border, or background of the cell using the **Custom Format** option. For this example, choose a green background on the Fill tab. When you are finished, click **OK**.
- To add another condition, simply repeat the process. As another example, make it greater than 40, with a font of white, bolded on a red background.

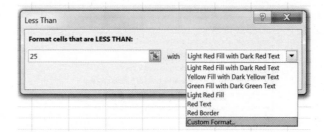

Your worksheet should now look like the one shown. If none of the specified conditions are TRUE, the cells keep their existing formats.

	A	B	C
1			
2			
3			
4	20	30	50
5			
6			

EXAMPLE 10-8

When building a computer network designed to run engineering systems that are used for design or manufacturing purposes, an important area of concern is the overall performance and health of the physical infrastructure. It could be catastrophic for a company that relies on technology to suffer from technology outages. To prevent these types of events from occurring, many companies rely on proactive measures to monitor the health of their network and continually measure certain parameters about the servers, applications, routers, wiring, and physical security that build their critical systems. One such parameter on servers is the ping time, which is the time it takes one computer on a network to communicate with another computer. When ping times are high, it implies there might be issues on the network or server that could eventually lead to unplanned outages.

Let us assume that we want to conditionally format a snapshot of data that contains the ping times measured in milliseconds for 10 different servers on a network. In order to visually inspect the data, we want to categorize the values as follows:

For all ping times greater than or equal to 1000 milliseconds, we want to highlight the value with a red background and white text.

For all ping times greater than or equal to 500 milliseconds, but less than 1000 milliseconds, we want to highlight the text with a yellow background with black bold text.

For all ping times less than 500 milliseconds, we want to leave the default formatting for the text.

To set up the conditional formatting described in this scenario, we will first need to select all of the ping times (range B5:B14) and then select the **Home > Styles > Conditional Formatting > Manage Rules** *window in Microsoft Excel. When the Rules Manager displays, click the button to create a* **New Rule***:*

- **New Formatting Rule > Format only cells that contain.** *First, we want to format all cells with ping times greater than 1000 milliseconds with a red background and white font. On the formatting rule screen, click the* **Format** *button to change the background and font color. When complete, click* **OK***.*

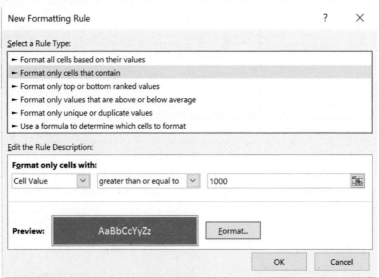

- **New Formatting Rule > Format only cells that contain.** *Next, we want to format the cells with ping times between 500 and 1000 milliseconds with a yellow background and bold black text. On the Rules Manager screen, click New Rule to create a new, additional rule for the selected cells. When complete, click OK on the New Rule screen and click OK on the Formatting Rules Manager screen to see the results of the applied rules.*

The final worksheet should appear as shown.

	A	B	C	D	E	F	G
1							
2	Server Response Times						
3							
4	Server	Ping (t) [ms]					
5	SRV-1	100					
6	SRV-2	25					
7	SRV-3	55					
8	SRV-4	5					
9	SRV-5	10					
10	SRV-6	1125					
11	SRV-7	770					
12	SRV-8	100					
13	SRV-9	55					
14	SRV-10	35					
15							

Comprehension Check | **10-11**

This is a continuation of the worksheet you created in Example 10-3. Modify it to highlight all heights greater than 100 meters with a light blue background and all heights less than 25 meters with a dark blue background with a white font.

10.6 Sorting and Filters

LEARN TO:	Use Excel to sort data with multiple levels of sorting
	Use Excel to filter data based on specified criteria
	Use the SUBTOTAL function to analyze filtered data

Excel provides a number of built-in tools for sorting and filtering data in a worksheet. This section describes how to use these tools effectively without causing unintended side effects.

Each year, the federal government publishes a list of fuel economy values. The complete lists for recent years can be found at http://www.fueleconomy.gov/feg. A partial list of 2013 vehicles is shown here. In the table, MPG = miles per gallon.

Make	Model	MPG City	MPG Highway	Annual Fuel Cost
Jeep	Grand Cherokee 4WD	16	23	$2,900
BMW	X5 xDrive 35i	16	23	$3,100
Honda	Civic Hybrid	44	44	$1,250
Volkswagen	Jetta 2.5L	24	31	$2,100
Ford	Mustang	15	26	$2,900
Bentley	Continental GTC	14	24	$3,450
Honda	Fit	28	35	$1,800

Given this information, assume you are to present it with some sort of order. What if you want to sort the data on text values (Make or Model) or numerical values (MPG City, MPG Highway, Annual Fuel Cost), or what if you want to view only certain vehicles that meet a certain condition?

Sorting Data in a Worksheet

- Select the cells to be sorted. You can select cells in a single column or row or in a rectangular group of cells.

- Select **Home > Editing > Sort & Filter**. By default, two commonly used sorting tools (Sort A to Z and Sort Z to A) appear, in addition to a button for Custom Sort. With a group of cells selected, the common sorting tools will sort according to the values in the leftmost column. If the leftmost column contained numerical values, the options would have read Sort Smallest to Largest/Largest to Smallest. Since it is often desired to involve multiple sorting conditions, click **Custom Sort**.

- The sorting wizard is displayed as shown below. If your selected group of cells had a header row (a row that displays the names of the columns and not actual data) the "My data has headers" checkbox should be selected.

  By default, Excel automatically detects whether the top row of your selected data is a header or a data row. Since you selected the data including the header rows, the "Sort by" drop-down menu will contain the header names. If you had not included the header row, the "Sort by" drop-down menu would have shown the column identifiers as options. It is good practice to select the headers in addition to the data to make sorting easier to understand.

- Assume you want to sort the list alphabetically (A to Z) by the make, then by smallest to largest annual fuel cost. Click the **Add Level** button to add two levels of sorting since there are two conditions. In the sorting wizard, the topmost sorting level will be the sort applied first, and then the next level will sort each data group that forms from the first sort. In the example, there is more than one Honda vehicle, so the second level will place the Civic Hybrid above the Fit, since the Civic Hybrid has a smaller annual fuel cost.

The resulting sorted data appear as shown.

	A	B	C	D	E
1	Fuel Economy of Vehicles				
2	This worksheet demonstrates the use of sorting and filtering in Excel				
3					
4	**Make**	**Model**	**MPG City**	**MPG Highway**	**Annual Fuel Cost**
5	Bentley	Continental GTC	14	24	$3,450
6	BMW	X5 xDrive 35i	16	23	$3,100
7	Ford	Mustang	15	26	$2,900
8	Honda	Civic Hybrid	44	44	$1,250
9	Honda	Fit	28	35	$1,800
10	Jeep	Grand Cherokee 4WD	16	23	$2,900
11	Volkswagen	Jetta 2.5L	24	31	$2,100

NOTE

To "undo" a sort, either choose the "Undo" arrow button on the top menu or use CTRL + Z.

It is important to be sure to select all of the data when using the sort functions because it is possible to corrupt your data set. To demonstrate, select only the first three columns (Make, Model, MPG City) and sort the data smallest to largest on the MPG City column.

Notice after sorting that the last two columns (MPG Highway, Annual Fuel Cost) are not the correct values for the vehicle. There is no way to recover the original association if you were to save the file and open it at a later time, so it is critical that when using the built-in sorting functions, you verify the correctness of your data before saving your workbook. In this case, you can click Excel's Undo button or CTRL + Z to unapply the last sort.

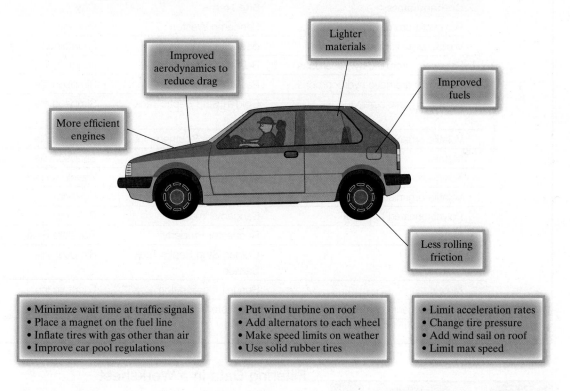

Improving automotive gas mileage, while keeping costs under control, is a complex puzzle, involving many different types of engineers. Shown here are some ways to possibly improve fuel efficiency. Some really work, some are false claims, and some are fictitious. Can you tell the difference? What other ways can you think of to improve today's automobiles?

Comprehension Check   10-12

In 1980, the Environmental Protection Agency (EPA) began the Superfund Program to help clean up highly polluted areas of the environment. There are over 1300 Superfund sites across the country. Not all Superfund sites are from deliberate pollution. Some sites are old factories, where chemicals were dumped on the ground; landfills where garbage was dumped along with other poisonous waste; remote places where people secretly dumped hazardous waste because they did not know what to do with it; or old coal, iron ore, or silver mines.

According to the EPA (http://www.epa.gov/superfund/index.htm), the following groundwater contaminants were found in South Carolina Superfund sites in Greenville, Pickens, Oconee, and Anderson counties.

- Sort by city in ascending order. Examine the result: Which city appears first?
- Sort again: first by city in descending order, then by site name in descending order. Examine the results: Which site name now appears first?
- Sort again by contaminant in ascending order, then by site name in ascending order. Examine the results: Which site name appears last?

Contaminants	Site Name	City
Polycyclic aromatic hydrocarbons	Sangamo Weston	Pickens
Volatile organic compounds	Beaunit Corporation	Fountain Inn
Polycyclic aromatic hydrocarbons	Beaunit Corporation	Fountain Inn
Polycyclic aromatic hydrocarbons	Para-Chem Southern, Inc.	Simpsonville
Volatile organic compounds	Golden Strip Septic Tank Service	Simpsonville
Volatile organic compounds	Para-Chem Southern, Inc.	Simpsonville
Metals	Para-Chem Southern, Inc.	Simpsonville
Polycyclic aromatic hydrocarbons	Rochester Property	Travelers Rest
Volatile organic compounds	Sangamo Weston	Pickens
Polychlorinated biphenyl	Sangamo Weston	Pickens
Metals	Rochester Property	Travelers Rest
Metals	Golden Strip Septic Tank Service	Simpsonville
Metals	Beaunit Corporation	Fountain Inn
Volatile organic compounds	Rochester Property	Travelers Rest

Filtering Data in a Worksheet

Assume you want to look only at a specific portion of the data set and hide all the other rows of data. For example, you might want to look only at Honda vehicles or all vehicles that have an MPG City rating between 10 and 15 MPG. Excel has a built-in filtering capability by which you can conditionally display rows in a data set.

- Select the header row for a data set and click the **Sort & Filter** button in the **Home > Editing** ribbon. Click the **Filter** option to enable filtering for each column of data. Each column label contains a drop-down menu with various sorting options, as well as a number of different approaches for filtering.

  - For data sets that contain a small number of options, use the checkboxes in the drop-down filter to manually check certain options to display.
  - For numerical values, use the Number Filters submenu to filter on certain conditional expressions. The Custom Filter option in the Number Filters submenu lets you combine up to two logical expressions to filter a single column of data.

Assume you want to revisit your fuel economy data set and add in a number of statistical functions to assist in analysis.

	A	B	C	D	E
1	Fuel Economy of Vehicles				
2	This worksheet demonstrates the use of sorting and filtering in Excel				
3					
4	Make	Model	MPG City	MPG Highway	Annual Fuel Cost
5	Jeep	Grand Cherokee 4WD	16	23	$2,900
6	BMW	X5 xDrive 35i	16	23	$3,100
7	Honda	Civic Hybrid	44	44	$1,250
8	Volkswagen	Jetta 2.5L	24	31	$2,100
9	Ford	Mustang	15	26	$2,900
10	Bentley	Continental GTC	14	24	$3,450
11	Honda	Fit	28	35	$1,800
12					
13		Average	22	29	$2,500
14		Min	14	23	$1,250
15		Max	44	44	$3,450

Suppose you filter the data set to look only at the Honda vehicles.

	A	B	C	D	E
1	Fuel Economy of Vehicles				
2	This worksheet demonstrates the use of sorting and filtering in Excel				
3					
4	Make	Model	MPG City	MPG Highway	Annual Fuel Cost
7	Honda	Civic Hybrid	44	44	$1,250
11	Honda	Fit	28	35	$1,800
12					
13		Average	22	29	$2,500
14		Min	14	23	$1,250
15		Max	44	44	$3,450

Notice that the statistical calculations at the bottom are still referencing the entire data set, even though, because of the filter, only a subset of the data is displayed. For data comparisons, this will be a valuable side effect; however, if you want the calculations to apply only to the visible data, you will need to use built-in functions other than the traditional functions (AVERAGE, MIN, MAX).

Using the SUBTOTAL Function

The **SUBTOTAL** function allows the worksheet to dynamically recalculate expressions generated with a filtered list. In the example where only Honda vehicles are selected, only the two visible vehicles will be used in the calculations, if you modify your worksheet to use the SUBTOTAL function instead of the traditional statistical functions. To use the SUBTOTAL function, pass in two different arguments:

= SUBTOTAL(function_num, range)

- The *function_num* argument is a number associated with various built-in Excel functions. Table 10-7 lists the available functions for use with the SUBTOTAL function.
- The *range* argument is the range of cells to which the function should be applied.

Table 10-7 Available functions in SUBTOTAL

function_num	Function	Definition
1	AVERAGE	Computes the average value of the range
2	COUNT	Counts the number of cells in the range that contain numbers
3	COUNTA	Counts the number of nonempty cells in the range
4	MAX	Calculates the maximum value of the range
5	MIN	Calculates the minimum value of the range
6	PRODUCT	Calculates the product of each number in the range
7	STDEVP	Calculates the standard deviation of the numbers in the range
8	SUM	Calculates the sum of all of the numbers in the range
9	VARP	Calculates the variance of the numbers in the range

In the example, use the following calculation in cell C13 to calculate the average of MPG City:

$$= \text{AVERAGE(C5:C11)}$$

The AVERAGE function corresponds to function_num 1, so the resulting calculation in cell C13 using the SUBTOTAL function would appear as follows:

$$= \text{SUBTOTAL(1, C5:C11)}$$

After you modified all of the statistical calculations in the worksheet to use the SUBTOTAL function, the sheet should appear as shown in the examples that follow. Note that the values recalculate automatically according to the filtered data.

Filter on Make: Honda Only

	A	B	C	D	E
1	**Fuel Economy of Vehicles**				
2	This worksheet demonstrates the use of sorting and filtering in Excel				
3					
4	**Make**	**Model**	**MPG City**	**MPG Highway**	**Annual Fuel Cost**
7	Honda	Civic Hybrid	44	44	$1,250
11	Honda	Fit	28	35	$1,800
12					
13		**Average**	36	40	$1,525
14		**Min**	28	35	$1,250
15		**Max**	44	44	$1,800

Filter on Annual Fuel Cost: Less than $3000

	A	B	C	D	E
1	Fuel Economy of Vehicles				
2	This worksheet demonstrates the use of sorting and filtering in Excel				
3					
4	**Make**	**Model**	**MPG City**	**MPG Highway**	**Annual Fuel Cost**
5	Jeep	Grand Cherokee 4WD	16	23	$2,900
7	Honda	Civic Hybrid	44	44	$1,250
8	Volkswagen	Jetta 2.5L	24	31	$2,100
9	Ford	Mustang	15	26	$2,900
11	Honda	Fit	28	35	$1,800
12					
13		Average	25	32	$2,190
14		Min	15	23	$1,250
15		Max	44	44	$2,900

IN-CLASS ACTIVITIES

ICA 10-1

▲	A	B	C	D	E	F
1						
2	45					
3	meters					
4	horses					
5	canoes					
6	99					
7	four					
8	equator					
9	Dabo					
10	33.3333					
11						

The worksheet displayed was designed to reflect on our understanding of the use of cell references. Answer the following questions using the worksheet.

(a) What is the cell address of the word "Dabo"?

(b) Using cell concatenation with the ampersand operator, what formula would you type into a cell to create the word "fourequator"?

(c) Using cell concatenation with the CONCAT function, what formula would you type into a cell to create the word "Dabocanoes"?

ICA 10-2

▲	A	B	C	D	E	F
1						
2	45					
3	meters					
4	horses					
5	canoes					
6	99					
7	four					
8	equator					
9	Dabo					
10	33.3333					
11						

The worksheet displayed was designed to reflect on our understanding of the use of cell references. Answer the following questions using the worksheet.

(a) If we type =MAX(A2:A10) into cell B2, what value displays?

(b) If we type =COUNT(A2:A10) into cell B3, what value displays?

(c) If we type =COUNTA(A2:A10) into cell B4, what value displays?

ICA 10-3

	A	B	C	D	E	F
1						
2	45					
3	meters					
4	horses					
5	canoes					
6	99					
7	four					
8	equator					
9	Dabo					
10	33.3333					
11						

The worksheet displayed was designed to reflect on our understanding of the use of cell references. Answer the following questions using the worksheet.

(a) If we want to calculate the average of all of the cells in the range A2:A10, what do we type into cell B5?

(b) If we want to calculate the sum of each **individual** numeric value (by individual cell, not range), what do we type into cell B6?

(c) If we want to concatenate every cell that contains text (in order from lowest numbered row to highest numbered row) in the range A2:A10, what is the resulting string of text that would appear in cell B7?

(d) If we want to concatenate every cell that contains text (in order from highest numbered row to lowest numbered row) in the range A2:A10, what would we write in cell B8 to generate that string of text?

ICA 10-4

	A	B	C	D	E	F
1						
2	45					
3	meters					
4	horses					
5	canoes					
6	99					
7	four					
8	equator					
9	Dabo					
10	33.3333					
11						

The worksheet displayed was designed to reflect on our understanding of the use of cell references. Answer the following questions using the worksheet.

(a) What displays in B9 if we type =A2+A3 into the cell?

(b) What displays in B10 if we type =A10/0 into the cell?

(c) What displays in B11 if we type =B11/0 into the cell? What is it called when the formula in cell B11 also uses B11 as a variable in the formula?

ICA 10-5

	A	B	C	D
1				
2				
3				
4	Surface Pressure	(Psurface)	2	[atm]
5	Specific Gravity of Fluid	(SG)	1.26	[–]
6	Gravity	(g)	9.8	[m / s^2]
7				
8	Surface Pressure	(Psurface)		[Pa]
9	Density of Fluid	(ρ)		[kg / m^3]
10				
11	Depth (H) [ft]	Depth (H) [m]	Total Pressure (P) [atm]	
12	1			
13	10			
14	20			
15	30			
16	40			
17	50			
18	60			
19	70			
20	80			
21	90			
22	100			

The worksheet shown here was designed to calculate the total pressure felt by an object submerged in a fluid as a function of the depth to which the object is submerged. The user will enter the surface pressure (in units of atmospheres), specific gravity of the fluid, and the gravity of the planet (in units of meters per second squared). All user input is shown in red. The worksheet will calculate the surface pressure in units of pascals, the density of the fluid in kilograms per cubic meter, and depth in units of feet. All conversions are shown in orange. Finally, the worksheet will calculate the total pressure in units of atmospheres.

(a) What formula should be typed in cell C8 to convert the surface pressure in cell C4 from atmospheres to pascals?

(b) What formula should be typed in cell C9 to determine the density in units of kilograms per cubic meter?

(c) What formula should be typed into cell B12 that can then be copied down column B to convert the depth from units of feet to units of meters?

(d) What formula should be typed into cell C12 that can then be copied down column C to calculate the total pressure in units of atmospheres?

ICA 10-6

The worksheet provided was designed to calculate the total pressure felt by an object submerged in a fluid as a function of the depth to which the object is submerged.

The user will enter the surface pressure (in units of atmospheres), specific gravity of the fluid, and the gravity of the planet (in units of meters per second squared). All user input is shown in red.

The worksheet will calculate the surface pressure in units of pascals, density of the fluid in kilograms per cubic meter, and depth in units of feet. All conversions are shown in orange. Format the pressure and density to a whole number, and the height in meters to three decimal places.

Finally, the worksheet will calculate the total pressure in units of atmospheres; format to two decimal places.

Complete the starting Excel file to meet these criteria. The following sample worksheet is shown for comparison.

	A	B	C	D
1				
2				
3				
4	Surface Pressure	(Psurface)	2	[atm]
5	Specific Gravity of Fluid	(SG)	1.26	[–]
6	Gravity	(g)	9.8	[m / s^2]
7				
8	Surface Pressure	(Psurface)	202650	[Pa]
9	Density of Fluid	(ρ)	1260	[kg / m^3]
10				
11	Depth (H) [ft]	Depth (H) [m]	Total Pressure (P) [atm]	
12	1	0.30	2.04	
13	10	3.05	2.37	
14	20	6.10	2.74	
15	30	9.15	3.11	
16	40	12.20	3.49	
17	50	15.24	3.86	
18	60	18.29	4.23	
19	70	21.34	4.60	
20	80	24.39	4.97	
21	90	27.44	5.34	
22	100	30.49	5.72	

ICA 10-7

Some alternate energy technologies, such as wind and solar, produce more energy than needed during peak production times (windy and sunny days) but produce insufficient energy at other times (calm days and nighttime). Many schemes have been concocted to store the surplus energy generated during peak times for later use when generation decreases. One scheme is to use the energy to spin a massive flywheel at very high speeds, then use the rotational kinetic energy stored to power an electric generator later.

The following worksheet was designed to calculate how much energy is stored in flywheels of various sizes. The speed of the flywheel (revolutions per minute) is to be entered in cell B2 and the density of the flywheel in cell B4. A formula in cell B3 converts the speed into units of radians per second. There are 2π radians per revolution of the wheel.

To simplify the computations, the stored energy was calculated in three steps. The first table calculates the volumes of the flywheels, the second table uses these volumes to calculate the masses of the flywheels, and the third table uses these masses to determine the stored rotational kinetic energy.

Note that in all cases, changing the values in cells B2 and/or B4 should cause all appropriate values to be automatically recalculated.

	A	B	C	D	E	F	G	H	I
1									
2	Speed (v) [rpm]	15,000		Volume (V) [m³]		Height (H) [m]			
3	Speed (ω) [rad/s]	1571		Diameter (D) [m]	0.3	0.6	0.9	1.2	1.5
4	Density (ρ) [kg/m³]	8000		0.2	0.009	0.019	0.028	0.038	0.047
5				0.4	0.038	0.075	0.113	0.151	0.188
6				0.6	0.085	0.170	0.254	0.339	0.424
7				0.8	0.151	0.302	0.452	0.603	0.754
8				1.0	0.236	0.471	0.707	0.942	1.178
9									
10				Mass (m) [kg]		Height (H) [m]			
11				Diameter (D) [m]	0.3	0.6	0.9	1.2	1.5
12				0.2	75	151	226	302	377
13				0.4	302	603	905	1206	1508
14				0.6	679	1357	2036	2714	3393
15				0.8	1206	2413	3619	4825	6032
16				1.0	1885	3770	5655	7540	9425
17									
18				Kinetic Energy (KE) [J]		Height (H) [m]			
19				Diameter (D) [m]	0.3	0.6	0.9	1.2	1.5
20				0.2	4.65E+05	9.30E+05	1.40E+06	1.86E+06	2.33E+06
21				0.4	7.44E+06	1.49E+07	2.23E+07	2.98E+07	3.72E+07
22				0.6	3.77E+07	7.53E+07	1.13E+08	1.51E+08	1.88E+08
23				0.8	1.19E+08	2.38E+08	3.57E+08	4.76E+08	5.95E+08
24				1.0	2.91E+08	5.81E+08	8.72E+08	1.16E+09	1.45E+09
25				Average KE [J]	9.11E+07	1.82E+08	2.73E+08	3.64E+08	4.55E+08
26				Max - Min KE [J]	1.08E+09				

(a) What should be typed in cell B3 to convert revolutions per minute in cell B2 into radians per second?

(b) What should be typed into cell E4 that can then be copied through the rest of the first table to calculate the flywheel volumes? Assume the shape of the flywheel to be a cylinder.

(c) What should be typed into cell E12 that can then be copied through the rest of the second table to calculate the flywheel masses?

(d) What should be typed into cell E20 that can then be copied through the rest of the third table to calculate the kinetic energies stored in the flywheels? The rotational kinetic energy is given by the formula: $KE_{Rot} = (I\omega^2) / 2 = (mr^2\omega^2)/4$

(e) What should be typed into cell E25 that can then be copied through row 25 to determine the average kinetic energy at each height (in each column)?

(f) What should be typed into cell E26 to determine the difference between the maximum kinetic energy and 800 times the minimum kinetic energy given in the table?

ICA 10-8

The worksheet shown was designed to calculate the cost of material that must be purchased to produce a given number of parts. The user will enter the specific gravity of the material, the diameter of the cylindrical part in units of inches, the cost of the raw material in dollars per pound-mass, and the number of parts to be manufactured. All user input is shown in red. The worksheet will calculate the radius of the cylindrical part in units of centimeters and the density of the fluid in grams per cubic centimeter. All conversions are shown in orange.

The worksheet will determine the volume and mass of a single part for a given height. Finally, the worksheet will determine the total mass of material needed to produce the desired number of parts in units of pounds-mass, and the total material cost.

The total material cost appears twice. In cells E13 to E20, a formula is written to determine the cost. In cells B26 to B33, the cells simply reference the corresponding cell in the table above. For example, in cell B26 the formula = E13 appears.

In the bottom table, the total cost for N parts is determined by the formula:

Total Cost = Total Material Cost + (Energy Cost + Labor Cost) × Number of Parts

	A	B	C	D	E	F
4	Specific Gravity of Material	(SG)	1.50	[--]		
5	Diameter	(D)	4.0	[in]		
6	Cost of Raw Material	(C)	$2.25	[$ / lbm]		
7	Number of Parts	(N)	150	[--]		
9	Radius	(R)		[cm]		
10	Density of Fluid	(ρ)		[g / cm^3]		
12	Height (H) [cm]	Volume (V) [cm^3]	Mass (m) [g]	Mass (M_N) [lbm] for N parts	Total Material Cost (MC) [$]	
13	1					
14	2					
15	3					
16	5					
17	6					
18	8					
19	10					
20	12					
22	Labor Cost	(LC)	$1.50	[$ / part]		
24		Total Cost of N Parts		Energy Cost (EC) [$ / part]		
25	Height (H) [cm]	Total Material Cost (MC) [$]	$0.05	$0.10	$0.20	$0.40
26	1					
27	2					
28	3					
29	5					
30	6					
31	8					
32	10					
33	12					

(a) What should be typed into cell C9 to determine the radius in the correct units?

(b) What should be typed into cell B13 that can then be copied down column B to determine the volume of a cylindrical part in units of cubic centimeters?

(c) What should be typed into cell C13 that can then be copied through down column C to calculate the mass of each part in unit of grams?

(d) What should be typed into cell D13 that can then be copied down column D to calculate the total mass needed to produce *N* parts in units of pounds-mass?

(e) What should be typed into cell E13 that can then be copied down column E to calculate the total material cost?

(f) What should be typed into cell C26 that can then be copied from C26 to F33 given the energy cost in row 25 and labor cost in cell C22 to calculate the total cost of producing *N* parts?

ICA 10-9

The worksheet shown was designed to calculate the cost of material that must be purchased to produce a given number of parts. The user will enter the specific gravity of the material, the diameter of the cylindrical part in units of inches, the cost of the raw material in dollars per pound-mass, and the number of parts to be manufactured. All user input is shown in red. The worksheet will calculate the radius of the cylindrical part in units of centimeters and the density of the fluid in grams per cubic centimeter. All conversions are shown in orange.

The worksheet will determine the volume and mass of a single part for a given height. Finally, the worksheet will determine the total mass of material needed to produce the desired number of parts in units of pounds-mass, and the total material cost.

The total material cost appears twice. In cells E13 to E20, a formula is written to determine the cost. In cells B26 to B33, the cells simply reference the corresponding cell in the table above. For example, in cell B26 the formula = E13 appears.

In the bottom table, the total cost for *N* parts is determined by the formula:

Total Cost = Total Material Cost + (Energy Cost + Labor Cost) × Number of Parts

The following sample worksheet is shown for comparison.

	A	B	C	D	E	F
1						
2						
3						
4	Specific Gravity of Material	(SG)	1.50	[--]		
5	Diameter	(D)	4.0	[in]		
6	Cost of Raw Material	(C)	$2.25	[$ / lbm]		
7	Number of Parts	(N)	150	[--]		
8						
9	Radius	(R)	5.08	[cm]		
10	Density of Fluid	(ρ)	1.5	[g / cm^3]		
11						
12	Height (H) [cm]	Volume (V) [cm^3]	Mass (m) [g]	Mass (M_N) [lbm] for N parts	Total Material Cost (MC) [$]	
13	1	81.07	121.61	40	$90.50	
14	2	162.15	243.22	80	$181.00	
15	3	243.22	364.83	121	$271.50	
16	5	405.37	608.05	201	$452.50	
17	6	486.44	729.66	241	$543.00	
18	8	648.59	972.88	322	$724.00	
19	10	810.73	1216.10	402	$905.00	
20	12	972.88	1459.32	483	$1,086.01	
21						
22	Labor Cost	(LC)	$1.75	[$ / part]		
23						
24		Total Cost of N Parts		Energy Cost (EC) [$ / part]		
25	Height (H) [cm]	Total Material Cost (MC) [$]	$0.05	$0.10	$0.20	$0.40
26	1	$90.50	$360.50	$368.00	$383.00	$413.00
27	2	$181.00	$451.00	$458.50	$473.50	$503.50
28	3	$271.50	$541.50	$549.00	$564.00	$594.00
29	5	$452.50	$722.50	$730.00	$745.00	$775.00
30	6	$543.00	$813.00	$820.50	$835.50	$865.50
31	8	$724.00	$994.00	$1,001.50	$1,016.50	$1,046.50
32	10	$905.00	$1,175.00	$1,182.50	$1,197.50	$1,227.50
33	12	$1,086.01	$1,356.01	$1,363.51	$1,378.51	$1,408.51

ICA 10-10

Refer to the following worksheet. The following expressions are typed into the Excel cells indicated. Write the answer that appears in the cell listed. If the cell will be blank, write "BLANK" in the answer space. If the cell will return an error message, write "ERROR" in the answer space.

	A	B	C	D	E	F	G	H
1								
2								
3	Fluid Type	Benzene			Fluid Type	Olive Oil		
4	Density (ρ)	0.879	[g / cm^3]		Density (ρ)	0.703	[g / cm^3]	
5	Viscosity (μ)	6.47E-03	[g / (cm s)]		Viscosity (μ)	1.01	[g / (cm s)]	
6								
7	Velocity (v)	15	[cm / s]		Velocity (v)	50	[cm / s]	
8								
9	Pipe Diameter	Reynolds Number			Pipe Diameter	Reynolds Number		
10	(D) [cm]	(Re) [--]			(D) [cm]	(Re) [--]		
11	1.27	2,588			1.27	44		
12	2.54	5,176			2.54	88		
13	3.81	7,764			3.81	133		
14	5.08	10,352			5.08	177		
15	6.35	12,940			6.35	221		
16	7.62	15,529			7.62	265		

	Expression	Typed into Cell
(a)	= IF(B4 > F4, B3, "F3")	D4
(b)	= IF(B7/2 > F7/10, " ", B7*2)	H7
(c)	= IF(B11 < F11, "B11", IF(B11 > F11, SUM(B11, F11), F11))	D11
(d)	= IF(AND(B4 < F4,B5 < F5), B3, MAX(F11:F16))	D9
(e)	= IF(OR(E16/2^2 > E15*2,E11+E12 < E14),F4*62.4,F4*1000)	H16

ICA 10-11

Write the output value that would appear in a cell if the equation was executed in Excel. You should answer these questions WITHOUT actually using Excel, as practice for the exam. If the cell will appear blank, write "BLANK" in the space provided.

= IF(AND(A1/A2 > 2, A2 > 3), A1, A2)		**Output**
(a)	A1 = 30 A2 = 5	
(b)	A1 = 5 A2 = 1	

= IF(SIN(A1*B1/180) < 0.5, PI(), IF(SIN(A1*B1/180) > 1, 180/A1,""))		**Output**
(c)	A1 = 30 B1 = PI()	
(d)	A1 = 5 B1 = PI()	

ICA 10-12

Write the output value that would appear in a cell if the equation were executed in Excel. You should answer these questions WITHOUT actually using Excel, as practice for the exam. If the cell will appear blank, write "BLANK" in the space provided.

= IF(OR(C1 > D3, D3 < E1), "YES", "NO")			**Output**
(a)	C1 = 10	E1 = −5	D3 = 0.1*C1^(−5*E1)
(b)	C1 = 10	E1 = 5	D3 = 0.1*C1^(−5*E1)

= IF(AND(G4/H3 > 2, H3 > 3), G4, MAX(2, G4, H3, 5*J2-10))			**Output**
(c)	G4 = 30	H3 = 5	J2 = 2
(d)	G4 = 10	H3 = 8	J2 = 10

ICA 10-13

Refer to the following worksheet. In all questions, give the requested answers in Excel notation, indicating EXACTLY what you would type into the cell given to properly execute the required procedures.

	A	B	C	D	E	F	G
1							
2							
3	Height (H) [ft]	5			Width (W) [ft]		
4					1	1.5	2
5							
6	Volume (V)	Radius (r)	Area (A)		Length [ft]		
7	[ft³]	[ft]	[ft²]	[cm²]	(L1)	(L2)	(L3)
8	79		70.5				
9	1		7.9				
10	55		58.8				
11	13		28.6				
12	39		49.5				
13	9		23.8				
14	63		62.9				
15	23		38.0				
16	72		67.3				
17	27		41.2				
18	67		64.9				

(a) In column B, you wish to determine the radius of a cylinder. The volume (column A) and height (cell B3) have been provided. Recall that the volume of a cylinder is given by $V = \pi r^2 H$. Assume you will write the formula in cell B8 and copy it down the column to cell B100. In the expression, fill in the blanks with any Excel functions and fill in the boxes with any dollar signs necessary for relative, mixed, or absolute references.

$$=\underline{\quad}(\Box A\Box 8 / (\underline{\quad}*\Box B\Box 3)\underline{\quad}$$

(b) In column C, the area of a cylinder corresponding to the radius (in column B) and the height (cell B3) has been determined in units of square feet. In column D, you wish to express these values in units of square centimeters. Fill in any Excel mathematical operators or parenthesis for the expression to correctly complete this conversion.

$$= \text{ C8}\underline{\quad}2.54\underline{\quad}2\underline{\quad}12\underline{\quad}2$$

(c) In columns E–G, we wish to determine the dimensions of a rectangular container with the same volume as the cylinders given in column A. The rectangle will be the same height as the cylinder (cell B3) but will have three possible widths (contained in cells E_4–G_4). Fill in the following boxes with any dollar signs necessary for relative, mixed, or absolute references to allow the expression to determine the length in cell E8 and be copied across to columns F and G, then down all three columns to row 100.

$$=\Box A\Box 8 / (\Box B\Box 3*\Box E\Box 4)$$

(d) In column H, we wish to tell the user how the length and radius of the different containers compare. Complete the following IF statement for cell H8 to display the maximum value of the length calculations (cells E8 through G8) if the maximum value of the length calculations is greater than the corresponding radius calculation; otherwise display the letter R.

$$= \text{ IF}(\underline{\quad}(1)\underline{\quad},\underline{\quad}(2)\underline{\quad},\underline{\quad}(3)\underline{\quad})$$

(e) Fill in the following IF statement for cell J8 to display the sum of Length 1 and Length 2 if the sum of these lengths is greater than Length 3; otherwise, leave it blank.

$$\text{IF}(\underline{\quad}(1)\underline{\quad}, \underline{\quad}(2)\underline{\quad}, \underline{\quad}(3)\underline{\quad})$$

ICA 10-14

Give all answers in EXACT Excel notation, as if you were instructing someone EXACTLY what to type into Excel. Be sure to use the values given in the worksheet as cell references and not actual numerical values in the formula. Use absolute, mixed, or relative addressing as required.

	A	B	C	D	E	F	G	H
1								
2	Ideal Gas Constant (R)		8,314	[(Pa L)/(K mol)]				
3	Amount of substance (n)		2	[mol]				
4	Molecular weight (MW)		28	[g / mol]				
5	Mass of substance (m)			[g]				
6								
7		Temperature (T) [°F]	Temperature (T) [K]	Volume (V) [ft³]	Volume (V) [L]	Pressure (P) [Pa]	Pressure Warning	Volume Warning
8		25		1				
9		30		1.2				
10		35		1.4				
11		40		0.84				
12		45		0.75				

(a) What would you type into cell C5 to calculate the mass of gas in the container?

(b) What would you type into cell C8 so that you could copy the cell down to cell C12 to calculate all corresponding values of temperature, converting the temperatures given in column B from units of degrees Fahrenheit to units of kelvins?

(c) What would you type into cell E8 so that you could copy the cell down to cell E12 to calculate all corresponding values of volume, converting the volumes given in column D from units of cubic feet to units of liters?

(d) What would you type into cell F8 so that you could copy the cell down to cell F12 to calculate all corresponding values of pressure using the ideal gas law, solving for pressure in units of pascals?

(e) What conditional statement would you type into cell G8 so that you could copy the cell down to cell G12 to display the words "Too High" if the pressure from the ideal gas calculation is equal to or greater than 500,000 pascals? If the pressure is less than this value, the cell should remain blank.

(f) What conditional statement would you type in cell H8 so that you could copy the cell down to cell H12 to display the words "Bigger" if the corresponding value in column E is greater than 5 gallons, "Smaller" if the value in column E is less than 1 gallon, or display the actual value of the volume, in units of gallons, if the value is between 1 and 5 gallons?

ICA 10-15

A bioengineer conducts clinical trials on stressed-out college students to see if a sleep aid will help them fall asleep faster. She begins the study by having 20 students take a sleep aid for seven days and records through biofeedback the time when they fall asleep. To analyze the data, she sets up the following worksheet. Evaluate the expressions that follow; state what will appear in the cell when the command is executed. Column I contains the average time each student took to fall asleep during the seven-day trial. Column J contains any adverse reactions the students experienced (H = headache; N = nausea).

(a) Column K will contain the rating of the time it took the student to fall asleep compared with the control group, who did not take the medication. The statement as it appears in cell K14 is given below. What will appear in cell K14 when this statement is executed?

=IF>(I14>$I$2+$I$3, "MORE", IF(I14<$I$2−$I$3, "LESS", " "))

(b) Column L groups the participants into three groups according to their reaction to the drug and the time it took them to fall asleep. Assume the statement for part (a) is executed in column K. The statement as it appears in cell L7 is given below. What will appear in cell L7 when this statement is executed?

= IF(AND(K7 = "MORE", J7 = "H"), "MH",
IF(AND(K7 = "MORE", J7 = "N"), "MN", ""))

(c) Suppose the formula in column L was changed to regroup the participants. The statement as it appears in cell L9 is given below. In Excel, this statement would appear as a continuous line, but because of space restrictions it is shown here on two lines. What will appear in cell L9 when this statement is executed?

= IF(AND(K9 = "MORE", OR(J9 = "H", J9 = "N")), "SEVERE",
IF(OR(J9 = "H", J9 = "N"), "MILD", IF(K9 = "LESS", "HELPFUL", "")))

(d) Suppose the formula in part (c) was copied into cell L16. What will appear in cell L16 when this statement is executed?

(e) Suppose the formula in part (c) was copied into cell L18. What will appear in cell L18 when this statement is executed?

	A	B	C	D	E	F	G	H	I	J	K	L
1						Control Group Data						
2						Overall Average			35	[min]		
3						Standard Deviation			4	[min]		
4												
5						Number of Minutes to Fall Asleep						
6	Patient	Day 1	Day 2	Day 3	Day 4	Day 5	Day 6	Day 7	Average	Reaction	Time	Group
7	A	45	39	83	47	39	25	42	46	H		
8	B	35	75	15	36	42	12	29	35			
9	C	42	32	63	45	37	34	31	41	N		
10	D	14	25	65	38	53	33	32	37	H		
11	E	14	71	48	18	29	14	24	31			
12	F	14	25	29	24	18	24	15	21	H N		
13	G	31	14	42	19	28	17	21	25			
14	H	12	24	32	42	51	12	16	27	H N		
15	I	28	29	44	15	43	15	22	28	N		
16	J	21	19	35	41	34	25	18	28	H		
17	K	44	36	51	39	30	26	25	36			
18	L	38	43	36	59	14	34	18	35	N		
19	M	19	15	63	50	55	27	31	37	H		

ICA 10-16

Refer to the worksheet shown, set up to calculate the displacement of a spring. Hooke's law states that the force (F, in newtons) applied to a spring is equal to the stiffness of the spring (k, in newtons per meter) times the displacement (x, in meters): $F = kx$.

	A	B	C	D	E	F	G	H
1								
2	Spring Code	Stiffness [N/m]	Maximum Displacement [mm]			Spring Code	Stiffness [N / m]	Maximum Displacement [mm]
3	3-Blue	50	20			1-Blue	10	40
4						1-Black	25	60
5	Mass [g]	Displacement [cm]	Warning			2-Blue	30	25
6	25	0.49				2-Black	40	60
7	50	0.98				2-Red	20	30
8	75	1.47				3-Blue	50	20
9	100	1.96				3-Red	40	30
10	125	2.45	Too Much Mass			3-Green	60	10
11	150	2.94	Too Much Mass					
12	175	3.43	Too Much Mass					
13	200	3.92	Too Much Mass					
14	225	4.41	Too Much Mass					
15	250	4.90	Too Much Mass					
16	275	5.39	Too Much Mass					
17	300	5.88	Too Much Mass					
18								

Cell A3 contains a data validation list of springs. The stiffness (cell B3) and maximum displacement (cell C3) values are found using a VLOOKUP function linked to the table shown at the right side of the worksheet. These data are then used to determine the displacement of the spring at various mass values. A warning is issued if the displacement determined is greater than the maximum displacement for the spring. Use this information to determine the answers to the following questions.

(a) Write the expression, in Excel notation, that you would type into cell B6 to determine the displacement of the spring. Assume you will copy this expression to cells B7 to B17.

(b) Fill in the following information in the VLOOKUP function used to determine the maximum displacement in cell C3 based on the choice of spring in cell A3.

$$= \text{VLOOKUP}(\_\_(1)\_\_, \_\_(2)\_\_, \_\_(3)\_\_, \_\_(4)\_\_)$$

(c) Fill in the following information in the IF function used to determine the warning given in cell C6, using the maximum displacement in cell C3. Assume you will copy this expression to cells C7 to C17.

$$= IF(\_\_(1)\_\_, \_\_(2)\_\_, \_\_(3)\_\_)$$

ICA 10-17

You are interested in analyzing different implant parts being made in a bioengineering production facility. The company can make nine different parts for shoulder, knee, or hip replacement.

On the worksheet shown, you have created a place for the user to choose the body location (shoulder, knee, or hip) in cell B5 using a data validation list. Once the body location is set, a list of material choices will appear in cells D5 to F5. The user can choose a material in cell B6 using a data validation list. If the material chosen does not match one of the possible choices in cells D5 to F5, a warning will be issued for the user to choose another material.

In cell B9, the user will choose if the part is size small (S), medium (M), or large (L) using a data validation list. Based upon body location and size, the part number will adjust automatically using a VLOOKUP function. After the part number has been determined, the material weight (cell B11) and part volume (cell B12) will adjust automatically using a VLOOKUP function.

The user will enter the number of desired parts in cell B14. If the user requests more than 250 parts, a warning of "Too Many" will be issued; if the user requests fewer than 20 parts, a warning of "Too Small" will be issued in cell C14.

The amount of material to be ordered will be determined in cell B16 by multiplying the number of parts and the material weight. The cost of the material to be ordered will be determined in cell B17. If the order cost is greater than $1000, a request to "Check with Purchasing" will appear; otherwise, the cost of the order will appear. Finally, in cell B18 the amount of boxes needed for shipping will appear determined by the number of parts requested and the number of parts per box, based on the part number chosen in cell B10.

Lookup functions in Excel contain four parts.

$$= VLOOKUP(\_\_(1)\_\_,\_\_(2)\_\_,\_\_(3)\_\_,\_\_(4)\_\_)$$

(a) Fill in the following information in the VLOOKUP function used to determine the third possible material choice in cell F5 based on the choice of body location in cell B5.

An IF statement in Excel contains three parts. Fill in the following information in the IF function used to determine the following conditions:

$$= IF(\_\_(1)\_\_,\_\_(2)\_\_,\_\_(3)\_\_)$$

	A	B	C	D	E	F	G	H	I	J	K	L	M	N
1								Body Location	Part Size	Part Number	Material Weight [lbm]	Number / Box	Part Volume [cin]	Energy Cost / Part
2								Shoulder	S	JB2	0.45	78	5.5	0.1
3								Shoulder	M	JB3	0.15	24	3.5	0.04
4								Shoulder	L	JB5	0.15	55	3.5	0.09
5	Body Location	Shoulder	Material Choices:	AuZn-4	WC-2	CuAg-5		Hip	L	KA9	0.05	64	1.5	0.02
6	MATERIAL	PdSi-3	Wrong Material Match / Choose Again					Hip	M	KA11	0.1	82	2.5	0.05
7	Cost / lbm of material	$ 22.00						Hip	S	KA2	0.4	36	6	0.02
8								Knee	M	DS3	0.5	65	1.5	0.07
9	Part Size	L						Knee	S	DS7	0.3	98	1.5	0.05
10	Part Number	JB5						Knee	L	DS8	0.4	93	5	0.04
11	Material Weight [lbm]	0.15												
12	Part Volume [cin]	3.5						Material	Material Cost / lbm			Body Location		Material Choice	
13								AuZn-4	$ 5.00			Shoulder	AuZn-4	WC-2	CuAg-5
14	Number of Parts	300	Too Many					WC-2	$ 17.00			Knee	ZnCd-2	CoNi-7	PtC-9
15								CuAg-5	$ 7.00			Hip	PtZn-4	PtC-9	MnPd-8
16	Amount of Mat'l to Order [lbm]	45						PdSi-3	$ 22.00						
17	Cost of Mat'l to Order [$]	$990.00						PdSi-5	$ 2.00						
18	Amount of Boxes Needed	6						ZnCd-2	$ 18.00						
19								CoNi-7	$ 24.00						
20								CoAg-12	$ 8.00						
21								CdAl-2	$ 13.00						
22								PtZn-4	$ 19.00						
23								PtC-9	$ 6.00						
24								MnPd-8	$ 25.00						
25								WTi-3	$ 3.00						
26								ScCo-4	$ 6.00						
27								ZrW-8	$ 7.00						
28								MnRh-5	$ 13.00						
29								PdCd-7	$ 11.00						

(b) In cell C6, a warning is issued to the user if the material chosen in cell B6 does not match the list of materials provided in cells D5 to F5. Fill in the IF statement used to create this error message, containing a complex IF test using AND or OR.

A nested IF statement in Excel contains three parts per IF statement. Fill in the following information in the IF function used to determine the following conditions:

$$= \text{IF}(\_\_(1a)\_\_, \_\_(2a)\_\_, \text{IF}(\_\_(1b)\_\_,\_\_(2b)\_\_,\_\_(3b)\_\_))$$

(c) In cell B14, the user can enter the number of parts needed in production. If this value is more than 250 parts, a warning will appear in cell C14 telling the user the quantity is too high; if the value is less than 20, a warning will tell the user the quantity is too small; otherwise, the cell remains blank.

(d) In order for Excel to display the correct number of boxes needed, the following functions are tried. Which one will correctly display the number of boxes needed to ship the parts?

(A) = B14/VLOOKUP(B10, J1:L10, 3, FALSE)
(B) = ROUND(B14/VLOOKUP(B10, J1:L10, 3, FALSE), 0)
(C) = ROUNDDOWN(B14/VLOOKUP(B10, J1:L10, 3, FALSE), 0)
(D) = ROUNDUP(B14/VLOOKUP(B10, J1:L10, 3, FALSE), 0)
(E) = TRUNC(B14/VLOOKUP(B10, J1:L10, 3, FALSE), 0)

ICA 10-18

You have a large stock of several values of inductors and capacitors, and you are investigating how many possible combinations of a single capacitor and a single inductor chosen from the ones you have in stock will give a resonant frequency between specified limits.

Create two cells to hold a minimum and maximum frequency the user can enter.

Incorrect Data:	Allowable Range		Correct Data:	Allowable Range	
	f_{min} [Hz]	f_{max} [Hz]		f_{min} [Hz]	f_{max} [Hz]
	2500	1000		2500	7777

Calculate the resonant frequency (f_R) for all possible combinations of one inductor and one capacitor, rounded to the nearest integer. For a resonant inductor/capacitor circuit, the resonant frequency in hertz [Hz] is calculated by

$$f_R = \frac{1}{2\pi\sqrt{LC}}$$

In this equation, L is the inductance in units of henry [H] and C is the capacitance in units of farads [F]. Note that the capacitance values in the table are given in microfarads. Automatically format each result to indicate its relation to the following minimum and maximum frequency values.

- $f_R > f_{MAX}$: The cell should be shaded white with light gray text and no border.
- $f_R < f_{MIN}$: The cell should be shaded light gray with dark gray text and no border.
- $f_{MIN} < f_R < f_{MAX}$: The cell should be shaded white with bold black text and a black border.

If done properly, the table should appear similar to the table that follows for $f_{MIN} = 2500$ and $f_{MAX} = 7777$.

After you have this working properly, modify the frequency input cells to use data validation to warn the user of an invalid value entry.

Resonant Frequency (fR) [Hz]	Capacitance (C) [µF]							
Inductance (L) [H]	0.0022	0.0082	0.05	0.47	0.82	1.5	3.3	10
0.0005	151748	78601	31831	10382	7860	5812	3918	2251
0.002	75874	39301	15915	5191	3930	2906	1959	1125
0.01	33932	17576	7118	2322	1758	1299	876	503
0.05	15175	7860	3183	1038	786	581	392	225
0.068	13012	6740	2729	890	674	498	336	193
0.22	7234	3747	1517	495	375	277	187	107
0.75	3918	2029	822	268	203	150	101	58

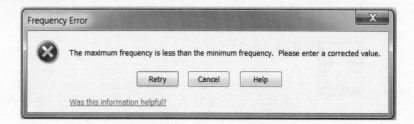

Frequency Error

The maximum frequency is less than the minimum frequency. Please enter a corrected value.

Retry   Cancel   Help

Was this information helpful?

ICA 10-19

We accidentally drop a tomato from the balcony of a high-rise apartment building. As it falls, the tomato has time to ponder some physics and says, "You know, the distance I have fallen equals one-half gravity times the time I have fallen squared." Create a worksheet to solve the question of when the tomato goes splat.

- The user will input the initial balcony height in units of feet. Use data validation to set a limit for the height of 200 feet.
- Place the acceleration due to gravity in a cell under the balcony height and not within the formulas themselves. *Be sure to watch the units for this problem!*
- Column A will be the distance the tomato falls, starting at a distance of zero up to a distance of 200 feet, in 5-foot increments.
- Column B will show the calculated time elapsed at each distance fallen.
- Column C will display the status of the tomato as it falls.
    - If the tomato is still falling, the cell should display the distance the tomato still has to fall.
    - If the tomato hits the ground, the cell should display "SPLAT" on a red background.
    - SPLAT should appear once; the cells below are blank.

Test your worksheet using the following conditions:

I.   At a balcony height of 200 feet, the tomato should splat at a time of 3.52 seconds.
II. At a balcony height of 50 feet, the tomato should splat at a time of 1.76 seconds.

ICA 10-20

You are interested in calculating the best place to stand to look at a statue. Where should you stand so that the angle subtended by the statue is the largest?

At the top of the worksheet, input the pedestal height (P) and the statue height (S).

In column A, create a series of distances (d) from the foot of the statue, from 2 feet to 40 feet by 2-foot increments.

In column B, calculate the subtended angle in radians using the following equation:

$$\theta = \tan^{-1}\left(\frac{P+S}{d}\right) - \tan^{-1}\left(\frac{P}{d}\right)$$

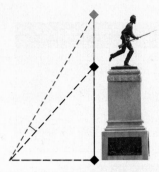

Photo courtesy of E. Stephan

In column C, write a function to change the angles in column B from radians to degrees. At the bottom of column C, insert a function to display the maximum value of all the angles.

In column D, use a conditional statement whose output is blank except at the single distance where the angle is a maximum; at the maximum, print "Stand Here." This font should be in the color of your choice, not the default black text.

Test your worksheet using the following conditions:

I. At a pedestal height of 20 feet and a statue height of 10 feet, the subtended angle is 11.5 degrees and you should stand 24 feet from the statue.
II. At a pedestal height of 30 feet and a statue height of 20 feet, the subtended angle is 14.5 degrees and you should stand 38 feet from the statue.

ICA 10-21

Many college students have compact refrigerator-freezers in their dorm rooms. The data set provided is a partial list of energy-efficient models less than 3.6 cubic feet [cft], according to the American Council for an Energy Efficient Economy (www.aceee.org). Complete the following analysis.

We would like to compute the cost to run each model for a year. Assume that it costs $0.106 per kilowatt-hour [kWh]. Create a new column, "Annual Energy Cost [$/year]," that calculates the annual energy cost for each refrigerator.

(a) Sort the first table by energy usage, with the model with the highest kilowatt-hour rating listed first. Which model appears first?
(b) Sort by the volume in ascending order and the annual energy cost in ascending order. Which model appears first?
(c) Assume we want to restrict our selection to refrigerators that can contain more than 2.0 cubic feet. Which models appear in the list?
(d) Assume we want to restrict our selection to refrigerators that can contain more than 2.0 cubic feet and only require between 0 and 300 kilowatt-hours per year. Which models appear in the list?

ICA 10-22

The complexity of video gaming consoles has evolved over the years. The data set provided is a list of energy usage data on recent video gaming consoles, according to the Sust-It consumer energy report data (http://www.sust-it.net).

Compute the cost to run each gaming console for a year, including the purchase price. Assume that it costs $0.086 per kilowatt-hour [kWh]. Create a new column, "Cost + Energy [$/yr]," that calculates the total (base + energy) cost for each gaming console.

On average, a consumer will own and operate a video gaming console for four years. Calculate the total carbon emission [kilograms of carbon dioxide, or kg CO_2] for each gaming console over the average lifespan; put the result in a column labeled "Average Life Carbon Emission [kg CO_2]." If these steps are completed correctly, the first-year cost for the Microsoft Xbox 360 should be $410.36 and the Average Life Carbon Emissions should be 207.88 kilograms of carbon dioxide.

(a) Sort the table by total cost, with the console with the highest total cost listed first. Which console appears first?
(b) Sort by the original cost in ascending order and the average life carbon emission in ascending order. Which console appears last?
(c) Restrict your selection to video game consoles that originally cost $300. Which models appear in the list?
(d) Restrict your selection to video game consoles that originally cost less than or equal to $300 and have an average life carbon emission less than or equal to 25 kg CO_2. Which models appear in the list?
(e) In the filtered selection in part (d), use the SUBTOTAL function to determine the average cost of the filtered models.

CHAPTER 10

REVIEW QUESTIONS

1. A history major of your acquaintance is studying agricultural commerce in nineteenth century Wales. He has encountered many references to "hobbits" of grain, and thinking that this must be some type of unit similar to a bushel (rather than a diminutive inhabitant of Middle Earth), he has sought your advice because he knows you are studying unit conversions in your engineering class.

He provides a worksheet containing yearly records for the total number of hobbits of three commodities sold by a Mr. Thomas between 1817 and 1824, and he has asked you to convert these to not only cubic meters, but also both U.S. and imperial bushels.

	A	B	C	D	E	F	G	H	I	J	K	L	M
1													
2													
3													
4													
5		Barley				Wheat				Oats			
6	Year	Hobbits	Imp. Bushels	US Bushels	Cubic Meters	Hobbits	Imp. Bushels	US Bushels	Cubic Meters	Hobbits	Imp. Bushels	US Bushels	Cubic Meters
7	1817	106				154				203			
8	1818	118				145				187			
9	1819	98				167				167			
10	1820	137				124				199			
11	1821	102				105				210			
12	1822	142				168				147			
13	1823	93				132				186			
14	1824	117				136				193			
15													
16		Hobbit	Imp. Bushels	US Bushels	Cubic Meters								
17		1											

After a little research, you find that the hobbit was equal to two and a half imperial bushels, the imperial bushel equals 2219 cubic inches, and the U.S. bushel equals 2150 cubic inches.

First, you create a table showing the conversion factors from hobbits to the other units, including comments documenting the conversion. You then use these calculated conversion factors to create the rest of the table.

2. You want to set up a worksheet to investigate the oscillatory response of an electrical circuit. Create a worksheet similar to the one shown, including the proper header information.

4			
5			
6			
7	Neper Frequency (α_0)	25	[rad/s]
8	Resonant Frequency (ω_0)	400	[rad/s]
9	Initial Voltage (V_0)	15	[V]
10			
11	Damped Frequency (ω_d)		[rad/s]
12			
13			
14	Time (t) [s]	Voltage (V) [V]	
15			

First, calculate another constant, the damped frequency ω_d, which is a function of the neper frequency (α_0) and the resonant frequency (ω_0). This can be calculated with the formula

$$\omega_d = \sqrt{\omega_0^2 - \alpha_0^2}$$

Next, create a column of times (beginning in A15) used to calculate the voltage response, ranging from 0 to 0.002 seconds at an increment of 0.0002 seconds.

In column B, calculate the voltage response with the following equation, formatted to one decimal place:

$$V = V_0\, e^{-\alpha_0 t} \cos(\omega_d t)$$

Test Cases: Use the following to test your worksheet.

I. Change neper frequency to 200 radians per second, resonant frequency to 800 radians per second, and initial voltage to 100 volts. At a time of 0.0008 seconds, the voltage should be 69.4 V.

II. Change neper frequency to 100 radians per second, resonant frequency to 600 radians per second, and initial voltage to 100 volts. At a time of 0.0008 seconds, the voltage should be 82.2 V.

3. A phase diagram for carbon and platinum is shown. Assuming the lines shown are linear, we can say the mixture has the following characteristics:

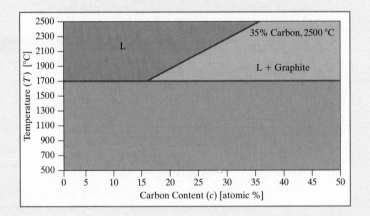

- Below 1700°C, it is a mixture of solid platinum and graphite.
- Above 1700°C, there are two possible phases: a liquid (L) phase and a liquid (L) + graphite phase. The endpoints of the division line between these two phases are labeled on the diagram.

Use the workbook provided to determine the phase of a mixture, given the temperature and carbon content.

	A	B	C	D	E
1					
2					
3	Maximum Temperature for Pt + G			1700	[°C]
4					
5					
6					
7	Temperature	Carbon Content	Temp between	Phase	
8	(T) [°C]	(c) [%]	L & L+G		
9	854	42			
10	564	20			
11	965	25			

(a) Write the equation to describe the temperature of the dividing line between the liquid (L) region and the liquid (L) + graphite region in column C. Reference the carbon content found in column B as needed. Add any absolute reference cells you feel are needed to complete this calculation.

(b) Write the conditional statement to determine the phase in column D. For simplicity, call the phases Pt + G, L, and L + G. For points on the line, YOU can decide which phase they are included in.

4. A simplified phase diagram for cobalt and nickel is shown. Assuming the lines shown are linear, we can say the mixture has the following characteristics:

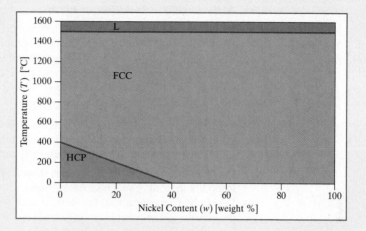

- Above 1500°C, it is a liquid.
- Below 1500°C, there are two possible phases: face-centered cubic (FCC) phase and hexagonal close-packed (HCP) phase.

Use the workbook provided to determine the phase of a mixture, given the temperature and nickel content.

	A	B	C	D	E
1					
2					
3	Minimum Temperature for Liquid			1500	[°C]
4					
5					
6					
7	Temperature	Nickel Content	Temp Between	Phase	
8	(T) [°C]	(w) [%]	HCP & FCC		
9	543	74			
10	1028	4			
11	1326	69			

(a) Write the mathematical equation to describe the dividing line between the HCP region and the FCC region in column C. Reference the nickel content found in column B as needed. Add any absolute reference cells you feel are needed to complete this calculation.

(b) Write the conditional statement to determine the phase in column D. For simplicity, call the phases HCP, FCC, and L. For points on the line, YOU can decide which phase they are included in.

(c) Use conditional formatting to indicate each phase. Provide a color key.

5. You enjoy drinking coffee but are particular about the temperature (T) of your coffee. If the temperature is greater than or equal to 70 degrees Celsius [°C], the coffee is too hot to drink; less than or equal to 45°C is too cold by your standards. Your coffee pot produces coffee at the initial temperature (T_0). The cooling of your coffee can be modeled by the following equation, where time (t) and the cooling factor (k) are in units per second:

$$T = T_0 e^{-kt}$$

(a) At the top of the worksheet, create an area where the user can modify four properties of the coffee. For a sample test case, enter the following data.

- Initial temperature (T_0); for the initial problem, set to 80°C.
- Cooling factor (k); set to 0.001 per second [s^{-1}].
- Temperature above which coffee is "Too Hot" to drink (T_{hot}); set to 70°C.
- Temperature below which coffee is "Too Cold" to drink (T_{cold}); set to 45°C.

(b) Create a temperature profile for the coffee:

- In column A, generate a time range of 0–300 seconds, in 15-second intervals.
- In column B, generate the temperature of the coffee, using the equation given and the input parameters set by the user (T_0 and k).

(c) In column B, the temperature values should appear on a red background if the coffee is too hot to drink, and a blue background if it is too cold using conditional formatting.

(d) In column C, create a warning next to each temperature that says "Do Not Drink" if the calculated temperature in column B is too hot or too cold in comparison with the temperature values the user enters.

A sample worksheet is shown here for the test case described in part (a).

Coffee Parameters		
Initial Temperature (T_0)	80	[°C]
Cooling Factor (k)	0.001	[1/s]
Too Hot Temperature (T_{hot})	70	[°C]
Too Cold Temperature (T_{cold})	45	[°C]

Time [s]	Temperature [deg C]	Warning!
0	80	Do not Drink
15	79	Do not Drink
30	78	Do not Drink
45	76	Do not Drink
60	75	Do not Drink
75	74	Do not Drink
90	73	Do not Drink
105	72	Do not Drink
120	71	Do not Drink
135	70	
150	69	
165	68	
180	67	

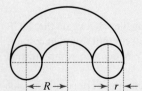

6. In the 1950s, a team at Los Alamos National Laboratories built several devices they called "Perhapsatrons," thinking that PERHAPS they might be able to create controllable nuclear fusion. After several years of experiments, they were never able to maintain a stable plasma and abandoned the project.

The perhapsatron used a toroidal (doughnut-shaped) plasma confinement chamber, similar to those used in more modern Tokamak fusion devices. You have taken a job at a fusion research lab, and your supervisor asks you to develop a simple spreadsheet to calculate the volume of a torus within which the plasma will be contained in a new experimental reactor.

(a) Create a simple calculator to allow the user to type in the radius of the tube (r) in meters and the radius of the torus (R) in meters and display the volume in cubic meters.

(b) Data validation should be used to assure that $R > r$ in part (a).

(c) Create a table that calculates the volumes of various toruses with specific values for r and R. The tube radii (r) should range from 5 centimeters to 100 centimeters in increments of 5 centimeters. The torus radii (R) should range from 1.5 meters to 3 meters in increments of 0.1 meters.

The volume of a torus can be determined using $V = 2\pi^2 R r^2$. A sample worksheet for parts (a) and (b) is shown here.

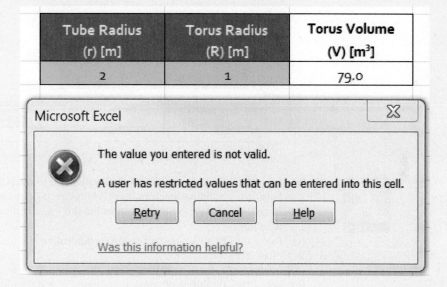

Tube Radius (r) [m]	Torus Radius (R) [m]	Torus Volume (V) [m³]
2	1	79.0

Use the following phase diagram for questions 7 and 8.

The following phase diagram for the processing of a polymer relates the applied pressure to the raw material porosity.

- Region A or B = porosity is too high or too low for the material to be usable.
- Region C = combinations in this region yield material with defects, such as cracking or flaking.
- Region D = below a pressure of 15 pound-force per square inch [psi] the polymer cannot be processed.
- Region E = optimum region to operate.

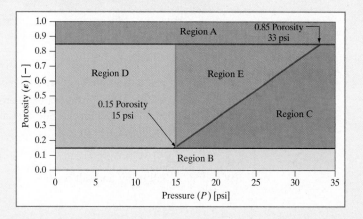

There are often multiple ways to solve the same problem; here we look at a few alternative ways to determine the phase of the material and the processability of the material.

7. (a) In column C, develop the equation for the line dividing the phases of Region E and Region C. Assume it was written in cell C9 and copied to column C.

(b) In column D, write an expression to determine the phase of the material (phase A–phase E).

(c) In column E, write an expression to determine if the material is processible.

(d) When the conditions of phase E are met, the cell should be highlighted by conditional formatting. Provide a color key.

	A	B	C	D	E
1					
2					
3	Porosity Upper Limit [%]			85	
4	Porosity Lower Limit [%]			15	
5	Pressure Limit [psi]			15	
6					
7	Pressure	Porosity	Porosity between	Phase	Is Material able to
8	(P) [psi]	(ε) [%]	C and E		be Processed?
9	16	91			
10	17	60			
11	18	20			

8. (a) In column A and column B, use data validation to restrict the user from entering values outside the valid parameter ranges—pressure: 0–35 psi and porosity: 0–100%.

(b) In column C, develop the equation for the line dividing the phases of Region E and Region C.

(c) In column D, write an expression to determine the phase of the material (phase A–phase E).

(d) In column E, write an expression to determine if the material is processible.

(e) When the conditions of phase E are met, the cell should be highlighted by conditional formatting.

(f) Write an expression in column F to tell the user why the material was rejected. For example, under the conditions of pressure = 25 psi and porosity = 40%, the statement might say "Porosity too low."

	A	B	C	D	E	F
1						
2						
3	Porosity Upper Limit [%]			85		
4	Porosity Lower Limit [%]			15		
5	Pressure Limit [psi]			15		
6						
7	Pressure	Porosity	Porosity between			
8	(P) [psi]	(ε) [%]	C and E			
9	50	68				
10						
11						
12						

Microsoft Excel dialog: The value you entered is not valid. A user has restricted values that can be entered into this cell. [Retry] [Cancel] [Help] Was this information helpful?

Use the following phase diagram for questions 9 and 10.

The following phase diagram is for salt water. There are four possible phases, which depend on the temperature and the sodium chloride content (NaCl).

- Ice and SC = Mixed ice and salt crystals.
- Ice and SW = Ice and salt water.
- SW = Salt water.
- SW and SC = Salt water and salt crystals.

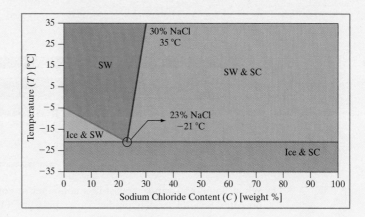

There are often multiple ways to solve the same problem; here we look at a few alternative ways to determine the phase of the mixture.

9. (a) In column C, develop the equation for the line dividing the phases of the ice–salt water mix and the salt water. Assume it was written in cell C11 and copied down.
   (b) In column D, develop the equation for the line dividing the phases of the salt water and the salt water–salt crystals mix. Assume it was written in cell D11 and copied down.
   (c) In column E, write an expression to determine the phase of the mixture.
   (d) Use conditional formatting to highlight the various phases. Provide a color key.

	A	B	C	D	E
1					
2					
3					
4					
5					
6					
7	Upper Limit of Mixed Ice and Salt Crystals			-21	[°C]
8					
9	NaCl [%]	Temp [°C]	Dividing Temp [° C]		Phase
10			Ice and SW to SW	SW to SW and SC	
11	84	31			
12	60	-17			
13	81	-17			
14	41	-17			

10. (a) In column A and column B, use data validation to restrict the user from entering values outside the valid parameter ranges: NaCl (%): 0–100%; Temp [°C]: −35°C to 35°C.
   (b) In column C, develop the equation for the line dividing the phases of the ice–salt water mix and the salt water.
   (c) In column D, develop the equation for the line dividing the phases of the salt water and the salt water–salt crystals mix.
   (d) In column E, write an expression to determine the phase of the mixture.
   (e) Use conditional formatting to highlight the various phases. Provide a color key.

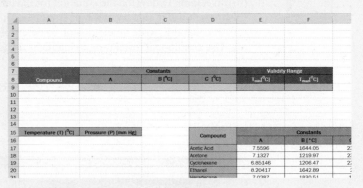

11. When liquid and vapor coexist in a container at equilibrium, the pressure is called vapor pressure. Several models predict vapor pressure. One, called the **Antoine equation**, first introduced by Ch. Antoine in 1888, yields vapor pressure in units of millimeters of mercury [mm Hg].

$$P = 10^{\left(A - \frac{B}{T+C}\right)}$$

The constants A, B, and C are called the *Antoine constants*; they depend on both fluid type and temperature. Note that B and C must be in the same units as temperature, and A is a dimensionless number, all determined by experiment.

Create a worksheet using the provided template. The Antoine constants, located in cells D17 to I24 of the workbook provided, should automatically fill in after the user selects one from a drop-down menu in cell A9 of the compounds shown below. (*Hint:* Use data validation and lookup expressions.)

Next, create a column of temperature (T) beginning at -100 degrees Celsius and increasing in increments of 5 degrees Celsius until a temperature of 400 degrees Celsius.

In column B, calculate the vapor pressure (P, in millimeters of mercury, [mm Hg]) using the Antoine equation, formatted to four decimal places. If the equation is outside the valid temperature range for the compound, the pressure column should be blank.

12. The ideal gas law assumes that molecules bounce around and have negligible volume themselves. This is not always true. To compensate for the simplifying assumptions of the ideal gas law, the Dutch scientist Johannes van der Waals developed a "real" gas law that uses several factors to account for molecular volume and intermolecular attraction. He was awarded the Nobel Prize in 1910 for his work. The *van der Waals equation* is as follows:

$$\left(P + \frac{an^2}{V^2}\right)(V - bn) = nRT$$

$P, V, n, R,$ and T are the same quantities as found in the ideal gas law. The constant a is a correction for intermolecular forces [atm L²/mol²], and the constant b accounts for molecular volume [L/mol]. Each of these factors must be determined by experiment.

	A	B	C	D	E	F
1						
2						
3						
4						
5						
6						
7	**Type of Gas**				**Compound**	**a**
8	**Quantity [g]**					**[atm L² / m**
9	**Temperature (T) [°C]**				Acetic Acid	17.587
10					Acetone	13.906
11					Ammonia	4.170
12	**Molecular Weight (MW) [g/mol]**				Argon	1.345
13	**vdW Constant "a" [atm L²/mol²]**				Benzene	18.001
14	**vdW Constant "b" [L/mol]**				Chlorobenzene	25.433
15					Diethyl ether	17.380
16	**Ideal Gas Constant (g) [(atm L)/mol K)]**		0.08206		Ethane	5.489
17					Ethanol	12.021
18					Hexane	24.387
19		**Ideal Gas**	**van der Waals**		Methanol	9.523
20	**Volume (V) [L]**	**Pressure (P) [atm]**			Neon	0.211
21					Oxygen	1.360

HINT

Use data validation and lookup expressions using the data found in the table located in E7 to H26 in the workbook provided.

NOTE

The astronomical unit (AU) is the average distance from Earth to the sun.

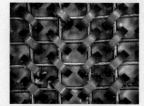

Photo courtesy of W. Park

Create a worksheet using the provided template. The molecular weight, a, and b should automatically fill in after the user selects the type of gas in cell B7. The user will also set the quantity of gas and the temperature of the system.

Next, create a column of volume beginning in A21 at 0.5 liters and increasing in increments of 0.1 liters to a volume of 5 liters.

In column B, calculate the pressure (P, in atmospheres [atm]) using the ideal gas law.

In column C, calculate the pressure (P, in atmospheres [atm]) using the van der Waals equation.

13. One of the NAE Grand Challenges for Engineering is **Engineering the Tools of Scientific Discovery**. According to the NAE website: "Grand experiments and missions of exploration always need engineering expertise to design the tools, instruments, and systems that make it possible to acquire new knowledge about the physical and biological worlds."

Solar sails are a means of interplanetary propulsion using the radiation pressure of the sun to accelerate a spacecraft. The table contained in the starting Excel file shows the radiation pressure at the orbits of the eight planets.

Create a table showing the area in units of square meters of a solar sail needed to achieve various accelerations for various spacecraft masses at the distances from the sun of the various planets. Your solution should use data validation and VLOOKUP to select a planet and the corresponding radiation pressure. The columns of your table should list masses of the spacecraft (including the mass of the sail) ranging from 100 to 1000 kilograms in increments of 100 kilograms. The rows should list accelerations from 0.0001 to 0.001 g in increments of 0.001 g, where g is the acceleration of Earth's gravity, 9.8 meters per second squared. All constants and conversion factors should be placed in individual cells using appropriate labels, and all formulae should reference these cells and should NOT be directly coded into the formulae. You should use absolute, relative, and mixed addressing as appropriate.

14. A hands-on technology museum has hired you to do background research on the feasibility of a new activity to allow visitors to assemble their own ferrite core memory device—a technology in common use until the 1970s, and in specialized applications after that. The computers onboard the early space shuttle flights used core memory due to their durability, nonvolatility, and resistance to radiation—core memory recovered from the wreck of the Challenger still functioned.

Ferrite core memory comprises numerous tiny ferrite rings ("cores") in a grid, each of which has either two or three wires threaded through it in a repeating pattern and can store a single bit, or binary digit—a 0 or a 1. Since the cores were typically on the order of one millimeter in diameter, workers had to assemble these under microscopes.

After investigating ferrite materials, you find several that would be suitable for fabrication of the cores. The museum staff has decided to have the visitors assemble a 4 × 4 array (16 cores—actual devices were MUCH larger) and anticipate that 2500 people will assemble one of these over the course of the project. Assuming that the cores are each cylindrical rings with a hole diameter half that of the outside diameter of the ring and a thickness one-fourth the outside diameter, you need to know how many grams of ferrite beads you need to purchase with 10% extra beyond the specified amount for various core diameters and ferrite materials. You also wish to know the total cost for the beads.

Using the provided online worksheet that includes a table of different ferrite material densities and costs, use data validation to select one of the materials from the list, then create a table showing the number of pounds of cores for core diameters of 1.2 to 0.7 millimeter in 0.1 millimeter increments as well as the total cost. For cores with a diameter less than 1 millimeter, there is a 50% manufacturing surcharge, thus the smallest cores cost more per gram. Include table entries for individual core volume and total volume of all cores. Your worksheet should resemble the following example.

	A	B	C	D	
1					
2					
3					
4			Cost of Ferrite Cores for Ha		
5					
6	Ferrite Compound	Specific Gravity	Cost per Gram [$/g]		Fer Comp
7	CMP C	3.73	$ 32.50		CM
8					CM
9	Sets needed	2500			CM
10	Cores per set	16			CM
11	Total Cores	40,000			CM
12	Total + 10%	44,000			CM
13					
14				Core Outs	
15			0.7	0.8	0
16	Volume of one core [mm^3]		0.0505	0.0754	0.1
17	Volume of all Cores [mm^3]		2,222	3,318	4.7
18	Mass of Cores [g]		8.29	12.37	17
19	Cost of Cores [$]		$ 404.11	$ 603.21	$ 8

15. Create an Excel worksheet that will allow the user to type in the radius of a sphere and select from a drop-down menu the standard abbreviation for the units used.

Standard Unit Abbreviations						
Unit	meter	centimeter	millimeter	yard	foot	inch
Abbreviation	m	cm	mm	yd	ft	in

The volume of the sphere should then be calculated and expressed by the following units: cubic meters, cubic centimeters, cubic millimeters, liters, gallons, cubic yards, cubic feet, and cubic inches. Your worksheet should appear similar to the sample shown below, although you will probably need additional information in the worksheet not shown here.

Enter Radius Value Here	Enter Radius Units Here	Volume							
		m^3	cm^3	mm^3	liters	gallons	yd^3	ft^3	in^3
1.7	in	3.372E-04	3.372E+02	3.372E+05	3.372E-01	8.910E-02	4.416E-04	1.191E-02	2.058E+01

16. Most resistors are so small that the actual value would be difficult to read if printed on the resistor. Instead, colored bands denote the value of resistance in ohms. Anyone involved in constructing electronic circuits must become familiar with the color code, and with practice, one can tell at a glance what value a specific set of colors means. For the novice, however, trying to read color codes can be a bit challenging.

Begin with the worksheet template provided. In the worksheet, the user will enter a resistance value as the first two digits and a multiplier, both selected using a drop-down menu created through data validation. The resistance should be calculated as the first two digits times the multiplier.

The worksheet should automatically determine the First Digit and the Second Digit of the value entered in cell E7, using the built-in functions LEFT and RIGHT. The number of zeros should be determined using the lookup function.

Finally, the worksheet should determine the corresponding resistance band color using the Color Code table. The cells should automatically change to the correct color when the digits or multiplier are changed using conditional formatting.

For example, a resistance of 4700 ohms [Ω] has first digit 4 (yellow), second digit 7 (violet), and two zeros following (red). A resistance of 56 ohms would be 5 (green), 6 (blue), and 0 zeros (black); 1,000,000 ohms is 1 (brown), 0 (black), and 5 zeros (green). Note that if the second digit is zero, it does not count in the multiplier value. There are many explanations of the color code on the Web if you need further information or examples.

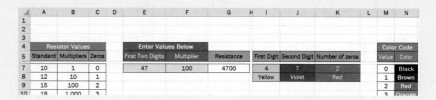

17. Download the starting file, and complete the following commands using the data provided.

   (a) Indicate the following using conditional formatting commands of your choice. Each of the following conditions should appear in a unique format.

   - Length shown in column B is greater than 6 inches or less than 4 inches.
   - Width shown in column C is less than 2.5 inches.
   - Inner radius shown in column D is above average for the inner radius values.
   - Outer radius shown in column E is below average for the outer radius values.
   - Volume shown in column F is less than 10 cubic inches or greater than 20 cubic inches.

   (b) For the following conditions, in column H use an IF statement to indicate the status:

   - If length is less than 4 inches or width is less than 2.5 inches, list the status as "Too Small."
   - Otherwise, if twice the inner radius is greater than the outer radius, list the status as "Off Center."
   - Otherwise, if the volume is greater than 20 cubic inches or the mass is greater than 3000 grams, list the status as "Too Large."
   - Otherwise, if none of these conditions are true, leave the cell blank.

   (c) For the following conditions, in column J use an IF statement to indicate the action code:

   - If the status is "Too Small" or "Too Large," list the action code as a numerical value of 1.
   - If the status is "Off Center," list the action code as a numerical value of 2.
   - If none of these conditions are met, list the action code as a numerical value of 3.

   (d) Use a conditional formatting icon set in column I to indicate the following:

   - Status as green for action code 3.
   - Status as yellow for action code 2.
   - Status as red for action code 1.

(e) Count the following items, showing the results somewhere above the data table. Be sure to indicate each counted item with an appropriate label.

- Indicate the number of items classified as each action code, such as how many items are listed as 1.
- Indicate number of parts when the length is greater than 6 inches.
- Indicate number of parts when the volume is less than 10 cubic inches or greater than 20 cubic inches. As a hint, use two COUNT functions and add them together.

(f) Sort the worksheet in the following order: Length, increasing and simultaneously then Outer Radius, decreasing. Be careful to select only the data and not the entire worksheet.

(g) Set the worksheet controls to be filtered in the header row. Filter the worksheet so only parts of length 2.80, 5.20, and 7.15 inches are shown.

18. Guitars are made in a way that allows musicians to place their fingers on different arrangements of strings to create chords. Chords are arrangements of three or more notes (or pitches) played at the same time. (There is some disagreement among musicologists concerning whether two notes played simultaneously should be classified as a chord.) On a standard electric guitar, there are typically six strings strung from the nut down the neck of the guitar with 21 frets set in the neck and terminating at the bridge on the guitar. On a six-string, 21-fret guitar, there are 126 different locations that can generate a pitch plus the six pitches generated by playing a string without pressing a fret. However, each fretted string on a guitar can take on one of 12 different semitones. As guitarist moves their finger from lower to higher frets on a string, each note change to increases by one semitone.

The 12-tone semitone transitions are shown in the chart below:

From	To		From	To
A	A#/Bb		D#/Eb	E
A#/Bb	B		E	F
B	C		F	F#/Gb
C	C#/Db		F#/Gb	G
C#/Db	D		G	G#/Ab
D	D#/Eb		G#/Ab	A

Source: Evgeny Guityaev/ Shutterstock

In other words, if we are on the fifth string playing a B on the 10th fret, moving to the 11th fret would play a C.

Fill in the provided Excel worksheet using LOOKUP statements to the semitone transition table for a standard tuning guitar (E-A-D-G-B-E) to show where each note will be located on the guitar fretboard. The worksheet should be set up so that it can be easily modified for different guitar tunings:

(a) What will the resulting notes on the fifth fret be in an "open tuning," typical of blues and slide guitar music, if the open G tuning were used (G-B-D-G-B-D)?

(b) What will the resulting notes on the eighth fret be if the D-A-D-G-A-D tuning were used (e.g., Led Zeppelin's "Kashmir")?

(c) Use conditional formatting to highlight the sharps and flats (notes with # or b) in light yellow.

(d) Use conditional formatting to highlight all entries with the same note name as one of the base tuning notes within the column by making the text white on a black background. For standard tuning, this would highlight all occurrences of E in the first column, A in the second column, etc.

(e) Use data validation to allow the user to choose the notes from a drop-down list.

Graphical Solutions

Often, the best way to present technical data is through a "picture." If not done properly, however, it is often the worst way to display information. As an engineer, you will have many opportunities to construct such pictures. If technical data are presented properly in a graph, it is often possible to explain a point in a concise and clear manner that is impossible any other way.

11.1 Graphing Terminology

LEARN TO: Identify the abscissa and the ordinate of a graph
Identify the independent and dependent variables in a problem

Abscissa is the horizontal axis; **ordinate** is the vertical axis. Until now, you have probably referred to these as x and y. This text uses the terms "abscissa" and "ordinate," or "horizontal" and "vertical," since x and y are only occasionally used as variables in engineering problems.

The **independent** variable is the parameter that is controlled or whose value is selected in the experiment; the **dependent** variable is the parameter that is measured corresponding to each set of selected values of the independent variable. Convention usually shows the independent variable on the abscissa and the dependent variable on the ordinate.

Data sets given in tabular form are commonly interpreted and graphed with the leftmost column or topmost row as the independent variable and the other columns or rows as the dependent variable(s). For the remainder of this text, if not specifically stated, assume that the abscissa variable is listed in the leftmost column or topmost row in a table of data values.

Time	Distance (d) [m]	
(t) [s]	Car 1	Car 2
Abscissa	Ordinate	Ordinate

Time (t) [s]		Abscissa
Distance	Car 1	Ordinate
(d) [m]	Car 2	Ordinate

11.2 Proper Plots

We call graphs constructed according to the following rules **proper plots**:

- **Label both axes clearly.** Three things are required unless the information is unavailable: category (e.g., Time), symbol used (t), and units [s]. Units should accompany all quantities when appropriate, enclosed in square brackets [ ].
- **Select scale increments (both axes) that are easy to read and interpolate between.** With a few exceptions, base your scale on increments of 1, 2, 2.5, and 5. You can scale each value by any power of 10 as necessary to fit the data. Avoid unusual increments (such as 3, 7, 15, or 6.5).

Increment	Sequence				
1	0	10	20	30	40
5	0.05	0.10	0.15	0.20	0.25
2.5	-2500	0	2500	5000	7500
2	6×10^{-5}	8×10^{-5}	1×10^{-4}	1.2×10^{-4}	1.4×10^{-4}

In this final case, reading is easier if the axis is labeled something like Time (t) [s] $\times 10^{-4}$ so that only the numbers 0.6, 0.8, 1.0, 1.2, and 1.4 show on the axis.

- **Provide horizontal and vertical gridlines** to make interpolation easier to aid the reader in determining actual numerical values from the graph.

  When minor gridlines are present, the reader should be able to easily determine the value of each minor increment. For example, examine the graphs shown in Figure 11-1. In which graph is it easier to determine the abscissa value for the blue point? In the graph on the left, the abscissa increment can easily be determined as 0.1 meters. In the graph on the right, it is more difficult to determine the increment as 0.08 meters.
- **Provide a clear legend** describing each data set of multiple data sets shown. Do not use a legend for a single data set. Legends may be shown in a standalone box or captioned next to the data set. Both methods are shown in Figure 11-2.

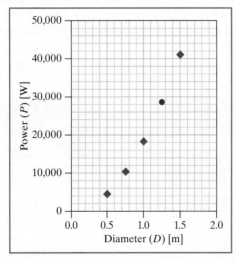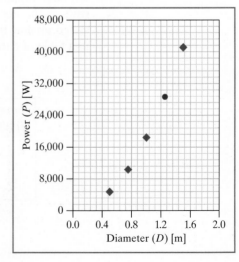

Figure 11-1 Example of importance of minor gridline spacing.

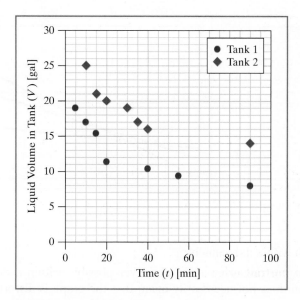
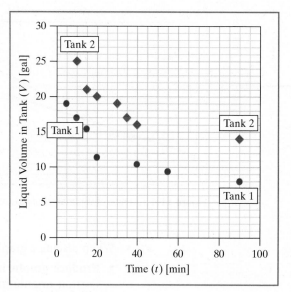

Figure 11-2 Options for displaying legends.

- **Show measurements as symbols. Show calculated or theoretical values as lines.** Do not display symbols for calculated or theoretical values. A symbol shown on a graph indicates that an experimental measurement has been made (see Figure 11-3).
- **Use a different symbol shape and color** for each experimental data set and a different line style and color for each theoretical data set. ***Never use yellow and other light pastel colors***. Remember that when graphs are photocopied, all colored lines become black lines. Some colors disappear when copied and are hard to see in a projected image. For example, in Figure 11-4, left, it is much easier to distinguish between the different lines than in the figure on the right.

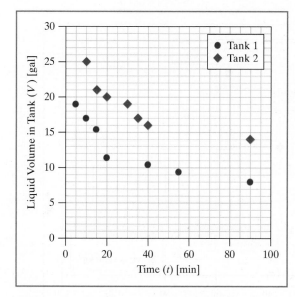
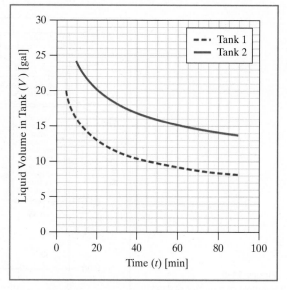

Figure 11-3 Illustration of experimental data (shown as points) versus theoretical data (shown as lines).

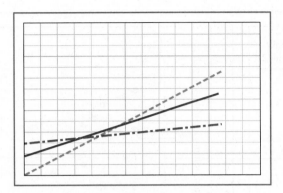
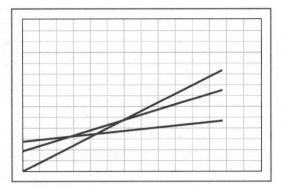

Figure 11-4 Example of importance of different line types.

- When placing a graph within a document:
  - **Produce graphs in portrait orientation** whenever possible within a document. Portrait orientation does not necessarily mean that the graph is distorted to be taller than it is wide; it means that readers can study the graph without turning the page sideways.
  - **Be sure the graph is large enough to be easily read.** The larger the graph, the more accurate the extracted information.
  - **Caption with a brief description.** The restating of "*d* versus *t*" or "distance versus time" or even "the relationship between distance and time" does not constitute a proper caption. The caption should give information about the graph to allow the graph to stand alone without further explanation. It should include information about the problem that does not appear elsewhere on the graph. For example, instead of stating "distance versus time," better choices would be "Lindbergh's Flight across the Atlantic," "The Flight of *Voyager I*," or "Walking between Classes across Campus, Fall 2008." When including a graph as part of a written report, place the caption below the graph.

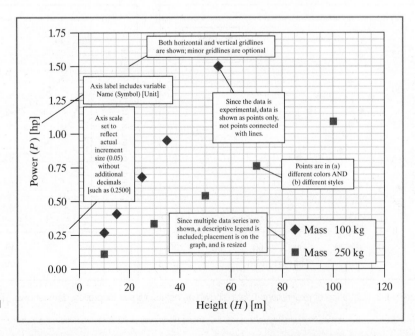

Figure 11-5

Example of a proper plot, showing multiple experimental data sets.

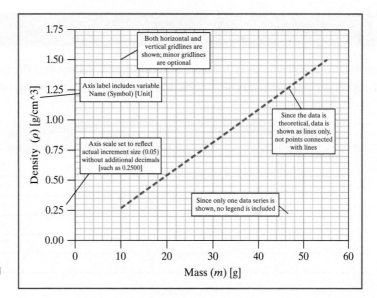

Figure 11-6
Example of a proper plot, showing a single theoretical data set.

Figure 11-7 is an example of a poorly constructed plot. Some problems with this plot are:

- It is a plot of distance versus time, but is it the distance of a car, a snail, or a rocket? What are the units of distance—inches, meters, or miles? What are the units of time—seconds, days, or years? Is time on the horizontal or vertical axis?
- Two data sets are shown, or are there three? Why is the one data set connected with a line? Is it a trendline? Is the same data set shown in the triangles? What do the shaded and open triangles represent—different objects, different trials of the same object, or modifications to the same object?
- Lack of gridlines and strange axis increments makes it difficult to interpolate between values. What is the location of the blue dot?

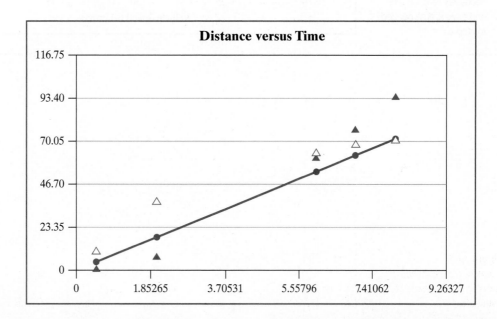

Figure 11-7
Example of a poorly constructed graph.

EXAMPLE **11-1**

When attempting to stop a car, a driver must consider both the reaction time and the braking time. The data are taken from http://www.highwaycode.gov.uk. Create a proper plot of these experimental data, with speed on the abscissa.

Vehicle Speed (v) [mph]	Distance	
	Reaction (d_r) [m]	Braking (d_b) [m]
20	6	6
30	9	14
40	12	24
50	15	38
60	18	55
70	21	75

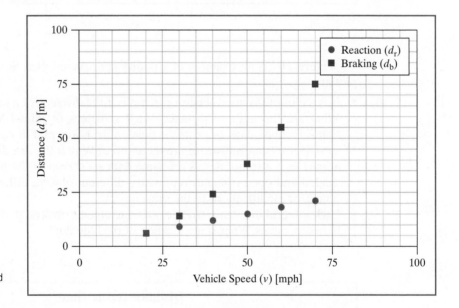

Figure 11-8
At various speeds, the necessary reaction time and braking time needed to stop a car.

EXAMPLE 11-2

Ohm's law describes the relationship between voltage, current, and resistance within an electrical circuit, given by the equation $V = IR$, where V is the voltage [V], I is the current [A], and R is the resistance [Ω]. Construct a proper plot of the theoretical voltage on the ordinate versus current, determined from the equation, for the following resistors: 3000 Ω, 2000 Ω, and 1000 Ω. Allow the current to vary from 0 to 0.05 A.

Note that while the lines were probably generated from several actual points along each line for each resistor, the points are not shown; only the resulting line is shown since the values were developed from theory and not from experiment (Figure 11-9). If you create a plot like this by hand, you would first put in a few points per data set, then draw the lines and erase the points so that they are not shown on the final graph.

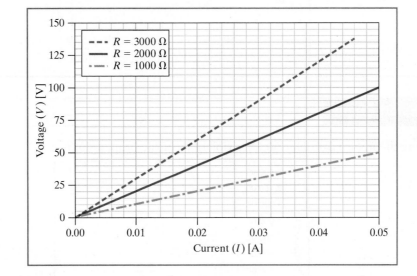

Figure 11-9
Ohm's law determined for a simple circuit to compare three resistor values

Comprehension Check 11-1

In the following experimental data plot, identify violations of the proper plot rules.

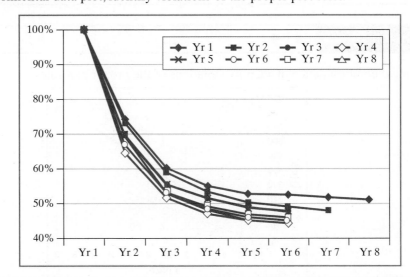

Comprehension Check | **11-2**

In the following experimental data plot, identify violations of the proper plot rules.

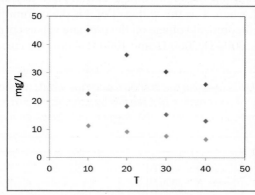

Comprehension Check | **11-3**

In the following theoretical data plot, identify violations of the proper plot rules.

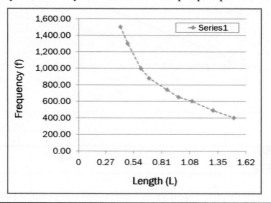

Comprehension Check | **11-4**

In the following theoretical data plot, identify violations of the proper plot rules.

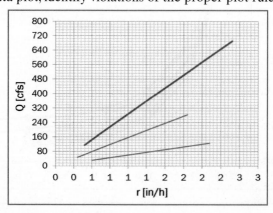

11.3 Available Graph Types in Excel

The following is an example of the level of knowledge of Excel needed to proceed. *If you are not able to quickly re-create the following exercise in Excel, please review graphing basics in Appendix C online before proceeding.*

Two graphs are given here; they describe the draining of tanks through an orifice in the bottom. When the tank contains a lot of liquid, the pressure on the bottom is large and the tank empties at a higher rate than when there is less liquid. The first graph shows actual data obtained from two different tanks. These data are given in the following table. The second plot shows curves (developed from theoretical equations) for two tanks. The equations for these curves are also given. Create these graphs exactly as shown, with matching legend, axis limits, gridlines, axis labels, symbol and line types, and colors.

Experimental data for first graph:

Time (*t*) [min]	5	10	15	20	40	55	90
Volume Tank 1 (*V1*) [gal]	19.0	17.0	15.5	11.5	10.5	9.5	8.0

Time (*t*) [min]	10	15	20	30	35	40	90
Volume Tank 2 (*V2*) [gal]	25	21	20	19	17	16	14

Theoretical equations for second graph (with *t* in minutes):

- Tank 1: Volume remaining in tank 1 [gal] $V = 33\,t^{-0.31}$
- Tank 2: Volume remaining in tank 2 [gal] $V = 44\,t^{-0.26}$

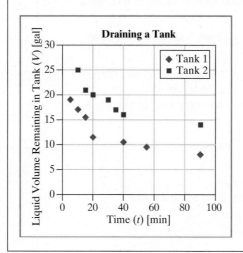

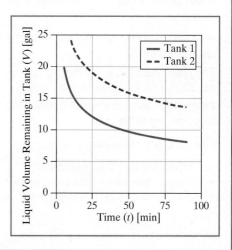

You can create many different types of charts in Excel. Usually, you will only be concerned with a few main types, shown in Table 11-1.

Table 11-1 Common chart types available in Excel

A **scatter plot** is a graph that numerically represents two-dimensional (2-D) theoretical or experimental data along the abscissa and ordinate of the graph. It is most commonly used with scientific data. To create a scatter plot, you specify each pair in the graph by selecting two identically sized columns or rows of data that represent the (x, y) values of each experimental symbol or point on a theoretical expression.

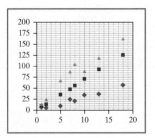

A scatter plot can be shown as discrete data points (used to show experimental data) or lines (used to show theoretical expressions). Excel will also show discrete data points connected by lines; the authors of this text do not find this type of chart particularly useful and do not discuss this type of chart.

The step size of both axes is evenly spaced as determined by the user and can be customized to show all or part of a data set plotted on a graph.

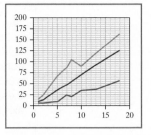

Use a scatter plot to visualize your data when you want to:

- Observe mathematical phenomena and relationships among different data sets
- Interpolate or extrapolate information from data sets
- Determine a mathematical model for a data set using trendlines

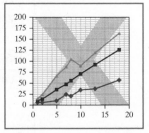

A **line plot** is a graph that visualizes a one-dimensional (1-D) set of theoretical or experimental data.

A line plot can be shown as points connected by lines, lines only, or in three dimensions (3-D).

The y-axis values of a line plot are spaced as determined by the user; however, the x-axis of a line plot is not. As shown in the graphs to the right, a line plot places each discrete element evenly along the x-axis regardless of the actual step-spacing of the data.

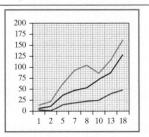

Use a line plot to visualize your data when you want to:

- Display any evenly spaced data
- Visualize time-series data taken at even intervals
- Display categorical data (e.g., years, months, days of the week)

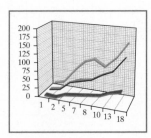

Table 11-1 Common chart types available in Excel (*continued*)

A **column graph** is used for displaying various types of categorical data.

The *y*-axis increments are spaced evenly, but the *x*-axis spacing has no meaning since the items are discrete categories. As a rule of thumb, a column graph can be used to represent the same information shown on a line plot.

A column plot can be shown as bars, cylinders, or cones; as a clustered group or stacked; or in 1-D or 3-D.

Use a column graph to visualize your data when you want to:

- Display any categorical data
- Observe differences between categories

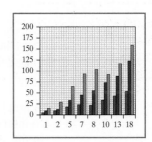

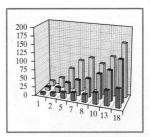

A **bar graph** is identical to a column graph, with the *x* and *y* categories reversed; the *x* category appears on the ordinate and the *y* category appears on the abscissa. Because of the similarity, only column graphs are covered in this text.

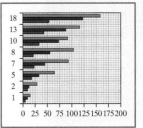

A **pie graph** is used on a single column or row of nonnegative numbers, graphed as a percentage of the whole. It is typically used for categorical data, with a maximum of seven categories possible.

A pie graph can be shown in 1-D or 3-D, with either the percentages or the raw data displayed with the category names.

Use a pie graph to visualize your data when you want to:

- Display categorical data as part of a whole
- Observe differences between categories

Pie charts are similar in form to column and bar charts; they are not covered in this text.

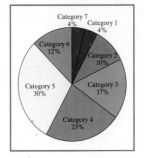

11.4 Graph Interpretation

LEARN TO: Calculate the area under a curve and describe its meaning
Calculate the slope of a line and describe its meaning
Understand the technical terms *derivative* and *integral* with respect to graph interpretation

A graph conveys a great deal of information in a small amount of space. By being able to interpret a graph, you can infer the story behind the lines. In addition to the value of the slope of the line, the shape of the line contains useful information.

EXAMPLE   11-3

Hourglass. Courtesy
of Thayer's Gifts,
Greenwood, SC.
Photo courtesy of
W. Park

Assume your company is designing a series of hourglasses for the novelty market, such as tourist attraction sales. You have determined that your prototype hourglass allows 275 cubic millimeters of sand to fall from the top to the bottom chamber each second. What volume of sand would be needed if the "hourglass" really measured a period of 10 minutes?

There are 60 seconds per minute, thus 10 minutes is 600 seconds. The sand flows at a rate of 275 cubic millimeters per second for 600 seconds, thus the total volume of sand is 165,000 cubic millimeters.

$$(275 \text{ mm}^3/\text{s}) (600 \text{ s}) = 165{,}000 \text{ mm}^3 \text{ or } 165 \text{ cm}^3$$

Let us consider the same problem graphically. Since the flow rate of sand is constant, a graph of flow rate with respect to time is simply a horizontal line. Now consider the area under the flow rate line. The area of a rectangle is simply the width times the height.

If we make a point of using the units on each axis as well as the numeric values, we get

$$((275 - 0) \text{ mm}^3/\text{s}) ((10 - 0) \text{ min} * 60 \text{ s/min}) = (275\text{mm}^3/\text{s}) (600 \text{ s}) = 165{,}000 \text{ mm}^3$$

This is exactly the same result we got above. In other words, the volume of sand is the area under the line (Figure 11-10).

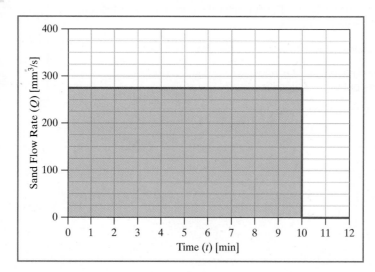

Figure 11-10
Sand in an hourglass.

This seems like much more effort than the straightforward calculation we did originally, so why should we bother with the graph? Let us look at a slightly more complicated situation.

EXAMPLE   11-4

Assume a container is being filled with sand. Initially, the sand enters the container at 100 grams per second, but the rate of filling decreases linearly for 20 seconds, then stops. The final rate of sand into the container just before it stops is 25 grams per second. How much sand enters the container during the 20 seconds involved?

NOTE

When discussing a rate of mass per time, such as grams per second, the quantity is referred to as the **mass flow rate**, symbolized by $\dot{m}$. When discussing a rate of volume per time, such as gallons per minute [gpm], this quantity is referred to as **volumetric flow rate**, symbolized by Q.

Let us compute the area under the line shown in Figure 11-11, being sure to include units, and see what we get. We can break this area into a rectangle and a triangle, which will make the calculation a bit easier.

- *The area of the rectangle at the base (below $\dot{m} = 25$) is ((25–0) grams/second) ((20–0) seconds) = 500 grams.*
- *The **slope** of the line on the triangle, which is the "rise" of the line divided by the "run" of the line, can be calculated to be ((100–25) grams per second)/((0–20) seconds) = 3.75 grams per second squared. In more technical terms, when two points on a line are described in terms of their x and y parameters, the rise is the change in the y parameter, and the run is the change in the x parameter.*
- *The area of the triangle is 0.5 ((100–25) grams per second) ((20–0) seconds) = 750 grams.*
- *The total area is 1250 grams, the total mass of sand in the container after 20 seconds.*

Again, many of you have realized that there is a much easier way to obtain this result. Simply find the average flow rate (in this case: 62.5 grams per second) and multiply by the total time.

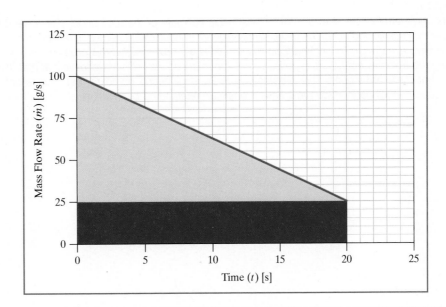

Figure 11-11

However, what if the parameter plotted on the vertical axis was not a simple straight line? Consider the following example.

EXAMPLE **11-5**

What if the parameter plotted on the vertical axis was not a simple straight line or straight-line segments? For example, the flow rate of liquid out of a pipe at the bottom of a cylindrical barrel follows an exponential relationship. Assume the flow rate out of a tank is given by $Q = 4\,e^{-t/8}$ gallons per minute. A graph of this is shown in Figure 11-12.

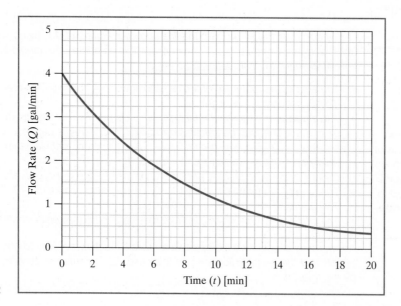

Figure 11-12

Although we might be able to make a reasonable estimate of the area under the curve (the total volume of water that has flowed out of the tank), simple algebra is insufficient to arrive at an accurate value. Those of you who have already studied integral calculus should know how to solve this problem. However, some students using this text may not have progressed this far in math, so we will have to leave it at that. It is enough to point out that there are innumerable problems in many engineering contexts that require calculus to solve. To succeed in engineering, you must have a basic understanding of calculus.

EXAMPLE   11-6

From the past experience of driving an automobile down a highway, you should understand the concepts relating acceleration, velocity, and distance. As you slowly press the gas pedal toward the floor, the car accelerates, causing both the speed and the distance to increase. Once you reach a cruising speed, you turn on the cruise control. Now, the car is no longer accelerating and travels at a constant velocity while increasing in distance. These quantities are related through the following equations:

$$\text{velocity} = (\text{acceleration})\,(\text{time}) \qquad v = (a)\,(t)$$
$$\text{distance} = (\text{velocity})\,(\text{time}) \qquad d = (v)\,(t)$$

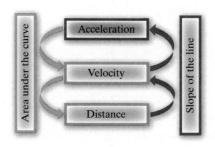

If we create a graph of velocity versus time, the form of the equation tells us that acceleration is the slope of the line. Likewise, a graph of distance versus time has velocity as the slope of the line.

However, if we have a graph of velocity versus time and we want to determine distance, how can we do this? The distance is determined by how fast we are traveling times how long we are traveling at that velocity; we can find this by determining the area under the curve of velocity versus time. Likewise, if we had a graph of acceleration versus time, we could determine the velocity from the area under the curve. In technical terms, the quantity determined by the slope is referred to as the **derivative**; the quantity determined by the area under the curve is referred to as the **integral**.

In the graph shown in Figure 11-13, we drive our car along the road at a constant velocity of 60 miles per hour [mph]. After 1.5 hours, how far have we traveled?

The area under the curve, shown by the rectangular box, is:

$$\text{Area of the rectangle} = (\text{height of rectangle})\,(\text{width of rectangle})$$
$$= (60 - 0\ mph)\,(1.5 - 0\ h)$$
$$= 90\ miles$$

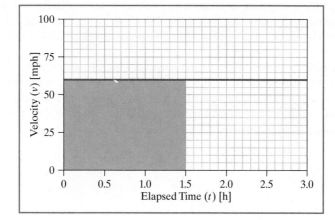

Figure 11-13
Example of distance calculation from area under velocity-versus-time graph.

Comprehension Check	11-5

Use the graph to answer the following questions.

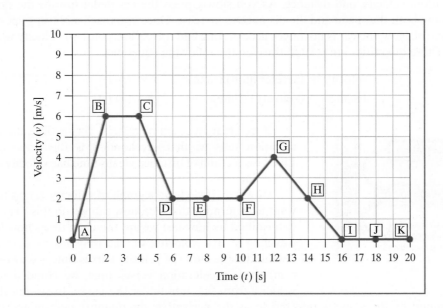

(a) What is the distance traveled by the vehicle when it reaches point C?
(b) What is the distance traveled by the vehicle when it reaches point F?

11.5 Meaning of Line Shapes

LEARN TO:	Recognize linear and nonlinear curves, and interpret the slope and area of the curve
	Understand the special linear cases, vertical lines, and horizontal lines
	Understand the physical meanings of the four combinations of curves with concavity

In addition to the value of the slope of the line, the shape of the line contains useful information. In Figure 11-13, the speed is shown as a horizontal line. This implies that it has a constant value; it is not changing over time. The slope of this line is zero, indicating that the acceleration is zero. Table 11-2 contains the various types of curve shapes and their physical meanings.

Table 11-2 What do the lines on a graph mean?

If the graph shows a . . .	. . . it means that the dependent variable . . .	Sketch
Horizontal line	The variable is not changing. The slope (the derivative) is zero. The area under the curve (the integral) is increasing at a constant rate.	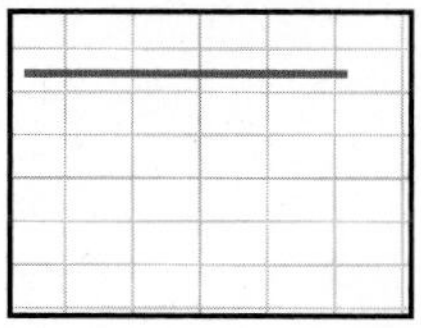
Vertical line	The variable has changed "instantaneously." The slope (the derivative) is "undefined" (infinite). The area under the curve is undefined (zero).	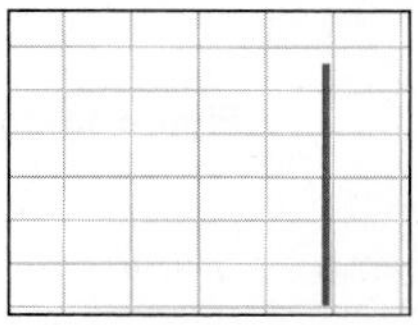
Straight line, positive or negative slope neither horizontal nor vertical	The variable is changing at a constant rate. The slope (the derivative) is constant and nonzero. The area under the line (the integral) is increasing. If the slope is positive, the rate of increase is increasing. If the slope is negative, the rate of increase is decreasing. If the negative slope line goes below zero, the area will begin to decrease.	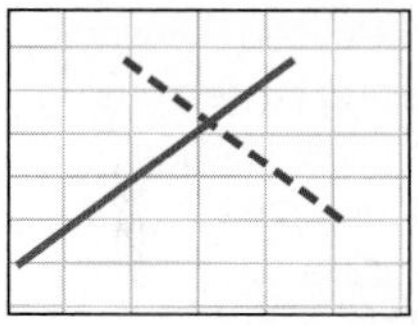
Curved line concave up, increasing trend	The variable is increasing at an increasing rate. The slope of the curve (the derivative) is positive and increasing. The area under the curve (the integral) is increasing at an increasing rate.	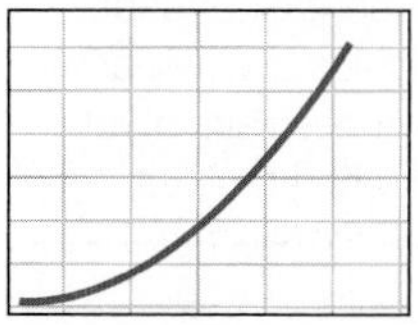
Curved line concave down, increasing trend	The variable is increasing at a decreasing rate. The slope of the curve (the derivative) is positive and decreasing. The area under the curve (the integral) is increasing at an increasing rate.	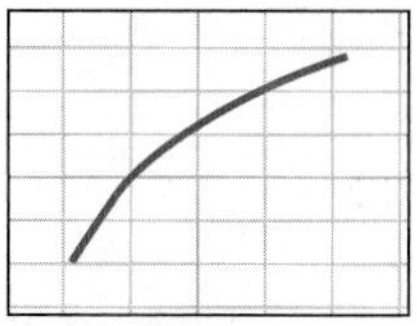
Curved line concave up, decreasing trend	The variable is decreasing at a decreasing rate. The slope of the curve (the derivative) is negative with a decreasing magnitude. The area under the curve (the integral) is increasing at a decreasing rate.	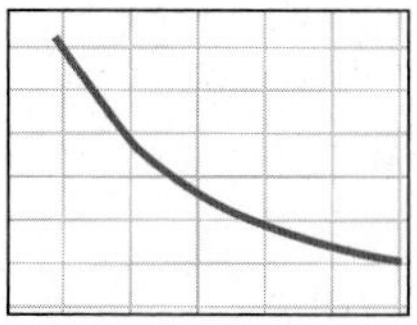
Curved line concave down, decreasing trend	The variable is decreasing at an increasing rate. The slope of the curve (the derivative) is negative with an increasing magnitude. The area under the curve (the integral) is increasing at a decreasing rate.	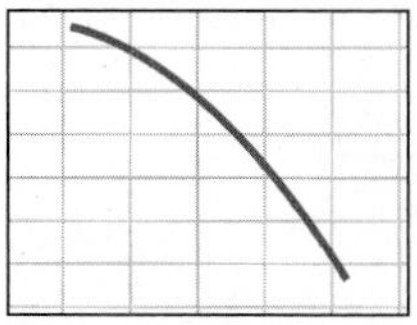

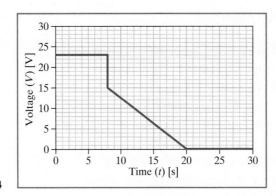

Figure 11-14

In Figure 11-14, the voltage is constant from time = 0 to 8 seconds, as indicated by the horizontal line at 23 volts. At time = 8 seconds, the voltage changes instantly to 15 volts, as indicated by the vertical line. Between time = 8 seconds and 20 seconds, the voltage decreases at a constant rate, as indicated by the straight line, and reaches 0 volts at time = 20 seconds, where it remains constant.

In Figure 11-15, the force on the spring increases at an increasing rate from time = 0 until 2 minutes, then remains constant for 1 minute, after which it increases at a decreasing rate until time = 5 minutes. After 5 minutes, the force remains constant at about 6.8 newtons.

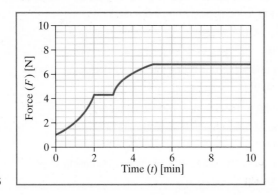

Figure 11-15

The height of a blimp is shown in Figure 11-16. The height decreases at an increasing rate for 5 minutes, then remains constant for 2 minutes. From time = 7 to 10 minutes, its height decreases at a decreasing rate. At time = 10 minutes, the height remains constant at 10 meters.

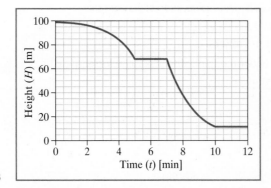

Figure 11-16

EXAMPLE   11-7

The Mars Rover travels slowly across the Martian terrain collecting data, yielding the velocity profile shown in Figure 11-17. Use this graph to answer the following questions.

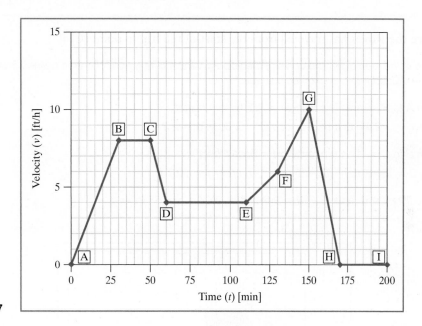

Figure 11-17

Between points D and E, the acceleration of the Rover is _____.

The velocity profile between points D and E is flat, indicating that the velocity is not changing. Acceleration is the derivative of velocity with respect to time, so acceleration is ZERO.

If the graph shows a . . .	. . . it means that the dependent variable . . .	Sketch
Horizontal line	The variable is not changing.	
	The slope (the derivative) is zero.	
	The area under the curve (the integral) is increasing at a constant rate.	

The value of acceleration of the Rover between points G and H is _____ ft/h^2.

Acceleration is the slope of the line of the velocity versus time graph. The slope between G and H is found by:

$$((10-0) \text{ ft/h})/(((170-150) \text{ min})*60 \text{ min/h}) = 30 \text{ ft/h}^2$$

Between points E and F, the distance traveled by the Rover is _____.

The velocity profile between points E and F is increasing at a constant rate. Distance is the integral of velocity with respect to time, so the distance is INCREASING at an INCREASING rate.

If the graph shows a . . .	. . . it means that the dependent variable . . .	Sketch
Straight line, positive or negative slope neither horizontal nor vertical	The variable is changing at a constant rate. The slope (the derivative) is constant and nonzero. The area under the line (the integral) is increasing. If the slope is positive, the rate of increase is increasing.	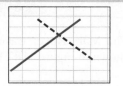

The distance the Rover has traveled from the start of the trip to point (C) is _____ ft.

Distance is the area under the curve of the velocity versus time graph. The area defined from point A to point C can be divided into two geometric shapes:

*From A to B = Area is a triangle = 1/2 base * height*
Area 1 = 1/2 (8−0) ft/h (30−0) min* 1h/60 min = 2 ft*

*From B to C = Area is a rectangle = base * height*
Area 2 = (8−0) ft/h (50−30) min * 1h/60 min = 2.67 ft*

Total distance = Area 1 + Area 2 = 2 ft + 2.67 ft = 4.67 ft

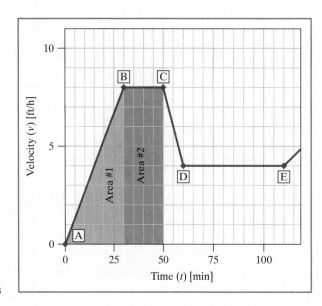

Figure 11-18

Comprehension Check   11-6

Use the graph to answer the following questions. Choose from the following answers:

1. Zero
2. Positive and constant
3. Positive and increasing
4. Positive and decreasing
5. Negative and constant
6. Negative with increasing magnitude
7. Negative with decreasing magnitude
8. Cannot be determined from information given

NOTE

The rate of change (derivative) of acceleration is called JERK.

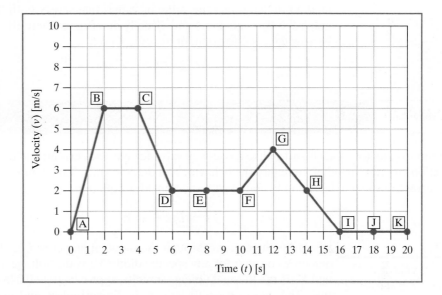

(a) Between points A and B, the acceleration is ___
(b) Between points B and C, the acceleration is ___
(c) Between points C and D, the acceleration is ___
(d) Between points D and E, the distance is ___
(e) Between points F and G, the distance is ___
(f) Between points G and H, the distance is ___

11.6 Graphical Solutions

LEARN TO: Use a graph of expressions to identify the overlapping points
Show fixed costs, variable cost, sales price, revenue, and profit graphically
Identify the breakeven point of an economic process (if one exists)

When you have two equations containing the same two variables, it is sometimes desirable to find values of the variables that satisfy both equations. Most of you have studied methods for solving simultaneous linear equations—however, most of these methods apply only to linear equations and do not work if one or both of the equations is nonlinear. It also becomes problematic if you are working with experimental data.

For systems of two equations, or data sets in two variables, you can use a graphical method to determine the value or values that satisfy both. The procedure is simply to graph the two equations and visually determine where the curves intersect. This may be nowhere, at one point, or at several points.

EXAMPLE   11-8

The semiconductor diode is sort of like a one-way valve for electric current: it allows current to flow in one direction, but not the other. In reality, the behavior of a diode is considerably more complicated. In general, the current through a diode can be found with the **Shockley equation**,

$$I = I_0(e^{\frac{V_D}{nV_T}} - 1)$$

NOTE

This example demonstrates the graphical solution of simultaneous equations when one of the equations is nonlinear. We do not expect you to know how to perform the involved circuit analyses. Those of you who eventually study electronics will learn these techniques in considerable detail.

where I is the current through the diode in amperes; I_0 is the saturation current in amperes, constant for any specific diode; V_D is the voltage across the diode in volts; and V_T is the thermal voltage in volts, approximately 0.026 volts at room temperature. The emission coefficient, n, is dimensionless and constant for any specific diode; it usually has a value between 1 and 2.

The simple circuit shown has a diode and resistor connected to a battery. For this circuit, the current through the resistor can be given by:

$$I = \frac{V - V_D}{R}$$

where I is the current through the resistor in milliamperes [mA], V is the battery voltage in volts [V], V_D is the voltage across the diode in volts, and R is the resistance in ohms [Ω].

In this circuit, the diode and resistor are in series, which implies that the current through them is the same. We have two equations for the same parameter (current), both of which are a function of the same parameter (diode voltage). We can find a solution to these two equations, and thus the current in the circuit, by graphing both equations and finding the point of intersection. For convenience of scale, the current is expressed in milliamperes rather than amperes.

Plot these two equations for the following values and determine the current.

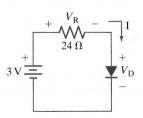

$$I_0 = 0.01 \text{ mA}$$
$$V = 3 \text{ V}$$
$$R = 24 \ \Omega$$
$$nV_T = 0.04 \text{ V}$$

The point of intersection shown in Figure 11-19 is at $V_D = 0.64$ V and $I = 100$ mA; thus, the current in the circuit is 100 mA or 0.1 A.

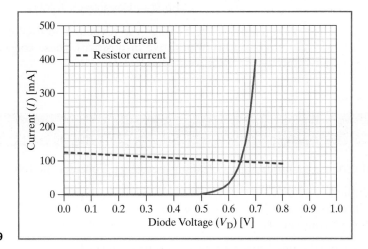

Figure 11-19

Comprehension Check	**11-7**

We assume that the current through two electromagnets is given by the following equations

Electromagnet C: $I = 7t + 8$

Electromagnet D: $I = 5t + 34$

We want to determine when the value of the current through the electromagnets is equal.

Using Graphs in Economic Analysis

Breakeven analysis determines the quantity of product a company must make before they begin to earn a profit. Two types of costs are associated with manufacturing: fixed and variable. **Fixed costs** include equipment purchases, nonhourly employee salaries, insurance, mortgage or rent on the building, etc., or "money we must spend just to open the doors."

Variable costs depend on the production volume, such as material costs, hourly employee salaries, and utility costs. The more product produced, the higher the variable costs become.

Total cost = Fixed cost + Variable cost * Amount produced

The product is sold at a **selling price**, creating **revenue**.

Revenue = Selling price * Amount sold

Any excess revenue remaining after all production costs have been paid is **profit**. Until the company reaches the breakeven point, they are operating at a **loss** (negative profit), where the money they are bringing in from sales does not cover their expenses.

Profit = Revenue − Total cost

The **breakeven point** occurs when the revenue and total cost lines cross, or the point where profit is zero (not negative or positive). These concepts are perhaps best illustrated through an example.

EXAMPLE | 11-9

Let the amount of product we produce be G [gallons per year]. Consider the following costs:

- Fixed cost: $1 million
- Variable cost: 10 cents/gallon of G
- Selling price: 25 cents/gallon of G

Plot the total cost and the revenue versus the quantity produced. Determine the amount of G that must be produced to break even. Assume we sell everything we make.

The plot of these two functions is shown in Figure 11-20. The breakeven point occurs when the two graphs cross, at a production capacity of 6.7 million gallons of G.

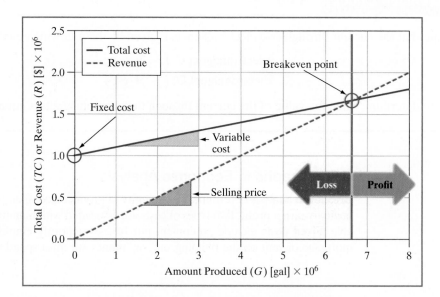

Figure 11-20
Breakeven analysis definitions.

EXAMPLE 11-10

In creating electrical parts for a Mars excursion module, you anticipate the costs of production shown in the graph. In your analysis, you assume the following costs of production:

- Labor cost = \$1.20/part
- Energy cost = \$0.60/part

Use Figure 11-21 to answer the following questions.

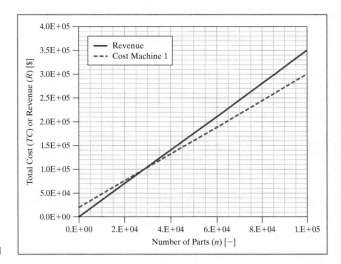

Figure 11-21

What is the material cost per part?

Variable cost is the slope of the total cost line.
  *Total cost = Fixed cost + Variable cost * Amount produced*

$$Slope\ total\ cost = \frac{3E5 - 2E4}{1E5 - 0} = \frac{\$2.80}{part}$$

The material cost is one of three costs that make up the variable cost.
Variable cost = Material cost + Labor cost + Energy cost
Solving for material cost:

$$Material\ cost = \frac{\$2.80}{part} - \frac{\$1.20}{part} - \frac{\$0.60}{part} = \frac{\$1.00}{part}$$

What is the selling price of each part?

Selling price is the slope of the revenue line.
  *Revenue = Selling price * Amount sold*

$$Slope\ revenue = \frac{3.5E5 - 0}{1E5 - 0} = \frac{\$3.50}{part}$$

You decide to consider a second option, with a fixed cost of $50,000 and a variable cost of $2.00 per part. Draw the total cost line for Machine 2 on the graph.

To draw the total cost line, two points are needed to draw a linear line.

At $n = 0$ parts, the total cost = fixed cost = $50,000.
At $n = 100,000$ parts,
the total cost = $50,000 + ($2.00/part)(100,000 parts) = $250,000.

To ensure the line is linear, it is a good idea to test at least one more point to make sure if falls along this line.

At $n = 40,000$ parts,
the total cost = $50,000 + ($2.00 / part)(40,000 parts) = $130,000.

Connecting a line through these points yields the green, dot-dash line shown in Figure 11-22.

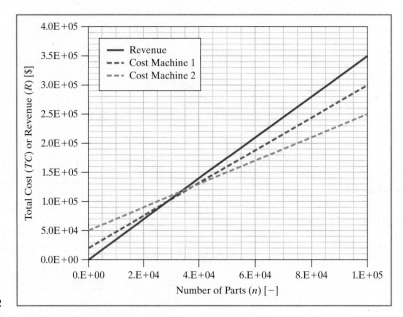

Figure 11-22

What is the profit of each machine at 80,000 parts?

Profit is the difference between the cost and revenue lines. If the cost line is above the revenue line, the process is operating at a loss. If the cost line is below the revenue line, the process is operating at a profit. Examining the graph at 80,000 parts (Figure 11-23), both cost lines are below the revenue line, so both machines are operating at a profit.

The difference for Machine 1 is four minor gridlines. Each ordinate minor gridline on the graph is $1E4. The profit for Machine 1 is $40,000.

The difference for Machine 2 is seven minor gridlines. The profit for Machine 2 is $70,000.

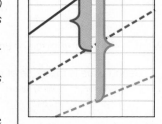

Figure 11-23

If the selling price is decreased by $0.50 per part, what will happen to the breakeven point for Machine 1?

(A) It will move to the left, indicating the breakeven will occur sooner than originally shown.

(B) It will move to the right, indicating the breakeven will occur later than originally shown.

(C) It will not change the breakeven point.

In the revenue line, the larger the slope, the higher the angle of line, the higher the selling price. A decrease in selling price translates graphically to a slope at a lower angle. If the slope of the revenue line decreases, the number of parts required to break even will increase, shifting the breakeven point to the right (Figure 11-24).

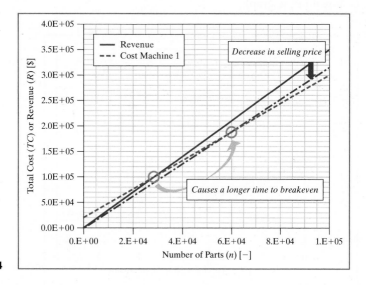

Figure 11-24

Comprehension Check 11-8

You are working for a tire manufacturer, producing wire to be used in the tire as a strengthening agent. You are considering implementing a new machining system, and you must present a breakeven analysis to your boss. You develop the graph, showing two possible machines that you can buy.

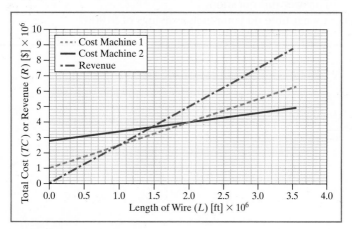

(a) Which machine has a higher fixed cost?
(b) Which machine has a lower variable cost?
(c) How much wire must be produced on Machine 1 to break even?
(d) If you make 3 million feet of wire, which machine will yield the higher profit?
(e) Which machine has the lower breakeven point?

Comprehension Check   11-9

You want to install a solar panel system on your home. According to one source, if you install a 40-square-foot system, the cost curve is shown in the graph.

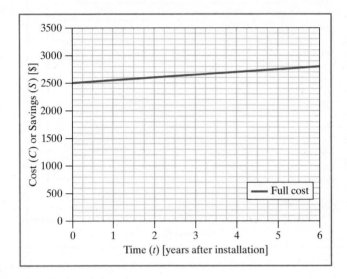

(a) List the fixed cost and the variable cost for this system.

(b) If the source claims that you can break even in 3.5 years, how much savings are you generating per year (or, what is the slope of the savings curve or the "revenue" that you generate by installing the system)? Draw the "revenue" curve on the graph and use it to answer this question.

(c) If you receive a federal tax credit for "going green," you can save 30% on the initial fixed cost. With this savings, how long does it take to break even? Draw this operating cost curve, labeled "Credit Cost" on the graph, and use it to answer this question.

(d) With the new tax credit, at what time do you reach a savings of $1000?

CHAPTER 11

IN-CLASS ACTIVITIES

For questions ICA 11-1 to ICA 11-9, your instructor will determine if you should complete this question by hand or using Excel. If you must complete this problem by hand, a blank graph has been provided online.

ICA 11-1

Joule's first law relates the heat generated to current flowing in a conductor. It is named after James Prescott Joule, the same person for whom the unit the joule is named. Use the following experimental data to create a scatter graph of the power (P, on the ordinate) and current (I, on the abscissa).

Current (I) [A]	0.50	1.25	1.50	2.25	3.00	3.20	3.50
Power (P) [W]	1.20	7.50	11.25	25.00	45.00	50.00	65.00

ICA 11-2

Data for a wind turbine is shown below. Use the following experimental data to create a scatter plot of the power (P, on the ordinate) and velocity (v, on the abscissa).

Velocity (v) [m/s]	5	8	12	15	19	23
Power (P) [W]	15	60	180	400	840	1500

ICA 11-3

There is a large push in the United States to convert from incandescent light bulbs to compact fluorescent bulbs (CFLs). The lumen [lm] is the SI unit of luminous flux (LF), a measure of the perceived power of light. To test the power usage, you run an experiment and measure the following data. Create a proper plot of these experimental data, with electrical consumption (EC) on the ordinate and LF on the abscissa.

	Electrical Consumption [W]	
Luminous Flux [lm]	Incandescent 120 V	Compact Fluorescent
80	16	
200		4
400	38	8
600	55	
750	68	13
1250		18
1400	105	19

ICA 11-4

Your team has designed three tennis ball launchers, and you have run tests to determine which launcher best meets the project criteria. Each launcher is set to three different launch angles, and the total distance the ball flies through the air is recorded. These experimental data are summarized in the table. Plot all three sets of data on a scatter plot, showing one data set for each of the three launchers on a single graph. Launch angle should be plotted on the horizontal axis.

Launcher 1		Launcher 2		Launcher 3	
Launch Angle (θ) [°]	Distance (d) [ft]	Launch Angle (θ) [°]	Distance (d) [ft]	Launch Angle (θ) [°]	Distance (d) [ft]
20	5	10	10	20	10
35	10	45	25	40	20
55	12	55	18	50	15

ICA 11-5

Plot the following pairs of functions on a single graph. The independent variable (angle) should vary from 0 to 360 degrees on the horizontal axis.

(a) $\sin \theta$, $-2 \sin \theta$
(b) $\sin \theta$, $\sin 2\theta$
(c) $\sin \theta$, $\sin \theta + 2$
(d) $\sin \theta$, $\sin (\theta + 90)$

ICA 11-6

Plot the following pairs of functions on a single graph. The independent variable (angle) should vary from 0 to 360 degrees on the horizontal axis.

(a) $\cos \theta$, $\cos 3\theta$
(b) $\cos \theta$, $\cos \theta - 3$
(c) $\cos \theta$, $\cos (2\theta) + 1$
(d) $\cos \theta$, $3 \cos (2\theta) - 2$

ICA 11-7

You need to create a graph showing the relationship of an ideal gas between pressure (P) and temperature (T). The ideal gas law relationship: $PV = nRT$. The ideal gas constant (R) is 0.08206 atmosphere liter per mole kelvin. Assume the tank has a volume (V) of 12 liters and is filled with nitrogen. The initial temperature (T) is 270 kelvin and the initial pressure (P) is 2.5 atmospheres. First, determine the number of moles of gas (n). Then, create a graph to model the gas as the temperature increases from 270 to 350 kelvin.

ICA 11-8

The decay of a radioactive isotope can be modeled using the following equation, where C_0 is the initial amount of the element at time zero and k is the half-life of the isotope. Create a graph of the decay of isotope A [k = 1.48 hours]. Allow time to vary on the abscissa from 0 to 5 hours with an initial concentration of 10 grams of isotope A.

$$C = C_0 e^{-t/k}$$

ICA 11-9

In researching alternate energies, you find that wind power is calculated by the following equation:

$$P = \frac{1}{2} A \rho v^3$$

where

- P = power [watts]
- A = sweep area (circular) of the blades [square meters]
- ρ = air density [kilograms per cubic meter]
- v = velocity [meters per second]

The specific gravity of air is 0.00123 and the velocity is typically 35 meters per second. Create a graph of the theoretical power (P, in units of watts) as a function of the blade diameter (D, in units of meters). Allow the diameter to be graphed on the abscissa and vary from 0.5 to 1.5 meters.

The following graph applies to ICA 11-10 to 11-15.

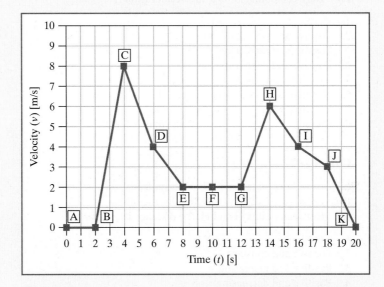

ICA 11-10

(a)   Between points A and B, the acceleration is
(b)   Between points C and D, the acceleration is
(c)   Between points G and H, the acceleration is

1. Zero
2. Positive and constant
3. Positive and increasing
4. Positive and decreasing
5. Negative and constant
6. Negative with increasing magnitude
7. Negative with decreasing magnitude
8. Cannot be determined from information given

ICA 11-11

(a) Between points B and C, the distance is
(b) Between points F and G, the distance is
(c) Between points I and J, the distance is

1. Zero
2. Positive and constant
3. Positive and increasing
4. Positive and decreasing
5. Negative and constant
6. Negative with increasing magnitude
7. Negative with decreasing magnitude
8. Cannot be determined from information given

ICA 11-12

(a) Between points B and C, the acceleration is
(b) Between points F and G, the acceleration is
(c) Between points I and J, the acceleration is

1. Zero
2. Positive and constant
3. Positive and increasing
4. Positive and decreasing
5. Negative and constant
6. Negative with increasing magnitude
7. Negative with decreasing magnitude
8. Cannot be determined from information given

ICA 11-13

(a) Between points A and B, the distance is
(b) Between points C and D, the distance is
(c) Between points G and H, the distance is

1. Zero
2. Positive and constant
3. Positive and increasing
4. Positive and decreasing
5. Negative and constant
6. Negative with increasing magnitude
7. Negative with decreasing magnitude
8. Cannot be determined from information given

ICA 11-14

(a) Use the graph to determine the numerical value and appropriate unit of the acceleration between points A and B.
(b) Use the graph to determine the numerical value and appropriate unit of the acceleration between points I and J.
(c) Use the graph to determine the numerical value and appropriate unit of the total distance traveled at point G.
(d) Use the graph to determine the numerical value and appropriate unit of the total distance traveled at point K.

ICA 11-15

(a) Use the graph to determine the numerical value and appropriate units of the acceleration between points C and D.
(b) Use the graph to determine the numerical value and appropriate units of the acceleration between points F and G.
(c) Use the graph to determine the numerical value and appropriate units of the total distance traveled at point E.
(d) Use the graph to determine the numerical value and appropriate units of the total distance traveled at point I.

ICA 11-16

Use the following graph to determine which statements about the two vehicles are true.

(a) At point B, the distance traveled by Vehicle 1 is equal to the distance traveled by Vehicle 2.
(b) At point B, the velocity of Vehicle 1 is equal to the velocity of Vehicle 2.
(c) The average acceleration of Vehicle 1 between points B and C is equal to the average acceleration of Vehicle 2 between points D and E.
(d) At point E, the distance traveled by Vehicle 1 is greater than the distance traveled by Vehicle 2.
(e) At point E, the velocity of Vehicle 1 is greater than the velocity of Vehicle 2.
(f) The average acceleration of Vehicle 1 between points E and F is greater than the average acceleration of Vehicle 2 between points E and F.

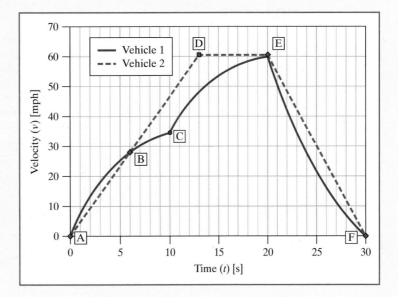

ICA 11-17

The following graph shows the power delivered to a motor over a period of 50 seconds. The power gradually increases to 200 watts and then remains constant until the power is turned off at 50 seconds.

(a) What is the total energy absorbed by the motor during the 50-second period shown?
(b) What is the rate of change of power delivery during the first 10 seconds?

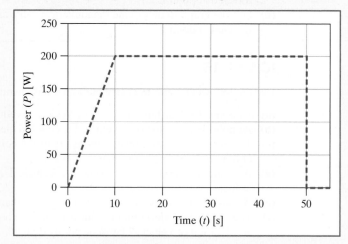

ICA 11-18

The music industry in the United States has had a great deal of fluctuation in profit over the past 20 years due to the advent of new technologies such as peer-to-peer file sharing and mobile devices such as the iPod and iPhone. The following graph displays data from a report published by eMarketer in 2009 about the amount U.S. consumers spend on digital music files and physical music formats (CDs, records, cassette tapes, etc.), where the values for 2009–2013 are reported as projections and for 2008 is reported using actual U.S. spending measurements.

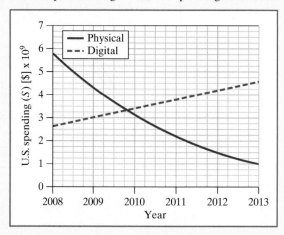

(a) According to the study, when will the sale of physical media be equivalent to the sale of digital audio files?
(b) When will the sales of digital audio files exceed that of physical media by $1 billion?
(c) If the physical media sales were $2 billion higher than the trend displayed on the graph, when would the sale of digital audio files exceed physical media?
(d) If the digital audio file sales were $1 billion lower than the trend displayed on the graph, when would the sale of digital audio files exceed physical media?

ICA 11-19

You are working for a chemical manufacturer, producing solvents used to clean lenses for microscopes. You are working on determining the properties of three different solvent blends. You develop the following chart, showing the evaporation of the three blends.

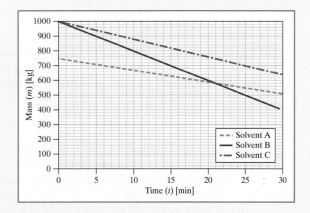

(a) Which solvent evaporates at the slowest rate?
(b) Which solvent evaporates at the fastest rate?
(c) What is the initial mass of Solvent A? Be sure to include units.
(d) What is the rate of solvent evaporation of Solvent A? Be sure to include units.

ICA 11-20

Use the accompanying graph to answer the following questions. Assume the company makes 30,000 parts per month of Product A and 17,500 parts per month of Product B.

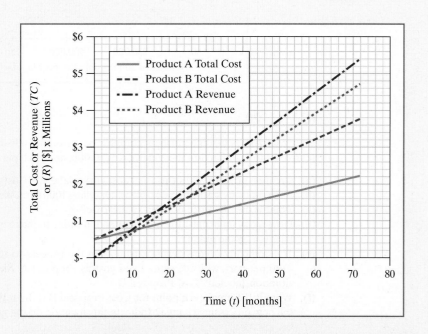

(a) Which product has the higher variable cost, and what is this value in units of dollars per part?
(b) Which product has the higher selling price, and what is this value in units of dollars per part?
(c) Which product has the faster breakeven time, and what is this value in units of months?
(d) At six years, which product makes more profit, and what is this value in units of dollars?
(e) If the fixed cost of product B is increased to $1,000,000 and the selling price is increased by $0.75 per part, what is the new breakeven point in units of months?

ICA 11-21

A company designs submersible robots with a new design for the robots that increases the rate of production. A new facility for manufacturing the submersible robots is constructed at a cost of $100,000,000. A contract is negotiated with a materials supplier (Supplier A) to provide all of the raw material and construction labor necessary for $250 per robot. The robots will be sold for $500 each.

(a) How many robots must be manufactured and sold to break even?
(b) How many robots must be manufactured and sold to make a profit of $100,000,000?
(c) An alternative materials supplier (Supplier B) comes along with a quote for the labor and material cost at $400 per robot, but only requires $50,000,000 to build a submersible robot construction facility. How many robots must be manufactured and sold to break even for this alternative supplier?
(d) Which supplier will generate a profit of $20,000,000 with fewer robots produced?

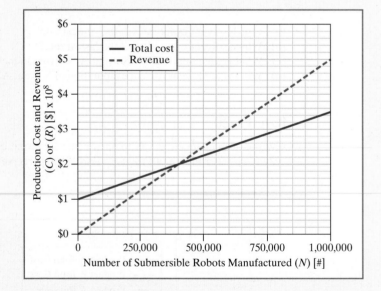

ICA 11-22

Your company is manufacturing a complex part from an advanced material. Assume the initial setup cost to manufacture these parts is $750,000, and each part costs $500 to make.

(a) Create a proper plot of this total cost curve, labeled "Cost Proposal A."
(b) If the company wishes to break even after selling 1000 parts, sketch the revenue curve on the graph.
(c) What is the sale price per unit in this case?
(d) How many units must the company sell in order to make a profit of $500,000? Indicate this location on the graph.
(e) The company is considering a change in the process to reduce the manufacturing cost by $100 per part, with the same fixed cost as Proposal A. Sketch the total cost curve for this situation, labeled "Cost Proposal B."
(f) What is the breakeven point for Cost Proposal B if the revenue curve remains the same for the new processing change? Indicate this location on the graph.

ICA 11-23

Using the list provided, you may be assigned a topic for which to create a graph. You must determine the parameters to graph and imagine a set of data to show on the chart. You may use grid paper or the grid below as directed by your instructor.

1. Air temperature
2. Airplane from airport to airport
3. Baking bread
4. Bird migration
5. Boiling water in a whistling teapot
6. Bouncing a basketball
7. Brushing your teeth
8. Burning a pile of leaves
9. Burning a candle
10. Climbing a mountain
11. Cooking a Thanksgiving turkey
12. Daily electric power consumption
13. Detecting a submarine by using sonar
14. Diving into a swimming pool
15. Drag racing
16. Driving home from work
17. Dropping ice in a tub of warm water
18. Engineer's salary
19. Exercising
20. Feedback from an audio system
21. Fishing
22. Flight of a hot air balloon
23. Football game crowd
24. Formation of an icicle
25. A glass of water in a moving vehicle
26. Hammering nails
27. Leaves on a tree
28. Letting go of a helium balloon

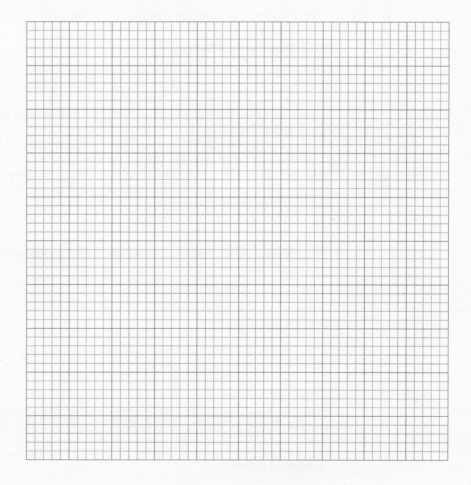

29. Marching band
30. Moving a desk down a staircase
31. Oak tree over the years
32. Oil supply
33. Person growing up
34. Playing with a yo-yo
35. Plume from a smokestack
36. Pony Express
37. Popping corn
38. Pouring water out of a bottle
39. Power usage on campus
40. Pumping air into a bicycle tire
41. Rain filling a pond
42. Recycling
43. River in a rainstorm
44. Skipping a stone on water
45. Sleeping
46. Snoring
47. Snow blowing over a roof
48. Solar eclipse
49. Sound echoing in a canyon
50. Space station
51. Spinning a hula-hoop
52. Student attention span during class
53. Studying for an exam
54. The moon
55. Throwing a ball
56. Thunderstorm
57. Traffic at intersections
58. Train passing through town
59. Using a toaster
60. Washing clothes

ICA 11-24

Materials

| Balloons (2) | Stopwatch (2) | String (40 inches) | Tape measure |

Part I: Blowing Up a Balloon

One team member is to inflate one balloon, a second team member is to time the inhalation stage (how long it takes to inhale a single breath), and a third team member is to time the exhalation stage (how long it takes to exhale a single breath into the balloon). A fourth team member is to measure the balloon size at the end of each inhale/exhale cycle, using the string to measure the balloon circumference.

Record the observations on a worksheet similar to the following one for three complete inhale/exhale cycles or until the balloon appears to be close to maximum volume, whichever occurs first. Repeat the entire balloon inflation process for a second balloon; average the times from the balloons to obtain the time spent at each stage and the average circumference at each stage. Calculate the balloon volume at each stage, assuming the balloon is a perfect sphere.

Balloon	Stage	Inhale Time	Exhale Time	Circumference
1	1			
	2			
	3			
2	1			
	2			
	3			

	Stage	Inhale Time	Exhale Time	Circumference	Volume
Average Balloon	1				
	2				
	3				

Part II: Analysis

Graph the balloon volume (V, ordinate) versus time (t, time). You may use grid paper or the grid below as directed by your instructor. Allow the process to be continuous, although in reality it was stopped at various intervals for measurements. The resulting graph should contain only the time elapsed in the process of inhaling and exhaling, not the time required for recording the balloon size. For this procedure, assume that the air enters the balloon at a constant rate and the balloon is a perfect sphere.

(a) What does the assumption of the air entering the balloon at a constant rate indicate about the slope?
(b) Calculate the following graphically.
  • The rate at which the air enters the balloon in the first stage.
  • The rate at which the air enters the balloon in the third stage.
(c) On the same graph, sketch the balloon volume (V, ordinate) versus time (t, time) if you were inflating a balloon that contained a pinhole leak.
(d) On the same graph, sketch the balloon volume (V, ordinate) versus time (t, time) if you were inflating a balloon from a helium tank.

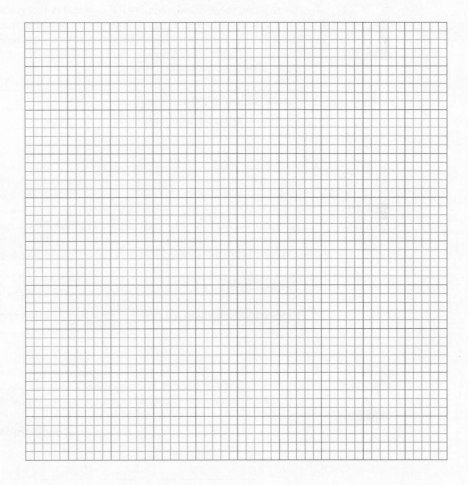

REVIEW QUESTIONS

For questions 1 through 10, your instructor will determine if you should complete this question by hand or using Excel. If you must complete this problem by hand, a blank graph has been provided online.

1. A computer engineer has measured the power dissipated as heat generated by a prototype microprocessor running at different clock speeds. Create a proper plot of the following experimental data set.

Speed (S) [GHz]	1.8	2.3	2.8	3.5	4.1
Power dissipated as heat (P) [W]	145	227	305	415	599

2. Due to increased demand, an industrial engineer is experimenting with increasing the speed (S) of a machine used in the production of widgets. The machine is normally rated to produce five widgets per second, and the engineer wants to know how many defective parts (D) are made at higher speeds, measured in defective parts per thousand. Create a proper plot of the following experimental data set.

Speed (S) [parts/min]	6.5	5.9	6.5	7.2	8.0
Defects in parts per thousand (D)	2	4	8	14	22

3. An engineer is conducting tests of two prototype toothbrush sanitizers that use ultraviolet radiation to kill pathogenic organisms while the toothbrush is stored. The engineer is trying to determine the minimum power needed to reliably kill pathogens on toothbrushes. Several toothbrushes are treated with a mix of bacteria, fungi, and viruses typically found in the human mouth, and then each is placed in one of the sanitizers for six hours at a specific power level (P). After six hours in the sanitizers, the viable pathogens remaining (R) on each toothbrush are assayed. Create a proper plot of the following experimental data set.

Power (P) [W]		15	28	35	50
Pathogens remaining (R) [%]	Sanitizer A	51	40	29	7
	Sanitizer B	59	42	22	8

4. Several reactions are carried out in a closed vessel. The following data are taken for the concentration (C) in units of grams per liter of solvent processed for compounds A and B as a function of time (t). Create a proper plot of the following experimental data set.

Time (t) [min]	Concentration [g/L]	
	A (C_A)	B (C_B)
26	0.135	0.170
55	0.110	0.165
90	0.090	0.160
150	0.070	0.10

5. The following experimental data are collected on the current (*I*, in units of milliamperes) in the positive direction and voltage (*V*, in units of volts) across the terminals of two different thermionic rectifiers. Create a proper plot of the following experimental data set.

Voltage (*V*) [V]	Current (*I*) [mA]	
	Rectifier A	Rectifier B
19	4	11
33	17	22
42	22	31
49	28	45

6. If an object is heated, the temperature of the body will increase. The energy (*Q*) associated with a change in temperature (ΔT) is a function of the mass of the object (*m*) and the specific heat (C_p). Specific heat is a material property, and values are available in literature. In an experiment, heat is applied to the end of an object, and the temperature change at the other end of the object is recorded. This leads to the theoretical relationship shown. An unknown material is tested in the lab, yielding the following results:

$$\Delta T = \frac{Q}{mC_p}$$

Heat applied (Q) [J]	13	19	28	44	52	60
Temp change (ΔT) [K]	1.50	2.25	3.25	5.50	6.25	7.25

Graph the experimental temperature change (ΔT, ordinate) versus the heat applied (*Q*).

7. Eutrophication is the result of excessive nutrients in a lake or other body of water, usually caused by runoff of nutrients (animal waste, fertilizers, and sewage) from the land, which causes a dense growth of plant life. The decomposition of the plants depletes the supply of oxygen, leading to the death of animal life. Sometimes, these excess nutrients cause algae blooms, or rapid growth of algae, which normally occur in small concentrations in the water body.

The following table contains data to illustrate the relationship between pressure (depth of fluid), the temperature of the water, and the solubility of oxygen in the water. Create a proper plot of the data.

Solubility of O₂ [mg/L]	Pressure (P) [mm Hg]		
Temperature (T) [°C]	760	1520	3040
10	11.3	22.6	45.1
20	9.1	18.2	36.4
30	7.6	15.2	30.3
40	6.5	12.9	25.9

8. In the 1950s, a team at Los Alamos National Laboratories built several devices called "Perhapsatrons," thinking that PERHAPS they might be able to create controllable nuclear fusion. After several years of experiments, they were never able to maintain stable plasma and abandoned the project.

The perhapsatron used a toroidal (doughnut-shaped) plasma confinement chamber, similar to those used in more modern Tokamak fusion devices. You have taken a job at a fusion research lab, and your supervisor asks you to develop a simple spreadsheet to calculate the volume of a torus within which the plasma will be contained in a new experimental reactor.

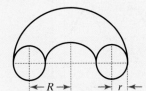

(a) Create a table that calculates the volumes of various toruses with specific values for r and R. The tube radii (r) should range from 10 to 100 centimeters in increments of 10 centimeters. The torus radii (R) should range from 1.5 to 3 meters in increments of 0.5 meters. The volume of a torus can be determined using $V = 2\pi^2 R r^2$.

(b) Using the table of volumes, create a graph showing the relationship between volume (ordinate) and tube radius (r) for torus radii (R) of 2 and 3 meters.

(c) Using the table of volumes, create a graph showing the relationship between volume (ordinate) and torus radius (R) for tube radii (r) of 40, 70, and 100 centimeters.

9. Generally, when a car door is opened, the interior lights come on and turn off again when the door is closed. Some cars turn the interior lights on and off gradually. Suppose that you have a car with 25 watts of interior lights. When a door is opened, the power to the lights increases linearly from 0 to 25 watts over 2 seconds. When the door is closed, the power is reduced to zero in a linear fashion over 5 seconds.

(a) Create a proper plot of power (P, on the ordinate) and time (t).

(b) Using the graph, determine the total energy delivered to the interior lights if the door to the car is opened and then closed 10 seconds later.

10. One of the 22 named, derived units in the metric system is the volt, which can be expressed as 1 joule per coulomb ($V = J/C$). A coulomb is the total electric charge on approximately 6.24×10^{18} electrons. The voltage on a capacitor is given by $V = \Delta Q/C + V_0$ volts, where ΔQ is the change in charge [coulombs] stored, V_0 is the initial voltage on the capacitor, and C is the capacitance [farads].

(a) Create a proper plot of voltage (V, on the ordinate) and total charge (ΔQ) for a 5-farad capacitor with an initial voltage of 5 volts for $0 < \Delta Q < 20$.

(b) Using the graph, determine the total energy stored in the capacitor for an addition of 15 coulombs.

11. The following is a graph of the vertical position of a person bungee jumping, in meters. A copy of this graph has been provided online; you may use one of these graphs, or use graph paper as directed by your instructor.

(a) What is the closest this person gets to the ground?

(b) When this person stops bouncing, how high off the ground will the person be?

(c) If the person has a mass of 70 kilograms, how would the graph change for a jumper of 50 kilograms? Approximately sketch the results on the graph.

(d) If the person has a mass of 70 kilograms, how would the graph change for a jumper of 80 kilograms? Approximately sketch the results on the graph.

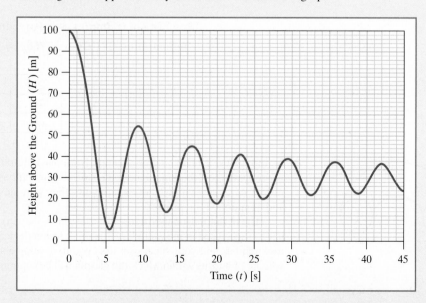

12. Shown are graphs of the altitude in meters, and velocity in meters per second, of a person skydiving. A copy of these graphs has been provided online; you may use one of these graphs, or use graph paper as directed by your instructor.

(a) When does the skydiver reach the ground?

(b) How fast is he moving when he reaches the ground?

(c) At what altitude does he open the parachute?

(d) Terminal velocity is the velocity at which the acceleration of gravity is exactly balanced by the drag force of air. How long does it take him to reach his terminal velocity without the parachute open?

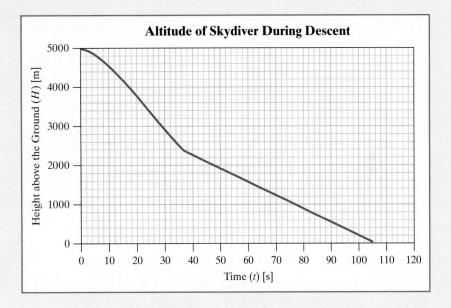

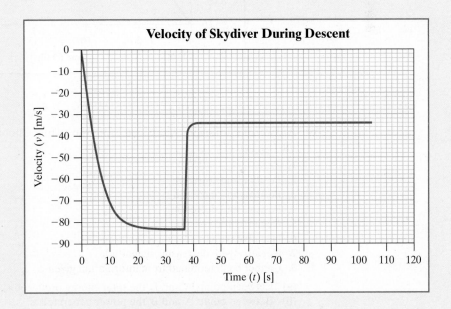

13. The following graph shows the current used to charge a capacitor over a period of 25 milliseconds [ms]. Choose from the following answers for (a)–(b).

1. Zero
2. Positive and constant
3. Positive and increasing
4. Positive and decreasing
5. Negative and constant
6. Negative with increasing magnitude
7. Negative with decreasing magnitude
8. Cannot be determined from information given

(a) At time $t = 10$ to 12 ms, classify the manner in which the **current** is changing.
(b) At time $t = 16$ to 18 ms, classify the manner in which the **charge** on the capacitor is changing.
(c) What is the total charge on the capacitor at time $t = 20$ ms?
(d) If the voltage on the capacitor at time 25 ms is 20 volts, what is the value of the capacitance? Express your answer using an appropriate prefix.

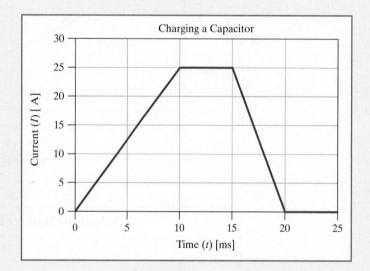

14. Answer the following questions using the graph. Choose from the following answers for (a)–(c):

1. Zero
2. Positive and constant
3. Positive and increasing
4. Positive and decreasing
5. Negative and constant
6. Negative with increasing magnitude
7. Negative with decreasing magnitude
8. Cannot be determined from information given

(a) Between points A and B, the total energy produced is:
(b) Between points A and B, the power generated is:
(c) Between points B and C, the power generated is:
(d) What is the power being generated at $t = 7$ minutes? State your answer in units of kilowatts.

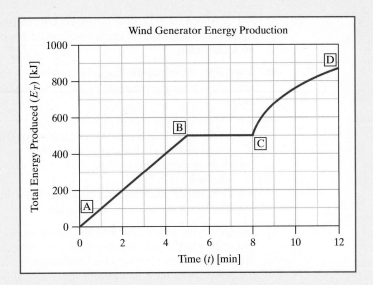

15. Answer the following questions using the graph. Choose from the following answers for (a)–(c):

1. Zero
2. Positive and constant
3. Positive and increasing
4. Positive and decreasing
5. Negative and constant
6. Negative with increasing magnitude
7. Negative with decreasing magnitude
8. Cannot be determined from information given

(a) For vehicle 2, between points A and D, the velocity is:
(b) For vehicle 2, between points D and E, the acceleration is:
(c) For vehicle 2, between points E and F, the distance is:
(d) What is the total distance traveled by vehicle 2 between points A and E? Give your answer in miles.
(e) Which vehicle travels the farthest distance between points A and F?

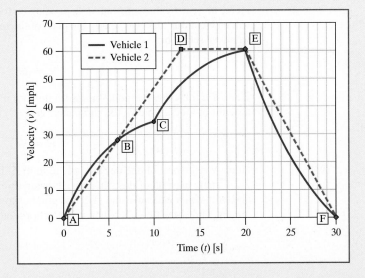

For questions 16 to 19, your instructor will determine if you should complete this question by hand or using Excel. If you must complete this problem by hand, a blank graph has been provided online.

16. In a simple electric circuit, the current (I) must remain below 40 milliamps ($I < 40$ mA) and must also satisfy the function $I > 10^{-6}\,e^{25}$V, where V is the voltage across a device called a diode.

   (a) Create a proper plot of these two inequalities with current on the ordinate. The values on the vertical axis should range from 0 to 50 milliamperes, and the values on the horizontal axis should range from 0 to 1 volt.

   (b) If graphing part (a) by hand, shade the region of the graph where *both* inequalities are satisfied.

   (c) Graphically determine the maximum allowable voltage across the diode. Indicate the location of this answer on your graph.

17. In a hard drive design, the faster the disk spins, the faster the information can be read from and written to the disk. In general, the more information to be stored on the disk, the larger the diameter of the disk must be. Unfortunately, the larger the disk, the lower the maximum rotational speed must be to avoid stress-related failures. Assume the minimum allowable rotational speed (S) of the hard drive is 6000 revolutions per minute [rpm], and the rotational speed must meet the criterion $S < 12{,}000 - 150\,D^2$, where D is the diameter of the disk in inches.

   (a) Create a proper plot of these two inequalities with rotational speed on the ordinate and diameter on the abscissa. The values on the vertical axis should range from 0 to 12,000 rpm, and the values on the horizontal axis should range from 0 to 7 inches.

   (b) If graphing part (a) by hand, shade the region of the graph where both inequalities are satisfied.

   (c) Graphically determine the range of allowable rotational speeds for a 4-inch-diameter disk. Indicate the location of this answer on your graph.

   (d) Graphically determine the largest diameter disk that meets the design criteria. Indicate the location of this answer on your graph.

18. We have decided to become entrepreneurs by raising turkeys for the Thanksgiving holiday. We already have purchased some land in the country with buildings on it, so that expense need not be a part of our analysis. A study of the way turkeys grow indicates that the mass of a turkey (m) from the time it hatches (at time zero) until it reaches maturity is:

$$m = K(1 - e^{-bt})$$

Here, we select values of K and b depending on the breed of turkey we decide to raise. The value (V) of our turkey is simply the mass of the turkey times the value per pound-mass (S) when we sell it, or:

$$V = Sm$$

Here, S is the value per pound-mass (in dollars). Finally, since we feed the turkey the same amount of food each day, the cumulative cost (C) to feed the bird is:

$$C = Nt$$

Here, N is the cost of one day's supply of food [\$/day].

Create a graph of this situation, showing three lines: cumulative food cost, bird value, and profit on a particular day. For the graph, show the point after which you begin to lose money, and show the time when it is most profitable to sell the bird, indicating the day on which that occurs. Use values of $K = 21$ pound-mass, $b = 0.03$ per day, $S = \$1$ per pound-mass, and $N = \$0.12$ per day.

19. As an engineer, suppose you are directed to design a pumping system to safely discharge a toxic industrial waste into a municipal reservoir. The concentrated wastewater from the plant will be mixed with freshwater from the lake, and this mixture is to be pumped into the center of the lake. You realize that the more water you mix with the waste, the more dilute it will be and thus the smaller the impact on the fish in the lake. On the other hand, the more water you use, the more it costs in electricity for pumping. Your objective is to determine the optimum amount of water to pump so the overall cost is a minimum.

    - Assume that the cost of pumping is given by the expression $C_{pump} = 10\,Q^2$. The cost C_{pump} [\$/day] depends on the pumping rate Q [gallons per minute, or gpm] of the water used to dilute the industrial waste.
    - Now, suppose that some biologists have found that as more and more water dilutes the waste, the fish loss C_{fish} [\$/day] can be expressed as $C_{fish} = 2250 - 150\,Q$.

    With this information, construct a graph, with pumping rate on the abscissa showing the pumping cost, the fish-loss cost, and total cost on the ordinate. For the scale, plot 0 to 15 gallons per minute for flow rate.

    Determine both the minimum cost and the corresponding flow rate. Indicate the location of this answer on your graph.

20. We have obtained a contract to construct metal boxes (square bottom, rectangular sides, no top) for storing sand. Each box is to contain a specified volume, and all edges are to be welded. Each box will require the following information: a volume (V, in units of cubic inches), the length of one side of the bottom (L, in units of inches), the box height (H, in units of inches), and the material cost (M, in units of dollars per square inch). To determine the total cost to manufacture a box, we must include not only the cost of the material, but also the cost of welding all the edges. Welding costs depend on the number of linear inches that are welded (W, in units of dollars per inch). The client does not care what the box looks like, but it should be constructed at the minimum cost possible.

    (a) Construct a worksheet that will depict the cost of the material for one box, the welding cost for one box, and the total cost for the box. First, create at the top of your worksheet a section to allow the user to specify as absolute references the variables V, M, and W. Next, create a column for length ranging from 2 to 20 inches in increments of 2 inches. Finally, determine the material cost per box, welding cost per box, and total cost.

    (b) Create a proper plot of the material cost, welding cost, and total cost (all shown as ordinate values) versus the box length.

    For the following values, use the graph to determine the box shape for minimum cost: $V = 500$ cubic inches, $M = \$1.00$ per square inch, and $W = \$3.00$ per inch. Indicate the location of this answer on your graph.

    (c) Below the table created in part (a), create a row to determine the minimum value for the material cost, the welding cost, and the total cost shown in the table. Use the information to create conditional formatting in the table to show the minimum values in the table as cells with a dark color background and white text. The highlighted cells should verify the solution found in part (b) using the graph.

21. Your company has developed a new high-mileage automobile. There are two options for manufacturing this new vehicle.

    - Process A: The factory can be completely retooled and workers trained to use the new equipment.
    - Process B: The old equipment can be modified.

A graph of the costs of each process and the revenues from sales of the vehicles is shown.

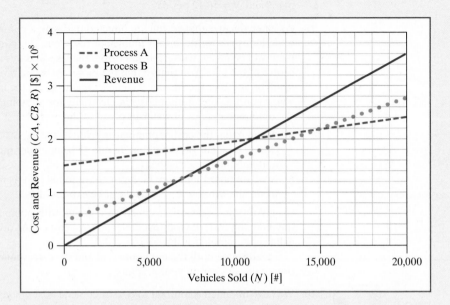

Use the chart to answer the following questions.

(a) What is the sale price per vehicle?

(b) What is the breakeven point (number of vehicles) for each of the two processes?

(c) Which process yields the most profit if 18,000 vehicles are sold? How much profit is made in this case?

(d) If the sale price per vehicle is reduced by $2000 with a rebate offer, what is the new breakeven point (number of vehicles) for each of the two processes?

22. One of the 14 Grand Challenges for Engineering as determined by a National Academy of Engineering committee is "Make Solar Energy Economical." According to the NAE website: The solar "share of the total energy market remains rather small, well below 1 percent of total energy consumption, compared with roughly 85 percent from oil, natural gas, and coal . . . today's commercial solar cells . . . typically convert sunlight into electricity with an efficiency of only 10 percent to 20 percent . . . Given their manufacturing costs, modules of today's cells . . . would produce electricity at a cost roughly 3 to 6 times higher than current prices . . . To make solar economically competitive, engineers must find ways to improve the efficiency of the cells and to lower their manufacturing costs."

The following graph shows a breakeven analysis for a company planning to manufacture modular photoelectric panels. A copy of this graph has been provided online; you may use one of these graphs, or use graph paper as directed by your instructor.

(a) What is the fixed cost incurred in manufacturing the photoelectric panels?

(b) How much does it cost to manufacture each photoelectric panel?

(c) What is the sale price of one photoelectric panel?

(d) If the company makes and sells 30,000 panels, is there a net loss or profit, and how much?

While the company is still in the planning stages, the government starts a program to stimulate the economy and encourage green technologies. In this case, the government agrees to reimburse the company $250 for each of the first 10,000 units sold.

(e) Sketch a modified revenue curve for this situation.

(f) Using this new revenue curve, how many units must the company make to break even? Be sure to clearly indicate this point on the graph.

(g) Also using the new revenue curve, how many units must the company make and sell to make a profit of $1,500,000? Be sure to clearly indicate this point on the following graph.

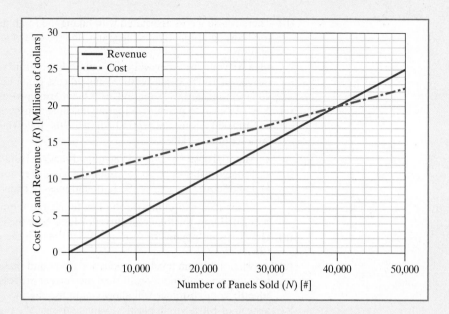

23. You are an engineer for a plastics manufacturing company. In examining cost-saving measures, your team has brainstormed the following ideas (labeled Idea A and Idea B). It is your responsibility to evaluate these ideas and recommend which one to pursue. You have been given a graph of the current process. A copy of this graph has been provided online; you may use one of these graphs, or use graph paper as directed by your instructor.

   (a) What is the selling price of the product?

   Current Cost: The current process has been running for a number of years, so there are no initial fixed costs to consider.
   In the operating costs, the process requires the following:

   - Material cost: $2.00/pound-mass of resin
   - Energy cost: $0.15/pound-mass of resin
   - Labor cost: $0.10/pound-mass of resin

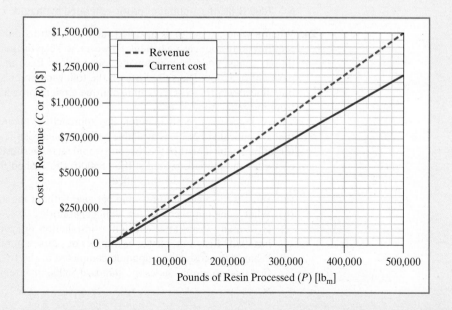

(b) There is also a cost associated with taking the scrap material to the landfill. Using the total cost determined from the graph, find the cost of landfill, in dollars per pound-mass of resin.

Idea A: Your customer will allow you to use regrind (reprocessed plastic) in the parts instead of 100% virgin plastic. Your process generates 10% scrap. Evaluate using all your scrap materials as regrind, with the regrind processed at your plant.

(c) You will need to purchase a regrind machine to process the plastic, estimated at a cost of $100,000. Using the regrind will alter the following costs, which account for using 10% scrap material:

- Material cost: $1.80/pound-mass of resin
- Energy cost: $0.16/pound-mass of resin
- Labor cost: $0.11/pound-mass of resin

This idea will eliminate the landfill charge required in the current process (see part [b]). Draw the total cost curve for Idea A on the graph or on a copy.

(d) How long (in pounds of resin processed) before the company reaches breakeven on Idea A?

(e) At what minimum level of production (in pound-mass of resin processed) will Idea A begin to generate more profit than the current process?

Idea B: Your customer will allow you to use regrind (reprocessed plastic) in the parts instead of 100% virgin plastic. Evaluate using 25% regrind purchased from an outside vendor.

(f) Using the regrind from the other company will alter the following costs, which account for using 25% scrap material purchased from the outside vendor:

- Material cost: $1.85/pound-mass of resin
- Energy cost: $0.15/pound-mass of resin
- Labor cost: $0.11/pound-mass of resin

This idea will eliminate the landfill charge required in the current process (see part [b]) and will not require the purchase of a regrind machine as discussed in Idea A. Draw the total cost curve for Idea B on the graph or on a copy.

(g) At what minimum level of production (in pound-mass of resin processed) will Idea B begin to generate more profit than the current process?

(h) At a production level of 500,000 pound-mass of resin, which Idea (A, B, or neither) gives the most profit over the current process?

(i) If the answer to part (h) is neither machine, list the amount of profit generated by the current process at 500,000 pound-mass of resin. If the answer to part (h) is Idea A or Idea B, list the amount of profit generated by that idea at 500,000 pound-mass of resin.

24. When a wind generator is installed there is a substantial initial cost, but daily operation requires no further cash payment. However, to keep the generator in proper operating condition, it must undergo maintenance once a year. Each maintenance cycle requires a cash payment of $5000. The solid lines on the following graph show this situation. The stepped blue line shows the cost over time and the straight brown line shows the revenue derived from the generator.

As the second yearly maintenance approaches, you are informed by the manufacturer that a significant upgrade is available for additional cost. The upgrade will make the generator far more efficient, thus the revenue would increase substantially. The yearly maintenance cost after the upgrade would still be $5000. The dashed lines show the cost and revenue projections if the upgrade is installed.

(a) What is the amount of revenue per year without the upgrade?

(b) What is the initial cost of the wind generator?

(c) How many years after the initial installation do you break even if the upgrade is installed? List your answer as number of years + number of months.

(d) What is the cost of the upgrade completed at the two-year maintenance cycle? Note that this amount includes the standard $5000 maintenance fee.

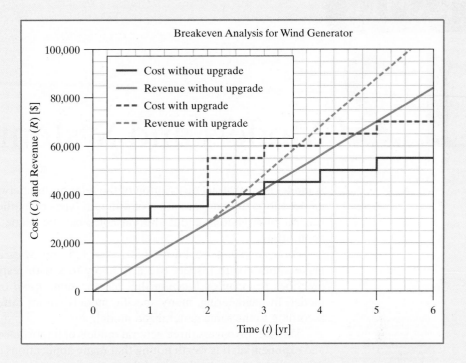

(e) How many years after the initial installation would the profit be the same whether you upgrade or not? List your answer as number of years + number of months.

(f) How many years after the initial installation will you have made a profit of $25,000 if the upgrade is NOT installed? List your answer as number of years + number of months.

(g) If the upgrade results in increased reliability, thus increasing the maintenance interval to two years, though still at a cost of $5000 per maintenance, how many years after the initial installation will you break even after the upgrade? List your answer as number of years + number of months.

12 Models and Systems

A **model** is an abstract description of the relationship between variables in a system. A model allows the categorization of different types of mathematical phenomena so that general observations about the variables can be made for use in any number of applications.

For example, if we know that $t = v + 5$ and $M = z + 5$, any observations we make about v with respect to t also apply to z with respect to M. A specific model describes a *system* or *function* that has the same *trend* or *behavior* as a generalized model. In engineering, many specific models within different subdisciplines behave according to the same generalized model.

This section covers three general models of importance to engineers: **linear**, **power**, and **exponential**. It is worth noting that many applications of models within these three categories contain identical math but apply to significantly different disciplines.

Linear models occur when the dependent variable changes in direct relationship to changes in the independent variable. We discuss such systems, including springs, resistive circuits, fluid flow, and elastic materials, in this chapter by relating each model to Newton's generalized law of motion.

Power law systems occur when the independent variable has an exponent not equal to 1 or 0. We discuss these models by addressing integer and rational real exponents.

Exponential models are used in all engineering disciplines in a variety of applications. We discuss these models by examining the similarities between growth and decay models.

The following is an example of the level of knowledge of Excel needed to proceed. *If you are not able to quickly re-create the following exercise in Excel, including trendlines and formatting, please review trendline basics in Appendix D online before proceeding.*

Energy (E) stored in an **inductor** is related to its inductance (L) and the current (I) passing through it by the following equation:

$$E = \frac{1}{2}LI^2$$

The SI unit of inductance, **henry** [H], is named for Joseph Henry (1797–1878), credited with the discovery of self-inductance of electromagnets.

Three inductors were tested and the results are given here. Create a proper plot of the data and add a properly formatted power law trendline to each data set.

Current (I) [A]	2	6	10	14	16
Energy of Inductor 1 ($E1$) [J]	0.002	0.016	0.050	0.095	0.125
Energy of Inductor 2 ($E2$) [J]	0.010	0.085	0.250	0.510	0.675
Energy of Inductor 3 ($E3$) [J]	0.005	0.045	0.125	0.250	0.310

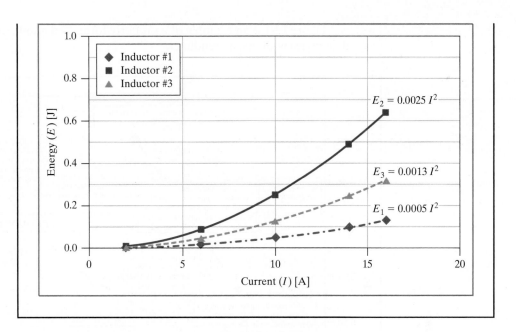

12.1 Proper Plot Rules for Trendlines

LEARN TO: Create a proper plot of experimental data containing trendlines

When creating a proper plot containing experimental data fitted to the appropriate trendline, it is important to follow several essential proper plotting rules. Note that several of the rules in the following outlined are repeated rules that apply to all graphs, or all graphs containing experimental data.

- Axis labels should include the name of the variable, the symbol used to represent the axis, as well as the units represented on the axis.
- Both vertical and horizontal major gridlines should be added at a minimum, with minor gridlines added on a case-by-case basis based on look and feel of the visual.
- For experimental data, data points should be represented as symbols only—not symbols connected by lines.
- Trendlines and symbols for each data set represented on a graph should differ in line type, symbol, and color—no data sets should repeat the same line type or symbols.
- If more than one set of data is represented on a graph, a legend should be included to assist in identification of the different experimental data sets.
- Trendline equation placement should be as close to the trendline as possible, as well as possibly formatting in the same color in order to reduce confusion in trendline association.
- Trendline equations should contain values that are formatted to be reasonable (not using the default values) and symbols in the trendline equation should match the symbols used on the horizontal and vertical axes (not using the default x, y symbols).

Figure 12-1 is an example of a properly formatted graph, showing an experimental data series with linear trendlines.

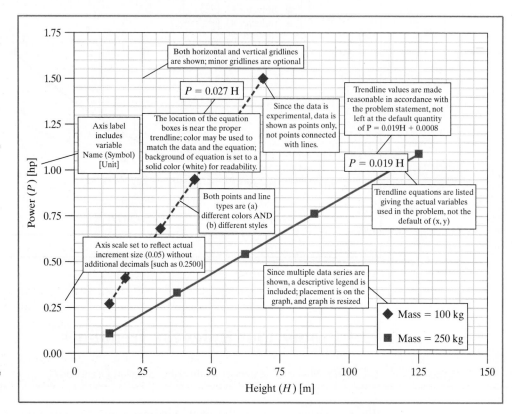

Figure 12-1
Example of a proper plot, showing multiple experimental data sets with linear trendlines.

12.2 Linear Functions

LEARN TO: Recognize the shape and boundaries of a linear function shown graphically
Recognize when an equation is a linear model
Determine the physical meaning and units of parameters of a linear function

Trend	Equation	Data Form	Graphical Example
Linear	$y = mx + c$	Defined value at $x = 0$ ($y = c$) ⎯⎯⎯⎯ Data appears as a linear (straight) line	Positive value of m / Negative value of m

One of the most common models is **linear**, taking the form $y = mx + c$, where the ordinate value (y) is a function of the abscissa value (x) and a constant factor called the **slope** (m). At an initial value of the abscissa ($x = 0$), the ordinate value is equal to the **intercept** (c). Examples include

- Distance (d) traveled at constant velocity (v) over time (t) from initial position (d_0):

$$d = vt + d_0$$

- Total pressure (P_{total}), relating density (ρ), gravity (g), liquid height (H), and the pressure above the surface ($P_{surface}$):

$$P_{total} = \rho g H + P_{surface}$$

- Newton's second law, relating force (F), mass (m), and acceleration (a):

$$F = ma$$

Note that the intercept value (c) is zero in the last example.

General Model Rules

Given a linear system of the form $y = mx + c$ and assuming $x \geq 0$:

- When $m = 1$, the function is equal to $x + c$.
- When $m = 0$, $y = c$, regardless of the value of x (y never changes).
- When $m > 0$, as x increases, y increases, regardless of the value of c.
- When $m < 0$, as x increases, y decreases, regardless of the value of c.

EXAMPLE 12-1

We want to determine the effect of depth of a fluid on the total pressure felt by a submerged object. Recall that the total pressure is

$$P_{total} = P_{surface} + P_{hydro} = P_{surface} + \rho g H$$

where P_{total} = total pressure [atm]; $P_{surface}$ = pressure at the surface [atm]; ρ = density [kg/m³]; g = gravity [m/s²]; H = depth [m]. We enter the lab, take data, and create the following chart.

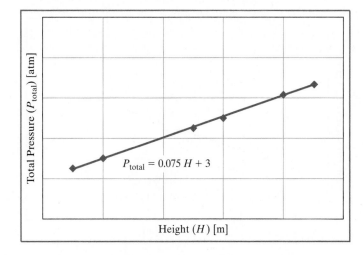

$P_{total} = 0.075\ H + 3$

Total Pressure (P_{total}) [atm]

Height (H) [m]

Determine the density of the fluid, in units of kilograms per cubic meter.

*We can determine the parameters by matching the trendline generated in Excel with the theoretical expression. In theory: total pressure = density * gravity * height of fluid + pressure on top of the fluid*

*From graph: total pressure = 0.075 * height + 3*

*By comparison: density * gravity = 0.075 [atm/m]*

$$\frac{0.075 \text{ atm}}{m} \left| \frac{101{,}325 \text{ Pa}}{1 \text{ atm}} \right| \frac{1 \frac{kg}{ms^2}}{1 \text{ Pa}} = \rho \left(\frac{9.8 \text{ m}}{s^2} \right)$$

$$\frac{7{,}600 \text{ kg}}{m^2 s^2} = \rho \left(\frac{9.8 \text{ m}}{s^2} \right)$$

$$\rho = \frac{7{,}600 \text{ kg}}{m^2 s^2} \left| \frac{s^2}{9.8 \text{ m}} = \frac{775 \text{ kg}}{m^3} \right.$$

Determine if the tank is open to the atmosphere or pressurized, and determine the pressure on the top of the fluid in units of atmospheres.

*Once again, we can compare the Excel trendline to the theoretical expression. In theory: total pressure = density * gravity * height of fluid + pressure on top of the fluid*

*From graph: total pressure = 0.075 * height + 3*

By comparison, the top of the tank is pressurized at 3 atm.

Comprehension Check   12-1

The graph shows the ideal gas law relationship ($PV = nRT$) between pressure (P) and temperature (T).

(a) What are the units of the slope (0.0087)?
(b) If the tank has a volume of 12 liters and is filled with nitrogen (formula, N_2; molecular weight, 28 grams per mole), what is the mass of gas in the tank in units of grams?
(c) If the tank is filled with 48 grams of oxygen (formula, O_2; molecular weight, 32 grams per mole), what is the volume of the tank in units of liters?

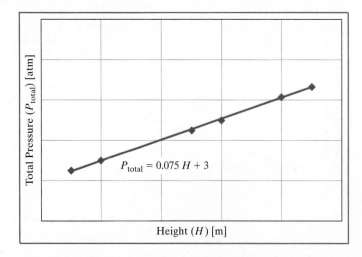

$P_{total} = 0.075\,H + 3$

Total Pressure (P_{total}) [atm]

Height (H) [m]

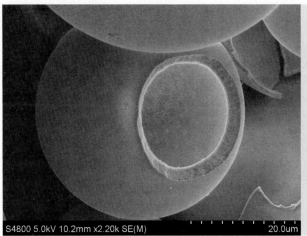

Increasingly, engineers are working at smaller and smaller scales. Tiny beads made of glass are on the order of 50 micrometers in diameter. They are manufactured so that they become hollow, allowing the wall thickness to be a few nanometers. The compositions of the glass were engineered, so when processed correctly they would sustain a hollow structure and the glass walls would be infiltrated with hundreds of thousands of nanometer-sized pores. These beads can possibly revolutionize the way fluids and gases are stored for use. The pores are small enough that fluids and even gases could be contained under normal conditions. However, if activated properly, the pores would allow a path for a gas to exit the "container" when it is ready to be used.

Photo courtesy of K. Richardson

S4800 5.0kV 10.2mm x2.20k SE(M)   20.0um

12.3 Linear Relationships

LEARN TO: Identify linear systems that are analogous to Newton's second law
Define dynamic and kinematic viscosity; identify units of centipoise and stokes

Most physics textbooks begin the study of motion ignoring how that object came to be moving in the first place. This is appropriate to the way physicists study the world, by observing the world as it is. Engineering is about changing the way things are. The fact that "engineer" is a verb as well as a noun is a reminder of this. As a result, engineers are concerned with forces and the changes those forces cause. While physicists study how far a car travels through the air when hit by a truck, engineers focus on stopping the truck before it hits the car or on designing an air-bag system or crush-proof doors. Engineering has many diverse branches because of the many different kinds of forces and ways to apply them.

Another Way of Looking at Newton's Laws

Newton's first law is given as "An object at rest remains at rest and an object in motion will continue in motion with a constant velocity unless it experiences a net external force." As we consider variables other than motion, we want to expand this definition: **A system keeps doing what it is doing unless the forces acting on the system change**.

Newton's second law is given as "The acceleration of an object is directly proportional to the net force acting on it and inversely proportional to its mass." This can be interpreted as follows: When an external force acts on a system to cause acceleration, the system resists that acceleration according to its mass. Expanding Newton's second law, we can generalize it for use with variables other than motion: **When a force influences a change to a system parameter, the system opposes the change according to its internal resistance**.

In generalizing these relationships, we can start to establish a pattern observed in a wide variety of phenomena, summarized in Table 12-1.

Table 12-1 Generalized Newton's second law

When a system . . .	. . . is acted upon by a force. . .	. . . to change a parameter . . .	. . . the system opposes the change by a resistance	Equation
Physical object	External push or pull (F)	Acceleration (a)	Object mass (m)	$F = ma$

Springs

When an external force (F), such as a weight, is applied to a spring, it will cause the spring to stretch a distance (x), according to the following expression:

$$F = kx$$

This equation is called **Hooke's law**, named for Robert Hooke (1635–1703), an English scientist. Among other things, he is credited with creating the biological term "cell." The comparison of Hooke's law and Newton's second law is shown in Table 12-2.

Table 12-2 Generalized second law applied to springs

When a system . . .	. . . is acted upon by a force . . .	. . . to change a parameter . . .	. . . the system opposes the change by a resistance	Equation
Physical object	External push or pull (F)	Acceleration (a)	Object mass (m)	$F = ma$
Spring	External push or pull (F)	Elongation (x)	Spring stiffness (k)	$F = kx$

The variable k is the **spring constant**, a measure of the stiffness of the spring. Stiff springs are hard to stretch and have high k values; springs with low k values are easy to stretch. The constant k is a material property of the spring, determined by how it is made and what material it is made from. The spring constant has units of force per distance, typically reported in newtons per meter.

EXAMPLE 12-2

Two springs were tested; a weight was hung on one end and the resulting displacement measured. The results were graphed. Using the following graph, give the spring constant of each spring and determine which spring is stiffer.

Spring 1 has a linear trendline of F = 66x. The slope of the line is the spring constant:

$$k_1 = 66\,N/m$$

Spring 2 has a linear trendline of F = 8x, which corresponds to:

$$k_2 = 8\,N/m$$

Spring 1 is stiffer since it has a higher spring constant.

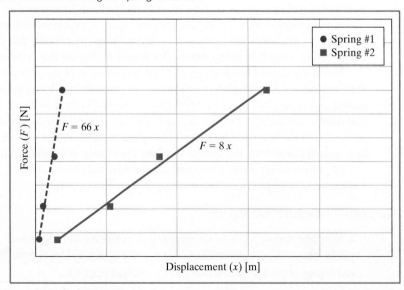

Electric Circuits

Electric current (I) is a measure of how many charges (normally electrons) flow through a wire or component in a given amount of time. This is analogous to measuring water flowing through a pipe as amount per time, whether the units are tons per hour, gallons per minute, or molecules per second. For additional information on electric circuits, refer to Section 8–11 for detailed discussion on various concepts and properties associated with electronics, electronic elements, and the mathematical foundation of how electricity works in a system.

NOTE

A now outdated term for voltage was actually "electromotive force" or EMF.

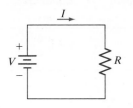

Voltage (V) is the "force" that pushes the electrons around. Although its effects on charged particles are similar to those of a true force, voltage is quite different dimensionally. The unit of voltage is the **volt** [V], described as the potential difference (voltage) across a conductor when a current of one ampere dissipates one watt of power.

Resistance (R) is a measure of how difficult it is to push electrons through a substance or device. When a voltage is applied to a circuit, a current is generated. This current depends on the equivalent resistance of the circuit. Resistance has units of volts per ampere, which is given the special name **ohm** [Ω]. It is named for Georg Ohm (1789–1854), the German physicist who developed the theory, called **Ohm's law**, to explain the relationship between voltage, current, and resistance. The similarities between Ohm's law and Newton's second law are given in Table 12-3.

$$V = IR$$

Table 12-3 Generalized second law applied to circuits

When a system . . .	. . . is acted upon by a force . . .	. . . to change a parameter . . .	. . . the system opposes the change by a resistance	Equation
Physical object	External push or pull (F)	Acceleration (a)	Object mass (m)	$F = ma$
Electrical circuit	Circuit voltage (V) (electromotive force)	Circuit current (I)	Circuit resistance (R)	$V = IR$

Fluid Flow

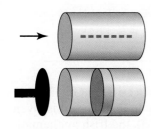

To create motion in a solid object, we can apply a force to that object by pushing on it. Imagine you have a small cube on the desk in front of you. If you take your pencil and push on that object at a single point, the entire object will move. For motion in a fluid to be created, a force must be applied over an area of the fluid. While both liquids and gases can be defined as fluids, we focus on liquids in this section. Imagine a section of fluid-filled pipe placed on the desk in front of you. If we apply a force at a single point in the fluid, only the particles at that point will move. To move the entire fluid uniformly, we must apply the force at all points at the pipe entrance simultaneously. Applying a force over the cross-sectional area of the pipe results in the application of a pressure to the fluid. The pressure that results in fluid flow has a special name: **shear stress** (τ, Greek letter tau).

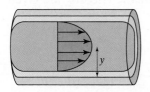

As the fluid moves, we find that the fluid molecules in contact with the wall adhere to the wall and do not move. The motion of the fluid can be visualized as occurring in layers; as the distance from the wall increases, the fluid moves faster. The fluid moves fastest at the farthest point from the wall, which is the center of the pipe. Since the velocity changes depend on the location in the pipe from the wall, the parameter we are changing cannot be expressed as a simple velocity, but rather as a **velocity gradient**, given as ($\Delta v/\Delta y$ or $\dot{\gamma}$). This is sometimes called the **shear rate** or **strain rate**.

Not all fluids respond equally to an applied pressure. The fluid property that represents the resistance of a fluid against flow is called the **dynamic viscosity** (μ, Greek letter mu). The relationship between shear stress and the velocity profile of a fluid is called **Newton's law of viscosity**, named after Isaac Newton. Fluids that behave in this way are called **Newtonian fluids** (e.g., water and oil). The comparison between Newton's law of viscosity and Newton's second law is given in Table 12-4.

$$\tau = \mu \frac{\Delta v}{\Delta y}$$

Table 12-4 Generalized second law applied to fluid flow

When a system . . .	. . . is acted upon by a force . . .	. . . to change a parameter . . .	. . . the system opposes the change by a resistance	Equation
Physical object	External push or pull (F)	Acceleration (a)	Object mass (m)	$F = ma$
Fluid	Shear stress (τ)	Shear rate ($\Delta v/\Delta y$)	Dynamic viscosity (μ)	$\tau = \mu \dfrac{\Delta v}{\Delta y}$

Sometimes, a fluid must have a certain amount of stress (called the **yield stress** τ_0) applied before it will begin to move like a Newtonian fluid. These fluids are called **Bingham plastics**, named after Eugene Bingham, a chemist who made many contributions to the field of **rheology** (the science of deformation and flow of matter, a term he, along with Markus Reiner, is credited in creating). Examples of Bingham plastics include toothpaste and slurries.

$$\tau = \mu \frac{\Delta v}{\Delta y} + \tau_0$$

Common units of dynamic viscosity are **centipoise** [cP], named after the French physician Jean Louis Poiseuille (1799–1869) who studied the flow of blood in tubes. Dynamic viscosity is a function of temperature. In most instances, viscosity decreases with increasing temperature; as the fluid heats up, it becomes easier to move.

Property	Symbol	Typical Units	Equivalent Units
Dynamic viscosity	μ	cP	1 P = 1 g/(cm s)
Kinematic viscosity	v	St	1 St = 1 cm^2/s

Another useful term in describing a fluid is **kinematic viscosity** (v, Greek letter nu). The kinematic viscosity is the ratio of dynamic viscosity to density and is given the unit of **stokes** [St], named after George Stokes (1819–1903), the Irish mathematician and physicist who made important contributions to science, including Stokes' law, optics, and physics. Several values of dynamic and kinematic viscosity are given in Table 12-5.

$$v = \frac{\mu}{\rho}$$

Table 12-5 Summary of material properties for several liquids

Liquid	Specific Gravity	Dynamic Viscosity (μ) [cP]	Kinematic Viscosity (v) [cSt]
Acetone	0.791	0.331	0.419
Corn syrup	1.36	1380	1015
Ethanol	0.789	1.194	1.513
Glycerin	1.260	1490	1183
Honey	1.36	5000	3676
Mercury	13.600	1.547	0.114
Molasses	1.400	8000	5714
Olive oil	0.703	101	143
SAE 30W oil	0.891	290	325
Water	1.000	1.000	1.000

Comprehension Check   **12-2**

Fluid A has a dynamic viscosity of 0.5 centipoise and a specific gravity of 1.1. What is the density of fluid A in units of pound-mass per cubic foot?

Comprehension Check   **12-3**

Fluid A has a dynamic viscosity of 0.5 centipoise and a specific gravity of 1.1. What is the dynamic viscosity of fluid A in units of pound-mass per foot second?

Comprehension Check   **12-4**

Fluid A has a dynamic viscosity of 0.5 centipoise and a specific gravity of 1.1. What is the kinematic viscosity of fluid A in units of stokes?

Elastic Materials

Elasticity is the property of an object or material that causes it to be restored to its original shape after distortion. A rubber band is easy to stretch and snaps back to near its original length when released, but it is not as elastic as a piece of piano wire. The piano wire is harder to stretch, but would be said to be more elastic than the rubber band because of the precision of its return to its original length. The term elasticity is quantified by **Young's modulus** or **modulus of elasticity** (E), the amount of deformation resulting from an applied force. Young's modulus is named for Thomas Young (1773–1829), a British scientist, who contributed to several fields: material elongation theory; optics, with his "double slit" optical experiment that led to the deduction that light travels in waves; and fluids, with the theory of surface tension and capillary action.

Like fluids, elastic materials accept a force applied over a unit area rather than a point force. **Stress** (σ, Greek letter sigma) is the amount of force applied over a unit area of the material, which has units of pressure [Pa]. The **strain** (ε, Greek letter epsilon) is the ratio of the elongation to the original length, yielding a dimensionless number. Since the modulus values tend to be large, they are usually expressed in units of gigapascals [GPa]. The generalized second law as applied to an elastic material is shown in Table 12-6.

$$\sigma = E\varepsilon$$

Table 12-6 Generalized second law applied to elastic materials

When a system . . .	. . . is acted upon by a force . . .	. . . to change a parameter . . .	. . . the system opposes the change by a resistance	Equation
Physical object	External push or pull (F)	Acceleration (a)	Object mass (m)	$F = ma$
Elastic object	Stress (σ)	Strain (ε)	Modulus of elasticity (E)	$\sigma = E\varepsilon$

From this discussion, you can see examples from many areas of engineering that are similar to Newton's second law. We often want to change something and find that it resists this change; this relationship is often linear. In all of these situations, we discover a coefficient that depends on the material encountered in the particular situation (mass, spring stiffness, circuit resistance, fluid viscosity, or modulus of elasticity).

Many other examples are not discussed here, such as Fourier's law of heat transfer, Fick's law of diffusion, and Darcy's law of permeability. You can enhance your understanding of your coursework by attempting to generalize the knowledge presented in a single theory to other theories that may be presented in other courses. Many different disciplines of engineering are linked by common themes, and the more you can connect these theories across disciplines, the more meaningful your classes will become.

12.4 Combinations of Linear Relationships

LEARN TO: Determine equivalency in systems of springs, circuits, capacitors, and inductors

When connected, both springs and circuits form a resulting system that behaves like a single spring or single resistor. In a combination of springs, the system stiffness depends on the stiffness of each individual spring and on the configuration, referred to as the effective spring constant (k_{eff}). In a network of circuits, the system resistance depends on the value of the individual resistors and on the configuration, referred to as the effective resistance (R_{eff}).

Springs in Parallel

When springs are attached in *parallel*, they must *displace the same distance* even though they may have different spring constants. The following derivation shows how this leads to an effective spring constant that is the sum of the individual spring constants in the system. Each spring is responsible for supporting a proportional amount of the force.

NOTE

Springs in parallel both
displace the same dis-
tance.

Writing Hooke's law for two springs each displacing the same distance (x):

$$F_1 = k_1 x \qquad \text{(a)}$$
$$F_2 = k_2 x \qquad \text{(b)}$$

Solve for F_1 in terms of F_2 since the displacement is the same:

$$F_1 = k_1 \frac{F_2}{k_2} = F_2 \frac{k_1}{k_2} \qquad \text{(c)}$$

Writing Hooke's law as applied to the overall system:

$$F = k_{\text{eff}} x \qquad \text{(d)}$$

The total force applied to the configuration (F) is the sum of the force supported by each spring:

$$F = F_1 + F_2 \qquad \text{(e)}$$

Eliminating force (F) from Equation (e) with Equation (d):

$$k_{\text{eff}} x = F_1 + F_2 \qquad \text{(f)}$$

Eliminating displacement (x) with Equation (b):

$$k_{\text{eff}} \frac{F_2}{k_2} = F_1 + F_2 \qquad \text{(g)}$$

Substituting for F_1 with Equation (c):

$$k_{\text{eff}} \frac{F_2}{k_2} = F_2 \frac{k_1}{k_2} + F_2 \qquad \text{(h)}$$

NOTE

A system of two springs
in parallel will always be
stiffer than either spring
individually.

Dividing Equation (h) by F_2:

$$\frac{k_{\text{eff}}}{k_2} = \frac{k_1}{k_2} + 1 \qquad \text{(i)}$$

Multiplying Equation (i) by k_2 gives:

$$\boldsymbol{k_{\text{eff}} = k_1 + k_2} \qquad \text{(j)}$$

Springs in Series

When two springs are attached in *series*, the *force is the same for both springs*. The effective spring constant is derived below. The applied force affects each spring as though the other spring did not exist, and each spring can stretch a different amount.

Writing Hooke's law for two springs each under the same applied force (F):

$$F = k_1 x_1 \qquad \text{(k)}$$
$$F = k_2 x_2 \qquad \text{(l)}$$

Solve for x_1 in terms of x_2 since the force is the same:

$$x_1 = \frac{k_2}{k_1} x_2 \qquad \text{(m)}$$

Writing Hooke's law as applied to the overall system:

$$F = k_{eff}x \tag{n}$$

The total distance stretched by the configuration (x) is the sum of the distance stretched by each spring:

$$x = x_1 + x_2 \tag{o}$$

Eliminating force (F) from Equation (n) with Equation (l):

$$k_2x_2 = k_{eff}x \tag{p}$$

Eliminating displacement (x) with Equation (o):

$$k_2x_2 = k_{eff}(x_1 + x_2) \tag{q}$$

Substituting for x_1 with Equation (m):

$$k_2x_2 = k_{eff}\left(\frac{k_2}{k_1}x_2 + x_2\right) \tag{r}$$

Dividing Equation (r) by x_2:

$$k_2 = k_{eff}\left(\frac{k_2}{k_1} + 1\right) \tag{s}$$

Dividing Equation (s) by k_2 gives:

$$1 = k_{eff}\left(\frac{1}{k_1} + \frac{1}{k_2}\right) \tag{t}$$

Thus,

$$\boldsymbol{k_{eff}} = \frac{1}{\left(\frac{1}{k_1} + \frac{1}{k_2}\right)} = \left(\frac{1}{k_1} + \frac{1}{k_2}\right)^{-1} \tag{u}$$

These equations for two springs connected in parallel and series generalize to any number of springs. For N springs in parallel, the effective spring constant is

$$k_{eff} = k_1 + k_2 + \cdots + k_{N-1} + k_N \tag{v}$$

For N springs in series, the effective spring constant is

$$k_{eff} = \left(\frac{1}{k_1} + \frac{1}{k_2} + \cdots + \frac{1}{k_{N-1}} + \frac{1}{k_N}\right)^{-1} \tag{w}$$

EXAMPLE   12-3

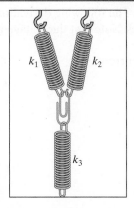

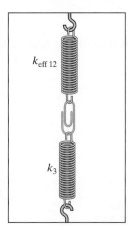

Find the displacement (x) in the spring combination shown, where Spring 1 (with a stiffness k_1) and Spring 2 (with a stiffness k_2) are connected in parallel, and the combination is then connected to Spring 3 (with a stiffness k_3) in series. Use the following values:

$$F = 0.2 \text{ N}$$
$$k_1 = 10 \text{ N/m}$$
$$k_2 = 5 \text{ N/m}$$
$$k_3 = 8 \text{ N/m}$$

First, we recognize that k_1 and k_2 are in parallel, so we can solve for an effective spring constant, using Equation (j).

$$k_{\text{eff12}} = k_1 + k_2 = 10 \text{ N/m} + 5 \text{ N/m} = 15 \text{ N/m}$$

The combination can then be redrawn to show k_{eff12} and k_3 in series.

Next, we solve for the effective spring constant using Equation (u).

$$k_{\text{eff}} = \left(\frac{1}{k_{\text{eff12}}} + \frac{1}{k_3} \right)^{-1} = \left(\frac{1}{15 \text{ N/m}} + \frac{1}{8 \text{ N/m}} \right)^{-1} = 5.2 \text{ N/m}$$

We can now solve for the displacement, using Hooke's law:

$$F = k_{\text{eff}} x$$
$$x = \frac{F}{k_{\text{eff}}} = \frac{0.2 \text{ N}}{5.2 \text{ N/m}} = 0.04 \text{ m} = 4 \text{ cm}$$

Comprehension Check   12-5

You have three springs, with stiffness 1, 2, and 3 newtons per meter [N/m], respectively. How many unique spring stiffnesses can be formed with these springs? Consider each spring alone, pairs of springs in both parallel and series, and all springs used at once.

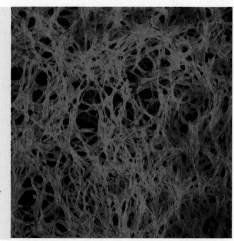

Human fibroblasts are connective tissue cells present in organs throughout the body. In this image, these cells can be seen spreading within a semi-interpenetrating network made of a polymer called polyethylene glycol diacrylate-hyaluronic acid (Pegda-HA). This material can be injected in a minimally invasive manner and cross-linked inside the body to form an insoluble gel with mechanical properties similar to many soft tissues in the human body. Such materials are being widely studied as "scaffolds" for cell transplantation in tissue engineering and regenerative medicine. The material degrades within 4–6 weeks, yielding physiological metabolites and water-soluble polymers that are readily excreted through the kidneys.

Photo courtesy of K. Webb and J. Kutty

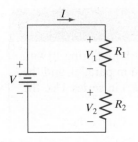

Resistors in Series

When two resistors are connected in *series*, the *current through both of the resistors is the same*, even though the value of each resistor may be different. The following derivation shows the effective resistance of two resistors connected in series. Note that the voltage is applied to the entire system.

Writing Ohm's law for two resistors each with the same current:

$$V_1 = IR_1 \tag{A}$$

$$V_2 = IR_2 \tag{B}$$

Solving for V_1 in terms of V_2 since the current is the same:

$$V_1 = R_1 \frac{V_2}{R_2} = V_2 \frac{R_1}{R_2} \tag{C}$$

NOTE

Resistors in series have the same current through both of the resistors.

Writing Ohm's law as applied to the overall system:

$$V = IR_{\text{eff}} \tag{D}$$

The total voltage applied to the configuration (V) is the sum of the voltage applied to each resistor:

$$V = V_1 + V_2 \tag{E}$$

Eliminating voltage (V) from Equation (E) with Equation (D):

$$IR_{\text{eff}} = V_1 + V_2 \tag{F}$$

Eliminating current (I) from Equation (F) with Equation (B):

$$R_{\text{eff}} \frac{V_2}{R_2} = V_1 + V_2 \tag{G}$$

Substitution for V_1 with Equation (C):

$$R_{\text{eff}} \frac{V_2}{R_2} = V_2 \frac{R_1}{R_2} + V_2 \tag{H}$$

Dividing Equation (H) by V_2:

NOTE

A system of two resistors in series will always provide more resistance than either resistor individually.

$$R_{\text{eff}} \frac{1}{R_2} = \frac{R_1}{R_2} + 1 \tag{I}$$

Multiplying Equation (I) by R_2:

$$\boldsymbol{R_{\text{eff}} = R_1 + R_2} \tag{J}$$

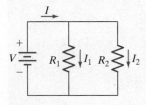

Resistors in Parallel

When two resistors are connected in *parallel*, the *voltage across both of the resistors is the same*. The current through each resistor may be different. The voltage is applied to the entire system.

Writing Ohm's law for two resistors each with the same voltage:

$$V = I_1 R_1 \tag{K}$$

$$V = I_2 R_2 \tag{L}$$

NOTE

Resistors in parallel have
the same voltage through
both of the resistors.

Solving for I_1 in terms of I_2 since the current is the same:

$$I_1 = \frac{R_2}{R_1}I_2 \tag{M}$$

Writing Ohm's law as applied to the overall system:

$$V = IR_{\text{eff}} \tag{N}$$

The total current (I) is the sum of the current flowing through each resistor:

$$I = I_1 + I_2 \tag{O}$$

Eliminating voltage (V) from Equation (N) using Equation (L):

$$I_2R_2 = IR_{\text{eff}} \tag{P}$$

Eliminating current (I) using Equation (O):

$$I_2R_2 = (I_1 + I_2)R_{\text{eff}} \tag{Q}$$

Substituting for I_1 using Equation (M):

$$I_2R_2 = \left(\frac{R_2}{R_1}I_2 + I_2\right)R_{\text{eff}} \tag{R}$$

NOTE

A system of two resistors
in parallel will always have
less resistance than either
resistor individually.

Dividing Equation (R) by I_2:

$$R_2 = R_{\text{eff}}\left(\frac{R_2}{R_1} + 1\right) \tag{S}$$

Dividing Equation (S) by R_2 gives:

$$1 = R_{\text{eff}}\left(\frac{1}{R_1} + \frac{1}{R_2}\right) \tag{T}$$

Thus,

$$R_{\text{eff}} = \frac{1}{\frac{1}{R_1} + \frac{1}{R_2}} = \left(\frac{1}{R_1} + \frac{1}{R_2}\right)^{-1} \tag{U}$$

These equations for two resistors connected in parallel and series generalize to any number of resistors. For N resistors in parallel, the effective resistance is

$$R_{\text{eff}} = \left(\frac{1}{R_1} + \frac{1}{R_2} + \cdots + \frac{1}{R_{N-1}} + \frac{1}{R_N}\right)^{-1} \tag{V}$$

This form, along with spring equation (w) shown earlier, is sometimes referred to as the "reciprocal of the sum of the reciprocals."

For N resistors in series, the effective resistance is

$$R_{\text{eff}} = R_1 + R_2 + \cdots + R_{N-1} + R_N \tag{W}$$

WARNING

Some of you may have seen a "simpler" form of the equation for two springs in series or two resistors in parallel.

For two springs in series:

$$k_{\text{eff}} = \frac{k_1k_2}{k_1 + k_2}$$

For two resistors in parallel:

$$R_{\text{eff}} = \frac{R_1 R_2}{R_1 + R_2}$$

These forms are sometimes referred to as "the product over the sum."

THESE FORMS *DO NOT GENERALIZE* TO MORE THAN TWO ELEMENTS

If you have three or more elements, then you must use the "reciprocal of the sum of the reciprocals" form given earlier.

EXAMPLE 12-4

Find the current (I) in the circuit shown. Resistor 1 (with resistance R_1) and Resistor 2 (with resistance R_2) are connected in series, and the combination is then connected to Resistor 3 (with resistance R_3) in parallel. Use the following values:

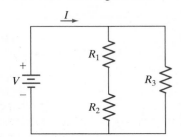

$V = 12$ V
$R_1 = 7.5$ kΩ
$R_2 = 2.5$ kΩ
$R_3 = 40$ kΩ

First, we recognize that R_1 and R_2 are in series, so we reduce R_1 and R_2 to a single effective resistor by using Equation (J).

$$R_{\text{eff12}} = R_1 + R_2 = 7.5 \text{ k}\Omega + 2.5 \text{ k}\Omega = 10 \text{ k}\Omega$$

Next, we can redraw the circuit so R_{eff12} and R_3 are in parallel. We solve for the effective resistance by using Equation (U).

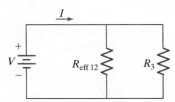

$$R_{\text{eff}} = \left(\frac{1}{R_{\text{eff12}}} + \frac{1}{R_3} \right)^{-1} = \left(\frac{1}{10 \text{ k}\Omega} + \frac{1}{40 \text{ k}\Omega} \right)^{-1} = 8 \text{ k}\Omega$$

We can now solve the problem with Ohm's law:

$$V = IR_{\text{eff}}$$

$$I = \frac{V}{R_{\text{eff}}} = \frac{12 \text{ V}}{8 \text{ k}\Omega} = 0.0015 \text{ A} = 1.5 \text{ mA}$$

Comprehension Check 12-6

You have three resistors with resistance 2, 2, and 3 ohms, respectively. How many unique resistances can be created with these resistors? Consider each resistor alone, pairs of resistors both in parallel and in series, and all resistors used at once.

When Are Components Connected in Series, Parallel, or Neither?

Note that in each diagram, the lines with one end loose indicate where the circuit or spring configuration is connected to other things.

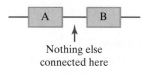

Series

When one end (but not both) of each of two components is connected together with nothing else connected at that point, they are in series.

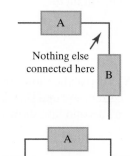

Note that they do not necessarily have to be in a straight line as shown. Electrical components can be physically mounted in any position relative to one another, and as long as a wire connects one end of each together (with nothing else connected there), they would be in series. Two springs can be connected by a string, so that the string makes a right angle direction change over a pulley, and the two springs would be in series.

Parallel

When each end of one component is connected to each of the two ends of another component, they are in parallel.

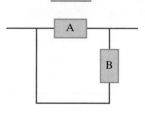

Similar to the series connection, the components do not have to be physically mounted parallel to each other or side by side, as long both ends are connected directly together with no intervening components. This is simple to do with electrical components since wire can be easily connected between any two points. Can you determine a method to physically connect two springs in parallel so that one is vertical and the other horizontal?

Sample Combinations

In the figure at left, B and C are in series. A is neither in series nor parallel with B or C since the lines extending to the left and right indicate connection to other stuff. A is, however, in parallel with the series combination of B and C.

In the figure at left, no components are in series or parallel with anything since the lines extending to the left and right indicate connection to other stuff. Note the extra line at lower right.

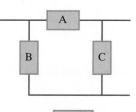

In the figure at left, A and B are in parallel. C is neither in series nor parallel with A or B. C is, however, in series with the parallel combination of A and B.

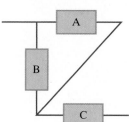

Capacitors and Inductors

In a capacitor, the **voltage** depends not only on the total charge stored, but also on the physical construction of the device, particularly the surface area of the plates. The charge

(Q) stored in a capacitor is proportional to the voltage (V) across it, where C is the proportionality constant.

$$Q = CV$$

Note that C must have units of coulombs per volt and is called **capacitance**. Capacitance is measured in units of **farads** [F], where one farad equals one coulomb per volt, or $1\,\text{F} = 1\,\text{C}/\text{V}$.

In its simplest form, an **inductor** is just a coil of wire. **Inductance** (L) is measured in units of **henrys** [H]. The voltage across an inductor is equal to the inductance of the device times the instantaneous rate of change of current through the inductor

$$V = L\frac{dI}{dt}$$

Dimensionally, the henry is one volt second per ampere [V s/A]. This can be shown to be dimensionally equal to resistance times time [Ω s] or energy per current squared [J/A²].

For both systems, we can write a generalized form of Newton's second law, shown in Table 12-7.

Table 12-7 Generalized second law applied to capacitors and inductors

When a system . . .	. . . is acted upon by a force . . .	. . . to change a parameter . . .	. . . the system opposes the change by a resistance	Equation
Physical object	External push or pull (F)	Acceleration (a)	Object mass (m)	$F = ma$
Capacitor	Charge (Q)	Voltage (V)	Capacitance (C)	$Q = CV$
Inductor	Voltage (V)	Rate of change of current (dI/dt)	Inductance (L)	$V = L\dfrac{dI}{dt}$

Combining Capacitors and Inductors

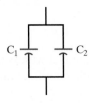

Mathematically, capacitors in series or parallel combine like springs, and inductors combine like resistors.

For two capacitors, C_1 and C_2 in **parallel**, the equivalent capacitance is given by

$$C_{\text{eq}} = C_1 + C_2$$

In general, for any number of capacitors in parallel, the equivalent capacitance is

$$C_{\text{eq}} = \sum_{i=1}^{N} C_i$$

On the other hand, two capacitors in **series** combine as the reciprocal of the sum of the reciprocals, given by

$$C_{\text{eq}} = \frac{1}{\dfrac{1}{C_1} + \dfrac{1}{C_2}}$$

or in general for any number of series capacitors

$$C_{\text{eq}} = \frac{1}{\displaystyle\sum_{i=1}^{N} \frac{1}{C_i}}$$

To help you remember which configuration matches which mathematical form, consider that the larger the area of the plates of the capacitor, the larger the capacitance. If capacitors are connected in parallel, the total plate area connected to each terminal is greater, thus the capacitance increases. This is represented by the sum, not the reciprocal of the sum of the reciprocals.

Comprehension Check   12-7

You have four 60-nanofarad [nF] capacitors. Using two or more of these capacitors in parallel or series, how many different equivalent capacitances can you form that are greater than 110 nF? Show the circuits for each such connection and list the resulting capacitances.

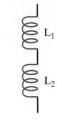

For two inductors, L_1 and L_2 in **series**, the equivalent inductance is given by

$$L_{eq} = L_1 + L_2$$

In general, for any number of inductors in series, the equivalent inductance is

$$L_{eq} = \sum_{i=1}^{N} L_i$$

On the other hand, two inductors in **parallel** combine as the reciprocal of the sum of the reciprocals, given by

$$L_{eq} = \frac{1}{\dfrac{1}{L_1} + \dfrac{1}{L_2}}$$

or in general for any number of parallel inductors

$$L_{eq} = \frac{1}{\displaystyle\sum_{i=1}^{N} \frac{1}{L_i}}$$

To help you remember which configuration matches which mathematical form, consider that the more turns of wire the current has to go through in an inductor, the larger the inductance. If inductors are connected in series, the total number of turns of wire the current must go through is larger, thus the inductance is larger, so this must be the sum, not the reciprocal of the sum of the reciprocals.

Comprehension Check   12-8

You have three 120 millihenry [mH] inductors. Can you connect two or three of these in a way that will yield an equivalent inductance of 180 mH? If not, what is the closest equivalent inductance to 180 mH you can achieve without going over 180 mH? Show the resulting connection and list the resulting inductance.

12.5 Power Functions

LEARN TO: Recognize the shape and boundaries of a power function shown graphically
Recognize when an equation is a power model
Determine the physical meaning and units of parameters of a power function

Trend	Equation	Data Form	Graphical Example
Power	$y = bx^m$	Positive m Value of zero at $x = 0$ _____ Negative m Value of infinity at $x = 0$	

Generalized power models take the form $y = bx^m + c$. One example:

- One expression for the volume (V) of a conical frustrum with base radius (r) is

$$V = \frac{\pi(H + h)}{3}r^2 - V_T$$

where H is the height of the frustrum, h is the height of the missing conical top, and V_T is the volume of the top part of the cone that is missing. In this case, $b = \frac{\pi(H + h)}{3}$ and $c = -V_T$.

In this chapter, we will only consider power law models, where c is zero. In the next chapter we will discuss ways of dealing with data when the value of c is nonzero. Examples of a power model where $c = 0$:

- Many geometric formulae involving areas, volumes, etc., such as the volume of a sphere (V) as a function of radius (r):

$$V = 4/3\pi r^3$$

- Distance (d) traveled by a body undergoing constant acceleration (a) over time (t), starting from rest:

$$d = at^2$$

- Energy calculations in a variety of contexts, both mechanical and electrical, such as the kinetic energy (KE) of an object as a function of the object's velocity (v), where the constant (k) depends upon the object shape and type of motion:

$$KE = kmv^2$$

- Ideal gas law relationships, such as Boyle's law, relating volume (V) and pressure (P) of an ideal gas, holding temperature (T) and quantity of gas (n) constant:

$$V = (nRT)P^{-1}$$

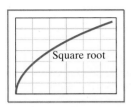

Square root

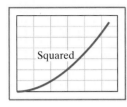

Squared

General Model Rules

Given a power system of the form $y = bx^m + c$, assuming $x \geq 0$:

- When $m = 1$, the model is a linear function.
- When $m = 0$, $y = b + c$, regardless of the value of x (y never changes).
- When m is rational, the function will contain a rational exponent or may be described with a radical symbol ($\sqrt{}$). Certain rational exponents have special names ($1/2$ is square root, $1/3$ is cube root).
- When m is an integer, the function will contain an integer exponent on the independent variable. Certain exponents have special names (2 is squared, 3 is cubed).
- When $0 < |m| < 1$ and $x < 0$, the function may contain complex values.

EXAMPLE 12-5

The volume (V) of a cone is calculated in terms of the radius (r) and height (H) of the cone. The relationship is described by the following equation:

NOTE

With a positive integer exponent, the dependent variable (volume) increases as the independent variable (radius) increases. This observation is true with any power model with a positive integer exponent.

$$V = \frac{\pi r^2 H}{3}$$

Given a height of 10 centimeters, calculate the volume of the cone when the radius is 3 centimeters.

$$V = \frac{\pi(3 \text{ cm})^2(10 \text{ cm})}{3} \approx 94.2 \text{ cm}^3$$

What is the volume of the cone when the radius is 8 centimeters?

$$V = \frac{\pi(8 \text{ cm})^2(10 \text{ cm})}{3} \approx 670 \text{ cm}^3$$

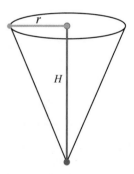

EXAMPLE 12-6

The resistance (R [g/(cm⁴s)]) of blood flow in an artery or vein depends upon the radius (r [cm]), as described by **Poiseuille's equation**:

$$R = \frac{8\mu L}{\pi} r^{-4}$$

The dynamic viscosity of blood (μ [g/(cm s)]) and length of the artery or vein (L [cm]) are constants in the system. In studying the effects of a cholesterol-lowering drug, you mimic the constricting of an artery being clogged with cholesterol, shown in the illustration. You use the data you collect to create the following graph.

NOTE

With a negative integer exponent, the dependent variable (resistance) decreases as the independent variable (radius) increases. This trend is true for any power model with a negative integer exponent.

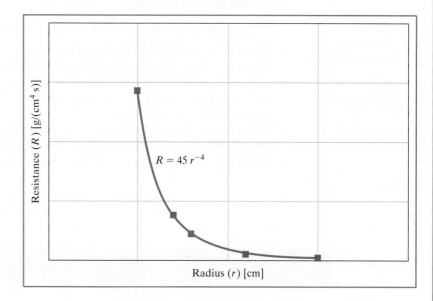

If the length of the artificial artery tested was 505 centimeters, what is the dynamic viscosity of the sample used to mimic blood, in units of grams per centimeter second [g/(cm s)]?

The constant 45 has physical meaning, found by comparison to the theoretical expression.

In theory: $R = \dfrac{8\mu L}{\pi} r^{-4}$ *and from graph:* $R = 45 r^{-4}$

By comparison:

$$45\frac{g}{s} = \frac{8\,\mu L}{\pi} = \frac{8\,\mu(505\ \text{cm})}{\pi}$$
$$\mu = 0.035\ \text{g}/(\text{cm s})$$

Comprehension Check | 12-9

The graph shows the ideal gas law relationship ($PV = nRT$) between pressure (P) and volume (V). If the tank is at a temperature of 300 kelvins and is filled with nitrogen (formula, N_2; molecular weight, 28 grams per mole), what is the mass of gas in the tank in units of grams?

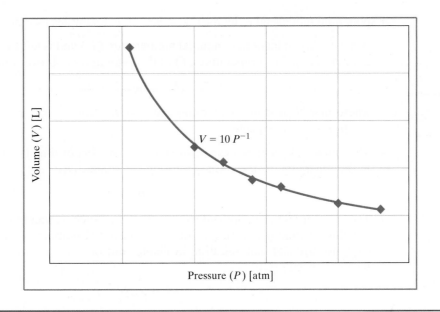

The preceding graph shows the ideal gas law relationship ($PV = nRT$) between pressure (P) and volume (V). If the tank is filled with 10 grams of oxygen (formula, O_2; molecular weight, 32 grams per mole), what is the temperature of the tank in units of degrees Celsius?

12.6 Exponential Functions

LEARN TO:	Recognize the shape and boundaries of an exponential function shown graphically
	Recognize when an equation is an exponential model
	Determine the physical meaning and units of parameters of an experimental function

Trend	Equation	Data Form	Graphical Example
Exponential	$y = be^{mx} + c$	Defined value at $x = 0$ ($y = b + c$)	
		Positive m: asymptotic to c for large negative values of x	Positive value of m / Negative value of m
		Negative m: asymptotic to c at large positive values of x	

Exponential models take the form $y = be^{mx} + c$. Examples include

- A newly forged ingot has an initial temperature (T_0) and is left to cool at room temperature (T_R). The temperature (T) of the ingot as it cools over time (t) is given by

$$T = (T_0 - T_R)\,e^{mt} + T_R$$

where m will be a negative value and $c = T_R$. Note that $b = T_0 - T_R$, so that at $t = 0, T = T_0$ as expected.

- The voltage (V) across a capacitor (C) as a function of time (t), with initial voltage (V_0) discharging its stored charge through resistance (R):

$$V = V_0 e^{-t/(RC)}$$

- The number (N) of people infected with a virus such as smallpox or H1N1 flu as a function of time (t), given the following: an initial number of infected individuals (N_0), no artificial immunization available, and dependence on contact conditions between species (C):

$$N = N_0 e^{Ct}$$

- The transmissivity (T) of light through a gas as a function of path length (L), given an absorption cross section (s) and density of absorbers (N):

$$T = e^{-sNL}$$

- The growth of bacteria (C) as a function of time (t), given an initial concentration of bacteria (C_0) and depending on growth conditions (g):

$$C = C_0 e^{gt}$$

Note that all exponents must be dimensionless, and thus unitless. For example, in the first equation, the quantity m must have units of inverse time so that the quantity mt will be unitless.

Note that the intercept value (c) is zero in all of the preceding examples except the first one.

General Model Rules

Given an exponential system of the form $y = be^{mx} + c$:

- When $m = 0, y = b + c$ regardless of the value of x (y never changes).
- When $m > 0$, the model is a **growth function**. The minimum value of the growth model for $x \geq 0$ is $b + c$. As x approaches infinity, y approaches infinity.
- When $m < 0$, the model is a **decay function**. The value of the decay model approaches c as x approaches infinity. When $x = 0, y = b + c$.

NOTE

An irrational number is a real number that cannot be expressed as the ratio of two integers. Pi (π) is an example.

What Is "e"?

The **exponential constant** "e" is a transcendental number, thus also an irrational number, that can be rounded to 2.71828. It is defined as the base of the natural logarithm function. Sometimes, e is referred to as **Euler's number** or the **Napier constant**. The reference to Euler comes from the Swiss mathematician Leonhard Euler (pronounced "oiler," 1707–1783), who made vast contributions to calculus, including the notation and terminology used today. John Napier (1550–1617) was a Scottish mathematician credited with inventing logarithms and popularizing the use of the decimal point.

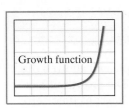

Growth function

Growth Functions

An exponential **growth function** is a type of function that increases without bound as the independent variable increases. For a system to be considered an exponential growth function, the exponential growth model ($y = be^{mx} + c$) with m is a positive value that represents the **growth rate**.

A more general exponential growth function can be formed by replacing the Napier constant with an arbitrary constant, or $y = ba^{mx} + c$. In general, a must be greater than 1 for the system to be a growth function. The value of a is referred to as the *base*, m is the *growth rate*, b is the *initial value*, and c is a *vertical shift*. Note that when $a = 1$ or $m = 0$, the system is reduced to $y = b + c$, which is a constant.

EXAMPLE 12-7

In 1965, Gordon E. Moore, cofounder of Intel Corporation, claimed in a paper that the number of transistors on an integrated circuit would double every two years. This idea by Moore was later referred to as **Moore's law**. The Intel 4004 CPU was released in 1971 as the first commercially available microprocessor. The Intel 4004 CPU contained 2300 transistors. This system can be modeled with the following growth function:

$$T = T_0 2^{t/2}$$

In the equation, T_0 represents the initial number of transistors, and t is the number of years since T_0 transistors were observed on an integrated circuit. Predict the number of transistors on an integrated circuit in 1974 using the Intel 4004 CPU as the initial condition.

$$t = 1974 - 1971 = 3 \text{ years}$$
$$T = T_0 2^{t/2} = 2300 \left(2^{3/2} \right) = 2300(2^{1.5}) \approx 6505 \text{ transistors}$$

In 1974, the Intel 8080 processor came out with 4500 transistors on the circuit.

Predict the number of transistors on integrated circuits in 1982 using the Intel 4004 CPU as the initial condition.

$$t = 1982 - 1971 = 11 \text{ years}$$
$$T = T_0 2^{t/2} = 2300 \left(2^{11/2} \right) = 2300(2^{5.5}) \approx 104{,}087 \text{ transistors}$$

In 1982, the Intel 286 microprocessor came out with 134000 transistors in the CPU.

Predict the number of transistors on integrated circuits in 2007 using the Intel 4004 CPU as the initial condition.

$$t = 2007 - 1971 = 36 \text{ years}$$
$$T = T_0 2^{t/2} = 2300 \left(2^{36/2} \right) = 2300(2^{18}) \approx 603{,}000{,}000 \text{ transistors}$$

In 2007, the NVIDIA G80 came out with 681,000,000 transistors in the CPU.

No one really knows how long Moore's law will hold up. It is perhaps interesting to note that claims have consistently been made for the past 30 years that Moore's law will only hold up for another 10 years. Although many prognosticators are still saying this, some are not. There is, however, a limit to how small a transistor can be made. Any structure has to be at least one atom wide, for example, and as they become ever smaller, quantum effects will probably wreak havoc. Of course, chips can be made larger, multilayer structures can be built, new technologies may be developed (the first functional memristor array on a CMOS chip was built in 2012), and so forth.

EXAMPLE   12-8

An environmental engineer has obtained a bacteria culture from a municipal water sample and allowed the bacteria to grow. After several hours of data collection, the following graph is created. The growth of bacteria is modeled by the following equation, where B_0 is the initial concentration of bacteria at time zero, and g is the growth constant.

$$B = B_0 e^{gt}$$

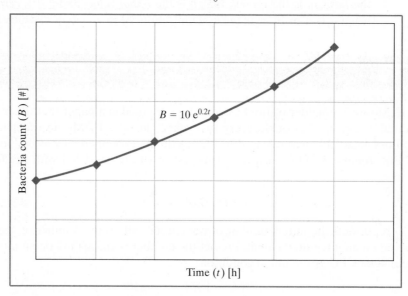

What was the initial concentration of bacteria?

In theory: $B = B_0 e^{gt}$ and from graph: $B = 10e^{0.2t}$

By comparison: $B_0 = 10$ bacteria

What was the growth constant (g) of this bacteria strain?

In theory: $B = B_0 e^{gt}$ and from graph: $B = 10e^{0.2t}$

By comparison: $g = 0.2$ per hour. Recall that exponents must be unitless, so the quantity of ($g\,t$) must be a unitless group. To be unitless, g must have units of inverse time.

The engineer wants to know how long it will take for the bacterial culture population to grow to 30,000.

To calculate the amount of time, plug in 30,000 for B and solve for t:

$$30{,}000 = 10e^{0.2t}$$

$$3{,}000 = e^{0.2t}$$

$$\ln(3{,}000) = \ln(e^{0.2t}) = 0.2t$$

$$t = \frac{\ln(3{,}000)}{0.2\left[\frac{1}{h}\right]} = 40 \text{ h}$$

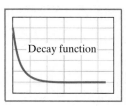

Decay function

Decay Functions

A **decay function** is a type of function that decreases and asymptotically approaches a value as the independent variable increases. In the exponential decay model ($y = be^{-mx} + c$), m is a positive value that represents the **decay rate**.

EXAMPLE 12-9

An electrical engineer wants to determine how long it will take for a particular capacitor in a circuit to discharge. The engineer wired a voltage source across a capacitor (C, farads) and a resistor (R, ohms) connected in series. After the capacitor is fully charged, the circuit is completed between the capacitor and resistor, and the voltage source is removed from the circuit. The product of R and C in a circuit like this is called the time constant and is usually denoted by the Greek letter tau ($\tau = RC$).

The following equation can be used to calculate the voltage across a discharging capacitor at a particular time:

NOTE

Exponential models are often given in the form $y = be^{-t/\tau} + c$, where t is time; thus τ also has units of time. In this case, the constant τ is often called the **time constant**.

Basically, the time constant is a measure of the time required for the response of the system to go approximately two-thirds of the way from its initial value to its final value, as t approaches infinity. The exact value is not two-thirds but $1 - e^{-1} \approx 0.632$ or 63.2%.

$$V = V_0 e^{-\frac{t}{\tau}} = V_0 e^{-\frac{t}{RC}}$$

Assuming a resistance of 100 kiloohms [kΩ], a capacitance of 100 microfarads [μF], and an initial voltage (V_0) of 20 volts [V], determine the voltage across the capacitor after 10 seconds.

$$V = 20\ [V]\ e^{-\frac{10\,s}{(100\,k\Omega)(100\,\mu F)}}$$

$$= 20\ [V]\ e^{-\frac{10\,s}{(100 \times 10^3 \Omega)(100 \times 10^{-6} F)}} \approx 7.36\ V$$

Assuming a resistance of 200 kiloohms [kΩ], a capacitance of 100 microfarads [μF], and an initial voltage (V_0) of 20 volts [V], determine the voltage across the capacitor after 20 seconds.

$$V = 20[V]e^{-\frac{20\,s}{(200\,k\Omega)(100\,\mu F)}} \approx 7.36\ V$$

Note that doubling the resistance in the circuit doubles the amount of time required to discharge the capacitor. In RC circuits, it is easy to increase the discharge time of a capacitor by increasing the resistance in the circuit.

Comprehension Check 12-11

The decay of a radioactive isotope was tracked over a number of hours, resulting in the following data. The decay of a radioactive element is modeled by the following equation, where C_0 is the initial amount of the element at time zero, and k is the decay constant of the isotope.

$$C = C_0 e^{-kt}$$

Determine the initial concentration and decay constant of the isotope, including value and units.

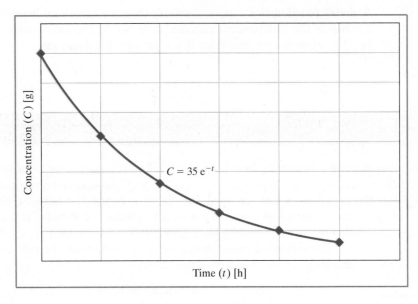

$$C = 35\,e^{-t}$$

Picture of a single mortar shot. The creation of fireworks involves knowledge of chemistry (what materials to include to get the desired colors), physics and dynamics (what amounts of combustible charge should be included to launch the object properly), and artistry (what colors, shapes, patterns, and sounds the firework should emit such that it is enjoyable to watch). This picture is a close-up of the instant when a firework is detonating.

Photo courtesy of E. Fenimore

CHAPTER 12

IN-CLASS ACTIVITIES

ICA 12-1

The graph shows the ideal gas law relationship ($PV = nRT$) between volume (V) and temperature (T).

(a) What are the units of the slope (0.0175)?

(b) If the tank has a pressure of 2.4 atmospheres and is filled with nitrogen (formula, N_2; molecular weight, 28 grams per mole), what is the mass of gas in the tank in units of grams?

(c) If the tank is filled with 20 grams of oxygen (formula, O_2; molecular weight, 32 grams per mole), what is the pressure of the tank (P) in units of atmospheres?

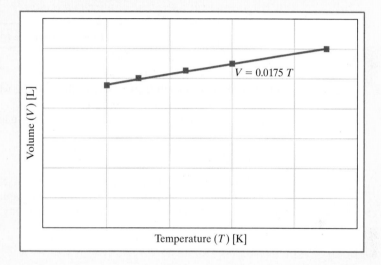

ICA 12-2

An inductor is an electrical device that can store energy in the form of a magnetic field. In the simplest form, an inductor is a cylindrical coil of wire, and its inductance (L), measured in henrys [H], can be calculated by

$$L = \frac{\mu_0 n^2 A}{\ell}$$

where

μ_0 = permeability of free space = $4\pi \times 10^{-7}$ [newtons per ampere squared, N/A^2]
n = number of turns of wire [dimensionless]
A = cross-sectional area of coil [square meters, m^2]
ℓ = length of coil [meters, m]
L = inductance [henrys, H] = [J/A^2]

Several inductors were fabricated with the same number of turns of wire (n) and the same length (ℓ), but with different diameters, thus different cross-sectional areas (A). The inductances were measured and plotted as a function of cross-sectional area, and a mathematical model was developed to describe the relationship, as shown on the following graph.

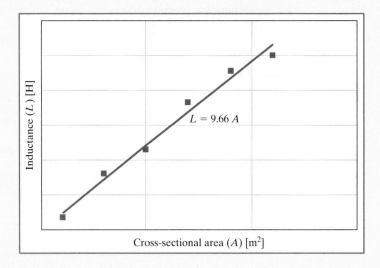

(a) What are the units of the slope (9.66)?

(b) For an inductor fabricated as described above, what is its diameter if its inductance is 0.4 henrys? Give your answer in centimeters.

(c) If the length of the coil (ℓ) equals 0.2 meter, how many turns of wire (n) are in the inductor?

ICA 12-3

Solid objects, such as your desk or a rod of aluminum, can conduct heat. The magnitude of the thermal diffusivity of the material determines how quickly the heat moves through a given amount of material. The equation for thermal diffusivity (α) is given by:

$$\alpha = \frac{k}{\rho \, C_p}$$

Experiments are conducted to change the thermal conductivity (k) of the material while holding the specific heat (C_p) and the density (ρ) constant. The results are shown graphically.

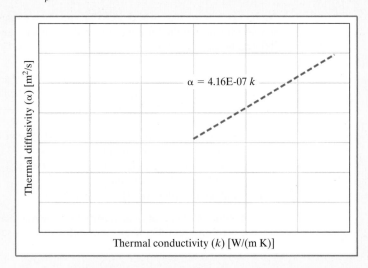

(a) What are the units of the constant 4.16×10^{-7}? Simplify your answer.
(b) If the specific heat of the material is 780 joules per kilogram kelvin, what is the density of the material?
(c) If the material has a density of 5400 kilograms per cubic meter, what is the specific heat of the material in units of joules per kilogram kelvin?

ICA 12-4

Mercury has a dynamic viscosity of 1.55 centipoises and a specific gravity of 13.6.

(a) What is the density of mercury in units of kilograms per cubic meter?
(b) What is the dynamic viscosity of mercury in units of pound-mass per foot second?
(c) What is the dynamic viscosity of mercury in units of pascal seconds?
(d) What is the kinematic viscosity of mercury in units of stokes?

ICA 12-5

SAE 99W10, a brand new type of motor oil has a dynamic viscosity of 0.28 kilograms per meter second and a specific gravity of 0.986.

(a) What is the density of the motor oil in units of kilograms per cubic meter?
(b) What is the dynamic viscosity of the motor oil in units of pound-mass per foot second?
(c) What is the dynamic viscosity of the motor oil in units of centipoise?
(d) What is the kinematic viscosity of the motor oil in units of stokes?

ICA 12-6

You have two springs each of stiffness 2 newton per meter [N/m] and one spring of stiffness 3 newtons per meter [N/m].

(a) There are ___ configurations possible, with ___ unique combinations, resulting in ___ different stiffness values.
     A "configuration" is a way of combining the springs. For example, two springs in parallel is one configuration; two springs in series is a second configuration.
     A "combination" is the specific way of combining given springs to form an effective spring constant. For example, combining springs 1 and 2 in parallel is one combination; combining springs 1 and 3 in parallel is a second combination. These combinations may or may not result in a unique effective spring constant.
(b) What is the stiffest combination, and what is the spring constant of this combination?
(c) What is the least stiff combination, and what is the spring constant of this combination?

ICA 12-7

You have three resistors of resistance 30 ohms [Ω].

(a) There are ___ configurations possible, with ___ unique combinations, resulting in ___ different resistance values.
     A "configuration" is a way of combining the resistors. For example, two resistors in parallel is one configuration; two resistors in series is a second configuration.
     A "combination" is the specific way of combining given resistors to form an effective resistance. For example, combining resistors 1 and 2 in parallel is one combination; combining resistors 1 and 3 in parallel is a second combination. These combinations may or may not result in a unique effective resistance.

(b) What is the greatest resistance that can be made from a combination of resistors, and what is the effective resistance of this combination?

(c) What is the least resistance that can be made from a combination of resistors, and what is the effective resistance of this combination?

ICA 12-8

Four springs were tested, with the results shown graphically below. Use the graph to answer the following questions.

(a) Which spring is the stiffest?

(b) Which spring, if placed in parallel with Spring C, would yield the stiffest combination?

(c) Which spring, if placed in series with Spring C, would yield the stiffest combination?

(d) Rank the following combinations in order of stiffness:

Spring A and Spring D are hooked in parallel

Spring B and Spring C are hooked in series, then connected with Spring D in parallel

Spring A

Spring D

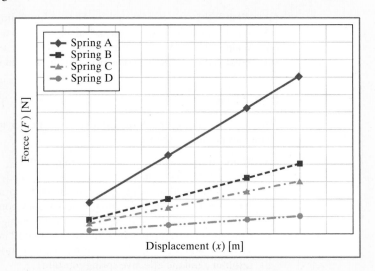

ICA 12-9

Four circuits were tested, with the results shown graphically below. Use the graph to answer the following questions.

(a) Which resistor gives the most resistance?

(b) What is the resistance of Resistor A?

(c) Which resistor, if placed in parallel with Resistor C, would yield the highest resistance?

(d) Which resistor, if placed in series with Resistor C, would yield the highest resistance?

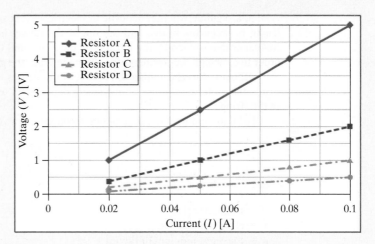

ICA 12-10

Assume you have an unlimited number of inductors all with the same inductance L.

(a) How would you connect four of these inductors so that the equivalent inductance equals L?

(b) How would you connect N^2 of these inductors so that the equivalent inductance equals L?

ICA 12-11

(a) The equivalent capacitance of the circuit shown is 6 nF. Determine the value of C.

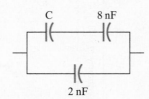

(b) The equivalent capacitance of the circuit shown is 5 nF. Determine the value of C.

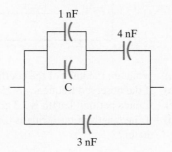

ICA 12-12

A standard guitar, whether acoustic or electric, has six strings, all with essentially the same total length between the bridge and the nut at the tuning head. Each string vibrates at a different frequency determined by the tension on the string and the mass per unit length of the string. In order to create pitches (notes) other than these six, the guitarist presses the strings down against the fretboard, thus shortening the length of the strings and changing their frequencies. In other words, the vibrating frequency of a string depends on tension, length, and mass per unit length of the string.

The equation for the fundamental frequency of a vibrating string is given by

$$f = \frac{\sqrt{T/\mu}}{2L}$$

where

f = frequency [Hz]
T = string tension [N]
μ = mass per unit length [kg/m]
L = string length [m]

Many electric guitars have a device often called a "whammy" bar or a "tremolo" bar that allows the guitarist to change the tension on the strings quickly and easily, thus changing the frequency of the strings. (Think of Jimi Hendrix simulating "the rockets' red glare, the bombs bursting in air" in his rendition of *The Star Spangled Banner*—a true *tour de force*.) In designing a new whammy bar, we test our design by collecting data using a single string on the guitar and creating a graph of the observed frequency at different string tensions as shown.

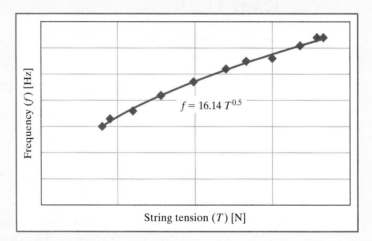

$f = 16.14\ T^{0.5}$

(a) What are the units of the coefficient (16.14)?
(b) If the observed frequency is 150 hertz, what is the string tension in newtons?
(c) If mass per unit length is 2.3 grams per meter, what is the length of the string in meters?
(d) If the length of the string is 0.67 meters, what is the mass per unit length in kilograms per meter?

ICA 12-13

The vibrating frequency of a guitar string depends on tension, length, and mass per unit length of the string. The equation for the fundamental frequency of a vibrating string is given by

$$f = \frac{\sqrt{T/\mu}}{2L}$$

where

f = frequency [Hz]
T = string tension [N]
μ = mass per unit length [kg/m]
L = string length [m]

Many electric guitars have a device often called a "whammy" bar or a "tremolo" bar that allows the guitarist to change the tension on the strings quickly and easily, thus changing the frequency of the strings. (Think of Jimi Hendrix simulating "the rockets red glare, the bombs bursting in air" in his rendition of *The Star Spangled Banner*—a true *tour de force*.) In designing a new whammy bar, we test our design by collecting data using a single string on the guitar and creating a graph of the observed frequency at different string lengths as shown.

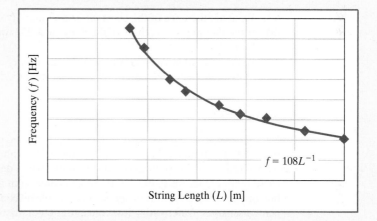

$f = 108L^{-1}$

String Length (L) [m]

(a) Is the relationship between frequency and length linear, power, or exponential?
(b) What are the units of the coefficient (108)?
(c) If the tension on the string is 135 newtons, what is the mass per unit length in grams per meter?
(d) If the mass per length of the string is 3.5 grams per meter, what is the tension in newtons?

ICA 12-14

Solid objects, such as your desk or a rod of aluminum, can conduct heat. The magnitude of the thermal diffusivity of the material determines how quickly the heat moves through a given amount of material. The equation for thermal diffusivity (α) is given by:

$$\alpha = \frac{k}{\rho\, C_p}$$

Experiments are conducted to change the specific heat (C_p) of the material while holding the thermal conductivity (k) and the density (ρ) constant. The results are shown graphically.

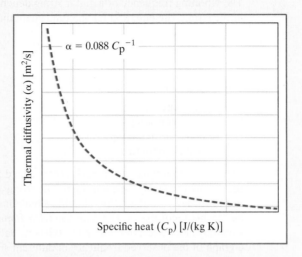

(a) What are the units of the constant 0.088? Simplify your answer.
(b) If the thermal conductivity of the material is 237 watts per meter kelvin, what is the density of the material?
(c) If the material has a density of 4500 kilograms per cubic meter, what is the thermal conductivity of the material in units of watts per meter kelvin?

ICA 12-15

Eutrophication is a process whereby lakes, estuaries, or slow-moving streams receive excess nutrients that stimulate excessive plant growth. This enhanced plant growth, often called an algal bloom, reduces dissolved oxygen in the water when dead plant material decomposes and can cause other organisms to die. Nutrients can come from many sources, such as fertilizers; deposition of nitrogen from the atmosphere; erosion of soil containing nutrients; and sewage treatment plant discharges. Water with a low concentration of dissolved oxygen is called hypoxic. A biosystems engineering models the algae growth in a lake. The concentration of algae (C), measured in grams per milliliter [g/mL], can be calculated by

$$C = C_0 e^{\left(\frac{kt}{r}\right)}$$

where

C_0 = initial concentration of algae [?]
k = multiplication rate of the algae [?]
r = estimated nutrient supply amount [mg of nutrient per mL of sample water]
t = time [days]

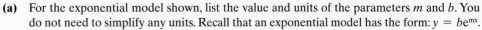

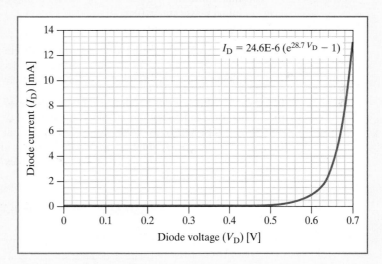

$$C = 42\, e^{(0.15\, t)}$$

(a) For the exponential model shown, list the value and units of the parameters m and b. You do not need to simplify any units. Recall that an exponential model has the form: $y = b e^{mx}$.

(b) What are the units on the multiplication rate of the algae (k)?

(c) If the algae are allowed to grow for 10 days with an estimated nutrient supply of 3 milligrams of nutrient per milliliter of water sample, what is the multiplication rate of the algae (k)?

ICA 12-16

The following graph shows the relationship between current and voltage in a 1N4148 small signal diode (a semiconductor device that allows current to flow in one direction but not the other).

$$I_D = 24.6\text{E-}6\,(e^{28.7\, V_D} - 1)$$

Semiconductor diodes can be characterized by the Shockley equation:

$$I_D = I_0\!\left(e^{\frac{qV_D}{nkT}} - 1\right)$$

where

I_D is the diode current [amperes]
I_0 is the reverse saturation current, constant for any specific diode
q is the charge on a single electron, 1.602×10^{-19} coulombs
V_D is the voltage across the diode [volts]
n is the emission coefficient, having a numerical value typically between 1 and 2, and constant for any specific device.
k is Boltzmann's constant, 1.381×10^{-23} joules per kelvin
T is the temperature of the device [kelvin]

(a) What are the units of the 1 following the exponential term? Justify your answer.
(b) If the device temperature is 100 degrees Fahrenheit, what are the units of the emission coefficient, n, and what is its numerical value? (*Hint: Electrical power [W] equals a volt times an ampere: P = VI. One ampere equals one coulomb per second.*)
(c) What is the numerical value and units of the reverse saturation current, I_0? Use an appropriate metric prefix in your final answer.

ICA 12-17

The total quantity (mass) of a radioactive substance decreases (decays) with time as

$$m = m_0 e^{-\frac{t}{\tau}}$$

where

t = time [days]
τ = time constant
m_0 = initial mass (at $t = 0$)
m = mass at time t [mg]

A few milligrams each of three different isotopes of uranium were assayed for isotopic composition over a period of several days to determine the decay rate of each. The data were graphed and a mathematical model derived to describe the decay of each isotope.

(a) What are the units of τ if time is measured in days?
(b) What is the initial amount of each isotope at $t = 0$?
(c) When will 1 milligram of the original isotope remain in each sample?
(d) Four isotopes of uranium are shown in the following table with their half-lives. Which isotope most likely matches each of the three samples? Note that one isotope does not have a match on the graph.

Isotope	Half-life [days]
230U	20.8
231U	4.2
237U	6.75
240U	0.59

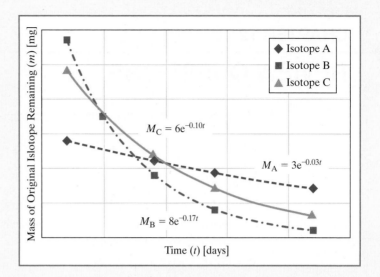

ICA 12-18

Match the data series from the options shown on the graph to the following model types. You may assume that power and exponential models do not have a constant offset. You may also assume that only positive values are shown on the two axes. For each match, write "Series X," where X is the appropriate letter, A through F. If no curve matches the specified criterion, write "No Match." If more than one curve matches a given specification, list both series.

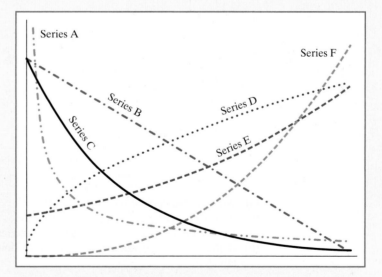

(a) Exponential, negative numeric value in exponent
(b) Power, negative numeric value in exponent
(c) Linear, negative slope
(d) Exponential, positive numeric value in exponent
(e) Power, positive numeric value in exponent

CHAPTER 12

REVIEW QUESTIONS

1. For a simple capacitor with two flat plates, the capacitance (C) [F] can be calculated by

$$C = \frac{\varepsilon_r \varepsilon_0 A}{d}$$

where

$\varepsilon_0 = 8.854 \times 10^{-12}$ [F/m] (the permittivity of free space in farads per meter)
ε_r = relative static permittivity, a property of the insulator [dimensionless]
A = area of overlap of the plates [m^2]
d = distance between the plates [m]

   Several experimental capacitors were fabricated with different plate areas (A), but with the same inter-plate distance $(d = 1.2\ mm)$ and the same insulating material, and thus the same relative static permittivity (ε_r). The capacitance of each device was measured and plotted versus the plate area. The graph and trendline follow. The numeric scales were deliberately omitted.

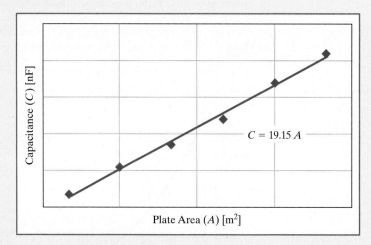

   (a) What are the units of the slope (19.15)?
   (b) If the capacitance is 3 nanofarads [nF], what is the area (A) of the plates?
   (c) What is the relative static permittivity of the insulating layer?
   (d) If the distance between the plates were doubled, how would the capacitance be affected?

2. When we wish to generate hydroelectric power, we build a dam to back up the water in a river. If the water has a height $(H$, in units of feet) above the downstream discharge, and we can discharge water through the turbines at a rate $(Q$, in units of cubic feet per second [cfs]), the maximum power $(P$, in units of kilowatts) we can expect to generate is:

$$P = CHQ$$

For a small "run of the river" hydroelectric facility, we have obtained the following data.

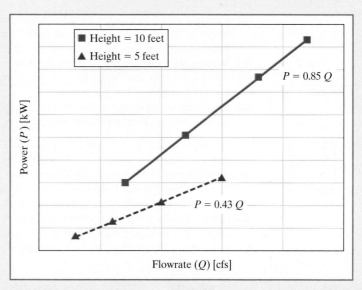

(a) Using the trendline results, and examining the preceding general equation, determine the value and units of the coefficient C for a height of 20 feet.

(b) If the flowrate was 30 cubic feet per second and the height is 3 meters, what would the power output be in units of horsepower?

(c) If the flowrate was 20 cubic feet per second and the height is 8 meters, what would the power output be in units of horsepower?

3. When rain falls over an area for a sufficiently long time, it will run off and collect at the bottom of hills and eventually find its way into creeks and rivers. A simple way to estimate the maximum discharge flowrate Q (in units of cubic feet per second [cfs]) from a watershed of area A (in units of acres) with a rainfall intensity i (in units of inches per hour) is given by an expression commonly called the Rational Method, as

$$Q = CiA$$

Values of C vary between about 0 (for flat rural areas) to almost 1 (in urban areas with a large amount of paved area).

A survey of a number of rainfall events was made over a 10-year period for three different watersheds. The data that resulted is given in the following table. Watershed A is 120 acres, B is 316 acres, and C is 574 acres.

Storm event	Watershed	Rainfall Intensity (i) [in/h]	Maximum Runoff (Q) [cfs]
1	A	0.5	30
2	A	1.1	66
3	A	1.6	96
4	A	2.1	126
5	B	0.3	47
6	B	0.7	110
7	B	1.2	188
8	B	1.8	283
9	C	0.4	115
10	C	1	287
11	C	1.5	430
12	C	2.4	690

(a) Create a graph containing all three watersheds, with flowrate on the ordinate, and fit linear trendlines to obtain a simple model for each watershed.

From the information given and the trendline model obtained, answer the following:

(b) What is the value and units of the coefficient C?

(c) What would the maximum flowrate be from a watershed of 200 acres if the rainfall intensity was 1.8 inches per hour?

(d) How long would it take at this flowrate to fill an Olympic-sized swimming pool that is 60 meters long, 30 meters wide, and 2 meters deep?

4. You are experimenting with several liquid metal alloys to find a suitable replacement for the mercury used in thermometers. You have attached capillary tubes with a circular cross section and an inside diameter of 0.3 millimeters to reservoirs containing 5 cubic centimeters of each alloy. You mark the position of the liquid in each capillary tube when the temperature is 20 degrees Celsius, systematically change the temperature, and measure the distance the liquid moves in the tube as it expands or contracts with changes in temperature. Note that negative values correspond to contraction of the material due to lower temperatures. The data you collected for four different alloys is shown in the following table.

Alloy G1		Alloy G2		Alloy G3		Alloy G4	
Temperature (T) [°C]	Distance (d) [cm]	Temperature (T) [°C]	Distance (d) [cm]	Temperature (T) [°C]	Distance (d) [cm]	Temperature (T) [°C]	Distance (d) [cm]
22	1.05	21	0.95	24	2.9	25	5.1
27	3.05	29	7.65	30	7.2	33	13.8
34	6.95	33	10.6	34	9.8	16	−4.3
14	−3.5	17	−2.6	19	−0.6	13	−7.05
9	−5.1	3	−14.8	12	−6.15	6	−14.65
2	−8.7	−2	−19.8	4	−11.5	−2	−22.15
−5	−11.7	−8	−25.4	−5	−18.55	−6	−26.3
−11	−15.5					−12	−32.4

(a) In Excel, create two new columns for each compound to calculate the change in temperature (ΔT) relative to 20 °C (for example, 25 °C gives $\Delta T = 5$ °C) and the corresponding change in volume (ΔV).

Plot the change in volume versus the change in temperature; fit a linear trendline to each data set.

(b) From the trendline equations, determine the value and units of the coefficient of thermal expansion, β, for each alloy. Note that $\Delta V = \beta V \Delta T$, where V is the initial volume.

(c) There is a small constant offset (C) in each trendline equation ($\Delta V = \beta V \Delta T + C$). What is the physical origin of this constant term? Can it be safely ignored? In other words, is its effect on the determination of β negligible?

5. The resistance of a wire (R [ohm]) is a function of the wire dimensions (A = cross-sectional area, L = length) and material (ρ = resistivity) according to the relationship

$$R = \frac{\rho L}{A}$$

The resistance of three wires was tested. All wires had the same cross-sectional area.

Length (L) [m]	0.01	0.1	0.25	0.4	0.5	0.6
Resistance Wire 1 (R1) [Ω]	8.00E-05	8.00E-04	2.00E-03	3.50E-03	4.00E-03	4.75E-03
Resistance Wire 2 (R2) [Ω]	4.75E-05	4.80E-04	1.00E-03	2.00E-03	2.50E-03	3.00E-03
Resistance Wire 3 (R3) [Ω]	1.50E-04	1.70E-03	4.25E-03	7.00E-03	8.50E-03	1.00E-02

(a) Plot the data and fit a linear trendline model to each wire.
(b) From the following chart, match each wire (1, 2, and 3) with the correct material according to the results of the resistivity determined from the trendlines, assuming a 0.2-centimeter diameter wire was used.

Material	Resistivity (ρ)[Ωm] $\times$ 10^{-8}
Aluminum	2.65
Copper	1.68
Iron	9.71
Silver	1.59
Tungsten	5.60

6. Use the figure shown to answer the following questions.

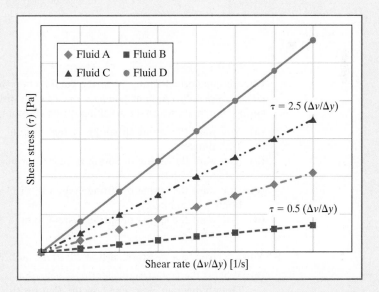

(a) Which fluid has the lowest dynamic viscosity?
(b) What is the dynamic viscosity of Fluid B in units of centipoise?
(c) If the specific gravity of Fluid C is 0.8, what is the kinematic viscosity of Fluid C in units of stokes?

7. You are given four springs, one each of 3.5, 6, 8.5, and 11 newtons per meter [N/m].

   (a) What is the largest equivalent stiffness that can be made using these four springs? Draw a diagram indicating how the four springs are connected.

   (b) What is the smallest equivalent stiffness that can be made using only three of these springs? Draw a diagram indicating how the three springs are connected.

   (c) How close an equivalent stiffness to the average of the four springs (6.25 newtons per meter) can you make using only these springs? You may use all four springs to do this, but you may use less if that will yield an equivalent stiffness closer to the average.

8. You have three springs. You conduct several tests and determine the following data.

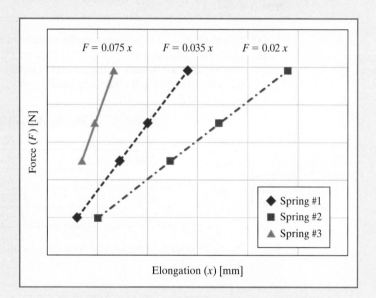

Choose one correct spring or spring combination that will meet the following criteria as closely as possible. Assume you have one of each spring available for use. List the spring or spring combination and the resulting spring constant.

   (a) You want the spring or spring system to hold 95 grams and displace approximately 1 centimeter.

   (b) You want the spring or spring system to displace approximately 4 centimeters when holding 50 grams.

   (c) You want the spring or spring system to displace approximately 5 millimeters when holding 75 grams.

   (d) You want the spring or spring system to hold 20 grams and displace approximately 1 centimeter.

9. You are given four resistors, each of 7.5, 10, 15, and 20 kiloohms [kΩ].

   (a) What is the largest equivalent resistance that can be made using these four resistors? Draw a diagram indicating how the four resistors are connected.

   (b) What is the smallest equivalent resistance that can be made using only three of these resistors? Draw a diagram indicating how the three resistors are connected.

   (c) How close an equivalent resistance to the average of the four resistors (13.125 kΩ) can you make using only these resistors? You may use all four resistors to do this, but you may use less if that will yield an equivalent resistance closer to the average.

10. You have three resistors. You conduct several tests and determine the following data.

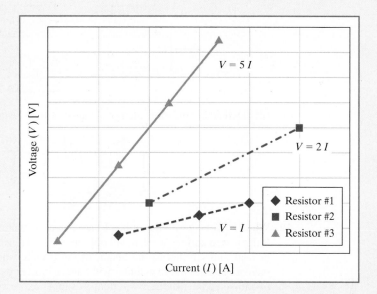

Choose one correct resistor or resistor combination that will meet the following criteria as closely as possible. Assume you have one of each resistor available for use. List the resistor or resistor combination and the resulting resistor constant.

(a) You want the resistor or resistor system to provide approximately 20 amperes when met with 120 volts.

(b) You want the resistor or resistor system to provide approximately 46 amperes when met with 30 volts.

(c) You want the resistor or resistor system to provide approximately 15 amperes when met with 120 volts.

(d) You want the resistor or resistor system to provide approximately 33 amperes when met with 45 volts.

11. Use the diagrams shown to answer the following questions.

(a) Determine the equivalent stiffness of four springs connected as shown.

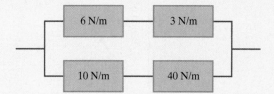

(b) Determine the equivalent stiffness of four springs connected as shown.

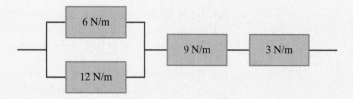

(c) Determine the equivalent resistance of four resistors connected as shown.

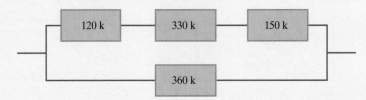

(d) Determine the equivalent resistance of four resistors connected as shown.

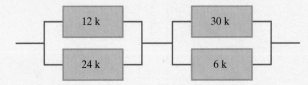

12. When a buoyant cylinder of height H, such as a fishing cork, is placed in a liquid and the top is depressed and released, it will bob up and down with a period T. We can conduct a series of tests and see that as the height of the cylinder increases, the period of oscillation also increases. A less dense cylinder will have a shorter period than a denser cylinder, assuming of course all the cylinders will float. A simple expression for the period is:

$$T = 2\pi \sqrt{\frac{\rho_{cylinder}}{\rho_{liquid}} \frac{H}{g}}$$

where g is the acceleration due to gravity, $\rho_{cylinder}$ is the density of the material, and ρ_{liquid} is the density of the fluid. By testing cylinders of differing heights, we wish to develop a model for the oscillation period, shown in the following graph.

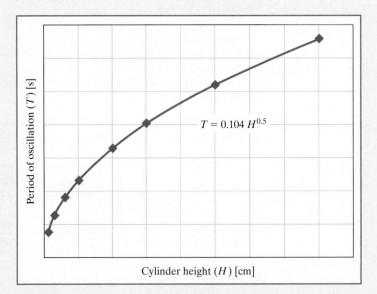

(a) What are the units of the coefficient (0.104) shown in the model?
(b) What is the oscillation period in units of seconds of a cylinder that is 4 inches tall?
(c) If the oscillation period is 0.2 seconds, what is the height of the cylinder in units of inches?
(d) We will conduct a series of tests with a new plastic (polystretchypropylene) that has a specific gravity of 0.6. What is the specific gravity of the fluid?

13. It is difficult to bring the Internet to some remote parts of the world. This can be inexpensively done by installing antennas tethered to large helium balloons. To help analyze the situation, assume we have inflated a large spherical balloon. The pressure on the inside of the balloon is balanced by the elastic force exerted by the rubberized material. Since we are dealing with a gas in an enclosed space, the ideal gas law will be applicable.

$$PV = nRT$$

where

P = pressure [atm]
V = volume [L]
n = quantity of gas [moles]
R = ideal gas constant [0.08206 (atm L)/(mol K)]
T = temperature [K]

If the temperature increases, the balloon will expand and/or the pressure will increase to maintain the equality. As it turns out, the increase in volume is the dominant effect, so we will treat the change in pressure as negligible.

The circumference of an inflated spherical balloon is measured at various temperatures; the resulting data are shown in the following graph.

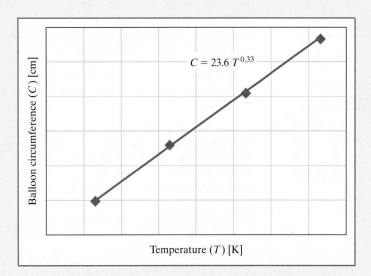

(a) What are the units of the constant 0.33?
(b) What are the units of the constant 23.6?
(c) What would the temperature of the balloon be if the circumference was 162 centimeters?
(d) If a circle with an area of 100 square centimeters is drawn on the balloon at 20 degrees Celsius, what would the area be at a temperature of 100 degrees Celsius?
(e) If the pressure inside the balloon is 1.2 atmospheres, how many moles of gas does it contain?

14. The data shown in the following graph was collected during testing of an electromagnetic mass driver. The energy to energize the electromagnets was obtained from a bank of capacitors. The capacitor bank was charged to various voltages, and for each voltage, the exit velocity of the projectile was measured when the mass driver was activated.

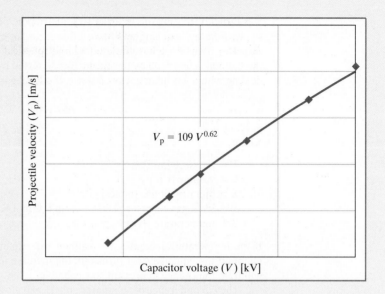

$$V_p = 109 \, V^{0.62}$$

Projectile velocity (V_p) [m/s] vs Capacitor voltage (V) [kV]

NOTE

Due to several complicated nonlinear losses in the system that are far beyond the scope of this course, this is a case of a power model in which the exponent does not come out to be an integer or simple fraction, so rounding to two significant figures is appropriate. In fact, this model is only a first approximation—a really accurate model would be considerably more complicated.

(a) What would the velocity be if the capacitors were charged to 100,000 volts?

(b) What voltage would be necessary to accelerate the projectile to 1000 meters per second?

(c) Assume that the total capacitance is 5 farads. If the capacitors are initially charged to 10,000 volts and are discharged to 2000 volts during the launch of a projectile, what is the mass of the projectile if the overall conversion of energy stored in the capacitors to kinetic energy in the projectile has an efficiency of 20%? Recall that the energy stored in a capacitor is given by $E = 0.5 \, CV^2$, where C is capacitance in farads and V is voltage in volts.

15. A standard guitar, whether acoustic or electric, has six strings, all with essentially the same total length between the bridge and the nut at the tuning head. Each string vibrates at a different frequency determined by the tension on the string and the mass per unit length of the string. In order to create pitches (notes) other than these six, the guitarist presses the strings down against the fretboard, thus shortening the length of the strings and changing their frequencies. In other words, the vibrating frequency of a string depends on tension, length, and mass per unit length of the string.

The equation for the fundamental frequency of a vibrating string is given by

$$f = \frac{\sqrt{T/\mu}}{2L}$$

where

f = frequency [Hz] T = string tension [N]
μ = mass per unit length [kg/m] L = string length [m]

Many electric guitars have a device often called a "whammy" bar or a "tremolo" bar that allows the guitarist to change the tension on the strings quickly and easily, thus changing the frequency of the strings. (Think of Jimi Hendrix simulating "the rockets red glare, the bombs bursting in air" in his rendition of *The Star Spangled Banner*—a true *tour de force*.) In designing a new whammy bar, we test our design by collecting data on a single string of the observed frequency at different string lengths (using the fretboard) with a specific setting of the whammy bar.

Length (L) [m]	0.25	0.28	0.32	0.36	0.40	0.45	0.51	0.57	0.64
Frequency (f) [Hz]	292	241	231	205	171	165	136	129	112

(a) Create a graph of the observed frequency data, including the power trendline and equation generated by Excel.

(b) If the tension was reduced to half of its original value, would the frequency increase or decrease and by what percentage of the original values?

(c) If the tension on the string is 125 newtons, what is the mass per unit length in grams per meter?

(d) If the mass per length of the string is 3 grams per meter, what is the tension in newtons?

16. Your supervisor has assigned you the task of designing a set of measuring spoons with a "futuristic" shape. After considerable effort, you have come up with two geometric shapes that you believe are really interesting.

You make prototypes of five spoons for each shape with different depths and measure the volume each will hold. The following table shows the data you collected.

Depth (d) [cm]	Volume (V_A) [mL] Shape A	Volume (V_B) [mL] Shape B
0.5	1	1.2
0.9	2.5	3.3
1.3	4	6.4
1.4	5	7.7
1.7	7	11

Use Excel to plot and determine appropriate power models for this data. Use the resulting models to determine the depths of a set of measuring spoons comprising the following volumes for each of the two designs:

Volume Needed (V) [tsp or tbsp]	Depth of Design A (d_A) [cm]	Depth of Design B (d_B) [cm]
1/4 tsp		
1/2 tsp		
3/4 tsp		
1 tsp		
1 tbsp		

17. One of the NAE Grand Challenges for Engineering is **Engineering the Tools of Scientific Discovery**. According to the NAE website: "Grand experiments and missions of exploration always need engineering expertise to design the tools, instruments, and systems that make it possible to acquire new knowledge about the physical and biological worlds."

Solar sails are a means of interplanetary propulsion using the radiation pressure of the sun to accelerate a spacecraft. The following table shows the radiation pressure at the orbits of several planets.

Planet	Distance from Sun (d) [AU]	Radiation Pressure (P) [μPa]
Mercury	0.46	43.3
Venus	0.72	17.7
Earth	1	9.15
Mars	1.5	3.96
Jupiter	5.2	0.34

(a) Plot this data and determine the power law model for radiation pressure as a function of distance from the sun.
(b) What are the units of the exponent in the trendline?
(c) What are the units of the other constant in the trendline?
(d) What is the radiation pressure at Uranus (19.2 AU from sun)?
(e) At what distance from the sun is the radiation pressure 5 μPa?

18. When volunteers build a Habitat for Humanity house, it is found that the more houses that are completed, the faster each one can be finished since the volunteers become better trained and more efficient. A model that relates the building time and the number of homes completed can generally be given by

$$t = t_0 \, e^{-N/\nu} + t_M$$

where

t = time required to construct one house [days]
t_0 = a constant related to (but not equal) the time required to build the first house
N = the number of houses completed [dimensionless]
ν = a constant related to the decrease in construction time as N increases
t_M = another constant related to construction time

A team of volunteers has built several houses, and their construction time was recorded for four of those houses. The construction time was then plotted as a function of number of previously built houses and a mathematical model derived as shown in the following graph. Using this information, answer the following questions:

(a) What are the units of the constants 8.2, 3, and 2.8?
(b) If the same group continues building houses, what is the minimum time to construct one house that they can expect to achieve?
(c) How long did it take for them to construct the first house?
(d) How many days (total) were required to build the first five houses?

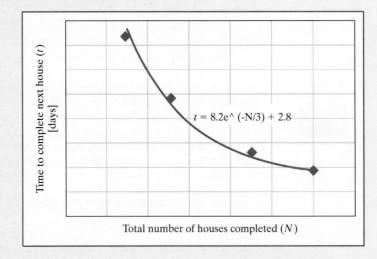

$t = 8.2e^{\wedge} (-N/3) + 2.8$

Time to complete next house (t) [days]

Total number of houses completed (N)

19. As part of an electronic music synthesizer, you need to build a gizmo to convert a linear voltage to an exponentially related current. You build three prototype circuits, make several measurements of voltage and current in each, and graph the results as shown in the following graph.

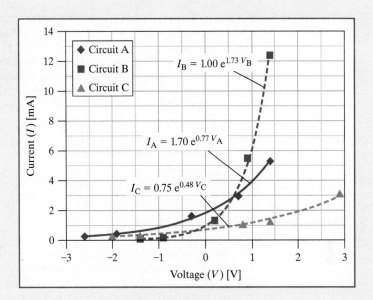

Assume that each circuit is modeled by the equation

$$I_X = A_X e^{(R_M/(R_X V_T))^{V_X}}$$

where

I_X is the current in circuit X [milliamperers, mA]
A_X is a scaling factor associated with circuit X
R_M is a master resistor, and has the same value in all circuits [ohms, Ω]
R_X is a resistor in circuit X whose value is different in each circuit [ohms, Ω]
V_T is the thermal voltage, and has a value of 25.7 volts
V_X is the voltage in circuit X [volts, V]

(a) What are the units of A_X?
(b) If you wish $I_X = 1$ mA when $V_X = 0$, what should the value of A_X be?
(c) Using the trendline models, if $R_M = 10$ kΩ, what is the value of R_A?

20. Essentially all manufactured items are made to some "tolerance," or how close the actual product is to the nominal specifications. For example, if a company manufactures hammers, one customer might specify that the hammers should weigh 16 ounces. With rounding, this means that the actual weight of each hammer meets the specification if it weighs between 15.5 and 16.5 ounces. Such a hammer might cost $10. However, if the U.S. military, in its quest for perfection, specifies that an essentially identical hammer should have a weight of 16.000 ounces, then in order to meet specifications, the hammer must weigh between 15.9995 and 16.0005 ounces. In other words, the weight must fall within a range of one-thousandth of an ounce. Such a hammer might cost $1000.

You have purchased a "grab bag" of 100 supposedly identical capacitors. You got a really good price, but there are no markings on the capacitors. All you know is that they are all the same nominal value. You wish to discover not only the nominal value, but the tolerance: are they within 5% of the nominal value, or within 20%? You set up a simple circuit with a known resistor and each of the unknown capacitors. You charge each capacitor to 10 volts, and then use an oscilloscope to time how long it takes for each capacitor to discharge to 2 volts. In a simple RC (resistor–capacitor) circuit, the voltage (VC) across a capacitor (C) discharging through a resistor (R) is given by:

$$V_C = V_0 e^{-t/(RC)}$$

where t is time in seconds and V_0 is the initial voltage across the capacitor.

After measuring the time for each capacitor to discharge to from 10 to 2 volts, you scan the list of times and find the fastest and slowest. Since the resistor is the same in all cases, the fastest time corresponds to the smallest capacitor in the lot, and the slowest time to the largest. The fastest time was 3.3 microseconds and the slowest was 3.7 microseconds. For the two capacitors, you have the two pairs of data points.

(a) Enter these points into a worksheet, then plot these points in Excel, the pair for C_1 and the pair for C_2, on the same graph, using time as the independent variable. Fit exponential trendlines to the data.

Time for C_1 (s)	Voltage of C_1	Time for C_2 (s)	Voltage of C_2
0	10	0	10
3.3×10^{-6}	2	3.7×10^{-6}	2

(b) Assuming you chose a precision resistor for these measurements that had a value of $R = 1000.0$ ohms, determine the capacitance of the largest and smallest capacitors.

(c) You selected the fastest and slowest discharge times from a set of 100 samples. Since you had a fairly large sample set, it is not a bad assumption, according to the laws of large numbers, that these two selected data sets represent capacitors near the lower and higher end of the range of values within the tolerance of the devices. Assuming the nominal value is the average of the minimum and maximum allowable values, what is the nominal value of the set of capacitors?

(d) What is the tolerance, in percent, of these devices? As an example, if a nominal 1 μF (microfarad) capacitor had an allowable range of 0.95 $\mu F < C < 1.05$ μF, the tolerance would be 5%.

If standard tolerances of capacitors are 5%, 10%, and 20%, to which of the standard tolerances do you think these capacitors were manufactured? If you pick a smaller tolerance than you calculated, justify your selection. If you picked a higher tolerance, explain why the tolerance is so much larger than the measured value.

Mathematical Models

As we have already seen, a large number of phenomena in the physical world obey one of the three basic mathematical models.

- Linear: $y = mx + b$
- Power: $y = bx^m + c$
- Exponential: $y = be^{mx} + c$

As we have mentioned previously, Excel can determine a mathematical model (trendline equation) for data conforming to all three of these model types, with the restriction that the constant c in the power and exponential models must be 0.

Here, we consider how to determine the best model type for a specific data set, as well as learn methods of dealing with data that fit a power or exponential model best but have a nonzero value of c.

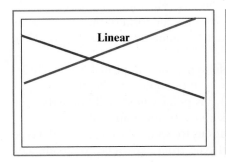

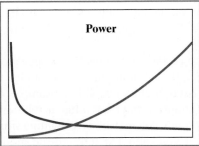

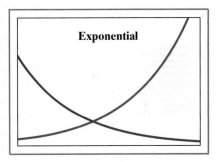

Except as otherwise noted, the entire discussion in this chapter assumes that the data fits one of the three trendlines models: linear, power, or exponential. You should always keep this in mind when using the techniques discussed here.

13.1 Selecting a Trendline Type

> **LEARN TO:** Evaluate the functional relationship between paired data sets using Excel trendlines
> Utilize boundary limits to determine whether a chosen model is appropriate
> Utilize linearization of data and/or of graph axes to determine whether a model is appropriate

When you determine a trendline to fit a set of data, in general you want the line, which may be straight or curved, to be as close as is reasonable to most of the data points.

The objective is not to ensure that the curve passes through every point.

To determine an appropriate model for a given situation, we use five guidelines, presented in general order of importance:

1. Do we already know the model type that the data will fit?
2. What do we know about the behavior of the process under consideration, including initial and final conditions?
3. What do the data look like when plotted on graphs with logarithmic scales?
4. How well does the model fit the data?
5. Can we consider other model types?

Guideline 1: Determine if the Model Type Is Known

If you are investigating a phenomenon that has already been studied by others, you may already know which model is correct or perhaps you can learn how the system behaves by looking in appropriate technical literature. In this case, all you need are the specific values for the model parameters since you already know the form of the equation. As we have seen, Excel is quite adept at churning out the numerical values for trendline equations.

If you are certain you know the proper model type, you can probably skip guidelines 2 and 3, although it might be a good idea to quantify how well the model fits the data as discussed in guideline 4. For example, at this point you should know that the extension of simple springs has a linear relationship to the force applied.

At other times, you may be investigating situations for which the correct model type is unknown. If you cannot determine the model type from experience or references, continue to guideline 2.

Guideline 2: Evaluate What Is Known About the System Behavior

The most important thing to consider when selecting a model type is whether the model makes sense in light of your understanding of behavior the physical system or process being investigated. Since there may still be innumerable things with which you are unfamiliar, this may seem like an unreasonable expectation. However, by applying what you *do* know to the problem at hand, you can often make an appropriate choice without difficulty.

When investigating an unknown phenomenon, we typically know the answer to at least one of three questions:

1. How does the process behave in the initial state?
2. How does the process behave in the final state?
3. What happens to the process between the initial and final states—if we sketch the process, what does it look like? Does the parameter of interest increase or decrease? Is the parameter asymptotic to some value horizontally or vertically?

EXAMPLE   13-1

Suppose we do not know Hooke's law and would like to study the behavior of a spring. We hang the spring from a hook, pull downward on the bottom of the spring with varying forces, and observe its behavior. We know initially the spring will stretch a little under its own weight even before we start pulling on it, although in most cases this is small or negligible. As an extreme case, however, consider what would happen if you hang one end of a Slinky® from the ceiling, letting the other end fall as it will.

As we pull on the spring, we realize the harder we pull, the more the spring stretches. In fact, we might assume that in a simple world, if we pull twice as hard, the spring will stretch twice as far, although that might not be as obvious. In words we might say,

The distance the spring stretches (x) is directly proportional to the pulling force (F), *or we might express the behavior as an equation:*

$$x = kF + b$$

where b is the amount of stretch when the spring is hanging under its own weight. This is what we mean by using an "expected" form. Always remember, however, that what you "expect" to happen may be in error.

In addition, suppose we had tested this spring by hanging five different weights on it and measuring the stretch each time. After plotting the data, we realize there is a general trend that as the weight (force) increases, the stretch increases, but the data points do not lie exactly on a straight line. We have two options:

- *If we think our assumption of linear behavior may be in error, we can try nonlinear models.*
- *Or we can use a linear model, although the fit may not be as good as one or more of the nonlinear models.*

In order to answer the three questions presented in Guideline 3, we should ask the following sequence of questions about the model or system:

Is the system linear?

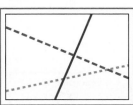

Linear systems have the following characteristics. If any of these is not true, then the system is not linear.

1. As the independent variable gets larger, the dependent variable continues to increase (positive slope) or decrease (negative slope) without limit. (See item 4.)
2. If the independent variable becomes negative, as it continues negative, the dependent variable continues to decrease (positive slope) or increase (negative slope) without limit unless one of the variables is constant. (See item 4.)
3. The rate of increase or decrease is constant; in other words, it will not curve upward or downward but is a straight line.
4. There are no horizontal or vertical asymptotes unless the dependent variable is defined for only *one* value of the independent variable *or* if the dependent variable is the same value for *all* values of the independent variable.

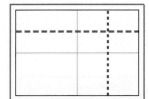

Examples illustrating if a system is or is not linear:

- You are driving your car at a constant speed of 45 miles per hour [mph]. The longer you drive, the farther you go, without limit. In addition, your distance increases by the same amount each hour, regardless of total time elapsed. This is a linear system.
- You observe the temperature of the brake disks on your car to be slowly decreasing. If it continued to decrease without limit, the temperature would eventually be less than absolute zero; thus, it is not linear. The temperature will eventually approach the surrounding air temperature; thus, there is a horizontal asymptote.

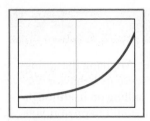

If the system is not linear, is there a vertical asymptote?

If there *is* a vertical asymptote, it will also have a horizontal asymptote. This is a power law model with a negative exponent. REMEMBER: We are assuming that our data fit one of the three models being considered here, and the previous statement is certainly not true for all other models. For example, $y = \tan x$ has multiple vertical asymptotes, but no horizontal asymptote.

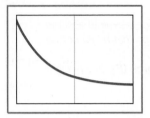

If there is not a vertical asymptote, is there a horizontal asymptote?

If there is a horizontal asymptote (but not a vertical one), then the model is exponential. If the horizontal asymptote occurs for positive values of the independent variable, then the exponent is negative. If the horizontal asymptote occurs for negative values of the independent variable, then the exponent is positive.

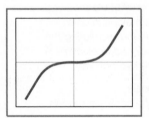

What if there is not a horizontal asymptote or a vertical asymptote?

It is a power law model with a positive exponent. Such models can have a variety of shapes.

This sequence of questions can be represented by the flow chart below. Remember, this is only valid if we assume the data fits one of the three models being discussed.

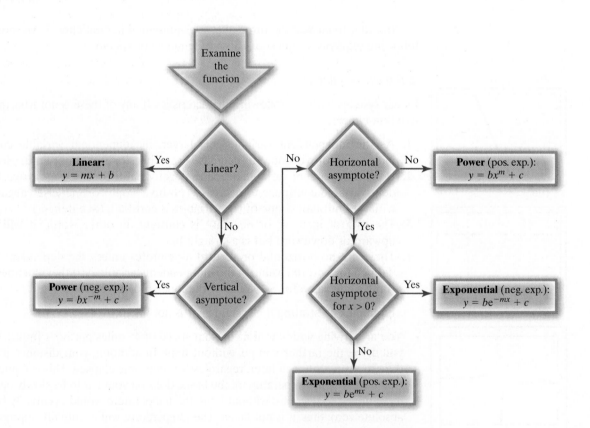

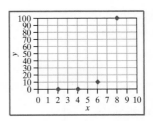

Guideline 3: Convert Axes to a Logarithmic Scale

If the logarithm of the dependent or independent variable is plotted instead of the variable itself, do the modified data points appear to lie on a straight line?

To see how logarithmic axes are constructed, let us consider a simple case. Plotting the data points below gives the graph shown to the left.

x	2	4	6	8
y	0.1	1	10	100

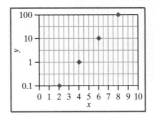

One way to linearize the data is to take the logarithm (base 10) of the independent variable and plot the results of log (y).

x	2	4	6	8
y	0.1	1	10	100
log y	−1	0	1	2

Another method of linearization is to take the logarithm (base 10) of the axis values and plot the original y values on this altered axis.

A note about the use of logarithmic scales:

- The original data would fit an exponential model ($y = 0.01e^{1.15x}$), and when plotted on a logarithmic vertical axis, the data points appear in a straight line.
- The logarithmic axis allows us to more easily distinguish between the values of the two lowest data points, even though the data range covers three orders of magnitude. On the original graph, 0.1 and 1 were almost in the same vertical position.
- Note that you *do not* have to calculate the logarithms of the data points. You simply plot the actual values on a logarithmic scale.

Logarithm graphs are discussed in more detail in the next section. We can use logarithmic axes to help us determine an appropriate model type using the following process:

1. Plot the data using normal (linear) scales for both axes. If the data appear to lie more or less in a straight line, a linear model is likely to be a good choice.
2. Plot the data on a logarithmic vertical scale and a normal (linear) horizontal scale. If the data then appear to lie more or less in a straight line, an exponential model is likely to be a good choice.
3. Plot the data with logarithmic scales for *both* axes. If the data then appear to lie more or less in a straight line, a power law model is likely to be a good choice.
4. Although not covered in this course, you could plot the data on a logarithmic horizontal scale and a normal (linear) vertical scale. If the data then appear to lie more or less in a straight line, a logarithmic model is likely to be a good choice.

REMEMBER, this is only valid if we assume the data fits one of the three models being discussed. This process is summarized in the following flow chart.

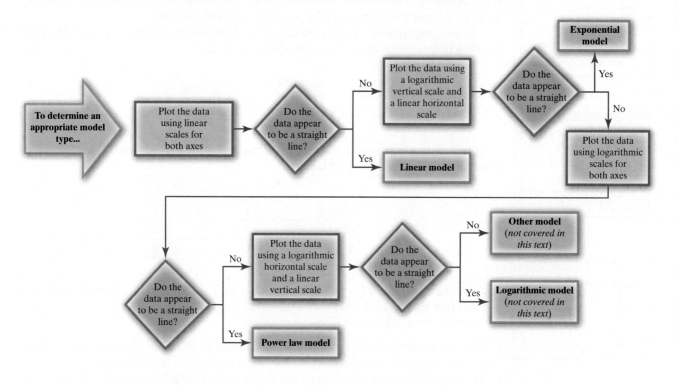

Guideline 4: Consider the R^2 Value

When a trendline is generated in Excel, the program can automatically calculate an **R^2 value**, sometimes called the **coefficient of determination**. The R^2 value is an indication of the variation of the actual data from the equation generated—in other words, it is a measure of how well the trendline fits the data. The value of R^2 varies between 0 and 1. If the value of R^2 is exactly equal to 1, a perfect correlation exists between the data and the trendline, meaning that the curve passes exactly through all data points. The farther R^2 is from 1, the less confidence we have in the accuracy of the model generated. When fitting a trendline to a data set, we always report the R^2 value to indicate how well the fit correlates with the data.

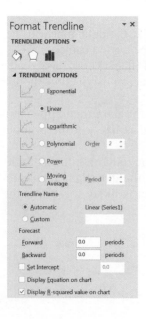

In reality, a fit of $R^2 = 1$ is rare, since experimental data are imprecise in nature. Human error, imprecision in instrumentation, fluctuations in testing conditions, and natural specimen variation are among the factors that contribute to a less-than-perfect fit. **The best R^2 value is not necessarily associated with the best model and should be used as a guide only**. Once again, making such decisions becomes easier with experience.

When displaying the equation corresponding to a trendline, you may have already noticed how to display the R^2 value.

✎ *To display an R^2 value:*

- Right-click or double-click on the trendline, or select the trendline then choose **Design > Add Chart Element > Trendline > More Trendline Options . . .**
- In the **Format Trendline** palette that opens, from the **Trendline** Options tab, check the box for **Display *R*-squared value on chart**. Click the "X" to close.

⌘ **Mac OS:** To show the R^2 value on a Mac, double-click the trendline. In the window that opens, click **Options** and select **Display *R*-squared value**. Click **OK**.

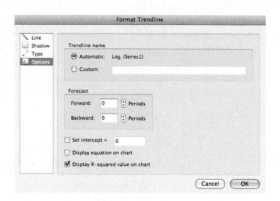

Try different models and compare the R^2 values.

- If one of the R^2 values is considerably smaller than the others, say, more than 0.2 less, then that model very likely can be eliminated.
- If one of the R^2 values is considerably larger than the others, say, more than 0.2 greater, then that model very likely is the correct one.

In any case, you should always consider guidelines 1 through 3 above to minimize the likelihood of error.

WARNING

While practicing with trendlines in the preceding chapters, you may have noticed a choice for polynomial models. Only rarely would this be the proper choice, but we mention it here for one specific reason—a polynomial model can always be found that will perfectly fit any data set. In general, if there are N data points, a polynomial of order $N - 1$ can be found that goes exactly through all N points. Excel can only calculate polynomials up to sixth order. For example, a data set with five data points is plotted in the following chart. A fourth-order polynomial can be found that perfectly fits the data. Let us consider a simple spring stretching example to illustrate why a perfect fit to the data is not necessarily the correct model.

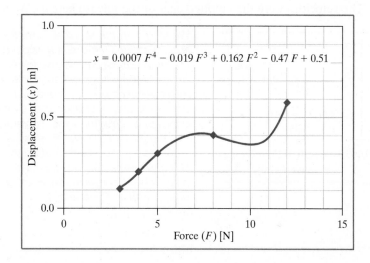

The graph shows the five data points for spring displacement as a function of force. As force increases, displacement increases, but the points are certainly not in a straight line. Also shown is a fourth-order polynomial model that goes through every point—a perfect fit. This, however, is a terrible model.

Presumably you agree that as force increases, displacement *must* increase as well. The polynomial trendline, however, suggests that as force increases from about 7 to 10 newtons, the displacement *decreases*.

Always ask yourself if the model you have chosen is obviously incorrect, as in this case. We do not use polynomial models in this book, and so we will discuss them no further.

THE THEORY OF OCCAM'S RAZOR

It is vain to do with more what can be done with less.

or

Entities are not to be multiplied beyond necessity.

—William of Occam

It is probably appropriate to mention Occam's razor at this point. Those who choose to pursue scientific and technical disciplines should keep the concept of Occam's razor firmly in mind. **Occam's razor refers to the concept that the simplest explanation or model to describe a given situation is usually the correct one.** It is named for William of Occam, who lived in the first half of the fourteenth century and was a theologian and philosopher.

EXAMPLE   13-2

The velocity of a ball was recorded as it rolled across a floor after being released from a ramp at various heights. The velocities were then plotted against the release heights. We want to fit a trendline to the data.

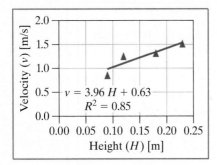

$v = 3.96\,H + 0.63$
$R^2 = 0.85$

We start with the simplest form, a linear fit, shown on the left. We know that if the ramp is at a height of zero, the ball will not roll down the ramp without any external forces. The linear fit yields an intercept value of 0.6, indicating that the ball will have an initial velocity of 0.6 meter per second when the ramp is horizontal, which we know to be untrue. It seems unlikely that experimental variation alone would generate an error this large, so we try another model.

We choose a power fit, shown in the center. With an R^2 value of 0.86, the equation fits the data selection well, but is there a better fit? Using the same data, we try a third-order polynomial to describe the data. The polynomial model, which gives a perfect fit, is shown on the bottom with an R^2 value of 1.

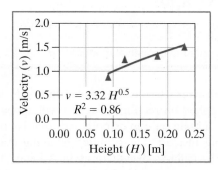

While the polynomial trendline gave the best fit, is this really the correct way to describe the data? Recall that in theory the potential energy of the ball is transformed into kinetic energy according to the conservation of energy law, written in general terms

$$PE_{initial} = KE_{final} \quad \text{or} \quad mgH = \frac{1}{2}mv^2$$

Therefore, the relationship between velocity and height is a relationship of the form

$$v = (2gH)^{1/2} = (2g)^{1/2}H^{1/2}$$

The relation between velocity and height is a power relationship; velocity varies as the square root of the height. The experimental error is responsible for the inaccurate trendline fit. In most instances, the polynomial trendline will give a precise fit but an inaccurate description of the phenomenon. **It is better to have an accurate interpretation of the experimental behavior than a perfect trendline fit!**

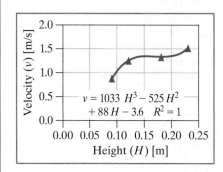

Guideline 5: Should We Consider Model Types Not Covered Here?

Many phenomena may be accurately characterized by a linear model, power law model, or exponential model. However, there are innumerable systems for which a different model type must be chosen. Many of these are relatively simple, but some are mind-bogglingly complicated. For example, modeling electromagnetic waves (used for television, cell phones, etc.) or a mass oscillating up and down while hanging from a spring requires the use of trigonometric functions.

You should always keep in mind that the system or phenomenon you are studying may not fit the three common models we have covered in this book.

NOTE ON ADVANCED MATH

Actually, sinusoids (sine or cosine) can be represented by exponential models through a mathematical trick first concocted by Leonhard Euler, so we now refer to it as Euler's identity. The problem is that the exponents are imaginary (some number times the square root of −1).

Euler's identity comes up in the study of calculus, and often in the study of electrical or computer engineering, early in the study of electric circuits. Euler's identity can be expressed in several different forms. The basic identity can be stated as the following equation, where i is the square root of −1:

$$e^{i\pi} = -1$$

Another form often used in electrical engineering is

$$\cos \theta = 0.5 \, (e^{i\theta} + e^{-i\theta})$$

13.2 Interpreting Logarithmic Graphs

LEARN TO: Plot data using logarithmic axis to linearize the data
Interpret a graph using logarithmic scales to develop a mathematical model

A "regular" plot, shown on a graph with both axes at constant-spaced intervals, is called **rectilinear**. When a linear function is graphed on rectilinear axis, it will appear as a straight line. Often, it is convenient to use a scale on one or both axes that is not linear, where values are not equally spaced but instead "logarithmic," meaning that powers of 10 are equally spaced. Each set of 10 is called a **decade** or **cycle**. A logarithmic scale that ranges either from 10 to 1000 would be two cycles, 10–100 and 100–1000. Excel allows you to select a logarithmic scale for the abscissa, the ordinate, or both.

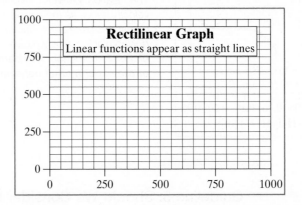

If one scale is logarithmic and the other linear, the plot is called **semilogarithmic** or **semilog**. Note in the following semilog graph that the abscissa has its values equally spaced and so is a linear scale. However, the ordinate has powers of 10 equally spaced and thus is a logarithmic scale.

If both scales are logarithmic, the plot is called **full logarithmic** or **log–log**. Note in the following log-log graph that both axes have powers of 10 equally spaced.

There are four different combinations of linear and logarithmic axes, each corresponding to one of four specific trendline types that will appear linear on that particular graph type. If the plotted data points are more or less in a straight line when plotted with a specific axis type, the corresponding trendline type is a likely candidate, as discussed earlier.

Once the data are plotted as logarithmic, how do you read data from this graph? This is perhaps best shown through examples.

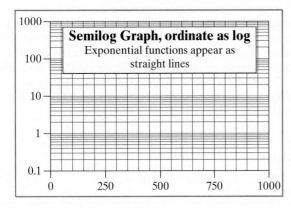

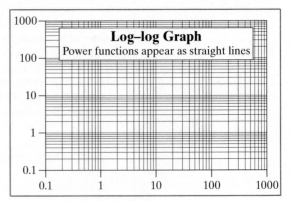

NOTE

For a refresher on logarithm rules, please see Appendix A online.

Derivation of Power Law Model

Consider a power law model:

$$y = bx^m$$

Now take the logarithm of both sides of the equation.

$$\log y = \log (bx^m) = \log b + \log x^m = \log b + m \log x$$

Using the commutative property of addition, you can write:

$$\log y = m \log x + \log b$$

Since b is a constant, $\log b$ is also a constant. Rename $\log b$ and call it b'. Since x and y are both variables, $\log x$ and $\log y$ are also variables. Call them x' and y', respectively.

Using the new names for the transformed variables and the constant b:

$$y' = mx' + b'$$

This is a linear model! Thus, if the data set can be described by a power law model and you plot the logarithms of both variables (instead of the variables themselves), the transformed data points will lie on a straight line. The slope of this line is m, although "slope" has a somewhat different meaning than in a linear model. The "intercept" value, b, occurs when $x = 1$, since $\log(1) = 0$.

EXAMPLE 13-3

When a body falls, it undergoes a constant acceleration. Using the figure, determine the mathematical equation for distance (d), in units of meters, of a falling object as a function of time (t), in units of seconds.

Since the graph appears linear on log–log paper, we can assume a power law relationship exists of the form:

$$d = bt^m$$

For illustration, a line has been sketched between the points for further clarification of function values.
   To establish the power of the function (m), we estimate the number of decades of "rise" (shown as vertical arrows) divided by the decades of "run" (horizontal arrow):

$$\text{Slope} = \frac{\text{Change in decades of distance}}{\text{Change in decades of time}} = \frac{2 \text{ decade}}{1 \text{ decade}} = 2$$

To establish the constant value (b), we estimate it as the ordinate value when the abscissa value is 1, shown in the shaded circle. When the time is 1 second, the distance is 5 meters.
   The resulting function:

$$d = 5t^2$$

This matches well with the established theory, which states

$$d = \frac{1}{2}gt^2$$

The value of ½ g is approximately 5 m/s².

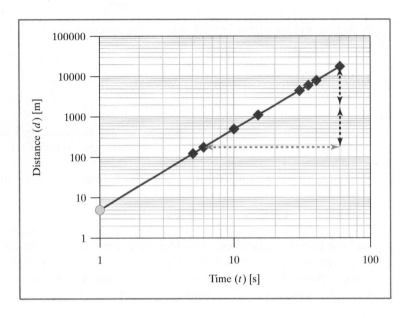

NOTE

For a refresher on logarithm rules, please see Appendix A online.

Derivation of Exponential Law Model

Consider an exponential model:

$$y = be^{mx}$$

Now take the logarithm of both sides of the equation.

$$\log y = \log (be^{mx}) = \log b + \log e^{mx} = \log b + (mx) \log e$$

Using the commutative property of addition, you can write:

$$\log y = m (\log e)x + \log b$$

Since b is a constant, $\log b$ is also a constant. Rename $\log b$ and call it b'. Since y is a variable, $\log y$ is also a variable; call it y'.

Using the new names for the transformed variable y and the constant b:

$$y' = m (\log e)x + b'$$

This is a linear model! Thus, if the data set can be described by an exponential law model, and you plot the logarithm of y (instead of y itself) versus x, the transformed data points will lie on a straight line. The slope of this line is $m(\log e)$, but again, "slope" has a somewhat different interpretation. The term $(\log e)$ is a number, approximately equal to 0.4343; the slope is 0.4343 m.

EXAMPLE 13-4

A chemical reaction is being carried out in a reactor; the results are shown graphically in the figure. Determine the mathematical equation that describes the reactor concentration (C), in units of moles per liter, as a function of time spent in the reactor (t), in units of seconds.

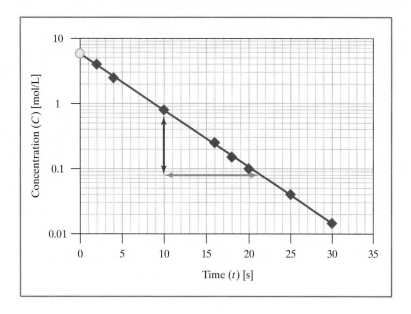

Since the graph appears linear on semilog paper where the ordinate is logarithmic, we can assume an exponential law relationship exists of the form:

$$C = be^{mt}$$

For illustration, a line has been sketched between the points for further clarification of function values.
Since this is an exponential function, to determine the value of m, we must first determine the slope:

$$\text{Slope} = \frac{\text{Change in decades of concentration}}{\text{Change in time}}$$

$$= \frac{-1 \text{ decade}}{21.5 \text{ s} - 10 \text{ s}} = -0.087 \text{ s}^{-1}$$

The value of m is then found from the relationship: slope = m(log e).

$$m = \frac{\text{slope}}{\log e} = \frac{-0.087 \text{ s}^{-1}}{0.4343} = -0.2 \text{ s}^{-1}$$

When time = 0 seconds, the constant (b) can be read directly and has a value of 6 [mol/L].
The resulting function:

$$C = 6e^{-0.2t}$$

Comprehension Check   **13-1**

An unknown amount of oxygen, kept in a piston-type container at a constant temperature, was subjected to increasing pressure (P), in units of atmospheres; as the pressure (P) was increased, the resulting volume (V) was recorded in units of liters. We have found that a log–log plot aligns the data in a straight line. Using the figure, determine the mathematical equation for volume (V) in units of liters, and of a piston filled with an ideal gas subjected to increasing pressure (P) in units of atmospheres.

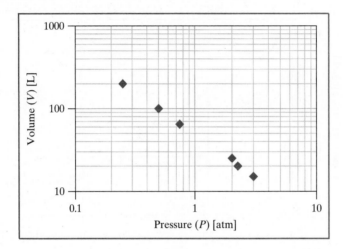

Comprehension Check   **13-2**

The data shown graphically in the figure describe the discharge of a capacitor through a resistor. Determine the mathematical equation that describes the voltage (V), in units of volts, as a function of time (t), in units of seconds.

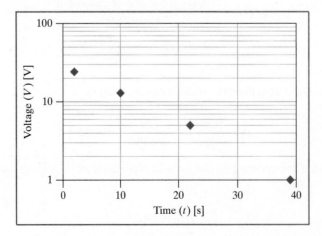

13.3 Proper Plot Rules for Log Plots

LEARN TO: Create a proper plot of experimental data containing logarithmic axes

When creating a proper plot containing experimental data fitted to logarithmic axes, it is important to follow several essential proper plotting rules. Note that several of the rules outlined below are repeated rules that apply to all graphs, or all graphs containing experimental data.

- Axis labels should include the name of the variable, the symbol used to represent the axis, and the units represented on the axis.
- Both vertical and horizontal major and minor gridlines should be added on a logarithmic graph to show the log scale on each axis. Note that minor gridlines are required only on logarithmic axes.
- The axis scale is represented with an appropriate number of decimal places (0.1 vs. 0.100) and moved to start at 1 or 0.1 (or other value that isn't the default value of 0).
- For experimental data, data points should be represented as symbols only—not symbols connected by lines.
- Trendlines and symbols for each data set represented on a graph should differ in line type, symbol, and color—no data sets should repeat the same line type or symbols.
- If more than one set of data is represented on a graph, a legend should be included to assist in identification of the different experimental data sets.
- Trendline equation placement should be as close to the trendline as possible, as well as possibly formatting in the same color in order to reduce confusion in trendline association.
- Trendline equations should contain values that are formatted to be reasonable (not using the default values) and symbols in the trendline equation should match the symbols used on the horizontal and vertical axes (not using the default x, y symbols).

Figure 13-1 is an example of a properly formatted graph, showing an experimental data series with trendlines and logarithmic axes.

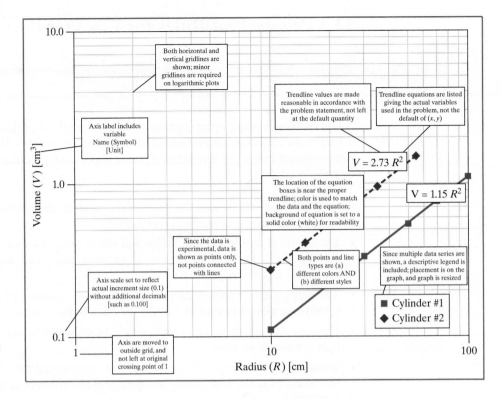

Figure 13-1

Example of a proper plot, showing multiple experimental data sets with trendlines and logarithmic axes.

13.4 Converting Scales to Log in Excel

LEARN TO: Use Excel to convert a graph into a logarithmic axis to make a data series appear linear

✏️ *To convert axis to logarithmic:*

- Right-click the axis > Format Axis or double-click on the axis. The **Format Axis** palette will appear.
- Click **Axis Options**, then check the box for **Logarithmic scale**. The Base should automatically appear as 10; this default value is correct.

- Click **Axis Options**, then check the box for **Logarithmic scale**.

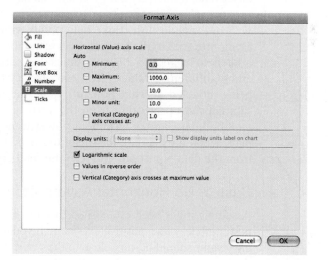

Alternatively:

- Click the chart. In the toolbar, select **Design > Add Chart Element > Axis > More Axis Options**.
- In the corresponding palette, select **Axis Options > Bar Graph Symbol**.
- ⌘ **Mac OS:** Double-click on the axis you want to convert to logarithmic. The Format Axis window will appear. Click **Scale** in the list on the left side of the window, and then click the checkbox near the bottom that says "Logarithmic scale."

Section 13.3 contains details on how to create a properly formatted graph, showing an experimental data series with power trendlines. The axes have been made logarithmic to allow the data series to appear linear.

13.5 Dealing with Limitations of Excel

LEARN TO: Understand the limitations in using Excel to model power or exponential data containing an offset

Determine appropriate steps to alter data using Excel if an offset is present

As we have mentioned earlier, Excel will not correctly calculate a trendline for a power or exponential model containing a vertical offset. In other words, it can calculate appropriate values for b and m in the forms

$$y = bx^m + c \text{ or } y = be^{mx} + c$$

only if $c = 0$. Note that if the data inherently have a vertical offset, Excel may actually calculate a trendline equation, but the values of b and m will not be accurate.

In addition, if any data value, dependent or independent, is less than or equal to zero, Excel cannot calculate a power law model. If one or more dependent variable data points are less than or equal to zero, exponential models are unavailable.

In the real world, there are many systems that are best modeled by either a power or an exponential model with a nonzero value of c or with negative values, so we need a method for handling such situations.

Case 1: Vertical Asymptote

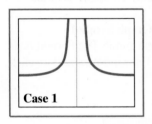

Case 1

Since a vertical asymptote implies a power model with a negative exponent, there will be a horizontal asymptote as well. If the horizontal asymptote is not the horizontal axis (implying a vertical offset), Excel will calculate the model incorrectly or not at all. The object here is to artificially move the asymptote to the horizontal axis by subtracting the offset value from every data point. If you have a sufficient range of data, you may be able to extract the offset from the data.

For example, if the three data points with the largest values of x (or smallest if the asymptote goes to the left) have corresponding y values of 5.1, 5.03, and 5.01, the offset is likely to be about 5. (This assumes there are other values in the data set with considerably different y values.) You can also try to determine from the physical situation being modeled at what nonzero value the asymptote occurs. In either case, simply subtract the offset value from the vertical component of *every* data point, plot this modified data, and determine a power trendline. Once the trendline equation is displayed, edit it by adding the offset to the power term. Note that you subtract the value from the data points but add it to the final equation.

Case 2: No Horizontal Asymptote

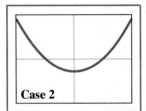

Case 2

Assuming it has been established that the model should not be linear, Case 2 implies a power model with a positive exponent. If you have a data point for $x = 0$, the corresponding y value should be very close to the vertical offset. Also, you may be able to determine the offset value by considering the physical situation. In either case, proceed as in Case 1, subtracting the offset from every data point.

Case 3: Horizontal Asymptote, No Vertical Asymptote

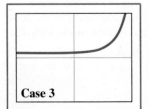

Case 3

Case 3 implies an exponential model. The object, as in Case 1, is to artificially move the asymptote to the horizontal axis. Also as in Case 1, you may be able to determine the offset by considering the physical system or by looking at the data points with the largest or smallest values. Again, subtract the determined value from every data point.

Case 4: A Few Values with Small Negative Value, Most Positive

In Case 4, the negative values may be a result of measurement inaccuracy. Either delete these points from the data set or change the negative values to a very small positive value.

Case 5: Many or All Data Points Negative

If the independent values are negative, try multiplying every independent value by -1. If this works, then make the calculated value of b negative after the trendline equation is calculated. You may have to apply some of the procedures in the previous cases after negating each data value.

Negative dependent values may simply be a negative offset to the data. If you can determine the asymptote value, ask if essentially all values are greater than the asymptote value. If so, it is probably just an offset. If not, then multiply every dependent data value by -1, and proceed in a manner similar to that described in the preceding paragraph.

EXAMPLE   13-5

The following data were collected in an experiment. We wish to determine an appropriate model for the data.

As the independent variable gets larger, the dependent variable appears to be approaching 10. This is even more apparent when graphed. Subtracting this assumed offset from every data point gives a new column of modified dependent data.

Since we subtracted 10 from every data point, we need to correct the equation by adding 10, giving $y = 14.3e^{-0.5x} + 10$.

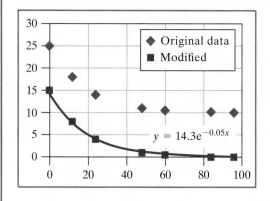

Independent	Dependent	Modified Dependent
0	25.0	15.0
12	18.0	8.0
24	14.0	4.0
48	11.0	1.0
60	10.5	0.5
84	10.2	0.2
96	10.1	0.1

If All Else Fails

If you are convinced that the model is exponential or power with an offset but you cannot determine its value, consider making further measurements for larger or smaller values of the independent variable. Particularly in data that has a horizontal asymptote, further measurements may make the value of the asymptote more obvious. In the first chart below, the value of the asymptote is not clear. By extending the measurements in the direction of the asymptote (positive in this case), it is clear that the asymptote has a value of 2. Note also that it becomes much clearer that the data are not linear.

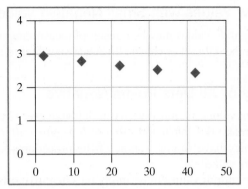 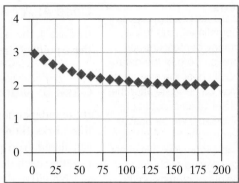

EXAMPLE   13-6

The data shown describe the discharge of a capacitor through a resistor. Before the advent of microprocessors, intermittent windshield wipers in automobiles often used such circuits to create the desired time delay. We wish to determine an appropriate model for the data.

Time (t) [s]	2	10	22	39
Voltage (V) [V]	24	13	5	1

- *Select the data series and create a linear trendline, being sure to display the equation and the R^2 value.*
- *Without deleting the first trendline, click one of the data points again—be sure you select the points and not the trendline—and again add a trendline, but this time choose a power trend.*
- *Repeat this process for an exponential trend.*

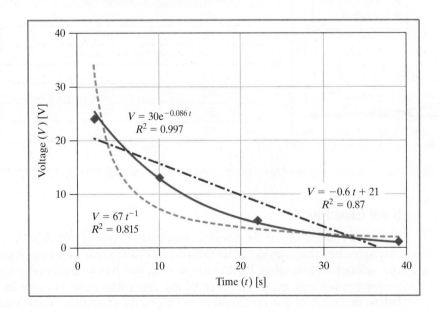

You should now have a chart with three trendlines. Things to note:

- *Neither the linear nor power trendlines are very good compared to the exponential line, and both have an R^2 value less than 0.9. These are probably not the best choice.*
- *The exponential model fits the data very closely and has an R^2 value greater than 0.95; thus, it is probably the best choice.*

As a model check, compare the graph by using logarithmic scales. If the model is exponential, the data should appear linear on a semilogarithmic plot with the ordinate shown as logarithmic.

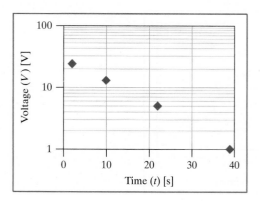
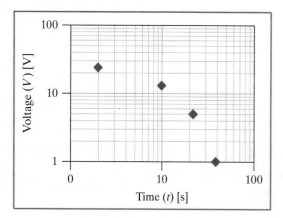

   Based on this analysis, you would choose the exponential model. As it turns out, the exponential model is indeed the correct one, being the solution to a differential equation describing the capacitor's behavior. Most students learn about this in second semester physics, and some study it in much more depth in electrical and computer engineering courses.

EXAMPLE   13-7

These data describe the temperature of antifreeze (ethylene glycol) in the radiator of a parked car. The temperature of the surrounding environment is -20 degrees Fahrenheit. The initial temperature (at $t = 0$) is unknown.

Time (t) [min]	10	18	25	33	41
Temperature (T) [°F]	4.5	1.0	-2.1	-4.6	-6.4

- Determine an appropriate model type for these data.
- Determine the vertical offset of the data.
- Plot the modified data and generate the correct trendline equation to describe these data.

   It seems reasonable that the temperature will be asymptotic to the surrounding temperature (-20 degrees Fahrenheit) as time goes on. Also, there is no known mechanism whereby the temperature could possibly go to infinity for any finite value of time, so there is no vertical asymptote. This indicates an exponential model with a negative exponent. Since the asymptote is at -20 degrees Fahrenheit, subtract -20 (i.e., add 20) from every data point before plotting.

Time (t) [min]	10	18	25	33	41
Temperature (T) [°F]	4.5	1.0	2.1	4.6	−6.4
Offset temperature (T_O) [°F]	24.5	21.0	17.9	15.4	13.6

Since you subtracted −20 from every data point, you should add −20 to the trendline equation, giving

$$T = 29.5e^{-0.019t} - 20$$

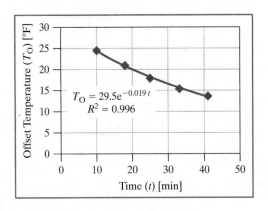

Comprehension Check   13-3

The following data were collected during an experiment. We wish to determine an appropriate model for the data.

Independent	Dependent
0	−25
10	−45
20	−85
25	−106
30	−154

Comprehension Check   13-4

Assume the car in Example 13-7 is cranked up and driven 50 feet into a garage. The temperature inside the garage is 5 degrees Fahrenheit. These data describe the temperature of antifreeze in the radiator after it is driven into the garage and the motor turned off.

- Determine an appropriate model type for these data.
- Determine the vertical offset of the data.
- Plot the modified data and generate the correct trendline equation to describe these data.

Time (t) [min]	5	13	25	34	51
Temperature (T) [°F]	−13.0	−10.0	−6.8	−4.5	−1.5

CHAPTER 13

IN-CLASS ACTIVITIES

ICA 13-1

Capillary action draws liquid up a narrow tube against the force of gravity as a result of surface tension. The height the liquid will move up the tube depends on the radius of the tube. The following data were collected for water in a glass tube in air at sea level. Show the resulting data and trendline, with equation and R^2 value, on the appropriate graph type (rectilinear, semilog, or log–log) to make the data appear linear.

Radius (r) [cm]	0.02	0.10	0.20	0.40	0.80	1.00
Height (H) [cm]	16.0	6.0	3.5	1.8	0.8	0.3

ICA 13-2

Several reactions are carried out in a closed vessel. The following data are taken for the concentration (C) of compounds A, B, and C [grams per liter] as a function of time (t) [minutes], from the start of the reaction. Show the resulting data and trendlines, with equation and R^2 value, on the appropriate graph type (rectilinear, semilog, or log–log) to make the data appear linear.

Time (t) [min]	3	6	9	16	21
Concentration of A (C_A) [g/L]	0.031	0.145	0.350	1.140	2.070
Concentration of B (C_B) [g/L]	0.042	0.222	0.570	1.826	3.425
Concentration of C (C_C) [g/L]	0.022	0.090	0.220	0.670	1.325

ICA 13-3

An environmental engineer has obtained a bacteria culture from a municipal water sample and allowed the bacteria to grow. The data are shown below. Show the resulting data and trendline, with equation and R^2 value, on the appropriate graph type (rectilinear, semilog, or log–log) to make the data appear linear.

Time (t) [h]	3	4	6	7	8	10	11
Concentration (C) [ppm]	27	50	117	159	209	324	391

ICA 13-4

In a turbine, a device used for mixing, the power requirement depends on the size and shape of the impeller. In the lab, you have collected the following data. Show the resulting data and trendline, with equation and R^2 value, on the appropriate graph type (xy scatter, semilog, or log–log) to make the data appear linear.

Diameter (D) [ft]	0.6	0.85	1.1	1.6	2.1	2.35	2.6	2.85
Power (P) [hp]	0.014	0.05	0.14	0.68	3.1	9.2	20.5	27

ICA 13-5

Being quite interested in obsolete electronics, Angus has purchased several electronic music synthesis modules dating from the early 1970s and is testing them to find out how they work. One module is a voltage-controlled amplifier (VCA) that changes the amplitude (loudness) of an audio signal by changing a control voltage into the VCA. All Angus knows is that the magnitude of the control voltage should be less than 5 volts. He sets the audio input signal to an amplitude of 1 volt, then measures the audio output amplitude for different control voltage values. The following table shows these data. Show the resulting data and trendline, with equation and R^2 value, on the appropriate graph type (*xy* scatter, semilog, or log–log) to make the data appear linear.

Control voltage (*V*) [V]	−3.5	−2.0	−0.5	0.5	1.5	3	4.5
Output amplitude (*A*) [V]	0.216	0.424	0.667	1.062	1.790	3.420	7.370

ICA 13-6

Referring to the previous ICA 13-5, Angus is also testing a voltage-controlled oscillator. In this case, a control voltage (also between −5 and +5 volts) changes the frequency of oscillation in order to generate different notes. The table below shows these measurements. Show the resulting data and trendline, with equation and R^2 value, on the appropriate graph type (*xy* scatter, semilog, or log–log) to make the data appear linear.

Control voltage (*V*) [V]	−5.0	−3.5	−2.0	0.0	2.0	3.5	5.0
Output frequency (*f*) [Hz]	38	102	237	549	999	3320	8450

The following instructions apply to ICA 13-7 to ICA 13-9. Examine the following models. Determine if the graph will appear linear on:

(A) Rectilinear axes
(B) Semilog, abscissa as logarithmic, axes
(C) Semilog, ordinate as logarithmic, axes
(D) Logarithmic (both) axes
(E) None of the above

ICA 13-7

Q	Model	Abscissa	Ordinate	Will appear linear on . . .				
				A	B	C	D	E
(a)		*F*	*L*					
(b)	$L = BF^{0.5}$	*L*	*B*					
(c)		*F*	*B*					

ICA 13-8

Q	Model	Abscissa	Ordinate	Will appear linear on . . .				
				A	B	C	D	E
(a)		*V*	*R*					
(b)		*L*	*R*					
(c)	$R = H^{0.5}V^{-2}L$	*V*	*L*					
(d)		*V*	*H*					

ICA 13-9

Q	Model	Abscissa	Ordinate	Will appear linear on . . .				
				A	B	C	D	E
(a)		S	M					
(b)	$M = \dfrac{W}{T} S^2 e^{-\frac{R}{L}}$	W	T					
(c)		1/L	M					
(d)		R	T					

ICA 13-10

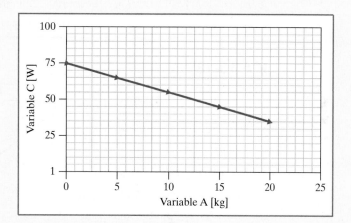

The following instructions will apply to ICA 13-10 to 13-21, for the preceding graph, identify:

- whether this model is linear, power, exponential, or none of those options,
- what is the equation of the line shown in the graph, and
- the units of each value represented in the equation of the line.

ICA 13-11

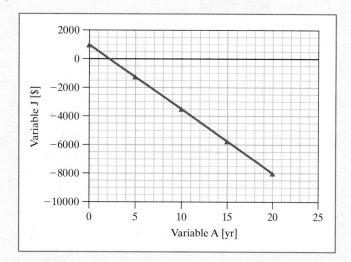

ICA 13-12

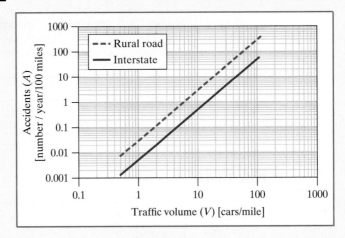

ICA 13-13

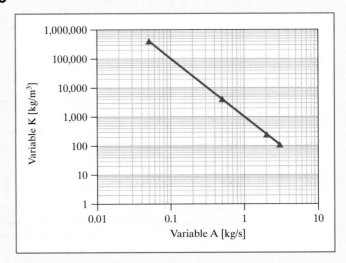

ICA 13-14

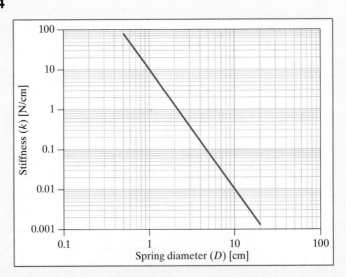

ICA 13-15

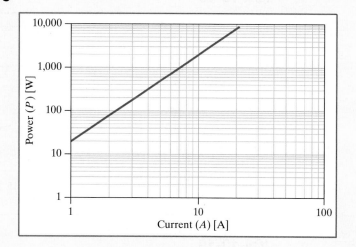

ICA 13-16

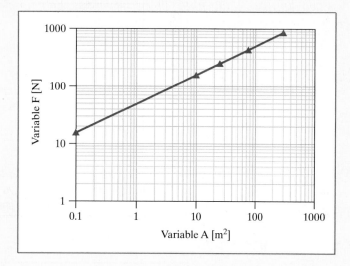

ICA 13-17

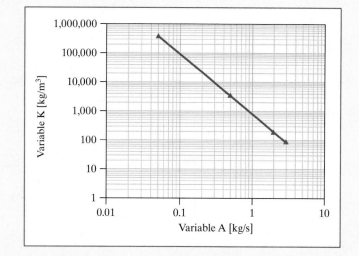

ICA 13-18

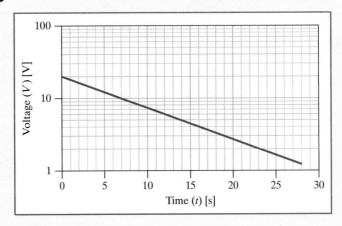

ICA 13-19

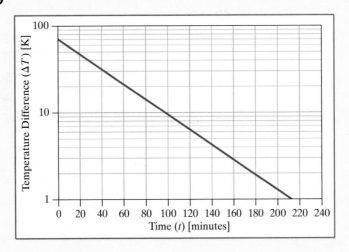

ICA 13-20

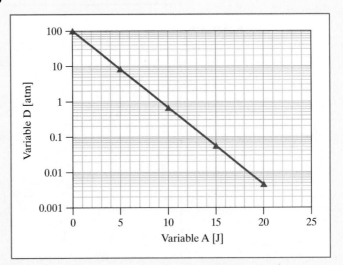

ICA 13-21

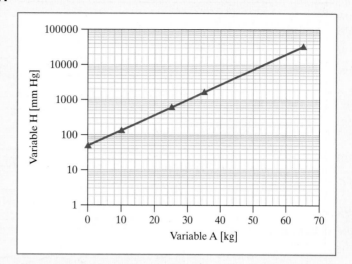

ICA 13-22

As a reminder, the Reynolds number is discussed in Chapter 9, Dimensionless Numbers.

When discussing the flow of a fluid through a piping system, we say that friction occurs between the fluid and the pipe wall due to viscous drag. The loss of energy due to the friction of fluid against the pipe wall is described by the friction factor. The **Darcy friction factor** (f) was developed by Henry Darcy (1803–1858), a French scientist who made several important contributions to the field of hydraulics. The friction factor depends on several other factors, including flow regime, Reynolds number, and pipe roughness. The friction factor can be determined in several ways, including from the Moody diagram (shown below).

Olive oil having a specific gravity of 0.914 and a viscosity of 100.8 centipoise is draining by gravity from the bottom of a tank. The drain line from the tank is a 4-inch-diameter pipe made of commercial steel (pipe roughness, $\varepsilon = 0.045$ millimeters). The velocity is 11 meters per second. Determine the friction factor for this system, using the following process:

Step 1: Determine the Reynolds number: $Re = \dfrac{\rho v D}{\mu}$.

Step 2: Determine flow regime.

- If the flow is laminar ($Re \leq 2000$), proceed to step 4.
- If the flow is turbulent or transitional ($Re > 2000$), continue with step 3.

Step 3: Determine the relative roughness ratio: (ε / D).
Step 4: Determine the Darcy friction factor (f) from the diagram.

ICA 13-23

Repeat ICA 13-22 with the following conditions:

Lactic acid, with a specific gravity of 1.249 and dynamic viscosity of 40.33 centipoise, is flowing in a 1½-inch-diameter galvanized iron pipe at a velocity of 1.5 meters per second. Assume the pipe roughness (ε) of galvanized iron is 0.006 inches. Determine the friction factor for this system.

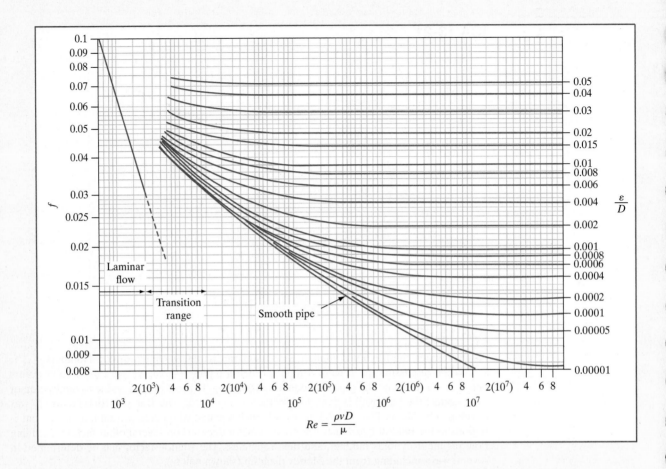

ICA 13-24

This activity requires data from ICA 8-41; the procedure is shown here for reference.

Materials

Bag of cylinders   Scale   Calipers   Ruler

Procedure

For each cylinder, record the mass, length, and diameter and/or width in an Excel workbook.

Analysis

- Using formulas in Excel, determine the volume and density for each cylinder.
- Use data from constant mass set, graph density (ordinate) versus length.
- Use data from constant volume set, graph density (ordinate) versus mass.
- Both graphs should be proper plots, with appropriate trendlines and logarithmic axes to prove your trendline choices by making the data appear linear.

REVIEW QUESTIONS

1. An environmental engineer has obtained a bacteria culture from a municipal water sample and has allowed the bacteria to grow.

Time (t) [min]	10	20	40	60	70	90	100
Concentration (C) [ppm]	18	30	64	127	200	430	649

(a) Show the resulting data and trendline, with equation and R^2 value, on the appropriate graph type (xy scatter, semilog, or log–log) to make the data appear linear.

(b) Assume the value of m in the resulting model is the growth constant. Use the trendline determined to find the value and units of the growth constant for these bacteria.

2. An environmental engineer has obtained a bacteria culture from a municipal water sample and allowed the bacteria to grow.

Time (t) [min]	10	20	40	60	70	90	100
Concentration (C) [ppm]	21.9	27.1	37.0	47.3	52.0	62.3	66.9

(a) Show the resulting data and trendline, with equation and R^2 value, on the appropriate graph type (xy scatter, semilog, or log–log) to make the data appear linear.

(b) Assume the value of m in the resulting model is the growth constant. Use the trendline determined to find the value and units of the growth constant for these bacteria.

3. An environmental engineer has obtained a bacteria culture from a municipal water sample and allowed the bacteria to grow.

Time (t) [min]	10	20	40	60	70	90	100
Concentration (C) [ppm]	1.0	8.4	65	215	343	701	970

(a) Show the resulting data and trendline, with equation and R^2 value, on the appropriate graph type (xy scatter, semilog, or log–log) to make the data appear linear.

(b) Assume the value of m in the resulting model is the growth constant. Use the trendline determined to find the value and units of the growth constant for these bacteria.

4. A growing field of inquiry that poses both great promise and great risk for humans is nano-technology, the construction of extremely small machines. Over the past couple of decades, the size that a working gear can be made has consistently gotten smaller. The table shows milestones along this path.

Years from 1967	0	4	8	17	24	32	39
Minimum gear size [mm]	0.9	0.3	0.1	0.07	0.003	1E-04	7E-06

(a) Show the resulting data and trendline, with equation and R^2 value, on the appropriate graph type (xy scatter, semilog, or log–log) to make the data appear linear.

(b) According to this model, how many years does it take (from any point in time) for the minimum size to be cut in half?

(c) According to the model, during what year will the smallest gear be one-tenth the size of the smallest gear in 2009?

5. If an object is heated, the temperature of the object will increase. The thermal energy (Q) associated with a change in temperature (ΔT) is a function of the mass of the object (m) and the specific heat (C_p). Specific heat is a material property, and values are available in literature. In an experiment, heat is applied to the end of an object, and the temperature change at the other end of the object is recorded. An unknown material is tested in the lab, yielding the following results.

Heat applied (Q) [J]	2	8	10	13	18	27
Temp change (ΔT) [K]	1.5	6.0	7.0	9.0	14.0	22.0

(a) Show the resulting data and trendline, with equation and R^2 value, on the appropriate graph type (xy scatter, semilog, or log–log) to make the data appear linear.
(b) If the material was titanium, what mass of sample was tested?
(c) If a 4-gram sample was used, which of the following materials was tested?

Material	Specific Heat Capacity (C_p) [J/(g K)]
Aluminum	0.91
Copper	0.39
Iron	0.44
Lead	0.13
Molybdenum	0.30
Titanium	0.54

6. The Volcanic Explosivity Index (*VEI*) is based primarily on the amount of material ejected from a volcano, although other factors play a role as well, such as height of plume in the atmosphere. The following table shows the number of volcanic eruptions (N) over the past 10,000 years having a VEI of between 2 and 7.

There are also VEI values of 0, 1, and 8. There is a level 0 volcano erupting somewhere on the Earth essentially all the time. There are one or more level 1 volcanoes essentially every day. The last known level 8 volcano was about 26,000 years ago.

Volcanic Explosivity Index (VEI) [−]	Number of Eruptions (N) [−]
2	3477
3	868
4	421
5	168
6	51
7	5

(a) Show the resulting data and trendline, with equation and R^2 value, on the appropriate graph type (xy scatter, semilog, or log–log) to make the data appear linear.
(b) How many level 1 volcanoes does the model predict should have occurred in the last 10,000 years?
(c) How many level 8 volcanoes does the model predict should have occurred in the last 10,000 years?

7. Biosystems engineers often need to understand how plant diseases spread in order to formulate effective control strategies. The rate of spread of some diseases is more or less linear, some increase exponentially, and some do not really fit any standard mathematical model.

Grey leaf spot of corn is a disease (caused by a fungus with the rather imposing name of *Cercospora zeae-maydis*) that causes chlorotic (lacking chlorophyll) lesions and eventually necrotic (dead) lesions on corn leaves, thus reducing total photosynthesis and yield. In extremely severe cases, loss of the entire crop can result.

During a study of this disease, the number of lesions per corn leaf was counted every 10 days following the initial observation of the disease, which we call day 0. At this time, there was an average of one lesion on every 20 leaves, or 0.05 lesions per leaf. The data collected during the growing season are tabulated.

(a) Show the resulting data and trendline, with equation and R^2 value, on the appropriate graph type (*xy* scatter, semilog, or log–log) to make the data appear linear.

(b) According to the model, how many lesions were there per leaf at the start of the survey?

(c) How many lesions are there per leaf after 97 days?

(d) If the model continued to be accurate, how many days would be required to reach 250 lesions per leaf?

Day	Lesions per Leaf	Day	Lesions per Leaf
0	0.05	110	4
20	0.10	120	6
30	0.20	140	17
40	0.26	150	20
60	0.60	170	40
80	1.30	190	112
90	2	200	151

8. A **pitot tube** is a device used to measure the velocity of a fluid, typically, the airspeed of an aircraft. The failure of a pitot tube is credited as the cause of Austral Líneas Aéreas flight 2553 crash in October 1997. The pitot tube had frozen, causing the instrument to give a false reading of slowing speed. As a result, the pilots thought the plane was slowing down, so they increased the speed and tried to maintain their altitude by lowering the wing slats. Actually, they were flying at such a high speed that one of the slats ripped off, causing the plane to nosedive; the plane crashed at a speed of 745 miles per hour.

In the pitot tube, as the fluid moves, the velocity creates a pressure difference between the ends of a small tube. The tubes are calibrated to relate the pressure measured to a specific velocity. This velocity is a function of the pressure difference (P, in units of pascals) and the density of the fluid (ρ in units of kilograms per cubic meter).

$$v = \left(\frac{2}{\rho}\right)^{0.5} P^m$$

Pressure (*P*) [Pa]	50,000	101,325	202,650	250,000	304,000	350,000	405,000	505,000
Velocity fluid A (*v*$_A$) [m/s]	11.25	16.00	23.00	25.00	28.00	30.00	32.00	35.75
Velocity fluid B (*v*$_B$) [m/s]	7.50	11.00	15.50	17.00	19.00	20.00	22.00	24.50

Fluid	Specific Gravity
Acetone	0.79
Citric acid	1.67
Glycerin	1.26
Mineral oil	0.90

(a) Show the resulting data and trendline, with equation and R^2 value, on the appropriate graph type (*xy* scatter, semilog, or log–log) to make the data appear linear.

(b) Determine the value and units of the density for each data set using the trendline equations.

(c) From the chart at left, match each data set (A, B) with the correct fluid name according to the results of the density determined from the trendlines.

9. As part of an electronic music synthesizer, you need to build a gizmo to convert a linear voltage to an exponentially related current. You build three prototype circuits and make several measurements of voltage and current in each. The collected data are given in the following table.

Circuit A		Circuit B		Circuit C	
Voltage (V_A) [V]	Current (I_A) [mA]	Voltage (V_B)[V]	Current (I_B) [mA]	Voltage (V_C) [V]	Current (I_C) [mA]
−2.7	0.28	−2.7	0.11	0	0.79
−0.4	1.05	−1.5	0.36	0.5	1.59
0	1.74	0	1.34	1.4	5.41
1.2	3.17	0.8	2.37	2.3	20.28
2.9	7.74	2.6	14.53	2.9	41.44

(a) Show the resulting data and trendline, with equation and R^2 value, on the appropriate graph type (*xy* scatter, semilog, or log–log) to make the data appear linear.

(b) Which of the three circuits comes the closest to doubling the current for an increase of 1 volt? Note that this doubling is independent of the actual values of voltage. Example: If the current was 0.3 mA at 2.7 volts, it should be 0.6 mA at 3.7 volts, 1.2 mA at 4.7 volts, 2.4 mA at 5.7 volts, etc.

(c) Calculate the value that should appear in the exponent if the current is to double with each increase of 1 volt. Note that you should perform this calculation without referring to the data, the plots, or the trendline equations. This is a purely theoretical calculation.

10. The following data were collected during testing of an electromagnetic mass driver. The energy to energize the electromagnets was obtained from a bank of capacitors. The capacitor bank was charged to various voltages, and for each voltage, the exit velocity of the projectile was measured when the mass driver was activated.

Voltage (V) [kV]	9	13	15	18	22	25
Velocity (v_p) [m/s]	430	530	580	650	740	810

(a) Show the resulting data and trendline, with equation and R^2 value, on the appropriate graph type (*xy* scatter, semilog, or log–log) to make the data appear linear.

(b) What would the velocity be if the capacitors were charged to 1000 volts?

(c) What voltage would be necessary to accelerate the projectile to 1000 meters per second?

(d) Assume that the total capacitance is 5 farads. If the capacitors are initially charged to 10,000 volts and are discharged to 2000 volts during the launch of a projectile, what is the mass of the projectile if the overall conversion of energy stored in the capacitors to kinetic energy in the projectile has an efficiency of 0.2? Recall that the energy stored in a capacitor is given by $E = 0.5\ CV^2$, where C is capacitance in farads and V is voltage in volts.

NOTE

Due to several complicated nonlinear losses in the system that are far beyond the scope of this course, this is a case of a model in which the exponent does not come out to be an integer or simple fraction, so rounding to two significant figures is appropriate. In fact, this model is only a first approximation—a really accurate model would be considerably more complicated.

11. The relationship of the power required by a propeller (shown as the power number, on the ordinate) and the Reynolds number (abscissa) is shown in the following graph. For a propeller, the Reynolds number (Re) is written slightly differently, as

$$Re = \frac{D^2 n \rho}{\mu}$$

where D is the blade diameter [meters] and n is the shaft speed [hertz]. The power number (N_p) is given by the following, where P is the power required [watts].

$$N_p = \frac{P}{\rho n^3 D^5}$$

Use the following chart to answer questions (a) through (d).

(a) If the Reynolds number is 500, what is the power number for a system described by curve A?

(b) If the power number (N_p) is 30, what is the Reynolds number for a system described by curve A?

(c) If the Reynolds number is 4000, what is the power (P) required in units of watts at a shaft speed (n) of 0.03 hertz? Assume the system contains acetone, with a kinematic viscosity of 0.419 stokes. The density of acetone is 0.785 grams per cubic centimeter. Use curve B in the graph to determine your answer. (*Hint:* Use the Reynolds number of the system to first calculate the diameter, then find the power number, and then calculate the power.)

(d) If the power number (N_p) is 5, what is the diameter (D) of the blade in units of centimeters at a shaft speed (n) of 0.02 hertz? Assume the system contains brine, with a kinematic viscosity of 0.0102 stokes. Use curve A in the graph to determine your answer. (*Hint:* Find the Reynolds number of the system first, and then calculate the diameter.)

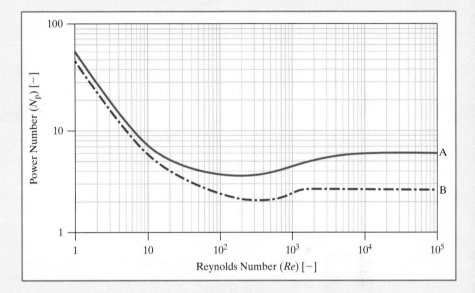

12. When a fluid flows around an object, it creates a force, called the drag force, that pulls on the object. The coefficient of drag (C_d) is a dimensionless number that describes the relationship between the force created and the fluid and object properties, given as

$$C_d = \frac{F_D}{\frac{1}{2}\rho v^2 A_p}$$

where F_D is the drag force, ρ is the fluid density, and v is the velocity of the object relative to the fluid. The area of the object the force acts upon is A_p, and for spheres is given by the area of a circle. The Reynolds number in this situation is written as

$$Re = \frac{D_p \rho v}{\mu}$$

where D_p is the diameter of the object the force acts upon. The following chart shows this relationship. The dashed lines show the predicted theories of Stokes and Newton compared to the solid line of actual results.

(a) If the Reynolds number is 500, what is the coefficient of drag?

(b) If the coefficient of drag is 2, what is the Reynolds number?

Ethylene glycol has a dynamic viscosity of 9.13 centipoise and a specific gravity of 1.109.

(c) If the fluid flows around a sphere of diameter 1 centimeter travelling at a velocity of 2.45 centimeters per second, determine the drag force on the particle in units of newtons. (*Hint:* First determine the Reynolds number.)

(d) If a coefficient of drag of 10 is produced, what is the diameter of the particle? Assume the fluid is moving at 1 centimeter per second. (*Hint:* First determine the Reynolds number.)

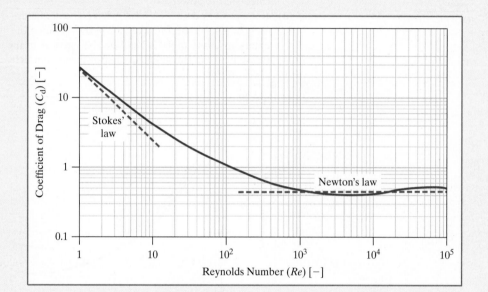

13. When discussing the flow of a fluid through a piping system, we say that friction occurs between the fluid and the pipe wall due to viscous drag. The loss of energy due to the friction of fluid against the pipe wall is described by the friction factor. The Darcy friction factor (*f*) was developed by Henry Darcy (1803–1858), a French scientist who made several important contributions to the field of hydraulics. The friction factor depends upon several other factors, including flow regime, Reynolds number, and pipe roughness. The friction factor can be determined in several ways, including the Moody diagram (discussed in ICA 13-22) and several mathematical approximations presented here.

In the laminar flow range, the *Darcy friction factor* can be determined by the following formula, shown as the linear line on the Moody diagram (see ICA 13-22 for the Moody diagram):

$$f = \frac{64}{Re}$$

In the turbulent range, the friction factor is a function of the Reynolds number and the roughness of the pipe (ε). For turbulent flow smooth pipes (where the relative roughness ratio (ε/D) is very small), the *Blasius formula* can be used to calculate an approximate value for the Darcy friction factor.

$$f = 0.316\,(Re)^{-1/4}$$

This simple formula was developed by Paul Richard Heinrich Blasius (1883–1970), a German fluid dynamics engineer. Later, a more accurate but more complex formula was developed in 1939 by C. F. Colebrook. Unlike the Blasius formula, the Colebrook formula directly takes into account the pipe roughness.

The *Colebrook formula* is shown below. Notice that both sides of the equation contain the friction factor, requiring an iterative solution.

$$\frac{1}{\sqrt{f}} = -2 \log\left(\frac{\varepsilon/D}{3.7} + \frac{2.51}{Re\sqrt{f}}\right)$$

To begin the iteration, the Colebrook calculation must have an initial value. Use the Blasius approximation as the first value for f, and determine the first iterative value of the Colebrook equation to use as your friction factor.

While this will only give us an approximation of the correct friction factor, a true solution requires using iteration. If you have covered iteration in Excel, your instructor may provide other instructions on how to determine f.

Prepare an Excel worksheet to compute the friction factor.

Sample Data						
Fluid	Water		Pipe Type	Commercial Steel		
Density (ρ) [g/cm^3]	Viscosity (μ) [cP]	Volumetric Flowrate (Q) [gpm]	Diameter (D) [in]	Roughness (ε) [mm]	Reynolds Number (Re) [-]	Flow Regime
1	1.002	500	6	0.045	263,020	Turbulent
Darcy Friction Factor (f) [-]			Initial f Value			
Laminar	Colebrook		Blasius		Moody Value	
	0.0174		0.0140		0.0185	

Input Parameters:

- Fluid: should be chosen from a drop-down list using the following material properties. Used with the lookup function to determine:
  - Density (ε) [grams per cubic centimeter]
  - Viscosity (μ) [centipoises]
  - Volumetric flow rate (Q) [gallons per minute]
  - Diameter (D) [inches]
- Type of pipe: should be chosen from a drop-down list using the following properties. Used with the lookup function to determine:
  - Pipe roughness (ε) [millimeters]

Output Parameters:

Be sure to include the appropriate unit conversions. You may add cells to the worksheet template to complete the necessary unit conversions.

- Reynolds number.
- Flow regime (laminar, transitional, or turbulent).
- Only the correct Darcy friction factor (one of these two values) should be displayed based on the flow regime.
  - For laminar flow, use the equation: $f = 64/Re$.
  - For turbulent flow determined with the Colebrook formula, use the Blasius equation as the initial f value.
- Determine the friction factor by hand from the Moody diagram (see ICA 13-22) and list the value found from the graph in the worksheet, as a comparison to your determined value.

Use the following parameters as a test case:

- Fluid = Acetone
- Pipe type = Cast iron
- Volumetric flowrate = 50 gpm
- Pipe diameter = 2 inches

CHAPTER

14 Statistics

Probability is associated with assessing the likelihood that an event will or will not occur. For example:

Airplane crash	River breaching a levee	Nuclear reactor accident
Tornado	Failure of equipment	Terminal cancer
Earthquake	Microprocessor failure	Space probe data reception

Statistics are used for design-concept evaluation because they provide quantitative measures to "things" that behave in a random manner. This evaluation helps us make rational decisions about everything from natural events to manufactured products. Statistics, as well as probability, use numerical evidence to aid decision making in the face of uncertainty. Roles of statistics in engineering include the following:

- Evaluation of new or alternative designs, concepts, and procedures
- Estimation of amount to bid on projects
- Management (human uncertainty, economic uncertainty, and others)
- Determination of degree of acceptable item-to-item variation (quality control)

Often, the best way to analyze an engineering problem is to conduct an experiment. When we take this approach, we face several questions:

- How many tests do we need?
- How confident are we in the results?
- Can we extrapolate the results to other conditions?
- Can we estimate how often the result will lie within a specified range?

There are many other related issues, but addressing all of them requires a separate book. Many readers will take or have taken an entire course in probability or statistics, so for now we just touch on some of the important fundamentals.

- **Repeated tests:** When a test is conducted multiple times, we will not get the same (exact) result each time. For example, use a ruler to measure the length of a particular brand of shoe manufactured by Company X. You can produce a table of values, all of which will be nearly, but not exactly, the same. This is because you make slight errors in measurement, so even if every shoe is identical, there will always be some errors in your measurement. Moreover, every shoe is not exactly identical.
- **Differences in a population:** What is the heart rate of all students in a class? Obviously, not everyone will have the same heart rate. We do expect, however, that everyone's rate will lie between, say, 40 and 140 beats per minute. Through measurement, we can determine this variation. In fact, we may find that the average rate for females and males differs. Statistical procedures help us analyze situations such as this.

- **Manufacturing errors:** Suppose you are manufacturing a run of widgets and a buyer wants all of them to be exactly alike. Obviously, this is impossible, but you can make them almost alike and then tell the buyer how much variation to expect. If you measure each widget as it comes off the assembly line, there will certainly be some variation.
- **Design criteria for products:** When you build a house, you would like it to stand safely for some period of time. For example, you might specify that the house be designed to withstand a windstorm that would occur, on average, every 50 years. For that case, you must be able to calculate the wind speeds associated with such a storm. Statistical methods allow you to do this.

14.1 Histograms

LEARN TO:	Create a histogram by hand given starting data
	Justify the choice of a reasonable bin size

To illustrate several common statistical concepts, we use data representing the height of several freshman engineering students. Table 14-1 shows the height, to the nearest inch, of each student in a typical class. Table 14-2 shows the same data, summarized by number of students at each height.

Table 14-1 Student height

Student ID	Height (*H*) [in]
A	67
B	73
C	71
D	69
E	68
F	64
G	70
H	72
I	67
J	71
K	70
L	68
M	66
N	71
O	74
P	71
Q	68
R	72
S	67
T	64
U	75
V	74
W	72

Table 14-2 Summary of height data

Height (*H*) [in]	Number of Students
62	0
63	0
64	2
65	0
66	1
67	3
68	3
69	1
70	2
71	4
72	3
73	1
74	2
75	1
76	0
77	0
Total	23

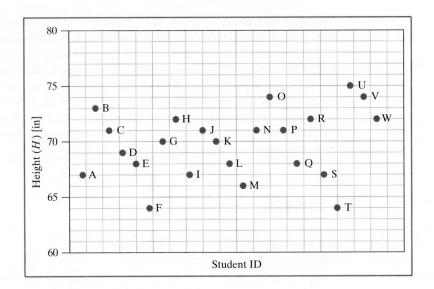

Figure 14-1

Example of student height, shown on scatter plot.

When we graph the values shown in Table 14-1, we end up with a scatter plot with data that is exactly the same: scattered, as shown in Figure 14-1.

Instead of using a scatter plot, we can group the data and plot the group values in a chart similar to a column chart, shown in Figure 14-2. Using the summarized data shown in Table 14-2, we will place two height ranges into a single column or **bin**. The first bin will contain all student-height values less than 63 inches. The next bin will contain student-height values of 64 and 65 inches. The next bin will contain student-height values of 66 and 67 inches, and so on. The abscissa of the graph is the height values; the ordinate is the number of students measured at each height range. Graphs of this nature are called **histograms**. By counting the number of blocks, we find the area under the curve represents the total number of samples taken, in this case, the total number of students (23) observed.

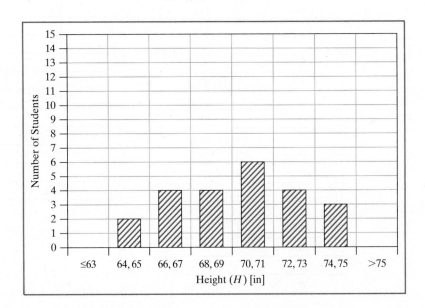

Figure 14-2

Example of student height, shown on a histogram.

Technically, before beginning this example we should have mathematically determined a bin size, rather than arbitrarily grouping the measurements in pairs (64 and 65 in one bin, 66 and 67 in the next bin, etc.). There are several ways to calculate the bin size that will best display the information; following is one method.

DETERMINATION OF BIN SIZE

Step One: Determine the number of bins needed.

Number of bins = Square root of number of data points, rounded to whole number

Step Two: Determine the range of the data.

Range = $X_{max} - X_{min}$

Step Three: Determine the number of items in each bin.

Bin size = Range divided by number of bins, rounded to whole number

---

Let us apply this to our example.

Step 1: As shown, we have a class of 23 students, so we would need five bins, since the square root of 23 is about 4.8, which rounds to 5. Four would probably also work fine, as would 6. Remember this is just a rule of thumb.

Step 2: The shortest person is 64 inches tall and the tallest is 75, so the range is 11 inches.

Step 3: Dividing the range determined in step 2 by the number of bins determined in step 1, we get 2.2, or about 2 inches per bin. On the other hand, we might instead decide to have four bins. If we divide the range by 4, we have 2.75 or 3 inches per bin.

Depending on the number of bins, we sometimes get two different, but acceptable, bin sizes. By changing the bin size, we can change the appearance of the data spread, or the data **distribution**.

What happens to the student height data if we alter the bin size? The plot on the preceding page shows a 2-inch bin interval, and Figure 14-3(a) shows a 3-inch interval.

In Figure 14-3(b), we have used a 4-inch bin interval, and while it is not what we obtained from the "rule of thumb" (2- or 3-inch intervals), it is still mathematically correct but not as informative as the other two.

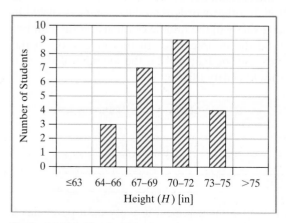

Figure 14-3(a) Bin size of 3.

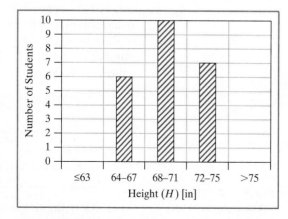

Figure 14-3(b) Bin size of 4.

| Comprehension Check | 14-1 |

The following table lists the number of computer chips rejected for defects during random testing over the course of a week on a manufacturing line. Four samples of 20 parts are pulled each day.

Use the following data to generate a histogram by hand.

1	1	8	0	2	0
0	2	10	1	3	2
0	1	12	0	2	1
1	6	15	0	0	
3	8	1	2	5	

14.2 Statistical Behavior

LEARN TO: Determine the mean, median, variance, and standard deviation by hand
Justify the choice of using mean or median in estimating central tendency
Define the relationship between variance and standard deviation

NOTE

Average or Mean = typical, expected value of the data set; sensitive to outliers.

Median = value representing the exact middle value of the list; typically unaffected by outliers. Data must be in ascending or descending order to determine median!

When we have gathered the data and plotted a distribution, the next step is to explain the outcome to others. For convenience, we identify a set of parameters to describe distributions.

One parameter of a distribution is the average value. The **average**, or **mean**, is an estimate of the value most representative of the population. This is often called the **central tendency** of the data. The computation of the mean ($\overline{X}$) of a data set containing N values is given in the following equation.

$$\text{Mean} = \overline{X} = \frac{1}{N}(X_1 + X_2 + \cdots + X_N) = \frac{1}{N}\sum_{i=1}^{N} X_i$$

In other words, the mean is the sum of all of the values divided by the total number of values.

The **median**, another measure of central tendency, is the value between the lower half and the upper half of the population. In other words, if all data points are listed in numerical order, the median is the value exactly in the middle of the list. If the number of data points is odd, the median will be the middle value of the population. If the number of data points is even, however, the median will be the average of the two values at the center. A few examples should clarify this.

Set	Data	Mean	Median
1	1, 2, 3, 4, 5, 6, 7	4	4
2	1, 50, 70, 100	55	60
3	5, 10, 20, 40, 80	31	20
4	50, 50, 50, 50, 50, 1000	208	50

Review the data shown in set 4. It would seem logical if every data point has a value of 50 except one, the average of the data should be about 50; instead, it is 208!

NOTE

Variance = measure of data scatter; has SQUARED UNITS of the original data set.

Standard deviation = square root of the variance; has units of the original data set.

This illustrates the sensitivity of the mean to **extreme values**, or **outliers**. Note that the median is unaffected or only slightly affected. It is for this reason that the mean is insufficient to describe the central tendency of all distributions.

Two other terms are useful in describing a distribution: **variance** and **standard deviation**. Both of these terms quantify how widely a set of values is scattered about the mean. To determine the variance (V_x^2), the difference between each point and the mean is determined, and each difference is squared to keep all terms positive. This sum is then divided by one less than the number of data points.

$$\text{Variance} = V_x^2 = \frac{1}{N-1}\left((\bar{X} - X_1)^2 + (\bar{X} - X_2)^2 + \cdots + (\bar{X} - X_N)^2\right)$$

$$= \frac{1}{N-1}\sum_i^N (\bar{X} - X_i)^2$$

The standard deviation (SD_x) is found by taking the square root of the variance:

$$\text{Standard deviation} = \text{SD}_x = \sqrt{V_x^2}$$

If we again examine the data found in Table 14-2, we can calculate the mean, median, variance, and standard deviation for our height data.

Height (H) [in]	Number of Students
62	0
63	0
64	2
65	0
66	1
67	3
68	3
69	1
70	2
71	4
72	3
73	1
74	2
75	1
76	0
77	0
Total	23

Calculation of the Mean:
Total number of points (N)
= 23 students

The sum of all heights
= (2 students *64 inches/student) + (1*66) + (3*67) + (3*68) + (1*69)
+ (2*70) + (4*71) + (3*72) + (1*73) + (2*74) + (1*75)
= 1604 inches

Mean
= 1604 inches/23 students
= 69.7 inches/student

Calculation of the Median:
Put data in order of value, listing each entry once
    64, 64, 66, 67, 67, 67, 68, 68, 68, 69, 70, 70, 71, 71, 71, 71, 72, 72, 72, 73, 74, 74, 75
Find the center value since the total number of students is odd
    64, 64, 66, 67, 67, 67, 68, 68, 68, 69, **70**, 70, 71, 71, 71, 71, 72, 72, 72, 73, 74, 74, 75
Median = 70 inches

Calculation of Variance: Note that the variance will have the same units as the variable in question squared, in this case, inches squared.

$$\text{Variance} = \frac{1}{23-1}\left((69.7 - 64)^2 + (69.7 - 64)^2 + (69.7 - 66)^2 + \cdots \right.$$
$$\left. + (69.7 - 75)^2\right) = 9.5 \text{ in}^2$$

Calculation of Standard Deviation: The standard deviation has the same units as the variable in question, in this case, inches.

$$\text{Standard deviation} = \sqrt{9.5} = 3.08 \text{ in}$$

EXAMPLE   14-1

Consider the following velocity data, listed in units of feet per second. Determine the mean, median, variance, and standard deviation of the data.

1	28	14	32	35	25	14	28	5
16	42	35	26	5	33	35	16	14

Calculation of the Mean:

Total number of points (N) = 18
Sum of all data $(\sum X_i) = (1) + (2*5) + (3*14) + \cdots + (42) = 404$
Mean $= 404/18 = 22.4$ feet per second

Calculation of the Median:

Put data in order, listing each entry once, including all duplicate values.
          1, 5, 5, 14, 14, 14, 16, 16, 25, 26, 28, 28, 32, 33, 35, 35, 35, 42
Find the center two values and average them, since total number of entries is even (18).
          *1, 5, 5, 14, 14, 14, 16, 16, **25**, **26**, 28, 28, 32, 33, 35, 35, 35, 42*
Median = (25 + 26)/2 = 25.5 feet per second

Calculation of Variance:

Variance $= \dfrac{1}{18 - 1}((22.4 - 1)^2 + \cdots + (22.4 - 42)^2) = 147$ (ft/s)^2

Calculation of Standard Deviation:

Standard deviation $= \sqrt{147} = 12.1$ feet per second

EXAMPLE   14-2

Consider the following energy data, given in units of joules. Determine the mean, median, variance, and standard deviation of the data.

159	837	618	208	971	571	379	220	31

Calculation of the Mean:

Total number of points (N) = 9
Sum of all data $(\sum X_i) = 159 + 837 + \cdots + 31 = 3,994$
Mean $= 3,994/9 = 443.7 = 444$ joules

Calculation of the Median:

Put data in order, listing each entry once.
                    31, 159, 208, 220, 379, 571, 618, 837, 971
Find the center value since the total number of students is odd (9; center value at entry 5).
                    *31, 159, 208, 220, **379**, 571, 618, 837, 971*
Median = 379 joules

Calculation of Variance:
Variance $= \dfrac{1}{9 - 1}[(444 - 31)^2 + \cdots + (444 - 971)^2] = 105,059$ joules^2

Calculation of Standard Deviation:

Standard deviation $= \sqrt{105,059} = 324$ joules

Comprehension Check	14-2

For the following mass data given in units of kilograms, determine the mean, median, variance, and standard deviation.

8	7	9	11	16
12	2	9	10	9

Comprehension Check	14-3

For the following temperature data given in units of degrees Celsius $[°C]$, determine the mean, median, variance, and standard deviation.

105	120	110	100	102
103	58	110	100	118

14.3 Distributions

LEARN TO:	Draw the expected distribution, given an expected change in baseline
	Assign a possible cause for change, given baseline distribution and new distribution
	Define a "normal" distribution

From Figure 14-2 and Figure 14-3, you can see a similarity in the histogram shape of all three plots. The values start small, increase in size, and then decrease again. In the case of student height, this means that a few people are short, most people have some "average" height, and a few people are tall. This same conclusion is true in many things we can measure. For example:

- If we weigh many standard-size watermelons (neither miniature nor giant), we will find that most weigh between 20 and 30 pounds. A few weigh less than 20 pounds, and a few weigh more than 30 pounds.
- As we look through a dictionary, we find that there are many words with between four and six letters. There are a few with one, two, or three letters and a few with more than six, but clearly most have between four and six letters.
- To improve efficiency in the office, we had an expert to monitor the length of phone calls made by the staff. The expert found that most of the time, phone calls lasted between 3 and 5 minutes, but a few were longer than 5 and some others lasted only a minute.

Normal Distributions

We wanted to know how many "flexes" it takes to cause a paper clip to fail, so we asked volunteers to test the bending performance of paper clips by doing the following:

- Unfold the paper clip at the center point so that the resulting wire forms an S shape.
- Bend the clip back and forth at the center point until it breaks.
- Record the number of flexes required to break the clip.

Using these data, we created Figure 14-4. This is the same as the earlier histogram, but the boxes are replaced by a smooth curve through the values.

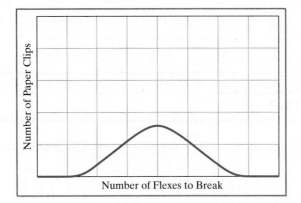

Figure 14-4
Distribution of paper clip failure.

When you are interested in the *shape* of the curve rather than the exact data values, you can replace the bars of the histogram with a smooth curve and rename the graph a **distribution**. A distribution is considered *normal* if the following rules hold true. This is known as the **68-95-99.7 rule**, shown in Figure 14-5.

- 68% of values are within one standard deviation (1σ) of the mean (μ).
- 95% of values are within two standard deviations (2σ).
- 99.7% of values are within three standard deviations (3σ).

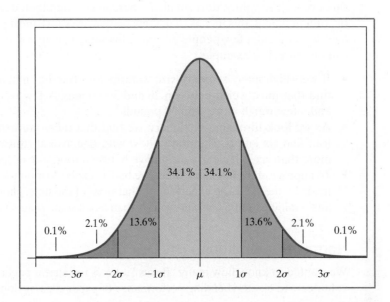

Figure 14-5
"Normal" distribution, showing the 68–95–99.7 rule.

EXAMPLE   14-3

Suppose we ask a class of students how many states they have visited. The results might appear as shown below.

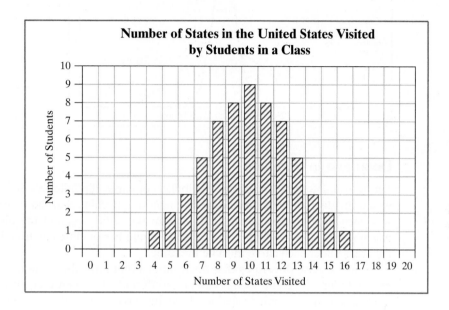

Number of States in the United States Visited by Students in a Class

Number of Students (y-axis, 0 to 10)

Number of States Visited (x-axis, 0 to 20)

It seems that most have visited between 8 and 12, and that as many have visited more than 10 as have visited fewer than 10. A few have visited as many as 16 states, and all the students have visited at least 4. Let us calculate some values pertinent to this situation.

How many students are there in the class?
To do this we simply add the number of students represented by each bar, or

$$1 + 2 + 3 + 5 + 7 + 8 + 9 + 8 + 7 + 5 + 3 + 2 + 1 = 61 \text{ students}$$

What is the cumulative number of state visits?
*We answer this by totaling the product of the bar height with the number of states represented by the bar. For example, 5 students have visited 7 states, so those 5 students have visited a total of 5 * 7 = 35 states. Or, 8 students have visited 11 states, so those students have visited a total of 88 states. We calculate*

$$1*4 + 2*5 + 3*6 \; + \; \cdots \; + 2*15 + 1*16 = 610 \text{ states}$$

What is the average number of states visited by a student?
Once we have the values from our first two answers, this is straightforward division: the total number of visits divided by the total number of students.

$$610/61 = 10 \text{ states per student}$$

Notice that the value 10 is in the center of the distribution. For distributions that are symmetrical (such as this one), the average value is the one in the center, the one represented by the largest number of occurrences.

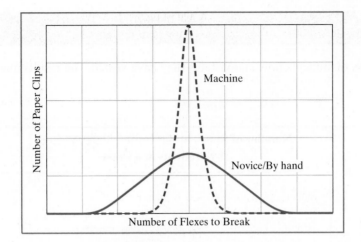

Figure 14-6

Distribution of paper clip failure after a decrease in variance.

Decrease in Variance

Let us examine the shape of the distribution of paper clip failures discussed earlier. How would the distribution change if we brought in a machine that did it "exactly" the same way each time? Both the distribution from the data class and the distribution of the machine are shown in Figure 14-6. The same number of clips was tested in each case, so the areas under each curve must be the same.

This exercise illustrates that distributions that have the same mean (and median) can look very different. In this case, the difference between these two distributions is in their *spread*, or their variation about the mean. The effect of using a machine to break the paper clips was a decrease in the variance. Similarly, many machine manufacturers may publish expected variance in results through illustrations like distribution graphs, but in general, the lower the variance, the higher the cost of the machine, and likewise, the higher the variance, the lower the cost.

Shift in Mean

Redraw the paper clip distribution; then on the same plot, sketch the distribution if each volunteer tested the same number of clips that were manufactured by the same manufacturer as before with the same variance, but were stronger and typically required 10 more

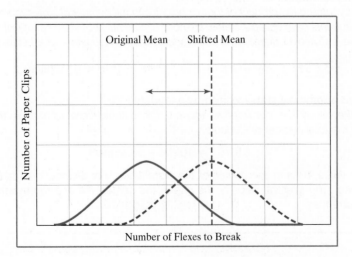

Figure 14-7

Distribution of paper clip failure with a shift.

flexes to fail. The result is shown in Figure 14-7. The stronger material caused the distribution to shift to the right. Since the variance and number of clips remained the same, the shape and size of the curve remains the same.

Skewed Data

It is often easy to place an upper or lower limit on the value of the possible outcome. In these cases, the distribution is no longer symmetric—it is **skewed**. A population is **positively skewed** if the mean has been pulled higher than the median, and **negatively skewed** if the mean has been pulled lower than the median (see Figure 14-8 for an example of a positively skewed graph). You have probably heard news reports that use the median to describe a distribution of income in the United States. The median is used in this case because the distribution is positively skewed. This skew is caused by two factors, the presence of extreme values (millionaires) and the **range restriction**, the latter because income cannot be lower than $0. The extreme values causing the positive skew are not shown on the graph. Most of these would be far off the page to the right.

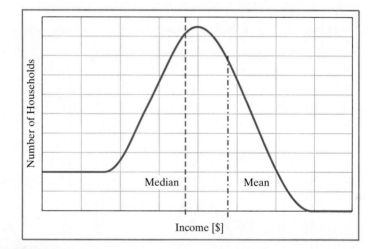

Figure 14-8
Distribution of positively skewed data.

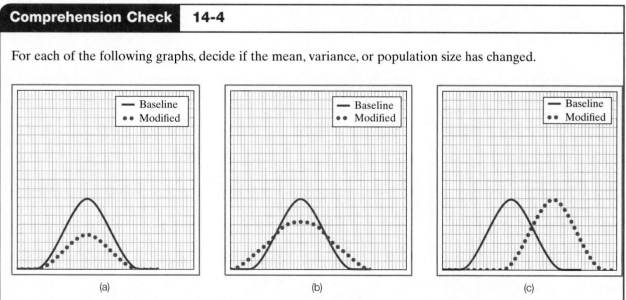

Comprehension Check   **14-4**

For each of the following graphs, decide if the mean, variance, or population size has changed.

(a)   (b)   (c)

EXAMPLE   14-4

For each scenario, identify one graph from the following set that best illustrates how the baseline curve would change under the conditions of that scenario. Each graph shows the usual distribution (labeled baseline) and the way the distribution would be modified from the baseline shape (labeled modified) under certain conditions.

The graphs show SAT composite (verbal + quantitative) scores, for which 400 is generally considered to be the minimum possible score and 1600 is considered to be the maximum possible score.

(a) The designers of the SAT inadvertently made the test more difficult, while the variance of the scores remains the same.

   This is shown by curve F: Area and variance the same; mean shifted to left.

(b) The variability of scores is reduced by switching to true/false questions, while the average remains the same.

   This is shown by curve D: Area and mean are the same; distribution is narrower.

(c) A population boom increases the number of students seeking college admission.

   This is shown by curve B: Area increases, distribution stays the same.

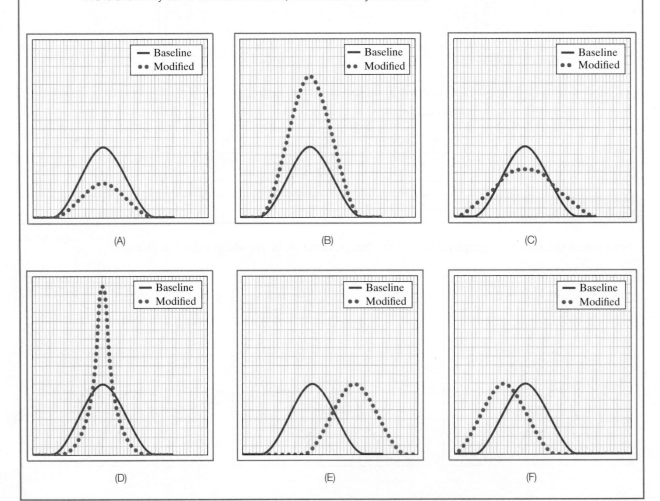

Comprehension Check | **14-5**

Use the scenario described in Example 14-4. For each scenario, identify one graph from the graph choices shown in Example 14-4 that best illustrates how the baseline curve would change under the conditions of that scenario.

(a) The economy declines, so more students decide to enter the workforce instead of attending college. The variance in the SAT scores remains the same.

(b) As a performance measure, all high school seniors are required to take the SAT; the variance remains the same.

(c) Due to mandatory test preparation courses, the mean of the SAT increases for the same number of students taking the exam. The variance remains the same.

14.4 Cumulative Distribution Functions

LEARN TO: Draw a CDF by hand given starting data or a histogram

For the earlier student height plot using two heights per bin, we will graph the bin data but now show the values on the ordinate as a fraction rather than a whole number. To do this, we divide the number of students in each bin of the histogram by the total number of students. If we now add the heights of all the bars in the new plot, they should equal 1. This is called a **normalized plot**, shown in Figure 14-9. We "normalized" the values by dividing by the total number of data points. This graph holds no new information; it is simply a rescaling of the histogram we drew earlier.

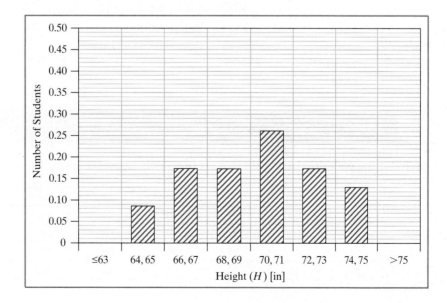

Figure 14-9

Normalized plot of student height, originally shown in Figure 14-2.

This plot can be used as an intermediate step to obtain a final plot called a **cumulative distribution function (CDF)**. We derive this plot by summing the values for each bin on the normalized plot from the first bin up to each individual bin. For example, suppose the values in the first three bins were 0, 0.08, and 0.18. By adding the values, we get new "cumulative" values: bin 1 = 0; bin 2 = 0 + 0.08 = 0.08; and bin 3 = 0 + 0.08 + 0.18 = 0.26. It should be obvious that we can obtain the CDF value for each bin by adding the normalized value of that bin to the CDF value of the bin before it. The CDF values are usually shown as percentages rather than fractions, for example, 50% instead of 0.5.

As we move across the plot, the values should go from 0 to 1. Using the height data in the normalized plot, we have produced a cumulative distribution shown in Figure 14-10. Sometimes, the CDF is shown as a continuous, curved line rather than a column chart. Both the original histogram and the cumulative distribution plot are useful tools in answering questions about the composition of a population.

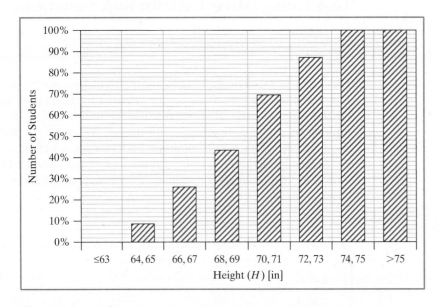

Figure 14-10
CDF of student height, originally shown in Figure 14-2.

EXAMPLE 14-5

Consider the following pressure data, given in units of pascals. Draw the histogram and CDF of the data.

36	9	33	11	23	3
34	39	56	51	39	1
27	25	2	1	53	32
14	41	55	28	29	19
51	15	25	10	35	38

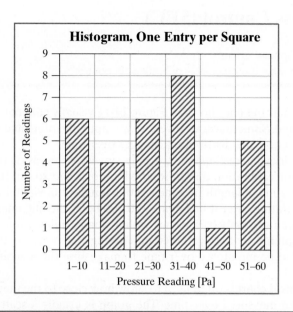

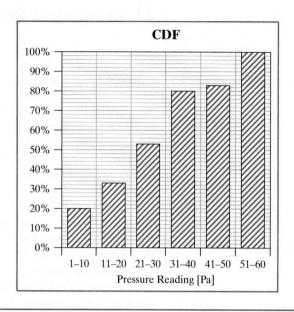

Comprehension Check   14-6

Consider the weight of shipping boxes sent down an assembly line, given in units of newtons. Draw the histogram and CDF of the data.

38	103	20	42	16
20	74	63	90	61
114	79	61	50	64

Comprehension Check   14-7

A team testing widget power usage in a trial of 25 experiments presented the results of their experiment as a CDF. Use the information from the CDF to create the histogram of their experiment.

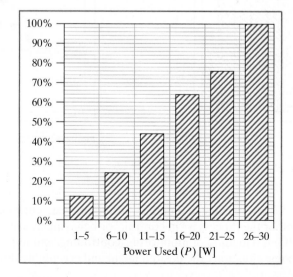

14.5 Statistical Process Control (SPC)

LEARN TO: Create a control chart of the appropriate "zones"
Use the eight Nelson rules to determine if a process is operating "in control"

We showed that a histogram such as the one shown in Figure 14-11 visually summarizes how a set of values is distributed. Sometimes, however, we are not only interested in the values themselves, but also in how the distribution changes over time.

For example, as a machine in a factory operates, it may slowly (or occasionally quickly) lose proper alignment or calibration due to wear, vibrations, and so on. If a machine was making bolts with a mean length of 1 inch and a standard deviation of 0.01 inch when it first began operating, after it had made 100,000 bolts, the alignment may have drifted so that the mean was only 0.95 inches with a standard deviation of 0.02 inch. This may be unacceptable to the customer purchasing the bolts, so the parameters of the process need to be monitored over time to make sure the machine is readjusted as necessary.

A graph called a **quality control chart** is often used to show how close to the mean the results of a process are when measured over time. The graph is usually a scatter plot, with the abscissa shown as time or another indicator that would change with time, such as batch number. Figure 14-12 shows a sample control chart, the mean, and the standard deviation.

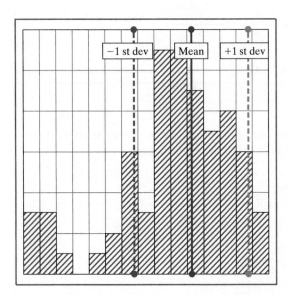

Figure 14-11 Sample histogram.

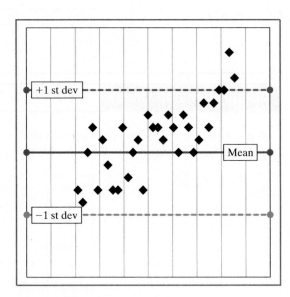

Figure 14-12 Sample quality chart.

When tracking a manufacturing process, engineers are often concerned with whether the process is "in control," or behaving as expected. **Statistical process control (SPC)** is a method of monitoring, controlling, and improving a manufacturing process. In some situations, the desired mean and acceptable deviation limits may be preset for a variety of reasons (chemistry, safety, etc.). Often, the upper or lower limits of control are determined by the desired end result. For example:

- The reactor temperature must not rise above 85 degrees Celsius or the reactant will vaporize.
- The injection pressure should be between 50 and 75 kilopascals to ensure that the part is molded properly.
- A bolt must be machined to ±0.02 inches to fit properly in a chair leg.

   An engineer will study how the process relates to the control limits and will make adjustments to the process accordingly. To discuss whether a process is in control, we can divide a chart into zones, shown in Figure 14-13, to create a control chart. The mean is determined either by the desired end result (the iron content of the product must average 84%) or by the process itself (the reactor temperature should average 70 degrees Fahrenheit for an optimum reaction to occur). The standard deviation is most often determined by experimentation. For example:

- The purity range for this product is ±0.0005%.
- The standard deviation for the reactor temperature must not exceed 5 degrees Fahrenheit or the reaction will create unwanted by-products.
- The standard deviation of the current gain of the transistors being produced must be less than 15.

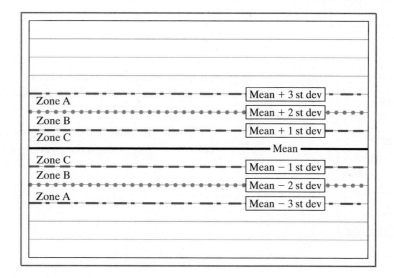

Figure 14-13
Standard deviation ranges.

Eight Ways to Be Out of Control

A variety of conditions can indicate that a process is out of control. First published by Lloyd S. Nelson in the October 1984 issue of the *Journal of Quality Technology*, the **Nelson rules** follow, with examples and graphs. In the graphs, solid points indicate the rule violations. The actual conditions may vary slightly from company to company, but most take the same standard form. For example, a company may operate with rule 3 stated as seven or eight points in a row instead of six.

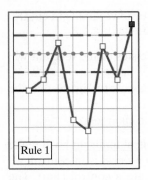

1. A point falls anywhere beyond Zone A. The value is more than three standard deviations away from the mean. May occur on either side of the mean.

Example: The mean temperature of a reactor is 85 degrees Celsius with a standard deviation of 5 degrees Celsius. If the temperature exceeds 100 degrees Celsius, the reactor vessel may explode. If the temperature falls below 70 degrees Celsius, the reaction cannot proceed properly.

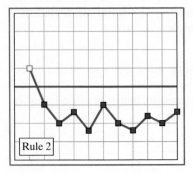

2. Nine points in a row occur on the same side of the mean. The actual value seems to be drifting away from the mean.

Example: The percentage of boron in a semiconductor should be 250 parts per billion. Nine consecutive samples have boron contents less than this value. The machine incorporating the boron into the semiconductor material may need to be cleaned or recalibrated.

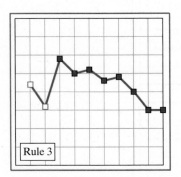

3. Six points occur with a consistently increasing or decreasing trend. If this pattern continues, the values will eventually become unacceptable.

Example: The shaft length of a part is increasing with each successive sample; perhaps the grinding wheel needs to be changed.

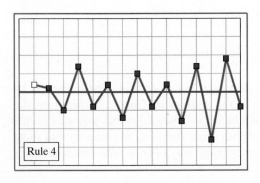

4. Fourteen points in a row alternate from one side of the mean to the other. The process is unstable.

Example: The control system for a crane errs from one side to the other. This may indicate a sensor failure or the need to reprogram the controller.

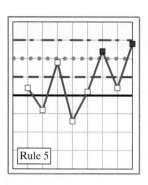

5. Two out of three points in a row occur in Zone A. The process is close to the upper limit; take preventive measures now.

Example: A robot that is spot-welding parts in an automobile is coming close to the edge of the material being welded. It probably needs attention.

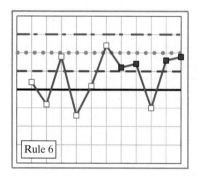

6. Four out of five points in a row occur in Zone B. The process is very close to the upper limit; take preventive measures now.

Example: Four out of five customers in the bank teller queue have waited more than one standard deviation to be helped. Perhaps another teller is needed.

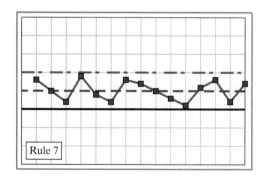

7. Fifteen points in a row occur in Zone C; points can occur on either side of the mean. The process is running too perfectly; in many applications the restrictions can be loosened to save time and money.

Example: The thickness of all washers being manufactured for quarter-inch bolts is within 0.0005 inch of the desired mean. Very few applications require washers with such close tolerances. Perhaps the process could be set to process the washers faster.

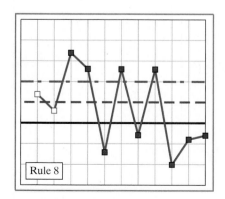

8. Eight points in a row occur beyond Zone C; points can occur on either side of the mean. The process does not run close enough to the mean; the parts are never quite on target; may indicate a need for a process adjustment.

Example: The postmark machine in a regional postal distribution center is stamping the envelopes too high or too low; it probably needs attention.

EXAMPLE 14-6

The data shown in the table were collected from a manufacturing process that makes bolts.

Assume the process specifies an average bolt length of 10 inches, with a standard deviation of 0.25 inches. Is this process under statistical control?

Part	Length (L) [in]	Part	Length (L) [in]
1	10.00	11	10.25
2	10.25	12	10.65
3	10.65	13	9.50
4	9.50	14	9.36
5	9.36	15	9.25
6	9.00	16	10.50
7	10.50	17	10.20
8	10.20	18	9.80
9	9.80	19	10.45
10	10.00	20	10.10

The control chart for these data is shown below. Rule 1 is violated since a point falls outside of Zone A (part 6) and rule 5 is violated as parts 14 and 15 fall inside Zone A. The process is not in statistical control.

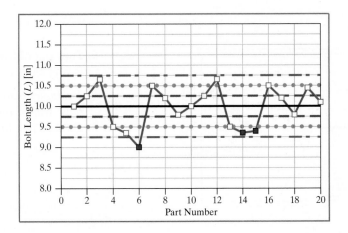

Comprehension Check 14-8

The pressure in a water filter is monitored in a chemical plant. The filter should operate at 18 pounds-force per square inch [psi], with a standard deviation of ±2 psi. Analyze the data shown to determine if the filter is behaving as expected (the process is in control; the filter does not require any attention) or if the filter needs attention (the process is out of control; the filter should be cleaned). Refer to the Nelson rules to explain your conclusion, and include the time the violations occur.

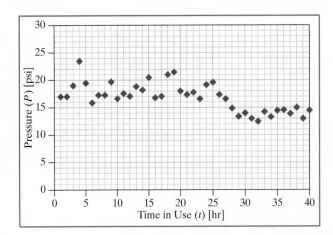

14.6 Statistics in Excel

LEARN TO: Create a histogram and CDF given a set of data in Excel
Determine the mean, median, variance, and standard deviation using Excel

In Chapter 10, some common built-in statistical functions were introduced. Table 10-3 is repeated here as Table 14-3. Please review Example 10-2 for a refresher on statistical functions in Excel.

Table 14-3 Statistical functions in Excel

Function as Written in Excel	Definition
AVERAGE (cells)	Finds the mean or average value of a list of cells
MAX (cells)	Finds the maximum value in a list of cells
MEDIAN (cells)	Finds the median value of a list of cells
MIN (cells)	Finds the minimum value in a list of cells
STDEV.P (cells)	Finds the standard deviation value of a list of cells
VAR.P (cells)	Finds the variance value of a list of cells

To create **histograms** and **CDFs** with Excel, you need to first activate the Analysis ToolPak in Microsoft Excel. Note that as of Microsoft Excel 2016, there is a built-in Histogram chart type, but this chart type does not provide key capabilities available in the Analysis ToolPak.

- In Excel, go to the Office button and click **Options**.
- Choose the **Add-Ins** tab on the left menu of the Excel Options window to display all the active add-in applications in Excel. Notice in our list that the Analysis ToolPak is listed as inactive.

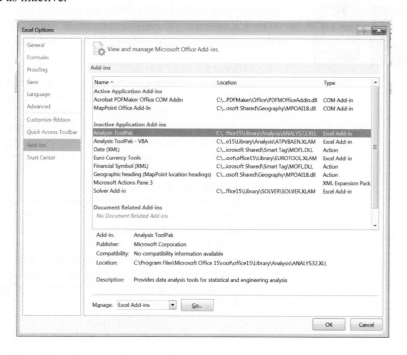

- At the bottom of the Excel Options window, select **Excel Add-Ins** in the **Manage** drop-down menu and click **Go**.
- In the Add-Ins window, check the **Analysis ToolPak** option and click **OK**.
- A prompt might pop up telling you to install the add-in—click **Yes** and finish the installation, using the Office Installer.

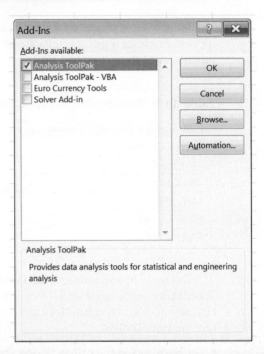

EXAMPLE   14-7

The following outline gives the steps necessary to use the data analysis tool in Excel for basic statistical analysis of a data set. This is presented with an example of the high and low temperatures during the month of October 2006.

- *If necessary, input the data; the data for this example have been provided online. Use column A to input an identifier for the data point, in this case, the date. Columns B and C will contain the actual high and low temperatures for each day, respectively.*
- *Next, decide on the bin range. This discussion focuses on the high temperatures but can easily be repeated with the low temperatures.*
  - *A rule of thumb is that the number of bins is approximately equal to the square root of the number of samples. While it is obvious in this example how many total samples are needed, the COUNT function is often very useful. October has 31 days, and the square root of 31 is 5.57; thus, you should choose either 5 or 6 bins.*
  - *Examine your data to determine the range of values. Using the MAX and MIN functions, you can determine that the highest high temperature during October was 86 degrees Fahrenheit and the lowest high temperature was 56 degrees Fahrenheit. Thus, your range is 86 − 56 = 30 °F.*
  - *Since 5.57 is closer to 6 than to 5, choose 6 bins. Remember, however, that you might want to try a different number of bins to see if that would result in a clearer representation of the data. With a range of 30 degrees Fahrenheit, 6 bins gives 30 °F/6 bins = 5 °F per bin.*

- *Type the range of values that will appear in each bin. For example, the first bin will contain temperatures 55, 56, 57, 58, and 59; the second bin will contain temperatures 60–64, and so on.*
- *In the adjacent column, type the corresponding upper value of temperature for each of the bins listed.*

🐾 *To create histograms and CDF charts:*

- *Go to **Data** > **Analysis** > **Data Analysis** and under Analysis Tools choose **Histogram**. Click **OK**.*

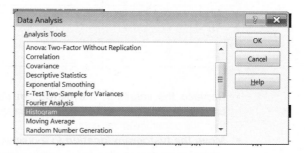

- *In the **Input Range**, click the icon at the right end of the blank box. You can then highlight the range (in this case, $B$6:$B$36). Close the box by clicking the icon at the right-hand end of this small box where the range is shown.*
- *Repeat this procedure for the **Bin Range**, highlighting the cells that contain the upper values.*
- *Next, for the **Output Range**, click the circle and identify a single cell to begin the placement of the output data.*
- *Finally, check the boxes to activate the options of **Cumulative Percentage** calculations and **Chart Output**.*
- *Click **OK**.*

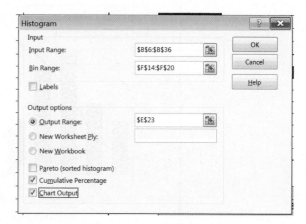

Your worksheet should now look like this:

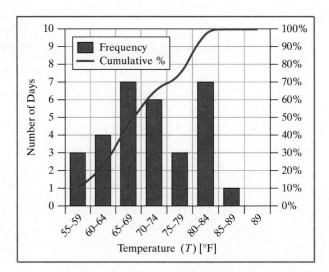

- ***Replace the values in the histogram data table for "Bin" with the "Bin Labels"*** *you entered earlier. This will change the axis labels to the range, rather than the upper value, for each bin.*
- ***Move the histogram location*** *to a new worksheet rather than imbedded in the original worksheet to allow the data to be seen clearly. After selecting the chart, use the* **Chart Tools > Design > Location > Move Chart** *option to select "As new sheet."*
- ***Modify the histogram to be a proper plot*** *just as you would with any other chart. The same rules for a "proper plot" apply to a histogram also, so make sure the background is white and alter the series colors, etc., as appropriate. The histogram generated with the preceding directions is shown below, properly formatted.*
- ***Change the vertical scale on the left axis*** *to be a multiple of 2, 5, or 10 to allow the cumulative percentages on the right axis to line up with the gridlines. This is important to do!*
- ***Change the vertical scale on the right axis*** *to be a maximum value of 100%. This is important to do. The resulting analysis should appear as follows:*

Comprehension Check | **14-9**

Repeat this analysis, using the daily low temperatures during October 2006.

⌘ Statistics on the Mac OS

Unfortunately, as this book goes to press, Microsoft has chosen not to include the histogram tool in Excel 2008 or Excel 2011 for the Mac OS. You have a few options.

- If you have an Intel-based Mac, you can use Excel 2007 or 2010 for Windows. If you do not know how to activate the Windows option on your machine, ask your friendly local Mac guru at your computer center.
- You can use Excel 2004 for the Mac OS, which did include a histogram tool.
- You can create the histogram manually according to the following instructions.

To create histograms and CDF charts using Mac OS Excel 2008 or 2011:

Create columns for the data, bin ranges, and upper value in each bin as described above.

Next, determine the number of data points in each bin. After doing a few, you will find it easy. Use the advanced Excel function, known as an array function, to accomplish this determination. A detailed explanation of array functions is beyond the scope of this book, but if you follow the instructions below *carefully*, you should not have any trouble. The specific array function to use is called FREQUENCY.

1. In the cell immediately to the right of the topmost "upper bin value" cell (this would be cell G14 in the example above) enter the formula

    = FREQUENCY(DataRange, UpperBinValueRange)

   and press Return. In the preceding example, this would be

    = FREQUENCY(B6:B36, F14:F20)

NOTE

Do not select these cells by clicking the bottom one and dragging up to the top one—it will not work correctly.

Do not use the replicate handle in the lower-right corner of the cell to drag down.

2. Click-and-hold the cell into which you entered the formula, then drag straight down to the cell in the row *following* the row containing the last "upper bin value." This would be cell G21 in the preceding example. Release the mouse button. At this point, you will have a vertical group of cells selected (G14:G21 in the example), the top cell will contain the number of data points in the first bin, and the rest of the selected cells *will be blank*. The formula you entered in the topmost of these cells will appear in the formula bar at the top of the window.
3. Click once in the formula in the formula bar. The top cell of the selected group will be highlighted.
4. Hold down the Command (⌘) key and press Return. The selected cells will now contain the number of data points in each bin immediately to the left. The bottommost selected cell will contain the number of data points larger than the upper bin value in the final bin. In the example, this "extra" cell should contain a 0, since no values are larger than those in the final bin. Note the formulae that appear in these cells are all identical—the cell references are exactly the same. This is normal for an array function.

Use these values to create the histogram.

1. Select the cells containing the bin ranges (E14:E20 in our example), then hold down the Command (⌘) key while you select the cells containing the number of data points

per bin. In our example, since the "extra" cell at the bottom contains a 0, you need not include it. If this were nonzero, you might want to add a cell at the bottom of the cell ranges that said something like >89. You should now have the two columns for bin ranges and number per bin selected (E14:E20 and G14:G20 in our example).

2. In the toolbar, select **Gallery > Charts > Column**. A row of column chart icons should appear.

3. Click the first icon, which shows pairs of columns. The chart that appears shows the histogram. Be sure to follow all appropriate proper plot rules for completing the histogram.

Finally, generate the CDF. If you have survived this far, you should be able to do this with minimal guidance. Create another column of values next to the column containing the number of data points per bin. In the cell next to the topmost bin cell, enter the number of data points in that bin. In the next cell down, enter a formula that will add the cell above to the cell beside it containing the number of data points in that bin. Replicate this formula down to the last bin. Each cell in the new column should now contain the sum of all data points in all bins to that point.

14.7 Statistics in MATLAB

LEARN TO: Create a histogram and CDF given a set of data in MATLAB
Determine the mean, median, variance, and standard deviation using MATLAB

Literally hundreds of functions are built into MATLAB. A few statistical functions, similar to Table 14-3 for Excel, are shown in Table 14-4.

Table 14-4 Common MATLAB statistical functions

MATLAB Function	Definition
`ceil(X)`	Rounds each element of X up to the next largest integer.
`fix(X)`	Rounds each element of X to the neighboring integer closest to 0.
`floor(X)`	Rounds each element of X down to the next smallest integer.
`length(X)`	If X is a vector, `length(X)` returns the number of elements in X. If X is a matrix, `length(X)` returns either the number of rows in X or the number of columns in X, whichever is larger.
`max(X)` and `min(X)`	Finds the maximum or minimum value of X. If X is a matrix, returns the maximum or minimum value of the elements of each column in X.
`mean(X)`	Finds the mean or average of the elements of X. If X is a matrix, `mean(X)` returns the mean of the elements in each column of X.
`median(X)`	Finds the median value of X. If X is a matrix, `median(X)` returns the median value of the elements of each column in X.
`round(X)`	Rounds each element of X to the nearest integer.
`size(X)`	Returns a vector of the number of rows and columns of X.
`std(X)`	Finds the standard deviation value of the elements of X. If X is a matrix, `std(X)` returns the standard deviation of each column of X.
`var(X)`	Finds the variance value of X. If X is a matrix, `var(X)` returns the variance of the elements of each column of X.

EXAMPLE   14-8

Week	Number of Fatal Accidents
A	190
B	202
C	179
D	211
E	160
F	185
G	172
H	205
I	177

You are studying the number of fatal accidents that occur during different times of the day. Using MATLAB and the data shown, determine the mean, median, variance, and standard deviation. The data represent the number of accidents between midnight and 6:00 a.m. for nine consecutive weeks.

Given the accident data:

```
>> accidents=[190 202 179 211 160 185 172 205 177];
```

Mean:

```
>> mean_accidents=ceil(mean(accidents))
mean_accidents =
   187
```

Median:

```
>> median_accidents=median(accidents)
median_accidents =
   185
```

Variance:

```
>> variance_accidents=var(accidents)
variance_accidents =
   281.9444
```

Standard deviation:

```
>> stdev_accidents=std(accidents)
stdev_accidents =
   16.7912
```

EXAMPLE   14-9

The following outline gives the steps necessary to use MATLAB for basic statistical analysis of a data set. This is presented as an example of the high and low temperatures during October 2006. The data are given in a starting MATLAB file in the online materials. This discussion focuses on the high temperatures, but would hold for the low temperatures as well.

1. **Input the data.** *The first column is simply an identifier (in this case, the date). The second and third columns contain the actual raw data of high and low temperatures for each day, respectively. This step has already been completed in the provided file.*

2. **Decide on the bin range.** *A rule of thumb is that the number of bins needed is approximately equal to the square root of the number of samples. While it is obvious in this example how many total samples are needed, the length function is often very useful.*

```
>> number_bins=round(sqrt(length(HighTemp)))
number_bins=
    6
```

The following histograms are created using four different types of data sets and the `histogram` function.

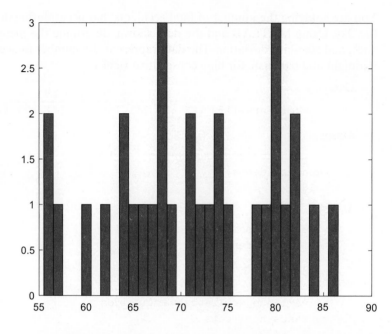

- **One argument specified (data set):** By default, `histogram` automatically determines what the appropriate size for each bin should be with a uniform width and displays the resulting distribution as a figure. Note that the figure shown is not a proper plot.

      >> histogram(HighTemp)

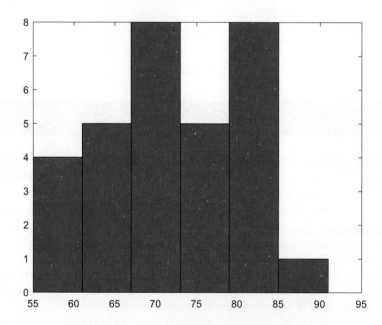

- **Two arguments specified (data set, number of bins):** `histogram` separates the data into the specified number of bins and displays the histogram in a figure. MATLAB divides the range between the minimum and maximum values of the data set into the number of bins specified by *the user. Note that MATLAB does not line up the bins on powers of 5, 10, 100, etc., in order to create a reasonable graphing axis. Note that the figure shown is not a proper plot.*

```
>> histogram(HighTemp, 6)
```

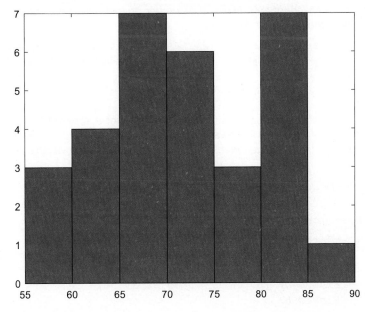

- **Two arguments specified (data set, edges of bins):** `histogram` *separates the data into bins specified by the edge values of each bin provided in the vector passed to the function and displays the histogram in a figure. This is the preferred method, since this result matches our results in Excel. Note that in the Edges variable, the list of numbers read left to right contain all of the "edges" of the bins, so the first bin contains the count of all temperatures in the 55 to 60 range, the next bin contains the count of all temperatures in the 60 to 65 range, and this continues until the final range of 85 to 90.*

```
Edges = [55 60 65 70 75 80 85 90];
>> histogram(HighTemp,Edges)
```

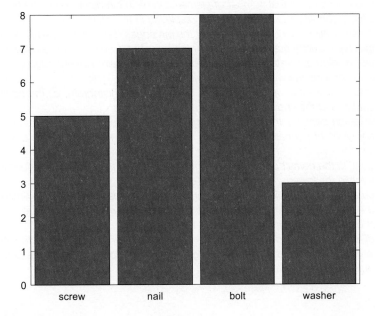

- **One argument specified (category data set):** `histogram` *offers the special capability of easily assessing the visualization of categorical data sets. For example, assume we have a data set containing a list of fasteners, where each fastener is represented by a different number—1 is screw, 2 is nail, 3 is bolt, 4 is washer. Given an array called fasteners, which is shown below, we could quickly count and display the number of each using the categories function in conjunction with the histogram function.*

```
>> fasteners = [1 2 3 3 3 2 1 2 4 2 3 1 4 3 2 1 1 4 3 3 2 3 2];
>> C=categorical(fasteners,[1 2 3 4],{'screw','nail','bolt',
'washer'});
>> histogram(C)
```

In our discussion of the `histogram` *function, we have not addressed what data the* `histogram` *function will return if we assign it to a variable. For example, if we save the result of* `histogram` *using the edges of each bin and save the result of the function call to the variable N, we see the following result.*

```
>> N=histogram(HighTemp,Edges)
N =
    Histogram with properties:
        Data: [1x31 double]
        Values: [3 4 7 6 3 7 1]
        NumBins: 7
        BinEdges: [55 60 65 70 75 80 85 90]
        BinWidth: 5
        BinLimits: [55 90]
    Normalization: 'count'
        FaceColor: 'auto'
        EdgeColor: [0 0 0]
```

The value returned in N is a MATLAB object that contains several attributes containing information about the visualization:

- *Data contains information about the data used in the creation of the histogram figure. In most situations, it will describe the number of elements and the data type of the values.*
- *Values contains the actual number of elements in each bin displayed in the histogram.*
- *NumOfBins contains the number of bins displayed in the histogram.*
- *BinEdges contains the numerical value of the bin edges displayed in the histogram. In our previous example, we explicitly stated the edges of the bins, but if we had used a different method for automatically binning our histogram, this attribute would contain different edge values.*
- *BinWidth contains the width of each bin in the histogram*
- *BinLimits contains the upper and lower bounds of the histogram, or the beginning value of the leftmost bin and the ending value of the rightmost bin.*
- *Normalization contains the type of histogram normalization used in the visualization. Many of the different types of normalization available are beyond the scope of this text, but values of interest include:*
  - *'count' is the default option; the height of each bar is the number of values in the data set that fall within the lower and upper values of the bin.*
  - *'countcum' is a cumulative count of the number of items in each range, where each bin contains the number of items in the range plus the count of items in the previous bin. The height of the last bin is the total number of items in the analysis.*
  - *'cdf' is a cumulative distribution function estimation, which is similar to the 'countcum' cumulative count, except rather than showing a cumulative count, it displays a percentage of the distribution from 0 to 1 (or 0% to 100%).*

- *'FaceColor' and 'EdgeColor' are both attributes that show the color of the bars and edges of the bars in the histogram.*

A key feature of the MATLAB object attributes on the histogram object is that the attributes are not read only—any parameter can be manipulated, and the corresponding figure updates in real time. For example, if we wanted to change the number of bins on the active histogram figure to use 10 bins, we could type the following expression, and the figure will update and all of the associated attributes for the histogram will automatically recalculate:

```
>> N.NumBins=10
N =

  Histogram with properties:

        Data: [1x31 double]
      Values: [3 1 4 5 4 4 2 6 2 0]
     NumBins: 10
    BinEdges: [55 58.5000 62 65.5000 69 72.5000 76 79.5000
      83 86.5000 90]
    BinWidth: 3.5000
   BinLimits: [55 90]
Normalization: 'count'
   FaceColor: 'auto'
   EdgeColor: [0 0 0]
```

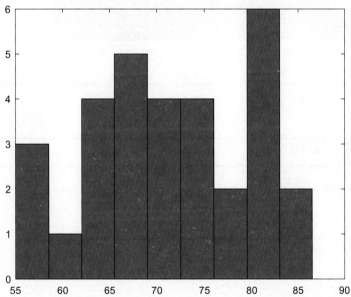

Note that all of the attributes like BinWidth, Edges, and Values all changed after we modified the total number of bins.

The cumulative distribution function (CDF) is created in MATLAB by the following procedure. For this situation, we will once again consider the histogram of high temperatures with manually specified bin edges.

```
Edges = [55 60 65 70 75 80 85 90];
>> histogram(HighTemp,Edges)

< >
```

1. *Create the cumulative sum of the histogram data using the '`cdf`' parameter.*

As we have seen previously, one approach we could take would be to save the result of the histogram function into a variable, and then change the Normalization parameter to 'cdf' to display the cumulative distribution function. MATLAB does provide a shortcut for manually specifying parameter values as arguments to the histogram function, as shown in the following code.

```
>> histogram(HighTemp,Edges,'Normalization','cdf')
```

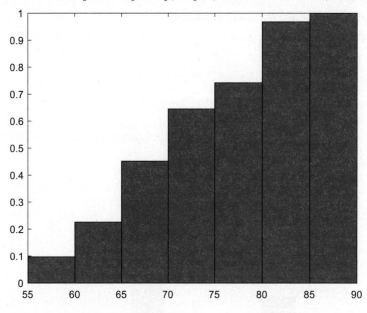

Comprehension Check	**14-10**

Repeat this analysis, using the daily low temperatures during October 2006.

IN-CLASS ACTIVITIES

ICA 14-1

This exercise includes the measurement of a distributed quantity and the graphical presentation of the results. You are to determine how many flexes it takes to cause a paper clip to fail.

Test the bending performance of 20 paper clips by doing the following:

- Unfold the paper clip at the center point so that the resulting wire forms an S shape.
- Bend the clip back and forth at the center point until it breaks.
- Record the number of flexes required to break the clip.

On a copy of the following table, record the raw data for the paper clips you break. Then, summarize the data for the team by adding up how many clips broke at each number of flexes. Each team member should contribute 20 data points unless otherwise instructed by your professor. Analyze the data using one of the methods (a) through (c), as specified by your instructor. Create a histogram with an appropriate bin size and a CDF, using the data collected and creating the graphs:

(a) by hand
(b) using Excel
(c) using MATLAB

Paper clip flexing data

Paper Clip	Flexes to Break
1	
2	
3	
4	
5	
6	
7	
8	
9	
10	
11	
12	
13	
14	
15	
16	
17	
18	
19	
20	

Summary of data

No. of Flexes	No. of Clips

ICA 14-2

For the following pressure data, recorded in units of pound-force per square inch, answer the following questions.

3	42	6	45	30
18	9	3	54	

(a) What is the mean of the data?
(b) What is the median of the data?
(c) What is the variance of the data?
(d) What is the standard deviation of the data?

ICA 14-3

A technician tested two temperature probes by inserting their probes in boiling water, recording the readings, removing and drying the probes, and repeating the process. The results follow, giving temperature reading in degrees Celsius.

(a) What is the mean of each probe?
(b) What is the median of each probe?
(c) What is the variance of each probe?
(d) What is the standard deviation of each probe?

Probe 1	90	89	90.5	92	89.5	91	91.5
Probe 2	98	102.5	104	100	93	94	106

ICA 14-4

One of the NAE Grand Challenges for Engineeering is **Develop Carbon Sequestration Methods**. According to the NAE website: "In pre-industrial times, every million molecules of air contained about 280 molecules of carbon dioxide. Today that proportion exceeds 380 molecules per million, and it continues to climb. Evidence is mounting that carbon dioxide's heat-trapping power has already started to boost average global temperatures. If carbon dioxide levels continue upward, further warming could have dire consequences, resulting from rising sea levels, agriculture disruptions, and stronger storms (e.g., hurricanes) striking more often."

The Mauna Loa Carbon Dioxide Record is the longest continuous record of atmospheric concentrations of carbon dioxide (CO_2), the chief greenhouse gas responsible for global climate warming. These data are modeled as the Keeling curve, a graph showing the variation in concentration of atmospheric CO_2 based on measurements taken at the Mauna Loa Observatory in Hawaii under the supervision of Charles David Keeling. It is often called the most important geophysical record on Earth and has been instrumental in showing that mankind is changing the composition of the atmosphere through the combustion of fossil fuels.

The Keeling curve also shows a cyclic variation in each year corresponding to the seasonal change in the uptake of CO_2 by the world's land vegetation. Most of this vegetation is in the northern hemisphere, where most of the land is located. The level decreases from northern spring onward as new plant growth takes CO_2 out of the atmosphere through photosynthesis and rises again in the northern fall as plants and leaves die off and decay to release the gas back into the atmosphere.

Data and wording for this problem set were obtained from www.esrl.noaa.gov/gmd/ccgg/trends/. Additional information on the Mauna Loa Observatory can be found at: http://scripps-co2.ucsd.edu/.

NOTE

In the graph, A and B are *not* drawn to scale, and the locations of C, D, and E are approximate. In other words, you cannot *guess* the value based upon the graph—you must *calculate* the value.

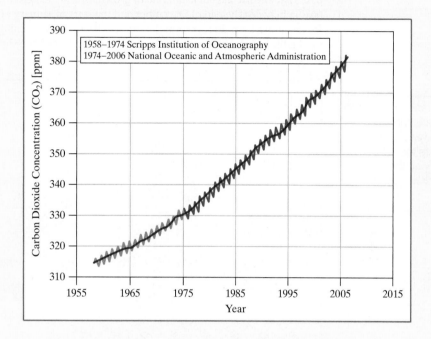

Examine the estimated increase in monthly CO_2 emissions for 2019, taken from the Mauna Loa data set. All values given are in parts per million [ppm] CO_2 as the difference between the December 2018 and the monthly 2019 reading.

2.38	3.64	−1.77	4.23	3.20	1.45
2.87	4.89	−2.16	4.92	−0.37	2.73

(a) What is the mean of these data?

(b) What is the median of these data?

(c) The variance of the data set shown here is 5.37 parts per million squared [ppm²]. What is the standard deviation of these data?

(d) The estimated annual growth rates for Mauna Loa are close, but not identical, to the global growth rates. The standard deviation of the differences is 0.76 parts per million per year [ppm/year]. What is the variance?

ICA 14-5

You use the data from the Mauna Loa observatory in the previous question to create the following histogram and CDF. These data reflect the observed yearly increase in CO_2 emissions for the past 51 years. The annual mean rate of growth of CO_2 in a given year is the difference in concentration between the end of December and the start of January of that year. If used as an average for the globe, it would represent the sum of all CO_2 added to, and removed from, the atmosphere during the year by human activities and by natural processes.

NOTE

In the graph, the locations of A–E are approximate. In other words, you cannot *guess* the value based upon the graph—you must *calculate* the value.

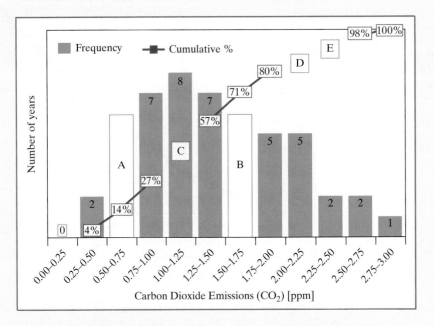

(a) What is the value of point A?
(b) What is the value of point B?
(c) What is the value of point C?
(d) What is the value of point D?
(e) What is the value of point E?

ICA 14-6

Polyetheretherketone (PEEK)™ polymers are resistant to both organic and aqueous environments; they are used in bearings, piston parts, and pumps. Several tests were conducted to determine the ultimate tensile strengths in units of megapascals [MPa]. The following CDF shows results from 320 points.

(a) What is the frequency value of A on the chart?
(b) What is the frequency value of B on the chart?
(c) What is the frequency value of C on the chart?
(d) What is the frequency value of D on the chart?
(e) What is the frequency value of E on the chart?

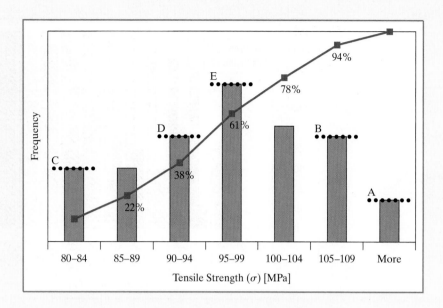

ICA 14-7

A technician tested a temperature probe by inserting it in boiling acetic acid (theoretical boiling point is 118 degrees Fahrenheit), recording the readings, removing and drying the probe, and repeating the process. The data are shown in the following table.

Temperature, Probe 1 (T) [°F]	122	120	107	115	107	122	122
Temperature, Probe 2 (T) [°F]	102	92	97	107	92	Missing Point	

(a) Determine the mean of probe 1.
(b) Determine the median of probe 1.
(c) A second probe was tested, yielding a mean of 97 degrees Fahrenheit and a median of 94.5 degrees Fahrenheit. If the data from probe 2 are as shown in the table, determine the missing data point.
(d) If a probe has a standard deviation of 10 degrees Fahrenheit, what is the variance of the probe?

ICA 14-8

A technician tested two temperature probes by inserting them in boiling water (theoretical boiling point is 100 degrees Celsius), recording the readings, removing and drying the probe, and repeating the process. The CDF for both probes is shown in the following graph.

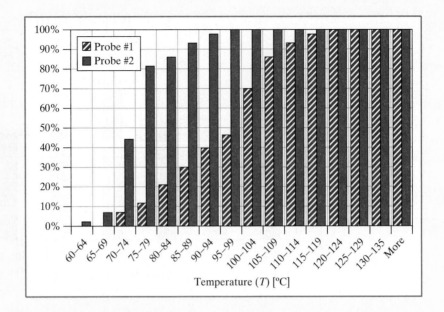

(a) Using this information, which probe would you choose to use? Explain why.
(b) How would you use the probe you chose to ensure that you found the correct boiling point?
(c) Which probe has the higher standard deviation?

ICA 14-9

During November, the heating system in your apartment appeared to be broken. To prove this, you record the following daily high temperatures in degrees Fahrenheit, taken every other day:

58	54	60	58	55	55	60	60
86	80	85	85	82	85	91	93

(a) Calculate the mean and median of the data.
(b) Draw the associated histogram and CDF for the data by hand; plot temperature on the abscissa and days on the ordinate.
(c) When you take your complaint to the apartment manager, he fails to see the problem; according to his heating bill, your apartment had an average temperature of 72 degrees Fahrenheit. Show both the mean and the median on the graph drawn for part (b). Which is a better presentation of the data, part (a) or part (b)? Justify.

ICA 14-10

You are assigned to inspect metal-composite beam trusses for a new bridge being built over a nearby lake. The manufacturer has run a prototype set of 500 beams and conducted strength tests, which you consider to be the baseline case, shown by the solid line in all graphs. Examine the graphs, and explain the changes to the baseline curve observed in the dashed line by choosing a cause from the following list. The strength of the beam is shown on the abscissa.

(a) The manufacturer tested 1000 beams instead of 500 beams.
(b) A reinforcing coating was used on a sample of 500 beams.
(c) The manufacturer upgraded processing equipment to lower the variability of the metal-composite strength.

Reading No.	Temperature (T) [°C]	Reading No.	Temperature (T) [°C]
1	100	11	103
2	105	12	101
3	106	13	100
4	97	14	98
5	98	15	97
6	95	16	96
7	101	17	104
8	100	18	102
9	96	19	95
10	105	20	101

ICA 14-16

The following data were collected from a manufacturing process involving reactor temperature measured in degrees Celsius. The following values are desired: average = 100 degrees Celsius; standard deviation = ±5 degrees Celsius.

Graph the data on a control chart. A blank grid has been provided online; you may use this grid, or use graph paper as directed by your instructor. Be sure to clearly indicate the zones of control.

Using the eight SPC rules, determine whether the process is in statistical control. If it is not in statistical control, indicate which rule or rules are violated and list the part numbers that violate those rules.

Reading No.	Temperature (T) [°C]	Reading No.	Temperature (T) [°C]
1	101.0	11	97.5
2	103.5	12	100.0
3	98.5	13	92.0
4	100.5	14	97.0
5	96.5	15	103.0
6	102.5	16	103.0
7	108.0	17	88.0
8	100.0	18	100.5
9	102.0	19	102.5
10	104.0	20	98.5

REVIEW QUESTIONS

1. The following table lists the number of resin patio furniture chairs rejected for defects during random testing over the course of a week on a manufacturing line. Four samples of 20 chairs are pulled each day. Use the following data to generate a histogram and CDF in Excel.

0	1	0	0	2	0
0	2	0	1	0	2
0	0	0	3	2	1
1	2	35	0	0	
0	0	1	2	1	

2. Repeat the analysis in Review Question 14-1, using MATLAB and the data available online.

3. An Excel worksheet, titled "Midterm Data," is available online. Our team is charged to analyze the actual impact of using a grading scale that isn't a traditional 10-point grading scale on a single midterm exam. Use the data provided to determine the following:

   (a) Class mean and median.
   (b) Class standard deviation.
   (c) We want to look at the use of a modified 5 letter scale, 8-point scale. Draw a histogram and CDF, based on the letter grade ranges given in the following table. After the histogram is created, change the bin labels to be the letter grade rather than the numerical value.

Grade	F	D	C	B	A
Minimum	0	69	77	85	93
Maximum	68	76	84	92	100

Use the charts to determine the following:

   (d) How many students received a C on the exam?
   (e) What percentage of students received an A on the exam?
   (f) What percentage of students received a passing grade (C, B, or A) on the exam?

4. Repeat the analysis in Review Question 14-3, using MATLAB and the data available online.

5. A company that fabricates small, custom machines has been asked to generate a machine that throws darts at a dart board as precisely and accurately as possible. To assess the precision and accuracy of each proposed design, the engineers build a model and record the distance from the bulls-eye of the dart board to the location of each dart thrown—both the straight-line distance (A) and the horizontal (B) and vertical (C) distances are recorded separately with regard to the bulls-eye, as demonstrated in the figure. The engineers throw 15 darts with their prototype machine and record the three data points for each dart.

   Using the data collected for a design in the starting workbook, create a histogram and a CDF in Excel for the straight-line distance (A), as well as the horizontal (B) and vertical (C) distances, and determine which graph or graphs are better for assessing the performance of the design if we were interested in (a) if the machine is throwing accurately to hit the bulls-eye or (b) if the machine needs to be calibrated (or adjusted) to correctly hit the bulls-eye. Justify your answer with a few sentences about why you selected the graph or graphs.

6. Repeat the analysis in Review Question 14-5, using MATLAB and the data available online.

7. This information was taken from the report of the EPA on the U.S. Greenhouse Gas Inventory (http://www.epa.gov).

"Greenhouse gas emission inventories are developed for a variety of reasons. Scientists use inventories of natural and anthropogenic emissions as tools when developing atmospheric models. Policy makers use inventories to develop strategies and policies for emission reductions and to track the progress of those policies. Regulatory agencies and corporations rely on inventories to establish compliance records with allowable emission rates.

In nature, carbon is cycled between various atmospheric, oceanic, biotic, and mineral reservoirs. In the atmosphere, carbon mainly exists in its oxidized form as CO_2. CO_2 is released into the atmosphere primarily as a result of the burning of fossil fuels (oil, natural gas, and coal) for power generation and in transportation. It is also emitted through various industrial processes, forest clearing, natural gas flaring, and biomass burning."

The EPA website provides data on emissions. The data found in the file online were taken from this website for the year 2001 for all 50 states and the District of Columbia.

(a) Use the data provided in the starting file to create a histogram with an appropriate bin size; use Excel.

(b) Determine the mean and median of the data.

(c) Which value more accurately describes the data? Indicate your choice (mean or median) and the value of your choice. Justify your answer.

8. Repeat the analysis in Review Question 14-7, using MATLAB and the data available online.

(a) Use the data provided in the starting file to write a program to determine an appropriate bin size. Use the bin size to continue the program and create a histogram of the data.

(b) Determine the mean and median of the data. Write a formatted output statement to the Command Window with this information.

(c) Which value more accurately describes the data? Indicate your choice (mean or median) and justify your answer using a comment statement at the end of your program file.

9. The Excel data provided online was collected by Ed Fuller of the NIST Ceramics Division in December 1993. The data represent the polished window strength, measured in units of kilopounds per square inch [ksi], and were used to predict the lifetime and confidence of airplane window design. Use the data set to generate a histogram and CDF in Excel (http://www.itl.nist.gov/div898/handbook/eda/section4/eda4291.htm).

10. Repeat the analysis in Review Question 14-9, using MATLAB and the data available online.

11. Choose *one* of the following options and collect the data required. For the data source you select, do the following using the analysis in Excel.

* Construct a histogram, including justification of bin size.
* Determine the mean, median, variance, and standard deviation values.
* Construct a cumulative distribution function.

(a) On a campus sidewalk, mark two locations 50 feet apart. As people walk along, count how many steps they take to go the 50 feet. Do this for 125 individuals.

(b) Select 250 words at random from a book (fiction). Record the number of letters in each word. Alternatively, you can count and record the words in 250 sentences.

(c) Go to one section of the library, and record the number of pages in 125 books in that same section.

(d) Interview 125 people to determine how far their home is, in miles, from the university.

12. Repeat the analysis in Review Question 14-11, using MATLAB.

PART 4

Programming Prowess

Chapter 15

MATLAB Basics

15.1 Variable Basics

15.2 Numeric Types and Scalars

15.3 Vectors

15.4 Matrices

15.5 Character Strings

15.6 Cell Arrays

15.7 Structure Arrays

Chapter 16

Algorithms, Programs, and Functions

16.1 Algorithms

16.2 Programs

16.3 Functions

16.4 Deriving Mathematical Models

16.5 Debugging MATLAB Code

Chapter 17

Input/Output in MATLAB

17.1 Input

17.2 Output

17.3 Plotting

17.4 Trendlines

17.5 Microsoft Excel I/O

Chapter 18

Logic and Conditionals

18.1 Algorithms Revisited—Representing Decisions

18.2 Relational and Logical Operators

18.3 Logical Variables

18.4 Conditional Statements in MATLAB

18.5 Application: Classification Diagrams

18.6 switch STATEMENTS

18.7 Errors and Warnings

Chapter 19

Looping Structures

19.1 Algorithms Revisited—Loops

19.2 while LOOPS

19.3 for LOOPS

Computers are controlled by software (programs) that can be designed in a variety of programming languages. Computer programs are a translation of what you want to accomplish into something the computer can understand, so the term "programming language" is particularly appropriate. Some computer programs are installed permanently or temporarily on computer chips, and others are installed on a variety of other media, such as hard drives.

Computers relentlessly produce a particular result given a particular set of input conditions. It can be frustrating when you make a simple mistake in a computer program—the computer will do exactly what you tell it to do, even if your mistake would be obvious to a person.

The biggest difference between a computer and a person is that you can ask a person open-ended questions—questions like design questions that can have many answers. Computers can only process questions that have a single correct answer.

This makes the process of programming a computer a bit like trying to ask another person to solve a problem when the person is on the other side of a wall, and you can communicate only by passing them slips of paper asking questions that can have only one correct answer and waiting for the person to pass back a slip of paper with the answer on it.

Learning Objectives

The overall learning objectives for this part include the following:

Chapter 15:

- Understand the various methods of storing information in MATLAB.
- Perform basic operations on data stored in MATLAB.

Chapter 16:

- Define the scope of a problem and create a written or graphical algorithm to solve the problem.
- Write MATLAB programs and/or functions to solve engineering problems.
- Read and interpret MATLAB programs written by others.
- Use MATLAB to determine mathematical models for experimental data.
- Debug a program to identify and correct different types of errors.

Chapter 17:

- Write input statements to allow the user to interact with the MATLAB environment.
- Write output statements to inform the user of program outcomes.
- Create graphs and use trendlines to enhance problem solving and communication of results.
- Transfer data between Microsoft Excel and MATLAB.

Chapter 18:

- Use conditional statements and `switch` statements to automate decision making.
- Use error and warning statements to aid the user in program execution.

Chapter 19:

- Use looping structures (`while` and `for`) to eliminate large blocks of repetitive code.

If a computer always produces the same result every time given the same input conditions, then why does my computer crash sometimes when I am doing something that should work?

The computers you use are simultaneously running a large number of complicated computer programs, including the operating system, background programs, and whatever programs you have started intentionally. Sometimes these programs compete for resources, causing a conflict. Other times, programs are complicated enough that the "input conditions," including the configuration of data in memory and on the hard disk, the time on the system clock, and other factors that change all the time while the computer is running, create a combination of circumstances that the programmers never anticipated and so did not include programming code to handle, and the system crashes.

**WISE WORDS: WOULD YOU CONSIDER YOUR CURRENT POSITION TO BE
"PURE ENGINEERING," A "BLEND OF ENGINEERING AND ANOTHER FIELD,"
OR "ANOTHER FIELD?"**

I always feel my work is not "pure engineering," but rather often a blend of engineering, sales, accounting, research, inspection, and maintenance.

E. Basta, Material Engineer

I would consider my career in another field from engineering, however, highly reliant on my engineering background. As a management consultant, I have to break down complex problems, develop hypotheses, collect data I believe will prove or disprove the hypotheses, and perform the analysis. My focus area is companies who develop highly engineered products.

M. Ciuca, ME

My position is mostly pure engineering.

E. D'Avignon, CpE

I work in a blend of engineering and business. I spend most of my time working on business-related activities—forecasting, variance reporting, and timing/work decisions—but I also have to work closely with our field engineers and understand our project scopes. I use both my business and engineering knowledge on a daily basis—without each, I would not be able to succeed at my job.

R. Holcomb, IE

It is definitely a blend of engineering and law with a heavy dose of technical writing. It takes the thinking of an engineer or scientist to truly comprehend the inventions and the skill of a writer to convey the inventor's ideas in written and image terms that others will understand (including juries of lay people). It takes the thinking of a lawyer to come up with creative strategies and solutions when faced with a certain set of facts.

M. Lauer, EnvE

My current position is definitely a blend of engineering and at least one other field, but more like five other fields. I definitely use my engineering background in the way I think, the way I analyze data, how I approach problems, and how I integrate seemingly unrelated information together. The project management skills that I learned in engineering are helpful, too.

B. Holloway, ME

Even though my boss calls Hydrology "Voodoo Engineering," it is pure engineering.

J. Meena, CE

A blend of mechanical/aerospace engineering and human factors engineering—and management.

R. Werneth, ME

Some Advantages of Computers

Given our description of how computers work, it may sound to some as if computers are too simple to be useful. The value of programming is linked to a few important characteristics of computers.

- *Calculation speed:* Although computers can only answer analytical questions, they can answer such questions very quickly—often in a fraction of a microsecond. Computer programs can therefore ask the computer a lot of questions in a short time, and thus we can find the answer to more complicated problems by breaking down the complicated question into a series of simple questions.
- *Information storage:* In "Memory: Science Achieves Important New Insights into the Mother of the Muses" (*Newsweek*, September 29, 1986), Sharon Begley estimates that the mind can store an estimated 100 trillion bits of information. The typical computer has a small amount of storage compared to that, but computers are gaining. Where computers have a bigger advantage is that new information can be incorporated in a fraction of the time it takes a human to learn it.
- *Information recall:* Computers have nearly 100% recall of information, limited only by media failures. The human brain can be challenged to recall information in exactly the same form as it was stored. A computer, for example, can store and "remember" the first million digits of pi faster than a human can recite the first 10 digits.

15 MATLAB Basics

At the most simplistic level, MATLAB can be used like a calculator on steroids. Not only can it do everything your fancy calculator can, it can also do far more with a single simple command than your calculator can as we will see in this chapter. First, we need to learn about two things:

- The MATLAB interface—what you see on the screen when you open MATLAB.
- The basic TYPES of data that MATLAB can deal with, since this encompasses considerably more than simple numeric values.

The Basic MATLAB Interface

When you launch MATLAB, you will see a window similar to the one that follows unless the default arrangement has been modified. You can return to the default layout by opening the Layout menu in the tool ribbon and selecting Default if your window looks a lot different.

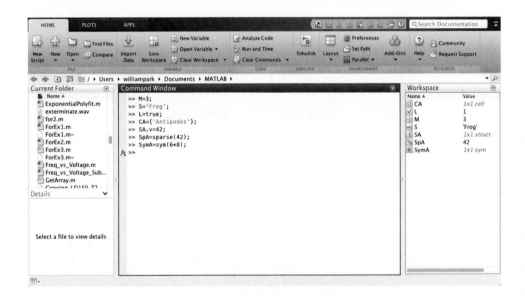

The MATLAB interface window is divided into several panes (or subwindows). These can be moved around, removed, others added, etc. We will discuss the use of the interface in more detail as we progress, but for now, we are going to focus on two of the panes.

The large pane in the middle is called the Command Window. This is where you can type instructions to tell MATLAB what you want it to do, and is where you will type instructions when you simply want to use MATLAB as a calculator, although you can do *many* other things in the Command Window. After MATLAB completes whatever task you told it to do, it displays the **Command Prompt**, two greater-than signs (>>). The Command Prompt is MATLAB's way of telling you, "OK, I am ready to accept a command." *If you do not see the prompt AFTER the last text in the Command Window, then either it is off the screen (often to the right) or MATLAB is busy doing something else such as running a really long program and cannot accept another command at that time.*

The smaller pane on the right is called the **workspace**. The workspace shows all of the variables (defined below) in use at the current time.

We will discuss both of these windows in more detail shortly, but first we will discuss a feature of considerable significance.

Using MATLAB to Learn MATLAB

The two functions `help` and `doc` can be invaluable in learning what various MATLAB functions do and how they are used (their syntax). They can also help you explore the often labyrinthine sets of built-in functions to discover new and useful features of the language.

To illustrate the use of the `help` function, at the prompt (>> in the Command Window) type `help sqrt`. A short explanation of `sqrt` appears, along with the syntax used. Near the end of the information is a section that says "See also" followed by a list of hyperlinks to related functions.

To illustrate the use of the `doc` function, at the prompt in the Command Window type `doc sqrt`. This launches the documentation browser and displays information on the specified function. Not only does the documentation browser give you much more information about a function than using `help` does, usually including examples, it also makes it relatively easy to explore MATLAB. In the following figure, note the following features:

- The **Main Information** window on the right. This area typically lists the function and its basic syntax, a description of what the function does, examples of using the function, and a few other items. Of particular note is See Also near the very bottom, which lists closely related functions.
- The **Contents** outline on the left. This shows an outline of where within the structure of the MATLAB documentation the particular item being viewed is found. If the Contents outline is not visible, click the little outline icon (three horizontal lines) near the top left to expand it. In the example shown for the square root function, it resides within the group for Exponents and Logarithms, which is within the Elementary Math group, which is in turn in the Mathematics group. By clicking on one of these more-inclusive topics, you can explore related functions. At the bottom of the Contents outline are hyperlinks to each area within the selected function displayed in the Main Information window. Of particular note are Examples, from which you can learn a lot about using the function, and See Also.
- The **Search Box** near the top right of the Documentation window. Here, you can type function names or keywords to search for functions that do particular types of tasks.

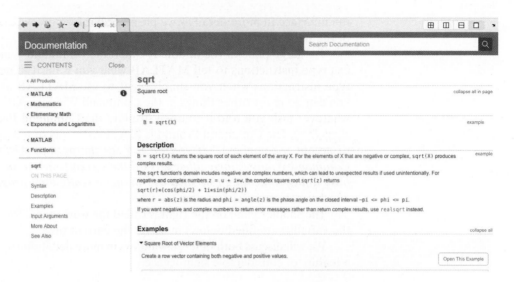

Table 15-1 lists a few functions that are useful in dealing with the MATLAB interface.

It is strongly recommended that you spend some time early in your study of MATLAB using the documentation browser to familiarize yourself with its organization. It will be really helpful in learning the language, but it takes practice to navigate.

Table 15-1 Useful MATLAB interface functions

Function	Description	Function	Description
clc	Clear the Command Window	help	Display basic information about a function in the Command Window
clear	Clear variables from the workspace	doc	Open the Documentation Browser
close	Close figure windows	save	Store a copy of workspace variables on the hard drive or other media
;	Do not display result of assignment operation in Command Window	load	Get stored variables from the hard drive or other media and place in the workspace

MATLAB Variables and Data Types

A variable in MATLAB is not like a variable in mathematics. In math, a variable is typically an unknown for which you wish to determine a value. In MATLAB, on the other hand, a variable is simply a label for a container (or location) in which information is stored. Variables can be classified according to the type of information they hold. In the workspace, the data type of each variable is represented by an icon shown in the leftmost column of the workspace beside the variable name. Table 15-2 illustrates the different variable classifications in MATLAB.

Table 15-2 Variable classifications in MATLAB

ICON	Data Type	Description
	Numeric array	Scalars, vectors, and matrices containing numbers
	Character array	Text strings. Each character occupies one element
	Logical array	0 means false; 1 means true Discussed in Chapter 18
	Cell array	Contains mixed data types. Indexed numerically.
	Structure array	Contains mixed data types. Data organized into named fields.
	Sparse array	Saves memory by storing only nonzero elements Not covered in this text
	Symbolic array	Allows symbolic mathematics within MATLAB Not covered in this text

15.1 Variable Basics

LEARN TO: Remember the naming rules for variables
Understand the use of = as the assignment operator

Every variable in MATLAB must have a name. This allows us to keep track of large amounts of information and gives a means whereby our code can use the stored information relatively easily. MATLAB names must follow certain rules to avoid unintended side effects.

NAMING RULES

- All names must consist only of letters, numbers, and the underscore character (displayed by holding Shift and typing a hyphen).
- Names can *only* begin with alphabetic letters (no numbers or special characters).
- Names cannot be longer than 63 characters.
- Names in MATLAB are case sensitive: `Frog` and `frog` are different.
- Names cannot have the same name as any other identifier (program name, function name, built-in function, built-in constant, reserved word, etc.), because MATLAB's order of execution may prevent the program from finding the correct item.

NOTE

A common error in naming variables (or other things) in MATLAB is to include a space character. MATLAB names cannot have a space in them—use an underscore instead.

These rules apply to *any* named item in MATLAB, not just variables. We will give further detail in the next chapter.

Always choose names that have meaning within the context of the program, not just random letters. We cannot overemphasize the importance of well-chosen variable names in creating self-documenting code: code that explains itself. For example, if a variable contained the diameter of a bolt, the name `BoltDiam` is immediately recognizable for what it contains; a randomly chosen variable such as `X1` or `Q` has no meaning in the context of the program.

Comprehension Check | **15-1**

Which of the following are valid MATLAB variable names? For those that are not valid, explain why they are invalid.

(a) `m`
(b) `mass6`
(c) `4mass`

(d) `m&m©`
(e) `Dog Breath`
(f) `MyFile-docx`

(g) `clear`
(h) `ReactorYield`

The Assignment Operator

Before beginning our discussion of data types in MATLAB, we need to introduce the concept of the **assignment operator**. The purpose of an assignment operator is to tell the computer to put a value into a variable. Perhaps unfortunately, the most commonly used assignment operator in modern programming languages is =, or the equal sign. We say unfortunately because this notation often confuses novices. In mathematics, a statement such as A=A + 1 is an equation and has exactly two solutions for A: infinity and negative infinity. However, in programming this means "add one to whatever is in variable A, then place the result back in A."

NOTE

To avoid the dual problem of mistaking = for "equals" and to explicitly indicate the direction of the movement of information, some languages use a left-pointing arrow (or other symbols) for the assignment operator.

It is important to note in MATLAB the = symbol should be read as "is assigned the value of." Thus, the programming statement V2=V2+4 is read " V2 is assigned the value of (the original value of) V2 plus four." Note that the original value of V2 is lost in this process.

It is also very important to note that the transfer of data in an assignment statement is always from right to left.

Consider the programming statement Ex1=V3−V4^P2.

When the computer executes this statement, it recognizes that it is an assignment statement. Therefore it first looks at the expression to the right of the equal sign and determines its value: Raise the value in V4 to the value in P2 and subtract that from the value in V3. Next, it looks to the left of the equal sign to find the destination, thus placing the calculated result into the variable Ex1.

It is also important to understand that whatever appears to the left of the assignment operator MUST be a variable, or as we will see later, a group of variables. For example, you will get an error if you type A+1=A since you cannot have a computation on the left of the equal sign. Also, any variables that appear on the right must already be defined before the statement is executed.

Comprehension Check | **15-2**

Which of the following assignment statements are valid? For those that are not valid, explain why. For those that are valid, show what value is placed into which variable. Assume the following variables (and no others) have already been defined: A=2 B=−1 C=3 D=2.5

Note that the priority of operators is the same as in Microsoft Excel.

(a) `X = A`

(b) `B = Y`

(c) `Z = A + 55`

(d) `W = A^B + C`

(e) `V = D − Q^2`

(f) `U + 13 = 65`

(g) `T^2 = 4`

(h) `Part#3 = C`

(i) `D = D^B`

(j) `A = 9999`

(k) `A + B = C − 5`

(l) `QPDoll = C^C`

Saving and Restoring Variables

For a variety of reasons, we sometimes would like to be able to save all or some of the variables in the workspace and then restore them later. For example, we might wish to archive results, we might want to use the same data in another program, or we might need to share a set of variables with colleagues. Some of the online files associated with this text, including some with the problems at the end of this chapter, were created using the following method.

Variables can also be saved as .txt files in ASCII format, but this is beyond the scope of this text. See the MATLAB documentation files for more information.

To save all variables in the workspace, type

```
save('filename')
```

In this case, the filename must have an extension of .mat on the end. If you do not include .mat, MATLAB will add it for you.

As an example, the command

```
save('TestVar.mat')
```

will save all of the variables currently in the workspace in a file named `TestVar.mat` placed in the current directory.

To place a set of stored variables into the workspace, type

```
load('filename')
```

Again, the filename must have an extension of .mat on the end. For example,

```
load('TestVar.mat')
```

will place all of the variables stored in `TestVar.mat` into the current directory.

To save only selected variables in a .mat file, use the format

```
save('filename', 'Var1', 'Var2', . . .)
```

As an example:

```
save('TV2.mat','Dens','Temp','Vol')
```

will store the three workspace variables `Dens`, `Temp`, and `Vol` in the file `TV2.mat`.

WARNING

If your workspace already has variables with the same names as any of the variables being loaded, they will be replaced, thus losing the contents in those variables prior to the load. No warning will be generated.

NOTE

The use of the single quotes around the filename and variable names will be made clearer in Section 15.5 discussing character strings.

You can also load only selected variables from a .mat file. The format is similar to the save format. For example, assume `TV2.mat` was created as in the previous example.

```
load('TV2.mat','Dens','Vol')
```

will place the two variables `Dens` and `Vol` from the file `TV2.mat` into the workspace.

The commands `save` and `load` also support wildcard variable names using the asterisk (*) symbol.

For example,

```
save('WC1.mat','Q*')
```

will save all workspace variables beginning with `Q` in the file `WC1.mat`.

```
load('WC2.mat','*3')
```

will load all variables stored in the file `WC2.mat` that end with `3`.

```
save('WC3.mat','*_*')
```

will save all workspace variables containing an underscore character in the file `WC3.mat`.

Comprehension Check | **15-3**

(a) Store all workspace variables in the file `TempData.mat`.
(b) Load all variables stored in `PressData.mat` into the workspace.
(c) Save the workspace variables `Pr1`, `Pr2`, and `Tmp3` in the file `PTData.mat`.
(d) Store all variables whose names end with "a1" in the file `a1Var.mat`.
(e) Load all variables stored in `VelData` whose names contain "XS" into the workspace.

15.2 Numeric Types and Scalars:

LEARN TO: Create numeric scalars in MATLAB
Perform basic scalar operations
Write mathematical expressions using scalar functions

Numeric data is exactly that—information consisting of one or more numbers. In MATLAB, as in most computer languages, there are actually several different types of numeric data. As an example, numbers can be classified as integers and nonintegers (floating-point numbers), and the two types are encoded differently using binary digits (1s and 0s) inside the computer. The major numeric types are listed in Table 15-3 and Table 15-4, but a complete discussion is beyond the scope of this introductory material, and these tables are merely included for completeness. For the moment, it is sufficient to note that the default numeric type in MATLAB is double precision, even if the value being stored has no fractional part.

Since the default numeric type in MATLAB is double precision, and doubles have the greatest range and, except for `int64` and `uint64`, also the greatest precision of all of the standard numeric types, why would you ever want to change to another format with less precision and/or range? Occasionally, when your application is processing gargantuan quantities of data, switching to either single precision or one of the integer formats will save memory space and/or increase execution speed. Since memory is cheap and processors are amazingly fast, this is seldom a significant factor in most engineering applications, and it is not worth dealing with the possible problems of increased round-off error, results beyond the range of the selected format, and other possible errors.

Table 15-3 Floating-point formats

Name	Bytes	Approximate Precision	Nonzero Minimum	Noninfinite Maximum
Double	8	16 decimal digits	$\pm 2.2251 \times 10^{-308}$	$\pm 1.7977 \times 10^{308}$
Single	4	7 decimal digits	$\pm 1.1755 \times 10^{-38}$	$\pm 3.4028 \times 10^{38}$

Table 15-4 Integer formats

	Signed			Unsigned		
Bytes	Name	Minimum	Maximum	Name	Minimum	Maximum
1	`int8`	−128	127	`uint8`	0	255
2	`int16`	−32,768	32,767	`uint16`	0	65,535
4	`int32`	−2,147,483,648	2,147,483,647	`uint32`	0	4,294,967,295
8	`int64`	-9.2234×10^{18}	9.2234×10^{18}	`uint64`	0	1.8447×10^{19}

A numeric variable can be categorized based on how many values are stored in that variable and how they are arranged. Most MATLAB variables containing only numeric data are considered to be matrices, hence the name MATrix LABoratory. We will separate these into three broad categories: scalars, vectors, and matrices. We will discuss vectors and matrices further in Section 15-3 and Section 15-4.

Scalars

NOTE

In MATLAB, a scalar is actually considered a 1×1 matrix

In this section we will discuss scalars and a variety of issues relating to scalars, although many of the concepts will carry over to our discussion of vectors and matrices.

A **scalar** is simply a single numeric value, such as 7, −35.12, 2.778×10^{15}, or $2.4 + 3.92i$.

In the simplest case, to define a scalar you simply use the assignment operator (=) to place the desired value into a variable. Remember that the location into which a value is being placed is ALWAYS to the left of the equal sign.

Examples of scalars include:

```
Num=34
Age=62
Root1=15.67−14.65i
BigNum=3.5E246
TinyNum=5.7E−105
```

COMPLEX NUMBERS

MATLAB has no difficulty dealing with imaginary and complex numbers. For example, if you type

`Q=5+sqrt(-9);`

Q will contain 5 + 3i where i = $\sqrt{-1}$

Predefined Constants

The following names are already defined as special constants in MATLAB. You should avoid using them for other purposes.

NOTE

j is used by electrical engineers to avoid confusion with electric current, since electric current usually uses *i* as the variable.

$$\texttt{pi} = \text{the constant pi } (\pi) \text{ to 15 decimal places.}$$
$$\texttt{i} = \sqrt{-1}$$
$$\texttt{j} = \sqrt{-1}$$
$$\texttt{inf} = \text{infinity}$$

Helpful Tip

If you wish to repeat a command previously typed in the Command Window, or make a small change to such a command, pressing the up arrow repeatedly will scroll back through the previous commands (called the Command History). Once the desired command is next to the prompt (>>), it may be edited if desired and re-executed.

Comprehension Check | **15-4**

Write MATLAB code to complete the following commands.

(a) Place the number 8 into the variable `Int2`.
(b) Place the number thirty five point seven into the variable `Real2`.
(c) Place the number 47.98×10^{56} into the variable `Big2`.
(d) Place the number 3×10^{-15} into the variable `Small2`.

Calculations with Scalars

For the most part, the mathematical operators used with scalars in MATLAB work the same way as they do in Microsoft Excel, including parentheses and priority of operators.

EXAMPLE | **15-1**

NOTE

Recall that variable names should have meaning in the context of their use. Here, DPM refers to Divide Plus Multiply, RTP refers to Raise To Power, and RTPP refers to Raise to Power with multiple elements in the Power.

What is stored in the variable when each line of the following MATLAB code is executed?

```
DPM=38/19 + 5*3;
```
    DPM *contains 17*
```
RTP=3^2 + 1;
```
    RTP *contains 10*
```
RTPP=3^(2 + 1);
```
    RTPP *contains 27*

Numeric Functions

There are hundreds of functions built into MATLAB for performing a wide variety of operations with numeric data. These are similar to the functions you learned about in your study of Excel, and in many cases even have the same name, such as `sin`, `sqrt`, or `round`. We will introduce quite a few functions throughout the subsequent sections. A small sample of common functions used with numeric values is shown in Table 15-5.

Table 15-5 Selected numeric functions

Function	Description	Function	Description
round	Round to specified decimal places or significant figures	sin	Sine of angle in radians
ceil	Round toward positive infinity	sind	Sine of angle in degrees
floor	Round toward negative infinity	acos	Inverse cosine in radians
fix	Round toward zero	acosd	Inverse cosine in degrees
sqrt	Determine the square root	deg2rad	Convert degrees to radians
nthroot	Determine the Nth root	exp	Exponential function
real	Real part of a complex number	log	Natural logarithm
imag	Imaginary part of a complex number	log10	Base 10 logarithm

NOTE

Even if a number has no fractional part, it will be stored as a double precision floating-point number. If you want a value to be stored in one of the integer formats, you must specifically tell MATLAB to do so.

Displaying Numbers

The default numeric display format in MATLAB is to show four decimal places unless the value has no fractional part. For numbers with fractional parts and a magnitude less than 1000 but greater than or equal to 0.001, MATLAB will display up to three digits to the left of the decimal point and four decimal places. For values outside of this range, MATLAB will display the value in scientific notation, with the most significant digit to the left of the decimal point and four decimal places.

100*pi	will display as 314.1593
1000*pi	will display as 3.1416e+03
pi/1000	will display as 0.0031
pi/10000	will display as 3.1416e−04

Values with no fractional part and a magnitude less than 10^{10} will display as integers. Larger values will display with five significant figures in scientific notation as above.

123123123	will display as 123123123
1231231231	will display as 1.2312e+09

You may change the default display format with the format function. For example, format long will change the default display format to show 15 decimal places instead of only four. More information about the format function can be found by typing help format or doc format in the Command Window.

Comprehension Check   **15-5**

(a) Calculate the two roots of the quadratic equation $3x^2 + 2x + 1$ using the quadratic formula and place them in R1 and R2. Note that MATLAB should do this calculation—do not calculate by hand and enter the resulting values.

(b) Calculate the tangent of 75 degrees and place the result in Trig2.

15.3 Vectors

A **vector** consists of numeric values organized into a single row or a single column. Examples include $[3.5 \;\; -38 \;\; 0 \;\; -1]$ and $\begin{bmatrix} -1 \\ 0 \\ 3.2 \end{bmatrix}$. **A vector is either an N × 1 matrix (column vector) or 1 × N matrix (row vector), where N is an integer greater than 1.**

Defining Vectors

As with scalars, the usual method for defining a vector is to use the assignment operator ($=$). There are a few immediate questions, however, since vectors have more than one value and may be organized as either rows or columns.

1. How do we tell MATLAB we are defining a vector and not a scalar?
2. How do we tell MATLAB whether we are creating a row or a column vector?
3. How do we distinguish the individual values from each other?

NOTE

A series of values separated by **spaces** or **commas** creates a row vector.

A series of values separated by **semicolons** creates a column vector.

First, when we wish to define a vector, the first character following the equal sign is [(open square bracket). Once we have entered all of the numbers, we end with] (close square bracket). It is also possible to define vectors using indexing, which is discussed later in this section.

The solution to the second question also solves the third: we separate the numbers by **delimiters**—either a space, a comma, or a semicolon. If we separate the individual values with spaces or commas, a row vector will be created. If we separate the values with semicolons, they will be placed in a column vector.

EXAMPLE   15-2

What is stored in the variable when each line of the following MATLAB code is executed?

```
Pair = [23 4.3];
```
    `Pair` *contains the row vector* [23 4.3]

```
Trio = [1, 4, 79857];
```
    `Trio` *contains the row vector* [1 4 79857]

```
CTrio = [0;0;97];
```
    `CTrio` *contains the column vector* $\begin{bmatrix} 0 \\ 0 \\ 97 \end{bmatrix}$

The Transpose (') Operator

If you need to convert a row vector to a column vector, or vice versa, use the apostrophe or single quote to transpose it.

Consider the following two lines of code:

```
V=[6, -9, 3];              VT = V';
```

V will contain $[6\ -9\ 3]$ VT will contain $\begin{bmatrix} 6 \\ -9 \\ 3 \end{bmatrix}$

Special Case: Linear Sequences

NOTE

A **colon** delimiter creates a sequence

It is fairly common to need a vector containing a set of equally spaced values, so MATLAB includes a shortcut method for creating such sequences.

If you wish to create a sequence of values in a row vector incrementing by one, such as 1, 2, 3, 4, 5, you can use the **colon operator** to separate the first and last values of the sequence.

EXAMPLE   15-3

What is stored in the variable when each line of the following MATLAB code is executed?

```
LSeq=1:5;
```

   LSeq *contains* [1 2 3 4 5]

   Note that the brackets are optional in this case. You could also enter LSeq=[1:5];

```
LSeq2=-2:4;
```

   LSeq2 *contains* [-2 -1 0 1 2 3 4]

If you wish a sequence that increments by a value other than 1, you may place the increment value between the first and last values, separated by colons.

EXAMPLE   15-4

What is stored in the variable when each line of the following MATLAB code is executed?

```
LseqA=5:15:80;
```

   LSeqA *contains* [5 20 35 50 65 80].

```
Frac=0.7:0.03:0.88;
```

   Frac *contains* [0.7 0.73 0.76 0.79 0.82 0.85 0.88]. *Fractional values are allowed.*

```
Rev=4:-3:-2;
```

   Rev *contains* [4 1 -2]. *Negative values are allowed.*

```
Miss=1:5:20;
```

   Miss *contains* [1 6 11 16]. *Note that the increment did not land on the final value (20). In cases like this, the sequence will stop on the last value BEFORE it goes beyond the stated end value.*

Comprehension Check   15-6

Each problem should be done with a single MATLAB statement:

(a) Create a column vector named `N4` containing the values 17, 34, −94, 16, and 0.
(b) Create a row vector named `Tiny` containing the values 3.4×10^{-14}, 9.02×10^{-23}, and 1.32×10^{-9}
(c) Create a row vector named `Ev` containing all even integers from 2 to 250.
(d) Create a column vector name `Tenths` containing the decreasing sequence $10, 9.9, \ldots 0.2, 0.1, 0$

NOTE

Whenever an error
message appears on the
screen, try to understand
what the message is
telling you. At first, some
of these messages will
seem cryptic and
incomprehensible, but
the more you practice
understanding them (ask
someone who knows
more than you!) the
easier your programming
tasks will become.

Calculations with Vectors

Basic calculations with vectors are somewhat more complicated than the same calculations with scalars.

Addition and Subtraction

When addition or subtraction involves a vector, there are two cases:

1. A scalar and a vector: The scalar is added or subtracted to each element of the vector.
2. Two equal size vectors: Corresponding elements of the two vectors are added or subtracted.

If you try to add two vectors that are not the same size, you will get angry red letters in the Command Window, indicating an error.

EXAMPLE   15-5

What is stored in the variable when each line of the following MATLAB code is executed?

Assume `V1=[2  5  9];    V2=[3  0];    V3=[4  -7  0];`

```
D1=V1-4;
D1 contains [-2  1  5].
S1=V1+V3;
S1 contains [6  -2  9].
E1=V1+V2;
```
Will return the message: `Error using +    Matrix dimensions must agree.`

Multiplication and Division

A complete discussion of vector multiplication will be deferred to the next section on matrices. For now, we will only consider cases similar to the two mentioned for addition and subtraction:

1. A vector multiplied or divided by a scalar: Each element of the vector is multiplied or divided by the scalar. NOTE: If you try to divide a scalar by a vector, you will get angry red letters.

NOTE

The **dot operator** indicates an element-wise operation

2. A vector multiplied or divided by an equal-length vector: Corresponding elements of the two vectors are multiplied or divided. NOTE: This requires the use of the **element-wise operator**, sometimes referred to as the **dot operator** since the actual operator is preceded by a period (or "dot") symbol, as shown below.

If you attempt an element-wise multiplication or division of two vectors that are not the same size, you will get angry red letters.

EXAMPLE 15-6

What is stored in the variable when each line of the following MATLAB code is executed?

Assume `V1=[2 5 9]; V2=[3 0]; V3=[4 -7 0];`

`Q1=V1/5;`

> `Q1` *contains* `[0.4 1 1.8]`.

`P1=V2*300;`

> `P1` *contains* `[900 0]`.

`P2=V1.*V3;`

> `P2` *contains* `[8 -35 0]`. *Note the dot before the asterisk (*`.*`* instead of *`*`*). This tells MATLAB that the multiplication is to be done on an element-wise basis.*

`Q2=V3./V1;`

> `Q2` *contains* `[2 -1.4 0]`. *Note the dot before the slash (*`./`* instead of *`/`*). This tells MATLAB that the division is to be done on an element-wise basis.*

`Q3=5/V1;`

> *Will return the message:* `Error using /    Matrix dimensions must agree.`

`P3=V1*V3;`

> *Will return the message:* `Error using *    Matrix dimensions must agree.`
>
> *The reasons for these errors will become clear when we discuss matrices in the next section.*

`Q4=V3/V1;`

> `Q4` *contains* −0.2445. *For an explanation of this seemingly mysterious result, refer to any general reference on matrix operations. For the moment just take note that vector division without the dot probably does not do what you expected.*

Powers

We will consider three cases here, all of which are element-wise operations.

1. A vector raised to a scalar.
2. A scalar raised to a vector.
3. A vector raised to a vector of equal size.

In all three cases, we will use `.^` (not simply `^`) to tell MATLAB to do an element-wise operation.

EXAMPLE   **15-7**

What is stored in the variable when each line of the following MATLAB code is executed?

Assume V1=[2 5 9];   V2=[3 0];   V3=[4 -7 0];

R1=V1.^3;

   R1 *contains* [8 125 729]   *Each element of* V1 *is cubed*

R2=2.^V3;

   R2 *contains* [16 0.0078 1]   *The number 2 is raised to each element of* V3

R3=V3.^V1;

   R3 *contains* [16 -16807 0]   *Raise each element of* V3 *to the corresponding element of* V1.

R4=V2.^V1;

   Will return the message: Error using .^   Matrix dimensions must agree.

Combined Operations

All of the standard rules, such as priority of operators and use of parentheses, apply to computations with vectors, the only difference being that both intermediate and final results may be vectors instead of scalars.

EXAMPLE   **15-8**

What is stored in the variable when the following line of MATLAB code is executed?

Assume V4=[2 1 -2]; V5=[3 -2 1.5];

R=(10-V4./V5).^2*3;

   R *contains* [261.33 330.75 385.33]

Sequence of operations:

1. V4 *is divided element-wise by* V5   [0.6667 −0.5 −1.3333]
2. *Each of these values is subtracted from 10*   [9.3333 10.5 11.3333]
3. *Each of those values is squared*   [87.1111 110.25 128.4444]
4. *Each of those values is multiplied by 3*   [261.33 330.75 385.33]

Comprehension Check   **15-7**

Assume a row vector named Vals has already been defined. Write a single MATLAB statement that will perform the following calculation using each element of Vals and leave the results in a vector named Comps: $C = (3V + 5)^4 - 16$ where V represents the individual values in Vals, and C represents the individual results in Comps. Example: if , Vals=[2 0 -1]; then Comps=[14625 609 0];

Functions Used with Vectors

All of the functions listed in Table 15-5 also work with vectors. In most cases, such as sqrt, sin, or round, these automatically perform the stated operation on each element individually. There are quite a few MATLAB functions that would seldom if ever be used with scalars but are very helpful in dealing with vectors. A few of these are listed in Table 15-6. You can obtain more information by typing either help or doc followed by the function name.

Table 15-6 Selected functions used with vectors

Function	Description
length	Find the number of elements in a vector. VERY useful!
find	Find the indices of values in a vector that meet some criterion. VERY useful.
sort	Sort elements in ascending or descending order.
rand	Create a vector of random numbers.
min	Find smallest value. Can also find location of smallest value in the vector.
max	Find largest value. Can also find location of largest value in the vector.
unique	List all values in a vector with no repetitions in result.
intersect	Find all values in both of two vectors, no repetitions in result.
union	Find all values in either of two vectors, no repetitions in result.

EXAMPLE 15-9

Assume V1=[2 4 0 7 0 -3 -5 0 -9 1] and V2=[5 -3 0 9 -2 -8 6 -1 4 1]

a. Place the indices of all nonzero elements of the vector V1 in the vector NotZero.
```
NotZero=find(V1) % find(X) finds non-zero elements in X
NotZero contains [1 2 4 6 7 9 10]
```

b. Place the indices of all elements in V1 that are greater than 2 in GT2.
```
GT2=find(V1>2) % finds cases where the inequality is true
GT2 contains [2 4]
```

c. Given two equal-length vectors V1 and V2, place the indices (n) of all elements for which V1(n) < V2(n) in V1Smaller
```
V1Smaller=find(V1<V2) % does an elementwise comparison
V1Smaller contains [1 4 7 9]
```

d. Place the values in V1 that are smaller than the corresponding values in V2 into Smaller.
```
Smaller=V1(V1Smaller); % use indices to acquire values
Smaller contains [2 7 -5 -9]   NOTE: Vector indexing will be discussed below.
```

Comprehension Check   15-8

a. Place the indices of all nonzero elements of the vector `TestVec1` into `NZLoc`.
b. Place the indices of all elements of the vector `TestVec1` that are less than −5 into `LTNeg5`.
c. Place the indices of all elements of the vector `TestVec1` that are less than the corresponding elements in `TestVec2` into `TV1_LT_TV2`. You may assume that `TestVec2` has the same dimensions as `TestVec1`.
d. For the elements found in part (c), replace all of these smaller elements in `TestVec1` with the larger values found in `TestVec2`. Example: `TestVec1=[2 3 7 5];` `TestVec2=[1 5 6 8];` then after this command, `TestVec1` would contain `[2 5 7 8]`. NOTE: You may wish to read ahead to the section on vector indexing to answer part d.

EXAMPLE   15-10

Random numbers are extremely useful for a variety of purposes such as code testing and simulations.

a. Create a row vector named `RVec42` containing 42 random numbers between 0 and 1.
`RVec42=rand(1,42);`

b. Create a column vector named `CVec500` containing 500 random numbers between 0 and 42.
`CVec500=42*rand(500,1);`

c. Create a row vector named `RVec1000` containing 1000 random numbers between −5 and 10.
`RVec1000=15*rand(1,1000)-5;`

d. Create a column vector named `CVecInt` containing 250 random integers between 1 and 75.
`CVecInt=randi(75,250,1);`

e. Create a row vector named `RVecInt` containing 10,000 random integers between −150 and 32.
`RVecInt=randi([-150 32],1,10000);`

Comprehension Check   15-9

(a) Create a column vector `CV1` containing 123 random numbers between 0 and 15.
(b) Create a row vector `RV1` containing 999 random numbers between −75 and 33.
(c) Create a row vector `RV2` containing 250,000 random integers between −75 and 33.

Other Notes on Creating Vectors

You can include calculations, functions, etc. inside the brackets when defining vectors.

EXAMPLE   15-11

NOTE

To **concatenate** means to combine end to end

1. Vectors can be combined (concatenated) by simply listing them in the desired order inside of square brackets. For row vectors, the individual items should be separated by spaces or commas; for column vectors they should be separated by semicolons. Use the single quote as needed to transpose a row to column before combining.

```
                    V1=[2 1 -2]; V6=[3;-2;1.5]; CV=[V4,V6'];
                    CV contains =[2 1 -2 3 -2 1.5]
```

2. You wish to create a two-element vector named MM containing the maximum and minimum values in vector V100. This can be done using

```
                       MM=[max(V100),min(V100)];
```

3. You wish to create a vector OddMM consisting of the odd integers from 1 to 50 followed by three times the square root of each element in MM from the previous example.

```
                     OddMM=[1:2:50, 3*sqrt(MM)];
```

Comprehension Check	**15-10**

(a) Assume you have four row vectors containing data on traffic flow named T1, T2, T3, and T4. Using a single MATLAB statement, combine these four, in that order, into a single column vector named TFC. Example:

```
              T1=[2 6 4]; T2=[0 -1]; T3=[9 -1 0];
              T4=[7 7]; TFC=[2;6;4;0;-1;9;-1;0;7;7];
```

(b) Using a single MATLAB statement, create a row vector named Rev that contains all even integers from 10 to 10,000 in ascending order followed by all integer multiples of 7 from 700 to 7 in descending order.

(c) Assume you have a row vector named RV5. Using a single MATLAB statement, create a new column vector named Pow that contains, in this order,

   1. The elements of RV5.
   2. The square roots of each element of RV5.
   3. The square of each element of RV5.

Vector Indexing

IMPORTANT CONCEPT

Index values MUST be integers and strictly positive. Talking about the 2.7^{th} element or the -3rd value does not make sense.

Often you will need to change a single element or group of elements within a vector, or perform some operation on only some of the elements within a vector. **Indexing** or **addressing** allows you to accomplish this. To access a specific element or elements within a vector, we place the numeric positions of the elements we wish to use in parentheses after the vector name. For example, in the vector V1=[2 5 9], the notation V1(2) indicates the second element in V1, or the value 5.

Creating Vectors Using Indexing

We have already seen that we can place a list of values in brackets to create a vector. We can also create vectors using indexing.

EXAMPLE   15-12

The vector `AVec` containing the values $2, 0,$ and 5 could be created by

```
AVec=[2  0  5];
```

This could also be accomplished using

```
AVec(1)=2;  % Define Element 1
AVec(2)=0;  % Define Element 2
AVec(3)=5;  % Define Element 3
```

Although this may look like a lot more work at first, as we will see later, this is the best way to define vectors in some situations.

What happens if you wish to specify some element of an undefined vector other than the first element?

In the Command Window, type this to be sure `try3` does not already exist:

```
clear try3
```

Now type

```
try3(3)=9
```

Do not place a semicolon at the end of the line so that `try3` will be echoed to the screen. What can you say about the first two elements, `try3(1)` and `try3(2)`, that were not defined?

Changing Individual Values of Vectors

You can change elements individually without modifying the others using vector indexing.

EXAMPLE   15-13

What is stored in the variable when each line of the following MATLAB code is executed?

Assume $V7 = \begin{bmatrix} 3 \\ -7 \\ 34 \\ 0 \\ 18 \end{bmatrix}$

```
V7(3)= -1;
```

$V7$ *will then contain* $\begin{bmatrix} 3 \\ -7 \\ -1 \\ 0 \\ 18 \end{bmatrix}$

Deleting Values from Vectors

Sometimes you need to simply delete elements from a vector—not replace them with zeros, but actually delete them making the vector shorter. This can be done by specifying the elements to be deleted to be the **empty element**, indicated by [ ].

EXAMPLE 15-14

What is stored in the variable when the following line of MATLAB code is executed?

Assume V8 contains [2 1 5 8 6 3].

```
V8(3)=[];
```

This will delete the third element, so V8 now contains [2 1 8 6 3].

Specifying More Than One Element Using Indexing

If you wish two or more elements of a vector to have the same value, you can list the indices of those values in the parentheses as a vector—enclosed in brackets and separated by commas or spaces.

EXAMPLE 15-15

What is stored in the variable when each line of the following MATLAB code is executed?

Assume V9 contains $[3 \ -2 \ 5 \ 0 \ -7]$

```
V9([2,4,5])=99;
```

This will replace the second, fourth, and fifth elements with the number 99, so V9 now contains [3 99 5 99 99].

Using Linear Sequences as Indices

You can use the colon operator to create a linear sequence of values to use as indices inside the parentheses.

EXAMPLE 15-16

What is stored in the variable when the following line of MATLAB code is executed?

Assume CV1 = $\begin{bmatrix} 3 \\ -7 \\ 34 \\ 0 \\ 18 \end{bmatrix}$

```
CV1(2:4)=99;
```

This will replace the second, third, and fourth elements with the number 99, so CV1 *now contains*

$$\begin{bmatrix} 3 \\ 99 \\ 99 \\ 99 \\ 18 \end{bmatrix}$$

EXAMPLE 15-17

What is stored in the variable when the following line of MATLAB code is executed?

Assume RV1 contains [3 6 −2 7 0 0 −4 −6 9 13]

```
RV1(2:3:10)=0;
```

This will replace every third element, starting with the second up to the tenth element (replacing the second, fifth, and eighth elements), so RV1 *now contains* [3 0 −2 7 0 0 −4 0 9 13].

Calculations with Part of a Vector

EXAMPLE 15-18

Determine the sine of the fifth element of vector T20, assuming the values in T20 are angles given in radians. Place the result in S5.

```
S5=sin(T20(5));
```

EXAMPLE 15-19

Assume you have a 100-element vector D1 containing data. Create a new vector RD1 containing 50 elements comprising the square roots of the even-numbered elements of D1.

```
RD1=sqrt(D1(2:2:100));
```

EXAMPLE 15-20

Assume you have a 1000-element vector `D2` containing data. Create a new vector `ZRD2` containing 1000 elements in which the odd-numbered elements are zero and the even-numbered elements are the squares of the corresponding even-numbered elements of `D2`.

```
ZRD2(2:2:1000)= D2(2:2:1000).^2;
```

Remember that when creating a new vector, any elements that are not specified default to zero.

EXAMPLE 15-21

Assume you have a vector `D3` containing an unknown number of elements. Modify `D3` so that all odd-numbered elements are replaced by the cosines of the odd-numbered values. All even-numbered values should be unchanged. Assume the odd-numbered elements contain angles in degrees.

```
D3(1:2:length(D3))=cosd(D3(1:2:length(D3)));
```

Comprehension Check 15-11

(a) Write a single MATLAB command that will create a 50,000-element vector named `Biggie` in which every tenth element equals −9999 and all other elements equal 0.

(b) Assume a vector `LV` has already been defined. Write a single MATLAB command that will delete all of the even-numbered elements, leaving `LV` with half as many elements.

(c) Assume a vector named `D` has an even number of elements. Write a single MATLAB command that will create a new vector named `DS` with half as many elements, in which each element is the sum of adjacent odd–even pairs in `D`.

Example:

If `D=[2 5 4 −7 3 0]`, then `DS` will equal `[7 −3 3]`, found by:
[D(1)+D(2) D(3)+D(4) D(5)+D(6)]

15.4 Matrices ⊞

LEARN TO: Create numeric matrices in MATLAB
Perform basic matrix operations
Write mathematical expressions using functions on an entire matrix

A **matrix** or **array** consists of numbers organized into rows and columns. The size of a matrix is given as R × C, where R is the number of rows and C is the number of columns.

NOTE

MATLAB can easily handle multidimensional arrays, but this text focuses on those with only two dimensions—conveniently referred to as rows and columns.

Examples include $\begin{bmatrix} 0.2 & 3 & 0 \\ -7 & 0.05 & 99 \end{bmatrix}$ which is a 2 × 3 matrix, and $\begin{bmatrix} 1 & 0 & 0 & 0 \\ 1 & 1 & 0 & 1 \\ 1 & 0 & 1 & 0 \\ 0 & 1 & 1 & 1 \end{bmatrix}$ which

is a 4 × 4 matrix. All numeric data in MATLAB, including scalars and vectors, are matrices. MATLAB is, after all, derived from the words **MAT**rix **LAB**oratory. A scalar is simply a 1 × 1 matrix, whereas a vector is either 1 × C (row) or R × 1 (column). Note that all numeric data, whether scalar, vector, or matrix, is represented in the workspace by a square icon divided into four smaller squares used to symbolize a matrix.

Defining Matrices

The basic way to define a matrix is to combine the methods used to create row and column vectors, but matrices are normally created row by row, not column by column—you specify the first row with elements separated by commas or spaces, then use a semicolon to move to the next row.

When defining matrices in this manner, interpret the meaning of commas, spaces, and semicolons as follows:

Comma or **space** means go to next element (column) in current row.
Semicolon means go to beginning of next row.

EXAMPLE   15-22

What matrix is formed by the following commands?

```
M1=[0 -4 3;-2 6 0.2];
```

Creates the 2 × 3 matrix M1= $\begin{bmatrix} 0 & -4 & 3 \\ -2 & 6 & 0.2 \end{bmatrix}$

```
M2=[0,-4;3,-2;6,0.2];
```

Creates the 3 × 2 matrix M2= $\begin{bmatrix} 0 & -4 \\ 3 & -2 \\ 6 & 0.2 \end{bmatrix}$

Creating Matrices Using Indexing

When using indexing with matrices, we need to specify both the row and column of the element in question. Instead of a single number, place two integers separated by a comma in the parentheses following the matrix name. **The first value is the row number, and the second value is the column number.**

EXAMPLE 15-23

NOTE

Note that when using indexing to create a matrix, the order in which the matrix is defined is not predefined. In this example, the matrix is defined column by column, not row by row.

What matrix is formed by the following commands?

```
M3(1,1)=5;
M3(2,1)=0;
M3(1,2)=1;
M3(2,2)=8;
```

Creates the 2 × 2 matrix M3 = $\begin{bmatrix} 5 & 1 \\ 0 & 8 \end{bmatrix}$

If you create an element of a matrix in such a way that for the row and column specified there are elements of the matrix that are unspecified, MATLAB will automatically make them zero.

EXAMPLE 15-24

What matrix is formed by the following commands?

```
M4(2,3)= 77;
```

Creates the 2 × 3 matrix M4 = $\begin{bmatrix} 0 & 0 & 0 \\ 0 & 0 & 77 \end{bmatrix}$

```
M5(2,3)=55;
M5(3,2)=88;
```

Creates the 3 × 3 matrix M5 = $\begin{bmatrix} 0 & 0 & 0 \\ 0 & 0 & 55 \\ 0 & 88 & 0 \end{bmatrix}$

Using the Colon Operator to Create Linear Sequences for Indexing Matrices

EXAMPLE   15-25

What matrix is formed by the following commands?

Assume M5 has already been defined as in the previous example.

```
M5(1,2:3)= -1;
```

This command states that Row 1, Columns 2 through 3 are set to −1.

$$\text{M5 } \textit{now contains } \text{M5}= \begin{bmatrix} 0 & -1 & -1 \\ 0 & 0 & 55 \\ 0 & 88 & 0 \end{bmatrix}$$

Defining a Matrix from Another Matrix

EXAMPLE   15-26

What matrix is formed by the following commands?

Assume M5 has been defined as in the previous example.

```
M6=M5(1:2,2:3);
```

This command states the variable M6 equals rows 1 and 2, columns 2 and 3 of M5.

$$\textit{This creates } \text{M6}= \begin{bmatrix} -1 & -1 \\ 0 & 55 \end{bmatrix}$$

NOTE

A **colon** used alone as an index indicates to use all rows or all columns in the matrix.

Accessing Entire Rows or Columns

Often you wish to specify an entire row or entire column or a matrix. MATLAB provides a shortcut notation for this situation. Instead of placing a 1 for the first row (or column) before the colon and the total number of rows (or columns) after the colon, simply type a colon by itself.

EXAMPLE   15-27

What matrix is formed by the following commands?

Assume M5 has been defined as above.

```
M7=M5(:,2);
```

This command states the variable M7 *equals all rows, column 2 of* M5.

This creates the 3 × 1 *matrix (or column vector)* M7 $= \begin{bmatrix} -1 \\ 0 \\ 88 \end{bmatrix}$

M8=M5(1:2:3,:);

This command states the variable M8 equals rows 1 and 3, all columns of M5.

This creates the 2 × 3 *matrix* M8 $= \begin{bmatrix} 0 & -1 & -1 \\ 0 & 88 & 0 \end{bmatrix}$

Creating a Column Vector (or an R × 1 Matrix)

You can use the colon to force a vector definition to be a column instead of the default row.

EXAMPLE   15-28

What matrix is formed by the following commands?

CV2(:,1)=[1 4 3 7];

This command states the variable CV2 *equals all rows, one column of values. This creates*

$$CV2 = \begin{bmatrix} 1 \\ 4 \\ 3 \\ 7 \end{bmatrix}$$

Deleting Rows or Columns from a Matrix

The colon operator in combination with the empty element [ ] can be used to delete entire rows or columns from a matrix.

EXAMPLE   15-29

What matrix is formed by the following commands?

Assume M9 $= \begin{bmatrix} 3 & -9 & 14 \\ 6 & 0.5 & -4 \\ 44 & 5 & 1 \end{bmatrix}$

M9(2,:)= [];

This command replaces row 2, all columns with an empty matrix, deleting row 2 from the matrix. Now,

$$M9 = \begin{bmatrix} 3 & -9 & 14 \\ 44 & 5 & 1 \end{bmatrix}$$

Deleting Individual Elements from a Matrix

EXAMPLE 15-30

What matrix is formed by the following commands?

Assume $M10 = \begin{bmatrix} -1 & -2 \\ 3 & 5 \end{bmatrix}$

```
M10(2,1)=[];
```

This command replaces row 2, the first column with an empty matrix. This will result in the error:
`Subscripted assignment dimension mismatch.`
The error occurs because in a matrix, all rows must have the same number of columns and all columns must have the same number of rows. Deleting a single element from a row or column would violate this rule. You can, however, replace elements with zeros.

```
M10(2,1)=0;
```

This command replaces the element in row 2, column 1 with a zero, so $M10 = \begin{bmatrix} -1 & -2 \\ 0 & 5 \end{bmatrix}$

Matrix Transpose

In the section on vectors, we saw that the single quote (') operator changes a row vector to a column vector, or vice versa. This is actually just a special case of a matrix transpose. The transpose of a matrix swaps rows and columns—the first row of the original matrix becomes the first column of the transposed matrix, the second row becomes the second column, etc.

EXAMPLE 15-31

What matrix is formed by the following command?

Assume $M1 = \begin{bmatrix} 0 & -4 & 3 \\ -2 & 6 & 0.2 \end{bmatrix}$

```
T1=M1';
```

This will transpose the M1 *matrix, and store it in* T1: $T1 = \begin{bmatrix} 0 & -2 \\ -4 & 6 \\ 3 & 0.2 \end{bmatrix}$

Comprehension Check   **15-12**

(a) Create the matrix CCM1=$\begin{bmatrix} 18 & 0.3 \\ -4.1 & -1 \\ 0 & 17 \end{bmatrix}$ using a single MATLAB command.

(b) Create the matrix CCM2=$\begin{bmatrix} 0 & 0 & 0 \\ 0 & 0 & 1 \times 10^{15} \\ 0 & 0 & 1 \times 10^{15} \end{bmatrix}$ using a single MATLAB command.

(c) Assume the 3 × 3 matrix CCM3 has already been defined as CCM3=$\begin{bmatrix} 3 & 9 & 14 \\ 6 & 0.5 & 4 \times 10^{-2} \\ 44 & 0 & 1 \times 10^{3} \end{bmatrix}$

Using a single MATLAB command, define a new 2×2 matrix named Corners that contains the corner elements of CCM3. Your code should still work correctly if the contents of CCM3 change but it remains a 3×3; in other words, DO NOT hard-code the four values.

Calculations with Matrices

For the basics of matrix operations, please refer to Appendix A.6 online. The following discussion assumes that you already understand things like a matrix transpose or matrix multiplication.

Addition and Subtraction

When addition or subtraction involves a matrix, there are two cases:

1. A scalar and a matrix: The scalar is added or subtracted with each element of the matrix.
2. Two equal-sized matrices: Corresponding elements of the two matrices are added or subtracted. If you try to add two matrices (both nonscalar) that are not the same size, you will get angry red letters.

EXAMPLE   **15-32**

What matrix is formed by the following commands?

Assume ExM1 = $\begin{bmatrix} 3 & 0.5 & 0 \\ 5 & -7 & 2 \end{bmatrix}$;   ExM2 = $\begin{bmatrix} 9 & -4 \\ 8 & 3.7 \end{bmatrix}$;   ExM3 = $\begin{bmatrix} 19 & 2 & 3.4 \\ -6 & 1 & -3 \end{bmatrix}$;

```
D1=7-ExM1;
```

D1 contains $\begin{bmatrix} 4 & 6.5 & 7 \\ 2 & 14 & 5 \end{bmatrix}$

```
S1=ExM1+ExM3;
```

$$\text{S1 } contains \begin{bmatrix} 22 & 2.5 & 3.4 \\ -1 & -6 & -1 \end{bmatrix}$$

```
E1=ExM1+ExM2;
```

Will return the message:
```
Error using +   Matrix dimensions must agree.
```

Comprehension Check   15-13

Write single MATLAB statements to perform each of the following tasks.

(a) Add 75 to each element of the third column of matrix `Tmp`.
(b) Assume `Press` is a 35×70 matrix. Replace the four corner elements of `Press` with their original values minus 100,000.

Multiplication

We will consider three cases of matrices involved in multiplication.

1. Multiplication of a scalar and a matrix: The scalar is multiplied by each element of the matrix.
2. Two equal-sized matrices multiplied on an element-by-element basis. Note that the two matrices MUST have the same dimensions. NOTE: This requires the use of the element-wise multiply operator (. *).
3. A true matrix multiplication (refer to Appendix A.6 online): In this case, the number of columns of the first matrix MUST equal the number of rows of the second matrix or the matrix multiplication is not defined.

EXAMPLE   15-33

What matrix is formed by the following commands?

$$\text{Assume ExM1=} \begin{bmatrix} 3 & 0.5 & 0 \\ 5 & -7 & 2 \end{bmatrix}; \text{ ExM2=} \begin{bmatrix} 9 & -4 \\ 8 & 3.7 \end{bmatrix}; \text{ ExM3=} \begin{bmatrix} 19 & 2 & 3.4 \\ -6 & 1 & -3 \end{bmatrix}$$

$$\text{RV=}[2\,5\,9]; \quad \text{CV=} \begin{bmatrix} 3 \\ -1 \\ 5 \end{bmatrix}$$

```
P1=3*ExM1;
```

$$\text{P1 } contains \begin{bmatrix} 9 & 1.5 & 0 \\ 15 & -21 & 6 \end{bmatrix}$$

```
P2=ExM1.*ExM3;
```

$$\text{P2 } contains \begin{bmatrix} 57 & 1 & 0 \\ -30 & -7 & -6 \end{bmatrix}$$ *Recall the symbol . * means element-wise multiplication*

```
P3=ExM1*ExM3;
```

> **Error using * Inner matrix dimensions must agree.**

```
P4=RV*CV;
```

> P4 *contains 46*

```
P5=CV*RV;
```

> P5 *contains* $\begin{bmatrix} 6 & 15 & 27 \\ -2 & -5 & -9 \\ 10 & 25 & 45 \end{bmatrix}$

> P4 *and* P5 *show that matrix multiplication is NOT commutative!*

```
P6=ExM2*ExM1;
```

> P6 *contains* $\begin{bmatrix} 7 & 32.5 & -8 \\ 42.5 & -21.9 & 7.4 \end{bmatrix}$

```
P7=ExM1*ExM2;
```

> **Error using * Inner matrix dimensions must agree.**

> P6 *and* P7 *show that commuted matrix multiplication is not necessarily even defined!*

Comprehension Check   15-14

Write single MATLAB statements to perform each of the following tasks.

(a) Multiply each element of the matrix Vel by the scalar value stored in T.
(b) Assume Voltage and Current are both N × M matrices. Determine the products of corresponding elements of Voltage and Current, placing the results in the N × M matrix Power.

Powers

We will consider four cases of matrices involved in powers.

1. Raising a matrix to a scalar power on an element-wise basis: Each element of the matrix is raised to the specified power. NOTE: This requires the use of the element-wise power operator (.^).
2. Raising a scalar to a matrix power on an element-wise basis: The scalar is raised to each element of the matrix. NOTE: This requires the use of the element-wise power operator (.^).
3. A matrix raised on an element-wise basis to another equal-sized matrix. NOTE: This requires the use of the element-wise power operator (.^) If the two matrices do not have the same dimensions, you will get angry red letters.
4. A square matrix raised to a strictly positive integer. Fractional and nonpositive powers will work with square matrices, but until you understand considerably more about matrix arithmetic, it would be wise to stick with strictly positive integer powers when using ^ . If the matrix is not square or the power is not a scalar, you will get angry red letters.

EXAMPLE   15-34

What matrix is formed by the following commands?

Assume $\text{ExM1} = \begin{bmatrix} 3 & 0.5 & 0 \\ 5 & -7 & 2 \end{bmatrix}$; $\text{ExM2} = \begin{bmatrix} 9 & -4 \\ 8 & 3.7 \end{bmatrix}$; $\text{ExM3} = \begin{bmatrix} 19 & 2 & 3.4 \\ -6 & 1 & -3 \end{bmatrix}$

`P8=ExM1.^2;`

> `P8` *contains* $\begin{bmatrix} 9 & 0.25 & 0 \\ 25 & 49 & 4 \end{bmatrix}$ *Recall using the symbol* `.^` *means element-wise power*

`P9=ExM1^2;`

> **Error using ^ Inputs must be a scalar and a square matrix.**

To compute elementwise POWER, use (`.^`) instead.

`P10=2.^ExM1;`

> `P10` *contains* $\begin{bmatrix} 8 & 1.4142 & 1 \\ 32 & 0.0078 & 4 \end{bmatrix}$

`P11=2^ExM1;`

> **Error using ^ Inputs must be a scalar and a square matrix.**

`P12=ExM3.^ExM1;`

> `P12` *contains* $\begin{bmatrix} 6859 & 1.4142 & 1 \\ -7776 & 1 & 9 \end{bmatrix}$

`P12=ExM3^ExM1;`

> **Error using ^ Inputs must be a scalar and a square matrix.**

`P13=ExM2^3;`

> `P13` *contains* $\begin{bmatrix} 34.6 & -383.96 \\ 767.9 & -474.147 \end{bmatrix}$ *Note that this is equivalent to* `ExM2*ExM2*ExM2` *using full matrix multiplication.*

Comprehension Check   15-15

Write single MATLAB statements to perform each of the following tasks.

(a) Cube all elements of the third column of matrix `Vol`.

(b) Create the matrix `Pof2` by raising the number 2 to each *element* in the matrix `P2`. `Pof2` will have the same dimensions as `P2`.

(c) Assume `D` is a matrix with N rows and `FracD` is a matrix with N columns. Modify `D` by raising each element in the first column of `D` to the corresponding elements in the first row of `FracD`.

Division

We will consider three cases of matrices involved in division.

1. Dividing a matrix by a scalar: Each element of the matrix is divided by the scalar.
2. Dividing a scalar by a matrix on an element-wise basis: The scalar is divided by each element of the matrix. NOTE: This requires the use of the element-wise divide operator (. /).
3. A matrix divided on an element-wise basis by another equal-sized matrix. NOTE: This requires the use of the element-wise divide operator (. /) If the two matrices do not have the same dimensions, you will get angry red letters.

Full matrix division is beyond the scope of this text. Refer to any general reference on matrix algebra for more information.

EXAMPLE 15-35

What matrix is formed by the following command?

Assume $\texttt{ExM1} = \begin{bmatrix} 3 & 0.5 & 0 \\ 5 & -7 & 2 \end{bmatrix}$; $\texttt{ExM2} = \begin{bmatrix} 9 & -4 \\ 8 & 3.7 \end{bmatrix}$; $\texttt{ExM3} = \begin{bmatrix} 19 & 2 & 3.4 \\ -6 & 1 & -3 \end{bmatrix}$

`Q1=ExM2/4;`

Q1 contains $\begin{bmatrix} 2.25 & -1 \\ 2 & 0.925 \end{bmatrix}$

`Q2=10./ExM1;`

Q2 contains $\begin{bmatrix} 3.3333 & 20 & Inf \\ 2 & -1.4286 & 5 \end{bmatrix}$

`Q3=10/ExM1;`

Error using /   Matrix Dimensions must agree.

`Q4=ExM3./ExM1;`

Q4 contains $\begin{bmatrix} 6.3333 & 4 & Inf \\ -1.2 & -0.1429 & -1.5 \end{bmatrix}$

Comprehension Check 15-16

Write single MATLAB statements to perform each of the following tasks.

(a) Divide all elements of the matrix `Circ` by pi.
(b) Create the matrix `UnitCost` by dividing the number 1500 by each element in the matrix `NumUnits`. `UnitCost` will have the same dimensions as `NumUnits`.
(c) Assume `Voltage` and `Current` are both $N \times M$ matrices. Determine the quotient of corresponding elements of `Voltage` and `Current` (`Voltage` over `Current`), placing the results in the $N \times M$ matrix `Resistance`.

Functions Used with Matrices

All of the functions listed in Table 15-5 and Table 15-6 also work with matrices. The functions in Table 15-5 automatically perform the stated operation on each element individually. The functions in Table 15-6 may behave a little differently than you would expect, however.

There are quite a few MATLAB functions that would seldom if ever be used with scalars and/or vectors but are very helpful in dealing with matrices. A few of these are listed in Table 15-7. You can obtain more information by typing either `help` or `doc` followed by the function name.

NOTE

All of the functions used with vectors in Table 15-6 can be used with matrices, but some do not do what you might expect. For example, `length` does not find the total number of elements in a matrix and `max` does not find the single largest element in the matrix.

Table 15-7 Selected functions used with matrices

Function	Description
size	Find the number of rows and columns in a matrix. VERY useful!
find	Find indices of all nonzero elements in a matrix. VERY useful.
zeros	Create a matrix of all zeros.
ones	Create a matrix of all ones.
eye	Create identity matrix with R rows and C columns, where all elements are zero except the main diagonal, which contains ones.
rand	Create a matrix of random numbers.
diag	Specify or extract diagonals of matrix.
trace	Sum of elements on the diagonal.
flipud	Flip matrix in up/down direction.
fliplr	Flip matrix in left/right direction.
rot90	Rotate matrix 90 degrees counterclockwise.

In the previous section, we saw that the `length` function will tell us the number of elements in a vector, so you might think that `length` tells you the number of elements in a matrix as well. **This is incorrect!** `length` will work with matrices, but what it returns is the larger of the number of rows or the number of columns. This is seldom what you want, so unless you are certain of what you are doing, do not use `length` with matrices that have at least two rows and at least two columns.

Normally what you are interested in is both the number of rows and the number of columns. For this purpose, use the `size` function. For a two-dimensional matrix, size returns a two-element vector: the first element contains the number of rows and the second element contains the number of columns.

EXAMPLE   15-36

What is placed in matrix S by the following command?

Assume $\text{ExM1}=\begin{bmatrix} 3 & 0.5 & 0 \\ 5 & -7 & 2 \end{bmatrix}$

`S=size(ExM1);`

S contains `[2  3]`

In this example, S(1) contains the number of rows in ExM1, and S(2) contains the number of columns in ExM1.

It would be convenient if the number of rows and columns could be placed in separate variables instead of having to index into a vector. We could, of course, set up two such variables as separate statements, such as R=S(1) and C=S(2), but there is a shortcut method for achieving this goal. We can explicitly name each element in the vector returned by the size function by placing the desired name for each inside square brackets. Example 15-35 would then become

```
[R,C]=size(ExM1)
```

Then R contains the number of rows in ExM1 and C contains the number of columns in ExM1.

EXAMPLE **15-37**

What variables are created by the following command and what do they contain?

Assume ExM2=$\begin{bmatrix} 6 & 4 & 9 \\ 3 & 4 & 8 \\ 0 & 3 & 9 \\ 1 & 0 & 2 \end{bmatrix}$

```
[rows,cols]=size(ExM2);
```

rows contains 4 and cols contains 3.

EXAMPLE **15-38**

What variables are created by the following commands and what do they contain?

Assume ExM1=$\begin{bmatrix} 3 & 0.5 & 0 \\ 5 & -7 & 2 \end{bmatrix}$

```
Ident=eye(3);
```

Ident *contains* $\begin{bmatrix} 1 & 0 & 0 \\ 0 & 1 & 0 \\ 0 & 0 & 1 \end{bmatrix}$

```
Mod2=fliplr(ExM1);
```

Mod2 *contains* $\begin{bmatrix} 0 & 0.5 & 3 \\ 2 & -7 & 5 \end{bmatrix}$

EXAMPLE   15-39

What variables are created by the following commands and what do they contain?

Assume $\text{ExM2} = \begin{bmatrix} 6 & 4 & 9 \\ 3 & 4 & 8 \\ 0 & 3 & 9 \\ 1 & 0 & 2 \end{bmatrix}$

```
[R,C]=size(ExM2);
```

```
NewM=rot90(ExM2(2:R,C-1:C));
```

`R` contains 4 and `C` contains 3.

`NewM` contains $\begin{bmatrix} 8 & 9 & 2 \\ 4 & 3 & 0 \end{bmatrix}$ *First, rows 2 through 4 of columns 2 and 3 are extracted, then rotated counterclockwise 90°.*

EXAMPLE   15-40

As we saw earlier, `find` can be used to find the indices of elements in a vector that meet certain criteria. `find` can also be used with matrices, but there are some differences.

Assume $M = \begin{bmatrix} 1 & -3 & 5 & -7 & 0 \\ -2 & 4 & -6 & 0 & 8 \end{bmatrix}$

What is placed in `MI` by the following command?

```
MI=find(M<0); % find indices of negative elements in M.
```

$$MI = \begin{bmatrix} 2 \\ 3 \\ 6 \\ 7 \end{bmatrix}$$

This is perhaps not what you were expecting. Since in English we read left to right, top to bottom, our natural inclination is to count across the top row, then continue with the second row, and so on. MATLAB, however, counts down the first column, then continues with the second column, and so on when indexing a matrix with a single value.

For many purposes, it may be preferable to have the `find` function return a row number and column number pair for each element it finds. The syntax for this is similar to what we saw with the `size` function earlier.

What is placed in `MR` and `MC` by the following command?

```
[MR,MC]=find(M<0); % find indices of negative elements in M.
```

$$MR = \begin{bmatrix} 2 \\ 1 \\ 2 \\ 1 \end{bmatrix} \quad MC = \begin{bmatrix} 1 \\ 2 \\ 3 \\ 4 \end{bmatrix}$$

Here the two vectors MR and MC contain the index PAIRS of the found elements: $(2,1)$; $(1,2)$; $(2,3)$; and $(1,4)$.

Comprehension Check   15-17

For each of the following questions, write a single MATLAB command that will achieve the stated goal.

(a) Create a matrix HSR defined as the square root of half of each element in matrix QPD.

(b) Replace the second row of matrix QPD with the cube of each of those elements divided by the natural logarithms of the corresponding elements in row 3 of QPD.

(c) Assume matrix M3R contains exactly three rows. Modify M3R so that the top and bottom rows are swapped.

(d) Assume MCC=[1 0 4 -5 -2;9 -4 0 -7 8;3 9 5 -1 0];

What is placed in MCCR and MCCC by the command

[MCCR,MCCC]=find(MCC<=-1);

15.5 Character Strings

LEARN TO: Create text strings in MATLAB
Combine existing text strings
Write expressions using special functions to modify character strings

Character Strings

A variety of data, such as names, part identification, or weekdays, requires alphabetic or alphanumeric data. Such information is stored in a **character string**, also called a text string or simply text, and is denoted in the workspace with a square icon containing the letters abc.

A character string is in one sense like a numeric vector, in that each character of the string is in a separate element of the variable where it is stored. For example, if the variable Z contains the text string 'Cat', the second element of Z contains the character 'a'. Examples of text strings include 'Ignatius J. Reilly', 'Part # 23B', and 'Wednesday'.

Note that punctuation marks and spaces each require a separate element within the variable. For example, 'Part # 23B' is a 10-element character string; elements 5 and 7 are the blank character, and element 6 is the number symbol #. Also, numerals in a text string each require a separate element, and they are encoded in binary very differently from the corresponding numeric value; in this example, elements 8 and 9 are the numerals 2 and 3 encoded as characters.

TEXT CREATION

We saw earlier that the single quote is used as the transpose operator. Single quotes are also used to define character strings. MATLAB is clever enough to understand the meaning of the quote from context: if a single quote immediately follows a matrix, it is the transpose operator; if a pair of single quotes surrounds characters, they are being used to indicate that those characters are simply text, not variables or calculations or MATLAB commands.

Defining Character Strings

When creating a character string using the assignment operator (=), simply enter the desired text enclosed in single quotes.

```
TS1='American Bison';
TS2='Tuesday, May 11, 868';
TS3='42';
```

Note that `TS3` is a text string, not a numeric value, and consists of two characters, `'4'` and `'2'`.

Including an Apostrophe in a Text String

If you wish to create a text string including an apostrophe, there is a minor problem since MATLAB will think the apostrophe is the end of the text string, not a character to be included in the text string. However, MATLAB also knows that if it is processing a text string and finds TWO single quotes in a row (2 single quotes, NOT one double quotation mark) it knows to place a single apostrophe in the text string, not to terminate the string. Thus, if you want to place the text "Willy's brain" into variable `WB`, you would type `WB='Willy''s brain';`

Combining Text Strings

Just like you can combine vectors by concatenation in square brackets, you can combine text strings.

```
W1='Willy''s';
W2='Brain';
TS=[W1,' Weird ',W2];
```

`TS` contains `Willy's Weird Brain`. Note the spaces in the single quotes before and after `Weird`.

Using Indexing with Character Strings

Character strings can be created or modified using indexing, and indexing can also be used to extract portions of a text string.

```
TS='Willy''s Weird Brain';
Word1=TS(1:7)
```
   `Word1` contains `Willy's`. Note the apostrophe is ONE character.
```
W2=TS(length(TS)-3:length(TS));
```
   `Word2` contains `rain`.
```
TS(9:14)=[];
```
   `TS` now contains `Willy's Brain`

NOTE

Some of these functions (e.g., `isletter`) return a logical array. Logical arrays will be covered in a later chapter.

Functions Used with Character Strings

Although many of the standard mathematical operators and functions will actually process character strings without errors, you may not get the results you expect, and you need to be certain that you understand what actually happens before trying to use these with text.

There are, however, numerous functions specifically for manipulating text, several of which are included in Table 15-8.

Table 15-8 Selected functions used with character strings

Function	Description
blanks	Create a string of blanks (compare the zeros function used with matrices)
isletter	Finds alphabetic characters in a string
isspace	Finds space characters in string
strfind	Find a string within another string
strrep	Replace substring within a string
strcmp	Compare two strings
strcmpi	Compare two strings ignoring case. VERY useful
lower	Convert string to all lowercase
upper	Convert string to all uppercase
num2str	Convert number to string
str2num	Convert string to number

Comprehension Check   **15-18**

(a) Create a variable named MTS containing the text: My hero's hat
(b) Create a variable named B containing 17 blanks followed by your name followed by 691 more blanks.
(c) Assume Avian='A wet bird never flies at night';
Modify Avian so that the word "wet" is replaced with the word "crepuscular" and the word "never" is replaced by "seldom". Note that you should actually modify ONLY the two words in question, not simply type in the complete new sentence.
(d) Assume Ultimate='The answer is 42';
Convert the substring "42" in Ultimate into its corresponding numeric value and place it in UltAns. Your solution should work as long as the "number" at the end of the string is exactly two digits. HINT: Look in Table 15-8.

15.6 Cell Arrays

LEARN TO: Create cell arrays in MATLAB
Perform basic operations on cell arrays

So far in our discussion of data types in MATLAB, all elements stored in a specific variable are the same type of thing: either all numeric or all text. Sometimes, however, it would be convenient to be able to store different types of information in a single variable. MATLAB provides two data types that allow us to store both numeric values and non-numeric values such as text within a single structure stored in a variable. The first of these structures is called a **cell array** or **cell matrix**.

One reason that we study cell arrays in this course has to do with Microsoft Excel. As we will see later, when we import data from an Excel worksheet into MATLAB, at least some of the data will be stored in a cell array. Similarly, when we want to send data from MATLAB to Excel, some of the data must be stored in a cell array prior to the transfer. We therefore need to understand how to use cell arrays if we wish to get MATLAB and Excel to communicate with each other.

Cell Arrays

The basic idea of the cell array is that each element in a cell array contains another MATLAB data structure. An element in a cell array might be a numeric matrix, whether a scalar, a vector or a matrix; it might be a text string; or it might be any of the other MATLAB data types, including cell arrays.

Defining Cell Arrays

Recall that when we define numeric variables, scalars are defined using a simple assignment statement such as `S1=57.43`, whereas vectors and matrices are defined by enclosing the multiple values in square brackets, such as `M1=[3.5,-9;13,-4.2]`. Text strings are defined using single quotes: `TS='ARF'`. We also saw that we could define individual elements of matrices or text strings using indexing. For example `M2(2,4)=42` places the value 42 into the second row, fourth column element of `M2` and `TS1(5:9)='BOOF!'` places the text string `BOOF!` into elements 5 through 9 of `TS1`.

Cell arrays are defined in a similar manner, with one key difference: the use of curly braces `{ }`.

This entire concept can be a little confusing at first, but as you work with cell arrays, take note of any errors or unexpected results you get, and it will eventually begin to congeal in your mind. To begin, let's look at different ways a cell array might be defined.

First type the following in the Command Window. Be sure NOT to include a semicolon so it will echo to the screen.

```
TS='test'
```

The echo to the screen is

```
TS =
test
```

This tells us that TS is a text string containing the word *test*.
Now let's simply enclose the right-hand side in curly braces.

```
CBRight={'test'}
```

The echo to the screen is

```
CBRight =
    'test'
```

Note that this time there are glitches around the text. The presence of the glitches tells us that this is a **cell array** containing a text string, not just a simple text string.

Now let's use curly braces on the left as an index into a variable.

```
CBLeft{1}='test'
```

The echo to the screen is

```
CBLeft =
    'test'
```

Once again, the glitches indicate that this is a **cell array** containing a text string.

We have thus seen that we can define a variable as a cell array either by enclosing what we wish to store in the variable in curly braces OR by indexing the variable with curly braces, writing the contents on the right as we normally would.

Now let's try it with curly braces on BOTH sides.

```
CBBoth{1}={'test'}
```

The echo to the screen is

```
CBBoth =
    {1X1 cell}
```

Note that now this is telling us that CBBoth is a cell array containing a cell array! The inner cell array contains the text string. Basically, the curly braces on the right say that the item on the right is a cell array. The curly braces on the left say that CBBoth is a cell array. Therefore, we are placing a cell array into a cell array!

EXAMPLE 15-41

What is stored in the variable by following commands?

Assume M3 = [3, 5; 1, 9] and TS3 = 'GRRR'
```
   CA1={TS3,42,M3,'REOWR'};
```

The curly braces say that the four items are in a cell array, thus CA1 is a cell array.

CA1 *now contains four elements in a single row:*
CA1{1} *The text string* GRRR
CA1{2} *The number* 42
CA1{3} *A 2×2 numeric matrix with the same elements as* M3
CA1{4} *The text string* REOWR
   Note that the individual elements are different types of things: a scalar, a matrix, and two text strings.
   Note the use of curly braces to index into the cell array.

We could also use indexing to define one or more elements of a cell array separately instead of all at once. The cell array CA1 above could also be created with the following code:

```
CA1(1)={'GRRR'};
```

The curly braces tell MATLAB that 'GRRR' is a cell array; thus when CA1 is created, it will be a cell array, and we are specifying the first element as the text string GRRR.

```
CA1(2:2:4)={42,'REOWR'}; % CA1(3) will be created empty
CA1(3)={[3,5;1,9]}; % Go back to specify CA1(3)
```

Note that the cell array indices are in parentheses, whereas the contents to be placed in the indexed values are in curly braces. This might be confusing at first, but perhaps the next example will help to clarify, along with the section on Extracting Data from Cell Arrays following Comprehension Check 15-19.

Yet another method to specify the same cell array is

```
CA1{1}='GRRR';
```

The braces used to index CA1 tell MATLAB that CA1 is a cell array into which a text string is placed

```
CA1{2}=42; % 2nd element of cell array is a scalar
CA1{3}=[3,5;1,9] % 3rd element of cell array is a matrix
CA1{4}='REOWR'; % 4th element is another text string
```

So, we have seen that curly braces on either the left or right of the assignment operator (=) specify that we are dealing with a cell array. As shown earlier, if we put the curly braces on both sides of the assignment operator, we would be putting cell arrays into the elements of a cell array. There is nothing specifically wrong with that, but it may not be what we want to do.

Cell arrays are not restricted to a single dimension. Just like matrices, they may have both rows and columns, and the notation used to separate columns and rows (commas, spaces, and semicolons) is basically the same.

EXAMPLE   15-42

What is stored in the variable by the following commands?

```
CA2D={56,'dogmatic';[3;5;-1],[3,-8;5,9]};
```

CA2D is a 2×2 cell array containing a scalar and a text string in the first row and a column vector and a 2×2 matrix in the second row.

Comprehension Check   15-19

For the following two problems, show two solutions for each, one that defines the entire cell array with one statement and a different solution that uses indexing to define the elements one at a time.

(a) Create a cell array named Cabinet with one row and three columns. The first cell should contain the number 77, the second cell should contain the text "Dr. Caligari", and the third cell should contain a 25×100 matrix filled with ones.

(b) Create a 2×2 cell array named Cyls containing the following:
   a. First row: "Diameter", "Length"
   b. Second row: 1.5, a column vector containing all integers from 1 to 25

Extracting Data from Cell Arrays

When we use indexing with parentheses to get information out of a numeric matrix, the result is another numeric matrix. If M4=[1 3 5;6 8 0] and we create another variable as M5=M4(:,2:3), then M5 is a 2×2 numeric matrix containing [3 5; 8 0].

When we use indexing with parentheses to get information out of a text string, the result is another text string. If TS4='Dogs vs Cats' and we create another variable as TS5=TS4(4:9), then TS5 is a text string containing s vs C.

In general, when we access some of the elements of a variable using index values in **parentheses**, the result is the **same variable type** as the original variable.

But what about cell arrays, since they often contain different variable types in the various elements?

This is perhaps best illustrated live at the computer, so we suggest that you fire up MATLAB and type the following simple commands in the Command Window.

First, we will set up three variables. Do not include a semicolon so each echoes to the screen.

Type:

```
S6=777 % define a scalar
```

The following appears on the screen:

```
S6=
   777
```

Next, type:

```
M6=[9 7 8;3 5 4] % define a 2×3 matrix
```

The following appears on the screen:

```
M6=
   9 7 8
   3 5 4
```

Finally, type:

```
TS6='apricot' % Define a text string
```

The following appears on the screen:

```
TS6=
apricot
```

As you will see in a moment, it is important that you note exactly how these values are echoed to the screen.

Next, we will create a cell array containing the scalar, the matrix, and the text string created above. Add a semicolon to suppress the output in this case, however.

```
CA6={S6,M6,TS6}; % Create cell array with three elements
```

If you look at CA6 in the workspace, it will say *1 × 3 cell*.

Now we will attempt to extract each of these three elements from the cell array using parentheses to get the original contents back. First, extract the scalar in element 1.

```
CA7=CA6(1) % Extract first element
```

Echoed to the screen is

```
CA7=
   [777]
```

Comparing this to the echoed value from the original scalar S6 above, we see there is a slight difference: there are brackets around the value!

Now do the same with the matrix in the second element of the cell array.

```
CA8=CA6(2) % Extract second element
```

Echoed to the screen is

```
CA8=
   [2×3 double]
```

This is even stranger. The original matrix echoed the individual values to the screen, but when we attempted to extract it from the cell array, it just gave us the matrix dimensions in brackets.

Finally, extract the third element, the text string.

```
CA9=CA6(3) % Extract third element
```

The echo to the screen is

```
CA9=
  'apricot'
```

Comparing this to the echo of the original text string, we see that the text is enclosed in single quotes, whereas in the original echo of TS6, there were no quotes.

Obviously something is different here. In the screen echoes, the presence of brackets or glitches is MATLAB's way of telling you that these are either matrices (if brackets) or text strings (if glitches) **stored in a cell array**.

If you were to try to use these extracted values in functions or computations that expected a scalar, a matrix, or a text string respectively, you would get an error message. For example, type

```
SUM=CA7+111
```

This seems perfectly reasonable, and should give an answer of 888. However, you get the error message

```
Undefined function 'plus' for input arguments of type 'cell'.
```

The point is that when parentheses are used to extract values from a numeric matrix, the result is a numeric matrix, from a text string the result is a text string, and **values extracted from a cell array using parentheses are a cell array**, thus they cannot be used where MATLAB is expecting numbers or text.

Obviously there must be a way to overcome this problem or cell arrays would be extremely limited in their utility. The answer is embodied in the well-known alliterative mnemonic: **Curly for Contents**.

With the cell array CA6 still in your workspace, try extracting the **contents** using curly braces instead of parentheses:

Type:

```
S7=CA6{1} % Extract contents first element
```

Echo:

```
S7=
   777
```

Type:

```
M7=CA6{2} % Extract contents second element
```

Echo:

```
M7=
   9 7 8
   3 5 4
```

Type:

```
TS7=CA6{3} % Extract contents third element
```

Echo:

```
TS7=
apricot
```

Note that the echoes to the screen are exactly the same as when we originally defined them. You may now use S7, M7, and TS7 anywhere you can use a scalar, a matrix, or a text string respectively.

We can also extract parts of these variables using indexing, just like we could with the original variables.

For example:

V=M7(2,3) places the value 4 into V

T=TS7(5:7) places the text string cot into T

What if we wanted to place the second column of the second element of the cell array CA6 into CV1?

We could use

```
CV1=CA6{2};
CV1=CV1(:,2);
```

or we can use a shortcut method, directly indexing into the extracted matrix:

```
CV1=CA6{2}(:,2);
```

Although all of the above examples have used a cell array with a single row thus we used a single index value in the curly braces, cell arrays can have 2 dimensions thus two index values as mentioned earlier.

Now let us look at a slightly more complicated example. Again it is recommended that you type this in on the computer.

We will set up a cell array CA12 containing two other cell arrays.

```
CA10={[8 6 4;3 7 0],'Occurrence'};
CA11={[2 3 4;7 6 5],'Then'};
CA12 = {CA10, CA11}; CA12={C10,CA11};
```

What if we wanted to add the two 2×3 matrices found in the two cell arrays of CA12? Try

```
Result=CA12{1}(1)+CA12{2}(1);
```

You might think that this will pull out the first elements of each of the two cell arrays in CA12 (the two 2×3 matrices), but instead, we get the error

Undefined function 'plus' for input arguments of type 'cell'.

The reason is that since CA12{1} refers to CA10, which is itself a cell array, when we use parentheses to index, we get another cell array, NOT the original matrix. In this case we have to extract the contents of the contents, so both indices must have curly braces.

```
Result=CA12{1}{1}+CA12{2}{1};
```

This will give us the correct sum of the two matrices.

A similar situation exists if we wanted to extract part of the text strings. To get characters 2, 3, and 4 from Then, we would use

```
T=CA12{2}{2}(2:4);
```

T now contains hen.

- The first {2} pulls out the contents of the second element in CA12: a cell array.
- The second {2} pulls out the contents of the second element of that inner cell array: Then.
- The final (2:4) pulls the 2nd through 4th elements from that text string, placing hen into T.

Comprehension Check 15-20

Assume a cell array named CA has three cells in a single row.

- The first cell contains a name in the format X. Y. Family. For example: M. V. Smith.
- The second cell contains a 2×3 matrix. For example: [1 3 5;9 6 3]
- The third cell contains a 3×2 matrix. For example: [4 7; −2 −6; 0 1]

   (a) Extract the family name (Smith in the example above) from the first cell and place it in column one of a newly created second row of CA. Your solution should work for family names of any length. Note that the first six characters of the name in the first cell will always be the two initials, each followed by a dot and a space.

   (b) Multiply the matrix in the second cell of row 1 by the matrix in the third cell of row 1 and place the result in the second row, second column of CA.

   (c) Calculate the element-wise product of the matrix in row 1, column 2 with the transpose of the matrix in row 1, column 3. Place the result in the third column of the second row of CA.

Functions Used with Cell Arrays

Some helpful functions for cell arrays are listed in Table 15-9. You can obtain more information by typing either help or doc followed by the function name.

Table 15-9 Selected functions used with cell arrays

Function	Description
iscell	Determines if a variable is a cell array
cell	Create a cell array filled with empty matrices
num2cell	Converts a numeric matrix into a cell array
mat2cell	Split matrix into sub-matrices and store as cell array
cell2mat	Convert cell array to matrix. All cells must be same type: numeric or text
cellfun	Apply a function to a cell array
celldisp	Display contents of cell array in Command Window

EXAMPLE 15-43

NOTE

iscell returns a logical variable. Logical variables will be covered in a later chapter.

What is stored in the variable by the following commands?

Assume $MI = \begin{bmatrix} 1 & 4 & -8 & 0 \\ 9 & 3 & 12 & 6 \\ 0 & 1 & -5 & 2 \end{bmatrix}$

```
T1='Zero';
S1=42;
CA1={M1,T1,S1};
IC1 = iscell(CA1);
```

   IC1 *contains 1 (true) since the variable* CA1 *is a cell array*

```
IC2 = iscell(M1);
```
   IC1 *contains 0 (false) since the variable* M1 *is a matrix, not a cell array*
```
CA2 = cell(4);
```
   CA2 *is a 4×4 cell array of empty matrices*
```
CA3 = cell(2,5);
```
   CA3 *is a 2×5 cell array of empty matrices*

Sometimes we would like to use a numeric or text function on the contents of part or all of a cell array. One solution is to first extract the contents of the desired cell array elements, then apply the function to the extracted values. However, the function cellfun gives us a somewhat more direct method to accomplish this.

EXAMPLE 15-44

NOTE

cellfun uses a handle (sort of like a pointer) to the function desired. The @ symbol before the function name creates the appropriate handle, telling MATLAB to use that function on the contents of the cell array.

Assume CA4 is defined by the following:
```
CA4={[2 3 5;0 4 6] 'TARDIS' [33;-99];'Dalek' 55 [3 5 4]'}
```
The following appears in the Command Window since there was no semicolon at the end
```
CA4=
[2×3 double]      'TARDIS'       [2×1 double]
'Dalek'           [ 55]         [3×1 double]
[rows,cols] = cellfun(@size,CA4)
```
The following appears on screen
```
rows=
2 1 2
1 1 3

cols=
3 6 1
5 1 1
```
NOTE: This gives us the sizes of the four matrices and the two text strings in CA4.

CA4{1,1} *is 2 × 3 (matrix)*
CA4{1,2} *is 1 × 6 (text string)*
CA4{1,3} *is 2 × 1 (matrix)*
CA4{2,1} *is 1 × 5 (text string)*
CA4{2,2} *is 1 × 1 (matrix)*
CA4{2,3} *is 3 × 1 (matrix)*

Comprehension Check	**15-21**

NOTE: You may want to use `help` or `doc` to investigate the operation of some of the functions in Table 15-9 to solve these problems.

(a) Convert an M × N matrix `BigMat` to an M × N cell array `BigCell` consisting of the scalar values from each corresponding element of `BigMat`.

Example:

If `BigMat` is defined by `BigMat=[2 6;5 8]`, your code would produce the same cell array as `Big-Cell={[2] [6];[5] [8]}`. Your solution must, of course, work for any matrix, whether or not you know its contents.

(b) Assume a cell array `VCell` contains only numeric vectors. Create a matrix `VCellMax` with the same dimensions as the cell array containing the maximum value of each vector in the corresponding location of the cell array.

Example:

If `VCell={[1 5 2] [-3 -5 -6 -1]};`, then `VCellMax=[5 -1]`

15.7 Structure Arrays   ⊟

LEARN TO:	Create structure arrays in MATLAB
	Perform basic operations on structure arrays

As we saw in the previous section, we can use cell arrays to store different types of things in one variable: matrices, text, other cell arrays, etc. To access the different elements of a cell array, we use standard indexing, usually with curly braces, to extract the content of a specific cell or cells.

Sometimes it would be convenient if we could refer to different cells or groups of cells using a descriptive name instead of a number. We could assign specific index values to named variables and use these variable names instead of the numbers they hold to access elements from a cell array. However, there is a more direct method of using names to access specific elements within a mixed-content variable. Such a variable is called a **structure array**, and it inherently uses names instead of numbers to access different parts of the structure.

Structure Arrays

A structure array consists of one or more **structures**, each containing the same **fields**. Each field contains a value or values relating to a specific aspect of the structure.

As an example, we might have a structure array named `MetalData` containing information about metals. `MetalData` might contain the fields `Type` (steel, brass, zinc, etc.), `SpecificGravity`, `ThermalConductivity`, and `Resistivity`.

Defining Structure Arrays

First, a quick review of defining the other data types we have considered.

When defining numeric variables, scalars are defined using a simple assignment statement such as `S1=-699`, whereas vectors and matrices are defined by enclosing the multiple values in square brackets, such as M1=[3.5,-9;13,-4.2]. Indexing can be used to define individual elements or groups of elements. For example, M2(4,3:5)=[9 6 3] places the three values 9, 6, and 3 into the third, fourth, and fifth columns of the fourth row of M2.

Text strings are defined using single quotes: TS1='Neon'. Individual elements or substrings of text variables can be defined using indexing. For example TS2(6:10)= 'Argon' places the text string Argon into elements 6 through 10 of TS2.

Cell arrays are defined in a similar manner, except curly braces are used to enclose the elements composing the cell array. These elements might themselves be scalars, matrices (using brackets), text (using single quotes), or even other cell arrays.

Structure arrays, on the other hand, use a somewhat different scheme, perhaps best explained by way of an example.

EXAMPLE 15-45

IMPORTANT CONCEPT

It may appear that these variables violate the naming rules (no periods allowed), but in the case of structure arrays, the period is not part of a name, but a separator between two names: the structure name and the field name.

IMPORTANT CONCEPT

We can create multiple rows in a structure array by using double index notation, such as MetalData(2,1).

Type the following lines into the Command Window:

```
MetalData.Type='Zinc';
MetalData.SpecGrav=7.14;
MetalData.ThermCond=116;
MetalData.Resistivity=59E-9;
```

Now the structure array MetalData contains a single structure comprising four fields.

We can add another similar structure to the MetalData structure array as follows:

```
MetalData(2).Type='Copper';
MetalData(2).SpecGrav=8.96;
MetalData(2).ThermCond=401;
MetalData(2).Resistivity=16.78E-9;
```

Now MetalData is a 1×2 structure array containing information about the two metals, zinc and copper.

We can add fields to the two structures within the MetalData in a similar manner.

```
MetalData(1).Symbol='Zn';
MetalData(1).Isotopes=[66 67 68];
MetalData(2).Symbol='Cu';
MetalData(2).Isotopes=[63 65];
```

Note that we had to use an index value of 1 to add fields to the first structure (Zinc), since there are now two structures in MetalData. When we first defined this single structure (for Zinc) there was no ambiguity, so an index was not needed, although it could have been used.

NOTE: We have used the field Isotopes to list only the stable isotopes. There are other isotopes of these metals with varying half-lives.

If we type MetalData in the Command Window, it will print the following to the screen:

```
MetalData=
1×2 struct array with fields:
Type
SpecGrav
```

```
        ThermCond
        Resistivity
        Symbol
        Isotopes
```

This tells us how many structures there are in `MetalData` and how they are organized (one row and two columns) as well as the names of the fields within the structures.

If we wish to know the contents of one of the structures within `MetalData`, we can type `MetalData(N)`, where N is the number of the structure for which we wish to see the contents. For example, typing `MetalData(2)` prints the following to the screen:

```
      Type: 'Copper'
   SpecGrav: 8.9600
  ThermCond: 401
 Resistivity: 1.6780e-08
     Symbol: 'Cu'
   Isotopes: [63 65]
```

Comprehension Check   **15-22**

Create a structure array named `Resistors` containing data on three resistors, including fields for Value, Power, Composition, and Tolerance. The three resistors should have the following specifications (enter numeric values only for Value, Power, and Tolerance, not the units):

- 100,000 ohms, ¼ watt, Metal Film, 0.1%
- 2,200,000 ohms, ½ watt, Carbon, 5%
- 15 ohms, 50 watts, Wire Wound, 10%

Extracting Data from Structure Arrays

Accessing the contents of a structure array is very straightforward. We do not have to worry about "contents" as we did with cell arrays since each field has its own data type. (The exception would be a field whose data type was a cell array!)

Using the previous `MetalData` example, if we wanted the type of the first metal, we would type

```
MetalData(1).Type
```

This returns the text string `Zinc`.
If we wanted the first two characters of the type of metal number 1, we would type

```
MetalData(1).Type(1:2)
```

This returns the text string `Zi`.
We can use calculations and functions directly with values or matrices extracted from structure arrays.

```
max(MetalData(1).Isotopes)
```

This returns 68, the mass number (number of protons and neutrons) of the heaviest stable isotope of the first metal, Zinc.

Comprehension Check **15-23**

Use the data stored in MetalData to answer the following questions:

(a) Write a single line of code that will determine the mass in kilograms of one cubic meter of zinc and place the result in MZn.

(b) Assume the number of one of the structures in MetalData is stored in a variable MNum. Write a single line of code that will place the mass number of the lightest isotope of the corresponding metal in the variable LtIso.

Functions Used with Structure Arrays

Table 15-10 lists several functions specifically designed to use with structure arrays.

Table 15-10 Selected functions used with structure arrays

Function	Description
isstruct	Determines if a variable is a structure array
isfield	Determines if a name is a field within a specific structure
struct	Another way to create structures
orderfields	Sorts field names in specified order
setfield	Another way to specify the content of fields
rmfield	Remove a field from a structure array
cell2struct	Convert cell array to structure array
struct2cell	Convert structure array to cell array

EXAMPLE **15-46**

NOTE

isfield returns a logical variable. Logical variables will be covered in a later chapter.

What is stored in the variable by following commands?

Assume MetalData has already been defined as above.

```
IsF1=isfield(MetalData,'Density');
```
   IsF1 *contains 0 (false), since* Density *is not a field defined in* MetalData

```
IsF2=isfield (MetalData,'SpecGrav');
```
   IsF2 *contains 1 (true), since* SpecGrav *is a field defined in* MetalData

---

To create a structure array named MetalSort with the same data but with the field names sorted in alphabetical order,

```
MetalSort = orderfields(MetalData);
```

If you now type **MetalSort** in the Command Window, the following will appear:

```
1X2 struct array with fields:
        Isotopes
        Resistivity
        SpecGrav
        Symbol
        ThermCond
        Type
```

Note that the original structure array `MetalData` remains unchanged. A sorted copy was created and placed in `MetalSort`. Fields do not have to be sorted in ASCII order; other types of sorts, including custom sorts, are possible.

Comprehension Check   15-24

The structure array named `Hdwr` has the following contents:

```
Hdwr=
    type: 'bolt'
    Diam: 1
    Material: 'Brass'
    price: 2.4000
    InStock: 387
    bin: '84B'
```

If the following command is issued, what will appear in the Command Window? (Note that there is no semicolon at the end of the command.)

```
SortH=orderfields(Hdwr)
```

Both cell arrays and structure arrays can contain any mixture of MATLAB data types. For example, we saw earlier that a cell array can contain other cell arrays. In a similar manner, stucture arrays can contain other structure arrays as elements.

EXAMPLE   15-47

Assume we are setting up a structure array to hold veterinary records. The structure will be named `Client`, which will have a field `Name` containing the Client's name. `Client` will also have a field `Pet`, which will be a structure array. This inner structure array will have data on the various pets of the specific client.

The the following code will set up the first client and that client's first pet.

```
Client.Name='Wild Bill';
Client.Pet.Name='Hero';
Client.Pet.Type='Canine';
Client.Pet.Breed='Poodle (min)';
Client.Pet.Birth='Jan. 2005';
```

Note that `Client.Pet` is followed in each case by a period and another field name. This establishes the field `Pet` as a structure array within the larger structure array `Client`.

To add a pet for `Client(1)`, we could use

```
Client.Pet(2).Name='Vegetable';
Client.Pet(2).Type='Feline';
Client.Pet(2).Color='Brown Mackerel Tabby';
Client.Pet(2).Birth='Mar. 1973';
```

This still refers to `Client(1)`, but sets up a second set of entries in `Pet`. To add a second client and a set of records for a pet of that client, we could use

```
Client(2).Name='Tom Terrific';
Client(2).Pet.Name='Mighty Manfred';
Client(2).Pet.Type='Canine';
Client(2).Pet.Breed='Mixed';
Client(2).Pet.Birth='Sep. 1957';
Client(2).Pet.Color='Black & White';
```

Note that not all of the defined fields in the various instances of `Pet` contain data. For example, if you access the color of the first pet of the first client, `Client(1).Pet(1).Color`, the result returned will be `[]`, an empty element.

IN-CLASS ACTIVITIES

ICA 15-1

Which of the following are not valid MATLAB variable names? Circle all that apply. For those that are invalid, state why.

(a) `BigNum.docx`
(b) `Six_Vals`
(c) `This is a very long MATLAB variable name`
(d) `2BR02B`
(e) `Is_This_Valid?`
(f) `Rocket=Space+100;`
(g) `Mult2`
(h) `log`
(i) `GClefSign`
(j) `AbcDEFGHijKLmnopqrstuvWXyz_Has_Lots_of_Characters`
(k) `DoNotPassGo_DoNotCollect$200`

ICA 15-2

Do the following scripted exercise. The >> prompt indicates what you should type into the MATLAB Command Window. Note that some of these commands may result in error messages.

`>>m=[10,20]*30`	% Multiply a row vector containing 10 and 20 by 30, store in variable `m`.
`>>m1=[10;20]*30`	% Multiply a column vector containing 10 and 20 by 30, store in variable `m1`.
`>>n=m+1`	% Add 1 to every element in variable `m`, store in variable `n`.
`>>r=m+m1`	% Add row vector in variable `m` to column vector in variable `m1`, store in `r`. Note that even though the two vectors contain the same number of elements, they are dimensionally inconsistent (one is a row, one is a column) thus the addition is not allowed.
`>>v=[2:2:6]`	% Create a row vector of numbers between 2 and 6, increasing by 2, store in `v`.
`>>u=[1:0.5:3]`	% Create a row vector of numbers between 1 and 3, increasing by 0.5, store in `u`.
`>>j=u+v`	% Add vector `v` to vector `u`. Why will this not work?
`>>a=sin(u)`	% Take the sine of each element in the vector `u`.
`>>b=sqrt(u)`	% Take the square root of each element in the vector `u`
`>>c=u^2`	% Square vector `u`. Why will this not work?
`>>d=u.^2`	% Square each element of vector `u`.
`>>e=[10,20;30,40]`	% Create a matrix of numbers 10 and 20 in the first row, 30 and 40 in the second row, store in the variable `e`.
`>>f=[1,2,3;4,5]`	% Create a matrix of numbers 1, 2, and 3 in the first row, 4 and 5 in the second row, store in the variable `f`. Why will this not work?

ICA 15-3

For the following questions, assume that the following scalar variables have already been defined:
`SG1=2.5;  SG2=0.87;  Vol2=35;  Vol3=273;  Mass1=392;  Mass3=42;`

Determine the contents of the variable defined by each of the following assignment statements:

(a) `SG3=Mass3/Vol3;`
(b) `Vol1=Mass1/SG1;`
(c) `Mass2=SG2*Vol2;`
(d) `SphRad2=(3*Vol2/4/pi)^(1/3);`
(e) `SphRad1=nthroot(3*Mass1/(4*pi*SG1),3);`

ICA 15-4

For each calculation described below, write a single MATLAB command that will perform the stated task.

(a) `L1` and `L2` are the lengths [cm] of two sides of a rectangular solid whose volume [cm³] is stored in `Vol1`. Calculate the length of the third side and place the result in `RS1`.
(b) `L2` and `L3` are the lengths [cm] of two sides of a rectangular solid whose mass [g] is stored in `Mass1` and whose specific gravity is stored in `SG1`. Calculate the length [cm] of the third side and place the result in `RS2`.
(c) The height [cm] of a cylinder is stored in `L2`, its radius [cm] is stored in `L1`, and its specific gravity is stored in `SG1`. Calculate the cylinder's mass [g] and place the result in `M1`.
(d) The diameter [cm] and height [cm] of a cylinder are equal and stored in `L2`, and the cylinder's mass [g] is stored in `Mass2`. Calculate the specific gravity of the cylinder and place the result in `SpG1`.
(e) The height [cm] of a cylinder is stored in `L3`, its specific gravity is stored in `SG2`, and its mass is stored in `Mass2`. Calculate the cylinder's diameter [cm] and place the result in `D1`.

ICA 15-5

For each calculation described below, write a single MATLAB command that will perform the stated task.

(a) Create a variable named `Voltage` that contains the result of the calculation $15-4\,e^{-0.34}$.
(b) Create a variable named `Rate` that contains the result of the calculation $3.7\,\text{Ln}(25)$.
(c) Create a variable named `Comp` that contains the result of the calculation

$$\sqrt[4]{\frac{14\tan(75°)}{\pi\log(42)}}$$

(d) Create a variable named `Wow` that contains the result of the calculation

$$\cos\left(\frac{\pi}{4}+5e^{2.7}\right)$$

(e) Place the value of `e` (the base of the natural logarithm) into the variable `e`. You MUST use MATLAB to determine e—DO NOT simply type in 2.71828...

ICA 15-6

For each of the vectors described below, write a single MATLAB statement that will create that variable. The results of the statements should not be echoed to the screen.

(a) VA=[18 34 0 0 0 −937 1]

(b) VB=$\begin{bmatrix} 9 \\ -8 \\ 0 \\ 44 \\ 42 \end{bmatrix}$

(c) VC is a row vector that contains all odd integers from 80 to 80,000.

(d) VD is a column vector that contains all even integers from 28 to −280,000.

(e) VE is a row vector that contains the sequence $0, 0.07, 0.14, 0.21, \dots, 139.93, 140$

(f) VF is a row vector containing the sequence $10^{22}, 2 \times 10^{22}, 3 \times 10^{22}, \dots, 9.9 \times 10^{23}, 10^{24}$.

(g) VG is a row vector containing the sequence $10^{-8}, 9.5 \times 10^{-9}, 9 \times 10^{-9}, \dots, -9.5 \times 10^{-9}, -10^{-8}$.

(h) VH is a row vector that contains all multiples of 3 from 0 to 300 followed by all multiples of 4 from 0 to 400 followed by all multiples of 5 from 0 to 500.

(i) VI is a 500-element column vector containing two hundred number seventeens followed by three hundred number negative elevens.

(j) VJ is a vector that contains 100 random values between −5 and 5.

(k) VK is a column vector that consists of the elements of VA in part (a) above followed by the elements of VB in part (b) above.

(l) VL is a row vector that consists of the elements of VB from part (b) above followed by the value 999,999,999,999 followed by all integers from 1 to 1000.

ICA 15-7

Assume the following vectors are already defined:

$$V_1 = [3 \ 0 \ -2] \qquad V_2 = \begin{bmatrix} -2 \\ 1 \\ -4 \end{bmatrix} \qquad V_3 = [5 \ 1 \ -3 \ -1] \qquad V_4 = \begin{bmatrix} 0.5 \\ -0.1 \\ 0.2 \\ -0.2 \end{bmatrix}$$

For each of the following operations that does not cause an error, determine the contents of the variable created by the statement. For those operations that will cause an error, explain why the error occurs.

(a) R1=55*V1+77;

(b) R2=V1^2;

(c) R3=V2.^2;

(d) R4=V3+V4;

(e) R5=V3'−V4;

(f) R6=V1.*V3;

(g) R7=V3./V4';

(h) R8=[V1,V3];

(i) R9=[V1;V2];

(j) R10=[V2;V4];

(k) R11=[V1(2:3),V3(1:2:3)];

(l) R12=V3(1:3).*V1;

(m) R13=V4−[5;V2];

(n) R14=V1(3)*V2(2)*V3(1)/V4(4);

ICA 15-8

For each of the following sequences, write a single MATLAB statement that will place that sequence in the stated variable.

(a) Place the sequence $10^{-150}, 10^{-147}, 10^{-144}, \dots, 10^{147}, 10^{150}$ into the row vector P10.

(b) Create a row vector P2 that contains all powers of 2 from $2^0 = 1$ to $2^{64} = 1.844674407370955 \times 10^{19}$.

(c) Create a column vector TanSeq that contains the tangents of all angles from 0 to 2π radians by increments of 0.01π radians.

ICA 15-9

Modify the following statements so that they are correct.

(a) An $N \times 1$ row vector contains N values.

(b) The column vector `CV1` containing $\begin{bmatrix} 0 \\ -5 \\ 2 \end{bmatrix}$ can be defined using the following three statements:

> CV1(1)=0;
> CV1(2)=-5;
> CV1(3)=2;

(c) In order to perform an element-wise addition of two vectors `V1` and `V2` containing the same number of elements, you would use `V1.+V2`, where you must place a dot (`.`) before the plus sign .

(d) The row vector `RV1` containing [8 4 2 1] can be defined using the statement

> RV1=[8;4;2;1];

(e) For any two vectors `V3` and `V4`, they may be combined into a single vector using [V3 V4].

(f) If `VP=[2 -1 3]*[4 -5 -6];` then `VP` contains [8 5 -18].

(g) Assuming `V5` and `V6` have the same number of elements, the following operation is defined only if `V5` is a row vector and `V6` is a column vector:

> V5'-V6.

ICA 15-10

Assume you have three equal-length row vectors, `Diam`, `Ht`, and `SG`, each with N elements. Corresponding elements in the three vectors are measurements of the diameter, height, and specific gravities of a variety of solid conical objects. All lengths (diameters and heights) are in units of centimeters.

(a) With a single MATLAB statement, create a new N element vector named `Vol`, in which each element is the volume [cc] of the object whose diameter, height, and specific gravity are in the same relative positions in `Diam`, `Ht`, and `SG`.

(b) With a single MATLAB statement, create a new N element vector named `Mass`, in which each element is the mass [kg] of the object whose parameters are in the same relative positions in `Diam`, `Ht`, `SG`, and `Vol`.

(c) Determine the average volume of all cones and place the result in `AvgCone`.

(d) Determine both the heaviest and lightest cone, placing the results in `HeavyCone` and `LightCone`.

(e) Determine the height of the heaviest cone and place the result in `HeavyHt`. Note that the heaviest cone is not necessarily the largest. HINT: Use `doc` to learn more about the `max` function.

ICA 15-11

Write the MATLAB code necessary to create the variables in (a) through (d) or calculate the vector computations in (e) through (q). If a calculation is not possible, explain why. You may assume that the variables created in parts (a) through (d) are available for the remaining computations in parts (e) through (q). For parts (e) through (q) when it is possible, determine the expected result of each computation by hand.

(a) Save vector [3 -2 5] in `Va`.

(b) Save vector $\begin{bmatrix} -1 \\ 0 \\ 4 \end{bmatrix}$ in `Vb`.

(c) Save vector [9 -4 6 -5] in `Vc`.

(d) Save vector $\begin{bmatrix} 7 \\ -3 \\ -4 \\ 8 \end{bmatrix}$ in Vd.

(e) Convert Vd to a row vector and store in variable Ve.

(f) Place the sum of the elements in Va in the variable S1.

(g) Place the product of the last three elements of Vd in the variable P1.

(h) Place the cosines of the elements of Vb in the variable C1. Assume the values in Vb are angles in radians.

(i) Create a new 14-element row vector V14 that contains all of the elements of the four original vectors Va, Vb, Vc, and Vd. The elements should be in the same order as in the original vectors, with elements from Va as the first three, the elements from Vb as the next three, and so forth.

(j) Create a two-element row vector V2 that contains the product of the first two elements of Vc as the first element and the product of the last two elements of Vc as the second element.

(k) Create a two-element column vector V2A that contains the sum of the odd-numbered elements of Vc as the first element and the sum of the even-numbered elements of Vc as the second element.

(l) Create a row vector ES1 that contains the element-wise sum of the corresponding values in Vc and Vd.

(m) Create a row vector DS9 that contains the element-wise sum of the elements of Vc with the square roots of the corresponding elements of Vd.

(n) Create a column vector EP1 that contains the element-wise product of the corresponding values in Va and Vb.

(o) Create a row vector ES2 that contains the element-wise sum of the elements in Vb with the last three elements in Vd.

(p) Create a variable S2 that contains the sum of the second elements from all four original vectors, Va, Vb, Vc, and Vd.

(q) Delete the third element of Vd, leaving the resulting three-element vector in Vd.

ICA 15-12

For each of the following problems, write a single MATLAB statement that will accomplish the stated purpose.

Assume the following vectors have already been defined.

```
V1 = [125, 367, 498, 24, 63, 0.25, 543, 89];
V2 = [75, 32, 0.67, 34];
V3 = [98, 56, 56, 1, 98, 56, 87, 98];
V4 = [5, 10, 100, 0.5, 67, 87];
```

(a) Place the largest value in V1 into BigVal.

(b) Find the length of the vector with the least elements from the vectors above and place the length of that vector into **MinL**. The CODE should determine the vector with the least elements, not you!

(c) Create a vector **UV** that contains one of each value that is stored in vector **V3**, eliminating duplicates.

```
Example: V3 = [98, 56, 542, 1, 98, 56, 87, 98]; UV = [98 56 1 87]
```

ICA 15-13

Assume you have two equal-length row vectors IV1 and IV2. For each of the following tasks, write a single MATLAB command to accomplish the desired goal.

(a) Place the indices of all negative elements of IV1 into N1.

(b) Place the values of all negative elements of IV1 and IV2 into a single row vector N2.

(c) Place the indices of all elements of `IV1` that are not greater than the corresponding elements of `IV2` into `NG`.

(d) Place the values of all elements in `IV1` that are equal to the corresponding elements in `IV2` into `EQ`. (HINT: in MATLAB, to ask if two things are equal, write the equal sign twice. Thus to ask if X is equal to Y, you write X==Y.

ICA 15-14

Write the MATLAB code necessary to create the variables in (a) through (d) or calculate the matrix computations in (e) through (m). If a calculation is not possible, explain why. You may assume that the variables created in parts (a) through (d) are available for the remaining computations in parts (e) through (m). For parts (e) through (m) when it is possible, determine the expected result of each computation by hand.

(a) Save matrix $\begin{bmatrix} 3 & 8 \\ 5 & 7 \end{bmatrix}$ in variable `W`

(b) Save matrix $\begin{bmatrix} 4 & 8 & 1 \\ 5 & 1 & 6 \end{bmatrix}$ in variable `X`

(c) Save matrix $\begin{bmatrix} 3 \\ 5 \end{bmatrix}$ in variable `Y`

(d) Save matrix $\begin{bmatrix} 4 & 2 \\ 2 & 9 \end{bmatrix}$ in variable `Z`

(e) Add `W+Z`, save in variable `A`
(f) Subtract `Y−Z`, save in variable `B`
(g) Multiply `W*X`, save in variable `C`
(h) Multiply `X*Z`, save in variable `D`
(i) Transpose `X`, save in variable `E`
(j) Multiply `W*Z` term-by-term, save in variable `F`
(k) Square `Z`, save in variable `G`
(l) Square `X`, save in variable `H`
(m) Add 20 to each element in `X`, save in variable `I`

ICA 15-15

Assuming `t=[9  10;11  12]` and `v=[2  4;6  8;10  12]` are currently stored in MATLAB's workspace, what is the output of each of the following statements? If an error will occur, explain why.

(a) `A=t(1,1)+v(1,1)`
(b) `B=t[1,1]+v[1,1]`
(c) `C=t*v`
(d) `D=v*t`
(e) `E=v(:,2)+t(:,2)`
(f) `F=v(2,:)+t(2,:)`
(g) `G=t(1,1)*v(3,:)`
(h) `H=t(1,:)*v(1,:)`
(i) `J=t(1,:).*v(1,:)`

ICA 15-16

Determine solutions to the following problems

(a) Assume the matrix `SpringData` has two columns. Each row (pair of numbers) represents data about a specific spring. The first column contains a force in newtons, and the second column contains a displacement in inches that the corresponding force caused in that specific spring. With a single MATLAB command, add a third column to `SpringData` that contains the corresponding spring constant [N/m] for each data pair.

(b) Assume two vectors containing the same number of elements have already been defined. `RV` is a row vector and `CV` is a column vector. Write a single MATLAB command that will calculate the sum of the products of the corresponding elements in `RV` and `CV` and place the result in `SOP`. For example, if `RV=[2 3 4]` and `CV=[5;-1;6]`, then `SOP=31`. Once you have a working solution, find a second completely different solution. HINTS: (a) consider how matrix multiplication works, (b) use `help` or `doc` to learn about the `sum` function.

(c) Assume you have a $2 \times N$ matrix OD. The first row contains measurements of the volumes of various objects [m^3]. The elements in the second row are measurements of the specific gravities of the corresponding objects.

Determine the total mass in kilograms of all of the objects. In other words, perform the calculation

`Totalmass=1000(OD(1,1)OD(2,1)+OD(1,2)OD(2,2)+ . . . +OD(1,N)OD(2,N))`

Determine two completely different one-statement solutions.

ICA 15-17

For each of the following problems except part (h), write a single MATLAB statement that will accomplish the stated purpose.

(a) Create a 150×150 matrix `N9999` that contains all zeros except for the major diagonal (upper left to lower right). All elements on the major diagonal should contain the value -9999.

(b) Create a 1200×1200 matrix `p75` that contains all zeros except for the minor diagonal (lower left to upper right). All elements on the minor diagonal should contain the value 75.

(c) Place the number of rows of matrix `Ma` into `MaRows` and the number of columns of matrix `Ma` into `MaCols`.

(d) Assume matrix `Mx` contains a single nonzero value. Place the row number containing this nonzero value into `NZR` and the column number into `NZC`.

(e) Assume matrix `Mz` has already been defined. Place the row numbers of all elements in `Mz` that are negative into vector `MzRN` and the corresponding column numbers into `MzCN`.

(f) Assume that matrix `MS` has already been defined. Place the elements on the major diagonal into row vector `MSD`.

(g) Create a 5×9 matrix `RM` that contains random numbers chosen between -50 and $+50$.

(h) Place the smallest value and the location within the matrix `RM` (created above) into `MinRM`. You may use more than one MATLAB statement to accomplish this task.

ICA 15-18

Assume you have an $N \times M$ matrix named `Gonzo`. For each of the following tasks, write a single MATLAB command that will accomplish the goal. You may use variables created in one part in order to solve subsequent parts.

(a) Place all values in `Gonzo` that are greater than or equal to 100 into the vector `GTE100`.

(b) Create pairs of row and column index values in the vectors `GR` and `GC` that correspond to all elements of `Gonzo` that are equal to 0.

(c) Place the number of zero elements in row 1 of `Gonzo` into the scalar `R1Z`.

(d) Replace all zero elements in `Gonzo` with the value `-inf` (negative infinity).

(e) Round all positive elements in Gonzo to one decimal place.

ICA 15-19

For each of the following tasks, write a single MATLAB command that will accomplish that goal.

(a) Place the text *All the king's horses* into the variable `Humpty`.

(b) Replace the word *horses* in `Humpty` with the word *men* and store the modified phrase in `Dumpty`. `Humpty` should remain unchanged.

(c) Combine the phrases in `Humpty` and `Dumpty` with the word *and* between the two phrases. Place the new combined phrase in `HD`.

(d) Extract appropriate individual letters from `HD` to form the word *egghead* and place this word in the variable `E`.

(e) Delete the word *king's* from `Humpty`. There should be a single space between *the* and *horses* after the deletion.

ICA 15-20

Determine the contents of the variables created or modified by the following MATLAB commands. Assume the following text strings have already been defined:

`AI` contains *Artificial Intelligence*
`Smell` contains *Dog breath*

(a) `Silly=[Smell(1:4) AI(1:3)];`
(b) `Time=Smell([5,7:9]);`
(c) `AI(1:11)=[];`
(d) `Double(2:2:2*length(Smell))=Smell;`
     `Double(1:2:2*length(Smell))=Smell;`

ICA 15-21

For each of the following tasks, write a single MATLAB statement to achieve that goal.

(a) Create a text string named `Greek` that begins with `Alpha`, ends with `Omega`, and has 250,000 blanks in between.

(b) Assume the text string `TS` contains a single word. Using only a single MATLAB command, create a text string named `Padded` that contains the same word stored in `TS`, but with a space between each character of the original word.

For example, `Rabbit` becomes `R a b b i t`.

(c) If `N1='5'` and `N2='8'`, then `N3=N1+N2` places the value 109 in `N3`. (Try it!) This is because you are actually adding the ASCII codes for the CHARACTERS '5' and '8', NOT the values 5 and 8. Write a statement that will correctly add the values represented by the characters in `N1` and `N2`, giving a result of 13 in this specific case. This should function correctly for any single digit, non-negative integers stored in `N1` and `N2`. HINT: Refer to Table 15-8.

(d) Assume that a text string `TS` already exists. Convert all letters in the first half of the text string to uppercase and leave the characters in the second half unchanged. You may assume there are an even number of characters in `TS`.

ICA 15-22

For each of the following problems, write a single MATLAB statement that will accomplish the stated purpose. Assume a text string `TS1` has already been defined.

(a) Create a vector `LetLoc` with the same number of elements as `TS1`. For each element of `TS1` that is alphabetic, the corresponding element of `LetLoc` should equal 1. All other elements of `LetLoc` should be 0. Example: `TS1='%@3Gb6'` returns `LetLoc=[0 0 0 1 1 0]`

(b) Create a vector `Alphas` that has the numeric positions of all alphabetic characters in `TS1`. Example: `TS1='%@3Gb6'` returns `Alphas=[4 5]`

(c) Place the numeric index of the first alphabetic character in `TS1` into `Alpha1`. Example: `TS1='%@3Gb6'` returns `Alpha1=4`

(d) Place the last alphabetic character in `TS1` into `LastLetter`. Example: `TS1='%@3Gb6'` `LastLetter=b`

(e) Find all occurrences of the string `@3G` in string `TS1` and place the locations of the first character of each such substring (`@`) into `Str3G`. Example: `TS1='%@3Gb6kl@3G9@33G'` returns `Str3G=[2 9]`

ICA 15-23

Each of the following questions contains a description of a desired result and code that was written to accomplish that goal.

For each item, state whether or not the code performs the desired task, and if not, show one way to modify the code so that it works correctly.

(a) Place the three words *The*, *red*, and *bee* into a 1×3 cell array `Words`.

```
Words{1:3}=['The','red','bee'];
```

(b) Create a 2×2 cell array `Nums` containing the values 15 and 18 in the first row, and the values −9 and −6 in the second row.

```
Nums={[15,18;-9,-6]};
```

(c) Place the entire lowercase English alphabet (a through z) into a one-element cell array `Alphabet`.

```
Alphabet{1}='a':'z';
```

(d) Replace the top right element in the 2×2 cell array `Nums` with the word *BOGUS*.

```
Nums(1,2)='BOGUS';
```

(e) Assume the 1×3 cell array `Stuff` contains the 2×3 matrix `[1 3 5;2 8 4]`, the text string *Goofball*, and a 1×1 cell array containing the number 42. You wish to divide the value stored in the cell array in the third element of `Stuff` (42) by 17 and place the result in the numeric scalar `R`.

```
R=Stuff{3}/17;
```

(f) Given the same cell array `Stuff` as described in the previous problem, you wish to delete the four letters *oofb* from the second element of `Stuff`.

```
Stuff{2}(2:5)=[];
```

(g) You wish to multiply the values in the top row of the first element of `Stuff` by the values in the second row of the first element of `Stuff` (element-wise multiply) and place the result in the numeric row vector `Prod`.

```
Prod=Stuff(1){1,:}*Stuff(1){2,:};
```

ICA 15-24

Header text in an Excel spreadsheet has been imported into MATLAB as a single-row cell array named `Headers`. The contents of `Headers` is

```
{'Temp (T) [°C]' 'Volume (V) [cc]' 'Amount (n) [mol]'}
```

You are going to process numerical data also imported from the spreadsheet (in a matrix), doing some unit conversions and creating some new data columns. You wish to modify the cell array `Headers` to reflect these changes and additions.

Show the code you would use to modify the individual elements as well as add two new elements to `Headers` so that its new contents is

```
{'Temp (T) [K]' 'Volume (V) [m^3]' 'Amount (n) [mol]'
 'Pressure (P) [atm]' 'Status'}
```

Show individual commands to modify each separate element; do not simply redefine `Headers` with a single statement.

ICA 15-25

A 4×2 cell array `RCData` contains data concerning a number of RC (resistor-capacitor) circuits. The first row contains `{'R [ohms]' 'C [farads]' 'Initial Vc' 'Final Vc'}`. Each element in the second row is an $N \times 1$ vector containing data referred to in the corresponding column of the first row. The first column vector has resistance values, the second column contains capacitance values, the third column contains the initial voltage for each RC pair in the first two columns, and the last column contains desired final voltages. The two voltages will always have the same sign (+ or −), and the magnitude of the final voltage will always be less than the initial voltage.

We wish to determine how long each circuit represented in these vectors will take for the capacitor to discharge from the initial voltage to the final voltage.

The final voltage can be calculated by

$$V_f = V_0 \, e^{\frac{-t}{RC}}$$

Add a fifth column to RCData. The first row element should be the text 'Time Required'. The second row element should be an $N \times 1$ vector containing the time required for the capacitor to discharge from V_0 to V_f (in seconds).

Example: If R = 120 KΩ, C = 50 nF, V_0 = 10 V, and V_f = 0.5 V, then t, the time required for the capacitor to discharge from V_0 to V_f, is 17.97 ms.

ICA 15-26

Assume a cell array CA1 has already been defined. The first row contains a text string and a scalar, and the second row contains a row vector and a matrix with at least two rows and at least two columns.

For each of the following problems, write a single MATLAB statement that will accomplish the stated purpose. You may use variables created in any problem to help solve problems farther down the list.

(a) Place the number of characters in the text string into LenTex.
(b) Place the number of alphabetic characters in the text string into NumChar.
(c) Place the number of nonalphabetic characters in the text string into NumNon.
(d) Place the number of values in the vector into LenVec.
(e) Place the sum of all elements in the vector except the last two into SumM2.
(f) Place the number of rows of the matrix into MR and the number of columns of the matrix into MC.
(g) Create a new 2x2 cell array CA1SubM from the matrix that contains the same contents as the matrix in CA1 but divided into four sections:
  • A scalar equal to the top left element from CA1
  • A row vector equal to the remainder of the first row
  • A column vector equal to the remainder of the first column
  • A matrix with all remaining elements.

ICA 15-27

You wish to design a structure array to maintain a list of small hardware needed for a variety of different products. The structure should include the following information:

• An alphanumeric product code for each product being produced
• For each product, the general types of hardware needed, such as bolt, screw, washer, nut, etc.
• For each specific hardware type, the specifications, including (as appropriate) length, diameter, material, subtype, thread type, etc. Note that there may be more than one specification for a given type of hardware, such as bolts with two different lengths and diameters.
• For each hardware item, the quantity required for the product in question.

Create a structure array containing the parts lists for the following two products.

Product 1: Product code WP42

• 4 – steel hex-head bolts, 3/4 inch long, 1/8 inch diameter
• 8 – steel nuts 1/8 inch diameter

Product 2: Product code ES65

• 16 – steel torx-head bolts, 1 inch long, 5/16 inch diameter
• 8 – steel lock washers 5/16 inch diameter
• 16 – steel nuts 5/16 inch diameter

ICA 15-28

You are setting up a structure array named `Client` to hold records of houses available for beach vacation rentals.

`Client` will contain a field `Name` containing the name of the person that owns the rental property. `Client` will also contain a structure array named `House` that contains details on each specific house available for rental.

Show the code to set up `Client` with the following data.

Client 1 is C. P. Zhang, who owns three rental houses as specified

House 1
Address: 123 Abalone Ave.
Number of bedrooms: 4
Maximum Occupancy: 12
Weekly Rental Rate: $2200

House 2
Address: 453 Cockle Ct.
Number of bedrooms: 6
Maximum Occupancy: 20
Weekly Rental Rate: $3500

House 3
Address: 989 Tuna Terrace
Number of bedrooms: 5
Maximum Occupancy: 18
Weekly Rental Rate: $2900

Client 2 is T. U. Desti, who owns one rental house
Address: 39 Whelk Way
Number of bedrooms: 2
Maximum Occupancy: 5
Weekly Rental Rate: $900

Client 3 is O. M. Geexer, who owns one rental house
Address: 14 Conch Circle
Number of bedrooms: 1
Maximum Occupancy: 2
Weekly Rental Rate: $350

CHAPTER 15

REVIEW QUESTIONS

1. You have three temperature values [°C] stored in the three scalar variables T1, T2, and T3.
   You have three pressure values [atm] stored in the three scalar variables P1, P2, and P3.
   You have a scalar variable n containing an amount of substance [mol].

   If these values represent parameters measured for gases in three reaction vessels, determine the volume of each reaction vessel in cubic centimeters. Place the three results in the scalar variables V1, V2, and V3. In the solution, you should set up a scalar variable R to hold an appropriate value for the ideal gas constant, then use that variable in any equations rather than hard-coding the value multiple times. If you need any conversion factors, these should also be placed in variables for use in the multiple equations.

2. You have N temperature values [°C] stored in the vector Temp.
   You have N pressure values [atm] stored in the vector Press.
   You have a scalar variable n containing an amount of substance [mol].

   If these values represent parameters measured for gases in N reaction vessels, determine the volume of each reaction vessel in cubic centimeters. Place the results in the variable Vol. In the solution, you should set up a scalar variable R to hold an appropriate value for the ideal gas constant, then use that variable in any equations where it is needed. If you need any conversion factors, these should also be placed in variables for use in the equations.

3. You have been collecting data on a nonlinear amplifier. Ideally, the output voltage [mV] should equal the input voltage [mV] squared. Thus an input of 5 mV should yield an output of 25 mV. You have measured the output at each integer value from 1 mV to N mV and recorded the outputs (in mV) in the vector SqOut. Note that the INDEX of each output value equals the input value in mV.

   Create a vector OutOfSpec that contains a list of all inputs that generated an output with difference from the ideal value of more than 1%. Note that the difference can be above or below the ideal value.

   Example: If SqOut contains [0.9985 4.052 8.973 15.81 25.15], OutOfSpec would contain [2, 4].

4. Assume four row vectors named Prod10, Prod11, Prod12, and Prod13 contain data on production of various electronic devices at your company during the four years 2010, 2011, 2012, and 2013, respectively. The corresponding elements in each represent the number of a specific part manufactured during that year. For example, the first element might contain the number of 2N3904 transistors produced during each year, whereas the fifth element might contain the number of IC555 timer chips produced. You may assume all four vectors contain the same number of elements, corresponding to the same produced items.

   Write a single line of code to answer each of the following questions. You may use the results of any question to answer subsequent questions if desired. Note that your solutions should work regardless of the number of elements in the four vectors.

   (a) Create a new vector TotalProd that contains the total number of each item produced during the four-year period. Note that TotalProd will have the same number of elements as the original four vectors.

   (b) Create a new vector AvgProd that contains the average number produced per year of each item during the four-year period. Note that AvgProd will have the same number of elements as the original four vectors.

   (c) Create a four-element column vector YearProd that contains the total number of all units produced during each year. 2010 production should be in the first (top) element.

   (d) Determine the maximum number of any type of device produced during each year and place the results in a four-element column vector MaxProd.

(e) Determine the maximum number of any device produced during any year and place the result in the scalar `OverallMax`.

(f) If your company makes a profit of one-fifth of one cent on each device produced, regardless of type, determine the total profit made during this four-year period and place the result in `Profit`. Your result should be in dollars.

5. You have a $2 \times N$ matrix named `GasData`.

    The first row of `GasData` contains N temperature values [°C].

    The second row contains N pressure values [atm].

    You also have a scalar variable n containing an amount of substance [mol].

    If these values represent parameters measured for gases in N reaction vessels, determine the volume of each reaction vessel in cubic centimeters. Create a third row of `GasData` to store the results. In the solution, you should set up a scalar variable R to hold an appropriate value for the ideal gas constant, then use that variable in any equations where it is needed. If you need any conversion factors, these should also be placed in variables for use in the equations.

6. Assume the matrix `M99` has at least two rows and at least two columns. Write no more than two MATLAB commands that, when executed in sequence, will create a 2×2 matrix `Corn` containing the four corner elements of `M99`, but with their positions swapped diagonally. For

    example, if `M99` had the contents $\begin{bmatrix} 6 & 0 & 3 & 9 \\ -7 & 4 & 5 & 8 \\ 1 & 3 & 9 & 2 \end{bmatrix}$, then `Corn` contains $\begin{bmatrix} 2 & 1 \\ 9 & 6 \end{bmatrix}$.

MATLAB
examples

7. Assume a matrix named `Prod` contains data on production of various electronic devices at your company during several years. Each row of the matrix contains production data for a single year. The first element in each row contains the year, e.g., 2007 or 2012. The remaining elements in each represent the number of a specific part manufactured during that year. For example, the second element might contain the number of 2N3904 transistors produced during each year, whereas the fifth column might contain the number of IC555 timer chips produced. You may assume that corresponding elements in each row contain production numbers for the same type of device.

    Write a single line of code to answer each of the following questions. You may use the results of any question to answer subsequent questions if desired.

    A sample `Prod` matrix is provided online. Note that your solution must work for any properly formatted matrix `Prod`.

    (a) Create a row vector `TotalProd` that contains the total number of years in the first element and the total number of each item produced during all listed years in the remaining elements. Note that `TotalProd` will have the same number of elements as the number of columns in the `Prod` matrix.

    (b) Create a row vector `AvgProd` that contains the total number of years in the first element and the average number of each item produced during all listed years in the remaining elements. Note that `TotalProd` will have the same number of elements as the number of columns in the `Prod` matrix.

    (c) Create a two-column matrix `YearProd`. The first column should contain the same years as those in the first column of `Prod,` and the second column should contain the total number of all units produced during each year.

    (d) Create a two-column matrix `MaxProd`. Determine the maximum number of any type of device produced during each year and place the results in the second column of the corresponding row in `MaxProd`. The first column should contain the years.

    (e) Determine the maximum number of any device produced during any year and place the result in the scalar `OverallMax`.

    (f) If your company makes a profit of one-fifth of one cent on each device produced, regardless of type, determine the total profit made during all listed years and place the result in `Profit`. Your result should be in dollars.

    (g) The solution to this problem is considerably more complicated than the corresponding problem using vectors (Review Question 4). What is the major advantage gained by this extra complexity?

8. You are studying the effects of climate change on the Arctic ice sheet. You have collected data at several locations on different dates. The data is stored in the matrix `Ice`. Each row of `Ice` contains the measurements made on a specific date. Each column of `Ice` contains the measurements made at a specific location. You may assume that `Ice` contains at least two rows and at least two columns, but may be considerably larger in either or both dimensions. As part of analyzing this data, write a single MATLAB command to achieve each goal below. The results from each step may be used in subsequent steps as necessary. HINT: Use `help` or `doc` to learn more about the functions `mean` and `min`.

(a) Place the number of rows in `Ice` in the variable `Dates` and the number of columns in `Ice` in the variable `Locs`.

(b) Add another row at the bottom of `Ice` that contains the averages of the values (ice thicknesses) in the corresponding columns.

(c) Add another column on the right side of the modified matrix `Ice` that contains the averages of the values (ice thicknesses) in the corresponding rows. Note that after this step, the lower right corner value will be an overall average of all data in the original matrix.

(d) Add another column on the right side of the modified matrix `Ice` that contains the minimum of the values (ice thicknesses) in the corresponding rows. Note that after this step, the lower right corner value will be the minimum of the average values that were added as a new bottom row in the second step above.

MATLAB *examples*

9. You are studying the properties of tiny spheres generated by an automated process. You collect data on a large number of the spheres and record this data in the two-row matrix `SphereData`. Each pair of values (each column) represents the data for a specific sphere. The first row of `SphereData` contains the diameters of the spheres in millimeters; the second row contains the masses of the corresponding spheres in micrograms. Write a single MATLAB command to achieve each of the following tasks.

(a) Create a third row in `SphereData` that contains the volumes [cubic millimeters] of the corresponding spheres represented in the first two rows of `SphereData`.

(b) Create a fourth row in `SphereData` that contains the specific gravities of the corresponding spheres represented in `SphereData`.

(c) Create a column vector `DiamData` that contains the following information:

First element: Average diameter of all spheres
Second element: Minimum diameter of all spheres
Third element: Maximum diameter of all spheres

MATLAB *examples*

10. Assume you have a three column matrix named `TempData` containing the daily minimum, average, and maximum temperatures at the geographic center of the United States (in western South Dakota at latitude 44° 58' 02.1" north, longitude 103° 46' 17.6" west) for an entire year. This matrix will have either 365 or 366 rows, depending on whether or not it is a leap year. The first column contains the daily minima, the second column the daily averages, and the third column the daily maxima.

You also have a three-element row vector `Overall` that contains long-term data for the same location prior to the year's data in `TempData`. The first element contains the minimum temperature ever recorded there, the second element contains the long-term average temperature, and the third element contains the maximum temperature ever recorded.

Perform the following calculations with this data.

(a) For each of the minimum temperatures in the first column of `TempData`, determine if it is less than the long-term minimum in `Overall`. Create a vector named `ColdDays` containing the indices of all such values.

(b) For each of the maximum temperatures in the third column of `TempData`, determine if it is greater than the long-term maximum in `Overall`. Create a vector named `HotDays` containing the indices of all such values.

(c) For each of the average temperatures in the second column of `TempData`, determine if it is greater than the long-term average in `Overall`. Create a vector named `OverAvg` containing the indices of all such values.

(d) Create a three-element vector named New. The first element contains the number of days in TempData that had a minimum temperature less than the previous overall minimum. The second and third elements contain the number of days in TempData that are greater than the previous long-term average and maximum, respectively.

(e) Create a three-element vector named Change. The first element contains the difference between the lowest temperature in TempData and the previous overall minimum. Note that if the minimum temperature in TempData is less than the previous overall minimum, this value will be negative. The second element of Change contains the differences between the average of the average temperatures in TempData and previous overall average. The third column contains a similar change value for the maximum temperatures.

(f) Assuming that half of all days in a year would typically be greater than the long-term average and half would be less than the long-term average, determine how many more days in TempData exceeded the long-term average than expected and store this value in Excess. For example, if TempData contains data for a leap year, you would expect 183 days to have an average temperature above the long-term average. If there were 191 such days, then Excess would contain the value 8. Note that if the year is not a leap year, you will probably get an answer that is not an integer. If there were no such days, then Excess will contain zero.

MATLAB *examples*

11. Assume you have a four-column matrix named BoltStock25 containing information about quarter-inch-diameter bolts your company has in stock. The first column contains the bolt length, the second column contains the number in stock, the third column contains the minimum stock that should be maintained (based on usage), and the fourth column contains the maximum number allowed in stock (based on storage space). You wish to determine how many bolts of each type to order according to the following guidelines:

- If the number in stock falls to less than the minimum stock desired plus 50 for any specific length, more of that length should be ordered.
- All orders must be in multiples of 100 of any specific length.
- The current stock plus the number ordered for a specific length must not exceed the maximum allowable stock.
- When an order is placed for a specific length, the total stock should then be as close as possible to the maximum allowed stock.
- You may assume that the maximum stock allowed is always at least 150 more than the minimum stock desired.

Create a new two-column matrix BoltOrder25. The first column should contain the lengths of all bolts that will be ordered. The second column will contain the number of each length to be ordered.

Example: One row of BoltStock25 contains [3, 219, 175, 900]. Since the number in stock (219) is less than 50 more than the minimum stock desired (175), more should be ordered to bring the total as close to 900 as possible. This means 600 should be ordered, bringing the total to 819.

12. One very old method of sending secret messages is to hide the message inside another (usually longer) message. A few possible cases are illustrated by the following questions.

(a) Assume Loud='THUNDERSTORM';
What is placed in Fate by the command Fate=Loud([5,10,10,12]);

(b) What is in Hidden after the following two MATLAB commands are executed?
En='Summer angst tells a sob story.';Hidden=En(4:4:length(En)−7);

(c) Assume Composer='Beethoven';
Replace a single letter in Composer with a blank and place the result in the variable Root_Crop_Cooker. The contents of this new variable should be appropriate for the variable name.

(d) Assume the following are defined:
```
Code='THREE BIRDS CAN EAT SUSHI';
Skip=-4;
Offset=4;
```
What is placed in Msg by the following command?
```
Msg=Code(length(Code)−Offset:Skip:1);
```

(e) You wish to send the secret message "April first" to a colleague. Conceal this message inside another text string stored in `Hide`. The string in `Hide` should be a complete sentence or phrase, not merely gibberish.

For your answer, show both the MATLAB statement that creates `Hide` and the MATLAB command that would be used to extract the message hidden within and place it in `Message`.

13. Create a two-row cell array named `IGP` that contains the initial data from Review Question 5. Each element of the first row should be a text string describing the contents of the element immediately below in row 2. The first element of row 2 will be the $2 \times N$ matrix that was originally stored in `GasData` in Review Question 5. The second element of row 2 should contain a scalar value representing an amount of substance [mol]. The third element of the second row should contain an appropriate value of the ideal gas constant. You should add any further columns to the cell array as needed to hold values for any conversion factors needed.

Add a third row to the matrix in element (2, 1) of `IGP` that will contain the volumes [cm^3] of each reaction vessel. (See Review Question 5 for further detail if this is unclear.)

14. Assume that a cell array `PrCA` already exists and contains a single item: the matrix `Prod` from Review Question 7. None of the other results required by that problem have been determined, however.

Augment `PrCA` so that it contains the original matrix `Prod` as well as all of the results required by Review Question 7. You may choose any organization of the cell array that seems logical to you. The information stored in each variable in Review Question 7 should be added to the cell array with a single line of code. You may NOT simply use an assignment statement with the results from Review Question 7.

A sample `PrCA` cell array is provided online. Note that your solution must work for any properly formatted initial matrix in `PrCA`.

The sample matrix provided has data for five different devices (columns 2 through 6). Add a five-element cell array to `PrCA` that contains the following part designations, corresponding to the five devices listed: `2N3904, 2N3906, 2N2222, IC555, IC741`. This should be the first element (first row, first column) of `PrCA`, thus the other entries will be shifted to other locations. Note that this is specific to the sample matrix provided. Other matrices would, of course, have different part designations.

15. (a) Create a one-row cell array named `Cylinder` containing one row with the following elements:
- The text string `Height(cm)` The value 12
- The text string `Diameter(cm)` The value 1.5
- The text string `Specific Gravity` The value 3.5

(b) Create a one-row cell array named `Pipe` containing the following elements:
- The text string `Height(cm)` The value 20
- The text string `OD/ID(cm)` A row vector containing the values 2.5 and 2.2

(c) Create a 2×3 cell array named `Parts` containing the following elements:
- The first row should contain the text string `Quantity`, the value 100 and the cell array `Cylinder`.
- The second row should contain the text string `Quantity`, the value 15, and the cell array `Pipe`.

(d) Using the `Parts` cell array, create the following variables. Note that your code should work if the contents of `Parts` changes—DO NOT hard-code any values.
- `NumCyl`, containing the number of cylinders (first row, second column of `Parts`)
- `VolCyl`, containing the volume (cm^3) of one cylinder ($V = \pi r^2 H$), where the diameter is in the fourth element of the cell array in the first row, third column of `Parts`, and the height is in the second element of that same cell array in `Parts`.
- `MassCyl`, containing the mass in grams of one cylinder. (You should be able to figure out where the specific gravity is by now.)

- NumPipe, containing the number of pipes (second row, second column of Parts)
- IVolPipe, containing the internal volume (cm³) of one cylinder ($V=\pi r^2 H$). You will have to determine where the values needed are stored using the explanations for NumCyl and VolCyl as a guide. Note that OD/ID means Outside Diameter/Inside Diameter.
- Add two columns to the cell array Parts. The fourth column of Parts should contain the text Total Mass(g) in the first row and Total Volume(cm³) in the second row. The fifth column of Parts should contain the corresponding TOTAL mass of all of the cylinders and the TOTAL internal volume of all of the pipes.

16. Assume you have a 2×1 cell array named Stocks. The top element is a cell array containing a single row of N text strings, each of which is the abbreviation for a stock on the New York Stock Exchange. The bottom element of Stocks is a $3 \times N$ matrix containing information about the stocks listed in the top row. Each column of this matrix (three values) represents the values of the stock in the corresponding position in text strings of the top row of Stocks at two different times as well as the number of shares owned. The top row of this matrix represents the original purchase price, the second row represents the current price, and the bottom row contains the number of shares. Answer the following questions. Each should be answered with a single MATLAB statement if possible.

(a) Place the abbreviation of the stock with the highest current price in CurrentHigh.

(b) Place the abbreviation of the stock with the lowest purchase price in Cheapest.

(c) Create a single row M element cell array named Best containing the abbreviations of all stocks that have increased in value by more than 25%.

(d) Create a 2×1 cell array named Profits. The top element contains the abbreviations of the stocks found in part (c) above in an M element single-row cell array, and the bottom element contains an M element row vector containing the amount of profit that would be made by selling each stock. (Calculated by (Current Price – Purchase Price) * Number of shares.)

(e) Place the total profit that would be made by selling all of the shares from the preceding question into the variable TotalProfits.

17. Assume that the matrix Prod from Review Question 7 already exists, but none of the other results required by that problem have been determined.

Create a single structure array named PrStr that contains the original matrix Prod as well as all of the results required by that problem. You may choose any organization of the structure array that seems logical to you as well as any field names that seem reasonable to you. You may NOT simply use the cell2struct function with the result from Review Question 14.

The sample matrix provided has data for five different devices (columns 2 through 6). Add a five element cell array to PrStr that contains the following part designations, corresponding to the five devices listed: 2N3904, 2N3906, 2N2222, IC555, IC741. You may choose any field name you deem appropriate. Note that this is specific to the sample matrix provided. Other matrices would, of course, have different part designations.

18. Refer to the specifications for Review Question 15.

Create a structure array to contain the same information given in parts (a) through (c) in Review Question 15. Note that it may be appropriate to incorporate the text values into the structure array as field names instead of data. You may use any organizational structure and field names that seem logical to you—there are various possibilities.

Add entries to the structure array corresponding to the calculated volumes and masses listed in part (d) of Review Question 15. Again, you may use any organizational structure or field names that seem logical to you.

Algorithms, Programs, and Functions

Learning to create effective algorithms is a crucial skill for any aspiring engineer. In general, an **algorithm** is a well-defined sequence of instructions that describe a process. Algorithms can be observed in everyday life through oral directions ("Simon says: raise your right hand"), written recipes ("Bake for 15 minutes at 350 degrees Fahrenheit"), graphical assembly instructions, or other graphical cues. As an engineer, writing any algorithm requires a complete understanding of all the necessary actions and decisions that must occur to complete a task.

16.1 Algorithms

> **LEARN TO:** Define the scope of a problem
> Define known and unknown quantities in a problem
> Document any assumptions necessary to solve a problem
> Create a written algorithm for a problem
> Create a graphical algorithm (flowchart) for a problem

When writing an algorithm, you must answer a few questions before attempting to design the process. To even begin thinking of a strategy to describe a process, you must have carefully defined the scope of the problem. The **scope** of an algorithm is the over-all perspective and result that the algorithm must include in its design.

For example, if we are required to "sum all numbers between 1 and 5," before thinking about an approach to solve the problem, we must first determine if the scope is properly defined. Does the word "between" imply that 1 and 5 are included in the sum? Do "numbers" include only the integer values? What about the irrational numerical values? Clearly, we observe that we cannot properly define the scope of the charge to add **all** numbers between 1 and 5 since there are an infinite number of such values.

Likewise, imagine you were charged to design a device that transports people from Atlanta, Georgia, to Los Angeles, California. How many people must the device transport? Does the device need to travel on land? Should it travel by air? Should it travel by water? Does the device require any human interaction?

This section covers two methods of defining a process: with **written algorithms** and with **graphical algorithms**. Both methods require properly identifying the scope of the problem and all of the necessary input and output of the process.

Scope

One of the most difficult steps in designing an algorithm is properly identifying the entire scope of the solution. Like solving a problem on paper involving unit conversions and equations, it is often necessary to state all of the known and unknown variables in order to determine a smart solution to the problem. If information is left out of the problem, it might be necessary to state an assumption in order to proceed with a solution. After all variables and assumptions about the problem have been identified, it is then possible to create a sequence of actions and decisions to solve the problem.

To clearly understand the scope of the problem, we often find it helpful to formally write out the known and unknown information, as well as state any assumptions necessary to solve the problem. In the following examples, notice that as the problem statements become more and more refined, the number of necessary assumptions decreases and eventually disappears.

EXAMPLE 16-1

For the problem statement, list all knowns, unknowns, and assumptions.
Problem: Sum all numbers between 1 and 10.

Known:

- The minimum value in the sum will be 1.
- The maximum value in the sum will be 10.

Unknown:

- The sum of the sequence of numbers.

Assumptions:

- We will only include the whole number values (e.g., 1, 2, 3, . . .) in the sum.
- The sum will include the starting value of 1 and the ending value of 10.

EXAMPLE 16-2

For the problem statement, list all knowns, unknowns, and assumptions.
Problem: Sum all numbers between (and including) 1 and 10.

Known:

- The minimum value in the sum will be 1.
- The maximum value in the sum will be 10.

Unknown:

- The sum of the sequence of numbers.

Assumptions:

- We will only include the whole number values (e.g., 1, 2, 3, . . .) in the sum.

EXAMPLE   16-3

For the problem statement, list all knowns, unknowns, and assumptions.
Problem: Sum all whole numbers between (and including) 1 and 10.

Known:

- The minimum value in the sum will be 1.
- The maximum value in the sum will be 10.

Unknown:

- The sum of the sequence of numbers.

Assumptions:

- [None]

Comprehension Check   16-1

For the problem statement, list all knowns, unknowns, and assumptions. Problem: Sum all even numbers between (and including) 2 and 20.

Comprehension Check   16-2

For the problem statement, list all knowns, unknowns, and assumptions. Problem: Multiply all powers of 5 between (and including) 5 and 50.

Written Algorithms

A written algorithm is a narrative set of instructions required to solve a problem. In everyday life, we encounter written algorithms in the form of oral instructions or written recipes. However, it is extremely common for humans to "fill in the blanks" on a poorly written algorithm. Imagine you are handed a strongly guarded family recipe for tacos. One of the steps in the archaic recipe is to "cook beef on low heat until done." To the veteran cook, it is apparent that this step requires cooking the prepared ground beef on a stovetop in a sauce pan for approximately 10 minutes on a burner setting of 2 to 3. To a first-time cook, the step is poorly defined and could result in potentially inedible taco meat.

Engineers and Written Algorithms

As an engineer, to write effective algorithms you must ensure that every step you include in a written algorithm must not be subject to misinterpretation. It is helpful to write an algorithm as if it were to be read by someone completely unfamiliar with the topic. Each step in the written algorithm should be written such that the stepwise scope is properly defined. The **stepwise scope** is all of the known and unknown information at that point in the procedure. If a step in an algorithm contains an assumption, you must formally declare it before proceeding with the next step. By ensuring that the stepwise scope is well defined, you ensure that your algorithm will not be subject to misinterpretation.

All written algorithms should be expressed sequentially. The most effective algorithms are written with many ordered steps, wherein each step contains one piece of information or procedure. While the author of an algorithm may consider each step in an algorithm to be "simple," it might not be trivial to an external interpreter. When writing an algorithm, it is helpful to assume that the reader of your algorithm can only perform small, simple tasks. Assume that your algorithm can be interpreted by a computer. A computer can execute small tasks efficiently and quickly, but unlike a human, a computer cannot fill in the blanks with information you intended the reader to assume.

Except for the simplest algorithms, most will involve either asking questions and taking different actions based on the answers, or repeating the same series of steps two or more times. We will introduce these types of algorithms in Chapter 18 and Chapter 19. In Chapter 16 and Chapter 17, all algorithms will consist of a series of steps that are done in sequential order exactly one time without skipping any steps.

Format of Written Algorithms

The first step in writing any algorithm is defining the scope of the problem. After you define the scope of the problem, create an *ordered* or *bulleted list* of actions and decisions. Imagine taking an English class and writing a research report on the influence of 19th-century writers on modern-day fiction authors. Before writing the paper, you would create an outline to ensure that your topics have connectivity and flow. Just like the outline of an English paper, an algorithm is best expressed as a sequential list rather than as complete paragraphs of information.

EXAMPLE   16-4

Determine the pressure [atm] in a sealed container with a volume of 5 gallons containing 25 moles of methane at a temperature of 155°F.

Known:

- Volume V = 5 gallons
- Temperature T = 155°F
- Amount of methane n = 25 moles

Unknown:

- Pressure [atm]

Assumptions:

- Ideal gas constant R = 8.314 L kPa $K^{-1}mol^{-1}$

Algorithm:

1. Convert volume to liters
2. Convert temperature to kelvin
3. Calculate pressure [kPa] using $P = nRT/V$
4. Convert pressure to atmospheres

Comprehension Check	16-3

The total distance traveled from position zero by an object undergoing constant acceleration can be determined by

$$d = v_0 t + \frac{1}{2}at^2$$

where v_0 is the initial velocity, t is time, and a is acceleration. Develop a written algorithm to calculate the total distance (in kilometers) traveled at time t in hours, assuming initial velocity is given in miles per hour and acceleration is given in meters per second squared.

Graphical Algorithms (Flowcharts)

To visualize a process, a graphical representation of algorithms is used instead of a written algorithm. A **flowchart** is a graphical representation of a written algorithm that describes the sequence of actions of a process. Designing a flowchart forces the author of the algorithm to create small steps that can be quickly evaluated by the interpreter and enforces a sequence of all actions. Flowcharts are used by many different disciplines of engineering to describe different types of processes, so learning to create and interpret flowcharts is a critical skill for a young engineer. In fact, in the United States, any engineer who discovers a new, innovative algorithm can submit the concept to a patent office by representing the process in terms of a flowchart.

Three different shapes are used in the creation of flowcharts in this book; a number of other widely used operators are encountered across the world. In this book, we describe all actions with rectangles, all decisions with diamonds, and all connections between shapes with directional arrows. The decision diamond will not be introduced until Chapter 18.

Rules for Creating a Proper Linear Flowchart

- The flowchart must contain a START rectangle to designate the beginning of a process.
- All actions must be contained within rectangles.
- All shapes must be connected by a one-way directional arrow.
- The flowchart must contain an END rectangle to designate the end of a process.

Actions

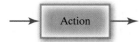

Actions are any executable steps in an algorithm that do not require a decision to be made. Based on this definition, any defined variables, calculations, and input or output commands (getting information from or sending information to the external world—to be covered in Chapter 17) would all be contained within action rectangles.

All simple actions are contained within a single rectangle on the flowchart. For each rectangle, two arrows are always associated with the shape, with two exceptions. The inward arrow to the rectangle represents the input to the action. It is assumed that any variables defined in the stepwise scope of an action rectangle are accessible and can be used in the action. The outward arrow from the rectangle represents the output of the action. If any new variables or calculations are performed within the rectangle, those values are passed along to the next shape's stepwise scope.

- Exception One: The START rectangle represents the beginning of the flowchart and does not contain an inward arrow. An oval shape is also commonly used to represent the start of an algorithm.
- Exception Two: The END rectangle represents the end of the flowchart and does not contain an outward output arrow. An oval shape is also commonly used to represent the end of an algorithm.

EXAMPLE   16-5

Create a flowchart for the problem of Example 16-4.

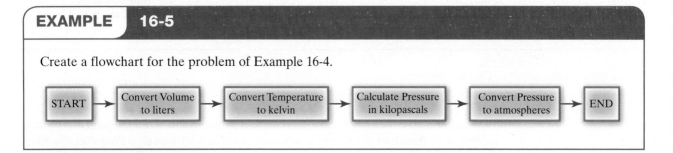

Comprehension Check   16-4

The total distance traveled from position zero by an object undergoing constant acceleration can be determined by

$$d = v_0 t + \frac{1}{2}at^2$$

where v_0 is the initial velocity, t is time, and a is acceleration. Create a flowchart to calculate the total distance (in kilometers) traveled at time t in hours, assuming initial velocity is given in miles per hour and acceleration is given in meters per second squared.

Algorithm Best Practices

The remaining part of this section details specifics on how to begin planning and writing algorithms from scratch. It is not intended to be a definitive resource on algorithm development, but it may provide guidance if you are struggling to break down a process into small, achievable steps.

Actions

In every action within an algorithm, there must be a key verb that defines the purpose of that step within an algorithm. The remaining subsections discuss different types of actions and list some of the common verbs associated with that category of action.

Establishing Variables and Constants

After defining the scope of a problem, it might become obvious that there are intermediate calculations or assumed constants that must be contained throughout the process. Along with the explicitly defined known values, these intermediate and constant values are referred to as variables (even if they are constant!). Algorithmic variables are different from the mathematic definition of a variable because algorithmic variables are treated more like containers to store known values and results of calculations rather than being some unknown entity in a mathematical expression. They are called variables because the stored value can be written, overwritten, and used by other actions or decisions in the algorithm.

Example			Action		
We assume the acceleration due to gravity is 9.8 meters per second squared.			Set variable g to be 9.8.		
Other Verbs					
Set	Define	Assign	Write	Store	Designate
Label	Name	Cast	Insert	Save	Initialize

User Interaction

It is often necessary to write algorithms that can be executed with prompts for input from the person using the algorithm, provide feedback on results, or display any error messages generated in the algorithm. This will be covered in more detail in Chapter 17.

User Input:

Example			Action		
We want the user of the algorithm to provide the amount of water in gallons.			**Input** the amount of water in gallons, save in variable W.		
Other Verbs					
Input	Ask	Load	Request	Query	Prompt

User Output:

Example			Action		
We want the algorithm to inform the user that the amount of water can't be negative.			**Display** error message to user "Warning: amount of water can't be negative!"		
Other Verbs					
Output	Display	Reveal	Write	Warn	

Calculations and Conversions

When algorithms involve calculating a value using an equation, it is helpful to write out the full equation and identify which variables in the algorithm correspond to the variables in the expression. For unit conversions, it is not necessary to write out the conversion factors since those are published standards that are readily available to anyone executing your algorithm. When using conversions, it is best to list them individually so they are easily recognizable to the user. For example, when converting from feet to centimeters, the expression $L = L/3.28 * 100$ is not difficult to recognize as the conversion from feet to meters, and then from meters to centimeters. It is harder to recognize the conversion of $L = L * 30.48$. Furthermore, it is easy to make a calculation error when lumping conversion factors together; it is easier to allow the program to calculate for you. When dealing with unit conversions, it is often best to save the converted value back into the original variable to reduce the number of variables you need to keep track of in your algorithm. We will discuss MATLAB's capabilities to handle this type of equation in later chapters.

Calculations:

Example	Action
We want to calculate the thermal energy of a substance using the expression $Q = m\, C_p\, \Delta T$, where m is the mass, C_p is the specific heat, and ΔT is the change in temperature.	**Compute** the thermal energy: $Q = m\, C_p\, \Delta T$ All variables should appear in the variable list.

Other Verbs					
Calculate	Adjust	Count	Measure	Add	Multiply
Subtract	Divide	Compute	Increment	Decrement	

Conversions:

Example	Action
We want to convert a variable t from minutes to seconds and save the result back in the variable t.	Convert t from minutes to seconds, save in t.

Other Verbs				
Convert	Change	Alter	Revise	Switch

Referencing Other Algorithms

When developing a large program, it is sometimes helpful to break that program into several smaller programs, and then reference the smaller programs within the large program. In MATLAB, these smaller programs are called **functions**. As a rule of thumb, each custom function you create should have its own separate algorithm. If you have separate algorithms for a program and the different functions referenced in the code, it makes the algorithms simpler to understand and easier to debug. When calling a function within an algorithm, it is critical to list the variables passed to the function and variables returned by the function. We often refer to this process of returning variables as "capturing" the variables. In general, the most common verb used with functions is "call." If you know the name you plan to use for your function, list it; otherwise, this can be set later.

Example	Action
We want to use a function named *Poltocar* that converts coordinates from polar to Cartesian. We will pass in the variable Z as the radius and the variable T as the angle. We will capture the x-coordinate in the variable X and the y-coordinate in the variable Y.	**Call** *Poltocar* In: Z, T Out: X, Y

Testing Your Algorithm

The last step in writing an algorithm is developing test cases that will reveal whether or not your algorithm behaves as expected. The key to writing test cases is figuring out how many test cases are necessary to confirm whether or not your code works. In general, there should be at least one or two test cases that will demonstrate the proper behavior of the algorithm given good input values. Not only should a test case include a list of all of the inputs used to generate the output, but you should also compute the expected output of the algorithm by hand in order to verify that the algorithm works. We will have more to say on this topic in Chapter 18 and Chapter 19, which involve conditional statements and loops, requiring more robust testing procedures.

To help you develop good algorithms, we have included an algorithm template online to help you document all of the variables, procedures, and test cases necessary to create correct and verifiable algorithms. Examples of how this template can be used are provided below and on select problems in the remaining chapters.

| EXAMPLE | 16-6 |

Create an algorithm to determine the volume and surface area of a cylinder, given the radius and height.

Known / Input:	Unknown / Output:	Assumptions:
Radius = r Height = H	Volume = V Surface area = SA	• All units will be in centimeters • Object is a cylinder. • Surface area does not include ends.

Algorithm: Written

1. Enter values of r and H
2. Calculate volume of a cylinder using $V = \pi r^2 H$
3. Calculate the surface area of a cylinder using $SA = 2\pi rH$

Algorithm: Flowchart

Start → Input radius (r) → Input height (H) → Calculate $V = \pi r^2 H$

END ← Calculate $SA = 2\pi rH$

Test Cases:

Input:	Output:	
r = 3 cm; h = 4 cm	$V = 113.1\ cm^3$	$SA = 75.4\ cm^2$

Once you have developed an algorithm for a problem, it is time to convert the algorithm into programming code. To some extent, each item in your algorithm will be translated into one programming statement, but this is not always a one-to-one correspondence.

Before we actually write our first programs, we need to address a few concepts involving programming in general and MATLAB in particular.

Programs and Functions

Programming is the process of expressing an algorithm in a language that a computer can interpret. To correctly automate a process on a computer, a programmer must be able to correctly speak the language both grammatically and semantically. In this text, we solve problems with a computer by using the MATLAB programming language. In Chapter 15, we looked at the process of creating variables containing different types of information in MATLAB, and in the first part of this chapter, we introduced the idea of creating algorithms to represent a process. Now we will combine these two ideas to create programs and functions.

16.2 Programs

LEARN TO:	Understand how to navigate the MATLAB interface
	Remember the naming rules for programs and functions
	Create, execute, and modify MATLAB programs

Certain principles apply to programming, regardless of what language is being used. Entire textbooks have been written on this subject, but a few important concepts are addressed here.

Notes About Programming in General

- **Programming style:** Just like technical reports should follow a particular format to ensure they can be understood by someone reading them, all programs should have a few common elements of programming style, particularly the inclusion of comments that help identify the source of the program and what it does. For any program, a proper header should be written describing the scope of the problem, including a problem statement and definition of all input and output variables used in the program. Properly commenting the source code is also critical for ensuring that someone else can follow your work on the program.

  As a rule of thumb, for each rectangle and diamond in a flowchart, there should be a few comments in the source code where that action occurs.

- **Program testing:** When you write a program, testing it is a critical step in making sure that it does what you expect. Lots of things can go wrong.

  - *Syntax errors:* These violate the spelling and grammar rules of the programming language. Compilers (programs that interpret your computer program) generally identify their location and nature before the code is actually executed.
  - *Runtime errors:* These occur when an inappropriate expression is evaluated during program execution. Such errors may occur only under certain circumstances. Examples of runtime errors are a variable not being defined before a line of code tries to use it; incorrect matrix dimensions, such as trying to add two matrices of different sizes; or the wrong kind of input, such as a user entering letters when numbers are expected. Many languages will report an error when an infinite result occurs in a calculation, such as division by zero or taking the logarithm of zero, but MATLAB will simply calculate this as inf (infinity) and continue without error. Computer programs should anticipate runtime errors and alert users to the error, rather than having the program print angry red letters on the screen and unexpectedly terminate.
  - *Formula coding errors:* It is not uncommon to enter a formula incorrectly in a form that does not create either a syntax or a runtime error. For example, if you wrote Q=X+1^2 with the intention of adding one to X, then squaring the result, no error would be generated, but the answer would not be correct due to priority of operators. Evaluating even a few simple calculations by hand and comparing them to the values calculated by the program will usually reveal this kind of error.
  - *Formula derivation errors:* This difficult error to discover in programming occurs when the computer is programmed correctly but the programming is based on a faulty conceptualization or erroneous formulae. Programmers generally assume that a program that executes without generating any errors is correctly written, and if the hand-verified sample calculations are based on the same misconceptions, this type of error can be very difficult to detect. This situation underscores the importance of checking a solution strategy before starting to write a computer program. For particularly complicated situations, it is helpful to have someone else check your approach to the solution.
- **Keeping track of units:** Some computer programs can be designed to have the user keep track of units as long as they use consistent units. Other computer programs require input data to be in a particular set of units. Regardless of which approach you use, the way in which units are managed must be clear to the user of the program and in the program's comments.
- **Documentation:** In any programming language, there is usually a special character or pair of characters that tell the program interpreter to ignore everything to the

right of the special character to allow the programmer to write human-readable notes within the code of the program. These notes are commonly referred to as "comments" within the code. A great deal of controversy surrounds how much documentation is necessary in any program, but this text proposes the following guideline to properly document a program:

For every numbered item in a written algorithm, you should include that item as a comment in your code.

In addition, at the top of each program, it is extremely helpful to at least write a problem statement and document any major variables used within the program. It is helpful to cluster your description of variables into three groups: input variables, output variables, and function variables.

Notes Particular to MATLAB

Before proceeding with the material on MATLAB, it would be wise to review the appendix material on matrices. Since variables in MATLAB are inherently matrices, it is critical that you understand not only some basic matrix concepts, but also the specific notation used in this text (as well as MATLAB) when referring to matrices. Certain details specific to MATLAB are important to note early on.

- **The MATLAB interface:** As we saw in the previous chapter, the MATLAB program window has a number of subwindows. The specific subwindows displayed and their organization can be customized using the Layout menu in the Home tab of the tool ribbon.

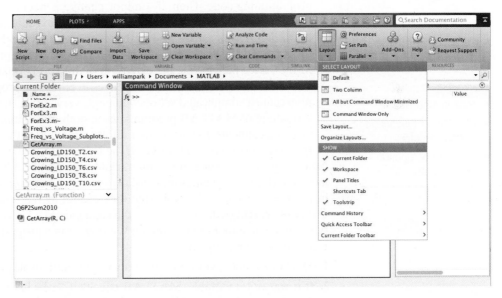

Individual subwindows can be *undocked* from the main window so that they appear in their own separate windows. The small arrow in a circle near the top right of each subwindow can be clicked and various actions involving that subwindow, including undock, may be selected.

- In Chapter 15, we typed individual commands in the Command Window. Sometimes we would type several commands in a sequence to achieve a larger purpose than was possible with a single command. If we wanted to repeat that same sequence of commands with perhaps one minor change in one of the commands, we could use the up arrow to select the previous commands in sequence

NOTE

`clc` = clear the Command Window

To clear all text from the Command Window, enter the special reserved MATLAB expression `clc` in the Command Window.

FILE TYPES

Word = .docx

Excel = .xlsx

Powerpoint = .pptx

MATLAB program
files = .m

MATLAB saved
variables = .mat

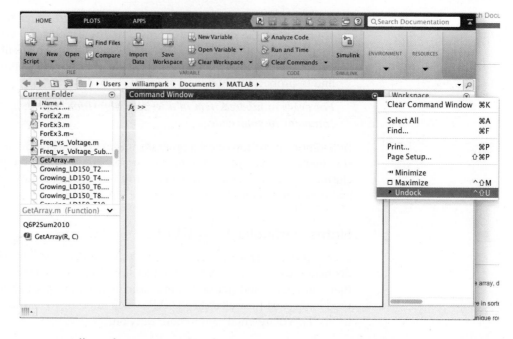

NOTE

Order of search:
- Variable
- Built-in function
- Program or function in current directory
- Program or function in current path

to edit and re-execute. For long sequences of commands, this method would become extremely cumbersome, however. MATLAB provides a special interface, called the Editor/Debugger, which is where you will normally write your programming code. Then, if a minor change is necessary, you only need to edit that one line of code and tell MATLAB to re-execute the entire sequence.

- The *Editor/Debugger window* is the interface of choice for developing MATLAB programs. The Editor stores your list of commands (your program) in what are called M-files, since they have a file extension of .m. This permanent record of your commands allows you to make small changes to either the program or the inputs and to re-execute a complicated set of commands very easily. The Editor also assists in the formatting of MATLAB programs, using both indentation and color to distinguish program elements. To launch the Editor window, press the **New Script** button on the **Home** ribbon or use CTRL+N. The Editor interface will launch in a new window.

- **Naming programs, functions, and variables in MATLAB:** In MATLAB, program names and function names must follow the same rules for naming variables that we introduced in Chapter 15. Note that they cannot include space characters.

- **Order of search:** Order of search is the sequence MATLAB goes through when you type a name in either the Command Window or in a program or function. If you type `Frog` in the Command Window:

  1. MATLAB first checks if `Frog` is a variable, and displays its value if it is.

  2. If it was not a variable, MATLAB checks if `Frog` is one of MATLAB's built-in functions, and executes that function if it is.

  3. If it was neither a variable nor a function, MATLAB checks if `Frog` is a program or function in its current directory, and executes the program or function if it is.

  4. If it still has not found `Frog`, MATLAB checks if `Frog` is a program anywhere in its path, and executes the program or function if it is. The location where you store your MATLAB programs or functions must be included in MATLAB's path. You can check where MATLAB is looking for programs and functions by typing `path` in the Command Window. Any M-files in the directories listed will be visible to MATLAB. If you try to execute an M-file outside any of the directories listed,

MATLAB will ask if you want to switch current directories or add that directory to the list of paths MATLAB can see.

5. If it cannot find Frog anywhere, it will send angry red letters to the screen stating that the name was an undefined function or variable.

- **what and who:** To display a list of all MATLAB programs in the current directory, type what into the Command Window. To display all variables currently in MATLAB's workspace, type who in the Command Window. Remember that all variables automatically clear after you exit MATLAB.

- **Comments:** For documenting or commenting any code in MATLAB, a special character is used to "comment" out text: the percent symbol (%). The comment character can appear anywhere in a MATLAB program or function, but preferably either at the beginning of a line or at the middle of a line of MATLAB code. Everything after the percent sign to the end of the line is ignored by MATLAB, so you as a programmer can "comment out" any broken lines of code during code testing in addition to providing comments. Program or function file headers will usually have 6–10 lines of comments at the top of every program and more throughout the rest of the program or function.

```
% This is an example of a comment line.
% Comment lines are shown in the Editor window in green.
```

A **block comment** is a block of nonexecuted text between the special block characters %{ and %} that allow the user to type full-length paragraphs as comments without needing to type the % symbol before each sentence. For example:

```
%{
This is an example of using
a block of comments in a code
%}
```

Program Structure and Use

A **program** is a set of instructions given to a computer to perform a specific task. In this section, we create MATLAB program files, or **M-files**, to automate our processes. To create a program in MATLAB, launch the MATLAB Editor. By default, the Editor will create a new, blank document named Untitled, or Untitled2, or Untitled3, and so on, depending on how many unsaved programs are open in the Editor at once. The following example provides a walkthrough for creating a basic program, as well as screenshots of the actions described in the text to help you to follow along while re-creating these steps on your own computer running MATLAB.

NOTE

In programming, the term Easter Egg refers to an unexpected "feature" that the programmers included just for fun. One Easter Egg in MATLAB is the why function. Type why in the Command window and see what appears. Type why again, and again, and again. . .

EXAMPLE   16-7

We want to create a MATLAB program to convert Cartesian coordinates (x, y) to polar coordinates (r, θ). Before we can even try to write the program, we need to recall some information about the coordinate systems.

We know the following:

- In the Cartesian coordinate system, the coordinates (x, y) represent where a particular data point is in relation to the origin $(0, 0)$ by specifying the horizontal and vertical distance with respect to the origin. The x-value represents the horizontal distance and the y-value represents the vertical distance.

- In the polar coordinate system, the coordinates (r, θ) represent the exact distance (r) from the point to the origin $(0, 0)$ as well as the angle of elevation (θ) with respect to the positive horizontal axis, typically in units of radians.

- The conversion from polar coordinates to Cartesian coordinates is

$$x = r \cos \theta$$
$$y = r \sin \theta$$

- The conversion from Cartesian coordinates to polar coordinates is

$$r = \sqrt{x^2 + y^2} \quad \theta = \tan^{-1}\left(\frac{y}{x}\right)$$

We now have the equations necessary to write a program to convert Cartesian coordinates to polar coordinates, and vice versa.

NOTE

CTRL + C = force quit a MATLAB program.

First, we need to learn how to ask MATLAB for more information on doing things like calculating a square root or calculating an inverse tangent of a value if we do not already know how to do these operations.

In MATLAB's Command Window, we can search for built-in functions to calculate a square root and an inverse tangent. The lookfor *function allows you to search MATLAB help documentation for certain key words. Note that if you type* lookfor square root *into the Command Window, MATLAB displays an error message. The* lookfor *command is expecting only one keyword following the command "lookfor," so we decide to try our search again, this time including only the word "square."*

Because MATLAB is searching through every built-in function, we need a way to tell MATLAB to stop, because when the function sqrt *appears, it seems to be the exact function we need. To terminate a MATLAB command early, we press* **CTRL + C** *(i.e., the CTRL key and the "C" key at the same time) on our keyboard, and the prompt will return to the Command Window.* CTRL + C *can also be used to stop a program caught in an infinite loop or that otherwise wandered away and got lost.*

To ask for more information on the function sqrt, *we can type* help sqrt *or* doc sqrt *to double-check that the function is going to perform our intended calculation as well as details of how to use it.*

```
Command Window                                                                    ⊙
  >> lookfor square
  hypot                            – Robust computation of the square root of the sum of squares
  realsqrt                         – Real square root.
  sqrt                             – Square root.
  magic                            – Magic square.
  lscov                            – Least squares with known covariance.
  sqrtm                            – Matrix square root.
  cgs                              – Conjugate Gradients Squared Method.
  spaugment                        – Form least squares augmented system.
  fifteen                          – A sliding puzzle of fifteen squares and sixteen slots.
  lsqnonneg                        – Linear least squares with nonnegativity constraints.
  slexSysIdentMATLABSystemLMSAdaptSysObj – Least mean squares (LMS) adaptive filtering.
  >> help sqrt
  sqrt   Square root.
     sqrt(X) is the square root of the elements of X. Complex
     results are produced if X is not positive.

     See also sqrtm, realsqrt, hypot.

     Reference page for sqrt
     Other functions named sqrt

fx >>
```

Repeating this process to find the inverse tangent function, we determine that atan *is the correct function necessary to calculate the inverse tangent. We note, from the help documentation, that the* atan *function will return the inverse tangent in radians.*

New Script

To create a program, we click the **New Script** button on MATLAB's Home tab. After creating a new blank M-file, we need to write out our problem statement as code comments. Comments are any non-code text that you would like to include in your code file. All comments start with the percent symbol (%), and everything to the right of the symbol will be ignored on a per-line basis. Comments can follow any executable statements in our code, but for now, we will add the first couple of lines in our MATLAB program by writing out a short problem statement with the % symbol before any text we write. We can also use block comments as mentioned previously.

Next, we start writing our code. In a MATLAB program, it is good practice to start the program by clearing all of the variables from the workspace (`clear`) and clearing the Command Window (`clc`). That way, we guarantee that if we run our program on a different computer or run it after restarting MATLAB, our program will create all the variables we need to solve the problem, and we need not rely on the workspace having variables predefined before code execution. Also, this will prevent values left over from a previous run of this or even a different program from affecting the results. Commands like this are often referred to as "housekeeping" commands.

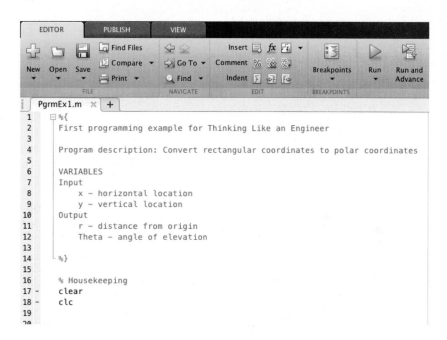

Note that all comments are shown in green, whereas our two program commands (`clear` and `clc`) are shown in black. As we will see later, MATLAB uses other colors for other specific types of things in your program. It is also helpful to use blank lines to visually separate different sections of your code.

Now we are ready to start our program. Let us look at the step-by-step written algorithm necessary to solve this problem:

1. Create a variable to contain the x-coordinate as a variable named x.
2. Create a variable to contain the y-coordinate as a variable named y.
3. Calculate the radius by taking the square root of the sum of the squares of x and y and store the result in a new variable called r.
4. Calculate the angle by taking the inverse tangent of the result of dividing y by x and store the result in a new variable called Theta.

In the written algorithm, the first thing we have to do is to manually assign the values of x and y. For this example, we assign x to contain the value of 5 and y to contain the value of 10, but in later chapters, we show how the user of a program can enter values.

Note that in addition to creating the variables x *and* y, *we have also added comments in the header of our program documenting what these variables represent.*

```
PgrmEx1.m  ×  +
 1    %{
 2      First programming example for Thinking Like an Engineer
 3
 4      Program description: Convert rectangular coordinates to polar coordinates
 5
 6      VARIABLES
 7      Input
 8          x - horizontal location
 9          y - vertical location
10      Output
11          r - distance from origin
12          Theta - angle of elevation
13
14    %}
15
16      % Housekeeping
17 -  clear
18 -  clc
19
20      % Define input parameters
21 -  x=5    % horizontal location
22 -  y=10   % vertical location
23    ⚠ Terminate statement with semicolon to suppress output (within a script). [ Details ▾ ] [ Fix ]
```

Note that MATLAB highlights the = *symbol in the Editor. When we move the cursor over the assignment operator* (=), *a pop-up message tells us that we can suppress the output of the assignment expression by using a semicolon. For now, we will omit the semicolons and observe MATLAB's behavior when we leave the code unsuppressed.*

Next, we type the equations to calculate the radius and the angle, using the built-in MATLAB functions we discovered with `lookup` *and* `help`. *In addition, we've added extra comments about the new variables representing the radius and angle of the polar coordinate.*

```
PgrmEx1.m  ×  +
 1    %{
 2      First programming example for Thinking Like an Engineer
 3
 4      Program description: Convert rectangular coordinates to polar coordinates
 5
 6      VARIABLES
 7      Input
 8          x - horizontal location
 9          y - vertical location
10      Output
11          r - distance from origin
12          Theta - angle of elevation
13
14    %}
15
16      % Housekeeping
17 -  clear
18 -  clc
19
20      % Define input parameters
21 -  x=5    % horizontal location
22 -  y=10   % vertical location
23
24      % Convert from rectangular to polar
25 -  r=sqrt(x^2+y^2)    % distance from origin (radius)
26 -  Theta=atan(y/x)    % elevation angle
27
```

NOTE

Remember, MATLAB programs and functions *cannot* have any spaces in their file names, so be sure that a file is named without spaces, for example, CartesianToPolar.m —*not* "Cartesian To Polar.m."

Next, we save the program in MATLAB's path with a file name that follows the program naming convention. We choose the name "CartesianToPolar" since that name nicely describes the purpose of the program.

*To save a program, we click the **Save** icon in the Editor window.*

⌘ *Mac OS: MATLAB is one of the few programs that does not place the main menu items (such as File and Edit) at the top of the screen. Instead, its main menu is at the top of the MATLAB window, just as in the Windows implementation. This is because MATLAB is actually running under an X Windowing System, similar to Linux or UNIX®.*

Running the Program

There are a number of different ways to run a MATLAB program located within MATLAB's path.

Method 1: *In the main MATLAB window (not the Editor), type the name of the program in the Command Window and press **Enter.***

Method 2: *In the main MATLAB window (not the Editor), right-click the program you want to run while in the Current Directory box and select **Run**.*

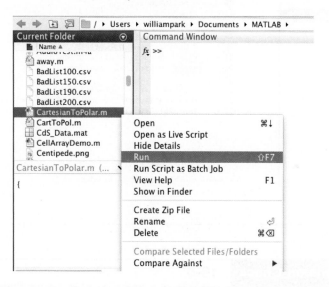

NOTE

To suppress output, end the code line with a semicolon (;).

Method 3: *With the program loaded in the Editor window, click the **Run** button in the toolbar.*

Looking in the Command Window, we see that MATLAB spits out the results of every assignment statement in the program.

Let us assume we want to show only the output from the calculations of r *and* Theta, *and not the output of the specification of* x *and* y. *To do this, we add a semicolon (;) at the end of each line of code we do not want showing up in the output of the program.*

```
CartesianToPolar.m  ✕  +
1   %{
2   First programming example for Thinking Like an Engineer
3
4   Program description: Convert rectangular coordinates to polar coordinates
5
6   VARIABLES
7   Input
8       x - horizontal location
9       y - vertical location
10  Output
11      r - distance from origin
12      Theta - angle of elevation
13
14  %}
15
16  % Housekeeping
17  clear
18  clc
19
20  % Define input parameters
21  x=5;    % horizontal location
22  y=10;   % vertical location
23
24  % Convert from rectangular to polar
25  r=sqrt(x^2+y^2)     % distance from origin (radius)
26  Theta=atan(y/x)     % elevation angle
27
```

Now when we save and run our modified program, we will see only the output from the calculations of r *and* Theta.

Comprehension Check | **16-5**

For the problem given in Comprehension Check 16-4, write a complete program to implement the flowchart you created. Be sure to include appropriate comments in your code. Suppress all assignment statements other than the final answer, which should appear on the screen. Use the following values to test your code: $V_0 = 10$ mph, $t = 0.02$ hr, $a = 3$ m/s^2.

Programming Features in Our Coverage of MATLAB

MATLAB has many advanced programming features that make sophisticated numerical analysis, including simulation, possible. In this text, you will learn basic programming structures to solve engineering analysis problems.

- **Data types:** Covered in the previous chapter, there are different data types that can be stored and manipulated by MATLAB. Some of the more common data types used by programs include numeric and string variables, but many applications also use vector and matrix variables for advanced calculations, as well as special types like cell arrays for storing a combination of different types within a single structure.

- **Input and output:** By gathering input from the external world, whether from a human user or some sort of sensor, the program can solve a different problem each time it is executed. Output statements allow you to produce formatted text to communicate with the user, as well as to plot graphs to clearly communicate data relationships. Likewise, being able to read and write Microsoft Excel files using MATLAB code allows your program to perform calculations on existing data without needing to enter it manually or to permanently save results to special MATLAB data files (.mat files). Input and Output will be covered in Chapter 17.
- **Conditional statements:** These allow a program to take different paths for various user inputs, intermediate calculations, or final results. Conditional statements allow the programmer to ask questions and have the code take different actions based on the answers. This is the topic of Chapter 18.
- **Looping:** Repetition is critical in breaking down complicated problems into many simple calculations. We will look at looping structures in Chapter 19.

16.3 Functions

> **LEARN TO:** Create, execute, and modify functions in MATLAB
> Understand the difference between local and global variables

A **function** is a special type of program that is not designed to be used by itself but to be used as an element in a larger program. The primary reason for using functions is to simplify and modularize programming by allowing commonly used code to be expressed as a single word instead of having to paste all of the code into the larger program. For example, when we need to perform a square root calculation, we do not have to write code to accomplish this, nor even paste the code that does the calculation into our program; we simply use the name sqrt followed by the values for which we wish the square roots. When we write something like Q=sqrt(Vals), the program knows to send the data in the variable Vals to a piece of code named sqrt that has already been written and debugged to do the calculation. This reduces the amount of redundant code and makes our programs cleaner and easier to read.

In general, functions require the user to pass input information to them—the data on which the function will perform its calculations and obtain the final results created by these calculations. However, some functions do not require any input to do their job. If you type RV=rand, for example, a single random number between 0 and 1 will be placed in RV. Some functions do not generate any data as output. An example would be pause(60), which simply stops program execution for one minute (60 seconds), after which it resumes. Some functions require neither input nor do they generate data as output. Make sure the volume on your computer is not muted and type beep in the Command Window, for example.

Sometimes our application requires a certain type of calculation that is not common enough to have a built-in function, but we need to use it in multiple places in our code. Rather than place the same code multiple times in our program, we can write our own function to accomplish the task, and once it is properly tested and debugged, we can simply use the name we give to our new function in the larger program. User-defined functions are similar to programs in that they contain a series of executable MATLAB statements in M-files that exist independently of any program we create in

MATLAB. To allow any MATLAB program to call (use) a user-defined function, however, we must follow a few guidelines to ensure that our function will run properly.

LAW OF ARGUMENTS

In computer programming, and often in math as well, the word "argument" does *not* refer to an altercation or vehement disagreement. In this context, it means the information that is given to a function as input to be processed. If you use the function `sqrt` to find the square root of 49, you would write `sqrt(49)`. In this case 49 is the argument of `sqrt`—the value upon which the `sqrt` function performs its calculations. Functions may have zero arguments (e.g., `Z=zeros` will place a single zero into `Z`—no input is required), one argument (e.g., `sqrt(4)`), or two or more arguments (e.g., `power(a,b)` raises a to the power b). Some functions may have different numbers of arguments depending on what you wish them to do. For example, `Ox=ones` will place a single 1 into `Ox`, `Ox=ones(5)` will place a 5 × 5 matrix filled with ones into `Ox`, and `Ox=ones(3,6)` will place a 3 × 6 matrix filled with ones into `Ox`.

Function Creation Guidelines

- The first line in a function other than comments **must** be a function definition line:

```
function[output_variables]=function_name(input_variables)
```

  - *output_variables:* A comma-separated list of output variables in square brackets.
  - *input_variables:* A comma-separated list of input variables in parentheses. The list of input variables to a function are commonly referred to as **arguments**.
  - *function_name:* Function names should not copy the names of built-in functions, user-created variables, or any reserved word in MATLAB. Avoid creating variables with the same name as a user-defined function because you will not be able to call your function.
- **The name of the function must be the same as the file name**. For example, if the function name is `sphereVolume`, the function must be saved in an M-file named `sphereVolume.m`, where `.m` is just the extension of the M-file. In addition to the aforementioned function-naming rules, the name of the function must also follow all program-naming rules, including not using spaces in the file name.
- To test a function, you should call it from MATLAB's Command Window or from a different program or function by passing in input variables and saving the resulting output into different variables. If you click the Run button in the Editor, a dialog box will appear asking you to enter necessary arguments, but then it will use those values as the default when testing in the future. This may or may not be something you wish to do. We will discuss this in more detail shortly.
- Do *not* use the built-in function `clc` within one of your own functions. The functions you are writing are intended to be general purpose and can be called from other programs (or functions), so if text is written to the screen prior to calling a function that contains the `clc` command, it will wipe the Command Window clean—losing all of the previous output on the screen.
- For a different reason, the `clear` command should not be used within a function since functions usually contain input arguments, so calling the `clear`

command would delete any input variables and their values. In general, the `clear` and `clc` commands should only appear at the top of programs, not functions!

- In general, all assignment statements in a function should have the screen echo suppressed by adding a semicolon at the end of each assignment statement. Otherwise, you are likely to fill the screen with unnecessary output. The main exception would be during debugging, where having key values echo to the screen may assist in finding errors. However, once the function is working properly, the screen echos should be suppressed before the final version is saved.

Function Structure and Use

The following examples demonstrate the structure of different functions and how MATLAB programs can use the functions to reduce code redundancy.

EXAMPLE   16-8

Assume we are required to create a function, `areaCircle`, to calculate the area of a circle given the radius. The function should accept one input (radius) and return one output (area).

Source Code

```
% Problem Statement: This function calculates the area of a
% circle
% Input:
% r — Radius [any length unit]
% Output:
% A — Area [units of input^2]
function [A]=areaCircle(r)
% Calculation of the area of a circle
A=pi*r^2;
```

Typical Usage (typed in Command Window or a Separate Program):

You can call the function by typing the name of the function and passing in a value for the radius. Note that the result is stored in a variable called `ans` when you do not specify a variable to capture the output.

```
>> areaCircle(3)
ans=
     28.2743
```

You can call the function by typing the name of the function and passing in a variable.

```
>> Rad=4;
>> CircleA=areaCircle(Rad)
CircleA=
   50.2655
```

Comprehension Check   16-6

What is the output when you "pass in" the value 10 to the function in Example 16-8?

Assume that you want to store the result of the previous function call in a variable named `Dogs`. What is the command you would type in MATLAB?

Comprehension Check   16-7

Write a function named `RAC` that will accept two parameters, `N` and `A`. The function should return a single value R calculated as `R = 3 sin(N^A)`. For this example, you do not need to include any comments with this code.

Testing Functions

MATLAB's Editor contains functionality to allow you to efficiently test functions that you write by allowing you to create different test scenarios and providing a shortcut to run those test cases you can create directly in the Editor window. Assume we have created and saved the following function in MATLAB:

```
function [out1,out2,out3]=myFunction(in1,in2)
out1 = in1 * 2;
out2 = out1 + in2;
out3 = [out1, out2]*3;
```

This function named `myFunction` accepts two input arguments (`in1`, `in2`) and returns three output variables: `out1` and `out2` are both scalars and `out3` is a vector. Assume we want to test to see what the resulting output variables will contain when the value 4 is used for `in1` and 7 is used for `in2`. To do this, we can use the same Run button we used for programs. The first time you press the Run button, an error message will appear in the screen indicating that the input variables are missing:

```
>> myFunction
Error using myFunction (line 2)
Not enough input arguments.
```

However, a little pop-up dialog appears in MATLAB's Editor window that actually allows you to type in values for the test case. Type in **4, 7** within the parentheses on the dialog, representing the values of in1 and in2 respectively, and press the Enter key. You'll notice that output has appeared in the Command Window:

```
>> myFunction(4,7)
ans =
     8
```

This output shows that the function was successfully executed with the value of 4 for in1 and 7 for in2, but since this function didn't specify any output variables, the value 8 is stored in the default variable ans, which is actually the value of the out1 variable calculated in the function.

By default, MATLAB requires you to provide values for all of the input variables, but it does not require you to capture all of the output variables with each call. If you do not specify any variable or variables in which to place the results, the first returned value in the list (out1 in this case) is placed in ans. However, the Editor is smart enough to be able to set up test cases where all of the output variables are assigned to variables when the function is called. To save all three output variables (for testing, we will save out1 into variable a, out2 into variable b, and out3 into variable c), select the Run drop-down menu in the Editor tool ribbon (NOT the Run icon), right-click the line that reads myFunction(4,7), which was stored by the Editor when you entered those values earlier, select Edit, and edit it to read:

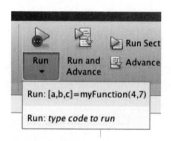

```
[a,b,c]=myFunction(4,7)
```

After we execute the function with our new test case, all three output variables are displayed in the Command Window:

```
>> [a,b,c]=myFunction(4,7)
a =
     8
b =
    15
c =
    24 45
```

You may have noticed that after you first entered the input values 4 and 7 that the Run icon in the Editor tool ribbon changed appearance: it now has a blue circle containing ellipses (three dots . . .). This indicates that there is at least one edited version of the function call stored. If you modify the function code, you may retest it with the same values by opening the Run menu and selecting the version you wish. To add another version with different input values, click "type code to run" in the Run menu, and that will be added to the selections in the future.

EXAMPLE   16-9

Assume we are required to create a function, `volumeCylinder`, to calculate the volume of a cylinder given the radius. The function should accept two inputs (radius, height) and return one output (volume). For this example, you may assume that the user will input the radius and height in the same units. As usual, any calculations in the functions must be suppressed.

Source Code

```
% Problem Statement: Calculate the volume of a cylinder.
% Input:
%    r — Radius [any length unit]
%    H — Height [same length unit as radius]
% Output:
%    V — Volume [units of input^3]
function [V]= volumeCylinder(r, H)
% Calculation of the volume of a cylinder
V=pi*H*r^2;
```

Typical Usage

```
>> Vol1 = volumeCylinder(1,2)
Vol1=
   6.2832
>> Vol2= volumeCylinder(2,1)
Vol2=
   12.5664
```

Note that the order of arguments passed in to the function matters. The order of variables expected by the function is based on the order they are listed in the function header. Therefore, the first input argument is the radius, followed by the height.

Comprehension Check   16-8

For the problem given in Comprehension Check 16-4, write a function to implement the flowchart you created. Be sure to include appropriate comments in your code. Remember to suppress all assignment statements. Test your code with at least three different sets of input values.

NOTE

Local variables are only visible to the program or function using them.

Global variables are visible within any program or function.

Local and Global Variables

One important fact to note about variables is that the variables in the workspace are only available within MATLAB's Command Window and to any running program. Functions create a private workspace and cannot access any of the variables available in MATLAB's main workspace. The only way to access any variables in MATLAB's main workspace is to "pass" the variable to the program as function input.

The reason for having "main workspace" variables inaccessible by functions is because functions exist within a different scope from the programs. This way, the programmer does not need to know the names of all variables used in all functions to avoid conflicts. For example, if the `sqrt` function defined a variable named `x1`, and your program also used `x1`, whenever the `sqrt` function was used, it would destroy the value of `x1` being used in your program.

The variables created within the "private workspace" of a function are commonly referred to as **local variables** since they are destroyed after the function executes and do not appear in the "main workspace" unless they are passed back as function output.

In some situations, this localization of variables could be very inconvenient if there were not a mechanism to circumvent it. For example, it would be a nuisance if every time you needed to use the predefined variable `pi` in a function, you had to include it in the input parameter list. Another case would be when both the main program and some of the functions it uses share a large number of variables, making the input parameter list excessively long.

In the first case, special constants like `pi` are predefined to exist within the scope of all programs and functions. Variables that are visible within any scope are commonly referred to as **global variables**. Global variables are created in functions using the reserved word **global** followed by a comma-separated list of the variables that should be considered global variables.

```
function [x] = functionName(y)
global z
z=30;
```

In addition, the same variables must be declared global in the main program and in any other functions that will rely on them.

In this example, after the function `functionName` is called, the value of `z` is accessible by other programs or functions that also include the statement `global z`.

On the left side of the following diagram, the inner box is a program (`Prgrm1`) and the outer box represents all of the variables in MATLAB's workspace. Since we do not include any commands to clear the variables from the workspace at the top of the program `Prgm1`, the program is able to "see" and use all of the variables within MATLAB's workspace. In addition, it will create a new variable in MATLAB's workspace (`c`) when we execute program `Prgm1`. The workspace then contains the three variables `a`, `b`, and `c`.

In the diagram on the right, again the outer box represents the workspace, which this time contains the variables `e` and `g`. The large box inside the outer box represents a program (`Prgrm2`) that uses a function (innermost box, `Func1`) to calculate a value for a new variable `h`. The function cannot "see" or use any of the variables in MATLAB's workspace and creates its own separate workspace to execute the code inside the function. In order to run the function, the programmer must provide the values of variables

MATLAB's Initial Workspace	MATLAB's Initial Workspace
<u>a</u>: 2 <u>b</u>: 3 Prgrm1: <u>c</u>=a+b;	<u>e</u>: 2 <u>g</u>: 5 Prgrm2: <u>h</u>=Func1(<u>e</u>,<u>g</u>); ——— e and g passed into x and y Func1: <u>function</u>[d]=Func1(<u>x,y</u>) <u>k</u>=2; <u>d</u>=k*x^y; <u>d</u> captured by h; d, k, x, and y discarded
Resulting Workspace	**Resulting Workspace**
<u>a</u>: 2 b: 3 c: 5	<u>e</u>: 2 <u>g</u>: 5 h: 64

x and y by "passing" the variables e and g into the function. The function code then defines a variable k, calculates the value of d, and returns d as function output. When the value of d is returned to Prgrm2, it is stored in the variable h. After the function has executed, the separate function workspace is destroyed, along with variables d, k, x, and y, which were never stored in MATLAB's main workspace.

EXAMPLE 16-10

Assume we are required to create a function, sphereCalculations, to calculate the volume and surface area of a sphere given the radius. The function should accept one input (radius) and return two outputs (volume, surface area). As usual, any calculations in the functions must be suppressed.

Source Code

```
% Problem Statement: This function calculates the volume and
% surface area of a sphere.
% Input:
%    r—Radius [any length unit]
% Output:
%    V—Volume [units of input^3]
%    SA—Surface Area [units of input^2]
function [V,SA]=sphereCalculations(r)
% Calculation of the volume of a sphere
V=4/3*pi*r^3;
% Calculation of the surface area of a sphere
SA=4*pi*r^2;
```

Typical Usage

```
>> [Vol,SurfArea]= sphereCalculations(7)
Vol=
   1.4368e+03
SurfArea=
   615.7522
```

EXAMPLE 16-11

Assume we are required to create a function, cylinderCalculations, to calculate the volume and lateral surface area of a cylinder given the radius and height. The function should accept two inputs (radius, height) and return two outputs (volume, surface area). For this example, you may assume that the user will input the radius and height in the same units. Any calculations in the functions must be suppressed.

Source Code

```
% Problem: Function calculates volume and surface area of
% cylinder.
% Input:
%    r—Radius [any length unit]
%    H—Height [same length unit as radius]
```

```
% Output:
%    V—Volume [units of input^3]
%    SA—Surface area [units of input^2]
function [V,SA]=cylinderCalculations(r,H)
% Calculation of the volume of cylinder
V=pi*H*r^2;
% Calculation of the surface area of a cylinder
SA=2*pi*r*H;
```

Typical Usage

```
>> [Volume,Surface]=cylinderCalculations(3,2)
Volume=
    56.5487
Surface=
    37.6991
```

EXAMPLE 16-12

Write a program that uses the two functions developed in Exercises 16-10 and 16-11 that will define a radius and a height, then place the volume and surface area of a sphere with the specified radius in SphVol and SphArea respectively, and the volume and surface area of a cylinder with the specified radius and height in CylVol and CylArea respectively.

```
% Problem: Calculate the surface area and volume of
% both a sphere and a cylinder.
% Input:
%    radius—Radius [any length unit]
%    ht—Height [same length unit as radius]
% Output:
%    SphVol—Volume of sphere [units of input^3]
%    SphArea—Surface area of sphere [units of input^2]
%    CylVol—Volume of cylinder [units of input^3]
%    CylArea—Surface area of cylinder [units of input^2]
clear
clc
radius=5; % define radius
ht=7; % define height
[SphVol,SphArea]=sphereCalculations(radius);
[CylVol,CylArea]=cylinderCalculations(radius,ht);
```

This yields the following values:

```
SphVol=
    523.5988
SphArea=
    314.1592
CylVol=
    549.7787
CylArea=
    219.9115
```

16.4 Deriving Mathematical Models

LEARN TO: Use `polyfit` to determine mathematical models to fit experimental data
Use the models to predict the behavior of the systems under study.

As a case study in writing simple programs, we will create a series of programs to acquire data from the user and create mathematical models to describe that data. Here we will consider the same three model types that were introduced in Chapter 13: linear, power, and exponential. In Excel, we used the trendline feature to determine these models. In MATLAB, we will use the function `polyfit`.

The **polyfit** function accepts two equal-sized vectors containing the data to be modeled. The first vector contains values of the independent variable, and the second vector contains the values of the dependent variable corresponding to the independent data measurements. Since polyfit is derived from "polynomial fit," it is easy to remember that it will determine the polynomial of order n that best fits the data contained in the vectors. We also must pass in a value for the desired order of the polynomial.

Consider the generic form of a polynomial:

$$p(x) = C_n X^n + C_{n-1} X^{n-1} + L + C_1 X + C_0$$

In the preceding equation, the coefficients $[C_n, C_{n-1}, \ldots, C_1, C_0]$ are returned by `polyfit` as a vector. Note that for an order n polynomial, there will be n + 1 coefficients.

The syntax of `polyfit` is:

```
C = polyfit (X, Y, n)
```

- **C**: Vector of the resulting coefficients. Note that `C(1)`, the first element of the vector C, contains the coefficient by the highest power of the variable listed in the equation, C_n.
- **X**: Vector of values that correspond to the abscissa
- **Y**: Vector (of equal length to X) of values that correspond to the ordinate
- **n**: The order of the polynomial.

Now, that may sound a bit complicated, but the ONLY case we will consider in this text is the linear case, where n = 1, and two coefficients are returned, C_1 and C_0. For a linear function, these are better known as m (slope) and b (intercept). Thus, `C=polyfit(X,Y,1)` returns the values of m and b in the two-element vector C.

WARNING

We have seen some cases where we can individually name the returned vector values. For example `Q=size([3,4,6]; will` return the vector `Q=[1, 3]`, or `[H,V]=size ([3,4,6]);` will return H=1 and V=3. This WILL NOT WORK with `poly-fit`. You cannot say `[m,b]=polyfit (X,Y,1)` and expect variables m and b to contain the slope and intercept. In this case, m would contain the values of both m and b in a single vector, and b would be a structure array containing other information that is beyond the scope of this discussion.

EXAMPLE 16-13

Given two row vectors, `Ext` and `Mass`, representing the extension of a spring [cm] and the mass [g] hung from that spring, write a program to determine a model to describe the spring. The spring constant [N/m] should be stored in the variable `SprConst` and the initial weight [N] with zero added weight should be in `InitialWt`. Use this model to determine the mass that would extend the spring 5 cm.

Assume `Ext` contains [3.4 7.2 11.9 14.2] and `Mass` contains [50 100 175 200].

```
% Housekeeping
clear
clc
```

```
% Set up data
g=9.801; % acceleration of gravity to calculate weight
Ext=[3.4 7.2 11.9 14.2]/100; %convert to meters
Mass=[50 100 175 200]/1000; % convert to kg
Wt=Mass*g; % calculate weight [N]

% Determine model coefficients
Coef=polyfit(Ext,Wt,1);
SprConst=Coef(1); % The slope is the spring constant [N/m]
InitialWt=Coef(2); % The intercept is the initial weight [N]

% Determine mass if extension is 5 cm
Ext5=5/100 % Convert to meters
W=SprConst*Ext5+InitialWt;
M=W/g*1000; % 1000 to convert to grams
```

After running the program, `SprConst = 13.95` *and* `InitialWt=0.006`*; thus the model for this spring is W = 13.95 E + 0.006. This yields a mass of 71.81 grams at an extension of 5 cm.*

Comprehension Check 16-9

The total distance [mi] a jetliner cruising at a constant speed has traveled is recorded every 15 minutes, beginning at 30 minutes, and the following values tabulated:

273   405   530   676   814   933   1070

At what time (in hours) will the distance traveled reach 2500 miles? All calculations must be done in MATLAB. Remember that total distance traveled by an object moving at a constant speed is $d = d_0 + vt$.

NOTE

You might at first think that if you know that the result will be a power model with a power of 2, you could just use `polyfit` with a 2 as the third parameter for the polynomial order. This WILL NOT return m and b for a power model, but the three coefficients of a quadratic equation.

Perhaps unfortunately, MATLAB does not have functions built in to directly calculate the parameters for power and exponential models. As we mentioned above, we will use only `polyfit`, and only first-order polynomials! At this point, some of you may be wondering how we can possibly do this. The key lies back in Chapter 13, where we first discussed these models.

First, we will consider the power model. As we saw in Chapter 13, for the power model, $y = bx^m$, if we take the base 10 logarithm of both sides, we get

$$\log (y) = m \log (x) + \log (b)$$

which is a linear form. The trick is that when we want to fit a power law model to a set of data using `polyfit`, we have to send it the base 10 logarithms of the independent and dependent data, and what it returns is the "slope" (the power of the independent variable) and the base 10 logarithm of the "intercept" (the scaling factor). In other words, we can use `polyfit` to create a linear fit between $\log(x)$ and $\log(y)$ to calculate the value of m and $\log(b)$. Thus when we get the results, we have to raise 10 to the second returned coefficient to get the value of b.

NOTE

Many computer applications and calculators use **log** for the base 10 logarithm and **ln** for the natural logarithm. It is easy to forget that MATLAB is different, so be careful!

Note that this calculation requires the use of the common (base 10) logarithm, not the natural logarithm. **In MATLAB, the base-10 logarithm function is `log10` and the natural logarithm function is `log`.**

Assuming abscissa and ordinate values are stored in variables X and Y respectively:

```
C = polyfit(log10(X),log10(Y),1)
m = C(1)
b = 10^C(2)
```

For power relationships, it is possible to use the natural logarithm instead of the common logarithm. The difference is how you calculate the values of b.

```
C = polyfit(log(X),log(Y),1)
m = C(1)
b = exp(C(2))
```

We encourage you to go through the derivation process using natural log to verify that this is correct. In summary, using either the common log or natural log for performing these calculations is correct as long as the calculation of b is handled appropriately.

EXAMPLE   16-14

You are making measurements of the electrical resistance [Ω] of a device as a function of temperature [K]. The results are stored in a 2 × N matrix named `TRData`. The first row of `TRData` contains the temperature measurements [K], and the second row has the corresponding resistance measurements [kΩ]. You suspect that the relationship between resistance and temperature follows a power law model. Determine the model parameters m and b in $R = bT^m$ and determine the resistance at a temperature of 1000 K.

Assume the temperature values are [250 290 315 350 425] and the corresponding resistance values are [1.25 1.35 1.51 1.61 1.92].

```
% Housekeeping
clear
clc

% Set up data. T in kelvins, R in ohms.
TRData=[250 290 315 350 425; 1250 1350 1510 1610 1920];

% Determine model coefficients
Coef=polyfit(log10(TRData(1,:)),log10(TRData(2,:)),1);
m=Coef(1);
b=10^Coef(2);

% Determine resistance at 1000 K
R1000=b*1000^m;
```

This yields b = 12.9 and m = 0.825, thus the model is R = 12.9 T$^{0.825}$. At 1000 K, the resistance is 3860 ohms.

Comprehension Check | 16-10

The height of liquid in a large, bowl-shaped container is recorded as measured amounts of water are added. The following data was recorded:

```
Heights (h) [ft]:    0.7    1.3     2.6    4.1
Volumes (V) [ft³]:  1.64   7.71    43.6   136
```

If the maximum depth of the container is 7.5 feet, what is the volume of liquid that it can contain? All calculations should be done in MATLAB, not manually. Assume the data fits a power model.

The final model we will consider is the exponential model: $y = b\,e^{mx}$. If we take the natural logarithm of both sides, we get

$$\ln(y) = mx + \ln(b)$$

This expression is essentially a linear model, so we can use `polyfit` to create a linear fit between x and $\ln(y)$ to calculate the value of m and $\ln(b)$. Recall that the natural logarithm function in MATLAB is `log`. To get the actual value of b, we will need to raise e to the value in the second coefficient returned by `polyfit`.

Assuming our abscissa and ordinate values are stored in variables X and Y, respectively:

```
C = polyfit(X,log(Y),1)
m = C(1)
b = exp(C(2))
```

EXAMPLE | 16-15

The pressure in a spacecraft has been slowly decreasing due to a small leak. The equipment on board has made measurements every hour for six hours and recorded the following pressures in pascals beginning at time = 0 (beginning time arbitrarily set to zero):

$$99{,}000 \quad 97{,}530 \quad 96{,}070 \quad 94{,}640 \quad 93{,}240 \quad 91{,}850 \quad 90{,}480$$

It is assumed that the pressure is decreasing according to an exponential model. At what time in hours measured from the arbitrary time 0 will the pressure reach 0.5 atmospheres?

```
% Housekeeping
clear; clc

% Set up data
Press=[99000,97530,96070,94640,93240,91850,90480];
Time=0:6;

% Determine model coefficients
Coef=polyfit(Time,log(Press),1);
m=Coef(1);  b=exp(Coef(2));

% Determine time to reach 0.5 atm
PAtm=0.5*101325; % convert to pascals
t=(log(PAtm)-log(b))/m;
```

This yields b = 99,000 and m = −0.015, thus the model is P = 99,000 e^(−0.015t). The pressure will reach one-half atmosphere at 44.7 hours.

Comprehension Check	**16-11**

The height of water slowly draining from a large barrel follows an exponential model: $h = h_0 e^{mt}$. The following three measurements of Height (h) [ft] versus time (t) [min] were recorded:

$$t = 4, h = 6.7; \quad t = 7, \quad h = 5.6; \quad t = 15, \quad h = 3.8$$

If a pump turns on to refill the barrel when the height drops below one foot, when [min] will the pump turn on? All calculation must be done in MATLAB, not by hand.

16.5 Debugging MATLAB Code

> **LEARN TO:**   Utilize the Debugger tool in MATLAB to eliminate errors
> Examine the variable values during execution of a program or function

NOTE

Why do we say that computers (and programs) have bugs? Grace Murray Hopper is quoted in the April 16, 1984, issue of *Time Magazine* as saying, "From then on, when anything went wrong with a computer, we said it had bugs in it"—referring to when a 2-inch-long moth was removed from an experimental computer at Harvard in 1947.

If a program has syntax errors, MATLAB (and other compilers) will give you useful feedback regarding the type and location of the error to help you fix, or **debug**, your program. If this information is not sufficient for you to understand the problem, check with others who might have seen this error previously—some errors are particularly common. If you still cannot identify the error, you might need to use a more formal process to study the error. For other kinds of errors, this formal process is essential for diagnosing the problems. When you first write a program, you must test the output of the program with a set of inputs for which you know the result. Such test cases would include:

- Simple cases for which you can quickly compute the expected output;
- Cases provided by an instructor or textbook with a published solution;
- Cases for which results have already been produced by a previously tested program that does the same thing;
- Test cases that are customarily used to test programs.

When the output of a program is different from what you expect, the debugging process begins. It is common to debug shorter programs simply by reading them and writing a few notes. Longer programs may require the use of the MATLAB Debugger, which is available as part of the MATLAB Editor. Whether you are using the MATLAB Debugger or are debugging "by hand" or in the Command Window, the same techniques apply.

Preparing for Debugging

When using the Editor/Debugger, open the program for debugging. If it is open, make sure changes are saved—otherwise MATLAB will run the most recently saved version without including changes made since the last time the program was saved. If you are debugging from a printed program and output, make sure that the printed output came from the version you are reading. In preparing to debug a program, it is critical to be able to reproduce the conditions that caused the bug in the first place. If the bug occurs in processing data from a large data set, you may have to split the data set to find the specific data that triggered the bug.

Setting Breakpoints

When using the Editor/Debugger, establish **breakpoints** that allow you to check your agreement with the program at various stages. A breakpoint stops the program at the specified location to allow you to examine the contents of variables and other elements before the program continues. This will help you find the specific location when the program does something you were not expecting. When debugging shorter programs, write down everything the program does or make all program results display in the Command Window by removing semicolons that were used to suppress output to the Command Window.

MATLAB has two main types of breakpoints: standard (set by location in the program) and conditional (triggered at a specified location by specified conditions). We focus on standard breakpoints in this section. Standard breakpoints are normally shown as little red dots next to the line of code where the executing should pause, and conditional breakpoints will appear as little yellow dots. If the breakpoint dot displays gray, either the file has not been saved since changes were made to it or there is a syntax error on the line or somewhere in the file. All breakpoints remain in a file until you clear (remove) them or until they are automatically cleared.

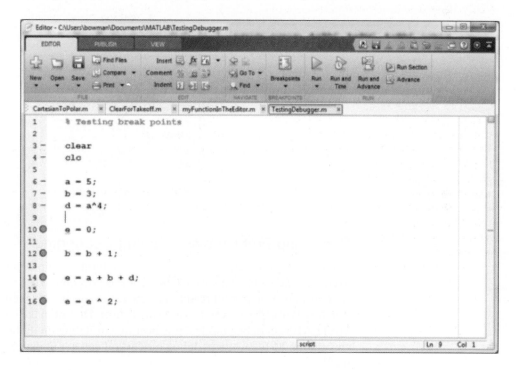

Stepping through a Program

When using the Editor/Debugger, you must start by running the program or function. You can choose to step through the program one line at a time or have it continue until a breakpoint is encountered. Typically, you will set a breakpoint and execute the program and the Editor will automatically stop before the line containing the breakpoint is executed. A green arrow will point to the program line when execution is paused. The controls in the Editor window will allow you to "step in" to functions you've written, or "step out" of functions you're currently in. Likewise, pressing the Continue button will resume execution—to completion or until the debugger encounters the next breakpoint.

Examining Values

When the program is paused at a breakpoint or when you are stepping through the program a line at a time, you can view the value of any variable by hovering your mouse over the variable to see whether a line of code has produced the expected result. If the result is as expected, continue running or step to the next line. If the result is not as expected, then that line, or a previous line, contains an error. You can also examine the value of a variable in the Workspace window in the main MATLAB program window.

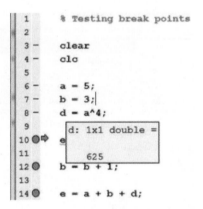

Correcting Problems and Ending Debugging

If a problem is difficult to diagnose, one method to help define or correct problems is to change the value of a variable in a paused program to see if continuing with the new value produces expected results. A new value can be assigned through the Command Window, Workspace browser, or Array Editor. Do not make changes to an M-file while MATLAB is in debug mode—it is best to exit debug mode before editing an M-file. If you edit an M-file while in debug mode, you can get unexpected results when you run the file.

Some time spent now familiarizing yourself with the debugger on simple programs will be well worth the effort when the programs become a lot more complicated.

IN-CLASS ACTIVITIES

For ICA 16-1 to ICA 16-9, create an algorithm (written and/or flowchart as specified by your instructor) to solve the following problems.

ICA 16-1

Your instructor will provide you with a picture of a structure created using K'Nex™ pieces. Describe the steps necessary to create the structure in the picture. When you are finished, hand your algorithm to the instructor and wait for further instruction.

ICA 16-2

Describe the steps necessary to create a paper airplane. You may assume that you are starting with a single sheet of 8½ × 11 inch paper. When you are finished, hand your algorithm to the instructor and wait for further instruction.

ICA 16-3

Describe the steps necessary to create a jelly sandwich. You may assume that you are starting with a loaf of bread, jar of jelly, a knife, and a plate on the table in front of you. When you are finished, hand your algorithm to the instructor and wait for further instruction.

ICA 16-4

Describe the steps necessary to cook your favorite meal, including a starting materials list.

ICA 16-5

Describe the steps necessary to walk from your home or dorm to your engineering classroom without using any street names.

ICA 16-6

An unmanned X-43A scramjet test vehicle has achieved a maximum speed of Mach number X.XX in a test flight over the Pacific Ocean, where X.XX is a positive value entered by the user. Mach number is defined as the speed of an object divided by the speed of sound. Assume the speed of sound is 343 meters per second. Determine the speed in units of miles per hour. For a test case, you may assume that the user provides the value of 9.68 for the Mach number.

ICA 16-7

Convert a temperature provided by the user in units of degrees Fahrenheit to units of kelvins. As a test case, you may assume the user provides the temperature of −129 degrees Fahrenheit, which is the lowest naturally occurring temperature ever recorded on Earth.

ICA 16-8

Determine the mass of oxygen gas (formula: O_2, molecular weight = 32 grams per mole) in a container, in units of grams. You may assume that the user will provide the volume of the container in units of gallons, the temperature in the container in degrees Celsius, and the pressure in the container in units of atmospheres. For your test case, you may assume that the user provides 1.25 gallons for the volume of the container, 125 degrees Celsius for the temperature, and 2.5 atmospheres for the pressure in the container.

ICA 16-9

Determine the length of one side of a cube of solid gold, in units of inches. You may assume that the specific gravity of gold is 19.3 and that the user will provide the mass of the cube in units of kilograms. As a test case, you can assume that the user has a 0.4 kilogram cube.

ICA 16-10

Which of the following are not valid program/function file names? Circle all that apply.

(A) `2b_solved.m`
(B) `calc_circum.m`
(C) `graph-data.m`
(D) `help4me.m`
(E) `MATLAB is fun.m`
(F) `matrix*matrix.m`
(G) `Mult2#s.m`
(H) `pi.m`
(I) `ReadFile.m`
(J) `SuperCaliFragiListicExpiAliDocious.m`

ICA 16-11

Without running these code segments in MATLAB, what will be the value stored in the specified variable after executing each program?

(a) What is stored in **Dogs**?

```
clear
clc
Greyhounds = 30;
G_Cost = 250;
Dalmations = 10;
D_Cost = 150;
Dogs = Greyhounds * G_Cost + Dalmations * D_Cost;
```

(b) What is stored in **W**?

```
clear
clc
X = 2;
Y = 4;
Z = 2;
W = (Y + Z/X + Z^1/2)
```

ICA 16-12

Without running these code segments in MATLAB, what will be the value stored in the specified variable after executing each program?

(a) What is stored in **X**?

```
clear
clc
A = 0;
B = 5;
X = A^B + B^A;
```

(b) What is stored in **Ramones**?

```
clear
clc
Joey = 1;
Johnny = 2;
DeeDee = 3;
Marky = 4;
Ramones = Joey * 2 + Johnny/2 + max([DeeDee, Marky]);
```

ICA 16-13

Write a program to store the following matrices into MATLAB. All variable assignments should be suppressed. If the variable assignment is not possible or causes an error message on the screen, write MATLAB comments explaining the problem.

$$A = \begin{bmatrix} 4 & 6 \\ -3 & 8 \end{bmatrix} \quad B = \begin{bmatrix} 3 & 5 \\ 5 & 5 \end{bmatrix} \quad C = \begin{bmatrix} 0 & 1 \end{bmatrix} \quad D = \begin{bmatrix} 1 \\ 1 \end{bmatrix} \quad E = \begin{bmatrix} 3 & 6 & 5 \\ 7 & 1 & 5 \end{bmatrix} \quad F = \begin{bmatrix} 2 & 8 \\ 4 & 3 \\ 1 & 5 \end{bmatrix}$$

$$G = \begin{bmatrix} 7 & 8 & 3 \\ 4 & 8 & 2 \\ 5 & 9 & 5 \end{bmatrix} \quad H = \begin{bmatrix} 1 & 2 & D & 299 & 300 \end{bmatrix} \quad I = \begin{bmatrix} 200 \\ 198 \\ D \\ -298 \\ -300 \end{bmatrix}$$

$$J = \begin{bmatrix} 1 & 2 & C & 299 & 300 \end{bmatrix} \quad K = \begin{bmatrix} 200 \\ 198 \\ C \\ -298 \\ -300 \end{bmatrix}$$

$$L = \begin{bmatrix} 20 & 40 & D & 180 & 200 \\ 10 & 20 & D & 90 & 100 \end{bmatrix} \quad M = \begin{bmatrix} 5 & 50 \\ 10 & 45 \\ D & D \\ 45 & -45 \\ 50 & -50 \end{bmatrix}$$

ICA 16-14

Write a program to perform the following calculations, assuming the variables are already defined as in ICA 16-13. Calculations should not be suppressed. If the calculation is not possible or causes an error message on the screen, write MATLAB comments explaining the problem.

The results of the calculations should be stored into variable names of your choosing. Each calculation should be stored in a different variable. Here, note that a superscripted "T" indicates the matrix should be transposed.

(a) A*B

(b) A + B

(c) A + C

(d) C − E

(e) F^T

(f) C*D

(g) $E^T + 30$

(h) E^2*50

(i) B^2*50

(j) E*F

(k) F*E

(l) G^2

(m) H + 35

(n) J − 100

(o) $I^T * I$

(p) $J^T * J$

ICA 16-15

Write a MATLAB program to evaluate the following mathematical expression. The equation should utilize variables for a, b, and c. Test the program with $a = 1$, $b = 2$, and $c = 3$.

$$x = \frac{b + \sqrt{b - 4a}}{c^7}$$

ICA 16-16

Write a MATLAB program to evaluate the following mathematical expression. The equation should utilize a variable for x. Test the program with $x = 30$.

$$A = \frac{x^2 \cos(2x + 1)}{(6x) \log(x)}$$

ICA 16-17

Write a MATLAB program to evaluate the following mathematical expression. The equation should utilize variables for x, μ, and σ. Test the program with $x = 0$, $\mu = 0$, and $\sigma = 1$.

$$P = \frac{1}{\sigma\sqrt{2\pi}} e^{-(x-\mu)^2/(2\sigma^2)}$$

This program calculates a *Gaussian normal distribution*.

ICA 16-18

The Shockley diode equation gives the relationship between the voltage (V) across a semiconductor junction and the current (I) through it.

$$I = I_0(e^{\left(\frac{V}{nV_T}\right)} - 1)$$

Assume V is a vector containing several voltage values. Write a MATLAB program that will calculate a vector I of the same length containing the current corresponding to each value in V.

Test your program with $I_0 = 2 \times 10^{-11}$, $n = 1$, $V_T = 25.85 \times 10^{-3}$, and $V = [0.4, 0.55, 0.65, -5, 0]$.

ICA 16-19

For each MATLAB code segment shown, write the function header necessary to convert the code segment into a function. Use the guidelines specified with each problem.

(a) The function will have one output, the variable X. The function will be stored in a file named `CandyCrush.m`.

```
X=(A+B+B^A)*pi;
```

(b) The function should have five output variables: the paintable area of each wall (all four walls) and the total paintable area. You may assume this function will be stored in a file named `FullHouse.m`.

```
P1 = Wall1 * RoomHeight;
P2 = Wall2 * RoomHeight;
P3 = Wall3 * RoomHeight;
P4 = Wall4 * RoomHeight;
PaintableArea = P1 + P2 + P3 + P4;
```

(c) The function should have three output variables: the minimum value, the maximum value, and the average value. You may assume this function will be stored in a file named MinMaxMean.m.

```
MyMin = min([D,D^2,D*10]);
MyMax = max([D^3,D*3,D + 3000]);
MyMean = mean([D/2,D/3,D/4]);
```

ICA 16-20

Assume you are given the following function:

```
function [status] = ClearForTakeoff(T)
status = [T + 2, T - 4; T*6, T/8];
```

(a) What is stored in the variable Z after executing the following lines of MATLAB code?

```
A = 2;
Z = ClearForTakeoff(A);
```

(b) What is stored in the variable E after executing the following lines of MATLAB code?

```
D = 0;
E = ClearForTakeoff (D);
```

ICA 16-21

A member of your team gives you the following MATLAB program. Your job is to create the functions used in the program.

```
A = 10; % area [cm^2]
H = 30; % height [cm]
V = 50; % volume [cm^3]
% calculates the radius [cm] of the circle
R1 = RadiusCircle(A);
% calculates the radius [cm] of a cone
R2 = RadiusCone(H,V);
```

ICA 16-22

Assume you are given the following function:

```
function[M_NEW,N] = SimpleCalculations(Wii)
M_NEW = Wii + 30;
N = M_NEW/2;
Wii = 35;
```

What is stored in the variables P, C, and r after executing the following lines of MATLAB code?

```
r=5;

[P,C]=SimpleCalculations(r);
```

ICA 16-23

Write a function named AddDiags that accepts two 3 × 3 matrices (A and B) as arguments, and returns a two-element vector S. The first element of S is the sum of the elements on the main diagonal of A, and the second element of S is the sum of the elements on the minor diagonal of B. You must calculate each term in S individually from terms in A and B—you may not use any built-in functions. The function is not expected to work properly for any other matrix dimensions. Use the following values to test the function:

```
A = [1 3 -4; -2 2 0;-3 6 5];
B = [-3 0 6;4 -1 -9;-5 0 7];
S = [8 0]
```

ICA 16-24

Write a function named `DiagCalcs` that accepts two 3×3 matrices (A and B) as arguments, and returns two variables, P and Q. P will return the sum of the elements on the main diagonal of A times the sum of the elements on the minor diagonal of B. Q will return the sum of the elements on the main diagonal of A divided by the sum of the elements on the minor diagonal of B. You MAY use the function `AddDiags` developed in the previous question, but you may not use any other functions. Use the following values to test the function:

```
A = [1 3 -4; -2 2 0;-3 6 5];
B = [-3 0 6;4 -1 -9;-5 0 7];
P = 0
Q = Inf
```

ICA 16-25

The mass of several different lengths of the same type of wire were measured and the following results recorded:

```
Length (L) [m]: 5      20      50       100
Mass (M) [g]: 700    2830    7070   14,200
```

Assume these data fit a linear model. Determine that model (mass as a function of length) and use it to predict the mass of one kilometer of the same wire. You must do all calculations in MATLAB.

ICA 16-26

The rotational kinetic energy of a spinning object is measured as a function of its angular velocity and the following data recorded.

```
Angular Velocity (ω) [rad/s]:  250    370    430     510     690
Rot. Kinetic Energy (E) [kJ]: 3130   6840   9250   13000   23800
```

Assume these data fit a power model. Determine that model and use it to predict the angular velocity of the object if its rotational kinetic energy is 10 megajoules. You must do all calculations in MATLAB.

ICA 16-27

Since the first exoplanet (a planet around a star other than our own sun) was discovered in 1988, the pace of new discoveries has increased roughly exponentially. The following table shows several dates and the total number of exoplanets discovered by the end of each year.

Year	1988	1992	1996	1999	2002	2005	2009	2012	2015
# Exoplanets	1	5	13	31	96	187	338	881	2078

Determine a mathematical model to describe the number of exoplanets discovered as a function of time, using 1988 as year zero. (Thus the first entry is year 0, the second entry is year 4, etc.)

If this trend continues, in what year will the one millionth exoplanet be discovered? You must do all calculations in MATLAB.

ICA 16-28

A novice programmer has attempted to write a program and two functions using global variables. These three files are provided for you with the online materials.

The purpose of the code is to determine the total pressure at the bottom of a cylindrical container filled with a liquid on an arbitrary planet, and to determine the potential energy of that cylinder at a given height above the planet's surface.

(a) Without running the code, determine what the values of the following variables will be following execution of the code: `g`, `atm`, `Radius`, `Depth`, `CylMass`, `rho`, `Height`, `H`, `Pressure`, `P`, `CylVol`, `ContentsMass`, `TotalMass`, `PE`

(b) Which of the variables in part (a) DO NOT appear in the main program workspace?

(c) For the values specified in the program, the correct answers are Pressure = 26,150 [Pa] and PE = 19,987 [J]. The program and associated functions calculate the pressure correctly, but the potential energy is wrong. Explain what went wrong and how to repair the problem.

ICA 16-29

Debug the MATLAB programs/functions provided for you with the online materials.

```
atm to Pa.m
ft to m.m
ICA #7.m
```

These files must be corrected to eliminate any syntax, runtime, formula coding, and formula derivation errors. You may assume that the header comments at the top of the program and functions are correct. In addition to the corrected files, you must submit an algorithm template, showing the main program and both functions.

ICA 16-30

Debug the MATLAB programs/functions provided for you with the online materials.

```
SG to rho.m
N_2_lbf.m
ICA S-D.m
```

These files must be corrected to eliminate any syntax, runtime, formula coding, and formula derivation errors. You may assume that the header comments at the top of the program and functions are correct. In addition to the corrected files, you must submit an algorithm template, showing the main program and both functions.

ICA 16-31

Consider the following MATLAB program and function, stored in MATLAB's Current Directory:

MyRadFunction.m

```
clear; clc
function [Out] = RadF (In1, In2, In3)
Out = (In1+In2)/2 + (In2+In3)/4;
```

MyRadProgram.m

```
clear; clc
InVar1=1; InVar2=3; InVar3=-1;
M=MyRadFunction(invar1), MyRadFunction(invar2), myradfunction(invar3);
```

Fix the program and function to eliminate all of the error messages. Note that for the variables provided in the program as InVar1, InVar2, and InVar3, the numerical result stored in M should be 2.5.

ICA 16-32

A novice MATLAB user created the following code with poor choice of variable names. Correct the errors and fill in the blanks provided to comment this code to determine the purpose of the code.

Main Program		Comments
`T = [30, 45, 120, 150];`	**(a)**	`% T =`
`Z = 2;`	**(b)**	`% Z =`
		`% The purpose of the function DTOR is . . .`
	(c)	`%`
`[W] = DTOR(T)`	**(d)**	`% W =`
		`% The purpose of the function PCAR is . . .`
`[X,Y] = PCAR(Z,W)`	**(e)**	`%`
	(f)	`% X =`
	(g)	`% Y =`
`function [A] = DTOR(T)`		
`A = T*2*pi/360;`		
`function (P,Q) = PCAR(M,N)`		
`P = M * cos(N);`		
`Q = M * sin(N);`		
	(h)	The output of this code, when run, is _____

REVIEW QUESTIONS

1. Create an algorithm to determine the weight of a rod in units of pounds-force on the surface of Callisto, Jupiter's moon. You may assume that the user of the algorithm will provide the volume of the rod in units of cubic meters. Your algorithm may assume that the specific gravity is 4.7 and the gravitational acceleration on Callisto is 1.25 meters per second squared. As a test case, you can assume the user has a rod with a volume of 0.3 cubic meters.

2. The Eco-Marathon is an annual competition sponsored by Shell Oil, in which participants build special vehicles to achieve the highest possible fuel efficiency. The Eco-Marathon is held around the world with events in the United Kingdom, Finland, France, Holland, Japan, and the United States.

   A world record was set in the Eco-Marathon by a French team in 2003 called Micro-joule with a performance of 10,705 miles per gallon. The Microjoule runs on ethanol. Create an algorithm to determine how far the Microjoule will travel in kilometers given a user-specified amount of ethanol, provided in units of grams. For your test case, you may assume that the user provides 100 grams of ethanol.

3. Create an algorithm to determine the density of tribromoethylene in units of kilograms per cubic meter in a cylindrical tank. Your algorithm should be written in such a way that the user provides the height measured in units of feet of the tribromoethylene in the tank, the surface pressure in units of atmospheres, and the total pressure at the bottom of the tank, also measured in atmospheres. For your test case, you may assume that the user provides a height of 25 feet for the tank, tank surface pressurization of 3 atmospheres, and a total pressure at the bottom of the tank of 5 atmospheres.

4. Create an algorithm to determine how long, in units of seconds, it will take a motor to raise a load into the air. Assume the user will specify the power of the motor in units of watts, the rated efficiency as a percentage (in whole number form—for example, 50 for 50%), the mass of the load in kilograms, and the height the load is raised in the air in units of meters. As a test case, you may assume the user provides 100 watts for the power of the motor, 60% for the efficiency of the motor, 100 kilograms for the mass of the load, and 5 meters for the height the load is raised.

5. The specific gravity of gold is 19.3. Write a MATLAB program that will determine the length of one side of a 0.4 kilogram cube of solid gold, in units of inches.

6. An unmanned X-43A scramjet test vehicle has achieved a maximum speed of Mach number 9.68 in a test flight over the Pacific Ocean. Mach number is defined as the speed of an object divided by the speed of sound. Assuming the speed of sound is 343 meters per second, write a MATLAB program to determine the record speed in units of miles per hour.

7. A rod on the surface of Jupiter's moon Callisto has a volume of 0.3 cubic meters. Write a MATLAB program that will determine the weight of the rod in units of pounds-force. The specific gravity is 4.7. Gravitational acceleration on Callisto is 1.25 meters per second squared.

8. The Eco-Marathon is an annual competition sponsored by Shell Oil, in which participants build special vehicles to achieve the highest possible fuel efficiency. The Eco-Marathon is held around the world with events in the United Kingdom, Finland, France, Holland, Japan, and the United States.

   A world record was set in the Eco-Marathon by a French team in 2003 called Micro-joule with a performance of 10,705 miles per gallon. The Microjoule runs on ethanol. Write a MATLAB program to determine how far the Microjoule will travel in kilometers given a user-specified amount of ethanol, provided in units of grams. For your test case, you may assume that the user provides 100 grams of ethanol. The specific gravity of ethanol is 0.789.

9. Write a program to determine the mass of oxygen gas (formula: O_2, molecular weight = 32 grams per mole) in units of grams in a container. You may assume that the user will provide the volume of the container in units of gallons, the temperature in the container in degrees Celsius, and the pressure in the container in units of atmospheres. For your test case, you may assume that the user provides 1.25 gallons for the volume of the container, 125 degrees Celsius for the temperature, and 2.5 atmospheres for the pressure in the container.

10. Write a program to convert a temperature provided by the user in units of Fahrenheit to units of kelvins. As a test case, you may assume the user provides the temperature of -129 degrees Fahrenheit, which is the world's lowest recorded temperature.

11. Write a program to determine how long, in units of seconds, it will take a motor to raise a load into the air. Assume the user will specify the power of the motor in units of watts, the rated efficiency as a percentage (in whole number form—for example, 50 for 50%), the mass of the load in kilograms, and the height the load is raised in the air in units of meters. As a test case, you may assume the user provides 100 watts for the power of the motor, 60% for the efficiency of the motor, 100 kilograms for the mass of the load, and 5 meters for the height the load is raised.

12. A cylindrical tank filled to a height of 25 feet with tribromoethylene has been pressurized to 3 atmospheres ($P_{surface} = 3$ atmospheres). The total pressure in at the bottom of the tank is 5 atmospheres. Write a MATLAB program to determine the density of tribromoethylene in units of kilograms per cubic meter.

13. Write a MATLAB program that implements the quadratic equation. Recall the quadratic equation:

$$r = \frac{-b \pm \sqrt{b^2 - 4ac}}{2a}$$

14. Write a MATLAB program that implements the Pythagorean theorem. Recall that the theorem states that the square of the hypotenuse (z) can be calculated by the sum of the squares of the adjacent sides (x and y), or

$$z = \sqrt{x^2 + y^2}$$

15. The specific gravity of gold is 19.3. Write a MATLAB function that will determine the length of one side of a cube of solid gold, in units of inches, provided the mass of the cube in kilograms.

16. An unmanned X-43A scramjet test vehicle has achieved a maximum speed of Mach number 9.68 in a test flight over the Pacific Ocean. Mach number is defined as the speed of an object divided by the speed of sound. Assuming the speed of sound is 343 meters per second, write a MATLAB function that will calculate the speed in units of miles per hour given the Mach number.

17. A rod on the surface of Jupiter's moon Callisto has a volume of 0.3 cubic meters. Write a MATLAB function that will determine the weight of the rod in units of pounds-force given the rod volume in cubic meters. The specific gravity is 4.7. Gravitational acceleration on Callisto is 1.25 meters per second squared.

18. A cylindrical tank filled to a height of 25 feet with tribromoethylene has been pressurized to 3 atmospheres ($P_{surface} = 3$ atmospheres) The total pressure in at the bottom of the tank is 5 atmospheres. Write a MATLAB function that will determine the density of tribromoethylene in units of kilograms per cubic meter given the height in units of feet and the surface and total pressures in units of atmospheres.

19. Write a function that implements the quadratic equation. Given three inputs (a, b, and c), calculate the roots ($r1$ and $r2$) of the quadratic formula. Recall the quadratic equation:

$$r = \frac{-b \pm \sqrt{b^2 - 4ac}}{2a}$$

20. Write a function that implements the Pythagorean theorem. Recall that the theorem states that the square of the hypotenuse (z) can be calculated by the sum of the squares of the adjacent sides (x and y), or

$$z = \sqrt{x^2 + y^2}$$

MATLAB examples

21. In the starting file provided on online, there are data sets from two different data collection sessions. In the first data collection session, a lab technician collected three different measurements and recorded mass in grams, height in feet, and time in minutes. However, in data collection session two, a different lab technician collected four different measurements and recorded mass in pounds-mass, height in centimeters, and time in hours.

Your job is to write a function that will calculate the potential energy in joules and power in watts for each data condition (seven total). In addition, you will need to write a program to call the function using the data sets provided in the starting file. Your function should only consider variables in SI units, so you will need to convert the vectors in the program before passing them in to your function. Note that all conversions must be done in MATLAB code—you may not hard-code any values you calculate by hand.

MATLAB examples

22. As part of a team investigating the effect of mass on the oscillation frequency of a spring, you obtain data from three different lab technicians, provided for you in the starter file online.

The data consists of frequency data on three different springs recording the amount of time it takes each spring with different masses attached to oscillate a certain number of times. In this experiment, each technician recorded the time it took for the spring to oscillate 25 times, stored in the variable N in the starter file.

As part of the analysis, you need to write a program containing the experimental data provided. The data should be converted to make all units consistent. The data should then be passed into a function.

The function should accept three inputs: a vector containing mass measurements (grams), a vector containing time measurements (seconds), and a variable containing the number of oscillations observed in the experiment.

The function should return two matrices. The first matrix should contain the mass measurements (kilograms) in the first column and the period (seconds), or the length of time required for one oscillation for each mass in the second column. The second matrix should contain a calculation of the force applied (newtons) by each mass in the first column and the frequency, the number of oscillations per second, or the inverse of the period (hertz) for each mass measured in the second column.

23. We have made many measurements of coffee cooling in a ceramic coffee cup. We realize that as the coffee cools, it gradually reaches room temperature. Consequently, we report the value of the coffee temperature in degrees above room temperature (so after a long time, the temperature rise will be equal to 0). Also, we realize that the hotter the coffee is initially (above room temperature), the longer it will take to cool. The values presented here are in degrees Fahrenheit.

Temperature Rise (T) [°F]						
Initial Temp Rise (T_0) [°F]	Cooling Time Elapsed (t) [min]					
	0	10	20	30	40	50
20	20	13	9	6	4	3
40	40	27	18	12	8	5
60	60	40	27	18	12	8
80	80	54	36	24	16	11

Write a MATLAB function that will perform a single interpolation given five numbers as input arguments and return the interpolated value as the only function output.

Write a MATLAB program that will calculate the following scenarios. Store each part in a different variable (e.g., part (a) should be stored in a variable named `PartA`, part (b) should be stored in a variable named `PartB`, etc.).

(a) What is the temperature (rise) of the cup of coffee after 37 minutes if the initial rise of temperature is 40 degrees Fahrenheit?

(b) If the coffee cools for 30 minutes and has risen 14 degrees Fahrenheit at that time, what was the initial temperature rise?

(c) Find the temperature rise of the coffee at 17 minutes if the initial rise is 53 degrees Fahrenheit.

24. In a factory, various metal pieces are forged and then plunged into a cool liquid to quickly cool the metal. The types of metals produced, as well as their specific heat capacity, are listed in the following table.

Material	Specific Heat [J/(g °C)]
Aluminum	0.897
Cadmium	0.231
Iron	0.450
Tungsten	0.134

The metal pieces vary in mass and are produced at a temperature of 300 degrees Celsius. The ideal process lowers the temperature of the material to 50 degrees Celsius. The liquid used to cool the metal is glycerol. The properties of glycerol are listed below.

Material Property	Value [Units]
Specific heat	2.4 J/(g °C)
Specific gravity	1.261
Initial temperature	25 °C

There are data sets from four different data collection sessions. In the first data collection session, a lab technician collected seven different measurements and recorded mass in grams of aluminum rods. The second data set contains cadmium rods; the third data set contains iron rods, and the fourth data set contains tungsten rods.

Mass of Object [g]			
Aluminum	**Cadmium**	**Iron**	**Tungsten**
2,000	3,000	2,500	4,800
2,500	4,000	3,500	6,400
3,000	6,500	4,500	10,400
4,000	8,000	5,000	12,800
5,500	10,000	5,500	16,000
7,500	11,000	7,500	17,600
8,000	15,000	9,000	24,000

Your job is to write two functions: (1) to calculate the thermal energy in joules that must be removed for each rod to cool it from 300 to 50 degrees Celsius for each mass and (2) to determine the volume of fluid needed in gallons to properly cool the rod for each mass. The result from the second function should be a matrix with the first column being the mass of the rod, and the second column the volume of fluid needed.

In addition, you will need to write a program to call the functions using the data sets provided in the final table. You will call each function four times, once for each material. Your function should only consider variables in SI units, so you will need to convert the vectors in the program before passing them to your function as necessary. Note that all conversions must be done in MATLAB code—you may not hard-code any values you calculate by hand.

25. You have been hired by the psychology department at your school to develop software to help analyze the progress of turtles on a track. Each turtle moves at a constant speed when it is not resting. Students monitoring the turtles record data for time and total distance traveled each time a turtle begins moving until it stops again. Write a function named `Turtle` to determine a model describing the movement of a turtle. It should also calculate how long it would take the turtle to reach specific distances. Recall that $d = d_0 + vt$

Specifications

Input:

- A $2 \times N$ matrix named `TurtleData`. The first row contains the times [s] at which the total distance traveled by the turtle was measured. The second row contains the corresponding total distances [ft] traveled by the turtle at the times stored in the first row. A new set of measurements will be started by resetting the time to zero whenever the turtle restarts after resting, so the total distance traveled at the instant the turtle starts will generally not be zero.
- A row vector `Dist` containing distances [ft] for which the required time is desired.

Output:

- A two-element vector `Model`, containing the speed of the turtle [ft/s] and the initial position [ft] of the turtle when it starts moving from rest at the arbitrary time zero.
- A row vector `Times` equal in length to `Dist` containing the times [s] necessary to reach each of the specified distances.

Test Case:

- `TurtleData = [2 15 35 72;42 51 63 88]`
- `Dist=[50 75 100]`
- `Model=[0.654 40.7]`
- `Times=[14.2 52.4 90.7]`

26. During the last three decades, the number of confirmed exoplanets (planets orbiting a star other than our own sun) has increased from none to over 3400, and the rate of discoveries is increasing. Write a function named `Planets` that will accept a set of measurements for the orbital periods of a group of planets orbiting another star and the distances of those planets from that star, and return the parameters (m and b) of a power model to describe the orbital period as a function of distance from the star. It should also determine how far a planet would have to be from that star if its orbital period was equal to one Earth year (365.24 Earth days). In addition, it should determine the mass of the star around which the planet orbits.

For systems where the mass of the star is considerably greater than that of the planet (the usual case), the orbital period can be given by

$$T = \frac{2\pi}{\sqrt{GM}} r^m$$

where r is the radius of the orbit [m], M is the mass of the star [kg], T is the orbital period [s], and G is the gravitational constant: 6.674×10^{-11} N m²/kg².

Specifications

Inputs:

- A $2 \times N$ matrix `ExoData`. The first row contains the orbital periods of planets circling the same star, and the second row contains the distances of those planets from that star.

Outputs:

- A two-element vector `SystemModel` containing the parameters (m and b) of the model of orbital period as a function of orbital radius.

- A scalar `Mass` containing the mass [kg] of the star
- A scalar `YrDist` containing the distance [m] of the orbit of a planet around that star with a period of one Earth year.

Test your function by using the orbital data for Mercury, Venus, and Mars.

- Mercury: Orbital radius: 57,909,000 km   Orbital period: 87.97 days
- Venus: Orbital radius: 108,208,000 km   Orbital period: 224.7 days
- Mars: Orbital radius: 227,939,000 km   Orbital period: 686.0 days
- Mass of sun: 1.989×10^{30} kg
- Orbital period of Earth: 31.56 Ms

27. You have to design a simple circuit to meet the following criteria: The voltage at time $t = 10$ ms should be 5 volts and the voltage at time $t = 100$ ms should be 1 volt. After $t = 100$ ms, the voltage should asymptotically approach zero. You know that the voltage across a simple circuit consisting of a resistor and a capacitor can be described by

$$V(t) = V_0 e^{-\frac{t}{RC}}$$

where V_0 is the initial voltage at $t = 0$, R is the resistance in ohms, and C is the capacitance in farads. Assume you have several different capacitor values stored in a vector named `Caps`. These values are 0.05, 0.1, 0.5, 2, and 5, all given in microfarads.

Write a program that will determine the parameters m and b for a standard exponential model to fit these data. Use that model to calculate the resistance values needed for each of the capacitor values to meet the circuit specifications. Store the resistance values corresponding to each capacitor in a vector named `Res`. Store the initial voltage in V_0. You should not do any calculations by hand—MATLAB should perform all necessary calculations.

28. Assume you are sent a file named `VolumeOfBox.m`, which is a function that calculates the volume of a box, given the length (`L`), width (`W`), and height (`H`) in nonspecific units. The assumption of the function provided to you is that the length, width, and height variables you provide the function are all measured in the same unit. When you attempt to call the MATLAB function, you notice that it crashes with the following error message:

```
Error using VolumeOfBox
Too many input arguments.
```

The contents of the `VolumeOfBox.m` file are shown below:

```
function [L,W,H] = VolumeOfBox(V)
% Calculate the volume of a box
V = L * W * H;
```

   (a) In your debugging, you have discovered that the error message on the screen means that the function header is incorrect. Write the correct function header.
   (b) Write a line of MATLAB code that will call the `VolumeOfBox` function with the value 3 for the length, 5 for the width, and 7 for the height variables and will store the volume calculation in a variable named `BoxVol`.

29. Assume you are sent a file named `GetCostByTime.m`, which is a function that calculates two different cost calculations given a time variable (`t`, non-unit specific) and a scaling factor, `S`. In the source code, the cost calculations are stored in two variables, `M` and `N`. When you attempt to call the MATLAB function, you notice that it crashes with the following error message:

```
Attempt to execute SCRIPT GetCostByTime as a function
```

The contents of the `GetCostByTime.m` file are:

```
t = t/s;
% Calculate the total cost
M = t^2 + 3*t + 10;
N = t^2 - 2*t + 30;
```

(a) In your debugging, you have discovered that the function header is missing from this file. Write the correct function header for the `GetCostByTime.m` file.

(b) Write a line of MATLAB code that will call the `GetCostByTime` function with the value 360 for the time variable and 60 as the scaling factor and will store the two different cost calculations in variables `C1` and `C2`.

30. You have been assigned to a new project at work. The previous engineer had created the following file, which your new boss claims does not work. Debug the program to correct all errors.

   These files must be corrected to eliminate any syntax, runtime, formula coding, and formula derivation errors. You may assume that the header comments at the top of the program and all comments throughout the program are correct.

```
% Problem Statement: A ball is thrown vertically into
% the air with an initial kinetic energy of 2,500 joules.
% As the ball rises, it gradually loses kinetic energy
% as its potential energy increases. At the top of
% its flight, when its vertical speed goes to zero, all of
% the kinetic energy has been converted into potential energy.
% Assume that no energy is lost to frictional drag, etc.
% How high does the ball rise in units of meters if it has
% a mass of 5 kilograms?
% Input Variables
% KE - Kinetic Energy [J]
% m - Mass of ball [kg]
% Output Variables
% H - Max height of ball [m]
% Other Variables
% g - Acceleration due to gravity [m/s^2]
% Assumptions
% No energy losses due to friction
% g = 9.8 m/s^2
clearscreen
clearworkspace
% Set initial variables
2500 = KE;
m = 5 kg;
gravity = 9.8;
%  Calculate height, assuming
%  Kinetic Energy = height * mass * -gravity
KE(2500) = H * m * g
```

31. You have been assigned to a new project at work. The previous engineer had created the following files, which your new boss claims do not work. Debug the program to correct all errors.

   These files must be corrected to eliminate any syntax, runtime, formula coding, and formula derivation errors. You may assume that the header comments at the top of the program and all comments throughout the program are correct.

```
% Problem Statement: A cylindrical tank filled to
% a height of 25 feet with tribromoethylene has been
% pressurized to 3 atmospheres (P_s = 3 atm).
% The total pressure (P_t) at the bottom of the tank is
% 5 atmospheres.
% This program will determine the density of the
% tribromoethylene in units of kilograms per cubic meter.
% Inputs:
%      Height H 25 ft
%      Surface Pressure P_s 3 atm
%      Total Pressure P_t 5 atm
```

```
% Outputs:
%      Density rho ? kg/m^3
% Assumptions:
%      Gravity g 9.8 m/s^2
% Define input variables
25 ft = H;
3 atm = P_s;
5 atm = P_t;
9.8 m/s^2 = g;
% Convert total and surface pressure from atm to Pa
Psurf  =  P*1,211,458;
P_t  =  P_t*14.7;
% Convert height from ft to m
H*3.28;
% Calculate density, knowing the calculation of total
% pressure using equation P_t = P_s + (rho)(g)(H)
P_t = P_s + (rho)(g)(H)
```

32. You have been assigned to a new project at work. The previous engineer had created the following files, which your new boss claims do not work. Debug the program to correct all errors.

   These files must be corrected to eliminate any syntax, runtime, formula coding, and formula derivation errors. You may assume that the header comments at the top of the program and all comments throughout the program are correct.

```
% Problem Statement: A rod on the surface of Jupiter's
% moon Callisto has a volume of 0.3 cubic meters and
% specific gravity of 4.7. The gravitational constant is
% 1.25 meters per second squared. Determine the weight of
% the rod in units of pounds-force.
% Inputs:
%      Specific Gravity SG 4.7 [-]
%      Volume V 0.3 m^3
% Outputs:
%      Weight w ??? lbf
% Other variables:
%      Density rho ??? kg/m^3
% Assumptions:
%      Gravity g 1.25 m/s^2
%      Density of water r_w 1000 kg/m^3
% Define input variables
25 = SG;
g = 1.25 m/s^2;
RHO_water = 1;
VOL = 0.3 m^3;
% Calculate rod density [kg/m^3] using specific gravity
SG=Rho*r_w/1000*62.4;
% Calculate weight of rod
w = (Rho / V) (g);
```

Input/Output in MATLAB

When writing a program, it is wise to make the program understandable and interactive. This chapter describes the following mechanisms in MATLAB:

- Allow the user of a program to input values or other information during runtime.
- Display clean, formatted output to the user so that the results can be read and interpreted.
- Generate graphs and charts.
- Exchange data between MATLAB and Microsoft Excel.

17.1 Input

LEARN TO: Accept numeric and character input with the `input` function
Create menus to allow interactive input
Use dialog pop-ups to get information from the user
Debug input statements when error messages appear on the screen

If all programs required the programmer to include the input quantities at the beginning of the program, everyone would need to be a computer programmer to use the program. Instead, MATLAB and other programming languages have ways of collecting input from the program's user at runtime. This section focuses on program-directed user input, using several different approaches.

Numerical Input Typed by the User

The **input** function allows the user to input numerical data into MATLAB's Command Window. The user is prompted to input values at the point where the input statement occurs in the code.

```
Variable = input ('String')
```

- **Variable**: The variable where the input value will be stored
- **'String'**: The text that will prompt the user to type a value

  It is important to note the difference between user input and functional input.

- Functional inputs are values passed in as arguments to a function before a function is executed. In the case of the `input` function, the functional input is the text string inside the parentheses. The `input` function displays the functional input on the screen as a prompt, then waits for the user to enter the value or values requested.

- User input is the value or values entered by the user in response to the prompt. Typically, user input occurs in programs instead of functions but certainly not exclusively. There are other forms in which user input can be obtained as we will see shortly.

EXAMPLE 17-1

Imagine you are writing a program to calculate the speed of a traveling rocket. Prompt the user to input the distance the rocket has traveled.

NOTE

It is usually a good idea to add a space at the end of the prompt so that the user's response is not jammed up against the last character of the prompt. There are other ways to accomplish this separation that we will see later.

NOTE

The use of a semicolon (;) suppresses output of an assignment statement to the Command Window.

Before writing an input statement, it is critical to remember that MATLAB cannot handle units, so it is up to the programmer to perform any unit conversions necessary in the algorithm. Imagine that the user wants to report the speed of the rocket in miles per hour. It would be wise to have the user input the distance traveled in miles and the travel time in hours.

```
Distance = input('Enter the total distance the rocket has
                        traveled [miles]: ')
Time = input('How long has the rocket been in the air [hours]? ')
Speed = Distance/Time
```

In the preceding code segment, the functions were executed without a semicolon being at the end of the input call. If you insert a semicolon at the end of the input call, output of the stored variables to the screen is suppressed, which allows for cleaner display in the Command Window. Note that the semicolon DOES NOT suppress the prompt displayed by input function, only the value or values that are assigned to the variable on the left.

```
% Variable output suppressed
Distance = input('Enter the total distance the rocket has
  traveled [miles]: ');
Time = input('How long has the rocket been in the air [hours]? ');
Speed = Distance/Time
```

Comprehension Check 17-1

(a) Write an input statement to ask for the user's height in inches.
(b) Write an input statement to ask the user to enter the temperature in degrees Fahrenheit.

Text Input Typed by the User

The input function also allows the user to input text data into MATLAB's Command Window. The programmer must include an additional argument ('s') to the input function to tell MATLAB to interpret the user input as text. The term 's' is used to indicate a string.

```
Variable = input ('String', 's')
```

- **Variable:** The variable where the input value will be stored
- **'String':** The text that will prompt the user to type information
- **'s':** Input type is text, not numeric

EXAMPLE   17-2

Imagine you are required to ask for the user's name and birth month. Suppress all extraneous output.

```
Name = input('Type your full name and press enter: ','s');
DataMonth = input('In what month were you born? ','s');
```

Comprehension Check   17-2

(a) Write an input statement to ask for the user's eye color.
(b) Write an input statement to ask the user to type the current month.

Debugging Input Statements

In the next chapter, you will learn how you can use conditional statements to prevent errors from occurring on the screen due to improper input from the user, but there are some errors that may appear during execution that will impact the flow of your program or function. Consider a simple line of MATLAB code that expects the user to type in a number:

```
Cats = input ('How many cats are outside? ')
```

The correct user behavior when this line of code is executed is to type a number and press Enter, but this is not always the case. Sometimes the user will accidentally include a typo in the input, or in nastier situations, the code might be facing a malicious user trying to break the code for a variety of reasons. Assume you type the input statement in a program. If the user types the number 4 and presses the Enter key, the variable M contains the number 4, as expected.

```
How many cats are outside? 4
Cats =
     4
```

However, what happens when the user types in a word or phrase? There are a few different error messages that may appear on the screen when invalid information is entered in response to the prompt. Consider the following error message sequence:

```
How many cats are outside? There are four cats outside
     Error: Unexpected MATLAB expression.
How many cats are outside? 4 cats
     Error: Unexpected MATLAB expression.
>>
```

On the user's first attempt, MATLAB simply did not recognize the sequence of nonnumeric characters, issued the error message, and redisplayed the prompt for the user to try again. On the second attempt, the user began with a digit, making MATLAB think that the input was valid. However, when the MATLAB input interpreter saw the text "cats" next, MATLAB became completely confused, issued the error message, terminated the program, and returned to the command prompt. Consider this error message and program termination as MATLAB's way of saying "I have no idea what's going on." If you are running an older version of MATLAB, the error display may be slightly different in the second case.

Consider the following error message sequence:

```
Cats = input ('How many cats are outside? ')
How many cats are outside? four
Error using input
Undefined function or variable 'four'.
How many cats are outside? four+cats
Error using input
Undefined function or variable 'four'.
How many cats are outside? [4; cats]
Error using input
Undefined function or variable 'cats'.
How many cats are outside? cats(4)
Error using input
Undefined function 'Cats' for input arguments of type 'double'.
```

In these error messages, the "**Undefined function or variable**" error message indicates that MATLAB is trying to use the user input as a MATLAB expression and treat some values as variables or function names. In the preceding error messages, the word in single quotes is the value MATLAB is trying to interpret as either a function or a variable name. None of these input attempts results in program termination, and the user is prompted again for input.

In addition, the `input` function will try to evaluate an expression typed by the user if there are valid variables referenced in the user input value. Assume the variable `rats` is defined in MATLAB's workspace and contains the value 10:

```
How many cats are outside? rats+5
Cats =
     15
```

This expression did not result in an error message because the variable `rats` existed in MATLAB's workspace, so MATLAB evaluated the expression and stored the result in the variable `Cats`.

```
How many cats are outside? cats=4
Error: The expression to the left of the equals sign is not a
valid target for an assignment.
```

In some error messages, MATLAB may report one problem but fail to report other issues. In the preceding error message, MATLAB is complaining about the expression to the left of the equal sign not being valid since you cannot write an assignment statement in response to an input statement. However, this error message fails to point out that `cats` is not a variable in MATLAB's workspace. As soon as MATLAB discovers the first problem, it quits looking for additional issues and displays the error on the screen. Removing the equal sign from the input will result in a different error message.

While some values typed as input into MATLAB when the interpreter is expecting numeric input will result in a nasty error message, sometimes no error or warning messages are generated, and the program will continue as if it has valid data. Consider the following scenario, using the same `input` expression, where MATLAB is expecting a number, but the user types a word as a string (enclosed by single quotes):

```
How many cats are outside? 'four cats'
Cats =
four cats
```

Note that the result of this action is the string `'four cats'` being successfully stored in the variable `Cats`. This is equivalent to using the `input` function with the extra argument denoting that MATLAB should expect character input, but that method does not require the user to type the input enclosed by single quotes. In the next chapter, you will see how functions like `isnumeric` or `ischar` can be used to determine if a variable is of a particular type, which will be necessary for assuring your program uses valid input of the correct data type.

In addition to allowing the user to type numeric input or a string, MATLAB will also allow the user to provide no input at all. Consider the following scenario, where the user only presses the Enter key without typing any other text after the input statement:

```
How many cats are outside?   (no text, the user only pressed
Enter)
Cats =
  []
```

In this situation, the value stored in the variable `Cats` is an empty matrix, which is different from an empty string or the number zero. The empty matrix in MATLAB is an absence of any data and is similar in nature to the concept of having a null value, which is common in other programming languages like C.

Note that all of the scenarios presented here discussed the input statement when it was expecting numeric input. The string input statement will not produce as many errors because the input provided by the user is written to a variable as a sequence of characters verbatim. Even in the scenario where the user presses the Enter key without typing any information, the result is an empty string.

```
M = input ('How many cats are outside? ','s')
How many cats are outside? four cats
M =
four cats
How many cats are outside?   (no text, the user only pressed
Enter)
M =
  ''
How many cats are outside? 4
M =
    4
```

In this final situation, the **numeral** 4 is stored in the variable `M` as a character. This could be used in a calculation without an error occurring, although the results may not be what you expect. For example, the statement N=M+1 will echo the following to the screen:

```
N =
    53
```

This has to do with the ASCII code that is used to encode text (including the 10 numerals) using binary digits.

Menu-Driven Input

In the preceding example, the user was required to type the name of the month he or she was born, but there is no good way to predict the way the user will actually type the name of the month into MATLAB. For example, the user may abbreviate, represent the month numerically, misspell the name of the month, or enter something totally unrelated, such as "Frog Breath." MATLAB has a built-in input function (**menu**) that allows the programmer to specify a list of options for user selection. The menu function will display a graphical prompt of a question and possible responses.

```
Variable = menu ('String', 'opt1','opt2', . . .)
```

NOTE

The menu function will also accept a cell array containing the individual text strings for the buttons in place of individual text strings.

- **Variable**: The variable where the ordinal value of the selection will be stored
- **'String'**: The text that will prompt the user to select an item in the menu
- **'optN'**: The text strings that appear as buttons in the menu

The first argument of the menu function is always the question to display ('String'), and the following arguments are the possible responses. The menu function does not store the actual text response, but rather the ordinal number of the response in the list. In other words, clicking the first button will return a 1, clicking the second button will return a 2, etc. An example will help clarify this idea.

EXAMPLE   17-3

NOTE

If you are running MATLAB under MacOS, menus and dialog boxes will look slightly different, particularly the controls at the very top.

Suppose we want to ask the user to input a favorite color. However, we want to force the user to choose from red, green, or blue. Suppress all extraneous output.

```
Color = menu('Favorite Color','Red','Green','Blue');
```

This command will create the menu shown. The value returned upon selection is an integer. For example, if the user were to select Red in the menu, the value stored in Color *would be 1 since Red is the first option available in the list.*

What if the user simply does not like any of the choices, or for any other reason closes the menu instead of making a selection of one of the buttons? In this case, a zero will be placed in the variable. This allows the programmer to include code to deal with this type of situation. We will consider this further in the next chapter.

Problem Issue: When I try to create a menu from a linear sequence, only the first entry shows up in the list.

Example:
```
A = 2:2:10;
MC1=menu('Select a number',A);
```

Solution:
```
ACA=num2cell(2:2:10));
MC2=menu('Select a number', ACA);
```

Menus can have LOTS of buttons. Type in the following command and observe the number of buttons in the menu.

```
Select_a_Prime=menu('Pick a Prime',num2cell(primes(1000)))
```

Comprehension Check   17-3

(a) Create a menu to ask the user to select the current month.
(b) Given the following menu statement, what is stored in variable C when the user clicks Red?

```
C = menu('My favorite color:','Green','Blue','Yellow','Black','Red')
```

Another function that limits the choices available to the user is **questdlg**. This will allow you to create a pop-up dialog that allows the user to select one of up to three different options for answers. The default buttons are Yes, No, and Cancel, although these can be customized by the programmer. The questdlg function returns the **text** shown on the button that was clicked, not the ordinal value.

```
Variable = questdlg ('qstring', 'title', 'opt1', 'opt2',
                     'opt3','Default')
```

- **Variable:** The variable where the selected button text will be stored
- **'qstring':** The question being asked
- **'title':** Creates a title in the dialog box title bar
- **'optN':** The options that appear as buttons. If nothing is listed, the options will be Yes, No, and Cancel. There may be UP TO THREE of these, no more.

- **'Default':** If more than two text strings are present in the argument list, the LAST one will always indicate the default selection. This is highlighted in the dialog box and will be the selection if the user presses Enter. If there are only one or two text strings, Yes is the default.

EXAMPLE   17-4

```
A = questdlg ('My favorite color is green', 'Color')
```

There are actually five user responses.

1. If the user chooses the Yes button, then the text string `'Yes'` will be stored in A.
2. If the user chooses the No button, then the text string `'No'` will be stored in A.
3. If the user chooses the Cancel button, then the text string `'Cancel'` will be stored in A.
4. Note that the Yes button is highlighted. This indicates the default choice. If the user simply presses the Enter (or Return) key, `'Yes'` will be placed in A.
5. Finally, perhaps the user does not like any of the options, so just closes the dialog box with the close button at the top. In this case, A will contain an empty text string.

Note that when there are only two text strings in the parentheses by `questdlg`, the first text string appears in the middle of the box, the second text string appears at the top, and the three buttons are Yes, No, and Cancel with Yes as the default.

NOTE

Question dialogs may have at most three buttons.

More than two text strings may be included in the parentheses, but this is a little more complicated. There may be a total of up to six text strings. If there are more than two, then the LAST one will denote the default button.

EXAMPLE   17-5

If you wanted to simply change the default button in the previous example to Cancel, you would type:

```
A = questdlg ('My favorite color is green', 'Color','Cancel')
```

(MacOS version shown)

You may have up to three additional text strings prior to the default choice at the end of the list. These specify the text to appear on the buttons.

EXAMPLE **17-6**

The following command generates the dialog shown:

```
KF = questdlg ('My favorite is', 'Klingon Food','Gagh','Bregit Lung','Bregit Lung')
```

EXAMPLE **17-7**

In the following example, the dialog box shown is created, but a warning is generated.

```
A = questdlg ('My favorite season is…', 'Season','Winter', 'Summer','Other')
```

The warning message is:

Warning: Default string does not match any button string name.

The dialog choice will work correctly, but there is no default choice, since the final text string ('Other') did not match any of the specified buttons.

Comprehension Check **17-4**

Write a question dialog statement to generate each of the following dialog boxes, and place the user's response in the variable `MyStatus`.

(a)

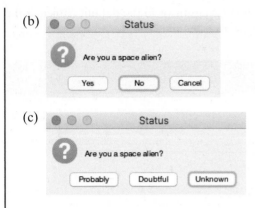

(b)

Status

Are you a space alien?

Yes   No   Cancel

(c)

Status

Are you a space alien?

Probably   Doubtful   Unknown

(d) In part (a), what is placed in `MyStatus` if the user clicks Cancel?
(e) In part (b), what is placed in `MyStatus` if the user closes the dialog box?
(f) In part (c), what is placed in `MyStatus` if the user presses Enter?

The final input function we will consider at this time is **inputdlg**. This function will create a pop-up dialog to allow users to type their responses in one or more text input boxes rather than typing them into the Command Window.

This is the most complicated of all the user input functions we will consider in this section, and we will introduce only the basics. For more information, refer to the MATLAB documentation on `inputdlg`.

In the simplest case, the form of `inputdlg` is

```
Variable=inputdlg(Instructions);
```

EXAMPLE   **17-8**

```
K = inputdlg('Enter your name')
```

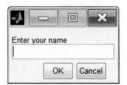

Enter your name

OK   Cancel

Assuming the user types something into the text box and clicks OK, whatever the user typed will be placed as a text string into the cell array K. If the user either clicks Cancel or closes the dialog box, K will be an empty cell array.

`Variable` is a **cell array** where the response will be stored, and `Instructions` is either a text string or a cell array containing a text string.

If you wish to have multiple text boxes in the dialog, Instructions must be a cell array containing the instructions for each desired text box as a text string. You may also add a title at the top of the dialog.

The following two lines will create the input dialog box shown.

```
Instr={'Enter Your Name','Enter your age','Enter your Gender'};
Resp=inputdlg(Instr,'Personal Data');
```

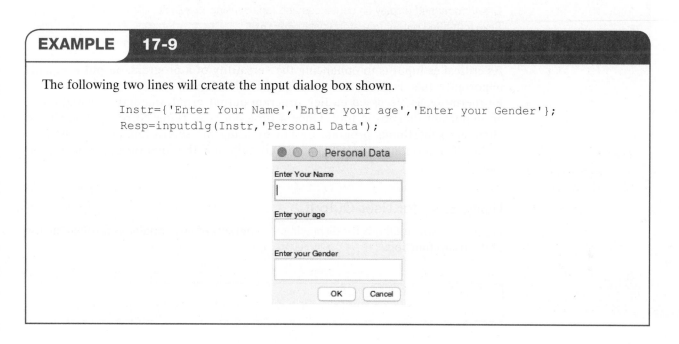

The user's responses will be stored as three text strings in the cell array Resp.

There are several other ways you can customize the input dialog, such as having multiline text boxes and having different length text boxes, but this is beyond the scope of our initial discussion here.

Comprehension Check   17-5

(a) Create an input dialog that asks users for their desired Username and places the response in UNReq.
(b) Create an input dialog that displays two text boxes asking users for their Username and Password and places the responses in LoginData.
(c) In part (b), if the user enters ClemKad as the username and Br@inLe55 as the password, what is stored in LoginData?

17.2 Output

As critical as input is to enhancing the versatility of a program, output is equally important—this is how the user gets feedback on the computer's solution. Furthermore, as discussed earlier, program output of intermediate calculations is an important diagnostic tool. Once a program is working correctly, you should eliminate superfluous program output by using the semicolon, as demonstrated earlier. The program user will also find it helpful if the final output is neatly formatted.

Using `disp` for User Output

One of the basic methods for displaying the contents of any variable is through the use of the **disp** function.

```
disp (Variable)
```

- **Variable:** The variable to be displayed on the screen

Given some variable x defined in MATLAB's workspace, `disp(x)` displays the value of x to the screen. However, the `disp` command cannot display a value and text on the same line. Furthermore, `disp` relies on using MATLAB's default numerical display format for displaying numbers, so it does not provide a capability for changing the form in which the number is displayed. For these reasons, `disp` is not generally recommended for use in generating the final output of a program. However, you might find it useful during debugging or for getting "quick and dirty" output during program development. It might also be used in cases where you need to simply place some text in the Command Window.

Formatting Output with `fprintf`

The formatted print command, **fprintf**, gives extensive control over the output format, including spacing. This is one of the major methods for displaying program results to the user.

NOTE

When the \ (backslash) character occurs in the formatting string inside fprintf or sprintf, it means "The next character is a special control character, not part of the text string itself."

```
fprintf ('String', Var1, Var2, . . .)
```

- **'String':** The formatted output string. This generally includes one or more insertion points, where values from variables will be inserted into the text string.
- **VarN:** The variables or values to be inserted into the formatted output string

The `fprintf` command can intersperse precisely formatted numbers within text, guarantee table alignment, align the decimal points of a column of numbers, etc. The `fprintf` command uses control and format codes to achieve this level of precision, as shown in Table 17-1 and Table 17-2.

Table 17-1 Output control with `fprintf`

Control Character	Use	Example
\n	Inserts a new line	`fprintf('Hello\n');`
\t	Inserts a tab	`fprintf('A\tB\tC\n');`
\\	Inserts a backslash	`fprintf('Hello\\World\n');`
'' (two single quotes)	Inserts a single quote mark	`fprintf('Bob''s car\n');`
%%	Inserts a percent symbol	`fprintf('25%%');`

NOTE

When the % character occurs in the formatting string inside `fprintf` or `sprintf`, it means "Insert a value here using the formatting code immediately following."

The biggest benefit of using `fprintf` is the ability to insert values contained in variables into formatted sentences. In addition to the formatting controls previously shown, MATLAB contains a number of special controls to represent different types of variables that might be displayed within a sentence. Inserting a variable control character (%) within the formatting string tells MATLAB to plug in the contents of a variable at that location in the string. Note that MATLAB's documentation on `fprintf` contains additional control characters not covered in this text that you may find useful in your programming.

Table 17-2 Variable (numeric) and TextVariable (text) control with `fprintf`

Control Character	Use	Example: `Var = 503/23;` `TVar='A wet bird';`
%f	Inserts a value shown to 6 decimal places	`fprintf('%f', Var)` 21.869565 is displayed
%0.Mf	Inserts a fixed point value with M decimal places	`fprintf('%0.1f',Var);` 21.9 is displayed
%N.Mf	Inserts a value with a field width of N (left padded with spaces), containing M decimal places	`fprintf('%8.2f',Var)` _ _ _ 21.87 is displayed, where _ indicates a blank space
%0N.Mf	Inserts a value with a field width of N (left padded with zeros), containing M decimal places	`fprintf('%08.2f',Var)` 00021.87 is displayed
%e or %E	Inserts a number in scientific notation	`fprintf('%e',Var);` 2.186957e+01 is displayed `fprintf('%0.3E',Var);` 2.187E+01 will display
%s	Inserts text	`fprintf('%s',TVar);` "A wet bird" is displayed

NOTE

You can display scientific notation using either lowercase e or uppercase E. To avoid confusion with the base of the natural logarithm e ≈2.71828, it may be best to use uppercase E. In either case, the letter E means "times 10 raised to the power."

EXAMPLE   17-10

Type each example into the Command Window and observe the output generated.
Define the following variables in MATLAB's workspace:

```
Age=20;
Average=79.939;
Food='pizza';
Postal=515;
```

Display the variables in formatted `fprintf` statements. At the end of each formatted output statement, insert a new line.

Since age is an integer, it is best to use the `%0.0f` control character:

```
fprintf('My age is %0.0f.\n',Age);
```

To display the average with two decimal places:

```
fprintf('%0.2f is the average.\n',Average);
```

Note that MATLAB rounds values when using `%f`.

To display a string within a sentence:

```
fprintf('Eating too much %s will make you fat.\n',Food);
```

To display a postal code, the field width of a postal code is typically five numbers long:

```
fprintf('My postal code is %05.0f.\n',Postal);
```

EXAMPLE   17-11

Type the following into the Command Window and observe the output. Assume the variables in Example 17-10 are still defined in the workspace.
Create a single formatted output statement that displays each sentence with a line break between each sentence.

```
fprintf('My age is %0.0f.\n%0.2f is the average.\n
        Eating too much %s will make you fat.\n
        My postal code is %05.0f.\n',Age,Average,Food,Postal);
```

In MATLAB, the previous command would be typed on a single line. Note that the order of the arguments to the function corresponds to the order of the inserted variables in the string.

Problem Issue: I want to round a number to the LEFT of the decimal place.

Example: Print 45678 as an integer rounded to three significant figures.

Solution: `fprintf('%0.0f\n',round(45678,-2))`

This will print 45700. The optional second argument in the round function specifies to what digit the first argument should be rounded relative to the decimal point. A negative second argument rounds to the left of the decimal point.

Comprehension Check | **17-6**

Assume that the variable M is stored in the workspace with the value 0.3539.

(a) What would be the control code used to display M with two decimal places?
(b) What would be the control code used to display M with three decimal places in scientific notation?
(c) What would be the control code used to display M if it should be padded with zeros to a field width of 10 characters, showing three decimal places?
(d) What would be the output of M if a control code is used to display it to two decimal places?
(e) What would be the output of M if a control code is used to display it to three decimal places in scientific notation?
(f) What would be the output of M if a control code is used to pad the field width to 10 characters, using zero padding and showing three decimal places?

Comprehension Check | **17-7**

Consider the following segment of code. What appears in the Command Window if this is executed?

```
clear
clc
S=[2012,27,17;2011,34,13;2010,29,7];
fprintf('Clemson-USC Football Rivalry:\n\n');
fprintf('Year\tUSC\t\tClemson\n');
fprintf('%0.0f\t%0.0f\t\t%0.0f\n',S(1,1),S(1,2),S(1,3));
fprintf('%0.0f\t%0.0f\t\t%0.0f\n',S(2,1),S(2,2),S(2,3));
fprintf('%0.0f\t%0.0f\t\t%0.0f\n',S(3,1),S(3,2),S(3,3));
```

Saving Formatted Output in a String: `sprintf`

The fprintf function is clearly the best tool available for displaying textual information in the Command Window, complete with handling the appropriate number of decimal places and plugging in values of different types of variables so the information can be read and interpreted by the user. In certain scenarios, we may wish to create a formatted string that is saved in a variable instead of printed on the screen.

The **sprintf** function uses the same syntax as the fprintf function to create formatted text, complete with formatting codes and variables, but the result of the sprintf function is stored in a variable that will contain the formatted text as a string variable. The contents of this variable could then be printed in the Command Window later, or as we will see shortly, used for other purposes, such as placing formatted text in graphs or labeling buttons in menus.

EXAMPLE | **17-12**

Assume that the following variables are defined in MATLAB's workspace:

```
Age = 20;
Average = 79.939;
Food = 'pizza';
Postal = 515;
```

To create strings of the formatted text displayed in the previous example, the `sprintf` function can be used instead of the `fprintf` function. Assume that it is desired to store the formatted strings into variables `Str1`, `Str2`, `Str3`, and `Str4`.

Since age is an integer, it is best to use the `%0.0f` control character:

```
Str1 = sprintf('My age is %0.0f\n',Age);
```

To store the average with two decimal places:

```
Str2 = sprintf('%0.2f is the average.\n',Average);
```

To store a string within a sentence:

```
Str3 = sprintf('Eating too much %s will make you fat.\n',Food);
```

To store a postal code, the field width of a postal code is typically five numbers long:

```
Str4 = sprintf('My postal code is %05.0f\n',Postal);
```

Comprehension Check   **17-8**

Assume the following variables have been defined:

```
Patient = 'Sarah';
Temp = 102.6;
SBP = 138;
DBP = 76;
Charge = 35;
```

Store a text string in `PatSum`, inserting values from the preceding variables as appropriate, that, if printed would appear as follows. Note that if the values in the variables change, the stored text string should reflect those changes.

```
Sarah had a temperature of 102.6 and a blood pressure of 138/76.
The charge for this visit is $35.00.
```

17.3 Plotting

LEARN TO: Create proper plots using **plot**
Create logarithmic plots with **loglog**, **semilogx**, and **semilogy**
Create multiple plots in a figure with **subplot**

MATLAB has many plotting options, not all of which concern us in this text. The simplest way to create plots of data in MATLAB is the **plot** command.

Creating Figure Windows in MATLAB

NOTE

The `figure` function creates a new figure window.

When MATLAB creates a new plot, it automatically draws the plot to a "figure" window. If no figure window has already been created, the command used to draw the plot (e.g., `plot` or `loglog`) will first create a figure window. If one or more figure windows are already open, MATLAB will normally **erase** whatever has already been drawn in the most recent figure window before drawing the new plot within that window. We will see later how to force MATLAB to **add** new plots to current plots without erasing the original first. For the moment, however, we will see how to create a new figure window whenever we want to plot something without losing the current plot. You simply type the command `figure` to create a new figure window before drawing your next plot. Add all titles, *x*- and *y*-axis labels, etc. for the previous plot **before** you initiate the `figure` command, because those functions normally apply to the most recently created figure. Once your new plot is created, you will have to include another set of commands for the axis labels, title, and so forth. If you need to reference a previous figure, you can type `figure(N)`, where N is the number of the figure. By default, MATLAB names the first figure created Figure 1 and increases figure number by 1 as new figures are created. Typing `figure(N)` makes Figure N the current figure, thus allowing you to go back and modify any figures that have already been created.

It is also possible to control the background color of the figure window using the `figure` command. This may be desirable if the figure will appear in another type of publication, such as a report or presentation where the default gray background would be inappropriate.

    figure('Color','White')

This will make the area surrounding the actual graphing area white instead of gray.

IMPORTANT CONCEPT

`'Color','White'` is an example of a "Property Name", "Property Value" pair. As we proceed through the course, you will encounter quite a few such pairs used as arguments in a variety of functions. These are used to do many kinds of things such as setting font characteristics like size or color, setting parameters such as size of data markers or width of lines, and specifying the locations of gridlines or legends.

Plotting Variables

If the vectors `time` and `distance` contain paired data, `plot(time, distance)` will plot the time data on the abscissa and the distance data on the ordinate.

In addition to time and distance, assume a second distance vector, `distance2`, contains additional distance recordings at the same time values in `time`. To plot both `distance` and `distance2` on the same graph, the plot command will contain both the `distance` and `distance2` vectors: `plot(time,distance,time, distance2)`.

```
plot (A,B,C, . . .)
```

- *A:* The horizontal axis values
- *B:* The vertical axis values
- *C:* Optional text string specifying the line type and/or symbol type and/or color. Type `help plot` to see a list of the linespec symbols or refer to the endpages in your text.
- *A,B,C:* Can be repeated for multiple data series (`plot(A1,B1,C1,A2,B2,C2)`)

Creating Proper Plots

A number of other functions allow the programmer to automatically insert information onto a graph generated in MATLAB. Many plot options are defined by separate commands, discussed below. These can be applied when the plot is created or afterwards, as long as the plot is unchanged.

- **close all:** A command to close all currently open figures. When writing a program that includes plotting, this would normally be included at the beginning with the other housekeeping commands `clear` and `clc`. This insures that any plots remaining from another program or a previous run of your own program are deleted to avoid confusion.

```
close all
```

- **xlabel**: Insert a label on the abscissa.

```
xlabel('Time (t) [s]')
```

- **ylabel**: Insert a label on the ordinate.

```
ylabel('Height (H) [m]')
```

- **title**: Insert a title. This is optional, and should not be used if the graph will be used in another presentation format (such as a report) and a caption will accompany the graph.

```
title('Flight Across The United States')
```

- **legend**: Insert a legend on the graph if more than one data series is present. The order of elements provided in the legend is based on the order in which they are passed into the plot function.

```
legend('Airplane 1','Airplane 2')
```

- **grid**: Allows the gridlines to be turned on or off.

```
grid on
```

- **axis**: Allows the setting of the minimum and maximum values on the abscissa and ordinate, in that order.

```
axis([0 50000 0 90000])
```

NOTE

When plotting multiple data sets with a single plot command, MATLAB will automatically make them different colors. You can specify any colors you wish, however.

Alternately, the functions `xlim` and `ylim` can be used to define the [minimum, maximum] limits on the abscissa and ordinate, respectively.

The following examples use several special graphing commands. For more information on graphing properties, refer to the MATLAB Graphing Properties, Linespec, and Special Character reference tables contained in the endpages of your text.

EXAMPLE 17-13

Create a graph of multiple experimental data series for the following data:

H1 = [10, 15, 25, 35, 55];   H2 = [10, 30, 50, 70, 100];
P1 = [0.27, 0.41, 0.68, 0.95, 1.5];   P2 = [0.11, 0.33, 0.54, 0.76, 1.09];

```
clear
clc
close all
H1=[10, 15, 25, 35, 55];
H2=[10, 30, 50, 70, 100];
P1=[0.27,0.41,0.68,0.95,1.5];
P2=[0.11,0.33,0.54,0.76,1.09];
plot(H2,P2,'+',H1,P1,'o');
axis([0 120 0 1.75]);
title('Experimental Data, Multiple Data Series Plot');
legend('Mass = 100 kg','Mass = 250 kg');
xlabel('Height (H) [m]');
ylabel('Power (P) [hp]');
grid on
```

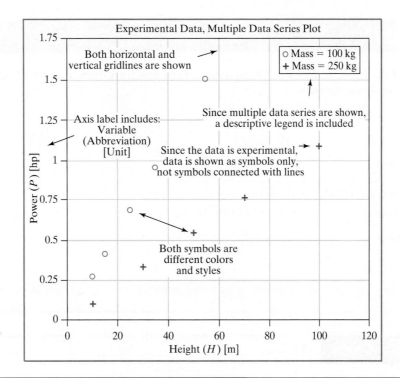

Figure 17-1

Example of a proper plot, showing multiple experimental data sets.

Plotting Theoretical Expressions

To draw a graph of a known equation, we first need to set up a series of data points to plot. We can then use the `plot` command to plot these points connected by straight-line segments. We will consider two general cases: Linear equations and nonlinear equations.

Plotting Linear Equations

You probably recall from earlier math courses that a straight line can be completely defined by only two points. Therefore, if we wish to draw a straight line on a graph, we only need to specify the two endpoints, and use the `plot` command to connect them with a straight line. You can, of course, have as many intermediate points as you wish, but these are not necessary in the case of a straight line.

EXAMPLE   **17-14**

Draw a graph of the function $y = 0.0273x + 0.05$ for values of t from 0 to 60.

```
% Housekeeping
clear
clc
close all

t=[0,60]; % Define the endpoints on the abscissa
d=0.0273*x+0.05; % calculate the corresponding ordinate values
plot(t,d) % The default is a solid blue line
```

For a proper plot, several more commands would need to be added for labels, axis limits, gridlines, etc. Remember that for theoretical plots, we draw lines with no data markers for the individual points. Since we did not give any line specifications in the plot command, it will draw a solid blue line.

EXAMPLE 17-15

Draw a graph of the function $F = kx + F_0$ as a dotted red line for values of x from 5 to 25. Values of k and F_0 will be obtained from the user.

```
% Housekeeping
clear
clc
close all
% Get model parameters from user
k=input('Enter spring constant [N/m]: ');
F0=input('Enter initial force (with no added mass) [N]: ');

x=[5,25]; % Define the endpoints on the abscissa
F=k*x+F0; % calculate the corresponding ordinate values
plot(x,F,':r') % ':r' makes it a dotted red line
```

For a proper plot, several more commands would need to be added for labels, axis limits, gridlines, etc.

Comprehension Check 17-9

Create a proper plot of the function $D = 50 - 3t$ for t ranging from 0 to 15. Draw the line as a dashed black line. Here D is depth in meters, and t is time in minutes.

Plotting Nonlinear Equations

Drawing graphs of nonlinear functions is really no harder than drawing straight lines, but we have to give a little attention to the set of points used to graph the function. Since the plot command connects points with straight-line segments, and nonlinear functions yield curved lines, we must specify some points between the two endpoints. The more pronounced the curvature, the closer together the points need to be. You should recall this concept from your study of graphing in Excel. Basically, the points need to be close enough together that the change in slope from one segment to the adjacent one is very small. You may have to experiment with the spacing a little to get a smooth curve. Fortunately, changing the spacing is actually easier to do in MATLAB than it was in Excel.

EXAMPLE 17-16

Draw a graph of the function $V = e^{-0.2t} \cos(2t)$ for $0 \le t \le 10$. Once again, we will not bother to add the commands to properly format the graph.

```
clear
clc
close all

t=0:.5:10; % Define sequence of independent values
V=exp(-0.2*t).*cos(3*t); % calculate dependent values
plot(t,V) % The default is a solid blue line
```

This gives a rather jagged plot, not a smooth curve, so we need to decrease the spacing between the points. This is easily done by changing the increment value in the definition of t.

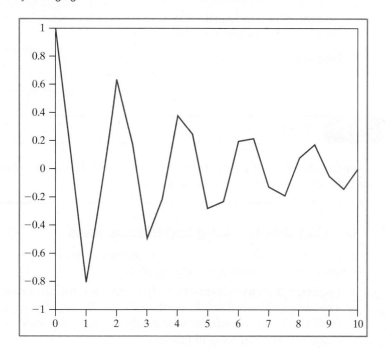

Let's change that one line to

```
t=0:.05:10; % Define sequence of independent values
```

and rerun the program with this change.
This yields an acceptably smooth curve.

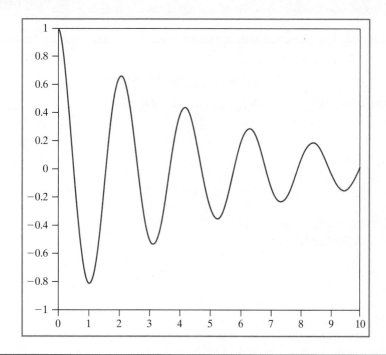

Problem Issue: I want to plot horizontal and vertical lines.

Example: Plot a red horizontal line at $y = 15$ for $-10 \le x \le 10$ and a blue vertical line at $x = -5$ for $0 \le y \le 20$.

Solution: `figure('Color','w')`
`        plot([-10,10],[15,15],'r',[-5,-5],[0,20],'b')`

Comprehension Check | 17-10

Create a proper plot of the function $I = 2.5 \cos(t) \cos(5t)$ for $-5 \le t \le 5$. Draw the curve as a solid red line. Here I is current in milliamperes, and t is time in microseconds.

Drawing Graphs Using Logarithmic Axes

The `plot` command uses standard linear axes to draw graphs. MATLAB also includes functions to draw semilog and log-log plots.

NOTE

When specifying axis limits on a logarithmic axis, it is often best to make the limits a power of 10, such as 0.01 or 1000.

- **loglog:** Both axes logarithmic. Typically used for power models.
- **semilogy:** Vertical axis logarithmic. Typically used for exponential models.
- **semilogx:** Horizontal axis logarithmic. Typically used for logarithmic models, which we do not discuss in this text.

Other than the names of the functions that create the plots, the other details for creating graphs remain the same as for linear plots.

EXAMPLE | 17-17

Create a graph of multiple experimental data series on logarithmic axes for the following data with R on the abscissa.

R1 = [10, 30, 50, 70, 100];   R2 = [10, 15, 25, 35, 55];
V1 = [0.11, 0.33, 0.54, 0.76, 1.09];   V2 = [0.27, 0.41, 0.68, 0.95, 1.5];

```
clear
clc
close all
R1=[10,30,50,70,100];
V1=[0.11,0.33,0.54,0.76,1.09];
R2=[10,15,25,35,55];
V2=[0.27,0.41,0.68,0.95,1.5];
loglog(R1,V1,'bs',R2,V2,'ro');
grid on
xlim([1 100]);
ylim([0.1 10]);
title('Logarithmic, Experimental, Multiple Data Series Plot');
xlabel('Radius (R) [cm]');
ylabel('Volume (V) [cm^3]');
legend('Cylinder #1','Cylinder #2')
```

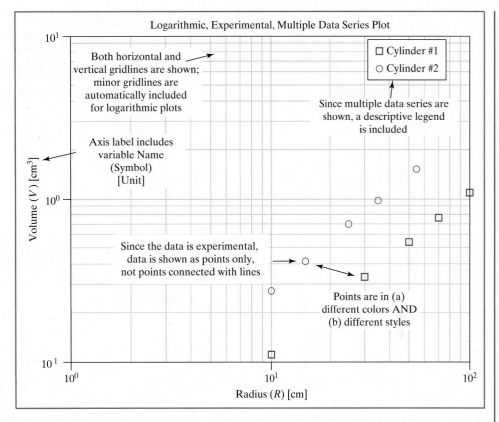

Figure 17-2
Example of a proper
plot, showing multiple
experimental data sets and
logarithmic axes.

EXAMPLE | **17-18**

Create a graph of multiple experimental data series on semilogarithmic axes (ordinate logarithmic) for the following data:

T1 = [2, 5, 6, 7, 10];   T2 = [1, 1.5, 4, 6.5, 9];
V1 = [30.8, 20.9, 18.3, 16.1, 10.9];   V2 = [9.8, 8.9, 5.4, 3.3, 2];

```
clear
clc
close all
T1 = [2, 5, 6, 7, 10];
V1=[30.8, 20.9, 18.3, 16.1, 10.9];
T2 = [1, 1.5, 4, 6.5, 9];
V2=[9.8, 8.9, 5.4, 3.3, 2];
semilogy(T1,V1,'ro',T2,V2,'bs');
grid on
axis([0 12 1 100])
title('Semilogarithmic, Experimental, Multiple Data Series Plot');
xlabel('Time (t) [ms]');
ylabel('Voltage (V) [V]');
legend('Resistor #1','Resistor #2')
```

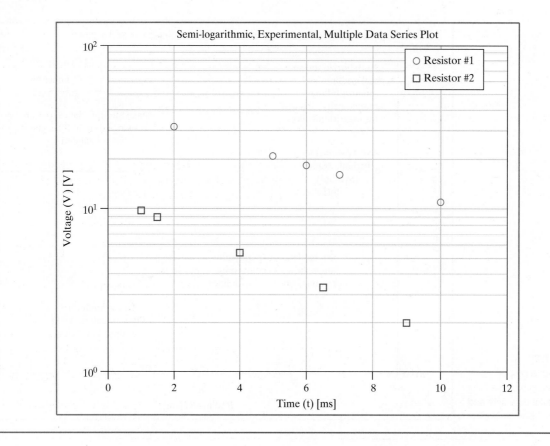

Using Subplots in MATLAB

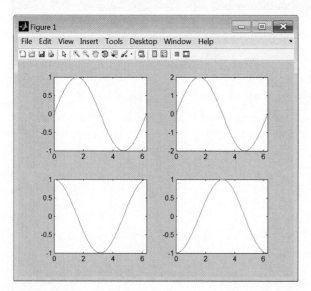

In addition to generating single plots, MATLAB can generate multiple plots, called subplots, in a single figure window. This can make it easier to compare the results of different cases or to study a process from a variety of perspectives simultaneously.

In the figure that follows, subplots have been used to study the effect of changing the input parameters to the general sine function $A \sin(Bx + C)$. The command to generate the grouping of plots is shown. Note that parameter A is amplitude, B is frequency, and C represents a phase shift.

Note that even plots in MATLAB are arranged in matrices. The **subplot** command is a void function that accepts three inputs but generates no output on the screen. The first two arguments specify the number of rows and columns of subplots, and the third argument specifies a specific subplot within that group. After a subplot command is executed, the next plotting command and related plotting functions will apply to that specific subplot until the next subplot command appears in the code.

```
subplot (R,C,N)
```

NOTE

The backslash character (\) can be used inside a text string to create special characters, such as Greek letters, at least in some contexts. These do not work in the Command Window, but they are often used to place special symbols in figure windows. See the endpages of your text for a list of special characters.

- **R**: The number of rows of plots to appear in the figure.
- **C**: The number of columns of plots to appear in the figure.
- **N**: The position in the subplot where the `plot` commands following the `subplot` command will apply.

The graph positions, N, are numbered starting in the top-left corner and going across each row, starting with the number 1 and ending with the value of R × C. The following program was used to create the subplot figure shown:

`x=0:0.01:2*pi;`	Sets up angle sequence
`subplot(2,2,1) plot(x,sin(x))`	Plots to top left corner of 2 × 2 group
`subplot(2,2,2) plot(x,2*sin(x))`	Plots to top right corner of 2 × 2 group
`subplot(2,2,3) plot(x,sin(x+pi/2))`	Plots to bottom left corner of 2 × 2 group
`subplot(2,2,4) plot(x,sin(x-pi/2))`	Plots to bottom right corner of 2 × 2 group

EXAMPLE 17-19

Create a graph with two subplots in a single column. The top graph should be a theoretical plot of the function $y = 0.0273x + 0.05$. The bottom plot should be the graph from Example 17-17.

```
clear
clc
close all
figure('color','white')
% Plot top graph
x=[0 60];
subplot(2,1,1);
plot(x,0.0273*x+0.05);
title('Theoretical Data, Single Data Series Plot');
xlabel('Mass (m) [g]');
ylabel('Density (\rho) [g/cm^3')
grid on
% Plot bottom graph
R1=[10,30,50,70,100];
V1=[0.11,0.33,0.54,0.76,1.09];
R2=[10,15,25,35,55];
V2=[0.27,0.41,0.68,0.95,1.5];
subplot(2,1,2);
loglog(R1,V1,'bs',R2,V2,'go');
grid on
xlim([1 100]);
ylim([0.1 10]);
title('Logarithmic, Experimental Data, Multiple Data Series Plot');
xlabel('Radius (R) [cm]');
ylabel('Volume (V) [cm^3]');
legend('Cylinder #1','Cylinder #2')
```

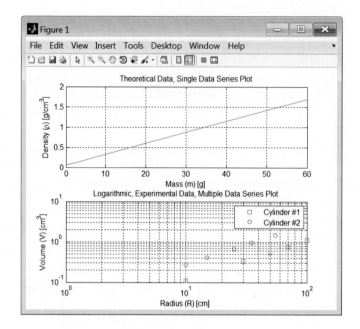

Figure 17-3

Example of a proper plot using subplots of the form two rows, one column.

JUST FOR FUN

Type in the following four commands and observe the pattern of subplots created. Normally, you would plot something to each subplot after it was created, of course.

```
subplot(3,2,1)
subplot(3,2,3)
subplot(3,2,5)
subplot(3,2,2:2:6)
```

Comprehension Check   17-11

In a single figure window, create three properly labeled subplots in a single column. Both axes in all three plots should range from 1 to 10. Assume the independent variable t is time in hours and the dependent variable F is flowrate in gallons per second.

(a) In the top subplot, plot the following experimental data using black diamonds on linear axes:
$t = [8, 6, 1, 3, 9]$; $F = [7, 6, 1, 8, 9]$

(b) In the center subplot, draw a theoretical plot of $F = 15\,e^{-0.4t}$, using a red dotted line and semilogarithmic axes (ordinate logarithmic).

(c) In the bottom subplot, plot the following experimental data using blue asterisks and logarithmic axes:
$t = [8, 4, 1, 3, 9]$; $F = [7, 2, 4, 1, 9]$

17.4 Trendlines

> **LEARN TO:**   Display trendlines using `polyfit` for linear, power, or exponential trends
> Place trendline equations on graphs

In the previous chapter, we learned how to determine the model parameters m and b for linear, power, and exponential models. We will now use these parameters to add a trendline that fits a set of experimental data.

Graphing a Trendline in MATLAB

1. Create a proper plot of the data (data points only!).
2. Determine if the form of the relationship is linear, power, or exponential.
3. Use the appropriate `polyfit` command based on step 2 to determine the parameters m and b for the trendline equation.
4. Create a theoretical data set from which the trendline will be plotted.
   (a) Set up a sequence of values for the independent variable (*x*). The range of these values should normally be similar to the range of the experimental values.
   (b) Using this sequence and the values of m and b determined in step 3, calculate corresponding values for the dependent variable (*y*).
   For example: `x = [1:1:10]; y = m*x + b;`
5. Add the theoretical data set (*x,y*) to the graph created in step 1.
6. Add the proper plot elements, such as the trendline equation, to complete the graph.

Linear Relationships

We can use `polyfit` to determine the parameters m and b of a linear model from a data set stored in X and Y using:

```
C = polyfit (X, Y, 1);
m = C(1);
b = C(2);
```

EXAMPLE   17-20

You are part of a firm designing nanoscale speedometers to measure speeds of small moving creatures like centipedes. To test your sensor, you place a centipede on a surface marked with a grid calibrated in millimeters and measure the following data, given in MATLAB notation. Use the data to create a graph and a mathematical model.

```
T=[0, 20, 40, 60, 80, 100, 120]; % time (t) [s]
D=[0, 105, 197, 310, 390, 502, 599]; % distance (d) [mm]
```

NOTE

The values for *x*- and *y*-coordinates use the actual coordinate values on the axes. The center-left end of the text string will be located at those coordinates.

The form of the `text` *function is*

```
text(x coordi-
nate,y coordi-
nate,text string)
```

In our specific case, we will use

```
text(35,450,TE)
```

Step 1. *Write the code to create a graph of the experimental data set [T, D]. To save space, most proper plot commands are not included here; see Example 17-13.*

```
plot(T,D,'rd')
```

Step 2. *This is a linear relationship.*

Step 3. *To determine the trendline parameters using the* `polyfit` *function:*

```
C=polyfit(T,D,1);
m=C(1);
b=C(2);
```

Note: the resulting elements in C ([4.9714, 2.1429]) *are the parameters* m *and* b *of the linear equation.*

Step 4. *Create a theoretical data set in MATLAB:*

```
T2=[0:5:120];
D2=m*T2+b;
```

Step 5. *Recall that we mentioned earlier that the plotting commands such as* plot *ERASE whatever is currently in the most recently used figure window, then draws the new graph. To prevent the erasure of the data points plotted earlier, you must include the* hold on *function before drawing the trendline.* hold on *tells MATLAB not to erase the current figure window, but to merely* **add** *the new information to what is already there. We will plot the trendline as a red solid line.*

```
hold on
plot(T2,D2,'r-');
```

Step 6. *To add a trendline equation, use the* sprintf *function to insert the numbers stored in variables* m *and* b *into a single string representing the equation. Here we will display m and b as integers.*

```
TE=sprintf('d = %0.0f t + %0.0f',m,b);
```

To add this expression on your graph near the trendline, you will need to provide coordinates where the equation should be shown using the text *function. The coordinates are determined by examining the resulting graph, and choosing a location that does not interfere with the data. Sometimes, it is easiest to create the graph, run the code, then alter the code to include the text location.*

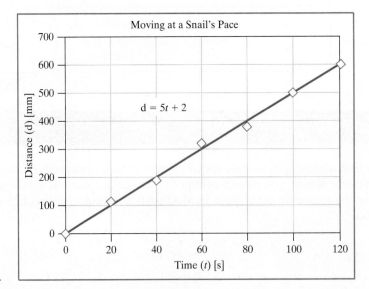

Figure 17-4

Example of a proper plot, showing single experimental data set with a linear trendline.

Comprehension Check | **17-12**

The following data represent the pressure [Pa] of a gas as a function of time [years]. Draw a proper plot of these data, and add a linear trendline and trendline equation to the plot.

$t = [0.1, 1.4, 3.6, 4.9, 8.0];$ $\qquad\qquad$ $P = [13800, 11100, 8900, 7300, 4000];$

MATLAB *examples*

Power Relationships

The parameters m and b in a power model can be determined from a data set stored in X and Y using:

```
C = polyfit(log10(X),log10(Y),1);
m = C(1);
b = 10^C(2);
```

EXAMPLE | **17-21**

Joule's first law, also known as the Joule effect, relates the heat generated to current flowing in a conductor. It is named for James Prescott Joule, the same person for whom the unit of the joule is named. The Joule effect states that the electric power (P) can be calculated as $P = I^2R$, where R is the resistance in ohms and I is the electrical current in amperes. The following data are collected in MATLAB notation. Use the data to create a graph and a mathematical model.

```
I=[0.50,1.25,1.50,2.05,2.25,3.00,3.20,3.50];    % Current (I) [A]
P=[1.2,7.5,11.25,20,25,45,50,65];               % Power (P) [W]
```

Step 1. *Write the code to create a graph of the experimental data set [I, P]. To save space, most proper plot commands are not included here; see Example 17-13.*

```
plot(I,P,'bs')
```

Step 2. *This is a power relationship.*

Step 3. *To determine the trendline parameters:*

```
C=polyfit(log10(I),log10(P),1);
m=C(1);
b=10^C(2);
```

Step 4. *Create a theoretical data set in MATLAB:*

```
I2=[0.5:0.1:3.5];
P2=b*I2.^m;
```

Step 5. *Plot the trendline as a blue solid line. Use the* `hold on` *function to prevent erasure of the data points plotted earlier:*

```
hold on
plot(I2,P2,'b-');
```

Step 6. *To add a trendline equation to your graph, use the* `sprintf` *function to convert the numbers stored in variables* n *and* b *into a single string variable:*

```
TE = sprintf('P = %3.1f I^%1.0f',b,m);
text(0.9,55,TE)
```

NOTE

When placing a text string on a figure, the caret (^) says to superscript the next single character or superscript the contents of braces { } immediately following.

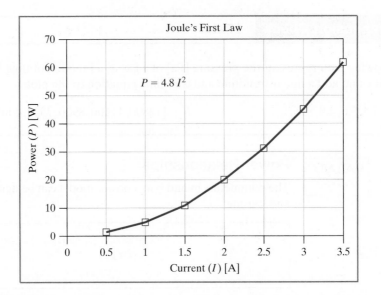

Comprehension Check 17-13

The following data represent the specific gravity of a gas as a function of volume [m³]. Draw a proper plot of these data, and add a power trendline and trendline equation to the plot.

$V = [0.7, 1.4, 3.6, 4.9, 8.0];$ $SG = [0.0081, 0.0039, 0.0014, 0.00099, 0.00052];$

Exponential Relationships

The parameters m and b in an exponential model can be determined from a data set stored in X and Y using:

```
C = polyfit(X, log(Y), 1);
m = C(1);
b = exp(C(2));
```

EXAMPLE 17-22

A reaction is carried out in a closed vessel. The following data are taken for the concentration of the organic material as a function of time from the start of the reaction. A proposed mechanism to predict the concentration at any given time is $C = C_0\, e^{-kt}$, where k is the reaction rate constant and C_0 is the initial concentration of the species at time zero. The following data are collected in MATLAB notation. Use the data to create a graph and a mathematical model.

```
T=[36,65,100,160];              % Time (t) [min]
C=[0.145,0.120,0.100,0.080];    % Concentration (C) [g / L]
```

Step 1. *Write the code to create a graph of the experimental data set [T, C]. To save space, most proper plot commands are not included here; see Example 17-13.*

```
plot(T,C,'g*')
```

Step 2. *This is an exponential relationship.*

Step 3. *To determine the trendline parameters:*

```
P=polyfit(T,log(C),1);
m=P(1);
b=exp(P(2));
```

Step 4. *Create a theoretical data set in MATLAB:*

```
T2=[20:1:160];
C2=b*exp(m*T2);
```

Step 5. *Plot the trendline*

```
hold on
plot(T2,C2,'-.g');
```

Step 6. *Add a trendline equation to the graph.*

```
TE=sprintf('C = %3.2f e^{%4.3ft}',b,m);
text(43,0.085,TE);
```

NOTE

When placing a text string on a figure, the caret (^) says to superscript the next single character or subscript the contents of braces { } immediately following.

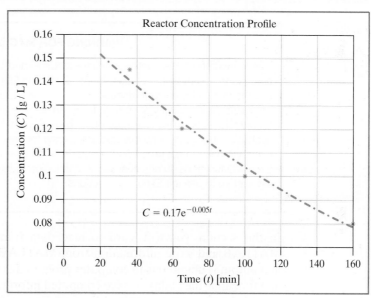

Problem Issue: Superscripts and subscripts do not work correctly when placing text on a figure.

Example: Trying to place $X^{4.25}$ on plot

```
text(5,10,'X^4.25')
```
generates $X^4.25$

Explanation: The caret (^) says superscript the next single item (4)

Solution: `text(5,10,'X^{4.25}')` generates $X^{4.25}$

Explanation: The braces ({}) effectively make the 4.25 a single entry in a cell array, thus the "next single item" is 4.25

Example: Place H_{FE} in the title of a plot

Solution: `title('H_{FE}')`

Explanation: The logic is the same as for superscripts, but using the underscore character (_) to subscript.

Comprehension Check	17-14

The following data represent the current [μA] into the base of a transistor as a function of the base-to emitter-voltage [mV]. Draw a proper plot of these data, and add an exponential trendline and trendline equation to the plot.

$V_{be} = [19, 27, 40, 57, 64];$ $I_b = [0.05, 0.21, 3.2, 51, 230];$

17.5 Microsoft Excel I/O

LEARN TO: Read basic Microsoft Excel workbook info with **xlsfinfo**
Read Microsoft Excel workbooks with **xlsread**
Write to Microsoft Excel workbooks with **xlswrite**

WARNING FOR MAC USERS!

Due to incompatibilities between Microsoft Excel and the MacOS version of MATLAB 2012 and earlier, you must run at least MATLAB version 2013 in Microsoft Windows in order to use the built-in Excel input function xlsread. MATLAB 2013 and later handles Excel input on both platforms, but writing data from MATLAB to Excel using xlswrite does not work on MacOS as of 2016. There are other ways to write MATLAB data to Excel in MacOS, such as using a comma separated value (csv) format, but we do not cover this here. Mac users will need to set up a Windows partition on their machines in order to fully utilize the Excel I/O functions. In any case, Microsoft Office must be installed on the matching operating system to use the Excel I/O functions.

In this section, you will learn to read data from Microsoft Excel workbooks into MATLAB and write information from MATLAB to an Excel workbook. File input and output is ubiquitous in computer programming if you want to give your programs and functions the ability to save computed information for use at a later time. We discuss file input and output using Microsoft Excel workbooks because it is a common file format used in many disciplines.

Reading Microsoft Excel Workbooks

For importing data from Excel files, there are two built-in functions in MATLAB that will allow us to read in all of the information we need from our Excel files; they are xlsfinfo and xlsread.

xlsfinfo

The **xlsfinfo** function in MATLAB allows us to extract information about Microsoft Excel files stored in MATLAB's current directory. The xlsfinfo function will return three outputs:

```
[FileType, Sheets, ExcelFormat] = xlsfinfo (filename)
```

where

- **FileType** is a variable containing a string that describes the type of the file. If we call `xlsfinfo` on a file that is not a Microsoft Excel workbook, this string will be empty (`' '`); otherwise, 'Microsoft Excel Spreadsheet' will be saved in the `FileType` variable. This is not particularly helpful information, but it does give us a mechanism for determining whether or not the file we are trying to open is a valid Microsoft Excel workbook.
- **Sheets** is a cell array variable containing the names of each worksheet within the Excel workbook. Each sheet name is stored as text within the cell array.
- **ExcelFormat** is a variable containing a string that describes the version of the Microsoft Excel workbook. In general, this information is not useful, but if the string is blank, it is an indicator that Microsoft Excel is not installed on your computer.
- **fileName** is the name of the Microsoft Excel file in which we are interested. The file name must be passed into `xlsfinfo` as a string and must contain the file extension of the Microsoft Excel workbook as part of the file name. The file extension of Microsoft Excel workbooks is usually .xls or .xlsx.

The following examples will use the starter file titled ClemsonWeather.xlsx found in the online materials. If you open the file, you will notice that there are three worksheets within the workbook, each containing day-by-day weather information for Clemson, SC, from 2008 to 2010.

For each day, we have the following values recorded: the high-, average-, and low-temperature readings measured in degrees Fahrenheit (`TempHighF`, `TempAvgF`, `TempLowF`).

xlsread

The **xlsread** function allows us to read in data from Microsoft Excel workbooks.

There are a number of different variations on how `xlsread` can be used, but we will only learn about a few of the variants involving different combinations of arguments and returned values.

If we want to tell `xlsread` to import numeric values from a particular worksheet, we can use the following syntax:

```
D = xlsread (filename, Sheet)
```

where

- **fileName** is the name of the Excel workbook including the file extension, as a string.
- **Sheet** is a string containing the name of the Excel worksheet we want to import into our variable D.

EXAMPLE   17-23

Use `xlsread` to import all of the weather data from 2008 from ClemsonWeather.xlsx.

```
WD08 = xlsread('ClemsonWeather.xlsx','2008');
```

If we open `WD08` *in MATLAB's Variable Editor, we can verify that our matrix contains all of the numeric values in the "2008" worksheet, beginning at cell* B2*. The first row (1) and the first column (A) were not imported since* *they contain only text data—no numbers. We will see how to acquire this information shortly. Notice that the dimension of* WD08 *is 366 × 15, which is what we expected since 2008 was a leap year.*

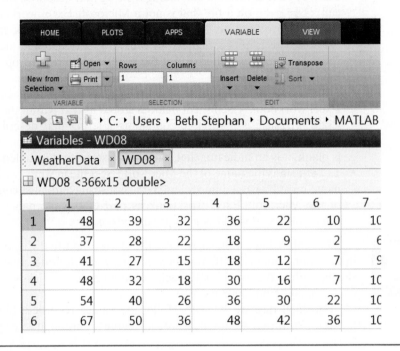

If we only want to import a particular range of an Excel file, we can add another argument into xlsread that will allow us to type in ranges of values. For example, if we only wanted to import all of the cells between B2 and B5, we could provide a range B2:B5 just like we would type into Microsoft Excel.

```
D = xlsread (filename, Sheet, Range)
```

where

- **fileName** is the name of the Excel workbook including the file extension, as a string.
- **Sheet** is a string containing the name of the Excel worksheet we want to import into our variable D.
- **Range** is a string containing the cell range written in Microsoft Excel notation.

EXAMPLE   17-24

Use xlsread to import the first 10 high temperatures from 2009 from file ClemsonWeather.xlsx. If we open ClemsonWeather.xlsx in Microsoft Excel, we notice that the first 10 values occur between B2 and B11.

```
WD09 = xlsread('ClemsonWeather.xlsx','2009','B2:B11');
```

MATLAB creates a 10 × 1 matrix containing a column of values representing the first 10 high temperatures recorded in 2009.

If we need to extract any of the text information from the Excel workbook, we need to modify our syntax to capture two outputs from `xlsread`. To extract nonnumeric values from an Excel workbook:

```
[D, T ] = xlsread (filename, Sheet)
```

where

- **fileName** is the name of the Excel workbook including the file extension, as a string.
- **Sheet** is a string containing the name of the Excel worksheet we want to import.
- **D** is a matrix of all of the numeric values in `Sheet`.
- **T** is a cell matrix of all nonnumeric (text, dates, etc.) values in the worksheet. If a cell contains a numeric value, that value will be stored as an empty string (`' '`) in `T`. In other words, matrices `D` and `T` will not necessarily have the same dimensions. For more information, refer to the examples that follow.

It is worth noting that even though we are using the syntax requiring a sheet name to save text values, any of the previously covered notations will allow you to capture text values in an Excel worksheet.

EXAMPLE   17-25

Use `xlsread` to import the first 10 high temperatures and the corresponding text values for the dates from 2009 from ClemsonWeather.xlsx. In the worksheet, the first 10 values occur between B2 and B11 and the dates are listed between A2 and A11, so the range we need to import is A2:B11.

```
[WD09,WT09] = xlsread('ClemsonWeather.xlsx','2009','A2:B11');
```

MATLAB creates two 10 × 1 matrices containing a column of values representing the first 10 high temperatures recorded in 2009 in WD09 and the corresponding text dates in a cell matrix WT09. In this special case, since the entire second column in WT09 was blank (all values were numeric), MATLAB automatically removed the second column so that WT09 is a 10 × 1 cell array.

EXAMPLE   17-26

Use `xlsread` to import the numeric and nonnumeric values in the 2009 sheet in ClemsonWeather.xlsx.

```
[D09,T09] = xlsread('ClemsonWeather.xlsx','2009');
```

If we look at the dimensions of D09 and T09, we note that D09 is a 365 × 15 matrix, but T09 is a 366 × 16 cell matrix. This is due to the fact that T09 contains the header row with all of the column labels, as well as the entire first column containing the corresponding dates. If we examine T09 in the Variable Editor, we see that the only values in the cell matrix are the first row and first column, and all of the other values in the cell matrix are empty strings (''). In light of these phenomena, special care must be taken when trying to write MATLAB code that tries to relate cell matrices and numeric data matrices imported using xlsread.

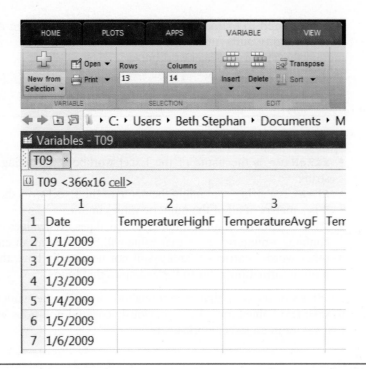

Writing Microsoft Excel Workbooks

MATLAB provides a built-in function, **xlswrite**, that enables MATLAB to export data by writing to Microsoft Excel workbooks.

xlswrite

Just like with xlsread, xlswrite accepts different syntax for different writing scenarios. For the sake of brevity, we will only cover the standard use of xlswrite, but refer to MATLAB's help and doc documentation for more information. The function xlswrite returns no output and only requires input arguments:

```
xlswrite (fileName, Matrix, Sheet, Cell)
```

where

- **fileName** is a string containing the file name you want to write to. If the Excel file does not exist, MATLAB will create a blank workbook with three blank worksheets ('Sheet1,' 'Sheet2,' and 'Sheet3') and display a warning "Warning: Added specified worksheet." This is common and is not considered an error message.
- **Matrix** is either a numeric or cell matrix containing the data you want to write to an Excel file.
- **Sheet** is a string that contains the sheet name where you want to write your data.
- **Cell** is a string that designates the top-left corner where MATLAB should start writing the data in Matrix in the Excel workbook. If Cell is not specified, MATLAB will default to writing with the top-left corner set to cell A1.

EXAMPLE   17-27

Use `xlsread` to import the numeric and nonnumeric values in the 2009 sheet in ClemsonWeather.xlsx, calculate all of the average high temperatures in units of degrees Celsius, and write the output to a new Excel file called newWeather.xlsx.

```
[D09,T09]=xlsread('ClemsonWeather.xlsx','2009');
T09{1,2}='TempHighC';
D09(:,1)=(D09(:,1)-32)/1.8;
xlswrite('newWeather.xlsx',T09,'2009');
xlswrite('newWeather.xlsx',D09,'2009','B2');
```

Note that we can use the `xlswrite` *function twice: first to write the headers and dates to the Excel file, followed by writing the matrix containing the new temperatures.*

Comprehension Check   17-15

Import all the data, including the header row, from the file **Widgets.xlsx**.

Assume each part is a cylinder. Determine the volume of each part in units of cubic inches [in³].

Write the part number and volume data to the Excel file. The data should write to a new worksheet in the **Widgets** file, containing the part name and the corresponding volume. Your worksheet should contain appropriate data column headings.

CHAPTER 17

IN-CLASS ACTIVITIES

ICA 17-1

Write a MATLAB statement that results in the input request shown in bold. The >> shows where your statement is typed, and the | shows where the cursor waits for input. The display must be correctly positioned. Each has a space before the cursor (shown as |). The input variable name and the variable type are shown at the right.

(a) >>

Enter the length of the bolt in inches: | (bolt, a number)

(b) >>

Enter the company's name: | (Company, text)

(c) >>

(In this statement, the window displayed will appear for the user to choose a color, and the result will be stored in the variable `LineColor`.)

ICA 17-2

You are writing code that is part of a purchasing system. For each of the following items, write a MATLAB statement to accomplish the task.

(a) Generate a menu that asks users what they wish to buy and gives them a choice of three items: Flange, Bracket, or Hinge. The user's choice should be placed in a variable named **PartType**.

(b) Ask users how many parts they wish to buy. Store the response in a variable named **NumParts**.

(c) Ask the user to enter his or her name. Store the response in a variable named **BuyerName**.

ICA 17-3

For each question dialog described below, write a MATLAB command to create the dialog.

(a) The title should be Banking, the question should be "What is your bank balance?", and the options should be Positive, Zero, and Negative, with Positive as the default. The response should be placed in `Bal`.

(b) The title should be Religion, the question should be "Which term describes your religious beliefs?", and the options should be Believer, Agnostic, and Atheist, with Agnostic as the default. The response should be stored in `Rel`.

(c) The title should be Dummy, the question should be "Are you one?" and there should be a single choice labeled "How should I know?" The response should be stored in the variable `Dummy`.

ICA 17-4

For each input dialog described below, write a MATLAB statement to create that dialog.

(a)   The title should be Vehicles, the question should be "How many vehicles do you own?" and the response should be stored in `Veh`.

(b)   The title should be Personal Information, the questions should be "What is your nickname?" "What is your gender?" "What is your age?" and "What is your marital status?" The responses should be stored in `PersInfo`.

ICA 17-5

For the following questions, assume $z = 100/810$. Write the MATLAB output that would result from each statement.

(a)   `>> disp(z)`
(b)   `>> fprintf('%f',z)`
(c)   `>> fprintf('%e',z)`
(d)   `>> fprintf('%0.4f',z)`
(e)   `>> fprintf('%0.8E',z)`

ICA 17-6

For the following questions, $z = 100/810$. Write the MATLAB output statement that displays z in the format shown in bold. DO NOT hard-code the values; use the value stored in `z` to print each one. In each case, the `>>` by the letter in parentheses shows where your statement is typed. In each case, the command prompt should appear on the line following the printed output.

(a)   `>>`
         0.1
(b)   `>>`
         0.123
(c)   `>>`
         1.235e-001
(d)   `>>`
         The value of z is 0.123.
(e)   `>>`
         z is 0.123, so 100,000 z is 1.23457E+04.

ICA 17-7

Assume that the following variables are stored in MATLAB's workspace:

```
MCost = 450;                % machine cost in $/day
WRate = 40;                 % widgets produced/day
OCost = 1150;               % operating cost in $/day
WPrice = 4787;              % sales price in $/widget
Days = 71;                  % number of production days
WName = 'Sonic Pliers';     % name of the widget produced
```

Determine the output displayed on the screen by the following code:

```
TCost=MCost+OCost;
fprintf('Total Cost per Day: $%0.2f\n',TCost)
NumW=WRate*Days;
fprintf('A total of %0.0f %s was produced in %0.0f days. \n',
NumW,WName,Days)
```

```
Income=NumW*WPrice;
fprintf('These will sell at $%0.2f each for a total of $%0.0f.\n',
WPrice,Income)
fprintf('This will make a profit of $%0.2f.\n',
Income-TCost*Days)
```

ICA 17-8

Assume that a three-element row vector V already exists. Write a single MATLAB statement that will print the contents of V diagonally from top left to bottom right. Display each value with three decimal places.

Example: V = [42, −17.9626, 0.03654]
Sample Output:
```
42.000
        −17.963
                0.037
```

ICA 17-9

The tiles on the space shuttle are constructed to withstand a temperature of 1950 kelvin. Write a MATLAB program that will display the temperature in the four temperature units [kelvins, degrees Celsius, degrees Fahrenheit, and degrees Rankine]. Each value should be incorporated into a sentence with appropriate text, and each sentence should appear on a new line. Format all values as integers.

ICA 17-10

The specific gravity of acetic acid (vinegar) is 1.049. Write a MATLAB program to display the density of acetic acid in units of pounds-mass per cubic foot, grams per cubic centimeter, kilograms per cubic meter, and slugs per liter. Incorporate each value into a sentence with appropriate text, each sentence on a new line. All numeric values should be given to two decimal places.

ICA 17-11

Write a MATLAB program that will allow a user to type the specific heat of a value in calories per gram degree Celsius. Display the converted value in units of British thermal units per pound-mass degree Fahrenheit in the Command Window using the following format with one decimal place: "The specific heat is ____ BTU / (lb_m*deg F)."

ICA 17-12

Write a program that will allow the user to type a liquid evaporation rate in units of kilograms per minute and display the value in units of pounds-mass per second, slugs per hour, and grams per second. Display the result in the Command Window using one decimal place: "The evaporation rate is _____ pounds-mass per second, ____ slugs per hour, or ____ grams per second."

ICA 17-13

In order to calculate the pressure in a flask, write a program that allows the user to type the volume of the flask in liters, the amount of an ideal gas in units of moles, and the temperature of the gas in kelvins. Display the result in the Command Window in the following format, using one decimal place: "The pressure is ___ atmospheres."

ICA 17-14

Write a program that will allow the user to type the input power of a motor in watts, the mass of an object in kilograms, the height to which the object will be raised by the motor in meters, and the time it took to raise the object in seconds. Use this information to determine the efficiency of the motor. Display the calculated value in the Command Window in the following format, using one decimal place: "The motor is ____% efficient."

ICA 17-15

Use a question dialog to determine whether the user's favorite type of meat is Fish, Bird, or Mammal. Then use an input dialog to ask the user how many servings of that type of meat they eat each week. Note that the question should include the name of the type of meat, such as "How many servings of Bird do you eat each week?" Next, use an input statement to ask the user how many servings of other types of meat they eat each week. Finally, you should print a report to the screen similar to the following example:

```
Your favorite type of meat is Bird.
Bird makes up 61% of your weekly meat consumption.
```

ICA 17-16

Write a MATLAB program that will ask users to enter their age, then ask them to enter the name of their best friend, and then ask them to enter their friend's age. Determine the difference in the ages, and express this value as a positive integer. Finally, a statement in the following format should be displayed in the Command Window:

```
My age is _____ years.
_____ is my best friend. My friend's age is _____ years.
The difference in our age is _____ years.
```

ICA 17-17

Write a program that asks a user to enter, one at a time, the four numbers to fill a 2 × 2 matrix. For each number entered, make sure the user knows the location in the matrix of the number being entered. When the program begins, the following information should display for the user before the prompts for the individual input values:

```
This program asks the user for a 2 × 2 matrix and displays the
result.
Please enter only integers, between −99,999 and 99,999.
```

Display the matrix in the Command Window as two rows and two columns, with the integers in the columns evenly spaced apart, using field width to control the column spacing.

ICA 17-18

You are writing a program to tabulate information about various spacecraft exploring other planets and moons in our solar system. A data file named PlanetData.mat is available online. Download this file and place it in your MATLAB directory. Write a program to accomplish the following goals.

Load the variables in PlanetData.mat into your workspace.

- PName is a **single-row cell array** containing the names of several planets and moons. A **partial** list of the contents is shown here.

Mercury	Venus	Luna	Mars	Phobos	Titan

- PData is a **three-row matrix** containing the same number of columns as PName. The first row contains the radii [kilometers, km] of the corresponding bodies in PName, the second row contains the masses [yottagrams, Yg], and the third row contains the gravitational acceleration on the surface of each planet or moon [meters per second squared].

2440	6051	1737	3390	11.3	2576
330	4870	73.4	624	1.07E-5	135
3.7	8.87	1.62	3.71	0.0057	1.35

Note that your program should be written so that if the number of columns in these two variables changes, your code will still work correctly.

The questions on this page concern a robotic spacecraft on the surface of one of the listed planets from the previous page.

Convert the radius values in row 1 of PData from kilometers [km] to meters [m]. Leave the results in the first row of PData.

Convert the mass values in row 2 of PData from yottagrams [Yg] to kilograms [kg]. Leave the results in the second row of PData.

Add a fourth row to PData that contains the specific gravities of each of the bodies given in PName. You may assume all are spherical. Recall that the volume of a sphere is given by

$$V_{sphere} = \frac{4}{3}\pi r^3$$

Determine which of the listed bodies has the lowest specific gravity and print a statement like the following:

```
The least dense body is Phobos with a specific gravity of 1.77.
```

Ask the user to enter the name of a spacecraft, storing the answer as a **text string** in the variable Craft.

Ask the user to enter the mass of the spacecraft in kilograms [kg], storing the answer as a scalar in the variable Mass.

Generate a menu similar to the partial one shown using the planet names stored in the cell array PName, and save the response in the variable Location.

You may assume that the user makes a valid choice from the menu.

Calculate the escape velocity in units of meters per second for the planet selected from the menu and store the velocity in the variable Escape.

Escape velocity can be calculated using $V_E = \sqrt{2gr}$ where r is the planet's radius and g is the planet's gravity.

Print in the Command Window statements similar to the ones shown. The sample output assumes the user specified the spacecraft Thumper with a mass of 512 kilograms located on Mercury.

```
Information on Thumper
    Spacecraft mass:    5.12E5 grams
    Location:           Mercury
    Escape Velocity:    4249 m/s
>>
```

ICA 17-19

Joule's first law, also known as the **Joule effect**, relates the heat generated to current flowing in a conductor. It is named for James Prescott Joule, the same person for whom the unit of the joule is named. Create a proper plot of the experimental data.

Current (*I*) [A]	0.50	1.25	1.50	2.25	3.00	3.20	3.50
Power (*P*) [W]	1.20	7.50	11.25	25.00	45.00	50.00	65.00

ICA 17-20

Create a proper plot of the following set of experimental data collected during the charging of a capacitor. In this plot, time should be on the abscissa.

Time (*t*) [s]	0.2	0.4	0.6	0.8	1.0
Voltage (*V*) [V]	75.9	103.8	114.0	117.8	119.2

ICA 17-21

There is currently an effort in the United States to replace incandescent light bulbs with more energy-efficient technologies, including compact fluorescent lights (CFLs) and light-emitting diodes (LEDs). The lumen [lm] is the SI unit of luminous flux, a measure of the perceived power of light. Luminous flux is adjusted to reflect the varying sensitivity of the human eye to different wavelengths of light.

To test the power usage, you run an experiment and measure the following data. Create a proper plot of these data, with electrical consumption (EC) on the ordinate.

Luminous Flux (LF) [lm]	Electrical Consumption (EC) [W]		
	Incandescent	CFL	LED
80	16		1.7
200		5	3
400	38	10	6
600	55		9
750	68	18	12
1250		27	
1400	105	33	

On the graph, use solid blue circles for incandescent bulbs, solid red squares for CLFs, and open black diamonds for LEDs.

ICA 17-22

You want to create a graph showing the theoretical relationship of an ideal gas between pressure (*P*) and temperature (*T*). Assume the tank has a volume of 12 liters and is filled with nitrogen (formula, N_2; molecular weight, 28 grams per mole). Allow the initial temperature to be 270 kelvin at a pressure of 2.5 atmospheres. Create a proper plot, showing the temperature on the abscissa from 260 to 360 kelvin.

ICA 17-23

In 1619, Johannes Kepler proposed his third law of planetary motion. This says that the ratio of the square of the orbital period (*P*) to the cube of the distance (*r*) from the sun is a constant (C): $P^2/r^3 = C$.

Given that the orbital period of the Earth is one year [yr], and its distance from the sun is one astronomical unit [au], create two subplots in a single figure window showing the theoretical relationship between orbital period and orbital radius. Place the radius on the abscissa over a range of 0.2 to 50 au, and the orbital period on the ordinate using units of years. One of the plots should use linear axes and the other should use logarithmic axes.

ICA 17-24

The decay of a radioactive isotope can be theoretically modeled with the following equation, where C_0 is the initial amount of the element at time zero and k is the decay rate of the isotope. Create a proper plot of the decay of isotope A [$k = 1.48$ hours]. Allow time to vary on the abscissa from 0 to 5 hours with an initial concentration of 10 grams of isotope A.

$$C = C_0 e^{-t/k}$$

ICA 17-25

Create a proper plot of the theoretical voltage decay of a resistor-capacitor circuit:

$$V(t) = V_0 e^{-\frac{t}{RC}}$$

You may assume that you have a capacitance (C) of 500 microfarads [μF], a resistance (R) of 0.5 ohms [Ω], and an initial voltage (V_0) of 10 volts [V]. The plot should start at time 0 seconds and increase by intervals of 1 microsecond to 600 microseconds.

ICA 17-26

Plot the following functions as assigned by your instructor using subplots, choosing an appropriate layout for the number of functions displayed. The independent variable (angle) should vary from 0 to 360 degrees.

(a) sin(u)
(b) 3 sin(u)
(c) sin(3u)
(d) sin(u) −3
(e) sin(u + 90)
(f) 3 sin(2u) − 2

ICA 17-27

Plot the following functions as assigned by your instructor using subplots, choosing an appropriate layout for the number of functions displayed. The independent variable (angle) should vary from 0 to 360 degrees.

(a) cos(u)
(b) −2 cos(u)
(c) cos(2u)
(d) cos(u) + 2
(e) cos(u − 45)
(f) 3 cos(2u) −2

ICA 17-28

Create a graph of $\cos(\theta)$, for angle values between 0° and 720°. The abscissa must be shown in units of degrees.

Ask the user to input four parameters (A, B, C, and D) into the generalized wave equation:

$$y = A \cos (B \theta + C) + D$$

Add the generalized wave function to the graph, using the input parameters from the user.

Add the equation to the graph as text in the upper left-hand corner of the graph. (See example.)

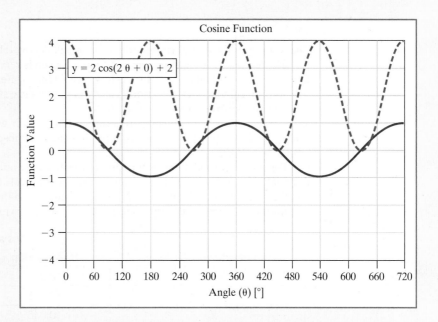

ICA 17-29

Enter the following data into MATLAB.

Diameter (D) [ft]	0.5	0.75	1	1.5	2	2.25	2.5	2.75
Power (P) [hp]	0.004	0.04	0.13	0.65	3	8	18	22

Graph the power (P, ordinate) versus the diameter (D, abscissa) assuming the data are experimental. Follow all proper plot rules! The title should be Turbine Power.

Try different combinations of linear and logarithmic axes to determine the proper `polyfit` form for the data set by answering the questions: Is it linear? Is it power? Is it exponential?

Use `polyfit` to determine a trendline for the data set. Add this trendline to the graph, including the trendline equation.

Show the axes in the necessary configuration so the data will appear linear.

ICA 17-30

If an object is heated, the temperature of the body will increase. The energy (Q) associated with a change in temperature (ΔT) is a function of the mass of the object (m) and the specific heat(C_p). In an experiment, heat is applied to the end of an object, and the temperature change at the other end of the object is recorded. This leads to the following equation.

$$\Delta T = \frac{Q}{mC_p}$$

An unknown material with a mass of 5 kilograms is tested in the lab, yielding the following results. Use the `polyfit` function to determine the specific heat of this material and store the final result in the variable C_p.

Create a proper plot of the data. Add a linear trendline, showing the resulting trendline equation, on the graph for a change in temperature over a range of energy from 5 joules to 70 joules.

Heat Energy Applied (Q) [J]	17	40	58
Temperature Change (ΔT) [K]	2	5	7

ICA 17-31

The resistance of a typical carbon film resistor will decrease by about 0.05% of its stated value for each degree Celsius increase in temperature. Silicon is very sensitive to temperature, decreasing its resistance by about 7% for each degree Celsius increase in temperature. This can be a serious problem in modern electronics and computers since silicon is the primary material from which many electronic devices are fabricated.

Create a proper plot to compare a carbon film resistor with a resistor fabricated from specially doped silicon ("doped" means impurities such as phosphorus or boron have been added to the silicon).

For relatively small temperature differences from the reference temperature, this process is essentially linear. Use `polyfit` to determine linear models for each data set. For each model, add the trendline and the associated trendline equation to the graph. Use an appropriate location for the equations to clearly associate them with the correct trendline.

Temperature (*T*)[°C]	Resistance (*R*) [Ω]	
	Carbon Film	**Doped Silicon**
15	10.050	10.15
20	10.048	9.85
25	10.045	9.48

ICA 17-32

Today, most traffic lights have a delayed green, meaning there is a short time delay between one light turning red and the light on the cross-street turning green. An industrial engineer has noticed that more people seem to run red lights that use delayed green. She conducts a study to determine the effect of delayed green on driver behavior. The following data were collected at several test intersections with different green delay times. These data represent only those drivers who continue through the intersection when the light turns red *before* they reach the limit line, defined as the line behind which a driver is supposed to stop. The data show the "violation time," defined as the average time between the light turning red and the vehicle crossing the limit line, as a function of how long the delayed green has been installed at that intersection.

Create a proper plot of the violation time (*V*, on the ordinate) and the time after installation (*t*, on the abscissa) for all three intersections on a single graph.

Use `polyfit` to determine linear models for each data set. For each model, add the trendline and the associated trendline equation to the graph. Use an appropriate location for the equations to clearly associate them with the correct trendline.

Time After Installation (*t*) [months]	Violation Time (*V*) [s]		
	Intersection 1	**Intersection 2**	**Intersection 3**
	1-Second Delay	**2-Second Delay**	**4-Second Delay**
2	0.05	0.1	0.5
5	0.1	0.5	1.5
8	0.3	1	2.5
11	0.4	1.3	3.1

ICA 17-33

Cadmium sulfide (CdS) is a semiconducting material with a pronounced sensitivity to light—as more light strikes it, its resistance goes down. In real devices, the resistance of a given device may vary over four orders of magnitude or more. An experiment was set up with a single light source in an otherwise dark room. The resistance of three different CdS photoresistors was measured when they were at various distances from the light source. The farther they were from the source, the dimmer the illumination on the photoresistor.

Create a proper plot of the data. Use `polyfit` to determine a power model for each data set. For each model, add the trendline and the associated trendline equation to the graph. Use an appropriate location for the equations to clearly associate them with the correct trendline.

Distance from Light (d) [m]	Resistance (R) [Ω]		
	A	B	C
1	79	150	460
3	400	840	2,500
6	1,100	2,500	6,900
10	2,500	4,900	15,000

ICA 17-34

Your supervisor has assigned you the task of designing a set of measuring spoons with a "futuristic" shape. After considerable effort, you have come up with two geometric shapes that you believe are really interesting.

You make prototypes of five spoons for each shape with different depths and measure the volume each will hold. The following table shows the data you collected.

Depth (d) [cm]	Volume of Shape	
	B (VA) [mL]	A (VB) [mL]
0.5	1	1.2
0.9	2.5	3.3
1.3	4	6.4
1.4	5	7.7
1.7	7	11

Create a proper plot of the data. Use `polyfit` to determine power models for each data set. For each model, add the trendline and the associated trendline equation to the graph. Use an appropriate location for the equations to clearly associate them with the correct trendline.

Use your models to determine the depths of a set of measuring spoons comprising the following volumes for each of the two designs. V = [¼ tsp, ½ tsp, ¾ tsp, 1 tsp, 1 Tbsp]. NOTE: 1 Tbsp = 3 tsp

Print the results to the Command Window in a table similar to the following format.

```
Volume Needed (V) [tsp]           0.25  0.5  0.75  1 3
Depth of Design A (dA)  [cm]
Depth of Design B (dB)  [cm]
```

ICA 17-35

Three different diodes were tested: a constant voltage (0.65 volts) was held across each diode while the current through each was measured at various temperatures. The following data were obtained.

Create a proper plot of the data. Use `polyfit` to determine exponential models for each data set. For each model, add the trendline and the associated trendline equation to the graph.

Temperature (T) [K]	Current (I_D) [mA]		
	Diode A	Diode B	Diode C
275	852	2086	264
281	523	1506	179
294	194	779	81
309	69	390	35
315	47	301	26

ICA 17-36

Design a program that will ask the user to enter two row vectors, `Ext` and `Wt`, representing the extension of a spring [cm] and the weight hung from that spring [N]. The program will then determine a model to describe the spring and print a report in the Command Window using the following format:

```
SPRING REPORT
Spring Constant: X.XX N/m
Initial Extension: X.X cm
>>
```

In addition, the original data should be plotted, a trendline added to the plot, and the equation for the trendline included on the plot.

ICA 17-37

If a hot liquid in a container is left to cool, its temperature $(T)\,[°C]$ will gradually approach room temperature. This model will have the form $T = A + B\,e^{mt}$. In this case, m will always be negative and its dimension will be inverse time. If it is not clear how to determine A and B, consider that $T = A + B$ when $t = 0$, and $T = A$ as t approaches infinity.

You are to write a function named `Cooling` that will accept two parameters:

- `Tr`: Room temperature in degrees C
- `TData`: A two-row matrix containing measurements of a cooling liquid at various times. The first row contains the times of the measurements in minutes, and the second row contains the corresponding temperature measurements in degrees C

The function must perform the following operations:

- Plot the measured points.
- Determine an exponential model to fit the data.
- Add the exponential trendline to the graph.
- Add the trendline equation for the exponential model to the graph.

Example:

Room Temperature: Tr = 30
Temperature Data: TData = [0.5 3 7 11; 85 48 35 33]

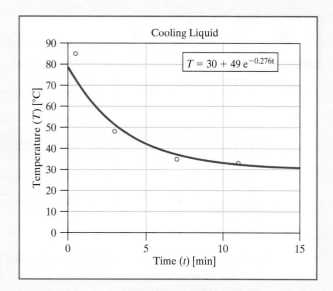

Cooling Liquid

$$T = 30 + 49\,e^{-0.276t}$$

ICA 17-38

You are an engineer working for M & M / Mars Corporation in the M&M plant. For Halloween, M&Ms are produced in "fun size." To help with quality control, you create the following worksheet. The factory workers will examine sample bags of M&Ms and enter the weight of the bag and the individual count of M&Ms contained in the bag.

Online, you have been given the following data in a Microsoft Excel workbook called **CandyCount.xlsx** with the data stored on a sheet named "M and M data." Only a portion of the actual data is shown.

	A	B	C	D	E	F	G	H
1	Bag Number	Mass of Bag (m) [g]	Red	Orange	Yellow	Green	Blue	Brown
2	1	15.0	3	1	5	4	4	1
3	2	13.9	3	2	3	5	1	2
4	3	14.7	1	1	9	2	3	1
5	4	14.3	2	5	4	2	1	3
6	5	14.7	2	3	6	3	2	2

Write a MATLAB program to read in the Excel data, calculate the data required as shown in the following output table, and write the results to a new Microsoft Excel file. Your output should appear as follows, where the highlighted portion is replaced by the values you calculate in your solution. Your program should also use formatted `fprintf` statements to display the same table in the Command Window, filling in the missing information (highlighted) and displaying the data using a reasonable number of decimal places.

The program should assume the user could enter more data to the original worksheet or modify the color names on the worksheet and the program would still run correctly.

	A	B	C	D	E	F	G	H
1	M&M Analysis by Color							
2		Overall	Red	Orange	Yellow	Green	Blue	Brown
3	Average per bag:							
4	Average mass per bag:							
5	Color that appears the most:							
6	Color that appears the least:							

ICA 17-39

Import all the data, including the header row, from the file **Voltage_Capacitance.xlsx**.

Note: This will require two input statements, capturing each data range. Data range 1 = columns A − C; data range 2 = columns E − I.

	A	B	C	D	E	F	G	H	I
1	Part No	Working Voltage [V]	Capacitance [microf]		Maximum Capacitor Voltage (Vmax) [V]				
2	C-6153	3	0.15			Resistor (R) [ohms]			
3	C-6473	3	0.47		Input Frequency (F) [Hz]	1000	2000	3000	4000
4	C-6225	5	0.22		100	1.18	0.59	0.39	0.30
5	C-6335	5	0.33		200	4.21	2.10	1.40	1.05
6	C-62210	10	0.22		300	28.61	14.31	9.54	7.15
7	C-63310	10	0.33		400	15.07	7.53	5.02	3.77
8	C-64710	10	0.47		500	7.86	3.93	2.62	1.97
9	C-66810	10	0.68		600	5.96	2.98	1.99	1.49
10	C-6471F	15	0.47		700	5.08	2.54	1.69	1.27

Using the data imported, ask the user to select the part number from a menu.

In a formatted output statement, list the voltage and capacitance for the part chosen, similar to the following statement.

```
Part number C-6153 has a voltage of 3 volts and a capacitance
of 0.15 microfarads.
```

Using the data imported, ask the user to select the input frequency from a menu.

Using the data imported, ask the user to select the resistance from a menu.

In a formatted output statement, list the maximum capacitor voltage for the chosen properties, similar to the following statement.

```
Input Frequency = 100 Hz
Resistor = 2000 Ohms
    Maximum capacitor voltage = 0.59 V
```

REVIEW QUESTIONS

1. The specific gravity of gold is 19.3. Write a MATLAB program that will ask the user to input the mass of a cube of solid gold in units of kilograms and display the length of a single side of the cube in units of inches. The output should display a sentence like the following one, with the length formatted to two decimal places.

   Sample Input/Output:

   ```
   Enter the mass of the cube [kilograms]: 0.4
   The length of one side of the cube is 1.08 inches.
   ```

2. An unmanned X-43A scramjet test vehicle has achieved a maximum speed of Mach number 9.68 in a test flight over the Pacific Ocean. Mach number is defined as the speed of an object divided by the speed of sound. Assuming the speed of sound is 343 meters per second, write a MATLAB program to determine speed in units of miles per hour. Your program should ask the user to provide the speed as Mach number and return the speed in miles per hour in a formatted sentence, displayed as an integer value, as shown in the following sample output.

   Sample Input/Output:

   ```
   Enter the speed as a Mach number: 9.68
   The speed of the plane is 7425 mph.
   ```

3. A rod on the surface of Jupiter's moon Callisto has a volume measured in cubic meters. Write a MATLAB program that will ask the user to type in the volume in cubic meters in order to determine the weight of the rod in units of pounds-force. The specific gravity is 4.7. Gravitational acceleration on Callisto is 1.25 meters per second squared. The input and output of the program should look similar to the output below. Be sure to report the weight as an integer value.

   Sample Input/Output:

   ```
   Enter the volume of the rod [cubic meters]: 0.3
   The weight of the rod is 397 pounds-force.
   ```

4. The Eco-Marathon is an annual competition sponsored by Shell Oil, in which participants build special vehicles to achieve the highest possible fuel efficiency. The Eco-Marathon is held around the world with events in the United Kingdom, Finland, France, Holland, Japan, and the United States.

   A world record was set in the Eco-Marathon by a French team in 2003 called Microjoule with a performance of 10,705 miles per gallon. The Microjoule runs on ethanol, which has a specific gravity of 0.789. Write a MATLAB program to determine how far the Microjoule will travel in kilometers given a user-specified amount of ethanol, provided in units of grams. Your program should ask for the mass using an input statement and display the distance in a formatted sentence similar to the output shown below.

   Sample Input/Output:

   ```
   Enter mass of ethanol [grams]: 100
   The distance the Microjoule traveled is 577 kilometers.
   ```

5. Your company wishes to write a computer program that will calculate the amount of heat needed to increase the temperature of a substance by a specified number of degrees. The company also wishes to know who ran the program and on what day of the week they ran it.

You are to write this program. Your program must allow the user to enter the following:
- The user's name *using a dialog box*
- Since the program will only be run either on Monday, Wednesday, or Friday, restrict the day of the week to these selections *using a question dialog*
- The mass of the substance *in units of grams*
- The specific heat of the substance *in units of J/(kg K)*
- The initial temperature of the substance *in units of degrees Fahrenheit*
- The final temperature of the substance *in units of degrees Fahrenheit*

    Your program should output the following formatted sentence:

```
On YYYY, the user XXXX entered a ZZ.ZZ gram mass with a specific
heat of AA.AA J/(kg K) that should be raised from BB.B degrees
Fahrenheit to CC.C degrees Fahrenheit.
This should require DD.D J of heat to be added to the substance.
```

Where:

- XXXX – User name
- YYYY – Day of the week
- ZZ.ZZ – Mass of the substance
- AA.AA – Specific heat of the substance
- BB.BB – Initial temperature
- CC.C – Final temperature
- DD.D – Required heat input

6. You are developing a program to record information about patients at a doctor's office. Write the code to accomplish the following goals.

Section 1: Patient Information

Create a single-column cell array named **Patients** containing the names shown in the menu.

Determine the number of names in **Patients** and place the result in **NumPat**. Your code must determine how many names there are in **Patients**; DO NOT hard-code this value. Your code should still work correctly if the number of names stored in **Patients** changes.

Ask the nurse to choose the patient name being seen during an office visit from a menu of patient names, located in the cell array **Patients**. A menu with the current five names is shown here as an example.

Ask the nurse to enter the selected patient's weight in pounds. This numeric value should be placed in a column vector **Wt** in the location corresponding to the location of that patient in the **Patients** cell array. For example, if the selected patient is Thomas Clemson, when the nurse types in his weight, the value will be placed in the fourth element of **Wt**. Note that the nurse does not enter the location in the vector, the program takes care of that.

Create a text string named **Gender** consisting of a series of the letter X. The length of this text string should equal the number of names in **Patients**. For example, with the values stored in **Patients** given above, **Gender** would contain **XXXXX**.

Ask the nurse to enter the patient's gender as a single letter chosen from M, F, T. (Some folks might argue vociferously concerning what categories should be included here, but this is just a homework assignment, not a philosophical discussion of gender identity and its classification, so chill out.) The letter typed by the nurse should REPLACE the corresponding X in the text string **Gender**. Thus if the nurse had selected Thomas Clemson and typed an M for gender, the text string **Gender** would then contain **XXXMX**, assuming no other patients' information had been previously entered. Again, the nurse does not specify the location in the text string, the code does that.

Ask the nurse to enter the selected patient's temperature [°F]. Place this numeric value in the corresponding element of a column vector named `Temp`. Thus if Thomas Clemson was selected and had a temperature of 98.7°F, the fourth element of `Temp` would then contain 98.7.

Section 2: Basic Report

Print a report on the screen similar to the following for the selected patient. Note that there is a blank line between the first line and the second line, and a blank line between the last line and the MATLAB prompt (>>). Also the last digit of the weight should be aligned beneath the "h" in "Weight", the gender identifier should be aligned under the letter d of "Gender", and the last digit of the temperature should be aligned under the "a" of "Temperature"

```
Patient: Thomas Clemson

Weight    Gender    Temperature
 195        M          98.7

>>
```

7. In a factory, various metal rods are forged and then plunged into a liquid to quickly cool the metal. The types of metals produced, as well as their specific heat capacity, are listed in Table A.

Table A Specific heat information for various metals

Material	Specific Heat [J/(kg K)]	Material	Specific Heat [J/(kg K)]
Aluminum	897	Iron	450
Cadmium	231	Tungsten	134

The metal rods may vary in mass and production (initial) temperature. The ideal process lowers the temperature of the material to 50 degrees Celsius [°C].

The liquid used to cool the metal is glycerol. The properties of glycerol are listed below. Initially, the glycerol is at room temperature, assumed to be 75 degrees Fahrenheit [°F].

Write a program containing the following elements.

- Enter the material names and specific heat information from Table A into a cell array.
- Ask the user to select a material from a menu.
- Ask the user to enter a 1×2 vector of the mass in grams [g] and the initial temperature in degrees Celsius [°C] of the material chosen from the menu. In the prompt to the user, state the material name. For example:
  Enter mass [g] and temperature [deg C] of Aluminum as [m T]:
- Call the function described below, sending any necessary information to the function as required by the solution. The function will return a single vector, described below.
- Write a formatted output statement to the user, similar to the final output shown below. The numerical formats and spacing should match exactly as shown.

Write a function to calculate the thermal energy in joules [J] that must be removed from the material and determine the volume of fluid needed in gallons [gal] to cool the rod to 50°C, assuming both the fluid and the rod are at this temperature at the end of the cooling process.

Table B Properties of glycerol

Specific heat (Cp)	2400 [J/(kg K)]	Specific gravity (SG)	1.261 [−]

- All properties of glycerol given in Table B should be entered into the function, not the program.
- Perform any necessary calculations and / or conversions. Be sure to track your units!

The function should return a 1×2 vector to the program, with the thermal energy required in joules [J] in the first element, and the volume required in gallons [gal] in the second element.

Sample Output:

```
Enter mass [g] and temperature [deg C] of Cadmium as [m T]: [4000 300]

For the 4.0 kg Cadmium rod with an initial temperature of 300 [deg C]
    Energy Removed [J]        2.3e+05
    Glycerol Volume [gal]     0.81
>>
```

8. Download the file `VI.mat` and use the `load` command to add the variables stored there to your workspace. The variables contain the information shown in the following table, organized as described.

Component	Voltage (V) [V]	5	7	10	12	15
Holtz100		128	142	165	180	212
Lever014	Current (I) [mA]	18	20	23	25	30
Dillard202		260	285	333	368	428

The three component names are stored in the single-row cell array `Name`.

The 4 × 5 matrix `CurData` contains the numeric values in columns three through seven of the preceding table.

Ask the user to choose a component from a menu created using the names stored in `Name`. You may assume the user will choose a button and not close the menu.

Calculate the power in units of watts [W] for all voltages for the component chosen by the user from the menu. Store the resulting vector in `Power`.

Recall that power (*P*) is determined by:

$$\mathbf{P} = \mathbf{VI}.$$

where:

V is the voltage in units of volts [V], given in row 1 of the matrix `CurData`.
I is the current in units of amperes [A], given in the row of the matrix `CurData` corresponding to the chosen component.
Be sure to note that the currents in `CurData` are given in mA.

Display for the user the choice of voltages given in row 1 of `CurData`. Note there is one tab between the numerical values. See example.

```
Voltage Choices:
5       7       10      12      15
```

Ask the user to enter a voltage value from the list shown. Store the scalar input as `Volt`.

You can assume users will enter a value from the list of values given and that they will not enter a number that is not shown.

Print a report similar to the following in the Command Window.

```
Component Holtz100
    Voltage = 10 V
    Current = 0.165 A
    Power = 1.65 W
>>
```

Next, the user will augment the data tables before performing another calculation. Ask the user to type the name of the new component. Place the text string response in a new column of `Name` at the end of the cell array, such that the cell array will now have one more column than the original array.

Ask the user to enter the current values corresponding to the five voltages for the new component in milliamperes [mA]. Place these values in a new row of `CurData` at the bottom of the matrix, such that the matrix will now have one more row than the original matrix.

Determine the maximum current value in units of milliamperes [mA] for all components shown in `CurData`. Store the resulting scalar in the variable `MaxCur`.

Determine the row and column location of the maximum current value in the matrix `CurData`. Store the resulting scalars in the variables `R` and `C`.

Print a report similar to the following in the Command Window.

```
Maximum Current = 428 mA
At voltage = 15 V
For Dillard202
```

9. Write a function and program to determine the mass of oxygen gas (formula O_2, molecular weight $= 32$) in a container in units of grams. The function accepts input arguments representing the volume of the container in units of gallons, the temperature in the container in degrees Celsius, and the pressure in the container in units of atmospheres. The function output should be a single formatted string that contains the mass of the oxygen gas formatted to 1 decimal place followed by the units. For example the strings '10.0 grams' or '13.3 grams' would be samples of the expected output of your function. Note that this function should not generate any output on the screen.

The program should prompt the user to input the volume in gallons, the temperature in degrees Celsius, and the pressure in atmospheres, and then call the function to create the formatted string containing the mass. Finally, your program should contain an output statement to display the mass of the oxygen gas in the container. For your test case, you may assume that the user provides 1.25 gallons for the volume of the container, 125 degrees Celsius for the temperature, and 2.5 atmospheres for the pressure in the container.

Sample Input/Output:

```
Enter the volume [gallons]: 1.25
Enter the temperature [deg C]: 125
Enter the pressure [atm]: 2.5
The mass of the oxygen gas in the container is 11.6 grams.
```

10. Write a program to calculate a temperature provided by the user into a different unit system. The program should ask the user to enter a temperature in units of degrees Fahrenheit. The program should then ask the user to choose a unit system they would like to have the value converted into, using a menu. The menu should display the choices "deg C", "K", and "deg R", which should be stored in a cell array. The program should display the output in a sentence formatted as shown below. The actual values in the sentence should change appropriately as the user input and menu selection changes.

Sample Input / Output:

Enter the temperature [deg F]: -129 Menu Choice: **K**
The equivalent temperature to -129 deg F is 184 K.

You are to write this program twice, using two different methods as indicated below. Note that in neither case may you use conditional statements, even if you know how to do so.

(a) The program should first convert the temperature in degrees Fahrenheit to all three other unit systems (degrees Celsius, kelvin, degrees Rankine) and store the results in a vector. It should then generate the report.

(b) The program should perform ONLY the specified calculation, not all three, then generate the report. *Hint:* Store the necessary conversion factors in a matrix.

11. You are part of an engineering firm on contract with the U.S. Department of Energy's Energy Efficiency and Renewable Energy task force to develop a program to help consumers measure the efficiency of their home appliances. Your job is to write a program that measures the efficiency of stove-top burners. Before using your program, the consumer will place a pan of room-temperature water on the stove (with 1 gallon of water), record the initial room temperature in units of degrees Fahrenheit, turn on the burner, and wait for it to boil. When the water begins to boil, the consumer will record the time in units of minutes it takes for the water to boil. Finally, the consumer will look up the power for the burner provided by the manufacturer.

The output of your program should look like the following output, where the highlighted values are example responses typed by the user into your program. Note that your code should line up the energy and power calculations as shown. In addition, your code must display the efficiency as a percentage with one decimal place and must include a percent symbol.

Sample Input/Output:

```
Household Appliance Efficiency Calculator: Stove

Type the initial room temperature of the water [deg F]: 68
Type the time it takes the water to boil [min]: 21
Type the brand name and model of your stove: Krispy 32-Z
Type the power of the stove-top burner [W]: 1200

Energy required:        1267909 J
Power used by burner:   1006 W

Burner efficiency for a Krispy 32-Z stove: 83.9%
```

Stove Model	Room Temp [°F]	Time to Boil [min]	Rated Burner Power [W]
Krispy 32-Z	68	21	1200
MegaCook 3000	71	25	1300
SmolderChef 20F	72	21	1500
Blaze 1400-T	68	26	1400
CharBake 5	69	18	1350

12. We want to conduct an analysis for a wooden baseball bat manufacturer that is interested in diversifying its product line by adding more materials and possibly upgrading the equipment. To help the manufacturer look at the different scenarios, write a program that includes the following:

- Load the information from the MAT file titled `BatCost.mat`. This file contains the following variables:
  - The bats can be produced from four different materials: ash, hickory, maple and pine. The material names are stored as a cell array in the variable `Materials`.
  - The estimated material cost to produce 25 bats from each material is saved in the vector `Cost`. To determine the material cost per bat, the value given in `Cost` will need to be divided by 25.
  - The labor and energy cost to produce one bat is estimated in the scalar variable `LECost`. This cost is the same regardless of the bat material.
- Using a menu, ask the user to select the material for the bat.
- Ask the user to enter, as a scalar, the selling price of a single bat made from the chosen material. The material name chosen by the user must appear in this statement.
- Ask the user to enter, as a vector, the total number of bats the manufacturer can produce per week without an upgrade, and the number of weeks the manufacturer plans to run the bat production machinery in a year; these values are the same regardless of the type of bat produced.

- Ask the user to enter a vector containing the ADDITIONAL number of bats that can be produced in a week if the equipment is upgraded, and the fixed cost for the upgrade.
- The variable cost of production will be the summation of the material cost, labor cost, and energy cost and is the same for a given material regardless of upgrades.

Your program should calculate and display the following information in the Command Window:

```
Selling Price per bat:              $##.##
Total Variable Cost per bat:        $##.##

NO UPGRADE
Producing ## MMM bats a week for ## weeks = ## total bats
        Profit:                     $##.#E##

WITH UPGRADE
Producing ## MMM bats a week for ## weeks = ## total bats
        Fixed Cost of upgrade:      $##.##
        Profit:                     $##.#E##
        Breakeven Point:            #### weeks
```

Formatting notes:

- The "#" character in the output displayed will be actual numbers. The word "**MMM**" should be replaced with the actual material chosen by the user.
- The number of bats per week, number of weeks per year, and total number of bats produced should be shown as integers.
- The selling price per bat, total variable cost per bat, and fixed cost of the upgrade should be displayed with two decimal places.
- The profit should be displayed in exponential notation with one decimal place.
- The lines for profit, fixed cost, and breakeven point should display as indented by a single tab from the left margin.
- There should be one blank line in the locations shown.
- The breakeven point should be shown as an integer and rounded to show two significant figures.

The program should generate a proper plot for the number of weeks bats are produced on the abscissa and the revenue and total cost on the ordinate.

- Cost curves for both scenarios (upgrade and no upgrade) should be shown along with the revenue curves.
- The breakeven point for the upgrade scenario should be clearly indicated on the graph with a solid black vertical line.
- The material name chosen by the user must appear in the title of the graph.
- The legend should appear in the upper-left corner of the graph.

13. You want to create a graph showing the relationship of an ideal gas between pressure (P) and temperature (T).

    Allow the initial temperature to be 270 kelvin. The range of temperatures to be modeled ranges from from 270 to 480 kelvin. Control the scale of the abscissa so the range is shown from 250 to 500 kelvin.

    Model two gases:

    - Nitrogen (formula, N_2; molecular weight, 28 grams per mole), using a 12-liter tank with an initial pressure of 2.5 atmospheres.
    - Oxygen (formula, O_2; molecular weight, 32 grams per mole), using a 15-liter tank with an initial pressure of 4 atmospheres.

    After drawing the graph for nitrogen and oxygen, model a third gas, with information entered by the user. Assume the volume is 12 liters.

    Ask the user to enter the name of the gas [Example: Chlorine]

    Ask the user to enter the initial pressure in the tank measured at 270 kelvin [Example: 3 atm]

Ask the user to enter the temperature of interest [Example: 400 kelvin]. The user is interested in a temperature in the range of 270 to 480 kelvin.

Output the following information to the Command Window:

`At a temperature of TTT kelvin for GGG, the pressure is P.P atm.`

where:

- TTT is the temperature of interest entered by the user;
- GGG is the name of the gas; and
- P.P is the pressure at that temperature shown to one decimal place.

Add the user-entered gas to the graph, showing the pressure over the range of temperatures from 270 to 480 kelvin.

In addition to the single graph with three data series, create an additional figure window with a set of subplots showing each data series as an individual subplot.

Your output should be similar to the following:

In the Command Window:

```
Enter the name of the gas: Chlorine
Enter the initial pressure [atm]: 3
Enter the temp of interest [K]: 400
At a temperature of 400 kelvin, chlorine has a pressure of 4.4 atm.
```

The figures should appear similar to the following. The colors and line types may vary.

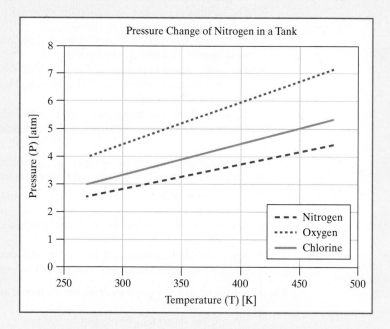

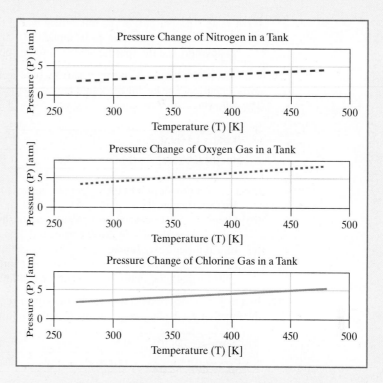

14. When one tries to stop a car, both the reaction time of the driver and the braking time must be considered. Create a proper plot of the following data.

Vehicle Speed (*v*) [mph]	Distance (*d*) [m]	
	Reaction (*d*$_r$)	Braking (*d*$_b$)
20	6	6
30	9	14
40	12	24
50	15	38
60	18	55
70	21	75

15. If an object is heated, the temperature of the body will increase. The energy (*Q*) associated with a change in temperature (ΔT) is a function of the mass of the object (*m*) and the specific heat (C_p). In an experiment, heat is applied to the end of an object, and the temperature change at the other end of the object is recorded. This leads to the theoretical relationship shown. An unknown material is tested in the lab, yielding the following results:

$$\Delta T = \frac{Q}{mC_p}$$

Heat applied (Q) [J]	12	17	25	40	50	58
Temp change (ΔT) [K]	1.50	2.00	3.25	5.00	6.25	7.00

(a) Create a proper plot of the experimental temperature change (ΔT, ordinate) versus the heat applied (*Q*, abscissa).

(b) Use `polyfit` to determine a linear relationship for the data set and graph the resulting trendline along with the experimental data.

(c) Create a proper plot of the theoretical model that represents the temperature change (ΔT, ordinate) versus the heat applied (Q, abscissa). Consider the mass (m) to be 5 kilograms and the specific heat (C_p) to be copper at 0.39 joules per gram kelvin.

16. Capillary action draws liquid up a narrow tube against the force of gravity as a result of surface tension. The height the liquid moves up the tube depends on the radius of the tube. The following data were collected for water in a glass tube in air at sea level.

Radius (r) [cm]	0.01	0.05	0.10	0.20	0.40	0.50
Height (H) [cm]	14.0	3.0	1.5	0.8	0.4	0.2

(a) Create a proper plot of the height (H, ordinate) versus the radius (r) assuming the data are experimental.
(b) Use `polyfit` to determine a power relationship for the data set and graph the resulting trendline along with the experimental data.
(c) In a different experiment with different data, you determine the relationship to be $H = 0.25 \ r^{-1}$. Create a proper plot of this expression on the same graph, assuming the units for both the radius and height are centimeters.

17. In a turbine, a device used for mixing, the power required depends on the size and shape of the impeller. In the lab, we have collected the following data:

Diameter (D) [ft]	0.5	0.75	1	1.5	2	2.25	2.5	2.75
Power (P) [hp]	0.004	0.04	0.13	0.65	3	8	18	22

(a) Create a proper plot of the power (P, ordinate) versus the diameter (D) assuming the data are experimental.
(b) Use `polyfit` to determine a power relationship for the data set and graph the resulting trendline along with the experimental data.
(c) In a different experiment with different data, you determine the relationship to be $P = 0.25 \ D^5$. Create a proper plot of this expression on the same graph, assuming the units for both the diameter and power are the same as shown in the table.

18. A pitot tube is a device that measures the velocity of a fluid, typically the airspeed of an aircraft. The failure of the pitot tube was credited as the cause of Austral Lineas Aéreas flight 2553's crash in October 1997. The pitot tube had frozen, causing the instrument to give a false reading of slowing speed. As a result, the pilots thought the plane was slowing down, so they increased the speed and tried to maintain their altitude by lowering the wing slats. Actually, they were flying at such a high speed that one of the slats ripped off, causing the plane to nosedive; the plane crashed at a speed of 745 miles per hour [mph].

In the pitot tube, as the fluid moves, the velocity creates a pressure difference between the ends of a small tube. The tubes are calibrated to relate the pressure measured to a specific velocity, using the speed as function of the pressure difference (P, in units of pascals) and the density of the fluid (ρ, in units of kilograms per cubic meter).

$$v = \left(\frac{2}{\rho}\right)^{0.5} P^{\mathrm{m}}$$

(a) Create a proper plot of the velocity (v, ordinate) versus the pressure (P) assuming the data are experimental.
(b) Use `polyfit` to determine the power relationships for the data sets and graph the resulting trendlines along with the experimental data.

Pressure (P) [Pa]	50,000	101,325	202,650	250,000	304,000	350,000	405,000	505,000
Velocity fluid A (v) [m/s]	11.25	16.00	23.00	25.00	28.00	30.00	32.00	35.75
Velocity fluid B (v) [m/s]	9.00	12.50	18.00	20.00	22.00	24.00	25.00	28.00
Velocity fluid C (v) [m/s]	7.50	11.00	15.50	17.00	19.00	20.00	22.00	24.50

19. A growing field of inquiry that both holds great promise and poses great risk for humans is nanotechnology, the construction of extremely small machines. Over the past couple of decades, the size that a working gear can be made has consistently gotten smaller. The table shows milestones along this path.

Years from 1967	0	5	7	16	25	31	37
Minimum gear size [mm]	0.8	0.4	0.2	0.09	0.007	2E-04	8E-06

(a) Create a proper plot of the gear size (ordinate) versus the number of years from 1967 assuming the data are experimental.

(b) Use `polyfit` to determine the exponential relationship for the data set and graph the resulting trendline along with the experimental data.

(c) In a different experiment with different data, you determine the relationship to be $MGS = 2.5\ e^{-0.5t}$. Create a proper plot of this expression on the same graph, assuming the units for both the time and gear size are the same between the two data sets.

20. Download the weekly retail gasoline and diesel prices Excel workbook Gasoline.xlsx associated with this review question and place it in your main MATLAB Current Directory. This data set is based on the data set available from the U.S. Energy Information Administration: http://www.eia.gov/dnav/pet/pet_pri_gnd_dcus_nus_w.htm. WARNING: Your system MAY import the first column (dates) as numeric date codes instead of ignoring them as text, so watch out.

(a) Calculate the average, minimum, and maximum retail fuel prices for each of the different types of fuel (regular, midgrade, premium, diesel) over the duration of the entire sample set. Your output should work in such a way that if the original Excel file were modified to include more weeks (rows), you would not need to change your MATLAB code. Your output should appear similar to the following format, where the blanks are replaced by the actual calculated values:

Average weekly retail gasoline and diesel prices

	Regular	Midgrade	Premium	Diesel
Min:				
Max:				
Average:				

Your code should calculate the values shown in the output—you should not hard-code the values in the output. Each value you display should appear with two decimal values. You may use any built-in MATLAB function, including functions that find the minimum, maximum, or average values.

(b) We have decided that we want to modify our previous analysis in part (a) to export the computed max, min, and average values to a new Microsoft Excel workbook. The data itself should be exported to a sheet named Fuel Price Analysis. Your data should appear similar to the worksheet that follows. Much like part (a), your code should be written in such a way that if the original Excel file were modified to include more weeks (rows), you would not need to change your MATLAB code.

	A	B	C	D	E	F
1	Average Weekly Retail Gasoline and Diesel Prices					
2		Regular	Midgrade	Premium	Diesel	
3	Min					
4	Max					
5	Average					
6						

21. A sample of the data provided in the Microsoft Excel file EnergyData.xlsx online follows. The file contains energy consumption data by energy source per year in the United States, measured in petaBTUs.

Year	Fossil Fuels	Elec. Net Imports	Nuclear	Renewable
2007	101.605	0.106	8.415	6.830
2006	99.861	0.063	8.214	6.922

(a) Write the MATLAB code necessary to read the Microsoft Excel file and store each column of data into different variables. Create the following:

- `Yr`: A vector of all of the years in the worksheet.
- `FF`: A vector of all of the fossil fuels for each year in `Yr`.
- `ENI`: A vector of all electric imports for each year in `Yr`.
- `Nuc`: A vector of all nuclear energy consumption for each year in `Yr`.
- `Ren`: A vector of all renewable energy consumption for each year in `Yr`.
- `Hdr`: A cell array of all of the headers in row 1.

You may not hard-code these variables—they should be imported from the Excel file.

(b) Create a new variable, **TotalConsumption**, which contains the sum of the four columns of energy consumption data for each year. In other words, since all five of the vectors (**Yr, FF, ENI, Nuc,** and **Ren**) we created in part (a) have the same length, the new variable, **TotalConsumption**, should have the same length. You may assume that you have correctly defined the variables in part (a).

(c) Calculate the average fossil fuel consumption in the entire data set. For this code, you may assume that you have correctly defined **FF**, the variable containing the fossil fuel consumption data, in part (a). Display the result of the calculation in the following format, where the number is shown to two decimal places:

```
The average fossil fuel consumption is _____ petaBTU.
```

22. An Excel file named DartTosses.xlsx has one worksheet named Darts containing the horizontal and vertical distance from the bull's-eye of a dart board from 20 different tosses of a dart. A portion of the data is shown below. Write the MATLAB code necessary to determine the darts that were the closest and the farthest from the bull's-eye. The program should tell the user in the Command Window, using formatted output, the darts that are closest and farthest from the bull's-eye.

Dart	X	Y
Dart 1	4.04	0.55
Dart 2	2.63	0.35
Dart 3	1.10	2.97
Dart 4	4.89	5.60

Logic and Conditionals

To this point, all steps in our algorithms, and all lines of code written to implement those algorithms, have been executed exactly once, following the sequence in which they were written. We never skipped a step or chose to do one step instead of another. That is about to change.

Until you can get the computer to ask questions and do different things based on the answers, the utility of computer programming is quite limited.

Outside the realm of computing, logic exists as a driving force for decision making. Logic transforms a list of arguments into outcomes based on a decision. Some examples of everyday decision making follow:

- If the milk has passed the expiration date, throw it out; otherwise, keep the milk.

  Argument: expiration date

  Decision: before or after?

  Outcomes: garbage, keep

- If the traffic light is red, stop. If the traffic light is green, go. If the traffic light is yellow, decide if you can safely stop before reaching the intersection.

  Argument: three traffic light colors

  Decision: which bulb lit?

  Outcomes: stop, go, ask can I stop safely and act accordingly

To bring decision making into our perspective on problem solving, we need to first understand how computers make decisions. **Boolean logic** exists to assist in the decision-making process, where each argument has a binary result and the overall outcome exhibits binary behavior. **Binary behavior**, depending on the application, is any sort of behavior that results in two and only two possible outcomes. Boolean logic, Boolean algebra, and related terms are named for George Boole (1815–1864), the English mathematician whose writings formed the basis of modern computer science.

18.1 Algorithms Revisited—Representing Decisions

Decision making can be expressed in a written algorithm. Assume you are designing a process to determine if the value read from a temperature sensor in a vehicle indicates it is unsafe for operation. To express the decision in a written algorithm, you should phrase your decisions in terms of questions that have a yes or no response.

When an algorithm requires that different actions be taken based on the answer to a question, it is helpful to indent the actions that might be taken relative to the question. This visually indicates that these actions are only associated with a particular answer to the question.

EXAMPLE   18-1

Create a written algorithm to express a temperature given in relative units [°F or °C] in the corresponding absolute units [K or °R].

Known:

- *Temperature in relative units (degrees Celsius or degrees Fahrenheit).*

Unknown:

- *Temperature in absolute units (kelvins or degrees Rankine).*

Assumptions:

- *Since the problem does not explicitly state the temperature of interest, assume that the interpreter of the algorithm will input the temperature and units.*

Algorithm:

1. *Input the numeric value of the temperature.*
2. *Input the units of the numeric value of the temperature.*
3. *Ask if the input unit is degrees Fahrenheit.*
   (a) *If yes, calculate the value in degrees Rankine.*
   (b) *If no, calculate the value in kelvins.*
4 *Display the new value and absolute unit.*
5. *End the process.*

Note that only one of the actions (a or b) under the question in item 3 will be completed depending on the answer.

Sometimes an action that is to be taken only if a question has a specific answer requires that another question be asked.

EXAMPLE   18-2

Create a written algorithm to express the decision process a person goes through when approaching a stoplight.

Known:

- *Status of stoplight, distance from light, speed of vehicle*

Unknown:

- *Vehicle motion: Stop or continue*

Assumptions:

- *This is not an emergency situation requiring violation of traffic laws*

Algorithm:

1. *Determine color of light.*
2. *Is the light red?*

    (A) *If yes, stop before reaching intersection.*
    (B) *If no, is the light yellow?*

        (i) *If yes, can I safely stop before the intersection?*

            (a) *If yes, stop.*
            (b) *If no, continue through intersection with caution.*

        (ii) *If no, continue through intersection.*

3. *End the process.*

Note that instead of asking a single question, "What color is the light?" this was broken down into questions that could be answered with yes or no. The reasons for this will become clearer as we proceed.

Also note that each time a question was asked, the actions to be taken based on the two possible answers were indented. We cannot overemphasize the importance of maintaining proper indentation, not only in algorithms, but also in the code that implements those algorithms.

Comprehension Check   **18-1**

Create a written algorithm that determines the dose of a medicine to be given based on whether the person is a child (≤ 14 years old) or an adult. Children get one pill, adults get two.

Comprehension Check   **18-2**

Create a written algorithm that determines the state of water given the temperature in °C.

To include decision making in flowcharts, we will introduce another symbol, the decision diamond. All decisions in a flowchart must be represented within a diamond shape.

There must be at least one arrow entering the decision diamond, although there may be more than one if different paths in the algorithm lead to the same question. The question itself is written inside the diamond. The decision diamond represents a binary question and thus has two arrows that leave it, one for each possible answer, typically labeled either Yes and No or True and False. If the answer to the question is yes, the algorithm follows the arrow labeled Yes, and similarly for No. By convention, arrows enter and leave decision diamonds at the vertices as shown. It is assumed that any variables defined in the stepwise scope of a decision diamond are accessible and can be used in the decision. Since no new

variables are created in a decision diamond, the stepwise scope that enters the decision diamond is passed on to the next shape of each conditional branch.

EXAMPLE 18-3

Create a flowchart to express a temperature given in relative units [°F or °C] in the corresponding absolute units [K or °R].

Known:

- *Temperature in relative units (degrees Celsius or degrees Fahrenheit).*

Unknown:

- *Temperature in absolute units (kelvins or degrees Rankine).*

Assumptions:

- *Since the problem does not explicitly state the temperature to be determined, the interpreter of the algorithm will ask for the temperature and units.*

Flowchart:

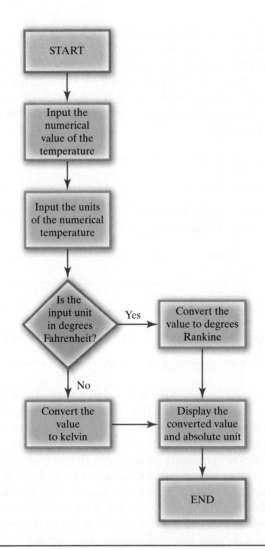

EXAMPLE   18-4

Create a flowchart to express the decision process a person goes through approaching a stoplight.

Known:
- *Status of stoplight, distance from light, speed of vehicle*

Unknown:
- *Vehicle motion: Stop or continue*

Assumptions:
- *This is not an emergency situation requiring violation of traffic laws*

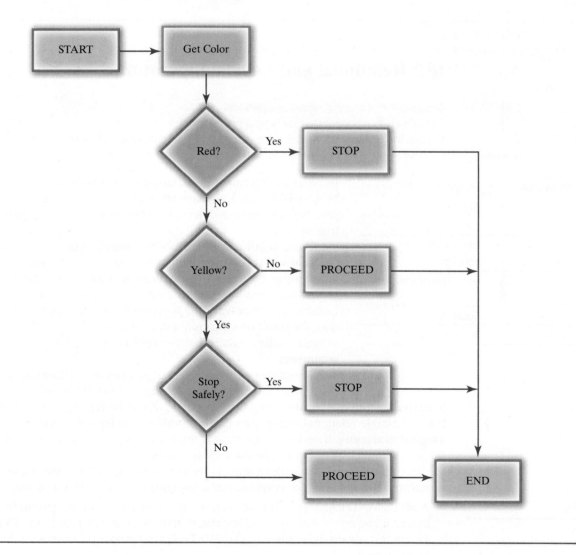

| **Comprehension Check** | **18-3** |

Create a flowchart that determines the dose of a medicine to be given based on whether the person is a child ($\leq$ 14 years old) or an adult. Children get one pill, adults get two.

| **Comprehension Check** | **18-4** |

Create a flowchart that determines the state of water given the temperature in °C.

18.2 Relational and Logical Operators

LEARN TO: Create decisions using relational operators
Create decisions connected by logical operators
Explain the difference between short-circuit and element-wise operators

Operator	Meaning
>	Greater than
<	Less than
>=	Greater than or equal to
<=	Less than or equal to
==	Equal to
~=	Not equal to

In order to determine the relationship between two values (numbers, variables, etc.), we have six operators with which we can compare two variables to determine whether or not the comparison is true or false.

These **relational operators** are usually placed between two different variables or mathematical expressions to determine the relationship between the two values. Examples would include `Var1<=17` and `4/3*pi*r^2>MaxVol`. This type of expression, variable–operator–variable is typically called a **relational expression**, the result of which is always either true or false. We will refer to such results—values that can only be true or false—as **Boolean arguments**.

Sometimes all we care about is a simple relationship between two things, for example, is A greater than B? In other cases, we need to ask more complicated questions; for example, is A both greater than B and less than C? Simple relational expressions can be combined by **logical operators** to create a **logical expression**. If no logical operator is required in a particular decision, then the single relational expression can be the logical expression.

To connect all of the Boolean arguments to make a logical decision, we use three operators with which we can combine our arguments to determine a final outcome.

NOTE

And: &&
Or: ||
Not: ~

- **And:** The AND logical operator connects two Boolean arguments and returns the result as true if and only if *both* Boolean arguments have the value of TRUE. In MATLAB, we use two ampersand symbols (`&&`) (SHIFT+7 on a keyboard) to represent the AND logical operator.

- **Or:** The OR logical operator connects two Boolean arguments and returns the result as true if *either one* (or both) of the Boolean arguments has the value of TRUE. In MATLAB, we use two pipe symbols (`||`) (SHIFT+\ on a keyboard) to represent the OR logical operator.

- **Not:** The NOT logical operator inverts the value of a single Boolean argument or the result of another Boolean operation. In MATLAB, we use the tilde (~) symbol (SHIFT+' on a keyboard) to represent the NOT logical operator. In other words, NOT True equals False, and vice versa.

EXAMPLE 18-5

Express the mathematical inequality in MATLAB code.

Mathematical Inequality	MATLAB Expression
$4 \leq X < 5$	`4<=X && X<5`
$10 < X \leq 20$	`10<X && X<=20`
$30 \leq X \leq 100$	`30<=X && X<=100`

In this example and others, MATLAB cannot operate with multiple relational operators in a single relational expression—a common situation in mathematical inequalities. Instead, we must separate out the two relations and combine the expressions with the And symbol.

Comprehension Check 18-5

What are the relational expressions for the following mathematical inequalities?

(a) $5 \leq t < 10$
(b) $-30 < M \leq 20$
(c) $Y \neq 100$

Express the mathematical inequality in MATLAB code.

Mathematical Inequality	MATLAB Expression
$X \leq 0$ or $X > 1000$	`X<=0 || X>1000`
X not between 0 and 1	`X<=0 || X>=1`
X not between 0 and 1	`~(X>0 && X<1)`

Be sure you understand the difference between the AND and OR logical operators. Also, note the two different solutions for X not between 0 and 1, and be sure you understand why both are correct. This pair of solutions illustrates DeMorgan's theorem, which many of you will study in other courses.

Comprehension Check 18-6

What are the relational expressions for the following mathematical inequalities?

(a) $Q < 17$ or $Q >= 42$
(b) Z not between -1 and $+1$ (give two different solutions)

(&&, &, and) + (||, |, or):   Say What?

By design, MATLAB contains multiple operators and functions for performing the `and` and `or` operations. The following set of guidelines will help you choose which operator will be the most appropriate choice for your code. In all of the following, assume the variable x contains the string `'robot'`.

And

- **&&:** The double-ampersand operator is often referred to in MATLAB documentation as a short-circuit AND operator because it will terminate the evaluation of a logical expression as soon as it encounters the first "false" in a logical expression. This is particularly useful when checking to see if a certain condition is false before continuing with a special computation that is part of the logical expression.

  Sample: `isnumeric(x) && x + 10 > 5`

In the preceding sample, the overall result will be false because the first computation (`isnumeric(x)`) was false, so the remainder of the expression was not evaluated.

- **&:** The single-ampersand operator is often referred to in MATLAB documentation as an element-wise AND operator because it will evaluate every element in a logical expression.

  Sample: `isnumeric (x) & x + 10 > 5`

In the preceding sample, the overall result will be an array of five false values. The reason the result is an array is because the first computation (`isnumeric(x)`) was false, so the expression will always be false, because FALSE & ANYTHING = FALSE. In this particular case, the value of x is a string, so the logical expression is computing the ASCII value of the letters r, o, b, o, and t, adding 10, checking to see if the result is greater than 5, and then using the AND result, which will be false, all five times.

- **and():** The `and` function behaves similarly to the AND function in Microsoft Excel. This function behaves in the same fashion as the single-ampersand element-wise operator.

  Sample: `and(isnumeric (x), x + 10 > 5)`

In the sample above, the overall result will be an array of five false values since the `and` function behaves identically to the element-wise operator.

Or

- **||:** The double-pipe operator is often referred to in MATLAB documentation as a short-circuit OR operator because it will terminate the evaluation of a logical expression as soon as it encounters the first True in a logical expression. This is particularly useful when checking to see if a certain condition is true before continuing with a special computation.

  Sample: `~isnumeric(x)||x + 5 > 0`

In the sample above, the overall result will be true because the first computation `~isnumeric(x)` was true, so the remainder of the expression was not evaluated.

- **|:** The single-pipe operator is often referred to in MATLAB documentation as an element-wise OR operator because it will evaluate every element in a logical expression.

Sample: `~isnumeric(x)|x + 5 > 0`

In the sample above, the overall result will be an array of five true values. The reason the result is an array is because the first computation (`~isnumeric(x)`) was true, and since the remainder of the expression (`x + 5 > 0`) will also be true for all five characters in the string `'robot'`, the result is an array of five true outcomes.

- **or():** The `or` function behaves similarly to the OR function in Microsoft Excel. This function behaves in the same fashion as the single-ampersand element-wise operator.

Sample: `or(~isnumeric(x),x+5>0)`

In the sample above, the overall result will be an array of five true values since the `or` function behaves identically to the element-wise operator.

IMPORTANT CONCEPT

The `&&` and `||` operators only work if the logical values being combined are logical SCALARS. If you wish an element-wise AND or OR of two matrices, you must use either `&` or `|` or use the `and`/`or` functions.

18.3 Logical Variables ☑

LEARN TO: Create and use logic variables in decisions
Define a logical matrix
Design logic within the constraints of operator priority

Logical Scalar Variables

When two numeric scalars or two single characters are compared using relational operators, the result is either a 1, representing true; or 0, representing false. In many cases, the results of such operations are used immediately by a MATLAB command and are not actually stored in a variable.

The results of such comparisons can, however, be assigned to a variable and used later. Although the `if` statement will not be introduced until the next section, the following example should be simple enough to understand.

EXAMPLE   18-6

Assume that A and B are scalars previously defined in the workspace. We wish to write a segment of MATLAB code to determine if variable A and variable B have different values.

```
% Option 1
if A ~= B
    fprintf('A and B are not equal')
end
```

The expression A `~=` B *is evaluated, giving a 1 if true or a 0 if false. This result is not stored in a variable but is immediately used by the* `if` *statement to decide whether or not to print the message. If we wanted to know what the result of the comparison was later, we would have to ask the question again, although A or B might have changed in the interim, so we might get a different result.*

```
% Option 2
A_B_Dif=A~=B; % A_B_Dif is a logical variable
```

```
if A_B_Dif
    fprintf('A and B are not equal')
end
```

The result of the relational question is stored in the logical variable `A_B_Dif` *and will have a value of 0 or 1. The icon shown beside* `A_B_Dif` *in the workspace will be a square with a checkmark, indicating it is a logical variable.*

The program will determine if the variable `A_B_Dif` *contains a 1 (true), and if so, then the message will be printed; if* `A_B_Dif` *contains a 0 (false), the message will not be printed. Note that we still have access to the result of the question later since we stored the result in* `A_B_Dif`*. We could, for example, use this in conjunction with asking the same question later to see if the status of the relationship between* `A` *and* `B` *had changed.*

Logical Matrices

Relational comparisons can be made between a matrix and a scalar. In this case, the result is a logical matrix with the same dimensions as the matrix being compared. Each element of this logical matrix contains the result (1 or 0) of the comparison between the scalar and the corresponding element of the matrix.

EXAMPLE   18-7

Determine the result of the following code, assuming $C = \begin{bmatrix} 1 & 0 & -1 \\ -2 & 2 & 3 \end{bmatrix}$

$$C0 = C <= 0;$$

The code will examine each element in the matrix and compare the value with zero. If the value is less than or equal to zero, the number 1 will be stored in the new `C0` *matrix in the same location as the element being compared. If the value is greater than zero, the number 0 will be stored. For example, if we compare* `C(1,1)` *– a value of 1 – to zero, this value is greater than zero, so the number 0 is stored in* `C0(1,1)`*.*

The resulting matrix will be: $C0 = \begin{bmatrix} 0 & 1 & 1 \\ 1 & 0 & 0 \end{bmatrix}$

Relational comparisons can also be made between two matrices, as long as they have the same dimensions. In this case, the result is a logical matrix with the same dimensions as the two matrices being compared. Each element of this logical matrix contains the result (1 or 0) of the element-wise comparisons between corresponding elements of the two matrices.

EXAMPLE   18-8

Determine the result of the following code, assuming

$$C = \begin{bmatrix} 1 & 0 & -1 \\ -2 & 2 & 3 \end{bmatrix} \text{ and } D = \begin{bmatrix} 2 & -2 & -1 \\ -1 & 2 & 1 \end{bmatrix}$$

$$CD = C <= D;$$

The code will examine each element in the matrix C *and compare the value to the same element position in matrix* D. *If the value in* C *is less than or equal the value in* D, *the number 1 will be stored in the new* CD *matrix, in the same location as the elements being compared. If the value in* C *is greater than the value in* D, *the number 0 will be stored.*

For example, if we compare the element C(1,1) *to* D(1,1), *the value of 1 is less than the value of 2, so the number 1 is stored in* CD(1,1).

The resulting matrix will be: $CD = \begin{bmatrix} 1 & 0 & 1 \\ 1 & 1 & 0 \end{bmatrix}$

Comprehension Check 18-7

Assume the following variables have been defined:

```
S = 5            V = [2 -4 6 -3]
```

$$M1 = \begin{bmatrix} 4 & -5 & 1 & 7 \\ -2 & 9 & 8 & -4 \end{bmatrix} \quad M2 = \begin{bmatrix} 0 & -2 & 4 & -7 \\ -2 & 4 & 1 & -1 \end{bmatrix}$$

What is placed in the variable *R* by each of the following statements? If an error will occur, explain why.

(a) R = V < S/2;
(b) R = V <= M1;
(c) R = M1 > M2;
(d) R = M1 < M2 || M1 > S;

NOTE

To determine if strings of possibly different lengths are the same, you can use the functions strcmp or strcmpi. In this case, the result is a logical scalar—true or false. We will discuss these functions in detail shortly.

Relational comparisons can be used with text strings as long as the strings have the same number of characters. Note that case matters: the lowercase letter "a" is not equal to the uppercase letter "A." Since text strings are essentially vectors containing text instead of numbers, the result of such a comparison is a logical vector containing the same number of elements as each of the two strings being compared.

Strings are compared based on the ASCII code used to represent them, so you need to know the codes assigned to the various characters, including punctuation, to understand the results of greater than or less than comparisons. The following table shows each ASCII character and the corresponding decimal value for each letter or number. Note that there are many more characters available in the full set of ASCII values (http://www.asciitable.com/). A comprehensive list can be found in the end-pages of this text.

Character	A	B	C	D	E	F	G	H	I	J	K	L	M
Value	65	66	67	68	69	70	71	72	73	74	75	76	77

Character	N	O	P	Q	R	S	T	U	V	W	X	Y	Z
Value	78	79	80	81	82	83	84	85	86	87	88	89	90

Character	a	b	c	d	e	f	g	h	i	j	k	l	m
Value	97	98	99	100	101	102	103	104	105	106	107	108	109

Character	n	o	p	q	r	s	t	u	v	w	x	y	z
Value	110	111	112	113	114	115	116	117	118	119	120	121	122

Character	0	1	2	3	4	5	6	7	8	9
Value	48	49	50	51	52	53	54	55	56	57

NOTE

You may have noticed that when variables are alphabetized in the workspace, those that begin with a capital letter come before those that begin with a lowercase letter.

If a comparison between two text strings involves greater than or less than (as opposed to just equal or not equal) the results may be surprising. For example `'a' < 'A'` is false, very likely not what you expected. On the other hand, alphabetic characters of the same case do compare as you probably expect: a is "less than" b and Y is "less than" Z. Also, non-alphanumeric characters can be problematic with greater than or less than.

EXAMPLE   18-9

Determine the result of the following code, assuming

```
T='AbCd 1; 2#3';        S='abCD 3:2, 1';
                TS=T<=S;
```

The code will examine each element in the vector `T` *and compare the value to the same element position in vector* `S`. *If the ASCII code value in* `T` *is less than or equal the ASCII code value in* `S`, *the number 1 will be stored in the new* `TS` *matrix, in the same location as the elements being compared. If the ASCII code value in* `T` *is greater than the ASCII code value in* `S`, *the number 0 will be stored.*

For example, if we compare the element `T(1)` *to* `S(1)`, *the text "A" has an ASCII code value of 65 and the text "a" has an ASCII code value of 97. Since 65 is less than 97, the number 1 is stored in* `TS(1)`.

The resulting matrix will be: `TS = [1 1 1 0 1 1 0 1 0 0 0]`

Note that in order to determine this manually, we would need to know the ASCII codes for the space character (32) as well as the various punctuation symbols that appear in the strings.

Comprehension Check   18-8

Assume `TS1` and `TS2` are defined by

```
TS1='AarDVArk 42' and TS2='AArdWoLf 33'
```

(a) Write a MATLAB command that will determine which elements in `TS1` are "greater than or equal to" the corresponding elements in `TS2` and place the results in `TSComp`.

(b) What is placed in `TSComp` by the statement you wrote in part (a)?

Calculations Using Logical Variables

Logical matrices allow you to do some moderately complicated operations very easily in MATLAB. Since logical matrices contain actual numeric values (although only zeros and ones), they can be used in computations with regular numeric matrices.

EXAMPLE   18-10

Assume $Q = \begin{bmatrix} 1 & -7 & 9 & -12 \\ -24 & 10 & 6 & 100 \end{bmatrix}$. Write a code segment to modify Q by dividing by 2 every element with a magnitude greater than or equal to 10 and leaving the other elements unchanged.

We can approach this problem by creating two matrices: one containing a value of 1 for all the elements with a magnitude greater than or equal to 10 [BigQ], and one containing a value of 1 for all the elements less than 10 [SmallQ]. Then, we can recombine these two matrices for our end result.

```
BigQ=abs(Q) >= 10;            % Find elements with magnitude ≥ 10
BigQAdjusted =0.5*BigQ.*Q;    % Divide big elements by 2. Others=0
SmallQ =~BigQ.*Q;             % Make big elements from Q = 0
Q = BigQAdjusted+SmallQ;      % Combine the two matrices
```

The result: $Q = \begin{bmatrix} 1 & -7 & 9 & -6 \\ -12 & 5 & 6 & 50 \end{bmatrix}$

Note that this will work with any size matrix: for a 1000 × 2000 matrix, the code would be the same.

Comprehension Check   18-9

IMPORTANT CONCEPT

If an arithmetic operation, such as dividing by 2 or adding 6, is performed on a logical matrix, the result is a numeric matrix.

Write a short section of MATLAB code that will replace all values in matrix Mat1 that are less than the value stored in scalar variable MIN with the value −9999. If you are clever, you can do this with a single statement.

Example:

BEFORE: $Mat1 = \begin{bmatrix} -54143 & 34 & 7 & 1950 & -2 \\ -777 & 0 & 9 & 5 & 42 \end{bmatrix}$ MIN = 7

AFTER: $Mat1 = \begin{bmatrix} -9999 & 34 & 7 & 1950 & -9999 \\ -9999 & -9999 & 9 & -9999 & 42 \end{bmatrix}$

Priority of Operators

The priority of operators must be observed carefully when mixing arithmetic, relational, and logical operators in the same expression. Particularly note that the arithmetic operators have priority over the relational operators, and the relational operators have priority over the logical operators. The one exception is **logical negation** ($\sim$ or not (A)), which has a priority equal to unary minus. A unary minus ($-x$) replaces the variable with the additive inverse; for example if $x = 3$, then $-x = -3$. The order of operator priorities is summarized in Table 18-1. Remember that parentheses have priority over everything.

Table 18-1 Priority of mixed operations

Priority	Operations
1	Transpose, power
2	Unary minus, logical negation
3	Multiplication, division
4	Addition, subtraction
5	Colon operator
6	Relational operators (`<`, `<=`, `>`, `>=`, `~=`, `==`)
7	`&`
8	`|`
9	`&&`
10	`||`

Comprehension Check   **18-10**

$$M1 = \begin{bmatrix} 6 & -3 \\ -7 & 0 \\ 2 & 10 \end{bmatrix} \text{ and } M2 = \begin{bmatrix} 9 & -5 \\ -4 & 1 \\ 2 & 0 \end{bmatrix}$$

(a) What is placed in R1 by the statement R1 = M1 <= M2?
(b) What is placed in R2 by the statement R2 = (M2 < M1).*M2?
(c) What is placed in R3 by the statement R3 = M2 < M1.*M2?

Comprehension Check   **18-11**

Assume M1 and M2 are defined as given in Comprehension Check 18-10. Write a short section of MATLAB code that will create a matrix R4 in which each element is the larger of the corresponding elements in M1 and M2.

For the values listed above, the result should be: $R4 = \begin{bmatrix} 9 & -3 \\ -4 & 1 \\ 2 & 10 \end{bmatrix}$

Functions Associated with Logical Variables

Many of the functions used with numeric matrices can also be used with logical matrices. Table 18-2 contains a few additional functions that you might find useful. For each function, examples are given using the following variables:

```
A = [1 3 5 7];   C = [2 4 5 7];   R = A~= B;     T = B == C;
B = [2 4 6 8];   D = [2 3 5 8];   S = A+1~= B;   U = C == D;
Canine1 = 'Dog';    Canine2 = 'Wolf';
```

Table 18-2 Selected functions used with logical variables

Function	Description
all	Are all elements in a given dimension of a matrix 1 (or nonzero)?
	Examples: H = all(R); % H contains 1
	I = all(T); % I contains 0
any	Are any elements in a given dimension of a matrix 1 (or nonzero)?
	Examples: J = any(S); % J contains 0
	K = any(T); % K contains 1
isequal	Determine if a group of matrices are all identical.
	Examples: F = isequal(B,C); % F contains 0
	G = isequal(A,B-1); % G contains 1
islogical	Determine if a variable is a logical variable.
	Examples: D = islogical(A); % D contains 0
	E = islogical(R); % E contains 1
strcmp	Case-sensitive string comparison; examples:
	CType1 = strcmp(Canine1,Canine2); % CType1 contains 0
	CType2 = strcmp(Canine1,'Dog'); % CType2 contains 1
	CType3 = strcmp(Canine1,'dog'); % CType3 contains 0
strcmpi	Case-insensitive string comparison; examples:
	CType4 = strcmpi(Canine1,Canine2); % CType4 contains 0
	CType5 = strcmpi(Canine1,'Dog'); % CType5 contains 1
	CType6 = strcmpi(Canine1,'dog'); % CType6 contains 1
xor	Exclusive OR: True if there is one 1 and one zero in corresponding elements.
	Examples: L = xor(T,U); % L contains [0 1 1 0]

EXAMPLE   **18-11**

For each statement shown, what would the user have to type for Q to contain the value `true`?

(a) `Q = strcmp('No',input('Please type Yes or No: '));`

The user would have to type No.

(b) `Q = strcmpi('No',input('Please type Yes or No: '));`

The user would have to type EITHER no, nO, No, or NO.

Comprehension Check | **18-12**

Assume a user has responded to the question "Do you want to continue (Yes or No)?" and the response has been stored in `UserResp`.

    If the user typed Yes, what is placed in `Cont` by each of the following statements? If an error is generated, write "error."

(a) `Cont = UserResp=='Yes';`
(b) `Cont = UserResp=='YES';`
(c) `Cont = UserResp=='No';`
(d) `Cont = strcmp('Yes',UserResp);`
(e) `Cont = strcmp('yes',UserResp);`
(f) `Cont = strcmp('No',UserResp);`
(g) `Cont = strcmpi('YES',UserResp);`

Comprehension Check | **18-13**

Assume a user has responded to the question "Do you want to continue (Yes or No)?" and the response has been stored in `UserResp`.

    Write a short section of MATLAB code (a single line if possible) that will return a value of TRUE in the variable `Cont` if the user either types the word "yes" (with any capitalization) or the user types the single letter "y" (upper or lowercase).

18.4 Conditional Statements in MATLAB

LEARN TO: Predict the number of possible outcomes given variability in a condition
                    Determine if two conditional statements are logically equivalent

Conditional statements are commands by which programmers give decision-making ability to the computer. Specifically, the programmer asks the computer a question framed in conditional statements, and the computer selects a path forward based on the answer to the question. Sample questions follow:

- If the water velocity is fast enough, switch to an equation for turbulent flow!
- If the temperature is high enough, reduce the allowable stress on this steel beam!
- If the pressure reading rises above the red line, issue a warning!
- If your grade is high enough on the test, state: You Passed!

    In these examples, the comma indicates the separation of the condition and the action that is to be taken if the condition is true. The exclamation point marks the end of the statement. Just as in language, more complex conditional statements can be crafted with the use of "else" and similar words. In these statements, the use of a semicolon introduces a new conditional clause. For example:

- If the collected data indicate the process is in control, continue taking data; otherwise, alert the operator.

- If the water temperature is at or less than 10 degrees Celsius, turn on the heater; or else if the water temperature is at or greater than 80 degrees Celsius, turn on the chiller; otherwise, take no action.

In the second example, the second conditional clause involves asking another question. This is known as a nested conditional statement.

Single Conditional Statements

In MATLAB, a single conditional statement involves the `if` and `end` commands. If two distinct actions must occur as a result of a condition, the programmer can use the `else` command to separate the two actions. The basic structure of a single `if` statement is as follows:

```
if Logical_Expression
        Actions if true
else
        Actions if false
end
```

Note that the *Logical_Expression*, as shown in the structure of a single `if` statement, is a logical expression as described previously.

EXAMPLE   18-12

Write a MATLAB statement to represent the following conditional statement.

If the water velocity (`v`) is less than 10 meters per second, determine the friction factor (`f`) of the piping system using laminar flow given by 64 divided by the Reynolds number (`Re`).

English (Pseudocode)	MATLAB Code
If the water velocity is slow enough,	`if v < 10`
use an equation for laminar flow	`    f = 64/Re;`
(period)	`end`

In this example, the comma indicates the separation of the condition and the action that is to be taken if the condition is true. The period marks the end of the statement.

EXAMPLE   18-13

Write a MATLAB statement to represent the following conditional statement.

If the speed of the vehicle is greater than 65 miles per hour, display "Speeding" to the Command Window, otherwise; display "OK."

English (Pseudocode)	MATLAB Code
If the speed is greater than 65,	`if Speed > 65`
display "Speeding";	`    fprintf('Speeding')`
Otherwise,	`else`
display "OK"	`    fprintf('OK')`
(period)	`end`

In this example, the comma indicates the separation of the condition and the action that is to be taken if the condition is true. The semicolon after "Speeding" marks the end of the true action and the beginning of the action to take if the answer is false. The period marks the end of the statement.

Comprehension Check   **18-14**

Write MATLAB code to represent the following conditional statements.

(a) If your time running a marathon (stored in `MTime`) is less than your previous fastest time (stored in `PersBest`), display the message "New personal best" and replace the value in `PersBest` with the value in `MTime`.

(b) If the scales indicate that the truck weighs more than its legal maximum weight, display the message "Reduce the load in the truck"; otherwise, display "Load acceptable."

Nested Conditional Statements

If more than two outcomes exist, the program can use an `if-elseif-else` command structure. This is similar to using the nested IF statements in Excel. Each time the word "`if`" appears, whether alone or as part of the `elseif` command, the program must ask a true/false question.

As a minimum, `if` statements can stand alone, without an `else` as seen in Example 18-12. As a maximum, the program requires one less `if` statement than the number of outcomes. For example, if you have four desired outcomes, the program will require three IF questions (`if-elseif-elseif`) and one else statement to determine the result.

EXAMPLE   **18-14**

Write a code segment to generate the menu shown and display the choice of the user in a sentence like: "You bought a Coke." Assume the result of clicking an option in the menu is saved in a variable called `soda` in the workspace. You can assume the user does not close the menu, but makes a valid selection from the choices listed.

Recall from our discussion of the menu function, the result from a menu selection by the user is the ordinal value of the choice. For example, if the user selects "Coke" from the menu shown, the variable soda would store the value 2.

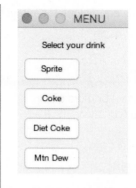

To decipher the choices into text, we can use an IF-ELSEIF-ELSE sequence:

```
soda = menu ('Select your drink','Sprite','Coke',
                                  'Diet Coke','Mtn Dew');
if soda == 1
        fprintf('You bought a Sprite\n');
elseif soda == 2
        fprintf('You bought a Coke\n');
elseif soda == 3
        fprintf('You bought a Diet Coke\n');
else
        fprintf('You bought a Mtn Dew\n');
end
```

Comprehension Check   18-15

Write MATLAB code to represent the following conditional statement.

If the water temperature (stored in variable TW) is at or less than zero degrees Celsius, the phase (stored in the text variable phase) is solid; otherwise, if the water temperature is at or greater than 100 degrees Celsius, the phase is vapor; otherwise, the phase is liquid.

Comprehension Check   18-16

Write a MATLAB function named SumItUp that will accept three input variables and return two output variables. If the sum of the first two input variables is greater than 100, the function should display the inputs showing two decimal places in the following format. Assume in this formatting example that the numbers 55, 66, and 77 are the input variables:

The inputs are 55.00, 66.00 and 77.00.

Otherwise, if the sum of the first two input variables is less than or equal to 100, the function should display the formatted output using no decimal places, for example:

The inputs are 25, 10 and 13.

The first variable the function returns should be the sum of the first and second variables. The second variable the function returns should be the sum of the second and third variables.

Equivalent Forms of Logic

Conditional statements are similar to traditional language in another way—there are many ways to express the same concept. The order, hierarchy (nesting), and choice of operators are flexible enough to allow the logic statement to be expressed in the way that makes the most sense. All of the following logic statements are equivalent for integer values of NumTeamMembers, but all read differently in English:

Conditional Statement	English Translation
| ```
if NumTeamMembers>=4
    if NumTeamMembers<=5
        fprintf('Team size OK.')
    end
end
``` | If the number of team members is 4 or more and if the number of team members is 5 or fewer, the team size is okay. |
| ```
if NumTeamMembers>=4 &&NumTeamMembers<=5
    fprintf('Team size OK.')
end
``` | If the number of team members is 4 or more and 5 or fewer, the team size is okay. |
| ```
If NumTeamMembers==4 ||NumTeamMembers==5
    fprintf('Team size OK.')
end
``` | If the number of team members is 4 or 5 the team size is okay. |

Comprehension Check   18-17

Which of the following expressions are equivalent to those shown above to determine if the status of NumTeamMembers is acceptable or needs to be adjusted?

(a)
```
if NumTeamMembers < 4 || NumTeamMembers ~= 5
    fprintf('Adjust team size.')
end
```

(b)
```
if NumTeamMembers >= 4 || NumTeamMembers <= 5
    fprintf('Team size OK.')
end
```

(c)
```
if NumTeamMembers ~= 4 && NumTeamMembers ~= 5
    fprintf('Adjust team size.')
end
```

(d)
```
if NumTeamMembers == 4 && NumTeamMembers == 5
    fprintf('Team size OK.')
end
```

18.5 Application: Classification Diagrams

During our study of Excel, we looked at classifying the phase of a material based on a phase diagram. A phase diagram is a specific type of a more general kind of graph that we will refer to as a classification diagram. Just as a phase diagram depicts the
```

boundaries between different phases of some substance based on values of two criteria, such as temperature and pressure, the classification diagram represents different conditions of some phenomenon based on two variables.

Our goal here is to understand how we can get MATLAB to automate this classification for us based on a classification diagram. MATLAB cannot "look" at the diagram to see in which region a particular pair of values lies, so we have to be able to express everything mathematically. This is probably best explained by way of an example, so we will jump right in.

EXAMPLE   18-15

A certain chemical reaction will only proceed normally under certain conditions of temperature and pressure as shown in the diagram below. Write a MATLAB program that will ask the user to enter a temperature [°C] and a pressure [atm] and print on the screen either "Acceptable" or "Not Acceptable." If the user enters a pressure less than 1 atm, the message should be "Pressure too low."

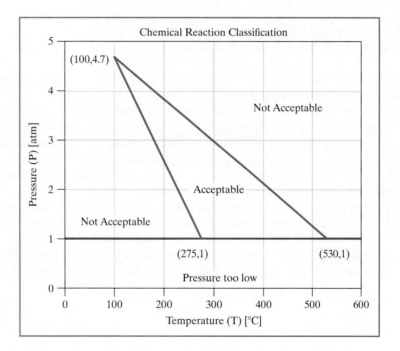

An explanation follows the code.

```
% Chemical Reaction Classification

% Housekeeping
clear
clc
```

```
% Obtain pressure and temperature from user.
Temp=input('Enter the temperature in degrees Celsius: ');
Pres=input('Enter the pressure in atmospheres: ');

% Classify the stated parameters
% First determine slope (m) and intercept (b) of the two non-horizontal lines
Low=polyfit([100 275],[4.7 1],1);     % Lower Line
mLow=Low(1);
bLow=Low(2);

High=polyfit([100 530],[4.7 1],1);     % Upper line
mHigh=High(1);
bHigh=High(2);

% Next calculate the pressure on each line at the entered temperature
PresLow=mLow*Temp+bLow;     % pressure on lower line at Temp
PresHigh=mHigh*Temp+bHigh;     % pressure on upper line at Temp

% Finally, do the classification
if Pres<1     % Determine if pressure too low
      fprintf('Pressure too low\n\n')
elseif Pres>=PresLow && Pres<=PresHigh     % is Pres between the two lines?
      fprintf('Acceptable\n\n')
else    % Not between lines if arrive here
      fprintf('Not Acceptable\n\n')
end
```

After getting the temperature and pressure from the user, we determine the equations (the slope and intercept values) for the two nonhorizontal lines. Once we have eliminated the case where the pressure is less than one atmosphere (the `if` statement), then we need to find out if the entered temperature and pressure values lie BETWEEN the two other lines. To do this, we determine the pressure ON each line corresponding to the entered temperature (the two lines before the `if` statement), then compare the entered pressure with the pressures on the two lines. If the entered pressure is greater than the pressure on the lower line AND less than the pressure on the upper line, then it lies between the two lines and is acceptable.

Note that if the temperature is greater than 530°C, ALL values of pressure greater than 1 atm would lie above the upper line (if you imagine extending it to higher temperatures), thus it cannot lie between the lines.

For temperatures below 100°C, the lines cross, so there is no point anywhere below 100°C that lies above the "lower" line and below the "upper" line since their relative positions have reversed.

EXAMPLE   18-16

The following diagram shows the boundary between acceptable power dissipation (OK) and excessive power dissipation (KABOOM!) in an electrical component given the voltage and current. Write a program that will ask the user to enter a voltage in volts and a current in milliamperes and determine if these parameters will exceed the rated power of the device. The curve follows a power law model.

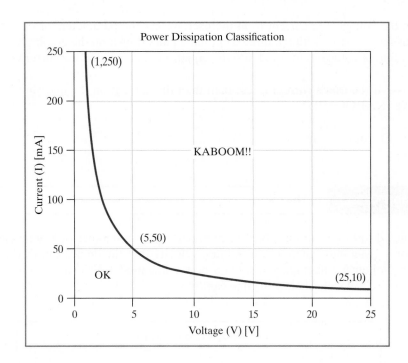

Explanation follows the code.

```
% Power Dissipation Classification

% Housekeeping
clear
clc

% Obtain pressure and temperature from user.
Volt=input('Enter the voltage in volts: ');
Curr=input('Enter the current in milliamperes: ');

% Classify the entered parameters
% First determine parameters (m and b) for the power model
Pow=polyfit(log10([1,5,25]),log10([250,50,10]),1);   % Lower Line
m=Pow(1);
b=10^Pow(2);

% Next calculate the current on the line at the entered voltage
LineCurr=b*Volt^m;   % pressure on lower line at Temp

% Finally, do the classification
if Curr<LineCurr   % Determine if current is below line
     fprintf('OK\n\n')
else   % Current is above line
     fprintf('KABOOM!!\n\n')
end
```

After obtaining the voltage and current from the user, we get the parameters of the power model for the boundary line using `polyfit` with the three coordinate pairs shown on the graph.

The user's value of voltage is inserted into the equation to determine the current on the boundary line at that voltage.

Finally, we ask if the user's current is less than than the current on the line, thus below the line (OK) or above the line (KABOOM!!).

Comprehension Check   18-18

A device constructed to throw various objects can impart up to 500 joules of kinetic energy to the object being thrown. For a given mass, there is a maximum velocity that the device can throw the object. This is represented in the following diagram, and it fits a power law model. Velocities above the line cannot be achieved by the device, velocities below the line can.

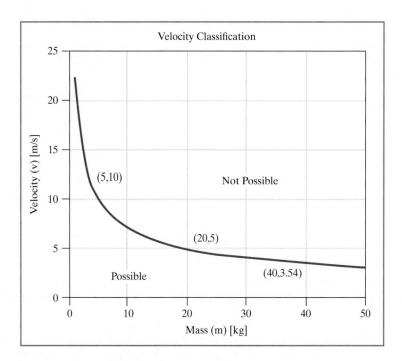

Write a MATLAB program that will accept a value of mass [kg] and desired velocity [m/s] from the user and classify them as "Possible" or "Not Possible." Three points on the line are labeled on the diagram. You must use `polyfit` to determine the equation of the line, and then use these values to do the classification.

18.6 `switch` Statements

A `switch-case` statement is another way to express `if-elseif-else` statements that test for equality. For certain types of questions, the `switch-case` statement is perhaps a little simpler to write and to understand.

The most important point about `switch-case` is that it ONLY tests for equality. If you need to make other types of relational tests (such as greater than), you should use `if` statements. Also, except in a very limited context, you cannot use logical operations (and, or, not) with `switch-case` statements.

The format of the `switch-case` structure is:

```
switch switch_expression
  case case_expression_1
        Execute code here if switch_expression==case _expression_1
  case case_expression_2
        Execute code here if switch_expression==case _expression_2
        .

        .

        .
  case case_expression_N
        Execute code here if switch_expression==case _expression_N
  otherwise
        Execute code here if switch_expression does not equal any
        of the cases
end
```

In general, the `switch-case` statement compares the value of a `switch` expression with different `case` expressions. When it finds a `case` expression that equals the `switch` expression, it executes the code associated with that case, then goes to the end of the `switch` structure and continues. Note that only one case will be executed (the first one that is equal) each time the `switch-case` is executed.

A common use of `switch-case` statements is processing menu selections.

EXAMPLE 18-17

Write a program that allows the user to enter an angle in degrees and then select a trigonometric function from a menu to calculate. Express the result in a sentence, such as:

```
The sine of 0.5236 radians is 0.5000.
```

In this case the user typed 30 for the angle in degrees.

One possible code:

```
Angle = input ('Enter angle in degrees: \n');
radAngle = Angle*pi/180;
trig = menu ('Select a function','sine','cosine','tangent');

switch trig
  case 1
     result = sin(radAngle);
     fprintf('The sine of %0.0f radians is %0.4f.\n', radAngle,result)
  case 2
     result = cos(radAngle);
     fprintf('The cosine of %0.0f radians is %0.4f.\n', radAngle,result)
  case 3
     result = tan(radAngle);
     fprintf('The tangent of %0.0f radians is %0.4f.\n', radAngle,result)
  otherwise
     fprintf('No selection made. Result set to NaN.\n')
     result = NaN;
end
```

If button number 2 (cosine) is selected, the cosine of the angle is displayed. Note that if the user closes the menu, the code between `otherwise` *and* `end` *is executed and the result is set to* `NaN` *(not a number).*

The expressions do not have to be simple variables or constants; they may also include calculations.

EXAMPLE 18-18

Write a code segment to accept from the user the nominal diameter of a bolt and the actual diameter of a specific bolt. The program should then classify the bolt diameter as within less than 1%, 2%, or 5% of the nominal value, and it should indicate if the bolt diameter is too far out of specifications when the nominal and actual diameters differ by 5% or more.

One possible code:

```
NomDiam = input ('Enter nominal diameter of bolt: \n');
BoltDiam = input ('Enter actual diameter of bolt: \n');
percent = BoltDiam/NomDiam*100-100;   % Calculate error

switch abs(fix(percent))    % round toward zero and make positive
  case 0
     fprintf('Diameter within < 1% of desired value.\n')
  case 1
     fprintf('Diameter between 1% and < 2% of desired value.\n')
  case {2,3,4}
     fprintf('Diameter between 2% and < 5% of desired value.\n')
```

```
    otherwise
        fprintf('Out of specifications (>= 5%% error).\n')
end
```

*The third case requires a bit of explanation. You can place multiple expressions (in this example, simple constants) in a cell array as the case expression. If **any** of the expressions in the cell array are equal to the* `switch` *expression, then the instructions associated with that* `case` *will be executed unless, of course, a prior case was true. This is the "limited use of logical operations" mentioned earlier, since it effectively implements an "or," although the "or" is not explicitly coded.*

Comprehension Check   18-19

Write MATLAB code using `switch-case` to allow users to choose from a menu the type of water sample they wish to prepare (solid, liquid, or vapor). Based on their choice, the program should provide guidelines for how to prepare the sample. For example, if the user selects "solid," the program should state:

```
For a solid, water must be at a temperature less than 0 degrees Celsius.
```

Text strings can also be used in `switch-case` structures. Note that this is case-sensitive, since it effectively implements a `strcmp`, not a `strcmpi`. If you need to test strings where case does not matter, you can use `if` statements in conjunction with `strcmpi`, or convert the strings to all upper or all lower case for use with `switch-case`.

EXAMPLE   18-19

Write a code segment to allow users to enter their favorite type of pet, then the program will describe the animal as the output is displayed to the user. If the user types a value that isn't one of the values specified, the code indicates that the user entry doesn't match any of the recognized animals.

One possible code:

```
pet = input ('What is your favorite pet? (Bird, Cat, or Dog)\n','s');

switch pet
    case 'Bird'
        fprintf('Your favorite pet flies through the air.\n')
    case 'Cat'
        fprintf('Your favorite pet makes a yowling noise. \n')
    case 'Dog'
        fprintf('Your favorite pet barks. \n')
    otherwise
        fprintf('Your entry does not match my database. \n')
end
```

Note that in this example, the first letter of the response must be capitalized to have one of the first three cases selected.

Write MATLAB code using `switch-case` to allow the user to type in the name of their favorite flower (Rose, Lily, Tulip, or Peony) and then display a message with information about that flower. You may look up your own facts or use the following: Roses have thorns, Lilies are popular at Easter, The Dutch produce about 3 billion tulips per year, Peonies abort their blooms if the weather is too hot.

18.7 Errors and Warnings

LEARN TO: Create a warning statement to display in the Command Window
Create an error expression to display in the Command Window
Design logic using **try-catch** statements to gracefully terminate the code

MATLAB contains built-in functions that allow you to create custom error and warning messages within your code, similar to those you encounter when you have a syntax error in your code or encounter a runtime error. You can use these functions, in conjunction with logical operators, to build error checking and prevention into your MATLAB code.

Error Messages

You might have noticed that when you run MATLAB code with syntax errors, the Command Window will display errors in red letters to report the error message rather than the standard black text font. In MATLAB, you can actually create your own custom error messages.

Error messages in MATLAB are shown in the Command Window in red letters:

NOTE

The `error` function displays the desired message, then terminates the program.

```
Undefined function or variable 'A'.
```

If you would like to build in your own error checking in MATLAB, you can use the `error` function to actually program custom error messages that may be more helpful than the default messages displayed by MATLAB.

EXAMPLE | **18-20**

We are given the weight (in variable w) in newtons of an object as well as the mass (in variable m) in kilograms and want to determine the gravity in meters per second squared. To solve this, we will use the equation $g = w/m$, but we want to wrap that expression inside a conditional just in case the user provides a zero mass, which does not make sense in our situation, but will not cause an error in MATLAB—if you divide a number by 0 in MATLAB, the value "Inf" is calculated, which represents infinity.

```
>> w=530;
>> m=0;
>> g=w/m
g=
   Inf
```

Instead, we could write the following segment of code to force MATLAB to throw an error message:

```
if m==0
    error('Error: The mass cannot be zero!');
else
    g=w/m;
end
fprintf('The gravity is %0.1f m/s^2\n',g);
```

When you run this code given the w and m above, the message below will display in the Command Window and your code will stop executing.

```
Error: The mass cannot be zero!
```

Warning Messages

If you want to provide a warning to the user but do not want MATLAB to stop running, you can use the `warning` function. This will allow the program to continue, while giving the user information about potential issues in the Command Window. Warning messages are displayed in orange.

EXAMPLE 18-21

Repeat Example 18-20, but this time if the user sets the mass variable to be zero, we want to display a warning message to the user and redefine the mass to be 3 kilograms to compute gravity.

```
w=530;
m=0;
if m==0
    warning('Mass=0! Using 3 kg for the mass');
    m=3;
end
g=w/m;
fprintf('The gravity is %0.1f m/s^2.\n',g);
```

When you run this code, the warning message will display in the Command Window and your code will continue executing. In this case, an `else` statement is unnecessary since the calculation will be done in either case.

```
Warning: Mass=0! Using 3 kg for the mass

The gravity is 176.7 m/s^2.
```

Comprehension Check 18-21

Write MATLAB code to modify the previous gravity calculation examples as follows: If the user enters a zero value for mass, an error message will be generated and the program will terminate; if the user enters a negative value for mass, the program will issue an appropriate warning message and calculate the gravity by using the absolute value of the mass; for strictly positive masses, the program will calculate gravity in the normal manner.

Try-Catch Statements

To customize how MATLAB deals with error messages, you can use `try-catch` statements to handle error exceptions. An exception, in MATLAB, is a structure that contains special information such as a description of the error, as well as information concerning where the error occurred in a particular file. The `try-catch` statement is formatted as follows:

```
try
        MATLAB code . . .
        goes here . . .
               .
               .
               .
catch e
        MATLAB code that executes on error occurs here.
end
```

Try-catch statements can be used over large blocks of code, even entire programs or functions, but their real value comes into play when used on smaller blocks of code to provide more meaningful feedback to the user running your program.

The variable `e` is a structure that contains information about the error that occurred (you "catch" that information). We will only consider the message field, `e.message`, in this text. `e.message` contains the message that would have appeared on the screen if we had not been using `try-catch`.

EXAMPLE 18-22

Write a code segment to ask the user to enter a matrix. If the matrix can be multiplied by a 2 × 4 matrix, the program will execute the operation and display "Finished!" when complete. If the multiplication is invalid, the program will produce a customized error message containing both the message we specify and the error message MATLAB generates.

```
try
    i = input ('Type a matrix: \n');
    m = i * [3,3,3,3;4,4,4,4];
    fprintf('Finished!\n')
catch MyError
    error('Houston, we have a problem:\n\n %s', MyError.message)
end
```

Sample Output:

```
Type a matrix: 3
Finished!
```

In the output above, there were no errors, so the catch statements did not execute.

```
Type a matrix: [1 1 2 3 5; 2 2 1 3 4; 3 3 3 3 3]
Houston, we have a problem:
Inner matrix dimensions must agree.
```

In the output above, there was an error, so the code terminated and the code in the error handler (the `catch` *block) executed. Note the* `fprintf` *in the* `try` *block did NOT execute.*

Comprehension Check   18-22

Write MATLAB code using `try-catch` to ask the user to enter two matrices of the same size. The two matrices will then be multiplied on an element-wise basis and the result stored in `MatProd`.

If the two matrices entered do not have the same dimensions, display a message informing users that they made a mistake, display the normal MATLAB error message for this situation, and terminate the program.

IN-CLASS ACTIVITIES

ICA 18-1

Create a written algorithm and flowchart to determine if a material is in the solid, liquid, or gaseous state given the temperature. The algorithm should acquire the name of the material, its freezing and boiling temperatures, and the actual temperature of the material from the user.

ICA 18-2

The Occupational Safety & Health Administration (OSHA) defines safety regulations on working environments to protect workers from unsafe conditions. The following flowchart shows how OSHA categorizes the safety level of the working temperature given the environmental temperature in degrees Fahrenheit. Given the flowchart, for what range of heat index (in degrees Fahrenheit) will the risk level be Lower (Caution), Moderate, High, or Very High to Extreme?

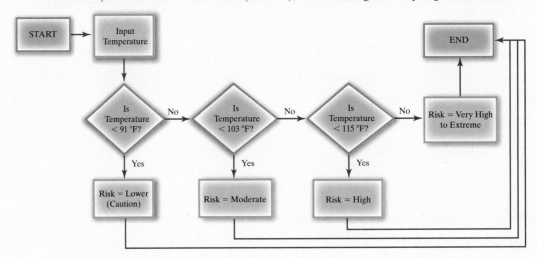

ICA 18-3

Given the following flowchart, for what range of values of pressure (in atmospheres) will the light be each of the colors Yellow, Blue, and Violet?

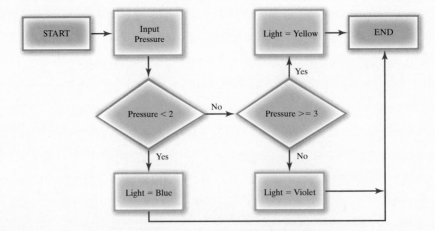

ICA 18-4

Create a flowchart that represents the following written algorithm.

- Input the height of person 1 [in units of feet] as P1
- Input the height of person 2 [in units of feet] as P2
- Is person 1 is taller than person 2
  - If yes, display "Person 1 is taller"
  - Otherwise, is person 2 is taller than person 1
    - If yes, display "Person 2 is taller"
    - Otherwise, display "They are the same height"

ICA 18-5

Create an algorithm to classify a given altitude [miles above sea level] entered by the user as in the troposphere, stratosphere, mesosphere, thermosphere, or exosphere. The algorithm should include a check to ensure the user entered a positive value. If a nonpositive value is entered, the program should warn the user and terminate. If the altitude entered is beyond the exosphere, the algorithm should report "Not in atmosphere." The atmospheric regions are defined by the following criteria:

Troposphere < 12 km
12 km $\leq$ stratosphere < 50 km
50 km $\leq$ mesosphere < 80 km
80 km $\leq$ thermosphere < 700 km
700 km $\leq$ exosphere $< 10,000$ km

ICA 18-6

Answer the following questions.

(a) For what integer values of D will the code between if and end execute?

```
if D >= -1 && 3 > D
    code
end
```

(b) For what values of F will the code between if and end execute?

```
if F >= 100 || F <= 5
    code
end
```

(c) What combinations of values for B and Q will cause a warning to be displayed?

```
if B>100 || (B<20 && Q~=0)
    fprintf('WARNING! Parameters out of bounds.\n')
end
```

ICA 18-7

Answer the following questions.

(a) For what integer values of G will the code between if and end execute?

```
if G < 2 && -5 <= G
    code
end
```

(b) For what values of R will the code between if and end execute?

```
if R < 75 || R >= 95
    code
end
```

(c) What combinations of values for W and M will cause a warning to be displayed?

```
if W<50 || (W>120 && M~=0)
    fprintf('WARNING! Parameters out of bounds.\n')
end
```

ICA 18-8

For each task listed, write a single MATLAB statement that will accomplish the stated goal.

(a) Replace each negative number in matrix M1 with zero.
(b) Given that a scalar value N is already defined, determine the sum of all elements in vector V1 that are greater than N. *HINT: Use the sum function.*

ICA 18-9

For each task listed, write a single MATLAB statement that will accomplish the stated goal.

(a) Replace each negative number in matrix M2 with –9999.
(b) Given that a scalar value N is already defined, determine the sum of all elements in matrix M3 that are less than or equal to N. *HINT: Use the sum function.*

ICA 18-10

What is stored in variable A after each of the following code segments is executed? If the program returns an error, indicate why.

(a)
```
A=5;
if A/2>2
  A = A*-1;
else
  A = A/2;
end
```

(c)
```
T=6; A=-1;
if T/3>2
  A = A*-1;
elseif T/3<=2
  A = A+2;
end
```

(b)
```
x=2; y=4; A=0;
if x+y<4
  A = A+2;
elseif x+y<6
  A = A+4;
end
```

(d)
```
x=2;y=0;A=10;
if x<=0&&y<=0
  A = A+2;
elseif x>=0&&y<0
  A = A/2;
elseif x>=2&&y<=0
  A = A*2;
end
```

ICA 18-11

What will be displayed by the following code in each of the following cases?

```
if flag==1 && alt>=30000
    fprintf('Normal operation at %0.0f feet.\n',alt)
elseif flag==0 || alt==0
    fprintf('On Ground')
elseif flag==2 && alt < 30000
    fprintf('Currently at %0.0f feet and climbing\n',alt)
elseif flag==3
    fprintf('Currently at %0.0f feet and descending\n',alt)
else
    fprintf('Status transitional')
end
```

(a) `flag = 1;` `alt = 25000`
(b) `flag = 0;` `alt = 7500`
(c) `flag = 2;` `alt = 10000`
(d) `flag = 1;` `alt = 37000`
(e) `flag = 3;` `alt = 35000`

ICA 18-12

Ask users to enter a matrix that could be any size but has a minimum of two columns. Check the matrix dimensions, and if there are fewer than two columns warn users of their error and ask them to enter the matrix again. If the user does not correctly enter the matrix a second time, use an error message to notify the user and terminate the program.

Using a matrix correctly entered by the user with two or more columns, create a new matrix containing the same number of rows and two columns. The first element in each row contains the smallest value in the odd-numbered columns of the corresponding row in the user matrix, and the second element in each row contains the largest value in the even-numbered columns of the corresponding row in the user matrix.

Use formatted output statements to print the results to the Command Window. This program should run correctly for any size matrix entered by the user.

Test Case and Sample Output:
```
Please enter a matrix with at least two columns, in the form
[X1 Y1; X2 Y2; . . . ]:
[0 5 9 8 -4 -5; 1 4 6 0 1 2; 8 -7 6 -2 3 -4]
```

The smallest value found in the odd-numbered columns:
Row 1 = −4
Row 2 = 1
Row 3 = 3

The largest value found in the even-numbered columns:
Row 1 = 8
Row 2 = 4
Row 3 = −2

ICA 18-13

A menu is generated using the following code:

```
Status =
    menu('Class','Freshman','Sophomore','Junior','Senior');
```

Write a segment of code using `if-elseif-else` that will classify the person making a menu selection as a new student (freshman) or a continuing student (sophomore, junior, or senior). The code should display one of the following messages:

```
You are a new student.
You are a continuing student.
You did not make a selection.
```

ICA 18-14

A menu is generated using the following code:

```
Status =
    menu('Class','Freshman','Sophomore','Junior','Senior');
```

Write a segment of code using `switch-case` that will classify the person making a menu selection as a new student (freshman) or a continuing student (sophomore, junior, or senior). The code should display one of the following messages:

```
You are a new student.
You are a continuing student.
You did not make a selection.
```

ICA 18-15

Write a program using `if-elseif-else` statements that displays a menu of traffic light colors (green, yellow, red). Depending on which button is pressed, the user should then be told to continue, slow, or stop. If the user closes the menu without making a selection, the user should be told to stop.

ICA 18-16

Write a program using `switch-case` statements that displays a menu of traffic light colors (green, yellow, red). Depending on which button is pressed, the user should then be told to continue, slow, or stop. If the user closes the menu without making a selection, the user should be told to stop.

ICA 18-17

In a middle-aged adult, high blood pressure is classified as a pressure reading greater than 140 millimeters of mercury [mm Hg] systolic or 90 millimeters of mercury [mm Hg] diastolic. For young adults, this range is classified as follows:

Age	Systolic Pressure [mm Hg]	Diastolic Pressure [mm Hg]
15–19	120	81
20–24	132	83

You are developing a blood pressure sensor that will take into account the user's age when making a diagnosis. Assume users will input their age and pressure readings in millimeters of mercury [mm hg]. Indicate as output if the user has high blood pressure or has normal blood pressure. If the age range entered is not between 15 and 24 years old, indicate the user should not use this method of diagnosis and terminate the program.

(a) Develop an algorithm (written or flowchart) for this problem.
(b) Write a MATLAB program to implement the algorithm

ICA 18-18

Write a program that asks the user to enter the length of the hypotenuse, opposite, and adjacent sides of a right triangle with respect to an angle (θ). After the user has entered all three values, a menu should pop up asking if the user wants to calculate $\sin(\theta)$, $\cos(\theta)$, $\tan(\theta)$, $\cot(\theta)$, $\csc(\theta)$, or $\sec(\theta)$. The program should display a final message, such as, "`For a right triangle of sides h, o and a, sin(`θ`) = 0.371,`" where h, o, and a are replaced by the actual lengths entered by the user, displayed to one decimal place. If the user enters lengths that do not work with a right triangle, an error message should appear and the program should terminate. If the user closes the menu instead of making a valid selection, an appropriate error message should appear and the program should terminate. You may use `if` statements and/or `switch-case` statements as appropriate.

ICA 18-19

Your boss hands you the following segment of MATLAB code that implements a safety control system in an automobile. Recently, the engineer who wrote the code was fired for inability to write efficient code. The control system takes 5 minutes to execute because it was not written using nested if statements, and worse, it does not work with multiple sensors because the former employee could not figure out how to make use of nested if statements! Your boss has instructed you to take the former employee's code and fix it, or you might meet the same fate as the original programmer!

```matlab
% Variable Definition:
% Input:
% Ignition: 0 if engine off, 1 if key in ignition, 2 if engine on
% Belt: 1 if buckled, 0 if unbuckled, -1 if sensor broken
% HLamp: 1 if all bulbs ok, 0 if 1 bulb out, -1 if 2 or more out
% TLamp: 1 if all bulbs ok, 0 if 1 bulb out, -1 if 2 or more out
% Light: 1 if sky is bright, 0 if sky is dark (need lamps)
% Output:
% Safety: 1 if safe, 0 if warn/caution, -1 if unsafe
function[Safety]=Vehicle(Ignition,Belt,HLamp,TLamp,Light)

if Ignition==1 || Ignition == 0
      Safety = 1;
end

if Ignition==2 && Belt == 1
      Safety = 1;
end

if Ignition==2 && Belt == 0
      Safety = 0;
end

if Ignition==2 && Belt == -1
      Safety = -1;
end

if Ignition==2 && Belt == 1 && HLamp == 1
      Safety = 1;
end

if Ignition==2 && Belt == 1 && HLamp == 0
      Safety = 0;
end

if Ignition==2 && Belt == 1 && HLamp == -1
      Safety = 0;
end

if Ignition==2 && Belt == 1 && HLamp == -1 && Light==0
      Safety = -1;
end

if Ignition==2 && Belt == 1 && TLamp == 1
      Safety = 1;
end

if Ignition==2 && Belt == 1 && TLamp == 0
      Safety = 0;
end
```

```
if Ignition==2 && Belt == 1 && TLamp == -1
      Safety = 0;
end
if Ignition==2 && Belt == 1 && TLamp == -1 && Light == 0
      Safety = -1;
end
```

ICA 18-20

Assume you are required to generate the menus shown below in a MATLAB program. Write a program to generate these menus and display the choice of the user in a sentence like: "You selected a car with automatic transmission." If the user clicks "Other," an input statement should allow the user to type a different vehicle type (as a string). For this program, use if-elseif statements instead of switch statements.

ICA 18-21

Assume you are required to generate the menus shown with ICA 18-20 in a MATLAB program. Write the program necessary to generate these menus and display the choice of the user in a sentence like "You selected a car with automatic transmission." If the user clicks "Other," an input statement should allow the user to type a different vehicle type (as a string). For this program, use switch statements instead of if-elseif statements.

ICA 18-22

Assume you are required to generate the menus shown below in a MATLAB program. Write a program to generate these menus and display the choice of the user in a sentence like: "You selected size 8 Nike shoes." If the user clicks "Other," an input statement should allow the user to type a different shoe size (as a number). For this program, use if-elseif statements instead of switch statements.

ICA 18-23

Assume you are required to generate the menus shown with ICA 18-22 in a MATLAB program. Write the program necessary to generate these menus and display the choice of the user in a sentence like "You selected size 8 Nike shoes." If the user clicks "Other," an input statement should allow the user to type a different shoe size (as a number). For this program, use `switch` statements instead of `if-elseif` statements.

ICA 18-24

We go to a state-of-the-art amusement park. All the rides in this amusement park contain biometric sensors that measure data about potential riders while they are standing in line. Assume the sensors can detect a rider's age, height, weight, heart problems, and possible pregnancy. Help the engineers write the conditional statement for each ride at the park based on their safety specifications.

	Variable Definitions:
A	% Age of the potential rider (as an integer)
H	% Height of the potential rider (as an integer)
HC	% Heart condition status ('yes' if the person has a heart condition, 'no' otherwise)
P	% Pregnancy status (1 if pregnant, 0 otherwise)

(a) The Spinning Beast: All riders must be 17 years or older and more than 62 inches tall and must not be pregnant or have a heart condition.

```
if _____
    fprintf ('Sorry, you cannot ride this ride');
end
```

(b) The Lame Train: All riders must be 8 years or younger and must not be taller than 40 inches.

```
if _____
    fprintf ('Sorry, you cannot ride this ride');
end
```

(c) The MATLAB House of Horror: All riders must be 17 years or older and must not have a heart condition.

```
if _____

    fprintf ('Welcome to the MATLAB House of Horror!');

end
```

(d) The Neck Snapper: All riders must be 16 years or older and more than 65 inches tall and must not be pregnant or have a heart condition.

```
if _____

    fprintf ('Welcome to The Neck Snapper!');

end
```

(e) The Bouncy Bunny: All riders must be between the ages of 3 and 6 (including those 3 and 6 years old)

```
if _____

    fprintf ('This ride is made just for you!');

end
```

ICA 18-25

A phase diagram for carbon and platinum is shown. It is assumed the lines shown are linear, and the mixture has the following characteristics. The endpoints of the division line between these two phases are labeled on the diagram.

- Below 1700 degrees Celsius, it is a mixture of solid platinum (Pt) and graphite (G).
- Above 1700 degrees Celsius, there are two possible phases: a Liquid (L) phase and a Liquid (L) + Graphite phase (G).

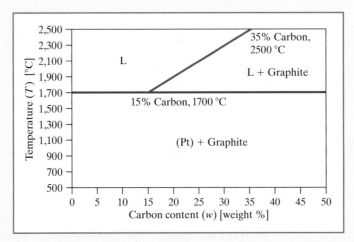

Write a program to determine the phase. The program should ask the user to enter the weight percent carbon and the temperature in degrees Celsius. Call the phases "Pt + G," "L," and "L + G," for simplicity. If the point falls directly on the $T = 1700°C$ line, include the point in the "Pt + G" phase. If the point falls directly on the L, L + G line, include the point in the "L + G" phase.

Store the phase as text in a variable. The equation of the line dividing the L and L + G phases must be found in the program using `polyfit`. The program should produce a formatted output statement to the command window, similar to "For X.XX weight percent carbon and a temperature of YYY degrees Celsius, the phase is PHASE," where X.XX, YYY and PHASE are replaced by the actual values formatted as shown.

The following partial code is designed to implement this program. You are to fill in the missing sections of code as appropriate to complete the program.

```
Q1.                    % Appropriate housekeeping commands
w = input ('Type the Carbon content [weight %%]: ');
T = input ('Type the temperature [deg C]: ');

% set parameters for line between L and L + G
Carbon1 = [  Q2.  ];
Temp1 = [ Q3.  ];

% set parameters for horizontal line separating Pt + G
Carbon2 = [  Q4.  ];
Temp2 = [  Q5.  ];

% determine polyfit parameters for L / L+G line
PhaseLine = polyfit(  Q6.          );

% determine temperature on L / L+G line for given w
TPhaseLine = Q7.          );

   if Q8.                  % Add appropriate question here
       Phase = 'Pt + G';
   elseif Q9.             % Is it a liquid?
       Q10.               % Add appropriate code here
Q11.               % First of two lines to complete if statement
Q12.               % Second of two lines to complete if statement
end

fprintf Q13.     % Add appropriate code
```

ICA 18-26

The graph shows a phase diagram for lead–tin solder. The most common alloy available commercially is 60% Sn (tin) and 40% Pb (lead), although eutectic solder (63% Sn and 37% Pb) is easily obtained and has some distinct advantages, primarily related to the lack of a "pasty" range, the range of temperatures in which the alloy is a mixture of solid within a liquid as it cools from liquid to solid. The gridlines have been removed from this graph to make it easier to read the phase locations.

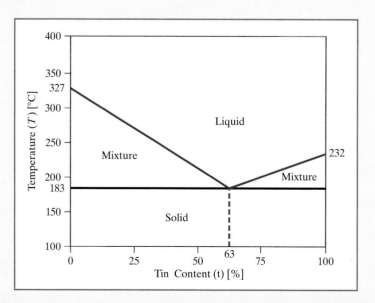

A function named `SnPbPhase` is to be written that will classify the phase of a Sn–Pb alloy given the percentage of tin and the temperature in units of degrees Celsius. The function should accept two scalars:

- `Temp`, containing the temperature in °C
- `PercSn`, the percentage of tin

The function should return a text string named `Phase` containing one of the following four values:

- Liquid
- Solid
- Mixture
- Eutectic point

If a specific combination of percentage and temperature is exactly on one of the lines (other than at the intersection of the eutectic point at 63% and 183°C), the classification should be with the region below the line. The equation of the lines dividing the phases must be found in the program using `polyfit`. The program should produce a formatted output statement to the command window, similar to "`For X.XX percentage of tin and a temperature of YYY deg C, the phase is PHASE,`" where `X.XX`, `YYY` and `PHASE` are replaced by the actual values formatted as shown.

The following partial code is designed to implement this function. You are to fill in the missing sections of code as appropriate to complete the function.

```
Q1. _____        % Declaration of function and passed parameters

% set parameters for Mixture / Liquid line when Tin < 63%
Tin1=[ Q2. ];
Temp1=[ Q3. ];
LoSn = polyfit( Q4. _____ );
TempLoSn = Q5. ;     % Temperature on LoSn line at given PercSn

% set parameters for Mixture / Liquid line when Tin > 63%
Tin2=[ Q6. ];
Temp2=[ Q7. ];
HiSn = polyfit( Q8. );
TempHiSn = Q9. ;     % Temperature on HiSn line at given PercSn

if Q10. _____        % Add appropriate question here
    Phase = 'Eutectic Point';
elseif Q11. _____        % Is it solid?
Q12. _____        % Add appropriate code here
elseif PercSn<=63 && Q13. _____        % Complete the question
    Phase='Liquid';
elseif Q14. _____        % Add appropriate question here
    Phase='Liquid';
Q15. _____        % First of two lines to complete if statement
Q16. _____        % Second of two lines to complete if statement
end
fprintf Q17. _____        % Add appropriate code
```

Upon reflection, you realize that the second and third `elseif` statements (the two that classify the material as liquid) can be written as a single `elseif` statement using a question with ONLY a SINGLE logical operator (`and` or `or`). What is the correct question to ask in a single `elseif` statement to correctly classify the material as liquid?

```
    elseif Q18. _____        % Add appropriate question here
        Phase='Liquid';
```

ICA 18-27

Radioactive decay follows an exponential model. The decay curve for a specific substance is shown here. Write a MATLAB program that will ask the user for a time [yr] and a radiation level in gigabecquerels [GBq], and then classify the decay rate indicated by these values as Faster, Slower, or Equal.

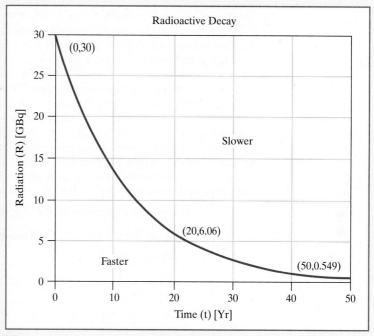

ICA 18-28

The following input statement is part of a silly program designed to flummox freshmen.

```
BS=input('How long is a battleship? (Type "true" or "false"): ','s');
```

Write a short segment of MATLAB code that will accomplish the following task:

- If the user typed either "true" or "false" with ANY combination of upper and lowercase characters, display the message "Surely you jest." in the command window.
- If the user typed anything other than the word "true" or the word "false," an error message stating "INVALID ENTRY" should appear and the program should terminate.

ICA 18-29

The following code will generate an error and terminate if the user does not enter scalar values as instructed.

```
Val=input('Enter a numeric scalar for comparison: ');
Lower=input('Enter a numeric scalar for the lower bound: ');
Upper=input('Enter a numeric scalar for the upper bound: ');
Status=Val>=Lower&&Val<=Upper;
```

Add appropriate code using `try-catch` so that the code will operate normally if the user enters scalars as requested but will terminate after generating a custom error message in addition to the standard error message as shown if the user does not enter scalars.

```
You did not enter scalar values as requested.
Operands to the || and && operators must be convertible to logical
scalar values.
```

REVIEW QUESTIONS

1. The Apple TV™ is a personal video device created by Apple, Inc. that attaches to high-definition televisions to present streaming content to a viewer on demand, in contrast to a standard source (cable/satellite) set-top box or digital video recorder (DVR) that presents recorded or live broadcast signals on a television. According to a study conducted by the Natural Resources Defense Council in June 2011, if consumers were to move from set-top boxes to smaller "thin" devices like the Apple TV, there would be an overall 70% reduction in annual energy costs resulting from the use of set-top boxes.

   The following data table shows the typical power consumption of the Apple TV in different power states (standby/off, idle—not playing any video content, playing via Ethernet, and playing via Wifi). Create an algorithm that allows a user to provide the power currently being used by their Apple TV in units of watts and determine the state of the device. If the user provides value for the power outside the range in the following table, your algorithm should report that the state of the device is unknown.

   ### Power Consumption of the Apple TV

State	Power Consumption [W]
Off/Standby	less than 0.5
Idle	0.5 to 1.5
Streaming via Ethernet	1.5 to 1.6
Streaming via Wifi	1.6 to 2

2. Humans can see electromagnetic radiation when the wavelength is within the spectrum of visible light. Create an algorithm to determine if a given wavelength [nanometer, nm] is one of the six spectral colors listed in the following chart. Your algorithm should ask the user to enter a wavelength, then indicate in which spectral color the given wavelength falls or provide a warning if it is not within the visible spectrum.

Color	Wavelength Interval [nm]
Red	700–635
Orange	635–590
Yellow	590–560
Green	560–490
Blue	490–450
Violet	450–400

3. Create an algorithm to display the letter grade in a course. A course grade is typically reported as a real number with values from 0 to 100. Your algorithm should ask the user to provide a grade in numerical form and display the letter grade corresponding to a numerical grade if the entered value is between 0 and 100 inclusive. If the value is outside this range, ask the user to enter the value again.

   Typical range of grades:

   A: $90 \leq$ grade
   B: $80 \leq$ grade < 90
   C: $70 \leq$ grade < 80
   D: $60 \leq$ grade < 70
   F: grade < 60

4. Create an algorithm to determine whether a given Mach number (actual speed divided by the speed of sound) is subsonic, transonic, supersonic, or hypersonic. Assume the user will enter the speed of the object in meters per second, and the program will calculate the Mach number and determine the appropriate Mach category. As output, the program will display the Mach category. The algorithm should include a check to ensure the user entered a positive value. If a nonpositive value is entered, the program should warn the user and terminate. The speed of sound is about 343 m/s. The classifications are as follows:

Subsonic < Mach 1
Transonic = Mach 1
Mach 1 < Supersonic < Mach 5
Hypersonic ≥ Mach 5

5. The specific gravity of gold is 19.3. Write a MATLAB program that will ask the user to input the mass of a cube of solid gold in units of kilograms and display the length of a single side of the cube in units of inches. The output should display a sentence like the one shown that follows, with the length formatted to two decimal places. If the user types a negative number or zero for the mass of the cube, your program should display an error message and terminate.

Sample Input/Output (two examples):

```
Enter the mass of the cube [kilograms]: -3
Error: Mass must be greater than zero grams.
Enter the mass of the cube [kilograms]: 0.4
The length of one side of the cube is 1.08 inches.
```

6. An unmanned X-43A scramjet test vehicle has achieved a maximum speed of Mach number 9.68 in a test flight over the Pacific Ocean. Mach number is defined as the speed of an object divided by the speed of sound. Assuming the speed of sound is 343 meters per second, write a MATLAB program that asks the user to provide a speed as a Mach number, calculates the corresponding speed in miles per hour, and displays the speed in a formatted sentence as shown in the following sample output. Note that the speed in mph should be shown as an integer. If the user provides a negative value for the Mach number, your program should display an error message and terminate.

Sample Input/Output (three examples):

```
Enter the speed as a Mach number: -2
Error: The Mach number must not be negative.
Enter the speed as a Mach number: 0
The speed of the plane is 0 mph.
Enter the speed as a Mach number: 9.68
The speed of the plane is 7425 mph.
```

7. A rod on the surface of Jupiter's moon Callisto has a volume measured in cubic meters. Write a MATLAB program that will ask the user to type in the volume in cubic meters in order to determine the weight of the rod in units of pounds-force. The specific gravity of the rod is 4.7. Gravitational acceleration on Callisto is 1.25 meters per second squared. The input and output of the program should look similar to the output that follows. Be sure to report the weight as an integer value. If the user types a negative number or a number greater than 500 cubic meters for the volume of the rod, your program should display an error message using the error function indicating that the provided input is outside of the desired range and terminate.

Sample Input/Output (three examples):

```
Enter the volume of the rod [cubic meters]: -10
Error: Volume must be between 0 and 500 cubic meters
Enter the volume of the rod [cubic meters]: 501
Error: Volume must be between 0 and 500 cubic meters
Enter the volume of the rod [cubic meters]: 0.3
The weight of the rod is 397 pounds-force.
```

8. The Eco-Marathon is an annual competition sponsored by Shell Oil, in which participants build special vehicles to achieve the highest possible fuel efficiency. The Eco-Marathon is held around the world with events in the United Kingdom, Finland, France, Holland, Japan, and the United States.

    A world record was set in the Eco-Marathon by a French team in 2003 called Microjoule with a performance of 10,705 miles per gallon. The Microjoule runs on ethanol. Write a MATLAB program to determine how far the Microjoule will travel in kilometers given a user-specified amount of ethanol, provided in units of grams. Your program should ask for the mass using an input statement and display the distance in a formatted sentence similar to the output that follows. If the user provides a mass less than zero or greater than 500 grams, your program should terminate after displaying an error message using the `error` function indicating that the provided mass of ethanol is outside of the desired range of input.

Sample Input/Output (three examples):

```
Enter mass of ethanol [grams]: -15
Error: Mass must be between 0 and 500 grams
Enter mass of ethanol [grams]: 1000
Error: Mass must be between 0 and 500 grams
Enter mass of ethanol [grams]: 100
The distance the Microjoule traveled is 577 kilometers.
```

9. Assume a variable R contains a single number. Write a short piece of MATLAB code that will:

   - Generate the message: `The square root of XXXX is YYYY`, if R is nonnegative.
   - If R is negative, the code should generate the message: `R is negative. The square root of XXXX is YYYYi`.

   The value of R should be substituted for XXXX, and the value of the square root of the magnitude of R should be substituted for YYYY. Both numbers should be displayed with three decimal places.

10. Write a program that gathers two data pairs in a 2×2 matrix and then asks the user to input another value of x for which a value of y will be interpolated or extrapolated. Report the value of y in the correct sentence below (with numbers in the blanks). Assume that the first column contains the x values and the second column contains the corresponding y values.

    ```
    Given x = _____, interpolation finds that y = _____ or
    Given x = _____, extrapolation finds that y = _____.
    ```

Sample Input/Output

Matrix [0 40; 10 70] is entered and the user chooses 5:
```
Given x = 5.00, interpolation finds that y = 55.00.
```

Matrix [0 40; 10 70] is entered and the user chooses 12:
```
Given x = 12.00, extrapolation finds that y = 76.00.
```

11. Create a program to determine whether a user-specified altitude [meters] is in the troposphere, lower stratosphere, or upper stratosphere. The program should include a check to ensure that the user entered a positive value less than 50,000. If a nonpositive value or a value of 50,000 or greater is entered, the program should inform the user of the error and terminate. If a positive value less than 50,000 is entered, the program should calculate and report the resulting temperature in units of degrees Celsius [°C] and pressure in units of kilopascals [kPa]. Use the following information where H is altitude in meters, T is temperature in °C, and P is pressure in kilopascals:

    Troposphere: Altitude $H < 11,000$ m

    $$T = 15.04 - 0.00649\,H$$

NOTE

Due to several complicated nonlinear phenomena in the system that are far beyond the scope of this course, these equations contain power models in which the exponent does not come out to be an integer or simple fraction. In fact, this model is only a first approximation—a really accurate model would be considerably more complicated.

$$P = 101.29\left(\frac{T + 273.1}{288.08}\right)^{5.256}.$$

Lower Stratosphere: Altitude 11,000 m $\leq H <$ 25,000 m
$T = -56.46.$

$$p = 22.65e^{(1.73 - 0.000157\,H)}$$

Upper Stratosphere: Altitude 25,000 m $\leq H <$ 50,000 m

$$T = -131.21 + 0.00299\,H.$$

$$P = 2.488\left(\frac{T + 273.1}{216.6}\right)^{-11.388}.$$

These formulae come from the atmosphere model provided by NASA at: http://www.grc.nasa.gov/WWW/K-12/airplane/atmosmet.html.

Sample Input/Output

Altitude = 500 meters.

```
An altitude of 500 is in the troposphere with a temperature of 12
degrees C and pressure of 96 kPa.
```

12. Create a program to determine whether a given Mach number is subsonic, transonic, supersonic, or hypersonic. Assume the user will enter the speed of the object in meters per second, and the program will determine the Mach number (actual speed divided by the speed of sound). As output, the program will display the Mach number to two decimal places. The program should include a check to ensure that the user entered a positive value. If a nonpositive value is entered, the program should inform the user of the error and terminate.

The speed of sound is about 343 m/s. The classifications are as follows:

Subsonic < Mach 1
Transonic = Mach 1
Mach 1 < Supersonic < Mach 5
Hypersonic ≥ Mach 5

Refer to the NASA page on Mach number:
http://www.grc.nasa.gov/WWW/K-12/airplane/mach.html

Sample Input/Output Speed = 100 m/s.

```
Subsonic, Mach number is 0.29.
```

13. Humans can see electromagnetic radiation when the wavelength is within the spectrum of visible light. Create a program to determine if a user-specified wavelength [nanometer, nm] is one of the six spectral colors listed in the following chart. Your program should ask the user to enter a wavelength, then indicate in which spectral color the given wavelength falls or state if it is not within the visible spectrum.

Color	Wavelength Interval
Red	700–635 nm
Orange	635–590 nm
Yellow	590–560 nm
Green	560–490 nm
Blue	490–450 nm
Violet	450–400 nm

14. For the protection of both the operator of a zero-turn radius mower and the mower itself, several safety interlocks must be implemented. These interlocks and the variables that will represent them are listed shown here:

Interlock Description	Variable Used	State
Brake switch	`Brake`	True if brake on
Operator seat switch	`Seat`	True if operator seated
Blade power switch	`Blades`	True if blades turning
Left guide bar neutral switch	`LeftNeutral`	True if in neutral
Right guide bar neutral switch	`RightNeutral`	True if in neutral
Ignition switch	`Ignition`	True if in run position
Motor power interlock	`Motor`	True if motor enabled

For the motor to be enabled (thus capable of running), all the following conditions must be true.

- The ignition switch must be set to Run.
- If the operator is not properly seated, both guide levers must be in the neutral position.
- If either guide lever is not in the neutral position, the brake must be off.
- If the blades are powered, the operator must be properly seated.

Write a function that will accept all of the preceding variables except `Motor`, decide whether the motor should be enabled or not, and place `true` or `false` in `Motor` as the returned variable.

Sample Input/Output

```
If the parameters are set to 0,0,0,0,0,1, the Motor output will be 0.
If the parameters are set to 0,1,0,1,1,1, the Motor output will be 1.
```

15. Most resistors are so small that the actual value would be difficult to read if printed numerically on the resistor. Instead, colored bands denote the value of resistance in ohms. Anyone involved in constructing electronic circuits must become familiar with the color code, and with practice, one can tell at a glance what value a specific set of colors means. For the novice, however, trying to read color codes can be a bit challenging.

You are to design a program that, when a user enters a resistance value, the program will display (as text) the color bands in the order they will appear on the resistor.

The resistance will be entered in two parts: the first two digits, and a power of 10 by which those digits will be multiplied. The user should be able to select each part from a menu. You should assume the ONLY values that can be selected by the user are the ones listed in Table 18-3 and Table 18-4.

As examples, a resistance of 4700 ohms has first digit 4 (yellow), second digit 7 (violet), and 2 zeros following (red). A resistance of 56 ohms would be 5 (green), 6 (blue), and 0 zeros (black); 1,000,000 ohms is 1 (brown), 0 (black), and 5 zeros (green). There are numerous explanations of the color code on the web if you want further information or examples.

Table 18-3 Standard resistor values

10	12	15	18	22	27	33	39	47	56	68	82

Table 18-4 Standard multipliers

1	10	100	1000	10000	100000	1000000

Table 18-5 Color codes

0	1	2	3	4	5	6	7	8	9
Black	Brown	Red	Orange	Yellow	Green	Blue	Violet	Gray	White

16. A phase diagram (from http://www.eng.ox.ac.uk/~ftgamk/engall_tu.pdf) for copper–nickel shows that three phases are possible: a solid alpha phase (a solid solution), a liquid solution, and a solid–liquid combination.

Write a program that takes as input a mass percent of nickel and a temperature in units of degrees Celsius and gives as output a message indicating what phases are present. Assume the region borders are linear, and determine a method for dealing with any points that lie directly on the lines. The equations of the lines dividing the phases must be found in the program using `polyfit`. The program should produce a formatted output statement to the Command Window, similar to "For w weight percent nickel and a temperature of T degrees Celsius, the phase is PHASE," where w, T, and Phase are replaced by the actual values.

The program should also reproduce the graph as shown here with all phase division lines and indicate the point entered by the user with a symbol.

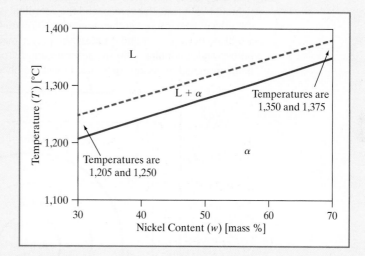

17. The heat index combines temperature and relative humidity in an attempt to describe how "hot" a person will feel in those conditions. NOAA issues safety guidelines based on the heat index, and a simplified chart of these recommendations is shown here.

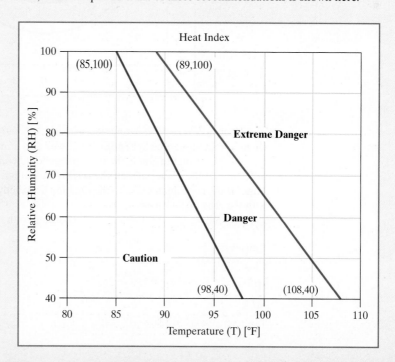

Write a MATLAB function named `HeatWarning` that will accept a temperature in °F and a relative humidity in percent. The function should return a text string containing the appropriate warning, "Caution," "Danger," or "Extreme Danger."

If the temperature passed to the function is less than 80°F, the returned text should be "Low temperature: not classified." If the temperature is greater than 110°F, the returned text should be "Extreme Danger."

Assuming the temperature is between 80°F and 110°F, if the relative humidity is either negative or greater than 100%, the returned text should be "Invalid Humidity Value."

Your function should use `polyfit` to determine the parameters of the boundary lines (assumed linear). The endpoints of the lines are noted on the graph. You should NOT calculate the slopes and intercepts manually.

18. You are part of a team designing a game to hit a target. A "gun" will fire a short pulse of light at a light-sensitive screen, and the coordinates of the location where the pulse hits the screen will be captured in the variables X (horizontal) and Y (vertical). The values in X and Y will be between −4 and 4 unless the player misses the screen altogether, in which case both X and Y contain −99. The software and hardware to accomplish all of the above are being designed by more experienced team members.

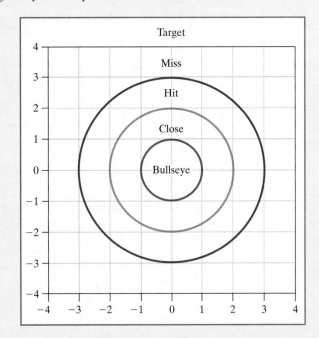

Your task is to write a function named `Classify` that returns a text string indicating where on the target the light pulse hit as well as a numeric score.

As shown on the graph, the text string should contain either "Bullseye," "Close," "Hit," or "Miss." If the player missed the screen altogether, the text string should contain "Total Miss." If the coordinates sent to the function lie exactly on a boundary, it should be classified with the area farther from the center.

The numeric score should be set according to the following list:

Bullseye:	4
Close:	2
Hit:	1
Miss:	0
Total Miss:	−1

Remember that a circle can be defined by

$$R^2 = X^2 + Y^2.$$

where R is the radius of the circle.

19. This generic phase diagram, based on the temperature and composition of elements A and B, is taken from http://www.soton.ac.uk/~pasr1/build.htm, where a description of how phase diagrams are constructed can also be found. The alpha and beta phases represent solid solutions of B in A and of A in B, respectively.

   The eutectic line represents the temperature below which the alloy will become completely solid if it is not in either the alpha or beta region. Below that line, the alloy is a solid mixture of alpha and beta. Above the eutectic line, the mixture is at least partially liquid, with partially solidified lumps of alpha or beta in the labeled regions.

   Assume the following:

   - The melting point of pure A is 700 degrees Celsius.
   - The melting point of pure B is 800 degrees Celsius.
   - The eutectic line is at 300 degrees Celsius, and the eutectic point occurs when the composition is 50% B.
   - The ends of the eutectic line are at 15% B and 85% B.
   - The lower left corner of the graph represents 0% B and 0 °C.

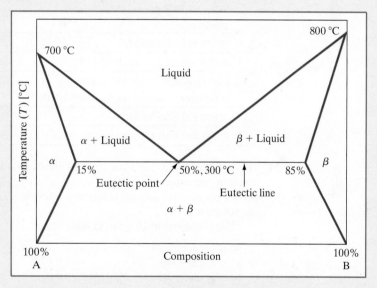

   Using the simplified phase diagram, which substitutes straight lines for the curved lines typical of phase diagrams, write a MATLAB program that takes as input the mass percent of B and the temperature in units of degrees Celsius and returns the phases that may exist under those conditions. The equation of the line dividing the phases must be found in the program using `polyfit`. The program should produce a formatted output statement to the Command Window, similar to "For the composition of x% A, y% B and a temperature of T degrees Celsius, the phase is PHASE," where x, y, T, and PHASE are replaced by the actual values.

   The program should also reproduce the graph as shown here with all phase division lines, and indicate the point entered by the user with a symbol.

   Your MATLAB program should include special notes if the provided conditions are on the eutectic line or at the eutectic point.

20. Soil texture can be classified with a moist soil sample and the questions in the following table from the Soil Texture Key found at http://www.pasture4horses.com/soils/hand-texturing .php/. Each answer results in either a soil classification or directions to continue. Write a

program that implements this decision tree. You may find it useful to download the key itself from the preceding address. You may also want to compare your results to an online implementation of the key found at the same website. Your program should include the use of menu windows to gather user responses.

Question	Yes	No
1. Does the soil feel or sound noticeably sandy?	Go to Q2	Go to Q6
2. Does the soil lack all cohesion?	SAND	Go to Q3
3. Is it difficult to roll the soil into a ball?	LOAMY SAND	Go to Q4
4. Does the soil feel smooth and silky as well as sandy?	SANDY SILT LOAM	Go to Q5
5. Does the soil mould to form a strong ball that smears without taking a polish?	SANDY CLAY LOAM	SANDY LOAM
6. Does the soil mould to form an easily deformed ball and feel smooth and silky?	SILT LOAM	Go to Q7
7. Does the soil mould to form a strong ball that smears without taking a polish?	Go to Q8	Go to Q10
8. Is the soil also sandy?	SANDY CLAY LOAM	Go to Q9
9. Is the soil also smooth and silky?	SILTY CLAY LOAM	CLAY LOAM
10. Does the soil mould like plasticine, polish, and feel very sticky when wetter?	Go to Q11	UNKNOWN SOIL
11. Is the soil also sandy?	SANDY CLAY	Go to Q12
12. Is the soil also smooth and buttery?	SILTY CLAY	CLAY

21. The variable `grade` can have any real number value from 0 to 100. Ask the user to enter a grade in numerical form. Write an `if-elseif-else` statement that displays the letter grade (any format) corresponding to a numerical grade in an appropriately formatted output statement.

   Use the standard 10-point grading scale:

A: $90 \le grade$
B: $80 \le grade < 90$
C: $70 \le grade < 80$
D: $60 \le grade < 70$
F: $grade < 60$

22. Write a MATLAB program that will allow the student to select the grade earned in a course from a menu (with the options of earning an A, B, C, D, or F) that will display the numerical range for the selected grade. The output of the program displayed on the screen should be formatted as shown in the sample output. In your program, you must use a `switch` statement instead of an `if-elseif` statement. Use the standard 10-point grading scale shown in the previous problem.

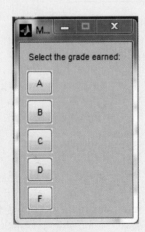

Sample Input/Output

If user selected A:

```
If you earn the letter grade A, your numeric grade is in the
range: 90 <= grade.
```

23. *One of the 14 NAE Grand Challenges is Engineering Better Medicines. Part of this Grand Challenge is creating individual medication plans for each patient, rather than sweeping recommendations based upon research using only one subset of the population. For example, in the past most heart attack protocols were created for white males. It has been discovered in recent years that women have different initial symptoms and respond better to a different treatment protocol than their male counterparts.*

To better tailor the medicine to the individual, you wish to write a computer program to allow the user to enter several variables. The program will then produce a suggested solution. Write the MATLAB code to:

(a) Ask the user to enter his or her name (text).

(b) Ask the user to type his or her weight (a number) in pounds-force.

(c) Ask the user to select his or her symptom from a menu, given the following choices.
- Cold
- Flu
- Migraine

(d) Based upon the answer selected to the menu, set the medicine name, volume [mL] and mass [g] of the medicine tablet.

Symptom	Medicine	Volume [mL]	Mass [g]
Cold	Achoo	3.5	9
Flu	Chill	5	16
Migraine	HAche	4	11

(e) Write (a) the function call from the program and (b) a function that will accept the mass and volume of the desired medicine from the main program and return the specific gravity of the desired medicine tablet to the program. Remember: The density of water is 1000 kilograms per cubic meter.

(f) Write (a) the function call from the program and (b) a function that will accept the weight of the person, specific gravity of the medicine, and tablet volume from the main program and return the number of tablets recommended for the person to the program. To determine the number of tablets needed, the following equation should be used and then rounded up to the next whole number of tablets. Any calculations or conversions necessary should appear in the function.

$$\text{Dose} = \text{Mass of person [kg]} \left| \frac{1.25 \frac{\text{ml}}{\text{kg}}}{2.5 * \text{SG}} \right| \frac{1 \text{ tablet}}{\text{Volume[mL]}}$$

(g) Write an output statement that appears as follows, where N is replaced by the name of the medicine and X.XXX is replaced by the calculated specific gravity formatted to show three decimal places: The specific gravity of N is X.XXX.

(h) Write an output statement that appears as follows, where AAAA is the name of the user, N is the name of the medicine, YYY is the name of their symptom, and Z is the number of tablets recommended:

AAAA, your recommended dosage of N to treat a YYY: Z tablets.

NOTE

Although you may use conditional statements to solve this problem, none are needed if you set up the medicine data table as a cell array.

24. In this problem, you will develop another medical application. Before you begin, download the two Excel files available online.

Main Program, Section 1: Patient Information

- Import the patient data from an Excel workbook titled "**PersonalMeds.xlsx**" on the worksheet "**PatientInfo**". *Note that there are several headings in the worksheet that are not currently used (Symptom, Medication, Dosage, Date). These will be utilized later in the program.*
- Determine the size of the imported data and text matrices for the patients. You must use the size limits in all code, not hard-coded numbers, to allow for patients to be added to

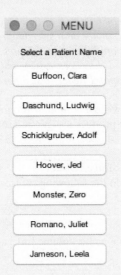

the worksheet. For example, the current worksheet contains seven names. The program should work correctly, without any changes required, if the number of patients is changed (to 30 names, for example).

- Ask the nurse to choose the patient name being seen during an office visit from a menu of patient names, located in column A of the worksheet. A menu with the current seven names included in the worksheet is shown here as an example.
- If the user closes the menu, ask if they wish to input a new name. The response to this input question must be a string entered by the user.
  - If the user answers "YES," the program should ask the user to enter the patient name (as a string).
    - The program should then ask the user to enter the patient weight in pound-force (as an integer), and then choose from a menu the patient gender as either "Male," "Female," or "Transgender." Based on the menu selection, the program should assign M, F, or T to the gender variable.
      - The program should create a 1×3 cell array containing the new information and then add this information to the next blank row, starting in column A, on the `PatientInfo` worksheet.
      - If the user closes this menu, the program should terminate after issuing an appropriate error message.
  - If the user answers "NO" to the menu regarding entering a new name, the program should terminate.
- Before the program continues to Section 2, the program should summarize the patient information for the nurse, similar to the following:

```
Patient Name: Calhoun, Anna    Gender: F
```

Main Program, Section 2: Symptom and Medication Information

- Import the medicine data from an Excel workbook titled "**PersonalMeds.xlsx**" on the worksheet "**MedicationInfo**". In the category "**Dose Type**", an "**L**" indicates the dosage comes in liquid form; a "**T**" indicates the dosage is given in a tablet form.
- Determine the size of the imported data and text matrices for the medication. You must use the size limits in all code, not hardcoded numbers, to allow for medications to be added to the worksheet. For example, the current worksheet contains 4 medications. The program should work correctly, without any changes required, if the number of medications was changed (to 30 medications, for example).
- Ask the nurse to select the patient's symptom from a menu, given the choices found in row 1 of the medication data. As a safety feature to make sure the nurse is selecting information for the correct patient, include the patient's name in the menu title.

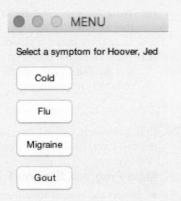

- Include an error message if the nurse closes the menu rather than selecting a symptom. The message should tell the user he or she has incorrectly selected a symptom, and the program should terminate.

Main Program, Section 3: Determine Medication Needed

- Write the function call from the program. The function will determine and return to the program the recommended dosage of the desired medicine. More information about the function is provided below.

Main Program, Section 4: Inform the User

- Write output statements that appear in the Command Window. Sample output is shown below. Here **AAA** is the name of the user, **YYY** is the name of their symptom, and **NNN** is the name of the medicine.

```
AAA Ailment: YYY
Medicine: NNN
```

- If the medication is in tablet form, the dosage should be listed as the number of tablets (**ZZ**):

```
Dosage: ZZ tablets
```

- If the medication is in liquid form, the dosage should be listed as follows. **M.M** is the total amount of liquid medicine, in milliliters; **Z.Z** is the dosage; and **V.V** is the volume of the medicine per dose.

```
Dosage: M.M mL [Z.Z doses, V.V mL/dose]
```

Main Program, Section 5: Record the information in the workbook

- Write the statements necessary to record the information on the ailment, medicine, dosage, and date to the Excel workbook **PersonalMeds.xlsx** on the worksheet **PatientInfo**. First, create a single cell array containing all information required. Then, write the single cell array to the correct columns, starting with column E, in the row corresponding to the patient selected.
  [**Hint**: To record the date, try the built-in function **date**]
  [**Hint**: To specify the starting location of the data in the Excel worksheet, use **sprintf** to create a variable which contains the column (**E**) and the row number.]

Function

You will need to determine what information is required from the program for the function to work correctly, and then appropriately pass this information from the program to the function. The function will return to the program the required medication dosage, determined by the following rules.

For medications requiring a tablet dosage, the following equation determines the number of tablets.

$$\mathbf{Dose_{tablet}} = \frac{\textbf{(Mass of person[kg])(SF)}}{\textbf{(3500)}\left(\textbf{Tablet Density}\left[\dfrac{\textbf{kg}}{\textbf{mL}}\right]\right)\left(\textbf{Tablet Volume}\left[\dfrac{\textbf{mL}}{\textbf{dose}}\right]\right)}.$$

Here, **SF** is a scaling factor depending the patient's gender. If the patient is male, the scale factor should be set to 1. If the patient is female, the scale factor should be set to 0.7. If the patient is transgender, the scale factor should be set to 0.85. The constant 3500 is an experimentally determined factor and is unitless.

For tablets, the actual dosage required should be rounded up to the nearest whole number. [**Hint**: There are four basic "rounding" functions: round, fix, floor, and ceil.]

For medications requiring a liquid dosage, the following equation determines the number of dosages.

$$\mathbf{Dose_{liquid}} = \frac{\textbf{Weight of person [lb}_\mathbf{f}\textbf{]}}{\left(\textbf{28}\dfrac{\textbf{lb}_\mathbf{f}}{\textbf{mL}}\right)\left(\textbf{Tablet Volume}\left[\dfrac{\textbf{mL}}{\textbf{dose}}\right]\right)}.$$

For the liquid dosage, there is no dependence upon gender. The constant 28 is an experimentally determined factor and has units of pounds-force per milliliter [lb_f/mL].

Testing

- Once you have thoroughly tested your program, alter your code to change the workbook name from **PersonalMeds.xlsx** to **PersonalMeds_List2.xlsx**. Your code should now have:
  - 15 patients, listed as LAST name, FIRST name in alphabetical order, given in the first column on the worksheet **PatientInfo**. The corresponding weight and gender of each person is given in columns 2 and 3, respectively.
  - 6 symptoms, listed in alphabetical order in row 2 on the worksheet **MedicationInfo**. The medication name, volume, mass, and dose type are given in rows 1, 3, 4, and 5.

  Run the code. If your code does not work with this new data without making any alterations other than changing the workbook name, you have HARD-CODED parts of your solution. Fix this before submitting.

25. Assume two matrices, M1 and M2, have already been defined and have the same dimensions. Create a matrix Comp1 with the same dimensions as M1 and M2. Each element of Comp1 should be an integer between −3 and 3 based on the relative values of the corresponding elements of M1 and M2 according to the following rules:

    - If the magnitudes (absolute values) of corresponding elements of M1 and M2 are equal, that element of Comp1 equals 0.
    - If the magnitude of the element in M1 is greater than that in M2, the corresponding element in Comp1 will be positive.
    - If the magnitude (absolute value) of the element in M1 is less than that in M2, the corresponding element in Comp1 will be negative.
    - If the values in both M1 and M2 are positive, the magnitude of the value in Comp1 is 1.
    - If the values in both M1 and M2 are negative, the magnitude of the value in Comp1 is 2.
    - If the values in M1 and M2 have opposite signs (assume zero is positive), the magnitude of the value in Comp1 is 3. (NOTE: If the magnitudes of the values in M1 and M2 are equal, that element of Comp1 will contain a zero.)

    ### Sample Input/Output

    $$M1 = \begin{bmatrix} 0 & 1 & 2 \\ -2 & -1 & 0 \\ 1 & -1 & -1 \end{bmatrix} \quad M2 = \begin{bmatrix} -1 & 1 & 0 \\ -1 & 2 & 2 \\ -1 & 0 & -2 \end{bmatrix} \quad Comp1 = \begin{bmatrix} -3 & 0 & 1 \\ 2 & -3 & -1 \\ 0 & 3 & -2 \end{bmatrix}$$

MATLAB examples

NOTE

Even if you know how to use programming loops, DO NOT do so for this program. The problem should be solved using logical arrays.

26. You are designing a vocabulary-building program for biological terms. Two cell arrays containing the same number of text strings have already been defined and are stored in the online file Critters.mat. The cell array ColAdj contains collateral adjectives for various groups of animals, and Common contains the common name of a representative animal in the corresponding group. For example, ColAdj{1} contains Canine and Common{1} contains Dog.

    Write a MATLAB program that will randomly choose one of the text strings in Common and ask the user to type in the collateral adjective of the group to which that animal belongs. For example,

    ```
    To what group do ostriches belong?
    ```

    (The correct answer is Ratite.)

    The user's response will be stored in the variable Answer. If the user's answer matches the corresponding group name stored in ColAdj (NOT case sensitive), a message similar to the following should be displayed:

```
CORRECT! The Ostrich is a Ratite.
```

If the user's answer did not match the correct group name in `ColAdj`, a message similar to the following should be displayed (assuming the user typed Corvine):

```
The Ostrich is not a Corvine.
The correct answer is Ratite.
```

Your program should work for any properly formatted cell arrays `ColAdj` and `Common`, whether they contain 5 entries or 500 entries (or any other number). Also note that when randomly choosing one of the text strings from `Common`, it must be possible for ANY of the entries to be chosen.

27. Consider the following section of code. Assume the matrix `IntMat` has already been defined.

```
[IntR,IntC]=size(IntMat);
Prompt=sprintf('Please enter a scalar or a matrix with %.0f
  rows.\n',IntC)
UserMat=input(Prompt);
Result=IntMat*UserMat;
```

If the user does not properly follow the instructions, an error will be generated. Add appropriate code using `try-catch` so that rather than terminating the program when the user enters an incorrectly sized matrix, a warning message similar to the following is generated, and `Result` set as described, after which the program continues. In this case, `IntMat` was a 3×7 matrix, and the user entered a matrix with five rows. Note that the column vector created when the user enters an incorrectly sized matrix has the same number of elements as the number of rows in `IntMat`.

```
The matrix entered should have had 7 rows.
Your matrix had 5 rows.
Result is being set to a 3 element column vector containing all
ones.
```

19 Looping Structures

Until this point, each line of code in the programs we have written was executed once at most each time the program was run. In the previous chapter, we learned how to choose which sections of code to execute and which to skip using conditional statements, but no statement was executed more than one time. It is extremely common, however, to want a group of lines in a program to be executed multiple times instead of just once, and the structures that allow us to do this are called loops. Typically, each time the code inside a loop is executed, it uses different values, generating different results. This ability to execute the same set of instructions many times extremely rapidly is perhaps the most important factor in making computers the powerful tools that they are.

In this chapter, we will learn about two types of looping structures:

- The `while` loop, that executes the code inside the loop as long as a condition is true.
- The `for` loop, that executes the code inside the loop a specified number of times.

First, however, we need to take another quick look at algorithms: specifically how to incorporate repetitive behavior.

19.1 Algorithms Revisited — Loops

LEARN TO: Design iterative algorithms

To incorporate looping in a written algorithm, we simply add an instruction that specifies to which step in the algorithm we wish to return and under what conditions.

EXAMPLE 19-1

Ask the user to make a selection from a menu. If the user did not make a correct choice but closed the menu instead, issue a warning message and ask the user to try again. This process should repeat until the user makes a correct selection, after which the selection is printed on the screen.

Known:

- *User's Choice*

Unknown:

- *None*

Assumptions:

- *None*

Algorithm:

1. *Display menu.*
2. *If the user closed the menu*
   (a) *Warn users that they must make a valid selection*
   (b) *Return to Step 1.*
3. *Display user's response on the screen*
4. *End the process.*

EXAMPLE   19-2

Create a written algorithm to calculate the sum of all integers between a lower bound and an upper bound.

Known:

- *Upper bound of whole number sequence.*
- *Lower bound of whole number sequence.*

Unknown:

- *Sum of all whole numbers between the upper and lower bound.*

Assumptions:

- *Since the problem does not explicitly state the upper and lower bounds, assume the program will ask for the values.*
- *Include the boundary values in the summation.*

Algorithm:

1. *Input the lower bound of the sequence.*
2. *Input the upper bound of the sequence.*
3. *If the lower bound is larger than the upper bound,*
   (a) *Warn the user that the input is invalid.*
   (b) *End the process.*
4. *If the upper bound is larger than the lower bound,*
   (a) *Create a variable to keep track of the sum (S).*
   (b) *Create a variable to keep track of the location in the sequence (L).*
   (c) *Set the initial value of S to be zero.*
   (d) *Set the initial value of L to be the lower bound.*
   (e) *If the value of L is less than or equal to the upper bound,*
      (i) *Add L to the current value of S.*
      (ii) *Add one to the current value of L.*
      (iii) *Return to step 4.e. and ask the question again.*
   (f) *Display the sum of the sequence (S).*
5. *End the process.*

In step 4.e.iii, we required that the interpreter return to an earlier step in the algorithm after changing the values of our variables. This allows us to create a **feedback loop** necessary to calculate the sequence of values. A feedback loop is a return to an earlier location in an algorithm with updated values of variables. It is important to note that if we failed to update the variables, the feedback loop will never terminate. A nonterminating feedback loop is also known as an **infinite feedback loop**.

Comprehension Check   19-1

Create a written algorithm to multiply all integer powers of 5, 5^x, for x between (and including) 5 and 50.

To implement looping in a flowchart, we simply add a directional arrow that eventually returns the flow of the chart back to a previous step. This is almost always associated with a decision of some form that answers the question, "Should this block of code repeat again?" The one notable exception is the rare case where you actually want an infinite loop that will never terminate.

EXAMPLE   19-3

Create a flowchart to ask the user to make a selection from a menu. If the user did not make a correct choice but closed the menu instead, issue a warning message and ask the user to try again. This process should repeat until the user makes a correct selection, after which the selection is printed on the screen.

Known:

- *User's Choice*

Unknown:

- *None*

Assumptions:

- *None*

Flowchart:

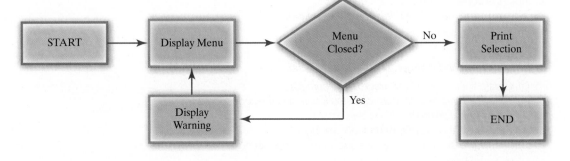

EXAMPLE   19-4

Create a flowchart to calculate the sum of a sequence of whole numbers, given the upper and lower bounds of the sequence.

Known:

- *Upper bound of whole number sequence.*
- *Lower bound of whole number sequence.*

Unknown:

- *Sum of all whole numbers between the upper and lower bound.*

Assumptions:

- *Since the problem does not explicitly state the upper and lower bounds, the interpreter will ask for the values. Include the boundary values in the summation.*

Flowchart:

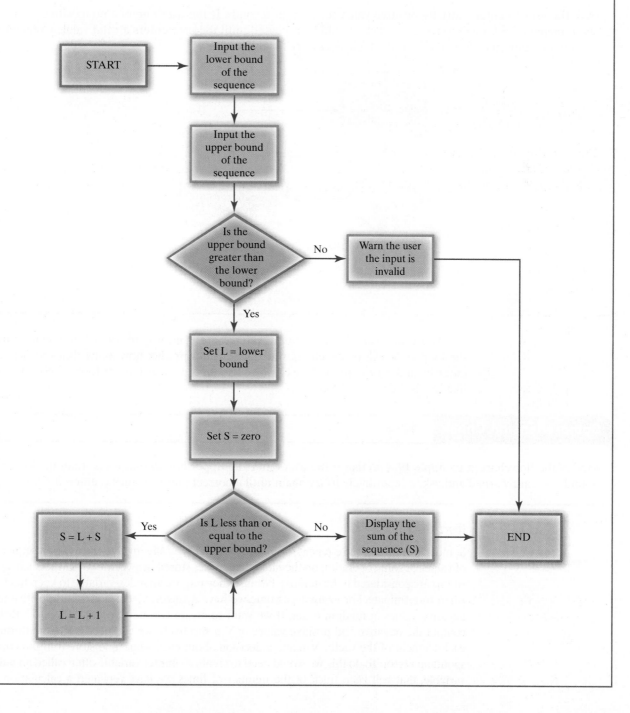

Comprehension Check   19-2

Create a graphical algorithm to multiply all integer powers of 5, 5^x, for x between (and including) 5 and 50.

EXAMPLE   19-5

Ask the user to enter a strictly positive value for time in seconds. If the user enters a nonpositive value, warn them and ask them to try again. This should be repeated until the user enters a valid value. Create a section of a flowchart to implement the logic necessary for this problem.

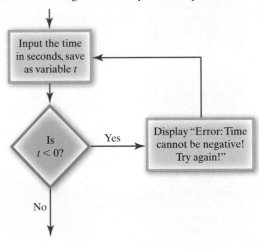

This is an example of a loop in which there is no way to predict how many times the loop code will be executed, since one cannot predict how many times a user will enter invalid information. As we will see soon, the usual type of loop to use in cases like this is the `while` loop.

Comprehension Check   19-3

Modify the flowchart in Example 19–4 so that if the user enters an upper bound that is less than the lower bound, they are warned and asked repetitively to try again until a correct pair of values is entered.

Iteration

Some algorithms require repeated calculations that typically involve the use of a sequence of values or some operation on the individual values stored in a vector or matrix. Such algorithms are considered to be iterative because they require an index variable to keep track of when to terminate. For example, assume we have a vector, V, which contains positive and negative values in random order. If we want to create two new vectors, VN and VP, that contain the negative and positive values of V respectively, we will need to iterate through each element of the vector V, make a decision about each value, and store it in the corresponding vector. To do this, we would need to create a **counter variable** often called an **index variable**, that will keep track of the number of times we have repeated a calculation or decision. If we create a counter variable, X, and initialize it to be the number 1, X will actually serve two purposes. In addition to keeping track of the number of times we have repeatedly made decisions and stored new values into VN and VP, it will also serve as the index variable into the V vector so that we can access element V(1), V(2), and so on until we reach the last element in V. Figure 15-3 demonstrates this scenario as a flowchart, including the iterated counter variable X.

Algorithms that will require iteration typically have a scenario where you have to repeat some decision or calculation "for each" or "for every" element or value within a sequence or vector. Since there is no "for each" or "for every" building block within an algorithm, this type of structure must be constructed out of the following steps:

- Initialization of a counter variable (e.g., Set X to be 1)
- A decision that involves the value of a counter variable (e.g., if X is less than or equal to the number of elements in V)
- Some action block that increments the counter variable (e.g., Set X equal to the current value of X plus 1)

EXAMPLE   **19-6**

Create a flowchart that implements the sorting of positive and negative values in a vector as previously described.

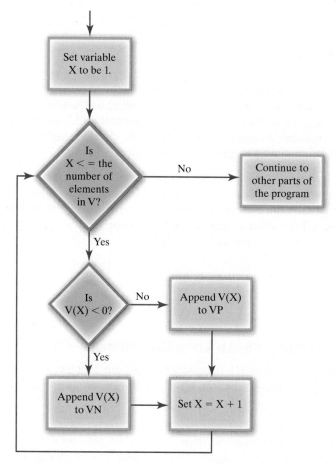

This is an example of a loop in which we can predict how many times the loop code will be executed, in this case the number of elements in the vector V. As we will see soon, the usual type of loop to use when the number of iterations is known or can be determined prior to executing the code is the `for` loop.

19.2 `while` Loops

We often want to repetitively execute a block of code as long as some condition remains true but there is no way to predict how many times the block of code will be executed without actually running the program. In many languages, including MATLAB, the best structure to use in cases like this is the `while` loop, which executes while a specified condition is true, or alternately stated, until that condition is false. The conditional part of the `while` statement has the same syntax as that of the `if` statement.

```
while Logical_Expression
      % executable statements
end
```

The big difference between the `while` and the `if` is as follows:

- After the logical statement of the **if** structure is evaluated and the correct block of code is completed, execution continues with the code below the corresponding `end` statement.
- After executing the code inside a **while** structure, execution returns from the `end` statement back to the `while` statement and asks the question again to determine whether to execute the code inside the loop again or to continue with the code immediately below the `end` statement.

As an example, some calculations cannot be expressed in closed form, and require an iterative solution to determine an approximate value. In this case, the value is estimated by some means, the error between the estimate and the actual value determined, then if the error is unacceptable, an improved estimate is made. This process repeats until the error is acceptably small.

```
while err >= 0.01   % require less than 1% error
   % perform another iteration of the estimation
   % recalculate error
end
```

Initializing `while` Loop Conditions

In a `while` loop, it is necessary to make sure that any variables used in the logical expression (the question) of the `while` statement are properly initialized before reaching the `while` command.

```
err = 1;
while err >= 0.01
   % Calculate new estimate and error
end
```

If err is not initialized, an error results the first time the while statement is executed since the variable err does not exist when MATLAB tries to compare it to the value 0.01. If err is initialized to a value less than 0.01, the loop never executes. In this case, the initial value is arbitrarily chosen to be 1, although we could have selected any number greater than or equal to 0.01 and the program would execute correctly.

EXAMPLE   19-7

We have a number stored in a variable T that we want to repeatedly divide by 17 until T is smaller than 3×10^{-6}. After determining the first value of T that meets the condition, it should be displayed in a formatted fprintf statement where the value of T is displayed in exponential notation. All extraneous output should be suppressed.

```
T=3993;
while T > 3E-6
        T=T/17;
end
fprintf('%E\n',T);
```

EXAMPLE   19-8

Given a vector of positive integers, V, we want to create two new vectors: vector Even will contain all of the even values of V; vector Odd will contain all of the odd values of V. To begin solving this problem, we need to seek out a MATLAB function that will help us to determine whether or not a given value is even or odd. The built-in function rem accepts two input arguments (Z = rem(X, Y)) such that Z is the remainder after dividing X by Y. For example, if we type rem(7, 2), the function would return 1 because 7/2 = 3 with a remainder of 1.

```
V=[2, 3, 91, 87, 5, 8];
Even=[];
Odd=[]; % initially, Even and Odd are both empty vectors
fprintf('\nV\t=\t%0.0f', V')
while length(V)>0
        if rem(V(1),2)==0        % is first element of V even?
                Even=[Even V(1)]; % place it in Even
        else                     % is first element of V odd?
                Odd=[Odd V(1)];  % place it in Odd
        end
        V(1)=[];                 % delete the first element of V
end
fprintf('\nEven\t=\t%0.0f', Even')
fprintf('\nOdd\t=\t%0.0f', Odd')
fprintf('\n')
```

The output of this code segment would be:

V	=	2	3	91	87	5	8
Even	=	2	8				
Odd	=	3	91	87	5		

Note that each time we "sort" a value from V into either Even or Odd, the value is then removed from V, and the process continues until V contains no elements. The original vector V is thus destroyed in the process. If we wanted to retain V intact, we could make a copy and destroy the copy instead. In Section 19–3, we will learn how keep V intact without the need for a duplicate.

Comprehension Check | **19-4**

Assume a vector V2 has already been defined and contains only values greater than 1. Write a while loop to calculate the product of the elements of V2 in sequence from the first element until the product is greater than 10^6 or all elements have been included in the product. The result should be stored in Prod.

Example: V2=[2 10 7 19];   Prod=2,660
Example: V2=[3 6 123 4 58 267 8 91 11]; Prod=137,144,016

Application: Repeating a Program under User Control

In some cases, it would be helpful to allow the user the option of having the program run again without shutting down and restarting it. A while loop can be a convenient mechanism for implementing this.

EXAMPLE | **19-9**

Assume you have a program to allow the user to choose a video game from a menu and then play that game as implemented by the following code. Note that PlayGame is a function that accepts the user's selection, runs the selected game, and returns the score.

```
clear
clc
Games={'Freakout','Chaoticon','Blimp'};
Sel=menu('Select a game',Games);
Score=PlayGame(Sel);
fprintf('Your %s score was %.0f.\n',Games{Sel},Score)
```

If you wished to give the user an option to play another game whenever they complete a game, this entire block of code (except possibly for the housekeeping and initialization commands) can be encapsulated in a while loop along with an appropriate request concerning whether they wish to play again or not.

```
clear
clc
Games={'Freakout','Chaoticon','Blimp'};
Again=1; % Initialization for first time
while Again==1 % Added to allow repetition
    Sel=menu('Select a game',Games);
    Score=PlayGame(Sel);
    fprintf('Your %s score was %.0f.\n',Games{Sel},Score)
    % Added to ask the user if they wish to continue playing
    Again=menu('Do you want to play another game?','YES','NO');
end
```

Note that a common method for writing a while loop is to write the code as though you were only going to execute it once, then add the while and end statements at the top and bottom of the code to be repeated, add whatever statements are necessary inside the loop to manage the criteria for repeating, and initialize any necessary values just before the while statement. Note that the housekeeping commands were not included in the loop, nor was the initialization of Games since these do not need to be repeated every time.

Comprehension Check 19-5

Given the following code, add the necessary statements to allow the user to repeat the program by answering the question, "Do you wish to calculate a new distance (Yes or No)?" The user is expected to type in either Yes or No as text, but the decision to repeat should not be case sensitive. You may assume the user actually types one of the two words, however.

```
LVel=input('Enter launch velocity [m/s]: ');
LAng=input('Enter launch angle [degrees]: ');
Dist=CalcDist(LVel,LAng);
fprintf('The projectile traveled %.1f meters.\n',Dist)
```

Application: Validating Input

When designing user interfaces, a fairly good rule to follow is "Never underestimate the incompetence of the user." All users make mistakes, even those who are extremely adept at interacting with computers. One way to mitigate user errors is to detect invalid input and require that the user try again. Since the programmer cannot predict how many attempts a user will require to enter valid information, a while loop is useful to repetitively inform the user they have erred and ask them to try again.

EXAMPLE 19-10

We want to input a number from the user that is between (and including) 5 and 10, but reject all other numbers. All extraneous output should be suppressed.

```
X=0;
while X < 5 || X > 10
    X=input('Enter a number between 5 and 10');
end
```

Note that X could be initially set to any value less than 5 or greater than 10.

Comprehension Check 19-6

Write a while loop that requires the user to input a number until a nonnegative number is entered.

In cases where a `while` loop is used to acquire valid user input, it is often desirable to warn the user when they enter an invalid value. The problem that arises is that we do not want to issue the warning prior to the first time the user attempts to specify valid input. There are two ways to handle this:

1. Ask for input before the `while` loop is reached, and then enter the loop only if the input was not acceptable. Inside the loop, the request for input is repeated following a warning statement.
2. The request for input occurs only inside the loop, and an `if` statement at the beginning of the loop code is used to selectively issue the warning only if the user has already entered invalid data at least one time.

There are other possible methods, but these are the most common.

EXAMPLE   19-11

Write a program to ask the user to make a selection from a menu. If the user did not make a correct choice but closed the menu instead, issue a warning message and ask the user to try again. This process should repeat until the user makes a correct selection, after which the selection is printed on the screen.

Version 1—display the menu both before the loop and inside the loop

```
Choices={'Choice A','Choice B','Choice C'};
Sel=menu('Choose one',Choices);
while Sel==0
    warning('You did not make a proper selection: Try again.')
    Sel=menu('Choose one',Choices);
end
fprintf('You selected %s.\n',Choices{Sel})
```

Version 2—display the menu only inside the loop

```
Choices={'Choice A','Choice B','Choice C'};
Sel=-1;
while Sel<=0
    if Sel==0
        warning('You did not make a proper selection: Try again.')
    end
    Sel=menu('Choose one',Choices);
end
fprintf('You selected %s.\n',Choices{Sel})
```

In version 2, note that `Sel` was initialized to –1, not 0, and the condition was `Sel <= 0` instead of `Sel ==0`. This allowed the loop to be entered the first time, but not display the "menu closed" warning prior to the first time the menu was displayed. `Sel` could have been initialized to any value less than zero in this case.

Comprehension Check   19-7

Write a program that asks the user to enter a positive integer that is evenly divisible by seven. If the user enters a value that does not meet this specification, warn the user and ask them to try again. This process should repeat until the user enters a valid value. Note that the warning should not be issued if the user enters a valid value on the first attempt.

Application: Estimating Intractable Functions

The solution to many equations can be easily calculated. For example, the quadratic formula gives us a simple way to find the two roots of a quadratic equation. Some equations, however, are extremely difficult or impossible to solve in a simple manner and require an iterative solution to obtain a close estimate. In cases like this, the solution is estimated by some means, this estimate is evaluated to see how close it is to the actual value, and then the estimate is improved based on the results.

EXAMPLE   **19-12**

We ask the user to enter a value greater than 1 to be stored in `Target`, and then find a value for X such that $Target = X^3 2^X \log X$. The final estimate should yield a value of the function within 0.001% of the correct value (`Target`). Note that for positive X, all three parts of the expression X^3, 2^X, and $\log X$, increase as X increases, so if our estimate is too low, we need to increase X, or if our estimate is too high, we need to decrease the estimate. We will proceed as follows:

- We will make two initial estimates: one that is guaranteed to be too high, and one that is guaranteed to be too low. We will then find the midpoint between the high and low estimates, use that as a new estimate, and if we are not close enough, we will replace either the high or low estimate with the new estimate as appropriate and find a new midpoint estimate. This will be continued until we are within 0.001% of the target value.
- Our initial low estimate will be `LoX=1`. Since $\log 1 = 0$, the function evaluates to 0 in this case. This estimate is thus guaranteed to be too low since `Target >1`.
- Our initial high estimate will be `HiX=10*Target`. The validity of this upper estimate follows from the fact that $\log 10 = 1$.

```
clear
clc
% We will assume the user inputs a value greater than 1
Target=input('Enter a target value greater than 1: ');
LoX=1; % Initial low X
HiX=10*Target; % Initial High X
Err=1; % Initialization to get into loop
while Err>0.00001 % aiming for <= 0.001%
    X=(LoX+HiX)/2; %get midpoint for new estimate
    EstVal=X^3*log10(X)*2^X; % Calculate function with estimate
    Err=abs(EstVal-Target)/Target; % Calculate error
    if EstVal>Target % Update High or Low X
        HiX=X;
    else
        LoX=X;
    end
end
fprintf('X=%.8f yields %.8E.\n',X,EstVal) % Display results
```

For example, if the user enters 42, the results are

```
X=2.57719240 yields 4.20000394E+01.
```

If the user enters 7879, the results are

```
X=5.76037774 yields 7.87899076E+03.
```

Comprehension Check 19-8

Given a value of F between 0 and 1 entered by the user, write a program to estimate the value of θ between 0 and $\pi/2$ that will satisfy the equation $F = e^{-\theta/10} \cos \theta$. Your final estimate should give a result accurate to within 0.0002% of the actual value of the function. *Hint*: Does F increase or decrease as θ increases over the range 0 to pi/2?

Converting "Until" Logic for Use in a `while` Statement

In some cases, it may make more sense to use the word *until* in phrasing a conditional loop, but MATLAB does not have such a structure. As a result, it is sometimes necessary to rephrase our conditions to fit the `while` structure.

"Until" Logic Condition	`while` Logic Translation

until cows == home
        party
end
	while cows ~= home
	        party
	end

until homework == done
        TVpower = off
end
	while homework ~= done
	        TVpower = off
	end

Comprehension Check 19-9

Write a program using a `while` loop that requires the user to input numbers until a negative number is entered. For each nonnegative number entered, a statement should be displayed in the command window in the following format:

```
You entered 39. The square root of 39 is 6.245.
```

When the user enters a negative value, the program should terminate after displaying the message:

```
You entered -0.42. Program terminated.
```

Tracking Number of Iterations in a `while` Loop

Sometimes, you wish to keep track of the number of times the code inside a `while` loop is executed. In general, the way to accomplish this is to set up a variable to use as a counter, and increment it each time the loop code is executed.

EXAMPLE 19-13

Write a program that will generate a random integer between 1 and 100. The user is then asked to guess the number. If the user fails to guess the random integer, the program prints a message specifying whether the guess was too high or too low, and asks the user to guess again. When the user guesses the correct value, a statement similar to the following is displayed on the screen:

```
You needed 7 attempts to guess 42.
```

```
Target=randi(100);
Guess=input('Try to guess the random integer between 1 and 100: ');
Tries=1;
while Guess~=Target
    if Guess<Target
        fprintf('%.0f is too low. Try again.\n',Guess)
    else
        fprintf('%.0f is too high. Try again.\n',Guess)
    end
    Guess=input('Try to guess the random integer between 1 and 100: ');
    Tries=Tries+1;
end
fprintf('You needed %.0f attempts to guess %.0f.\n',Tries,Target)
```

Comprehension Check   19-10

Write a section of code that will ask the user to enter a positive number P between 10 and 100. If the user fails to enter a valid value, issue a warning and ask them to try again. If the user has made 10 attempts without success, exit the loop and continue with the last value entered by the user.

19.3 for Loops

LEARN TO: Automate a segment of code using a `for` loop
Calculate the number of executions of a `for` loop
Convert an algorithm into code involving a `for` loop

We often need to execute a block of code when we know, or can calculate ahead of time, the number of times the code will be iterated. The looping structure that is most commonly used in cases like this is called in many languages, MATLAB included, a `for` loop. Contrast this to the `while` loop, which is typically used when we cannot determine the number of times the loop will iterate without actually running the code. With that said, anything you can do with a `while` loop can also be done with a `for` loop, and anything you can do with a `for` loop can be done with a `while` loop. The reason we have two different looping structures is that it can be extremely cumbersome and confusing to code `while` logic using a `for` structure and vice versa, so try to wisely choose the looping structure to fit the situation.

A few examples where a `for` loop would probably be the better choice:

- For every pressure sensor, record the pressure reading. (You know how many sensors there are.)
- For each employee, check the date of the employee's last safety training. (You know how many employees there are.)
- For every fifth data point, enter its value into an array for plotting. (You know how many data points there are, or can determine this ahead of time using `length` or `size`.)
- For every second counting down from 10 to 0 seconds, announce the time remaining until launch.

The most common syntax of a `for` loop is shown in the following illustration:

```
for index = start : step : finish
   % executable statements
end
```

- `index`: variable chosen to keep track of the number of times the loop has executed
- `start`: first value of the index variable in the loop
- `step`: incremental value to use to advance from start to finish
- `finish`: final value of the index variable in the loop

In addition to being merely a counter that determines how many times the loop code will execute, the index variable can be used in the loop to calculate other variables, to index an array, to set function parameters, etc.

A sample `for` loop follows. This loop will display on the screen the numbers from 1 to 5, each on a separate line.

```
for k=1:1:5
    fprintf('%0.0f\n',k)
end
```

NOTE

The variable k is defined in the header of the `for` loop. If you modify the value of k inside the `for` loop, it might lead to unintended results.

Understanding the `for` Loop

Most program statements do exactly the same type of thing every time, although they may be manipulating different information. The `for` loop is slightly different. A `for` statement can be arrived at during program execution in one of two ways:

1. It can be executed immediately following a statement that was outside the loop, typically the statement immediately above it.
2. It can be executed upon returning from the `end` statement that marks the end of the loop after the loop has executed one or more times.

What the `for` statement does in these two cases is slightly different and is crucial to understanding its operation.

1. When the `for` statement is arrived at from another statement *outside* the loop, it does the following:
   - Places the start value into the index variable.
   - Decides whether or not to execute the instructions inside the `for` loop.
2. When the `for` statement is arrived at from its corresponding `end` statement below, it does the following:
   - Adds the step value to the index variable.
   - Decides whether or not to execute the instructions inside the `for` loop.

NOTE

Since the loop can count either up or down, we use the word "beyond" to describe the comparison of the index variable and the finish value.

The second step is the same in both cases and requires a bit more explanation. To determine whether or not to execute the instructions inside the loop (those between the `for` and its corresponding `end`), the `for` statement asks the question, "Has the index variable gone beyond the finish value?"

- If the answer is no, execute the statements inside the loop.
- If the answer is yes, skip over the loop and continue with the first statement (if any) immediately following the corresponding `end` statement.

If the loop is counting up (step > 0), then we could ask, "Is the index variable greater than the finish value?" If the loop is counting down (step < 0), we could use "less than" instead. Several examples follow.

EXAMPLE 19-14

Determine how many times each `for` loop will execute, and what values the index variables will take on.

(a)
```
for n=1:1:4
    % loop statements
end
```
The statements in the loop will be executed four times, once each with $n = 1, 2, 3,$ and 4.

(b)
```
for time=1:2:4
    % loop statements
end
```
The statements in the loop will be executed two times, once each with `time` = 1 and 3. Note that `time` never actually equals the finish value in this case. When it is incremented to 5, it is beyond the finish value, so the loop only executes with `time` equal to 1 and 3.

(c)
```
for gleep=4:-2:0
    % loop statements
end
```
The statements in the loop will be executed three times, once each with `gleep` = 4, 2, and 0.

(d)
```
for raft=13:-6:-4
    % loop statements
end
```
The statements in the loop will be executed three times, once each with `raft` = 13, 7, and 1. When `raft` is decremented to -5 it is beyond the finish value.

(e)
```
for k=6:1:4
    % loop statements
end
```
The statements in the loop will not be executed. `k` is beyond the finish value when the loop is first entered.

(f)
```
S1=5;
S2=7;
for wolf=S1:S2:S1^2+S2+1
    % loop statements
end
```
The statements in the loop will be executed five times, once each with `wolf` = 5, 12, 19, 26, and 33.

As noted earlier, it is generally a bad idea to redefine the index variable inside of the loop. Similarly, redefining any variables from which the index variable is calculated is also strongly discouraged. For example, S1 and S2 should not be changed inside the loop.

Arithmetic Sequences

When dealing with **arithmetic sequences**—those that require only changing the **index variable** by adding or subtracting a number from it—there are two primary ways to proceed.

1. Set up the sequence explicitly, such as
```
for k=2:2:10
    % Use k as the sequence in whatever calculations
end
```

2. Set the index variable to count from 1 through the desired number of iterations, and then calculate the desired values from the index values, such as

```
for k=1:1:5
    k1=2*k;
    % Use k1 as the sequence in whatever calculations
end
```

In some cases, the direct sequence may be more intuitive, and in others the calculated equivalent makes more sense.

In each of the following examples, a sequence of five numbers is given, with two solutions shown to generate that sequence. The first code segment in each example uses the for loop to count from 1 to 5, and the values desired are calculated inside the loop. The second segment of code accomplishes the same goal by manipulating the start, step, and finish values instead. Although we are using print statements inside the example loops so that you can easily type these into MATLAB and see the sequence generated, you will seldom print the index values. Most of the time, the index values will be used in a calculation of some form, not just printed to the screen.

EXAMPLE   19-15

We desire the sequence to be 3, 6, 9, 12, 15.

```
for k=1:1:5
    fprintf('%0.0f\n',3*k)
end
```

Given that k goes through the sequence k = 1, 2, 3, 4, 5, the desired sequence can be calculated as 3*k.

```
for k=3:3:15
    fprintf('%0.0f\n',k)
end
```

This illustrates the idea that multiplication and division of a sequence by a constant to generate a new sequence affects all three loop control parameters: start, step, *and* finish.

EXAMPLE   19-16

We desire the sequence to be 3, 4, 5, 6, 7.

```
for k=1:1:5
    fprintf('%0.0f\n',k+2)
end
```

This can be calculated as k + 2

```
for k=3:1:7
    fprintf('%0.0f\n',k)
end
```

Adding and subtracting a constant to create a new sequence affects only the start *and* finish *values.*

EXAMPLE　19-17

We desire the sequence to be 4, 6, 8, 10, 12.

Note that the standard order of operations applies, and the multiplication must be done first.

```
for k=1:1:5
    fprintf('%0.0f\n',2*k+2)
end
```

This can be calculated as `2*k + 2`*.*

```
for k=4:2:12
    fprintf('%0.0f\n',k)
end
```

Note that all three loop parameters are affected: `start` *and* `finish` *are subject to both multiplication and addition, but* `step` *is only affected by multiplication.*

Comprehension Check　19-11

Write a `for` loop to display every even number from 2 to 20 on the screen.

Comprehension Check　19-12

Write a `for` loop to display every multiple of 5 from 5 to 50 on the screen.

Comprehension Check　19-13

Write a `for` loop to display every odd number from 13 to −11 on the screen.

NOTE

Although it is common in most languages to choose as a generic index variable name i or j, this is best avoided in MATLAB since both i and j are predefined to be the square root of negative one.

Using Variable Names to Clarify Loop Function

Just like other variables, you should strive to pick a meaningful name for the index variable of a `for` loop. If you are simply using it to index into a vector, you might use the variable name n. If you were using it to step through the columns of an array, you might choose C or col. In many cases, the index variable is setting values of specific types of parameters, such as pressure, in which case you might choose P or Press.

Using variables with meaningful names as loop controls makes it easier for those trying to interpret the program to keep track of what the loop is doing. The following loop keeps track of the position of an object moving at a constant speed. By defining the step distance as Speed*TimeStep, thus an increment in position, the number of steps in the loop is flexible as well. Note that these variables would need to be defined earlier in the program.

```
for Position=StartPosition:Speed*TimeStep:FinalPosition
    fprintf('%0.0f\n',Position)
end
```

Using a `for` Loop in Variable Recursion

An important use of loops is for **variable recursion**—passing information from one loop execution to the next. A simple form of recursion is to keep a running total. The following loop determines the total sales for all the vendors at a football game. The loop assumes that the vector `Sales` contains the total sales for each vendor. The number of vendors is computed when the loop begins, using the `length` command, and the total sales are accumulated in `TotalSales`. Note that `TotalSales` must be initialized before the loop is entered, since it appears on the right-hand side of the equation:

```
TotalSales = 0;
for Vendor=1:1:length(Sales)
    TotalSales = TotalSales + Sales(Vendor);
end
```

This code also introduces a very important use of `for` loops: to automatically step through the values in a vector or array. Note here that the index variable `Vendor` is used to sequentially access each element of the `Sales` vector.

Comprehension Check   19-14

Assume a vector `Vals` has already been defined. Write a `for` loop that will calculate the sum of all positive values in `Vals` and place the result in the variable `PosSum`.

Manipulating a `for` Loop Index

Many kinds of computations can be performed in a loop. Many of the previous examples of loops will output the value of `k` to the screen each time through the loop. Other sequences can be achieved by having an expression other than `k` in the executable part of the loop.

Comprehension Check   19-15

Consider the following table of values. Determine the formulae to represent columns A through J. In some cases, the formulae in rows 1–10 will be similar. In other cases, the entry in the first row will be a number rather than a formula.

Create a `for` loop to display the values; use tabs in a formatted `fprintf` statement to create each column.

k	A	B	C	D	E	F	G	H	I	J
1	2	1	1	1	7.25	10	1	2	1	1.00
2	4	3	4	3	7.50	9	−1	4	2	0.50
3	6	5	9	6	7.75	8	1	8	6	0.33
4	8	7	16	10	8.00	7	−1	16	24	0.25
5	10	9	25	15	8.25	6	1	32	120	0.20
6	12	11	36	21	8.50	5	−1	64	720	0.17
7	14	13	49	28	8.75	4	1	128	5020	0.14
8	16	15	64	36	9.00	3	−1	256	40320	0.13
9	18	17	81	45	9.25	2	1	512	362880	0.11
10	20	19	100	55	9.50	1	−1	1024	3628800	0.10

Using the Index Variable as an Array Index

The previous example used the index variable `Vendor` to access each element of the `Sales` vector in sequence to add them together. If we wanted to add only the odd-numbered elements, the only change necessary would be to change the step variable to 2. If we wanted every fourth element of the array beginning with the fourth element, the `for` statement would become

```
for Vendor=4:4:length(Sales)
```

and so forth.

To access elements in a matrix, we could use a single index since MATLAB will handle this, but we would have to be very careful to make certain which row and column is being accessed. For example, if A is a 3×2 matrix, $A(4) = A(1,2)$. Since this can be quite confusing, it is usually better to use the double index notation when accessing values in a matrix. One method to accomplish this would be to use nested `for` loops, that is, a `for` loop inside another `for` loop. Let us modify the previous football sales example to illustrate.

Assume we have a 2-D array named `Sales` where each row represents a specific vendor and each column represents a different category of item, such as drinks, hot dogs, hats, etc. Now if we want the total sales, we need to step through every element in every column (or every element in every row) and add them all up. To determine the number of vendors and items in the matrix, we use the `size` function, rather than the `length` function we used previously with the vector.

```
% Initialize the sum of all sales to 0
TotalSales = 0;
% Determine the number of vendors (rows) and items (columns)
[NumVendors, NumItems]=size(Sales)
for Vendor=1:1:NumVendors % this will index the rows
        for Item=1:1:NumItems % this will index the columns
                TotalSales = TotalSales + Sales(Vendor, Item);
        end
end
```

Note that the inner `for` loop (`Item` loop) goes through all items (columns) before exiting, at which point the outer `for` loop increments to the next vendor (row) and the inner loop resets to the first column and steps through all items again, but for a different vendor.

If there were four vendors (rows) and three items (columns), the order in which the entries in the `Sales` array would be added to `TotalSales` would be:

(1,1); (1,2); (1,3); (2,1); (2,2); (2,3); (3,1); (3,2); (3,3); (4,1); (4,2); (4,3)

You may recall from Chapter 15 that this problem could be solved more simply with a single line of code:

```
TotalSales=sum(sum(Sales));
```

There are many situations, however, where the built-in matrix operations will not accomplish the desired purpose and you *must* set up a pair of nested loops to step through the elements of the array.

Comprehension Check | **19-16**

Write two nested `for` loops to determine how many positive values (including zeros) are in each column of the matrix M1, which has already been defined. The results should be stored in a row vector PosNums in which each element contains the number of positive values found in the corresponding column of M1.

Example: `M1=[3 -6 0;-4 -8 2;8 -9 1];`   `PosNums=[2 0 3]`

EXAMPLE | **19-18**

Assume you're required to write a MATLAB program that will allow the user to generate a graph and plot a linear trendline of flowrate data manually provided by the user. Assume that your program will prompt the user to enter the initial time of data collection, the final time, and the interval between data samples, all in units of minutes. Your code should then iterate and prompt the user to record the corresponding flowrate sample value in units of gallons per minute at each required sample time.

```
clear
clc

AbsStart=input('What is the initial sample time [min]? ');
AbsEnd=input('What is the final sample time [min]? ');
AbsInc=input('How often should samples be taken [min]? ');

Abs = []; % create empty vectors to hold the user input
Ord = [];
cnt = 1; % cnt = count number of samples entered by user
for I=AbsStart:AbsInc:AbsEnd
    prompt=sprintf('Sample #%0.0f, at %0.2f min\nFlowrate (Q) [gpm]: ',cnt,I);
    Abs=[Abs I];
    x = input(prompt);
    Ord = [Ord x];
    cnt = cnt + 1;
end

% polyfit to calculate trend:
C = polyfit(Abs,Ord,1);
TrendAbs = [AbsStart:AbsInc/20:AbsEnd];
TrendOrd = C(1)*TrendAbs + C(2);

% generate plot
plot(Abs,Ord,'x',TrendAbs,TrendOrd,'-')
% other proper plot commands to show, for example, the axis labels, are
% entered here, but are not shown for space considerations
```

Sample Usage:

```
What is the initial sample time [min]?: 1
What is the final sample time [min]?: 6
How often should samples be taken [min]?: 2
Sample #1, at 1.00 min
```

```
Flowrate (Q) [gpm]: 7
Sample #2, at 3.00 min
Flowrate (Q) [gpm]: 17
Sample #3, at 5.00 min
Flowrate (Q) [gpm]: 27
```

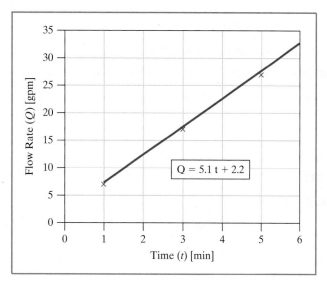

$Q = 5.1\,t + 2.2$

Comprehension Check   19-17

Assume a matrix `M2` has already been defined. Write two nested `for` loops to determine how many rows of the matrix `M2` contain only negative values. The results should be stored in `NegRows`.

Example: `M2=[3 -6 0;-4 -8 -2;8 -9 1];    NegRows=1`

Ending `for` Loops Early

In certain situations, it might be desirable to build in logic within a `for` loop that allows for the loop to terminate when a certain situation is encountered. The `break` command instructs MATLAB to exit the loop and continue as if the entire loop sequence was finished. It is important to realize that the `break` command does *not* terminate the function or program—it will only exit the loop code.

EXAMPLE   19-19

Consider the following segment of MATLAB code:

```
S = input('Type the starting point of the loop: ');
E = input('Type the ending point of the loop: ');
for n=S:1:E
```

```
        if 1/n == Inf
            fprintf('Uh oh! Division by zero!\n')
            break
        end
        fprintf('%0.1f\n',1/n)
    end
    fprintf('. . . ALL DONE!\n')
```

What is the outcome if the user enters S=1 and E=3?

When the user provides the value of 1 for the starting value and 3 for the ending value, everything runs smoothly:

```
Type the starting point of the loop: 1
Type the ending point of the loop: 3
1
0.5
0.3
. . . ALL DONE!
```

What is the outcome if the user enters S=−3 and E=3?

However, when the user provides the values of −3 for the starting value and 3 for the ending value, notice what happens:

```
Type the starting point of the loop: -3
Type the ending point of the loop: 3
-0.3
-0.5
-1
Uh oh! Division by zero!
. . . ALL DONE!
```

There are a few important concepts to note from the output of this execution of our code segment. First, the code after the for *loop, the output statement that displays ". . . ALL DONE!" is executed, so the* break *command will not skip any code after the* for *loop has terminated. Second, since the output of the number occurred after the check to see if a division by zero has occurred, the output command did not display* Inf *(1/0) because the* break *command terminated the loop, which means that no code after the* break *command within an iteration of a loop will execute. Last, it is important to note that the* break *command terminates the entire loop—not just the iteration, so the numbers 1 through 3 will not be displayed because the loop finished early.*

When a break command appears within a nested for loop, the break will apply to only the loop where the break occurs and will not terminate the entire nested loop.

EXAMPLE   19-20

Consider the following segment of MATLAB code:

```
Si = input('Type the starting point of the inner loop: ');
Ei = input('Type the ending point of the inner loop: ');
So = input('Type the starting point of the outer loop: ');
```

```
Eo = input('Type the ending point of the outer loop: ');
for n=So:1:Eo
    for m=Si:1:Ei
        if n == m
            fprintf('Same value. . . skipping!\n')
            break
        end
        fprintf('%0.0f\t%0.0f\n',n,m)
    end
end
fprintf('. . . ALL DONE!\n')
```

What is the outcome if the user enters 3, 5, 1, 2?

When the user provides the value of 3 for the starting value and 5 for the ending value of the inner loop and 1 for the starting value and 2 for the ending value of the outer loop, everything runs smoothly since these two sets of numbers will never intersect.

```
Type the starting point of the inner loop: 3
Type the ending point of the inner loop: 5
Type the starting point of the outer loop: 1
Type the ending point of the outer loop: 2
1    3
1    4
1    5
2    3
2    4
2    5
...ALL DONE!
```

What is the outcome if the user enters 3, 5, 3, 5?

However, when the user provides the values of 3 for the starting value and 5 for the ending value of both the inner and outer loops, notice what happens:

```
Type the starting point of the inner loop: 3
Type the ending point of the inner loop: 5
Type the starting point of the outer loop: 3
Type the ending point of the outer loop: 5
Same value ...skipping!
4    3
Same value ...skipping!
5    3
5    4
Same value ...skipping!
... ALL DONE!
```

Notice that the outer loop is unaffected by the `break` *command and continues looping over values 3 to 5, regardless of whether the inner loop encounters a* `break` *command.*

Comprehension Check   **19-18**

Determine the output of the following code:

```
Up=1;
for Down=10:-2:0
    if Down==Up
       fprintf('Program Terminated')
       break
    end
    fprintf('Up=%.0f\tDown=%.0f\n',Up,Down)
    Up=Up+1;
end
```

Ending `while` Loops Early

Just like the `for` loop, the `while` loop allows code to be written in such a way that the `break` command will terminate the loop early. Note that the behavior of the `break` command is the same with a `while` loop—it will only end the loop early and will *not* terminate a program or function. After encountering the `break` command, the code continues executing at the next line after the end of the loop.

EXAMPLE   **19-21**

NOTE

`true` and `false` are predefined logical constants with rather obvious values.

Consider the following `while` loop, where the loop will execute while "true." Note that this would normally create an infinite loop, but we will use the break command to exit under certain conditions. This is a common approach to designing algorithms where the looping logic is indeterminate or too complicated to efficiently implement with a pure `while` loop or `for` loop structure.

```
S = 0;
while true
    A = input('Type the alpha value: ');
    if A == 5
      break
    end
    S = A + S;
    B = input('Type the beta value: ');
    if B == 7
      break
    end
    S = 2*B + S;
end
fprintf('The final answer is: %0.0f\n',S)
```

Consider the following output sequence from the code segment shown.

```
Type the alpha value: 1
Type the beta value: 2
Type the alpha value: 4
Type the beta value: 5
Type the alpha value: 5
The final answer is: 19
```

NOTE

Remember that to escape from an infinite loop use CTRL+C.

Note the loop terminates before the alpha value of 5 can be added to the summation variable. Likewise, if the user never types the values that trigger the `break` commands (alpha value of 5 or beta value of 7), this code will loop "infinitely." In practice, this code will eventually terminate once the value stored in the variable is "overflown" with a value greater than the variable S can handle, but if the user continues manually typing in small numbers, this will not happen for a long time.

Comprehension Check | **19-19**

Determine the output of the following code:

```
Q=20;
R=75;
while R<100
    R=R-Q;
    if R<0
       fprintf('R Negative')
       break
    end
    fprintf('%.0f\n',R)
end
```

IN-CLASS ACTIVITIES

ICA 19-1

Create a written algorithm or flowchart as assigned to ask the user to draw a card from a deck of cards until the user pulls out the queen of hearts.

ICA 19-2

Create a written algorithm or flowchart as assigned to ask the user to enter the number of internet-enabled devices they own. If the user enters a negative number or a number greater than 50, the message "This is not reasonable. Try again." Should appear to the user and they should be asked to enter another value. The warning message should continue to appear each time they enter an invalid value. When the user provides an acceptable value, the message "Thank you" should be displayed and the algorithm should terminate.

ICA 19-3

Create a written algorithm or flowchart as assigned to determine the factorial value of an input integer between 1 and 20. Add a statement to check for input range.

ICA 19-4

Create a written algorithm or flowchart as assigned to ask the user to enter two numeric values in a vector Comp. The algorithm should then determine how many rows of the matrix M1 contain at least one value between (and including) the values entered by the user. Note that the user may enter the two values in either order: smallest first or largest first.

ICA 19-5

Create a written algorithm or flowchart as assigned to first ask the user to enter a positive number. After checking that the value entered is positive, determine all of the even numbers between 0 and that value that are also evenly divisible by both 3 and 5. If the user enters a nonpositive value, issue a warning and ask the user to enter a new value repetitively until a positive value is entered. At the end of your algorithm, your algorithm should display the total number of values that meet the specified criteria.

ICA 19-6

Create a written algorithm or flowchart as directed by your instructor for the following problem.

Ask the user to enter a 1×5 vector of numbers. Determine the size of the user entry. If the user enters a 5×1 vector, use a warning statement to inform the user of their error, and set the new input value to be the transpose of the user input value. If the user enters a vector that is not the correct size, ask the user to reenter the vector using the correct dimensions. The program should continuously check the user input each time it is entered. Continue to ask the user to input a new vector until the user has entered three tries. If the user fails to enter a vector correctly given three chances (including the first entry), use an error message to terminate the program. If the user enters a correct vector, or the vector is transposed, the program should restate the result for the user. An example is shown here:

```
The resulting 1 × 5 vector is:
    1 2 3 4 5
```

ICA 19-7

Create a written algorithm or flowchart as assigned to repetitively read a flowmeter monitoring the flow of fuel to an engine as well as a speedometer measuring the speed of the vehicle. If the speedometer reading is less than 10, no action should be taken other than to repeat the measurements. If the speedometer reading is greater than or equal to 10, then if the ratio of the flowmeter value to the speedometer reading is greater than or equal to 0.004 a warning message should be displayed on the screen stating "Excessive fuel use," and the readings should be taken again. The warning should be repeated each time the ratio is too high. When the speed drops below 10 or the ratio drops below 0.004, the message "Normal operation" should be displayed. This message should *not* be repeated each time unless the warning message appears again. If both the flowrate and the speed are zero, the program should terminate.

ICA 19-8

For each of the following code segments, determine the contents of the specified variable following execution. If an error will occur, write ERROR and explain what caused the error.

(a) What is stored in Q?

```
Q=[];
N=1;
while length(Q)<4
  Q=[Q,N]*2;
  N=N*3;
end
```

(b) What is stored in Q?

```
Q=[1 2 3 4];
N=1;
while N<length(Q)
  Q=[2*Q,N];
  N=N*2;
end
```

(c) What is stored in S and N?

```
Q=[4 -3 3 -3 5 -2 -3 -3 1 0];
S=0;
N=1;
while S>=0 && N<=length(Q)
  S=S+Q(N);
  N=N+1;
end
```

(d) What is stored in S and N?

```
Q=[4 -3 3 -3 5 -2 -3 -3 1 0];
S=0;
N=1;
while S>=0 && N<=length(Q)
  if Q(N)<0
    if S<0
      S=S-Q(N);
    else
      S=S+Q(N);
    end
```

```
    else
      if S>0
        S=S-Q(N);
      else
        S=S+Q(N);
      end
    end
    N=N+1;
end
```

ICA 19-9

(a) Assume `CA1` and `CA2` are cell arrays that both contain the same number of text strings in a single row. Explain what the following code does.

```
NumWords=length(CA1);
Point=1;
while Point<=NumWords && strcmpi(CA1{Point},CA2{Point})
   fprintf('Word %.0f: %s.\n',Point,upper(CA1{Point}))
   Point=Point+1;
end
if Point>NumWords
   fprintf('100%% match.\n')
else
   fprintf('\nWord %.0f mismatched: %s and %s.\n',...
           Point,upper(CA1{Point}),upper(CA2{Point}))
end
```

(b) What would appear on the screen if `CA1` and `CA2` are defined as follows?

```
CA1={'Art','BAG','Cat','dog','EgG','Fan','GAB','HOe','iCe','jar'};
CA2={'Art','Bag','CaT','DOG','eGG','Far','gab','Hot','ICe','ajar'};
```

ICA 19-10

Fill in the table with the values of `k`, `b`, `x`, and `y` that would be displayed as output each pass through the `while` loop of the following program. If the program terminates before it makes seven passes, leave any additional entries blank.

```
k = 1;
b = -2;
x = -1;
y = -2;
while k <= 3
  x = x + 1;
  k = k + 1;
  fprintf('%0.0f %0.0f %0.0f %0.0f\n',k,b,x,y)
  y = x^2 - 3;
  if y < b
    b = y;
  end
end
```

Pass	k	b	x	y
First				
Second				
Third				
Fourth				
Fifth				
Sixth				
Seventh				

ICA 19-11

Draw a grid similar to the example shown that has the number of rows and columns of subplots that will be generated by the following code. Indicate the location of the graphs created by shading the location of each subplot in a manner similar to the example.

```
a = 1;
X=a*[1 2 3;4 5 6];
Y=[1 2 3;7 8 9];
subplot(4,5,a)
plot(X,Y)
while a <= 20
  X=a*[1 2 3;4 5 6];
  Y=[1 2 3;7 8 9];
  subplot(4,5,a)
  plot(X,Y)
  a = a * 2;
end
```

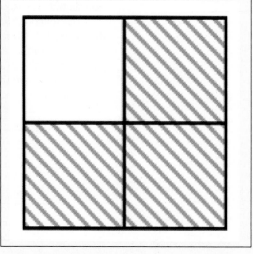

Example: If a 2 x 2 subplot grid was created and a graph would appear in positions 2, 3 and 4, then the answer would appear as:

How many data series are plotted on each subplot graph in the code?

ICA 19-12

Assume that a simple menu has been created by the following line of code:

```
SystemStatus=menu('System Status','ON','OFF');
```

Write a short section of code that will handle the situation if the user closes the menu instead of making a selection. If the menu is closed without a proper selection, the user should be told to try again, then the menu should be redisplayed. This should continue until the user makes a proper selection.

ICA 19-13

Write a MATLAB program to implement the algorithm described in ICA 19–2.

ICA 19-14

Write a program that will ask the user to input a single number N. If the number is nonzero, a one hundred element row vector named SEQ will be filled with the values 0.01*N, 0.02*N, . . . , 0.99*N, N. Note that this must work for both positive and negative values of N. After filling the vector, the program should ask (yes or no, not case sensitive) if this is correct, where "No" implies that the user wants to enter a new value for N. If the user answers "No," the user should be asked for the revised value of N, in which case it should refill SEQ with the corrected values. This should continue until the user answers "Yes" or enters a zero as the revised value, at which point the program terminates.

ICA 19-15

Write a MATLAB program to implement the algorithm described in ICA 19-6.

ICA 19-16

Write a function named `CountDown` that accepts a total time `T` in seconds and displays the time remaining after half of the previously displayed time has elapsed as described below. The function should also accept an integer `Steps` as also described below.

The function should immediately display the value of `T`, and then wait T/2 seconds before displaying T/2. It should then wait T/4 seconds and display T/4, etc. This should continue until only $1/2^{Steps}$ of the original time remains, at which point it waits for this remaining time, displays a zero, and emits an audio tone.

It is suggested that you look up the `pause` function and the `beep` function.

Example: `CountDown(60,6)` displays the following followed by an audio tone.

60	Displayed immediately
30	Displayed 30 seconds later
15	Displayed 15 seconds later
7.5000	Displayed 7.5 seconds later
3.7500	etc.
1.8750	etc.
0.9375	etc.
0	Displayed 0.9375 seconds later accompanied by a beep

ICA 19-17

You are writing the code to control a chemical reaction. Sensors are used to place values into the variables `Temp` (temperature in °C), `Pres` (pressure in atmospheres), `pH` (pH of solution), and `Time` (elapsed time in minutes since start of reaction). A function called `Update` reads the sensors and modifies the contents of the variables to indicate the current conditions. To use the `Update` function, you simply include the line

```
[Temp,Pres,pH,Time]=Update;
```

in your code as needed.

Write a program that will use the `Update` function once every minute to read new values from the sensors, and continue doing so as long as the pH has decreased by at least 0.02 since the last reading. (*Hint:* Remember the `pause` function – `pause(n)` pauses execution for n seconds.)

If at any time the pressure exceeds 5 atmospheres or the temperature exceeds 200 degrees Celsius, a message should be generated indicating which parameter is beyond the safe limits and the value of that parameter. In this case, the program should call the function `terminate` which will shut down the reaction and the program stops after printing a message to the screen such as

```
Reaction terminated after X minutes.
Final pH is x.xxx.
```

where X and x.xxx are replaced with the elapsed time and the final pH.

If at any point, the pressure exceeds 4 atmospheres or the temperature exceeds 175 degrees Celsius, an appropriate warning should be generated stating which parameter is too high and what its value is, but the reaction should continue and the program keep running.

When the pH has decreased by less than 0.02 from the previous reading, the reaction should be terminated, a message printed giving elapsed time and final pH as above, and the program should end.

You may assume that the initial pH is 8.0, the initial temperature is 30 degrees Celsius, the initial pressure is 1 atmospheres, and the initial time is 0.

ICA 19-18

While experimenting with coding sequences, you decide to try a modification of the factorial sequence by calculating the product of consecutive odd integers instead of all consecutive integers. You call this sequence `OFact`. The first four values of the `OFact` sequence are:

$$OFact(1) = 1$$
$$OFact(2) = 1*3 = 3$$
$$OFact(3) = 1*3*5 = 15$$
$$OFact(4) = 1*3*5*7 = 105$$

Write a program that will ask the user for a maximum sequence value desired, and then print the `OFact` sequence to the command window as long as the value calculated does not exceed the maximum value entered by the user. For example, if the user specifies a maximum of 100, it would print the first three values since $OFact(4) = 105 > 100$. This should be printed in two columns: `N` in the first column and `OFact(N)` in the second column. All values generated should be stored in the vector `OFact` for later use.

When the program reaches the maximum value, a message should be printed that states:

```
The desired maximum was XXX.
OFact (Y) = ZZZ is the closest to this value without exceeding it.
```

where XXX is the maximum entered by the user, Y is the sequence number of the calculated value, and ZZZ is the last calculated value.

ICA 19-19

Answer the question with each of the following code segments. If an error will occur, write ERROR and explain what caused the error.

(a) What is stored in `Count` by the following code?

```
Count=0;
for N=10:-0.2:8.5
   Count=Count+1;
end
```

(b) What is displayed on the screen by the following code?

```
for m=3:8
    fprintf('%0.0f\n',m^2-m)
end
```

(c) What is displayed on the screen by the following code?

```
for m=3:8
   M2=m^2-m;
end
fprintf('%0.0f\n',M2)
```

(d) What is stored in `P` and `C` after the following code executes?

```
S=3;
I=2;
P=1;
C=0;
for K=S:I:S^I
  P=P*K;
  C=C+1;
end
```

ICA 19-20

For each of the following code segments, determine the contents of the specified variable following execution. If an error will occur, write ERROR and explain what caused the error.

(a) What is stored in VC?

```
VC=0;
V1=[9 5 -3 6 -1 0];
for Val=1:length(V1)
   if V1(Val)<=0
      VC=VC+1;
   end
end
```

(b) What is stored in VS?

```
V2=[9 5 -3 6 -1 0];
for Val=2:length(V2)
   VS(Val-1)=V2(Val-1)+V2(Val);
end
```

(c) What is stored in VE?

```
VE=0;
V1=[9 -4 -3 3 6 -1 0 5 5 -2 -2 -3];
for Val=1:length(V1)
   if V1(Val)==round(V1(Val)/2)*2
      VE=VE+1;
   end
end
```

(d) What is stored in VD? *Hint:* Look up the `sign` function.

```
VD=0;
V1=[9 -4 -3 3 6 -1 0 5 5 -2 -2 -3];
for Val=2:length(V1)
   VD(Val)=1;
   if sign(V1(Val))~=sign(V1(Val-1))
      VD(Val)=-1;
   end
   if abs(V1(Val))==abs(V1(Val-1))
      VD(Val)=0;
   end
end
```

ICA 19-21

(a) Explain what the following function does, assuming `Grades` is a text string.

```
function[Fixed]=FixGrades(Grades)
NumGrades=length(Grades);
for G=NumGrades:-1:1
   if ~(strcmpi(Grades(G),'a')||...
        strcmpi(Grades(G),'b')||...
        strcmpi(Grades(G),'c')||...
        strcmpi(Grades(G),'d')||...
        strcmpi(Grades(G),'f'))
      Grades(G)=[];
   end
end
Fixed=Grades;
```

(b) If the following text string was sent to `FixGrades`, what would be returned in `Fixed`?

```
'AvBcDgEfAdbCAAacdbtYcdaNDCFAB'
```

(c) Explain why the `for` loop must count down instead of up.

ICA 19-22

For each of the following code segments, determine the contents of the specified variable follow-ing execution. If an error will occur, write ERROR and explain what caused the error.

(a) What is stored in `IT`?

```
IT=0;
for k=3:2:7
  C=k^2;
  while C>=0
    C=C-25;
    IT=IT+1;
  end
  IT=IT*2;
end
```

(b) What is stored in `R`?

```
R=1;
N=1;
while R>1E-6
  R=1;
  for k=1:N
    R=R/K
  end
  N=N+1;
end
```

ICA 19-23

For each of the following code segments, determine the contents of the specified variable follow-ing execution. If an error will occur, write ERROR and explain what caused the error.

(a) What is stored in `A`?

```
for m=1:3
  for n=1:4
    A(n,m)=m^2-n;
  end
end
```

(b) What is stored in `SP`?

```
M=[1 2 3;4 5 6];
[Rows,Cols]=size(M);
SP(1)=0;
for r=1:Rows
  P=1;
  for c=1:Cols
    P=P*M(r,c);
```

```
           end
         SP(r+1)=SP(r)+P;
      end
```

(c) What is stored in PR?

```
   M=[0 2 4 6;1 -1 2 -2;-3 0 -2 0];
   [R,C]=size(M);
   PR=zeros(R,C);
   for row=1:1:R
     for col=1:1:C
        PR(row,col)=(PR(row,col)+M(row,col))*max(M(row,:));
     end
   end
```

ICA 19-24

Write a MATLAB program to implement the algorithm described in ICA 19–3. You may *not* use the built-in `factorial` function. You must implement this using a `for` loop.

ICA 19-25

Write a MATLAB program to implement the algorithm described in ICA 19–4.

ICA 19-26

Write a MATLAB program to implement the algorithm described in ICA 19–5.

ICA 19-27

Write a function named `GetArray` that will accept two positive integers R and C representing the number of rows and columns of a matrix NA, and return the R × C matrix NA with values input by the user. For each value to be input a prompt should appear similar to the following:

```
     Please input the value for row 2 and column 4:
```

with, of course, the 2 and 4 replaced by the indices of the actual element being entered.

ICA 19-28

Assume the vector AM contains an even number of elements. Write a short section of code that will divide the product of the even elements by the sum of the odd elements and place the result in PDS. Note that this must work correctly for any vector as long as it contains an even number of elements. Write the code using two different methods:

(a) a `for` loop without using built-in functions like `sum` or `prod`
(b) direct matrix operations, no loops.

> EXAMPLE: AM =[9 3 5 0.5 10 4]
> PDS = 0.25   Detailed calculation: (3*0.5*4)/(9+5+10) = 0.25

ICA 19-29

Write a function named `EvenSum` that will accept a matrix of unknown size and return the sum of all elements at the intersection of even numbered rows and even numbered columns. Note that your function must work with any size matrix. Also note that if the matrix is a scalar or a vector, the function will return zero since there are no even numbered rows or columns in those cases. Accomplish this using two methods:

(a) a `for` loop without using built-in functions like `sum`
(b) direct matrix operations, no loops

> **EXAMPLE:** For the matrix shown, `A=EvenSum(Test);` will place the value 13 in `A`. The numbers added together are shown in **red**.

$$= \begin{bmatrix} 1 & 3 & 5 & 7 & 9 \\ 0 & 1 & 2 & 3 & 4 \\ 6 & 4 & 2 & 1 & 8 \\ 3 & 5 & 2 & 4 & 0 \end{bmatrix}$$

ICA 19-30

You wish to design a program to tabulate values for the function $V = e^{-t/\tau} \cos \theta t$, a common type of voltage response in electric circuits, over a specified range of times.

(a) The user should be asked to enter the following parameters:
- τ — time constant in seconds
- θ — frequency in radians per second
- t_0 — initial time in seconds
- t_f — final time in seconds
- *Steps* — number of divisions into which the range of time is to be divided.

(b) The program should print a statement similar to the following:

```
Calculations of the equation e^(-t/x.xxx) cos (x.xxx t) from
x.xxx to x.xxx seconds.
```

(c) The program should calculate the value of V [voltage] for the stated values of τ and θ over the range of t (time in seconds) from t_0 to t_f. The results should be printed in a neatly formatted table showing both time and voltage.

ICA 19-31

Write a function named `CLASS` that will accept a vector of unknown length called `INT` and return four variables containing the following information about the contents of the vector:

- `NotInt`—contains the number of values in `INT` that are not integers.
- `NotPos`—contains the number of values in `INT` that are zero or less.
- `TooBig`—contains the number of values in `INT` that are greater than 99. Note that `NotInt` has priority over `NotPos` and `TooBig`. For example, an element in `INT` equal to -9.43 would add one to `NotInt`, but *not* to `NotPos`.
- `Ints`—a 99 element vector in which each element contains the number of occurrences of the element's index value in `INT`. For example, if `INT` contains a total of fourteen 57s, `Ints` (57) will contain 14.

Example:
```
INT=[-3, -7.3, 0, 16, 2, 99, 11.3, -42, 298, 176.98, 16]
    NotInt=3
    NotPos=3
    TooBig=1
    Ints(2)=1
    Ints(16)=2
    Ints(99)=1
```

All other elements of `Ints` contain a zero.
Note that the sum of all elements in the returned variables should equal the number of elements in `INT`.

You *may not* use any built in functions to count the number of occurrences of each integer between 1 and 99, nor may you use any function whose sole purpose is to determine if a value is an integer or not. You *may* use other functions, however.

ICA 19-32

You are assessing the price of various components from different vendors and wish to find the least expensive vendor for each component. The prices of the parts from each vendor are stored in a matrix, VendCost. Each row corresponds to a specific vendor and each column corresponds to a specific component. If a specific part is not offered by a vendor, the corresponding entry will be −1.

Write a program that will determine which vendor offers the cheapest price for each component, and place the results in a two row matrix Cheapest with the same number of columns as there are columns in VendCost. Each entry in the first row of Cheapest should be an integer corresponding to the row number of the vendor with the cheapest price for the corresponding component, and the entries in row 2 should contain the lowest price for that component. You may assume that each part is available from at least one of the listed vendors. If two or more vendors offer a component at the same lowest price, you may choose either one.

You may not use the built-in min function or other similar functions to solve this problem. You may not use direct matrix operations to solve this problem; you must do it using for loops (in a meaningful way). Your solution must work for any number of vendors and any number of components.

Example:

VendCost =

4.97	8.54	2.04	0.44	13.55	−1.00
5.23	8.23	2.12	0.39	15.98	2.67
5.24	8.22	2.09	0.51	−1.00	2.76

Cheapest =

1	3	1	2	1	2
4.97	8.22	2.04	0.39	13.55	2.67

ICA 19-33

You are studying the number of defective parts produced each week by several machines to help adjust maintenance protocols.

Assume the rows of matrix Def represent different machines and all columns except the last represent weeks. The last column contains the long-term average of the number of defects per week produced by that machine. Write a short section of MATLAB code that will generate a new matrix Comp with the same number of rows but one fewer columns as described.

The code will compare each value in the matrix, except those in the last column, to the value in the last column of the same row to compare the number of defective parts produced by each machine each week with that machine's long-term defect rate.

- If the number of errors equals that machine's average, the corresponding element in the new matrix Comp will equal 0.
- If the number of errors is greater than that machine's average, the corresponding element in the new matrix Comp will equal 1.
- If the number of errors is less than that machine's average, the corresponding element in the new matrix Comp will equal −1.

Example:

$$
\text{Def} = \begin{bmatrix} 3 & 6 & 9 & 2 & 7 & 7 \\ 0 & 6 & 3 & 2 & 1 & 3 \\ 12 & 7 & 9 & 2 & 11 & 8 \end{bmatrix} \quad \text{Comp} = \begin{bmatrix} -1 & -1 & 1 & -1 & 0 \\ -1 & 1 & 0 & -1 & -1 \\ 1 & -1 & 1 & -1 & 1 \end{bmatrix}
$$

ICA 19-34

The Pascal triangle has an amazing number of uses, from linear algebra to design of fractal antennas. Write a program following the five steps listed that will create a Pascal triangle.

1. Prompt the user to enter an integer and place it in a variable N.
2. Create an N × N matrix named Pascal filled with zeros.
3. Modify the Pascal matrix so that the first row and first column contain all ones.
4. The value of every other element of the Pascal matrix should be changed to the sum of the element to the left and the element above. You should complete row 2, then row 3, etc. until all rows have been calculated. Example for N = 5 shown.

Note: You *may not* use the pascal function built in to MATLAB.

$$
\text{Pascal} = \begin{matrix} 1 & 1 & 1 & 1 & 1 \\ 1 & 2 & 3 & 4 & 5 \\ 1 & 3 & 6 & 10 & 15 \\ 1 & 4 & 10 & 20 & 35 \\ 1 & 5 & 15 & 35 & 70 \end{matrix}
$$

5. Reset all elements of the Pascal matrix below the minor diagonal (from lower left to upper right) to zero. Example for N = 5 shown.

$$
\text{Pascal} = \begin{matrix} 1 & 1 & 1 & 1 & 1 \\ 1 & 2 & 3 & 4 & 0 \\ 1 & 3 & 6 & 0 & 0 \\ 1 & 4 & 0 & 0 & 0 \\ 1 & 0 & 0 & 0 & 0 \end{matrix}
$$

ICA 19-35

Write a function named ProdStats that will accept a matrix ProdData that has at least one row and at least two columns. The rows each represent a different machine in a factory, and the columns represent successive days of production. The value in each element is the number of units produced by that machine on the given day.

ProdStats should return a new matrix Trend with the same dimensions as ProdData and a row vector TrendNum with three elements.

Each element of Trend indicates whether that day's production for that machine was less than, equal to, or greater than the previous day's production for that machine. Since the first day (first column) does not have previous data upon which to base the comparison, the first column of Trend will arbitrarily be set to all zeros. All other elements will be set to either -1 (lower production than previous day), 0 (equal production to previous day), or 1 (higher production than previous day).

The first element of TrendNum contains the total number of days that production decreased (in other words, the number of negative ones in Trend), the second element of TrendNum contains the number of days with no change in production (zeros in Trend, *not* counting the zeros on the first day), and the third element contains the number of days with higher production (ones in Trend).

Example:

$$\text{ProdData} = \begin{bmatrix} 1 & 2 & 4 & 4 & 5 & 4 \\ 7 & 8 & 9 & 0 & 1 & 5 \\ 9 & 8 & 5 & 6 & 7 & 7 \end{bmatrix}$$

$$\text{Trend} = \begin{bmatrix} 0 & 1 & 1 & 0 & 1 & -1 \\ 0 & 1 & 1 & -1 & 1 & 1 \\ 0 & -1 & -1 & 1 & 1 & 0 \end{bmatrix} \quad \text{TrendNum} = \begin{bmatrix} 4 & 2 & 9 \end{bmatrix}$$

ICA 19-36

You are studying the effect of experience on the productivity of workers.

For a group of new workers, you record how many units each person completes per day for several weeks. You make the assumption that when the average of three consecutive days is not greater than 2% more than the maximum daily production prior to those three days, then that worker has reached maximum productivity, and you do not need to continue looking at the data for subsequent days.

The provided file, `workers.mat`, contains an array with this productivity data. Each row represents a specific worker, and the columns represent successive days. The value in each element of the matrix represents the number of units produced by a specific worker on a specific day.

Your program should analyze the data for each worker and print the following message for each worker, one per line:

```
Worker # X reached maximum productivity after Y days.
```

where X is the worker number (same as row number) and Y is the number of days where the above mentioned criterion is first reached.

If a worker has not reached maximum productivity after the trial period represented by the data, the following message should be produced instead:

```
Worker # X did not reach maximum productivity within Y days.
```

where in this case Y is the total number of days in the trial period.

Note that although this problem can be solved using `for` loops and `if` statements without a `while` statement, you are expected to use a `while` loop to stop scanning each worker's record when the day of maximum productivity has been found. (You may also use `for` loops and `if` statements as necessary, of course.)

Your code should work for any number of workers and any number of days, not just the sample data provided.

REVIEW QUESTIONS

1. Design a flowchart for a program to help students understand damped sinusoids having the form

$$F = be^{-mt}\cos(\theta t).$$

The user should first be prompted to enter values for b (real number), m (positive number), and theta (positive number). If users enter invalid values for any of the parameters, they should be warned and asked to try again until they enter a valid value. After they have successfully entered a set of parameters, a graph of the function should be generated for $0 \le t \le 25$. After the graph is generated, a menu should appear asking if they want to enter another set of parameters. If they select Yes, the entire process repeats. If they select No, the program should terminate. If they close the menu, a warning message should be generated and the menu redisplayed. If they close the menu three times in a row, an error message should be displayed and the program terminated.

2. Create an algorithm to allow the user to enter a list of courses they have taken along with the grades received and the credit hours for each course, and then calculate the user's GPA. The program should first ask the user to type either Yes or No in response to the question, "Do you want to enter a grade?" If the answer is Yes, the user should be asked to enter the name of the course (text), the grade in the course (A, B, C, D, or F), and the number of credit hours for the course (a positive integer). If the user enters an invalid grade or an invalid number for the credit hours, he or she should be warned and asked to try again a maximum of three times, after which an appropriate error message should be generated and the program terminated. If they successfully enter the information for a course, the course name, the letter grade, and the credit hours should be displayed in a single row on the screen in an appropriate format, and then they should be asked again if they wish to enter a grade, and the entire process repeats if they answer Yes. If they answer No, the GPA should be calculated and displayed in an appropriate format, and then the program terminated unless no grades were entered (they typed No the first time) in which case an appropriate message should be generated and the program terminated.

3. Write a program to implement the algorithm described in Review Question 1.

4. A zombie picks up a calculator and starts adding odd whole numbers together, in order: $1 + 3 + 5 + \ldots$ etc. What will be the last number the zombie will add that will make the sum on his calculator greater than 10,000? Your task is to write the MATLAB code necessary to solve this problem for the zombie or he will eat your brain. The user should be asked to enter the target number (10,000), and your code should be written in such a way that if a target number other than 10,000 is entered, the correct answer for the value entered will be determined.

5. Write a function called `Balloon` that will accept a single variable named S. The function should replace S with the square of S and repeat this process until S is either greater than 10^{15} or less than 10^{-15}. The function should return a value Q containing the number of times S was squared during this process. If S equals 1, the function should display a warning message and return the value `Inf` in Q.

Examples:
- If $S = 100, Q = 3$ (S equals 10^{16} after the third square).
- If $S = 0.1, Q = 4$ (S equals 10^{-16} after the fourth square).
- If $S = 3, Q = 5$ (S equals 1.853×10^{15} after the fifth square).

MATLAB
examples

6. You are constructing a program to convert the specific heat of common materials from units of calories per gram degree Celsius to units of British thermal units per pound-mass degree Fahrenheit.

A file named SpecificHeat.mat is available online. Place this file into your MATLAB directory. Following the housekeeping commands, use the load command to load the contents of this file into your workspace [load('SpecificHeat.mat')]. This will place the variable SpHeat into your workspace. SpHeat is a two column cell array containing the following initial compounds and associated specific heat values: Aluminum, 0.22; Calcium, 0.15; Gold, 0.031; Silicon, 0.17; Zinc, 0.093.

Instruct the user to select a material from a menu or enter a new material name [see sample menu shown]. If the user closes the menu, issue a warning message and redisplay the menu until the user makes a correct choice.

If the user chooses to enter a new material, the user should be prompted to enter the material name and the specific heat of the material in units of calories per gram degree Celsius. This new material should be stored in a new row at the bottom of the cell array with the original data. Once the new material has been entered, save the revised cell array variable to "SpecificHeat.mat". [Refer to the documentation files on the save command if you do not know how to do this.] Note that if you run the program again after this that SpHeat will contain the newly added data when you load the variable from SpecificHeat.mat.

For the compound either chosen or entered by the user, convert the specific heat value to units of British thermal units per pound-mass degree Fahrenheit. Display in the Command Window both values similar to the output shown below, using three decimal places for the values.

```
The specific heat of Gold is 0.031 cal/(g deg C) = X.XXX BTU/
(lb_m deg F).
```

Ask the user if they wish to run the program again, using a menu. Repeat the program until the user chooses to terminate the program.

Test Case and Sample Output:

```
Enter a metal: Brass
Enter the specific heat capacity of Brass in cal/(g deg C): 0.09
The specific heat of Brass is 0.090 cal/(g deg C)=X.XXX BTU/(lb_m deg F).
```

Here, the value of X.XXX will be replaced with the actual problem results.

7. The relationship between time (t, in seconds) and the temperature of a cooling liquid (T, in degrees Celsius), is dependent upon the initial temperature of the liquid (T_0, in degrees Celsius), the ambient temperature of the surroundings (T_A, in degrees Celsius) and the cooling constant (k, in hertz); the relationship is given by:

$$T = T_A + (T_0 - T_A)e^{-kt}$$

You are to write a program to determine what the temperature of a fluid will be at a specific time. Begin by entering the following data into the program.

```
Kvalues = [0.01; 0.03; 0.02];
Fluids = {'ABC', 'FGH', 'MNO'};
```

Ask the user the following questions:

- From a menu entitled "Choose a fluid" the user can choose fluid **ABC**, **FGH**, or **MNO**. If the user closes the menu, issue a warning message and redisplay the menu until the user makes a proper selection.
- Enter the initial fluid temperature and ambient temperature, both in units of degrees Celsius.
  - If the user enters an ambient temperature greater than the initial temperature, ask the user to enter both values again. Continue asking until the user enters valid data.
  - If the user enters an ambient temperature greater than 35°C or less than 5°C, the program should tell the user what the problem is with the entry and ask the user to enter the ambient temperature value again until a valid entry is made.
- Enter the time at which the temperature is desired, in units of minutes.

Determine the cooling constant (k) based upon the fluid choice. The cooling constant matrix is in units of hertz, and the values are given in corresponding order to the fluid names in the cell array.

Determine the temperature of the fluid (T) at time t using the user input values (T_A, T_0, and t) and the cooling constant (k) for the fluid chosen using the equation above.

Create a formatted output statement for the user in the Command Window similar to the following. The decimal places must match.

```
ABC has temp 83.2 deg C after 1.4 min.
```

Once the temperature is determined, the program should ask the user (Yes or No—not case sensitive) if they want to calculate the temperature for another fluid, without using a menu. The program should continue to run until the user enters **No**; then the program should tell the user:

```
Thank you!!
```

Below is the output from a sample run of the program. The items listed in **BOLD** are the user's responses to the questions when prompted.

ABC selected from menu.
Enter initial fluid temperature in degrees Celsius: **100**
Enter the ambient temperature of the room in degrees Celsius: **42**
Warning: The room temperature must be between 5 and 35 degrees and less than 100
> In RQ_19_7 (line 38)
Enter the ambient temperature of the room in degrees Celsius: **12**
At what time [min] do you wish the fluid temperature? **2.3**
ABC has a temperature of 34.1 degrees C after 2.3 minutes.
Do you want to enter another set of data? (Yes or no): **YES**

FGH selected from menu.
Enter initial fluid temperature in degrees Celsius: **22**
Enter the ambient temperature of the room in degrees Celsius: **33**
Warning: Ambient temperature should be less than Fluid temperature.
> In RQ_19_7 (line 30)
Initial Temperature must be greater than or equal to 5.
Enter initial fluid temperature in degrees Celsius: **122**
Enter the ambient temperature of the room in degrees Celsius: **33**
At what time [min] do you wish the fluid temperature? **3.1**
FGH has a temperature of 33.3 degrees C after 3.1 minutes.
Do you want to enter another set of data? (Yes or no): **yes**

MNO selected from menu.
Enter initial fluid temperature in degrees Celsius: **254**
Enter the ambient temperature of the room in degrees Celsius: **10**
At what time [min] do you wish the fluid temperature? **0.7**
MNO has a temperature of 115.3 degrees C after 0.7 minutes.
Do you want to enter another set of data? (Yes or no): **no**

Thank you!!

---

8. You have written three functions for three different games. The names of the games (and the functions that implement them) are Dunko, Bouncer, and Munchies. Each function will accept an integer between one and three indicating the level of difficulty (1 = easy, 3 = hard), and when play is complete will return a text string, either won or lost, to the program that executed the function.

   Write a program that will use a menu to ask the user if he or she wants to play a game. If not, the program should terminate. If so, the program should generate another menu to allow the user to select one of the three games by name.

   After selecting a game, the program should display another menu asking the user for the desired level of difficulty (Easy, Moderate, or Hard), and the game should begin by calling the appropriate function.

   When the user has finished playing the selected game, a message should be displayed indicating whether the user won or lost the game. A menu should then be generated asking if the user wishes to repeat the game just played. If so, the difficulty level menu should be displayed again and the game repeated.

   If the user does not wish to repeat the same game, the program should display a menu asking if the user wants to play another game. If so, the game selection menu should be generated again, followed by level selection, etc. If the user does not wish to play another game, the program should display a message saying, "Thanks for playing" and terminate.

   If the user closes a menu rather than making a selection, a warning message should be displayed and the user given three more chances to make a selection from that menu. If they still close the menu on the fourth attempt, an error message should be displayed and the program terminated.

9. The Microsoft Excel file `EnergyData.xlsx` has been provided online. The file contains energy consumption data by energy source per year in the United States, measured in petaBTUs.

Year	Fossil Fuels	Elec. Net Imports	Nuclear	Renewable
2007	101.605	0.106	8.415	6.830
2006	99.861	0.063	8.214	6.922
2005	100.503	0.084	8.160	6.444
2004	100.351	0.039	8.222	6.261
2003	98.209	0.022	7.959	6.150

(a) Write the MATLAB code necessary to read the Microsoft Excel file and store each column of data into different variables. Create the following:

- `Yr`: A vector of all of the years in the worksheet.
- `FF`: A vector of all of the fossil fuels for each year in `Yr`.
- `ENI`: A vector of all electric imports for each year in `Yr`.
- `Nuc`: A vector of all nuclear energy consumption for each year in `Yr`.
- `Ren`: A vector of all renewable energy consumption for each year in `Yr`.
- `Hdr`: A cell array of all of the headers in row 1.

You may not hard-code these variables—they should be imported from the Excel file.

(b) Write the MATLAB code necessary to generate the following table of nuclear energy consumption by year using formatted output in the Command Window. You may assume that you have correctly defined the vector `Nuc` and cell array `Hdr` in part (a). Note the nuclear energy consumption should be displayed to two decimal places.

```
Nuclear Energy Consumption by Year [petaBTU]

    2007    2006    2005    2004    2003    ... etc
    8.42    8.21    8.16    8.22    7.96
```

(c) Add code to ask the user if they wish to select another type of fuel or terminate the program. If the user wants to select again, then part (b) should repeat using the chosen energy source.

10. You are to program part of the interface for a simple ATM. When the user inserts their card and types the correct PIN (you do NOT have to write this part of the program), the system will place the users' checking account balance in a variable `CBal` and the users' savings account balance in `SBal`.

You are to write a function that will accept `SBal` and `CBal` as inputs and return two variables `NewCBal` and `NewSBal` containing the checking and savings balances after the transaction is completed. The function should do the following:

- Display a menu titled "Main Menu" with the following three options.
  - Get cash
  - Get balance
  - Quit
- If "Get cash" is selected, another menu titled "Withdrawal amount" with the following four items is displayed:
  - $20
  - $60
  - $100
  - $200

- After selecting an amount, a menu titled "From which account?" should be displayed showing the following two options:
  - Checking
  - Savings
- At this point, the program should verify that the selected account contains sufficient funds for the requested withdrawal.
- If not, a message should be displayed that says, "Sorry. You do not have sufficient funds in your SSSS account to withdraw $XX" where SSSS is either Savings or Checking and $XX is the selected withdrawal amount.
- If funds are available, the program should call a function named Disp20(x), where x is the number of $20 bills to dispense. (See the following note about Disp20.) After that, the withdrawal amount should be subtracted from the appropriate balance.
- After processing the "Get cash" request, the program should return to the main menu.

  Note About Disp20(x): The purpose of this function is to dispense the requested number of $20 bills—that is, to shove x bills out of the slot in the ATM machine. This does not really exist, since we do not have an ATM machine to work with. Thus, if you try to run your code, you will get an error ("Undefined function …").

  In order to test your program, add the following function to your current path:

  ```
  function[]=Disp20(x)
  fprintf('%.0f $20 bills were dispensed.\n',x)
  ```

  where x is the number of bills to be dispensed.

  This allows you to know if the program reached the proper location in the code. It is fairly common in software development to use a "dummy" function in the place of a real one when the device to be controlled has not been completed or is not available in order to help verify whether the software is reaching the correct places in the code for various situations.

- If "Get Balance" is selected, another menu titled "Which account?" should appear with the two choices:
  - Checking
  - Savings

  and the program should then display "Your SSSS balance is $bb.bb.," where SSSS is either Savings or Checking and $bb.bb is the balance in the selected account.
- After processing the "Get balance" request, the program should return to the main menu.
  - If Quit is selected, the function should return to the calling program with the updated balances in NewCBal and NewSBal. Note that the new balances will be equal to the original balances if no money was drawn from an account, but they must still be returned in the two new balance variables.

  If the user closes a menu rather than making a selection, a warning message should be displayed and the user given two more chances to make a selection from that menu. If they still close the menu on the third attempt, an error message should be displayed and control should return to the main program after making sure that NewCBal and NewSBal were set as appropriate based on earlier transactions.

11. We want to conduct an analysis for a wooden baseball bat manufacturer who is interested in diversifying their product line. To help the manufacturer look at the different scenarios, write a program that includes the following features:

Section 1: User Input

- Load the information from the .mat file titled NewBatCost.mat. This file contains the following variables:
  - The bats can be produced from multiple different materials. The material names are stored as a cell array in the variable Materials.

- The estimated material cost to produce 10 bats is saved in the vector `Cost` in the location corresponding to the materials listed in `Materials`. To determine the material cost per bat, the value given in `Cost` will need to be divided by 10.
- The labor and energy cost to produce one bat is estimated in the scalar variable `LE-Cost`. This cost is the same regardless of the bat material.
- Using a menu, ask the user to select the material for the bat.

  - Include a warning message if the user chooses to close the menu without making a selection. Show the menu again until the user chooses a material from the list.
- Ask the user to enter, as a scalar, the selling price of a single bat made from the chosen material. The material name chosen by the user must appear in this statement.

  - Include a warning message if the user enters a negative value. Ask users to re-enter the value until they enter a positive value.
- Ask the user to enter, as a vector, the total number of bats the manufacturer can produce per week and the number of weeks the manufacturer plans to run their bat production machinery in a year; these values are the same regardless of the type of bat produced.

  - Include a warning message if the user enters either (a) zero or negative values for either value; (b) more than 52 for the number of weeks; or (c) a matrix that is not a 1×2 vector. Ask the user to reenter the values until all conditions are met. All conditions should be checked each time a vector is entered.
- Ask the user if they are evaluating upgrading the current equipment by responding Yes or No, saving the response as a text string. If the user responds Yes (not case sensitive), the user should be asked to enter a fixed cost for the upgrade. Otherwise, the fixed cost should be set to zero. (You may *not* use a menu to collect this information from the user).

Section 2: Breakeven Calculations

Your program should calculate the information necessary to display in Sections 3 and 4.

- The variable cost of production will be the summation of the material cost, labor cost, and energy cost.
- If the process will not produce a profit, include a warning message to tell the user the current values entered will not make any money. Ask users if they wish to alter any input values; you may choose the method of asking the user and storing the response. If the user indicates no changes will be made, terminate the program. Otherwise, restart the program by asking the user to choose a bat material.

Section 3: Output to Command Window

Your program should calculate and display the following information in the Command Window:

```
Producing ## MMM bats a week for ## weeks = ## total bats
     Selling Price per bat:           $##.##
     Total Variable Cost per bat:     $##.##
     Fixed Cost of upgrade:           $##.##


     Profit:   $##.#E##
Breakeven Point:    #### bats
```

Formatting notes:

- The # characters in the output displayed above will be actual numbers. The word "MMM" should be replaced with the actual material chosen by the user.
- The number of bats per week, number of weeks per year, and total number of bats produced should be shown as integers.
- The selling price per bat, total variable cost per bat, and fixed cost of the upgrade should be displayed with two decimal places. The fixed cost should only appear if the user is planning to upgrade the equipment and the fixed cost is greater than zero.
- The profit should be displayed in exponential notation with one decimal place.

- The lines for selling price, variable cost, fixed cost and profit should display as indented by a single tab from the left margin.
- There should be one blank line preceding the profit line.
- The breakeven point should be shown as an integer, and rounded to show two significant figures. The breakeven point should only appear if the user is planning to upgrade the equipment and the fixed cost is greater than zero.

Section 4: Output to Graph

The program should generate a proper breakeven plot with the number of bats produced on the abscissa and the revenue and total cost on the ordinate.

- If the user is planning to upgrade the equipment and the fixed cost is greater than zero:
  - The breakeven point should be clearly indicated on the graph with a black solid line
  - Add a textbox to the graph, in the upper-right corner, with the profit and breakeven point using the same display properties as given above for the Command Window. The format of this textbox should be a black background with white font.
- The material name chosen by the user must appear in the title of the graph. For example, the title might be: Maple Bat Production.
- The legend should appear in the upper-left corner of the graph.
- The axis limits of the graph should change with user input, with the abscissa limit set to the rounded up total number of bats produced, and the ordinate limit set to the rounded up revenue made at the total number of bats produced.

12. You are part of an engineering firm on contract by the U.S. Department of Energy's Energy Efficiency and Renewable Energy task force to develop a program to help laboratory technicians measure the efficiency of their lab equipment. Your job is to write a program that measures the efficiency of hot plates.

    The program will begin by suggesting four possible fluids for the technician to choose from using a menu: acetic acid, citric acid, glycerol, and olive oil. If the technician closes the menu without making a selection, a warning message is displayed and the menu reappears. If the technician fails to make a proper selection after five attempts, an error message should be generated and the program terminates.

    The technician is then prompted to enter the initial room temperature in units of degrees Fahrenheit, the brand name and model of the hot plate, and the theoretical power for the hot plate provided by the manufacturer. If the room temperature entered is less than 30 or greater than 100, display a warning and ask the technician to try again until an appropriate value is entered. If the technician enters a negative value for power, display a warning and ask the technician to try again until an appropriate value is entered.

    The program will then call a function, which will determine the following:

    - All fluid properties [Specific Gravity, Specific Heat] should be contained in your function, not in the main program.
    - The technician will use 2 liters of fluid.
    - The technician will then begin to heat the fluid. The technician should take 2 data points, one at 2 minutes and one at 5 minutes during the heating process. The program will prompt the technician to enter each data point at the time interval. The technician will record the temperature of the fluid [degrees Fahrenheit] for the two data measurements.
    - Once the final data point has been entered, the function will calculate the energy required to heat the fluid, in joules, and the power used to heat the fluid, in watts
    - The function will return to the program the time interval, the temperature readings, and the power used.
    - The program will then calculate the efficiency of the hot plate, in percentage.

    The output of your program should look like the following output displayed; where the highlighted blue values are example responses provided by the user (typed or by pressing a button depending on the requirements previously mentioned) into your program, and the highlighted yellow values are the calculated values that will change based upon your starting properties. Note that the data shown as results and on the graph are examples only and may not reflect actual calculations! The code should line up the output calculations. In addition, your code must display the efficiency rounded to the nearest integer and must include a percent symbol.

Finally, the program should produce a graph of the two experimental data points relating time and temperature and a trendline with a formatted equation found using `polyfit`.

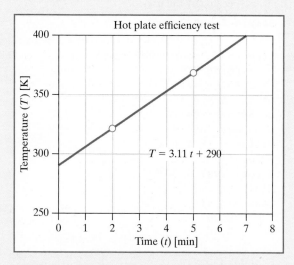

```
Lab Equipment Efficiency Calculator: Hot Plate

    Select a Fluid Type

    Acetic Acid

    Citric Acid

(Choose Acetic Acid)

Enter the initial temperature of the fluid [deg F]: 65
Enter the brand name and model of your hot plate: CU Tigers 2011
Enter the power rating of the hot plate [W]: 1040

Reading #1: 2 minutes
Temperature [deg F]: 73

Reading #2: 5 minutes
Temperature [deg F]: 86

Energy required to heat Acetic Acid:        1.1e+005 J
Power used by the hot plate:                222 W

Efficiency for a CU Tigers 2011 hot plate:  21%
```

Hot plate efficiency test

$$T = 3.11\,t + 290$$

Use the following fluid properties in your function:

Fluid	Specific Gravity [—]	Boiling Point [°C]	Specific Heat [J/(g K)]
Acetic acid	1.049	118	2.18
Citric acid	1.665	153	4
Glycerol	1.261	290	2.4
Olive oil	0.915	300	1.97

13. Write a function for finding the maximum value in a vector of data. It should receive an array as an argument and return the maximum of the values. You must use a loop to accomplish this. Do not use the `max` built-in function or any other functions related to ordering the magnitudes of elements.

14. A matrix named `mach` contains three columns of data concerning the energy output of several machines. The first column contains an ID code for a specific machine, the second column contains the total amount of energy produced by that machine in calories, and the third column contains the amount of time required by that machine to produce the energy listed in column 2 in hours.

Write a function named `MPower` to accept as input the matrix `mach` and return a new matrix named `P` containing two columns and the same number of rows as `mach`. The first column should contain the machine ID codes, and the second column should contain the average power generated by each machine in units of watts. You must use a loop to accomplish this. Do not use direct matrix operations.

15. You are working for a data analytics firm that has been asked to create a generic data collection tool that will provide basic statistics on input typed in by the user. This data collection tool should ask the user to type the number of desired data points they need to record, the number of decimal places for all of the numbers displayed in the final output, then the program should allow them to enter all of the data points, one by one, where the user presses the Enter key after each data point. You may assume that this program only needs to handle scalar numeric values as input and will not need to worry about strings, vectors, or other variable types.

As soon as the user inputs all of the values, your program should generate a string representation of the vector of data (such as `"Vector=[3 -6 1 7 87]"`) as well as provide basic statistics, including the number of negative values, number of positive values, the sum of all of the values, the mean, median, and standard deviation of the data set, and the minimum and maximum values in the data set. The output should appear similar to the following output. Note that the vector of data does not need to be displayed with the number of decimal places specified by the user.

Sample Input/Output:

```
Type the number of data points to record: 4
Type the number of decimal places to show in output: 2
Data Point #1: 8
Data Point #2: 2
Data Point #3: 6
Data Point #4: 4
Data Set Information:
Vector = [8 2 6 4]
# Negative: 0.00
# Positive: 4.00
Minimum: 2.00
Maximum: 8.00
Sum: 20.00
Mean: 5.00
Median: 5.00
Standard Deviation: 2.58
```

16. Write a function named `MatchCA` that will accept two cell arrays, `CA1` and `CA2`, each containing one or more text strings in a single row, and return a single cell array according to the following guidelines.

(a) If the two cell arrays contain the same number of text strings, successively compare the text strings in the corresponding locations of each to determine if they contain the same characters (*not* case sensitive).

- For each pair that matches, display a statement like `String 2 contains Frog.`

- For each pair that does not match, display a menu asking the user to choose the correct string. This menu should have three buttons. The top two buttons should contain the two different text strings in the corresponding locations of `CA1` and `CA2`, and the third button should read `Other`.
  - If the user selects one of the text strings, that string should replace the mismatched string in the cell arrays.
  - If the user selects `Other`, the user should be asked to type in the correct string, and both corresponding strings in `CA1` and `CA2` should be replaced with this new string.
  - If the user closes the menu, a warning should be issued and the menu redisplayed until the user makes a correct selection.
  - Display a statement like `String 2 now contains Dirge`.

(b) If the two cell arrays do not have the same number of text strings, proceed as above for the first *N* text strings, where *N* is the number of strings in the shorter cell array. For each element in the longer cell array that does not have a matching string in the shorter one, display a menu with two buttons, the first containing the string in the longer cell array and `Other`. Proceed in a similar manner to part (a), either adding the string in the longer cell array to the shorter one, or asking the user to type in the correct string and replacing both. Each time, display a message indicating the new contents of that element of the two cell arrays.

(c) The returned cell array may be either of the modified cell arrays since they should contain the same strings after processing.

17. You are programming the control system for a robot tractor that will be used to plant and till circular fields. (Circular fields are very common in the arid western United States, where they commonly use center-pivot irrigation, inherently giving a circular field.) To avoid having the tractor run over crops traveling directly from the outside to the center or vice versa, it will be programmed to do one double-wide spiral inward, followed by a second outward

spiral exactly halfway between the lines of the inward spiral to give the desired spacing. See diagram, inward in black, outward in red (or vice versa).

Write a program to generate this spiral pattern in a figure window.

- First, ask the user how many turns they wish for one arm of the spiral to have. In the example shown, the user entered 5, for a total of 10 turns, 5 inward and 5 outward.
- All turns should be evenly spaced as shown. You may assume any outside diameter that is convenient, and the values do not need to be marked on the axes.
- You can consider either spiral to be the starting (inward) spiral as long as you end up with the correct pattern.
- *Hint:* Recall the math of polar to rectangular conversions:
  - **X = R*cos(A)**
  - **Y = R*sin(A)**
- *Hint:* Make the radius (distance from center) dependent on the total angle through which the tractor has traveled.

18. A sample data set D is provided online in the file `Spacecraft.mat`. Use the `load` command to load this data into your workspace. A matrix D contains data on several experimental spacecraft engines. The matrix is organized in pairs of rows. Each pair of rows contains data measured during tests of the engines. For each pair of rows, the odd numbered row contains several measurements of the total energy used by the engine and the even numbered row contains the corresponding total kinetic energy imparted to the spacecraft. Recalling the equation for energy efficiency, $E_0 = \eta E_I$ we see that the efficiency of the engine is the slope of the line if the kinetic energy of the spacecraft (the output energy, E_0) is plotted versus energy input to the engine (E_I).

For each data set in matrix D:

(a) Determine the efficiency of the engine represented by each data set.

(b) For each data set with an efficiency of at least 0.8 (80%), plot the data (all data sets on the same graph) represented by **red diamonds**. Add a **solid red** trendline to each such data set, including a trendline equation.

(c) For each data set with an efficiency of at least 0.5 (50%) but less than 0.8 (80%), plot the data (all data sets on the same graph) represented by **blue circles.** Add a **dotted blue** trendline to each such data set, including a trendline equation.

(d) For each data set with an efficiency less than 0.5 (50%), plot the data (all data sets on the same graph) represented by **black X's**. These low-efficiency data sets will NOT have a trendline or equation added.

(e) The trendline equations should be in the form $E\#_0 = \eta E\#_I$ where # is the number of the dataset (half of the even-numbered row of the data set).

(f) The trendline equations should be positioned immediately to the right of the largest (rightmost) data point of the corresponding data set.

Note that you SHOULD NOT figure out the efficiencies and hard-code the line types, etc. Your program must do this automatically so it will work for any set of data without user intervention.

The graph should be properly labeled and formatted, but the colors and types of both the lines and data markers will not follow the standard proper plot rules since they are specified in the problem.

You may assume the largest value of E_I is 12 and the largest value of E_0 is 10.

For the sample dataset provided, your graph should be similar to the following.

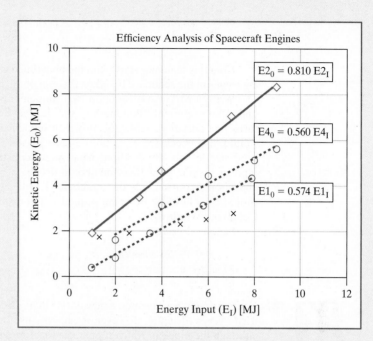

19. Write a program to implement the GPA calculator described in Review Question 2 with the following modifications. The answers "yes" and "no" as well as the letter grades should NOT be case sensitive. Rather than printing the course, grade, and credit hours each time one is entered, the course names should be stored in successive elements of a cell array, the letter grades should be stored in successive elements of a text string, and the credit hours should be stored in a vector. After entering one or more courses and stating that they wish to enter no more grades, a properly formatted table containing the course names, grades received, and credit hours should be printed, followed by the overall GPA.

20. The online Excel file `BaseballData.xlsx` contains salary data from major league baseball teams in 2005. Write a MATLAB program that uses this data to determine the following:
 (a) Salary mean, median, and standard deviation for all players.
 (b) Salary mean, median, and standard deviation for the Arizona Diamondbacks.
 (c) Salary mean, median, standard deviation for all pitchers.
 (d) Salary mean, median, standard deviation for all outfielders.
 (e) Which team had the highest average salary?
 (f) Which position had the lowest average salary?
 (g) What percentage of players earned more than $1 million?

21. The Fibonacci sequence is an integer sequence calculated by adding the previous two numbers together to calculate the next value. This is represented mathematically by saying that $F_n = F_{n-1} + F_{n-2}$ (where F_n is the nth value in the sequence, F) or:

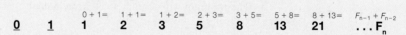

Note this sequence starts with the underlined values $(0, 1)$ and calculates the remaining values in the sequence based on the sum of the previous two values.

Professor Bowman found this sequence to be extremely insufficient and created the Bowman sequence, which is an integer sequence calculated by adding the previous three numbers together (instead of two like in the Fibonacci sequence) to calculate the next value. This is represented mathematically by saying that $F_n = F_{n-1} + F_{n-2} + F_{n-3}$ (where F_n is the nth value in the sequence, F) or:

$$\underline{0} \quad \underline{1} \quad \underline{2} \quad \overset{0+1+2=}{3} \quad \overset{1+2+3=}{6} \quad \overset{2+3+6=}{11} \quad \overset{3+6+11=}{20} \quad \overset{F_{n-3}+F_{n-2}+F_{n-1}=}{...F_n}$$

Note this sequence starts with the underlined values $(0, 1, 2)$ and calculates the remaining values in the sequence based on the sum of the previous three values.

Write a MATLAB function that implements the Bowman sequence that accepts one input argument, the length of the desired Bowman sequence to generate, and returns one output variable, the Bowman sequence stored in a row vector. This function should also check to see if the number passed in to the function is a valid Bowman sequence length (think about what might constitute valid sequence lengths!). If the input is invalid, your function should display a warning message and the output variable should contain only one number: –1. Otherwise if the input is valid, your function should calculate the Bowman sequence and display each value in the sequence in the Command Window.

Sample Output:

```
>> B=FUNCTIONNAME(8);
Bowman sequence:
0 1 2 3 6 11 20 37
```

22. Your company has invented a new device that can convert either sunlight or heat into electricity very efficiently. However, it can only use one source of energy at any given time, although the mode in which it operates (heat or light) can easily be changed by a voltage applied to a control terminal. In order to design an automated controller so that the device always uses the source that will provide the most energy at any given time, a graph is plotted showing the combinations of illumination and temperature for which

- heat conversion yields the most energy
- light conversion yields the most energy
- neither heat nor light are sufficient to maintain the control circuits without a net loss of energy

The program should begin by loading a matrix of temperature and illumination values from the Excel workbook `LightHeatEnergy.xlsx` provided online. Column 1 will contain the temperature (T) in units of Fahrenheit and column 2 will contain the illumination (E) in units of lux.

The program should ask the user to choose a temperature value they wish to display, giving the user a choice of temperatures found in Column 1 on a menu in increasing numerical order. Each button of the menu should be labeled with both the temperature [°F] and the illumination [lux] values, such as (53, 40505).

The program should check to ensure the temperature chosen is between 260 and 310 kelvins. If the user selects a value outside of this range, the program should issue a warning, telling the user the value chosen was outside the temperature limits of the program and ask the user to select another value. The menu should continue to appear until the user selects a value within the temperature limits of the device.

The program will contain a cell array named `Status` containing the following words as text strings:

HEAT   LIGHT   OFF

The program will correctly categorize the temperature selected by the user and matching illumination value as one of the categories contained in `Status`. The program should produce a formatted output statement to the Command Window, similar to the following, for the temperature value chosen:

```
Temperature of TF degrees Fahrenheit = TK kelvin.
For TK kelvin and E lux, the device is in mode Status.
```

In this statement, **TF**, **TK**, **E,** and **Status** are replaced by the actual values. Format the values of temperature and illumination as integer values. Use a blank line between user selections for visual clarity.

The program should ask the user by using a menu if they wish to display another temperature value and repeat this process of temperature selection and Command Window output until the user selects No.

The program should produce the graph as shown with all classification division lines, all classification names, and should indicate all points chosen by the user with a symbol of your choice. A sample graph is shown at the right.

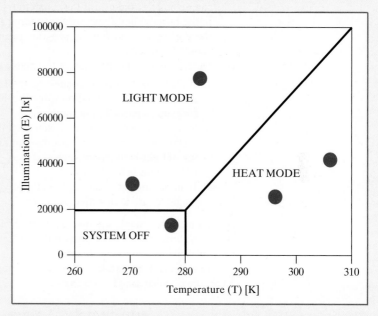

23. As early as 650 BC, mathematicians have been composing magic squares, a sequence of n numbers arranged in a square such that all rows, columns, and diagonals sum to the same constant. Used in China, India, and Arab countries for centuries, artist Albrecht Dürer's engraving *Melencolia I* (year: 1514) is considered the first time a magic square appears in European art. Each row, column, and diagonal of Dürer's magic square sums to 34. In addition, each quadrant, the center four squares, and the corner squares all sum to 34. The following is an example of a magic square:

16	3	2	13
5	10	11	8
9	6	7	12
4	15	14	1

Write a program to determine if a series of numbers is indeed a 4×4 magic square. Your program should complete the following steps, in this order:

(a) Ask the user to enter the proposed magic square in a single input statement (e.g., [1 2 3 4; 5 6 7 8; 9 10 11 12; 13 14 15 16]—note this example is a 4×4 matrix, but NOT a magic square). You may assume the user will enter whole numbers; they will not enter either decimal values or text.

(b) Check that all values are positive; ** **for-loop or nested `for`-loop required in the solution.** If one or more of the values in the matrix are negative or zero, issue a statement to the command window informing the user of the mistake and ask the user to enter another matrix. This check should be repeated until the user enters a matrix with positive values. This check should work even if the user does not enter a 4×4 matrix; it should work regardless of the size of matrix entered.

(c) Check for an arrangement of 4×4. If the matrix is not a 4×4, issue a statement to the command window informing the user of the mistake, and ask the user to enter another matrix. This check should be repeated until the user enters a 4×4 matrix. You may assume the reentered matrix contains only positive values; you do not need to recheck the new matrix for positive values, only for matrix dimensions.

(d) Determine if the matrix is a form of a magic square. The minimum requirement to be classified as a magic square is each row and column sums to the same value. ** **for-loop or nested `for`-loop required in the solution.** If this criterion is not met, issue a statement to the command window informing the user they have not entered a magic square and ask the user if they wish to try another magic square. This question can be posed using either a text answer entered by the user (Yes, No) or by using a menu. If the user chooses to run the program again, the entire program starting with step (a) should begin again.

(e) Determine the classification of the magic square using the following requirements:

1. If each row and column sums to the same value, the magic square is classified as semi-magic; the summation value is called the magic constant. ** **for-loop or nested `for`-loop required in the solution**.

2. If, in addition to criterion 1, each diagonal sums to the same value, the magic square is classified as normal; ** **for-loop or nested `for`-loop required in the solution. The use of built-in functions such as `diag`, `fliplr`, `rot90`, `trace` or similar built-in functions is forbidden.**

3. If, in addition to 1 and 2, the largest value in the magic square is equal to 16, the magic square is classified as perfect.

Format your magic square classification similar to the format shown below. You may choose to format your table differently, but each classification should contain a Yes or No next to each magic square category.

The magic constant for your magic square is 24. The classification for your magic square:

Semi-magic	Normal	Perfect
yes	yes	yes

After this table appears, ask users if they wish to try another magic square following the instructions given earlier.

A few test cases for you to consider:

- Albrecht Dürer magic square: [16, 3, 2, 13; 5, 10, 11, 8; 9, 6, 7, 12; 4, 15, 14, 1];
- Chautisa Yantra magic square: [7, 12, 1, 14; 2, 13, 8, 11; 16, 3, 10, 5; 9, 6, 15, 4];
- Sangrada Familia church, Barcelona magic square: [1, 14, 14, 4; 11, 7, 6, 9; 8, 10, 10, 5; 13, 2, 3, 15];
- Random magic square: [80, 15, 10, 65; 25, 50, 55, 40; 45, 30, 35, 60; 20, 75, 70, 5];
- Steve Wozniak's magic square: [8, 11, 22, 1; 21, 2, 7, 12; 3, 24, 9, 6; 10, 5, 4, 23].

24. Write a program to analyze the cooling of a cup of coffee. Start by asking the user to enter a matrix in the format [a, b; c, d]. The matrix will contain two columns; the first is time in minutes and the second is temperature in degrees Celsius. These pairs of values indicate the actual temperatures of the coffee at the specified times. If the matrix entered does not have at least two rows or does not have exactly two columns, warn the user and ask them to enter a matrix in the correct format until they do it correctly. Next ask the user to enter the room temperature [°C].

Create a table of output showing the time in hours and the temperature above room temperature in degrees Fahrenheit.

Time must be shown to two decimal places; temperature should be shown as an integer.

- The output statement must be written such that the number of entries in the matrix could change and the table would still be created correctly.

Using the data entered, create a graph with an exponential trendline to project the temperature above room temperature at 30 minutes on a graph.

- The axis limits, grid, axis labels, and title must match exactly to the graph shown. The title is shown in bold, red letters, size 20. Time should be in minutes and temperature above room temperature should be degrees Fahrenheit.
- The markers are open red circles, size 16.
- The line is red dashed format, line width 2.
- Background color on the trendline text is green; font size is 16.
- The background on the graph is white.

Example:

```
Enter a matrix of form [a,b; c,d] for time [min], temperature
[deg C] of a cup of coffee.
[1, 91; 5, 87.5; 8, 84.9];
Enter the temperature of the room [deg C]: 26
```

Time (t) [h]	Temp above Room (T) [deg F]
0.02	117
0.08	111
0.13	106

After the table and graph have been produced, use a menu to ask users if they wish to analyze another set of data, and either repeat or terminate the program as appropriate.

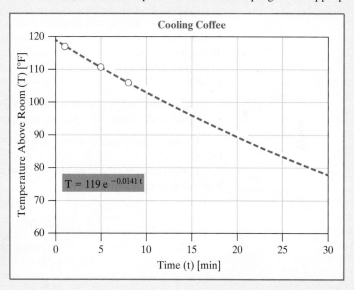

25. Your company manufactures a variety of products, each with a unique numeric part number. On any given day, only a few different parts are manufactured. On some days, nothing is manufactured, such days being devoted to product development and other tasks. As part of recordkeeping, an Excel worksheet is created in the following format. Each row of the worksheet corresponds to a day on which parts were manufactured. The first column of the worksheet lists the Ordinal Day for that year (the Ordinal Day for January 1 is 1, and for December 31 is 365 or 366 depending on leap year). The second column lists the month and date that correspond to the Ordinal Day. Then, the remaining odd columns list specific part numbers manufactured on the date listed in that row, and the value in the subsequent even column following each part number contains the number of that specific part manufactured on that date.

EXAMPLE: Assume the worksheet `Parts2013` contains the following data:

Ordinal Day	Date	First Part #	Number of Parts	Second Part #	Number of Parts	Third Part #	Number of Parts
13	T Jan 13	J109	100	R222	50	0	0
29	M Jan 29	G459	50	X565	25	P649	40
45	W Feb 14	H106	25	J109	50	R222	25
67	F March 8	R222	70	0	0	0	0

For example, on Ordinal Day 13, which was Tuesday, January 13, 2013, 100 of Part #J109 and 50 of Part #R222 were manufactured. Further, on Ordinal Day 45, which was Wednesday, February 14, 2013, 25 of Part #H106, 50 of Part #J109, and 25 of Part #R222 were manufactured.

Note the following characteristics of the worksheet that will be loaded:

- A valid worksheet will *always* contain an even number of columns: the first column is the Ordinal Day (in ascending order); the second column is the corresponding date; and then the columns appear in pairs, Part # first, then number of parts manufactured.
- The worksheet can contain any number of rows and any even number of columns as long as there are at least four columns.
- Unused columns for a given date contain zeros as placeholders.
- The part numbers are not sorted in any way, but any part number will appear only once on a given date.

As part of your assigned task, you must write MATLAB code to accomplish the goals specified below. You should suppress output to the screen except as shown.

All sample output shown below is based on the example worksheet `Parts2013` shown above, *not* the file `Parts2014.xlsx` that your program will actually use.

Load the data from the Excel file `Parts2014.xlsx` into your workspace. Please note the following:

- When writing your program, it should be written to work for any size data in the Excel file.
- The data will always be arranged as described in the preceding section.

Create a menu similar to the one shown to ask the user to select the type of data analysis to perform.

Note: The user should be able to continue to display summaries until "End Program" is selected or the menu is closed.

If the user selected **Part Summary**, ask the user to select a Part Number from a menu. Each part number should appear only once in the menu (or, only unique part numbers should appear in the menu). If the user closes the menu, a statement should be issued warning the user they have not selected a part and the Part Number menu should reappear until the user chooses a Part Number. Then, print a statement similar to that shown below showing the total number of the specified Part # that were manufactured according to the data file.

```
A total of 145 of Part #R222 were produced.
```

Note: There may be more than one day where the chosen part number was manufactured. The program should display the total number of the specified Part # manufactured for all days.

The program should then allow the user to choose from the menu another summary or end the program.

If the user selected **Day Summary**, ask the user to type in a specific Ordinal Day (between 1 and 366), as follows. You should check that the user enters a number between 1 and 366. If the number is outside of this range, the program should ask the user to enter a correct value, and keep checking until a valid number is entered or the user has tried four times. After the fourth invalid try, terminate the program.

```
For what ordinal date do you wish a summary?
25698
The number is out of range. Enter a value from 1 to 366.
For what ordinal date do you wish a summary?
29
```

After the user types a valid Ordinal Day, print a table similar to the following example. The table should show each Part # and the number of each part manufactured on that day. If there are zeroes used as placeholders, they should *not* appear.

```
The following parts were produced on M Jan 29.

          Part #          Number Produced
          G459                  50
          X565                  25
          P649                  40
```

If the specified date does not appear in the data file, the program should write an output statement stating no parts were produced on that day.

```
For what ordinal date do you wish a summary?
365
There was no production this day.
```

The program should then allow the user to choose from the menu another summary or end the program.

If the user selected **End Program, closes the data analysis menu, or attempts too many entries for the ordinal date**, the program should display the message similar to the following.

```
Thank you for using this program!
```

26. After numerous experiments with a circuit configuration, you determined that as long as the transistor chosen meets certain criteria, other components in the circuit can be adjusted to yield an ideal frequency response. In the classification diagram shown, there are two regions where the transistor is acceptable.

Zone A, shown in solid magenta lines, is defined as:

- a horizontal line at 7500 Hz for voltages from 0 to 3 V;
- a linear line from (3, 7500) to (6, 9000);
- A vertical line at 6 V for a frequency from 6000 to 9000 Hz; and
- a power law curve which extends from (0, 0) to (6, 6000), passing through the point (3, 4244).

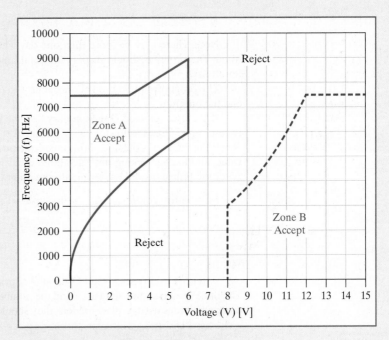

Zone B, shown in blue, dotted line segments, defined as:

- a horizontal line at 7500 Hz for voltages from 12 to 15 V;
- a vertical line at 8 V for a frequency from 0 to 3000 Hz; and
- an exponential curve which extends from $(8, 3000)$ to $(12, 7500)$, passing through the point $(11, 6000)$.

If a point falls on a line dividing the Accept and Reject regions, it is considered acceptable. Create the classification diagram, duplicating the diagram exactly as shown.

The user will be asked if they wish to enter a set of test values using a menu choice of Yes or No. If the user selects Yes, they will be allowed to specify a pair of experimental test values for a specific transistor and store the response in the two-element vector `Test`. The first value in `Test` is the voltage and the second value is the measured frequency. If the user closes the menu rather than making a selection, a warning message should be displayed and the menu redisplayed until the user makes a selection.

The program will check if the user has entered a two-element row vector, with values of voltage between 0 and 15 volts and values of frequency between 0 and 10,000 hertz. If the user has entered an incorrect number of elements or values outside of either or both allowable ranges, give the user two more chances to enter the correct information. If the user fails to enter a matrix correctly within three total tries (the first try + two additional tries), terminate the program.

As long as the user enters a two-element row vector and values are within the correct ranges, the program will follow these steps:

- The program will determine if the V, f pair entered by the user is Accepted or Rejected:
  - If the test value lies in an Accept region, the program will **save in the cell array** the text **Accept Device in Zone** # in a variable, where # is replaced by the letter **A** or **B** depending on the Zone the point lies in.
  - If the test value is anywhere outside an Accept region, the program will **save in the cell array** the text **Reject Device**.
  - **The program should also keep track, as running totals, of the number of points that have been accepted and the number rejected.**

- The point is added to the classification diagram as a size 20 red, solid circle. The center of the circle should contain the point number entered by the user (1 for first point, 2 for second point, etc.)
- The user is then prompted by the menu to select if they wish to enter another pair of test values.
- If the user selects No from the menu or closes the menu at any time, the program will end the input loop and show the results in the Command Window. If the user has entered at least one valid row vector and then enters a voltage value outside the acceptable range of 0 to 15 volts, the program will print the results in tabular format, similar to the following table. It will also report the number of points accepted and number of points rejected that were entered by the user.

Example showing tabular output to Command Window and the graph created.

```
Your tests resulted in:
   4 Accepted Devices
   6 Rejected Devices
```

Entry #	Voltage [V]	Frequency [Hz]	Status
1	2	1000	Reject Device
2	11	5000	Accept Device in Zone B
3	9	1000	Accept Device in Zone B
4	5	8000	Accept Device in Zone A
5	13	8000	Reject Device
6	4	9000	Reject Device
7	10	7000	Reject Device
8	1	5000	Accept Device in Zone A
9	6	5000	Reject Device
10	3	3000	Reject Device

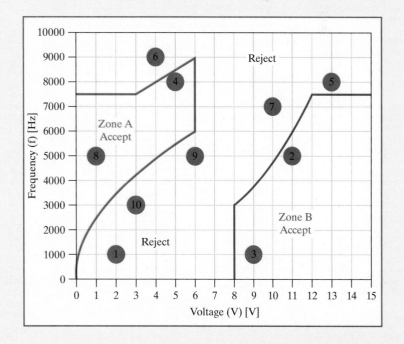

27. For this assignment, you will need the Cincinnati Reds player data file `CincinnatiReds2010 .mat`. The data in this file is a capture of the 2010 season on Baseball-Reference.com (http://www.baseball-reference.com/teams/CIN/2010.shtml).

In the MATLAB workspace provided, there are three variables of interest:

- `PlayerData`, a 37 × 9 matrix which contains:
  - Column 1: The age of the baseball player
  - Column 2: Games played or pitched
  - Column 3: At bats
  - Column 4: Number of runs scored/allowed
  - Column 5: Singles hit/allowed
  - Column 6: Doubles hit/allowed
  - Column 7: Triples hit/allowed
  - Column 8: Home runs
  - Column 9: Runs batted in

  Note that each row in `PlayerData` represents a different baseball player.

- `PlayerNames`, a 37 × 1 cell array that contains the names of each player in the `PlayerData` matrix.

- `PlayerPositions`, a 37 × 1 cell array that contains the position abbreviation of each player in the `PlayerData` matrix.

  (a) Create a function that will accept a single input, a cell array of player positions for an entire team—with our data set, this would be a variable like `PlayerPositions`. Your function must create a new cell array that contains only the unique positions, sorted alphabetically.

  (b) Create a program that will display the average age, at bats, home runs, and runs batted in for user-selected positions. In order to allow the user to select the positions, you must use the menu function to allow the user to click positions to include in the analysis, as well as a Done button to indicate that all desired selections have been made and to allow the code to display the output. Note that the menu will continue to be redisplayed until the user selects Done.

Sample Output: Assuming the user selects all positions

Position Stats for 2010 Cincinnati Reds

Pos	Ave(Age)	Ave(AB)	Ave(HR)	Ave(RBI)
1B	25	288	19	58
2B	29	626	18	59
3B	31	242	8	39
C	32	197	5	32
CF	25	514	22	77
IF	27	3	1	4
LF	29	338	11	52
MI	24	38	1	2
OF	27	123	4	11
P	27	21	0	1
RF	23	509	25	70
SS	31	347	5	34
UT	36	23	2	2

Sample Output: Assuming user presses 3B, OF, Done

```
Position Stats for 2010 Cincinnati Reds

Pos      Ave(Age)    Ave(AB)    Ave(HR)    Ave(RBI)
3B          31         242         8          39
OF          27         123         4          11
```

(c) MLB.com is consistently ranked one of the top 500 websites in the United States, bringing in millions of visitors a year. However, recent surveys have shown that http://MLB.com is one of the slowest top 500 websites and is looking for new solutions to speed up their website. After hearing about the baseball statistics programs we have developed in our class, MLB.com has contracted us to help them create a MATLAB program to help them improve the user experience of their website. Re-create the analysis in part (b), except this time export the output to a Microsoft Excel workbook.

Sample Output: Assuming the user selects all positions

	A	B	C	D	E	F
1	Position Stats for 2010 Cincinnati Reds					
2	Pos	Ave(Age)	Ave(AB)	Ave(HR)	Ave(RBI)	
3	1B	25	288	19	58	
4	2B	29	626	18	59	
5	3B	31	242	8	39	
6	C	32	197	5	32	
7	CF	25	514	22	77	
8	IF	27	3	1	4	
9	LF	29	338	11	52	
10	MI	24	38	1	2	
11	OF	27	123	4	11	
12	P	27	21	0	1	
13	RF	23	509	25	70	
14	SS	31	347	5	34	
15	UT	36	23	2	2	
16						

Sample Output: Assuming user presses 3B, OF, Done

	A	B	C	D	E	F
1	Position Stats for 2010 Cincinnati Reds					
2	Pos	Ave(Age)	Ave(AB)	Ave(HR)	Ave(RBI)	
3	3B	31	242	8	39	
4	OF	27	123	4	11	
5						

28. The graph shows a simplified plasma phase diagram. The gridlines have been removed from this graph to make it easier to read the phase locations.

The program should ask the user to enter a value for the log of temperature, and a value for the log of density. Write a program to classify the phase. The phase division information given in the following table should be entered into MATLAB using a cell array.

Phase Division	Log Density (ρ) [g/cc]		Log Temperature (T) [K]	
Plasma – Molecular fluid	−10	0	3.3	3.9
Plasma – Metallic fluid	0	2	3.9	5.6
Molecular fluid – Metallic fluid	0	0	2	3.9

The equation of the lines dividing the phases must be found in the program using `polyfit` within a `for` loop, such that only one set of `polyfit` commands are written within a loop to determine the model parameters for all dividing lines. The loop should work for any size cell array of phase data.

If a specific combination of density and temperature is exactly on one of the lines, the classification should be with the region *below* the line, or to the *left* in the case of the vertical line.

The program should produce a formatted output statement to the Command Window, similar to

```
For the log density x and log temperature y, the phase is PHASE
```

where x, y and Phase are replaced by the actual values. Format the values of x and y to one decimal place.

The program should also produce the graph as shown here with all phase division lines, phase names, and should indicate the point entered by the user with a symbol (shown here as a blue dot).

The information provided about the line endpoints (the T=5.6, T=3.9, and T=3.3 text with the arrow) does not need to be duplicated on the graph produced.

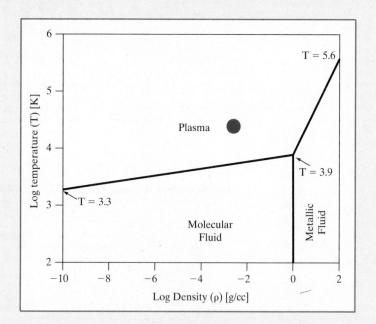

29. You are designing a voltage-controlled oscillator. Ideally, you want the frequency to double if the input voltage increases by 1 volt. This can be expressed mathematically as

$$F = f_i 2^V$$

where **F** is the output frequency, **V** is the input voltage, and $\mathbf{f_i}$ is the initial frequency when the input voltage is zero. You wish to graph the ideal frequency response along with measured data from the actual devices.

1. Create the theoretical data series by:

   A. Ask the user to enter a scalar variable `fi` to store the value of initial frequency. Check that the user enters a positive value, less than 20 for `fi`; if the user enters an incorrect value, continue to ask the user to enter a value until they enter a positive number that is less than 20.

   B. Create the row vector `V` containing a linear sequence of voltages from 1 to 10 volts by increments of 0.25 volts.

   C. Create a row vector named `F` containing the frequencies corresponding to the series of voltage values in `V`.

2. Load the initial data for the rest of the problems which is stored in `TranData .mat`. The variables in this file are

   - `TransNames`, which is a **3 × 2 cell array**, containing manufacturer (such as **Motorola**) in column 1 and the transistor name (such as **2N3904**) in column 2.
   - `ExpData`, which is a **4 × 4 matrix**, containing measurements from the circuits using the different transistors. The matrix `ExpData` will contain one more row than the number of rows in cell array `TransNames`. The first row of `ExpData` contains the voltages at which measurements were made. The remaining rows each contain the frequency measured for each transistor type corresponding to each voltage in row one. The second row has frequency data for the first transistor in `TransNames` (**2N3906**), the third row has frequency data for the second transistor (**SSM2212**), etc.

3. Create a menu to ask the user to select one of the tested transistor types and store the response in `MenuChoice`. You must use the variable `TransNames` to create the menu selections.

   The user will have three attempts to make a correct menu selection instead of closing the menu. If the user closes the menu instead of making a selection and has not already used three attempts, issue a warning message similar to that shown below, including the number of attempts remaining, and redisplay the menu. If the user closes the menu on the third attempt, issue an error message similar to that shown, and terminate the program.

Sample: Menu closed with less than three attempts:	Sample: Menu closed on third try:
`Warning: 2 remaining attempts`	`Error: No more attempts!`

4. Modify the menu in step 3 to include a bottom choice of **User Enters Info**, where the user will enter the information for a transistor. If the user selects this option, the program should prompt the user to enter a transistor manufacturer, and then a transistor name, both stored as text; then four frequencies as a row vector, each element being the frequency at the corresponding voltage values given in row 1 of `ExpData`. Verify that the user enters four elements for the frequency vector; if the user enters an incorrect number of elements, continue to ask the user to enter data until they enter a correctly sized row vector.

5. Each time the user enters a new transistor, save the new data as a new row in the `ExpData` matrix and add a new row to the `TransNames` cell array. Before terminating the program by any mechanism, these modified variables should be saved in `TranData.mat` so that they will be available the next time the program is run.

6. Create a proper plot of both the (a) theoretical data series given in Question 1 and (b) the experimental data series chosen by the user in question 4.

A sample graph is shown for the AD818 transistor experimental data. All axis labels, title, and graphing properties must match the graph shown. The title must include the name of the transistor chosen. This graph should correctly execute regardless of the menu choice by the user, even if the user enters additional data.

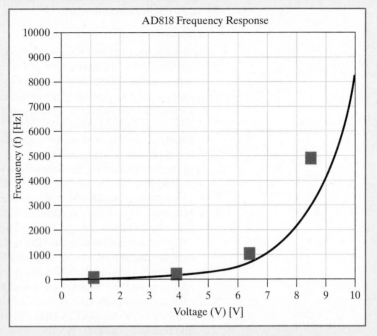

MATLAB
examples

30. Palindromes are sequences, typically of alphabetic characters, that are the same whether read from left to right or right to left, typically ignoring punctuation.

Well known examples are **Madam, I'm Adam**, and **A man, a plan, a canal: Panama**.

Palindromes can be numeric as well. Examples of numeric palindromes are 343 and 0845335480.

Write a MATLAB program to find the palindromic sequences in the first million decimal places of pi.

The file `Pi1M.mat` available online contains a text string with the first one million decimal places of pi.

This file contains the single vector `Pi1M` that contains the number pi to one million decimal places (thus 1,000,001 significant figures) represented as a text string, and includes the decimal point in element 2. In order to facilitate processing, it is useful to simply delete the decimal point by replacing it with nothing using the code

```
Pi1M(2)=[];
```

If you check your workspace following this command, you should see that `Pi1M` is a text string containing 1,000,001 characters.

First, prompt the user for the maximum length of palindromic sequences you wish to find. If the user enters 5, your program should find all palindromes of length 2, 3, 4, and 5. (Length 1 palindromes are a bit silly since a single character is the same read forward or backward.)

The number of palindromes of each length up to the entered maximum should be printed to the screen in a format similar to that shown below. (These *are* the correct answers, by the way.)

```
Maximum number of digits in the palindromes? 15

There are 100072 palindromes of length 2.
There are 99947 palindromes of length 3.
There are 10010 palindromes of length 4.
There are 10027 palindromes of length 5.
There are 992 palindromes of length 6.
There are 982 palindromes of length 7.
```

```
There are 104 palindromes of length 8.
There are 107 palindromes of length 9.
There are 10 palindromes of length 10.
There are 9 palindromes of length 11.
There are 1 palindromes of length 12.
There are 1 palindromes of length 13.
There are 0 palindromes of length 14.
There are 0 palindromes of length 14.
There are 0 palindromes of length 15.
```

You *may not* use the built-in palindromes function—this defeats the purpose of this problem.

Comments about probability theory:

The digits of pi illustrate many properties of an equally distributed random variable. This means that each of the 10 digits occurs about the same number of times, and the likelihood of a specific digit occurring at any given location is one-tenth. This means that it is easy to predict approximately how many palindromes of a given length should occur. For length 2, whatever the first digit is, the next digit must be the same. Since they are random and equally distributed, the probability that the second digit will be the same as the first is 0.1 or 10%. Thus, there should be about 100,000 length 2 palindromes in 1,000,000 digits. Similarly for length 3, the center digit does not have to be duplicated, only the first and third, so again the probability is 0.1. For length 4 and 5, two digits have to match, giving a probability of 0.01 or 1%. For one million digits, this yields about 10,000 palindromes. By the time you consider 12 or 13 digits, there is probably only one palindrome in one million digits. These values are borne out in the calculated results.

Other comments:

Note that other than length 2 and 3 palindromes, every palindrome contains a shorter palindrome—simply delete the character on each end and the rest is still a palindrome.

For one million digits, probability theory indicates that there should be only 0.1 palindromic sequence of length 14 or 15, thus these are unlikely. Indeed, the results show that there are no such sequences. Since any palindromes longer than 15 would *have* to include a palindrome of length 14 or 15, there are no palindromes longer than 13 in the first million digits of pi. It is therefore a waste of time to ask for sequences longer than 15. (If we had 10 million digits, there would likely be one palindrome of length 14 and one of length 15.)

Options for investigating the properties of the distribution of digits in pi:

A. Create a two-column cell array containing all palindromes found. The first column contains the character position where the palindrome begins, and the second column contains the palindrome itself. The first row will contain 25 and '33'. Row 100,073 contains 2 and '141'. (Remember that pi = 3.14159 . . . so digits 2, 3, and 4 are the palindromic sequence 141.) *Do not* try to print this cell array to the screen—it would take a *long* time since the cell array contains about 220,000 rows.

B. Do not include in the counts of palindromes, or in the cell array table, any palindromes that are within a longer palindrome. For example, the first length 5 palindrome is 46264 starting at location 20. With this option, the length 3 palindrome (626) inside this longer palindrome would not be counted as a separate palindrome, nor included in the cell array.

C. To verify that the digits are equally distributed, determine how many times each digit $(0, 1, 2, \ldots, 9)$ occurs. The number of each digit should be about 100,000, or one tenth of one million.

D. Determine the number of sequences of repeated digits for all lengths from 2 to 13. (You already know that there cannot be any longer than that since there are no palindromes of length greater than 13, and any repeated sequence is a palindrome.) Examples include '00' and '4444444'.

E. Determine the number of increasing sequences of various lengths assuming when 9 is reached, the next digit would be 0. Examples include '1234', '789012', and '01'. Do not include subsequences within longer sequences in the count.

F. Repeat the previous option using decreasing sequences.

MATLAB®
examples

31. Write a function named CAX to determine what type of data is stored in each element of a cell array CAIn, available online in the file CAIn.mat. CAIn may contain any number of elements, and those elements may be any combination of numeric matrices, text strings, and cell arrays. No other data types are stored in CAIn. In addition, CAIn will not have more than two dimensions.

The function should receive as input the cell array CAIn to be analyzed. It should return a cell array (which we will call CAOut) with the same number of elements as CAIn, and with the same basic structure (number of rows and columns). Note that your function should work correctly with any cell array, not just the sample in the file provided.

Each element of CAOut should itself be a cell array, the contents of which depends on the type of element—text, numeric, or cell array—in the corresponding element of CAIn.

- If the element of CAIn was a numeric matrix, the corresponding element in CAOut should contain six entries in a single row as follows:
  1. The text string *Matrix*.
  2. A text string indicating the subtype of matrix:
     - Scalar
     - Row vector
     - Column vector
     - 2-D matrix
  3. A two-element row vector containing the size (rows and columns) of the matrix.
  4. The largest value in the matrix
  5. The smallest value in the matrix
  6. The average of all values in the matrix

You may assume there are no matrices with three or more dimensions.

- If the element of CAIn was a text string, the corresponding element in CAOut should contain four entries in a single row as follows:
  1. The text string *Text*
  2. The total number of characters, including punctuation, spaces, special symbols, etc.
  3. The number of alphabetic characters
  4. The number of numeric digits
- If the element of CAIn was a cell array, the corresponding element in CAOut should contain two entries in a single row as follows:
  1. The text string *Cell Array*
  2. A two-element row vector containing the size (rows and columns) of the cell array.
- If the element of CAIn did not match one of these three data forms, the corresponding element in CAOut should contain a cell array with the single text string *Invalid*.

MATLAB®
examples

32. One type of simple electronic oscillator comprises three resistors, a capacitor, and an integrated circuit known as an operational amplifier. The design is somewhat simplified if one of the resistors (R_2) equals approximately 0.86 times one of the others (R_1). (The ideal value is actually $(e^1 - 1)/2 \cong 0.85914$.)

The problem is that resistors are only available in certain specific values. The file ResVals.mat available online contains two vectors. R5V contain the first two digits of all standard 5% resistance values. R1V contain the first three digits of all standard 1% resistance values. Each of these values can be multiplied by an appropriate power of 10 to yield a total range of values from about 1 ohm to about 10 megohms.

Write a MATLAB program to determine the pair of values chosen from the list specified by the user (5% or 1% list) such that $1.0 \text{ K}\Omega \leq R_1 < 9.9 \text{ k}\Omega$ and the ratio R_2/R_1 is as close to 0.85914 as possible. As examples, in the 5% series $R_1 = 2.7\text{K}\Omega$ and $R_2 = 2.4\text{K}\Omega$ gives a ratio of 0.88889 and $R_1 = 1.1\text{K}\Omega$ and $R_2 = 910\Omega$ gives a ratio of 0.82727, but there is probably a pair of values that will be closer than either of these.

CHAPTER 3

CC 3-1

Product	Computer	Automobile	Bookshelf
Inexpensive	Less than $300	Less than 1/5 the median annual U.S. family income	Less than $30
Small	Folds to the size of a DVD case	Two can fit in a standard parking space	Collapses to the size of a briefcase
Easy to assemble	No assembly; just turn on	All parts easily replaceable	Requires only a screwdriver
Aesthetically pleasing	Body color options available	Looks like the Batmobile©	Blends well with any décor
Lightweight	Less than one pound	Less than one ton	Less than 5% of the weight of the books it can hold
Safe	Immune to malware	Receives 5 star rating in NCAP crash tests	Stable even if top-loaded
Durable	Survives the "Frisbee® test"	200,000 mile warranty	Immune to cat claws
Environmentally friendly	Contains no heavy metals	Has an estimated MPG of at least 50	Made from recycled materials Low VOC finish

CC 3-2

- Decrease aerodynamic drag
  - Remove unnecessary protrusions
  - Redesign body based on wind-tunnel testing
- Decrease weight
  - Use carbon-composite materials
  - Remove unnecessary material
- Increase engine efficiency
  - Use ceramic parts (e.g., valves and pistons) so engine can run hotter, thus more efficiently
  - Change to hybrid design with regenerative braking
  - Limit maximum acceleration if consistent with safety
- Manage friction
  - Improved bearings in wheels, engine, etc.
  - Redesign tires for less slippage on road surface
- Encourage fuel-efficient driving
  - Include displays for real-time and cumulative fuel usage (MPG)

CC 3-3

PAT

Hindrance: Tended to slow the team down due to difficulty understanding things.

Helpful: Always available and willing to do assigned tasks. The individuals who helped explain things to Pat probably developed a deeper understanding of the material in the process, thus this was helpful to them personally, though not to the team directly. Probably the second most useful team member.

CHRIS

Hindrance: Often absent, seldom prepared, seldom contributes, offers excuses for poor performance. Definitely the worst team member.

Helpful: Teaches the other team members about the real world, and how to deal with slackards.

TERRY

Hindrance: Not a team player, impatient, not encouraging to others, wants to dominate the team. Despite cleverness, probably the second worst team member.

Helpful: Solves problems quickly.

ROBIN

Hindrance: No obvious hindrance.

Helpful: The real leader of the group, kept things going, encouraging to weaker members. The most useful team member.

CHAPTER 5

CC 5-1

Value	Significant Figures	Decimal Places
(a) 0.0050	2	4
(b) 3.00	3	2
(c) 447×10^9	3	0
(d) 75×10^{-3}	2	0
(e) 7,790,200	5	0
(f) 20.000	5	3

CC 5-2

(a) −58.9
(b) 247
(c) 2.47
(d) 0.497

CC 5-3

(a) Elevator capacity may require a conservative estimate of 180 pounds for safety limits.
(b) Serving sizes are usually measured in integer cups, so estimate at 1 cup; but to make sure the bowl is full, round up to 1.3 cups.
(c) If recording, conservative might be to round up to 33 seconds. If it doesn't matter, 30 seconds is fine.

CC 5-4

Value	Scientific Notation	Engineering Notation
(a) 58,093,099	5.809×10^7	58.083×10^6
(b) 0.00458097	4.581×10^{-3}	4.581×10^{-3}
(c) 42,677,000.99	4.268×10^7	42.677×10^6

CHAPTER 6

CC 6-1

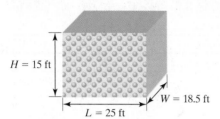

$H = 15$ ft
$W = 18.5$ ft
$L = 25$ ft

CC 6-2

Objective:

Determine the mass of the gravel in units of kilograms.

Observations:

- Assume the height of the container is 15 feet., because that is the depth the gravel fills
- Assume the gravel is flat on the top and not bumpy
- Neglect the space between the gravel pieces

CC 6-3

List Variables:

- ρ = density $[=]$ lb_m/ft^3 = 97 lb_m/ft^3 — given
- V = volume of container $[=]$ ft^3
- m = mass of gravel $[=]$ kg
- L = length $[=]$ ft   = 25 ft — given
- W = width $[=]$ ft   = 18.5 ft — given
- H = height $[=]$ ft   = 15 ft — assumed

CC 6-4

Equations:

1. Density in terms of mass and volume: $\rho = m/V$
2. Volume of a rectangle: $V = L \times W \times H$
3. Unit conversion: $1\ kg = 2.2\ lb_m$

CC 6-5

NOTE: All equation numbers refer to the previous Comprehension Check.

Manipulate:

- First, find the volume of the container; use equation (2).

$V = L \times W \times H$

- Substituting L, W, and H:

$V = (25\ \text{ft})(18.5\ \text{ft})(15\ \text{ft}) = 6{,}937.5\ ft^3$

- Second, find the mass of gravel; use equation (1).

$\rho = m/V$

- Substituting the density and the volume:

$(97\ lb_m/ft^3) = m(6{,}937.5\ ft^3)$ $m = 672{,}937.5\ lb_m$

- Convert from pound-mass to kilograms:

$$\frac{672{,}937.5\ lb_m}{} \left| \frac{1\ kg}{2.2\ lb_m} \right. = 305{,}881\ kg$$

CHAPTER 7

CC 7-1

Standard	With Prefix
(a) 3,100 joules [J]	3.1 kJ
(b) 26,510,000 watts [W]	26.51 MW
(c) 459,000 seconds [s]	459 ks
(d) 0.00000032 grams [g]	320 ng

CC 7-2

(a) Incorrect: 5 s
(b) Incorrect: 60 mm
(c) Incorrect: 3,800 µm OR 3.8 L

CC 7-3

5.5 mi

CC 7-4

1,800 min

CC 7-5

5×10^{-9} pL

CC 7-6

2 km/min

CC 7-7

26 flushes

CC 7-8

84 ft^3

CC 7-9

14 cubits

CC 7-10

10.3 m/s

CC 7-11

Quantity	Common Units	Dimensions						
		M	L	T	Θ	N	J	I
Density	lb$_m$/ft^3	1	−3	0	0	0	0	0
Evaporation	slug/h	1	0	−1	0	0	0	0
Flowrate	L/min	0	3	−1	0	0	0	0

CC 7-12

Quantity	Common Units	Dimensions						
		M	L	T	Θ	N	J	I
Energy	calories [cal]	1	2	−2	0	0	0	0
Power	horsepower [hp]	1	2	−3	0	0	0	0
Pressure	atmospheres [atm]	1	−1	−2	0	0	0	0

CC 7-13

B. Energy

CC 7-14

M	L	T	Θ	N	J	I
1	2	−3	0	0	0	−2

CC 7-15

No; the quantity on the left hand side {=}L/T and the quantity on the right hand side {=}L^2/T

CC 7-16

(a) 210 Tm
(b) 0.022 light years

CC 7-17

2.72 gal

CHAPTER 8

CC 8-1

$161,000\dfrac{mi}{h^2}$

CC 8-2

13.3 N

CC 8-3

2.2 N

CC 8-4

$3122\dfrac{lb_m}{ft^3}$

CC 8-5

1.53

CC 8-6

4.85 g

CC 8-7

0.03 moles

CC 8-8

194.5 K

CC 8-9

$0.000357 \dfrac{BTU}{g\,K} = 3.57 \times 10^{-4} \dfrac{BTU}{g\,K}$

CC 8-10

221 in Hg

CC 8-11

3.38 atm

CC 8-12

1.22 atm

CC 8-13

220 kPa

CC 8-14

2.4 L

CHAPTER 9

CC 9-1

B. $MT^{-3}\Theta^{-1}$

CC 9-2

$\varphi = \dfrac{M}{LT^2} = \text{pressure}$

CC 8-15

16.9 m

CC 8-16

20 K

CC 8-17

0.83 h

CC 8-18

1.04 h

CC 8-19

61.3 W

CC 8-20

0.3 V

CC 8-21

14.4 W or ~ 15 W

CC 8-22

(a) $-1.07\ \mu A$
(b) 28.3 V

CC 8-23

106 V

CC 8-24

$\dfrac{1}{S}$

CC 9-3

$M^{-1}L^3$

CC 9-4

$\dfrac{\Delta P}{\rho v^2}$

CHAPTER 10

CC 10-1

Relative addressing
Cell F23 displays 17

CC 10-2

Absolute addressing
Cell H26 displays 55

CC 10-3

Mixed addressing
Cell D30 displays 50
Cell F28 displays 10

CC 10-4

Mixed addressing
Cell G30 displays 5
Cell J28 displays 45

CC 10-5

In Cell . . .	Enter the Formula . . .	The Cell Will Display . . .
A1	= SQRT(169)	13
A2	= MAX(5, 8, 20/2, 5 + 7)	12
A3	= AVERAGE(15, SQRT(400), 25)	20
A4	= POWER(2, 5)	32
A5	= PI()	3.141593
A6	= PI	#NAME?
A7	= PRODUCT(2, 5, A2)	120
A8	= SUM(2 + 7, 3 * 2, A1:A3)	60
A9	= RADIANS(90)	1.570796
A10	= SIN(RADIANS(90))	1
A11	= SIN(90)	0.893887
A12	= ACOS(0.7071)	0.785408
A13	= DEGREES(ACOS (0.7071))	45.00055
A14	= CUBRT(27)	#NAME

CC 10-6

B =4/3*PI()*$A5^3*B$4

CC 10-7

32
The cell will be blank
B9

CC 10-8

A
C
B

CC 10-9

(A) YES
(B) No
(C) No
(D) YES
(E) No

CC 10-10

In cell E7:   =VLOOKUP(B7, G8:H17,2,FALSE)

This assumes the table of planet names and planetary gravities is in G7:H17, with header in row 7.

In cell B7, a data validation list using G8:G17 must be created to allow the user to select the name of the planet from a drop-down menu.

** This question has a VIDEO SOLUTION **

CC 10-11

** This question has a VIDEO SOLUTION **

CC 10-12

City ascending: Fountain Inn

City descending first, Site Name descending second:
Rochester Property

Contaminant ascending first, Site Name ascending second:
Sangamo Weston

CHAPTER 11

CC 11-1

- No title or caption describing the data
- No axis labels
- Is year 1 equal to 1993 or 2000 or . . . ?
- Need different line types
- The data is too crowded to read accurately

This may be better shown as a column chart instead of a line graph

CC 11-2

- No title or caption
- No units or description of axis quantities (both X and Y)
- Need different symbols
- No legend
- Need gridlines if wish to read data

CC 11-3

- Missing title or caption
- Missing units on both axis
- Too many decimal places on ordinate
- Poor choice of increment value on abscissa
- Missing vertical gridlines
- With theoretical data, should be line only and no data points
- With a single data series, should not include a graph

CC 11-4

- Missing title or caption
- Missing axis variable names
- Poor choice of increments on ordinate
- Missing decimal places on abscissa
- Differentiate data series using different line types
- With more than one data series, need to include a legend

CC 11-5

(a) Area can be divided into a triangle for A – B and a rectangle for B – C
$\frac{1}{2}(6-0)(2-0)+(6-0)(4-2)=18$ meters

(b) Area can be divided into the following areas:
- triangle for A – B
- rectangle for B – C
- triangle (from 2 – 6) + rectangle (from 0 – 2) for C – D
- rectangle for D – F
$\frac{1}{2}(6-0)(2-0)+(6-0)(4-2)+\frac{1}{2}(6-2)(6-4)+$
$(2-0)(6-4)+(10-6)(2-0)=34$ meters

CC 11-6

(a) ii. Positive and Constant
(b) i. Zero
(c) v. Negative and Constant

(d) iii. Positive and Increasing
(e) iii. Positive and Increasing
(f) iii. Positive and Increasing

CC 11-7

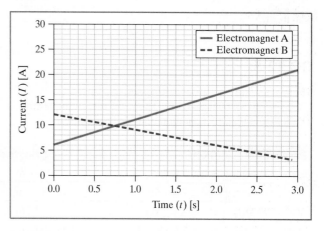

$I = 9.8$ A

CC 11-8

(a) Machine 2
(b) Machine 2
(c) 1×10^6 feet
(d) Machine 2
(e) Machine 1

CC 11-9

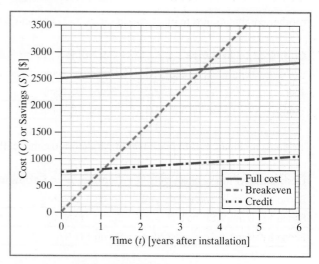

(a) Fixed: \$2500; Variable: \$50/yr
(b) Approximately \$750/yr
(c) 2.5 years to breakeven
(d) 4.0 years = \$1,000 profit

CHAPTER 12

CC 12-1

(a) atm/K   **(b)** M = 35.6 g   **(c)** V = 14.1 L

CC 12-2

68.6 lb$_m$/ft^3

CC 12-3

$3.36 \times 10^{-4} \dfrac{\text{lb}_m}{\text{ft s}}$

CC 12-4

4.55×10^{-3} St

CC 12-5

There are 17 unique spring stiffnesses.
Each spring can be used alone. There are three unique stiffnesses: 1 N/m, 2 N/m, 3 N/m
The following parallel combinations are possible:

Parallel Combination	Effective Spring Constant
Spring 1 (1 N/m) + Spring 2 (2 N/m)	3 N/m
Spring 1 (1 N/m) + Spring 2 (2 N/m)	4 N/m
Spring 2 (2 N/m) + Spring 3 (3 N/m)	5 N/m
Spring 1 (1 N/m) + Spring 2 (2 N/m) + Spring 3 (3 N/m)	6 N/m

The following series combinations are possible:

Series Combination	Effective Spring Constant
Spring 1 (1 N/m) + Spring 2 (2 N/m)	0.67 N/m
Spring 1 (1 N/m) + Spring 3 (3 N/m)	0.75 N/m
Spring 2 (2 N/m) + Spring 3 (3 N/m)	1.2 N/m
Spring 1 (1 N/m) + Spring 2 (2 N/m) + Spring 3 (3 N/m)	0.54 N/m

Another combination is to place two in series, and connect that combination to a third in parallel.

Combination in Series . . .	Connected in Parallel to. . .	Effective Spring Constant
Spring 1 (1 N/m) + Spring 2 (2 N/m)	Spring 3 (3 N/m)	3.67 N/m
Spring 1 (1 N/m) + Spring 3 (3 N/m)	Spring 2 (2 N/m)	2.75 N/m
Spring 2 (2 N/m) + Spring 3 (3 N/m)	Spring 1 (1 N/m)	2.2 N/m

The final combination is to place two in parallel, and connect that combination to a third in series.

Combination in Parallel . . .	Connected in Series to . . .	Effective Spring Constant
Spring 1 (1 N/m) + Spring 2 (2 N/m)	Spring 3 (3 N/m)	1.5 N/m
Spring 1 (1 N/m) + Spring 3 (3 N/m)	Spring 2 (2 N/m)	1.3 N/m
Spring 2 (2 N/m) + Spring 3 (3 N/m)	Spring 1 (1 N/m)	0.83 N/m

CC 12-6

There are 12 unique combinations.
Each resistor can be used alone. There are two unique resistances: $2\,\Omega, 3\,\Omega$
The following parallel combinations are possible:

Parallel Combination	Effective Spring Constant
Resistor #1 ($2\,\Omega$) or Resistor #2 ($2\,\Omega$) + Resistor #3 ($3\,\Omega$)	$1.2\,\Omega$
Resistor #1 ($2\,\Omega$) + Resistor #2 ($2\,\Omega$)	$1\,\Omega$
Resistor #1 ($2\,\Omega$) + Resistor #2 ($2\,\Omega$) + Resistor #3 ($3\,\Omega$)	$0.75\,\Omega$

The following series combinations are possible:

Series Combination	Effective Spring Constant
Resistor #1 ($2\,\Omega$) or Resistor #2 ($2\,\Omega$) + Resistor #3 ($3\,\Omega$)	$5\,\Omega$
Resistor #1 ($2\,\Omega$) + Resistor #2 ($2\,\Omega$)	$4\,\Omega$
Resistor #1 ($2\,\Omega$) + Resistor #2 ($2\,\Omega$) + Resistor #3 ($3\,\Omega$)	$7\,\Omega$

Another combination is to place two in series, and connect that combination to a third in parallel.

Combination in Series . . .	Connected in Parallel to . . .	Effective Spring Constant
Resistor #1 ($2\,\Omega$) + Resistor #3 ($3\,\Omega$) or Resistor #2 ($2\,\Omega$) + Resistor #3 ($3\,\Omega$)	Resistor #2 ($2\,\Omega$) or Resistor #1 ($2\,\Omega$)	$1.43\,\Omega$
Resistor #1 ($2\,\Omega$) + Resistor #2 ($2\,\Omega$)	Resistor #3 ($3\,\Omega$)	$1.72\,\Omega$

The final combination is to place two in parallel, and connect that combination to a third in series.

Combination in Parallel . . .	Connected in Series to . . .	Effective Spring Constant
Resistor #1 ($2\,\Omega$) + Resistor #3 ($3\,\Omega$) or Resistor #2 ($2\,\Omega$) + Resistor #3 ($3\,\Omega$)	Resistor #2 ($2\,\Omega$) or Resistor #1 ($2\,\Omega$)	$3.2\,\Omega$
Resistor #1 ($2\,\Omega$) + Resistor #2 ($2\,\Omega$)	Resistor #3 ($3\,\Omega$)	$4\,\Omega$

CC 12-7

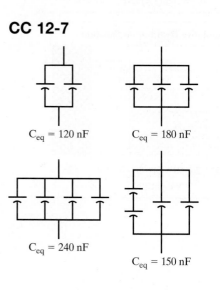

$C_{eq} = 120\ \text{nF}$ $C_{eq} = 180\ \text{nF}$

$C_{eq} = 240\ \text{nF}$ $C_{eq} = 150\ \text{nF}$

CC 12-8

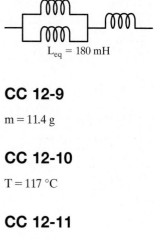

$L_{eq} = 180\ \text{mH}$

CC 12-9

$m = 11.4\ \text{g}$

CC 12-10

$T = 117\,°\text{C}$

CC 12-11

$C_0 = 35\ \text{g};\ k\ [\,=\,]\ 1/\text{h}$

CHAPTER 13

CC 13-1

$V = 50\,P^{-1}$

CC 13-2

$V = 30\,e^{-0.87t}$

CC 13-3

$y = -25.6\,e^{0.06\,x}$

CC 13-4

Data is exponential. The vertical offset is 5°F
$T = 5 - 20\,e^{-0.022\,t}$

CHAPTER 14

CC 14-1

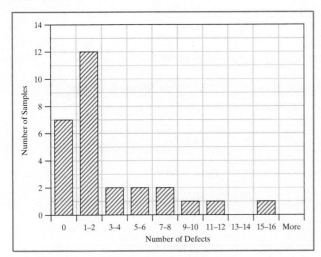

CC 14-2

Mean: 9.3 kg
Median: 9 kg
Variance: 12.9 kg^2
Standard Deviation: 3.6 kg

CC 14-3

Mean: 102.6 °C
Median: 104 °C
Variance: 295.4 °C^2
Standard Deviation: 17.2 °C

CC 14-4

(a) Population size has decreased
(b) Variance has increased
(c) Mean has increased

CC 14-5

(a) Curve (A) – Population size has decreased
(b) Curve (B) – Population size has increased
(c) Curve (E) – Mean has increased

CC 14-6

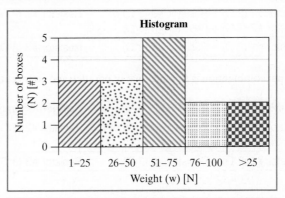

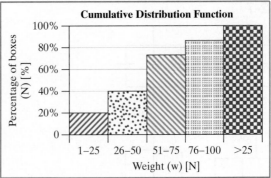

CC 14-7

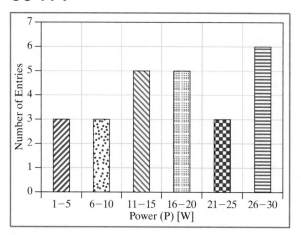

CC 14-8

Not in control

Rule 2 – More than nine consecutive points below mean after 26 hr

Rule 3 – Downward trend between 25 hr and 37 hr

Rule 5 – Two out of three in zone A – several after 29 hr

Rule 6 – Four out of five on same side of mean in zone B – start at 33 hr

Rule 8 – Eight consecutive points outside of zone C – after 28 hr

CC 14-9

** See video in online materials

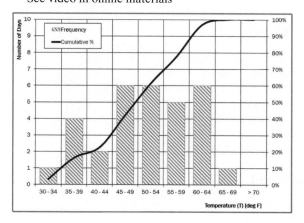

CC 14-10

** See video in online materials

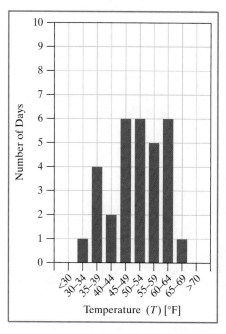

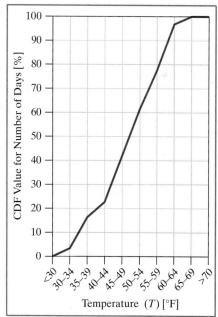

CHAPTER 15

CC 15-1

(a) m is OK
(b) mass6 is OK
(c) INVALID: cannot begin with number
(d) INVALID: cannot contain special characters (& and ©)
(e) INVALID: No spaces allowed
(f) INVALID: cannot contain special characters (-)
(g) INVALID: clear is the name of a common function
(h) ReactorYield is OK

CC 15-2

(a) X contains the value 2
(b) INVALID: Y not defined
(c) Z contains the value 57
(d) W contains the value 3.5
(e) INVALID: Q is not defined
(f) INVALID: Computation not allowed on left of =
(g) INVALID: Computation not allowed on left of =
(h) INVALID: Variable Name is invalid (cannot contain #)
(i) D contains 0.4
(j) A contains 9999
(k) INVALID: Computation not allowed on left of =
(l) QPDoll contains 27

CC 15-3

(a) `save('TempData.mat')`
(b) `load('PressData.mat')`
(c) `save('PTData.mat','Pr1','Pr2','Tmp3')`
(d) `save('a1Var.mat','*a1'))`
(e) `load('VelData.mat','*XS*')`

CC 15-4

(a) `Int2=8;`
(b) `Real2=35.7;`
(c) `Big2=47.98E56;`
(d) `Small2=3E-15;`

CC 15-5

(a)
```
% Solution hardcoding coefficients
R1=(-2+sqrt(2^2-4*3*1))/(2*3);
R2=(-2-sqrt(2^2-4*3*1))/(2*3);

% Better solution using variables
C2=3;
C1=2;
C0=1;
R1=(-C1+sqrt(C1^2-4*C2*C0))/(2*C2);
R1=(-C1-sqrt(C1^2-4*C2*C0))/(2*C2);
```

(b)
```
% Solution using tand
Trig2=tand(75);

% Solution using tan
Trig2=tan(75*pi/180);

% Solution illustrating one of the many
% built-in functions
Trig2=tan(deg2rad(75));
```

CC 15-6

(a) `N4=[17;34;-94;16;0];`
(b) `Tiny=[3.4E-14,9.02E-23,1.32E-9];`
(c) `Ev=2:2:250;`
(d) `Tenths=[10:-0.1:0]';`

CC 15-7

```
Comps=(3*Vals+5).^4-16;
```

CC 15-8

(a) `NZLoc=find(TestVec1~=0);`
(b) `LTNeg5=find(TestVec1<-5);`
(c) `TV1_LT_TV2=find(TestVec1<TestVec2);`
(d) `TestVec1(TV1_LT_TV2)=TestVec2(TV1_LT_TV2);`

CC 15-9

(a) `CV1=rand(123,1)*15;`
(b) `RV1=rand(1,999)*108-75;`
(c) `RV2=randi(109,1,250000)-76;`

CC 15-10

(a) `TFC=[T1,T2,T3,T4]';`
Alternate solution
`TFC=[T1';T2';T3';T4']`
(b) `Rev=[10:2:10000,700:-7:7];`
(c) `Pow=[RV5,sqrt(RV5),RV5.^2]';`
Alternate solution
`Pow=[RV5';sqrt(RV5');RV5'.^2];`

CC 15-11

(a) `Biggie(10:10:50000)=-9999;`
(b) Solution 1
`LV=LV(1:2:length(LV));`

Solution 2
`LV(2:2:length(LV))=[];`

(c) `DS=D(1:2:length(D))+D(2:2:length(D));`

CC 15-12

(a) `CCM1=[18 0.3;-4.1 -1;0 17];`
(b) `CCM2(2:3,3)=1E15;`
(c) `Corners=CCM3(1:2:3,1:2:3);`

CC 15-13

(a) `Tmp(:,3)=Tmp(:,3)+75;`
(b) `Press(1:34:35,1:69:70)=Press`
    `(1:34:35,1:69:70)-100000;`

CC 15-14

(a) `Vel=Vel*T;`
(b) `Power=Voltage.*Current;`

CC 15-15

(a) `Vol(:,3)=Vol(:,3).^3;`
(b) `Pof2=2.^P2;`
(c) `D(:,1)=D(:,1).^(FracD(1,:)');`
    OR
    `D(:,1)=FracD(1,:)'.^D(:,1)`
    NOTE: Transpose and power have the same precedence, thus the extra parentheses in the first solution.

CC 15-16

(a) `Circ=Circ/pi;`
(b) `UnitCost=1500./NumUnits;`
(c) `Resistance=Voltage./Current;`

CC 15-17

(a) `HSR=sqrt(QPD/2);`
(b) `QPD(2,:)=QPD(2,:).^3./log(QPD(3,:));`
(c) `M3R=flipud(M3R);`
(d) `MCCR=[2;1;2;3;1]`   `MCCC=[2;4;4;4;5]`

CC 15-18

(a) `MTS='My hero''s hat';`
(b) `B=[blanks(17),'Dr. Willy',blanks(691)];`
(c) `% Two different solutions`
    `% Two line solution`
    `Avian=strrep(Avian,'wet','crepuscular');`
    `Avian=strrep(Avian,'never','seldom');`
    `% One line solution`
    `Avian=strrep(strrep(Avian,'never',`
      `'seldom'),'wet','crepuscular');`

(d) `% Will work as long as the number is`
    `% two digits long.`
    `% To make this work with a variable number of`
    `% digits requires concepts not yet covered.`
    `UL=length(Ultimate);`
    `UltAns=str2num(Ultimate(UL-1:UL));`

CC 15-19

(a) `Cabinet={77,'Dr. Caligari',ones(25,100)};`
(b) `Cyls={'Diameter','Length';1.5,[1:25]'};`

CC 15-20

(a) `CA{2,1}=CA{1,1}(7:length(CA{1,1}));`
(b) `CA{2,2}=CA{1,2}*CA{1,3};`
(c) `CA{2,3}=CA{1,2}.*CA{1,3}';`

CC 15-21

(a) `BigCell=num2cell(BigMat);`
(b) `VCellMax=cellfun(@max,VCell);`

CC 15-22

```
Resistors(1).Value=100E3;
Resistors(1).Power=1/4;
Resistors(1).Composition='Metal Film';
Resistors(1).Tolerance=0.1;

Resistors(2).Value=2.2E6;
Resistors(2).Power=1/2;
Resistors(2).Composition='Carbon';
Resistors(2).Tolerance=5;

Resistors(3).Value=15;
Resistors(3).Power=50;
Resistors(3).Composition='Wire Wound';
Resistors(3).Tolerance=10;
```

CC 15-23

(a) `MZn=MetalData(1).SpecGrav*1000`
(b) `LtIso=min(MetalData(MNum).Isotopes)`

CC 15-24

```
SortH =
      Diam: 1
   InStock: 387
  Material: 'Brass'
       bin: '84B'
     price: 2.4000
      type: 'bolt'
```

CHAPTER 16

CC 16-1

Known:
- The minimum value in the sum will be 2
- The maximum value in the sum will be 20
- Only even numbers will be included

Unknown:
- The sum of the sequence of even numbers

Assumptions:
- [None]

CC 16-2

Known:
- The minimum value in the product of powers of 5 will be 5
- The maximum value in the product of powers of 5 will be 50

Unknown:
- The product of the sequence of powers

Assumptions:
- Only include integer values

Concerning the actual operation to be performed, there are two alternate assumptions, since the wording is intentionally unclear:

1. Multiply all values between 5 and 50 that are an integer power of 5. In other words, the product of 5 and 25.
2. Multiply all integer powers of 5 with a power between 5 and 50 inclusive. In other words $5^5 * 5^6 * 5^7 * \ldots * 5^{49} * 5^{50}$.

CC 16-3

Known:
- Initial velocity, v_o, in miles per hour
- Acceleration, a, in meters per second squared
- Time, t in hours

Unknown:
- Distance traveled, d, in kilometers

Assumptions:
- Initial position is zero

Algorithm:
1. Acquire values for v_o, a, and t.
2. Convert v_o to meters per second.
3. Convert t to seconds.
4. Calculate d in meters using specified formula.
5. Convert d to kilometers.

CC 16-4

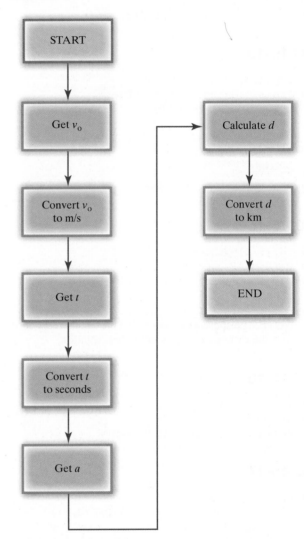

CC 16-5

```
%{
Comprehension Check 16-5

Program description: Calculate distance
traveled given initial velocity,
acceleration, and time

VARIABLES
Input
    vo - Initial velocity [mph]
    a - acceleration [m/s^2]
    t - time [hr]
Output
    d - distance traveled [km]

%}

% Housekeeping
clear
clc

% Define input parameters
vo=10;   % Initial velocity
a=3;     % Acceleration
t=0.02;  % Time traveled

% Define conversion factors
mi2km=1/0.621; % miles to km
hr2sec=60^2;   % hr to sec
km2m=1000;     % km to meters

% Conversions
vo=vo*mi2km*km2m/hr2sec; % mph to m/s
t=t*hr2sec;             % hr to sec

% Calculate distance traveled
d=vo*t+0.5*a*t^2; % distance in meters
d=d/km2m          % convert to km
```

CC 16-6

```
314.1593
Dogs=areaCircle(10);
```

CC 16-7

```
function[R]=RAC(N,A)
R=3*sin(N^A)
```

CC 16-8

```
%{
Comprehension Check 16-8

Function description: Calculate distance
traveled given initial velocity, acceleration,
and time

VARIABLES
Input
    vo - Initial velocity [mph]
    a - acceleration [m/s^2]
    t - time [hr]
Output
    d - distance traveled [km]

%}

function[d]=CC16_8(vo,a,t)

% Define conversion factors
mi2km=1/0.621; % miles to km
hr2sec=60^2;   % hr to sec
km2m=1000;     % km to meters

% Conversions
vo=vo*mi2km*km2m/hr2sec; % mph to m/s
t=t*hr2sec;             % hr to sec

% Calculate distance traveled
d=vo*t+0.5*a*t^2; % distance in meters
d=d/km2m;         % convert to km
```

CC 16-9

```
%{
Comprehension Check 16-9

Program description: Determine model for
travel of jetliner and determine
when it will have traveled 2500 miles

VARIABLES
Input
    dist - vector containing distances
           traveled [mi]
    time - vector containing times at which
           distances were measured [min]
```

```
Output
    m, b - parameters of linear model
    t2500 - time at which 2500 miles have
            been traveled

%}

% Housekeeping
clear
clc

% Define measured data
dist=[273 405 530 676 814 933 1070];
time=30:15:120;

% Determine model parameters
Params=polyfit(dist,time,1);
m=Params(1);
b=Params(2);

% Calculate time at 2500 miles
t2500=m*2500+b; % Time in minutes
t2500=t2500/60  % Convert to hours; Do not
                % suppress final calculation
```

Final answer is about 4.68 hours.

CC 16-10

```
%{
Comprehension Check 16-10

Program description: Determine model for
volume in a container

VARIABLES
Input
    Vol - vector containing measured volumes
          [ft^3] at specified heights
    Ht - vector containing heights [ft] at
          which volumes were measured

Output
    m, b - parameters of power model
    VMax - Volume at height of 7.5 feet

%}

% Housekeeping
clear
clc

% Define measured data
Vol=[1.64 7.71 43.6 136];
Ht=[0.7 1.3 2.6 4.1];
```

```
% Determine model parameters
Params=polyfit(log10(Ht),log10(Vol),1);
m=Params(1);
b=10^Params(2);

% Calculate volume at 7.5 feet
VMax=b*7.5^m     % Volume in ft^3. Do not
                 % suppress final calculation
```

Final answer is about 615.5 cubic feet.

CC 16-11

```
%{
Comprehension Check 16-11

Program description: Determine model for
water draining from a barrel

VARIABLES
Input
    Ht - vector containing measured heights
         [ft] at specified times
    t - vector containing times [min] at
         which heights were measured

Output
    m, b - parameters of power model
    t1ft - Time when water level drops to
           one foot

%}

% Housekeeping
clear
clc

% Define measured data
t=[4 7 15];
Ht=[6.7 5.6 3.8];

% Determine model parameters
Params=polyfit(Ht,log(t),1);
m=Params(1);
b=exp(Params(2));

% Calculate time to reach 1 foot
t1ft=b*exp(m*1)     % Time in minutes.
                    % Do not suppress final
                    % calculation
```

Final answer is about 54 minutes.

CHAPTER 17

CC 17-1

(a) Height=input('Please enter your height in inches: ');

(b) TempF=input('Enter the temperature in degrees Fahrenheit: ');

CC 17-2

(a) EyeColor=input('Enter the color of your eyes: ','s');

(b) Month=input('What is the current month? ','s');

CC 17-3

(a) Month=menu('What is the current month?', 'Jan','Feb','Mar','Apr','May', 'Jun','Jul','Aug','Sep','Oct', 'Nov,','Dec');

(b) C will contain 5

CC 17-4

(a) MyStatus=questdlg('Are you a space alien?','Status')

(b) MyStatus=questdlg('Are you a space alien?','Status','No')

(c) MyStatus=questdlg('Are you a space alien?','Status','Probably', 'Doubtful','Unknown','Unknown')

(d) MyStatus contains the text string 'Cancel'.

(e) MyStatus contains an empty string.

(f) MyStatus contains the text string 'Unknown'.

CC 17-5

(a) UNReq=inputdlg('What UserName would you like?');

(b) LoginData=inputdlg({'UserName', 'Password'});

(c) LoginData is a cell array containing {'ClemKad';'Br@inLe55'}

CC 17-6

(a) %0.2f

(b) %0.3E or %0.3e

(c) %010.3f

(d) 0.35

(e) 3.539E-01 or 3.539e-01

(f) 000000.354

CC 17-7

```
Clemson-USC Football Rivalry:
Year        USC        Clemson
2012        27          17
2011        34          13
2010        29           7
```

NOTE: Tab spacing may be slightly different on different platforms (Windows/Mac/Linux).

CC 17-8

PatSum=sprintf('%s had a temperature of %.1f and a blood pressure of %.0f/%.0f. \nThe charge for this visit is $%.2f. \n',Patient,Temp,SBP,DBP,Charge);

CC 17-9

```
%{
Comprehension Check 17-9

Program description: draw a plot of 50 - 3t
Input
    D - vector containing values of depth [m]
    t - vector containing times [min] at
        which depths occur
Output
    a plot
%}

% Housekeeping
clear
clc
close all
```

```
% Define data
t=[0,15];
D=50-3*t;

% Create plot and label
figure('Color','w')
plot(t,D,'k--')
title('Comprehension Check 17-9')
xlabel('Time (t) [min]')
ylabel('Depth (D) [m]')
grid on
```

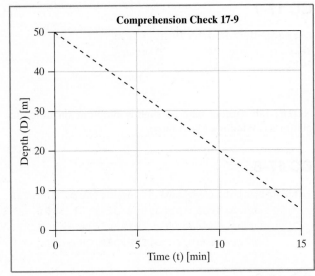

```
% Create plot and label
figure('Color','w');          plot(t,I,'r-')
title('Comprehension Check 17-10')
xlabel('Time (t) [\mus]')
ylabel('Current (I) [mA]')
grid on
```

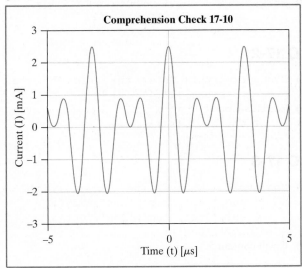

CC 17-10

```
%{
Comprehension Check 17-10

Program description: draw a plot of
2.5 cos(t)cos(5t)
Input
    I - vector containing values of current
       [mA]
    t - vector containing times
       [microseconds] at which currents occur
Output
    a plot
%}

% Housekeeping
clear
clc
close all

% Define data
t=-5:0.01:5;
I=2.5*cos(t).*cos(5*t);
```

CC 17-11

```
%{
Comprehension Check 17-11

Program description: plot three sets of
data using subplots in a single figure
Input
    ta,Fa - vectors containing values for
             first graph (experimental)
    tb,Fb - vectors containing values for
             second graph (theoretical)
    tc,Fc - vectors containing values for
             third graph (experimental)
Output
    Three subplots
%}

% Housekeeping
clear
clc
close all

% Define data
ta=[8,6,1,3,9];
Fa=[7,6,1,8,9];
tb=1:0.1:10;
Fb=15*exp(-0.4*tb);
tc=[8,4,1,3,9];
Fc=[7,2,4,1,9];
```

```
% Create plots and label
figure('Color','w')
subplot(3,1,1)
plot(ta,Fa,'kd')
axis([1 10 1 10])
title('Comprehension Check 17-11 (a)')
xlabel('Time (ta) [hr]')
ylabel('Flow Rate (Fa) [gal/s]')
grid on

subplot(3,1,2)
semilogy(tb,Fb,'r:')
axis([1 10 1 10])
title('Comprehension Check 17-11 (b)')
xlabel('Time (tb) [hr]')
ylabel('Flow Rate (Fb) [gal/s]')
grid on
```

```
subplot(3,1,3)
loglog(tc,Fc,'b*')
axis([1 10 1 10])
title('Comprehension Check 17-11 (c)')
xlabel('Time (tc) [hr]')
ylabel('Flow Rate (Fc) [gal/s]')
grid on
```

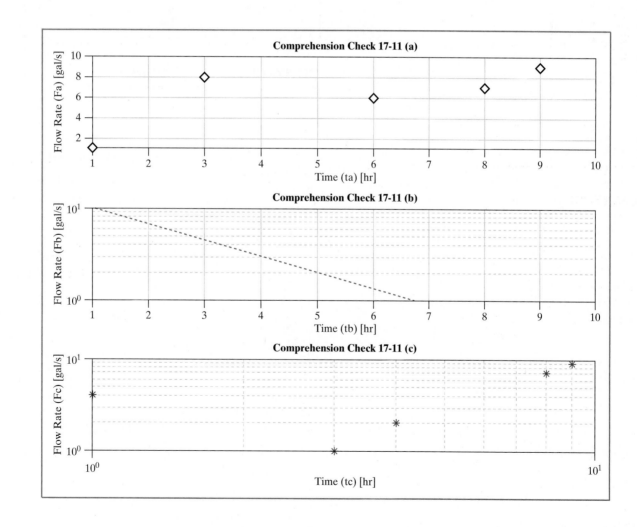

CC 17-12

```
%{
Comprehension Check 17-12

Program description: Plot experimental
data, determine mathematical model, and add
trnedline to graph.
Input
    P - vector containing values of pressure
        [Pa]
    t - vector containing times [yr] at
        which pressures occur
Output
    Plot of data and trendline
%}

% Housekeeping
clear
clc
close all

% Define data
t=[0.1,1.4,3.6,4.9,8.0];
P=[13800,11100,8900,7300,4000];

% Create plot and label
figure('Color','w')
plot(t,P,'b*')
axis([0 10 0 15000])
title('Comprehension Check 17-12')
xlabel('Time (t) [yr]')
ylabel('Pressure (P) [Pa]')
grid on
set(gca,'FontSize',18)

% Determine linear model
LP=polyfit(t,P,1);
m=LP(1);
b=LP(2);

% Plot trendline
hold on
TLt=[0,10];
TLP=m*TLt+b;
plot(TLt,TLP,'b-')

% Place trendline equation on graph.
TLEq=sprintf('P = %.0f t + %.0f',m,b);
```

```
text(1,6000,TLEq)
```
NOTE: A number of things were added to enhance the graph below, but the code will work correctly as shown.

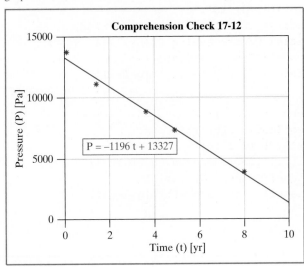

CC 17-13

```
%{
Comprehension Check 17-13

Program description: Plot experimental
data, determine mathematical model, and add
trendline to graph.
Input
    SG - vector containing values of
         specific gravities
    V - vector containing volumes [m^3] at
        which SG was measured
Output
    Plot of data and trendline
%}

% Housekeeping
clear
clc
close all

% Define data
V=[0.7,1.4,3.6,4.9,8.0];
SG=[0.0081,0.0039,0.0014,0.00099,0.00052];
```

```
% Create plot and label
figure('Color','w')
plot(V,SG,'rs')
axis([0 10 0 0.01])
title('Comprehension Check 17-13')
xlabel('Volume (V) [m^3]')
ylabel('Specific Gravity (SG) [-]')
grid on

% Determine power model
PP=polyfit(log10(V),log10(SG),1);
m=PP(1);
b=10^PP(2);

% Plot trendline
hold on
TLV=0:0.1:10;
TLSG=b*TLV.^m;
plot(TLV,TLSG,'r:')

% Place trendline equation on graph.
TLEq=sprintf('SG = %.5f V^{%.0f}',b,m);
text(2.4,0.005,TLEq)
```

NOTE: A number of things were added to enhance the graph below, but the code will work correctly as shown.

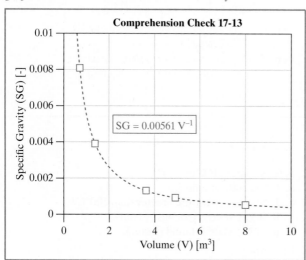

CC 17-14

```
%{
Comprehension Check 17-14

Program description: Plot experimental
data, determine mathematical model, and
add trendline to graph.

Input
    Ib - vector containing values of
        currents [microamperes]
    Vbe - vector containing voltages [mV] at
        which Ib was measured
Output
    Plot of data and trendline
%}

% Housekeeping
clear
clc
close all

% Define data
Vbe=[19,27,40,57,64];
Ib=[0.05,0.21,3.2,51,230];

% Create plot and label
figure('Color','w')
plot(Vbe,Ib,'kd','MarkerSize',12)
axis([0 70 0 250])
title('Comprehension Check 17-14',...
    'FontSize',18)
xlabel('Base to Emitter Voltage (V_{be})
    [mV]','FontSize',18)
ylabel('Base Current (I_b) [\muA]',...
    'FontSize',18)
grid on
set(gca,'FontSize',18)

% Determine exponential model
EP=polyfit(Vbe,log(Ib),1);
m=EP(1);
b=exp(EP(2));

% Plot trendline
hold on
TLVbe=0:70;
TLIb=b*exp(m*TLVbe);
plot(TLVbe,TLIb,'k--','LineWidth',2)

% Place trendline equation on graph.
TLEq=sprintf('I_b = %.5f e^{%6.3fV_
            {be}}',b,m);
text(25,75,TLEq,'Color','k','EdgeColor','k',
    'BackgroundColor','w','FontSize',18)
```

NOTE: The graph enhancements mentioned in CC's 17-12 and 17-13 have been included here. These are not required in your solution.

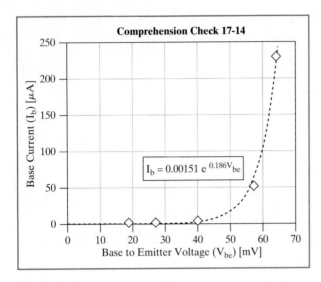

Comprehension Check 17-14

$$I_b = 0.00151\ e^{\,0.186 V_{be}}$$

CC 17-15

```
%{
Comprehension Check 17-15

Program description: Read Excel file,
perform calculations on data, write
results to new worksheet in original file
Input
    Data - matrix containing numeric values
           from worksheet
    Text - cell array containing text from
           worksheet
```

```
Output
    New worksheet in original workbook
    containing results of calculations
%}

% Housekeeping
clear
clc
close all

% Import data
[Data,Text]=xlsread('Widgets.xlsx');
% The following two lines are not
% necessary, but enhance readability
% of program in next step.
Radius=Data(:,2); Length=Data(:,1);

% Calculate volumes
Vol=pi*Radius.^2.*Length;

% Create cell array containing text for new
% worksheet
NewText=Text(:,1);        % Part numbers with
                          % header
NewText{1,2}='Volume';    % Header for volumes

% Write results to new worksheet
xlswrite('Widgets.xlsx',NewText,'Widget
        Volumes')         % Write text
xlswrite('Widgets.xlsx',Vol,'Widget
        Volumes','B2')    % Write volumes
```

CHAPTER 18

CC 18-1

1. Acquire person's age.
2. Is age 14 or less?
   a. If Yes, give one pill.
   b. If No, give two pills.
3. End process.

ALTERNATE SOLUTION:
1. Acquire person's age.
2. Is age greater than 14?
   a. If Yes, give two pills.
   b. If No, give one pill.
3. End process.

CC 18-2

1. Acquire temperature.

2. Is temperature 0°C or less?
   a. If Yes, state is solid (ice).
   b. If No, is temperature greater than 100°C?
      i. If Yes, state is gaseous (steam).
      ii. If No, state is liquid (water).
3. End process.

ALTERNATE SOLUTION:
1. Acquire temperature.
2. Is temperature greater than 100°C?
   a. If Yes, state is gaseous (steam).
   b. If No, is temperature greater than 0°C?
      i. If Yes, state is liquid (water).
      ii. If No, state is solid (ice).
3. End process.

There are numerous other possible solutions.

CC 18-3

One possible solution:

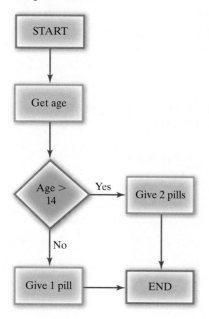

Another possible solution:

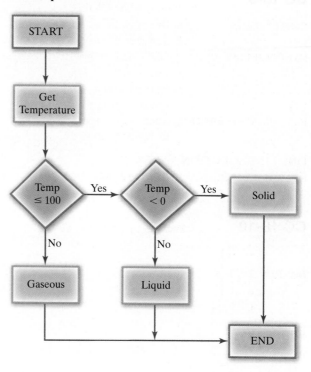

CC 18-4

One possible solution:

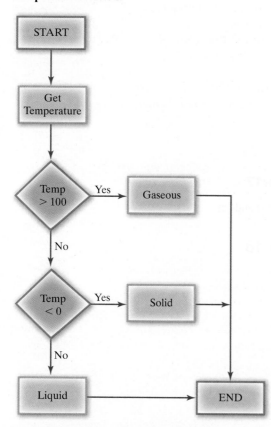

CC 18-5

(a) `5 <= t && t < 10`
(b) `-30 < M && M <= 20`
(c) `Y ~= 100`

CC 18-6

(a) `Q < 17 || Q >= 42`
(b) `Z <= -1 || Z >= 1`
OR
`~(Z > -1 && Z < 1)`

CC 18-7

(a) `R = [1 1 0 1]`
(b) ERROR: Dimensions of matrices must be the same
(c) $R = \begin{bmatrix} 1 & 0 & 0 & 1 \\ 0 & 1 & 1 & 0 \end{bmatrix}$
(d) ERROR: `||` only works with SCALARS. Both `M1<M2` and `M1>S` yield logical matrices. The elementwise operation indicated could be performed using
`M1 < M2 | M1 > S;`

CC 18-8

(a) `TSComp = TS1 >= TS2;`
(b) `[1 1 1 0 0 0 1 1 1 1 0]`

CC 18-9

```
C=Mat1<MIN;          % Determine elements that
                     % are less than MIN
Mat1=Mat1.*(~C);     % zero all elements less
                     % than MIN.
                     % Parentheses for clarity
                     % only - not necessary.
Mat1=Mat1-9999*C;    % subtract 9999 from the
                     % zeroed elements.
```

ONE LINE SOLUTION
```
Mat1=Mat1.*(Mat1>=MIN)-9999*(Mat1<MIN);
          % Parentheses are necessary here.
```

CC 18-10

(a) $R1 = \begin{bmatrix} 1 & 0 \\ 1 & 1 \\ 1 & 0 \end{bmatrix}$

(b) $R2 = \begin{bmatrix} 0 & -5 \\ 0 & 0 \\ 0 & 0 \end{bmatrix}$

(c) $R1 = \begin{bmatrix} 1 & 1 \\ 1 & 0 \\ 1 & 0 \end{bmatrix}$

CC 18-11

```
M1G=M1>M2;
R4=M1.*M1G+M2.*(~M1G);
```

ONE LINE SOLUTION
```
R4=M1.*(M1>M2)+M2.*(M2>=M1);
```
NOTE: One and only one of the inequalities must include the equality. WHY?

CC 18-12

(a) [1 1 1]
(b) [1 0 0]
(c) ERROR: String lengths must be the same.
(d) 1
(e) 0
(f) 0
(g) 1

CC 18-13

```
Cont=strcmpi(UserResp,'y')||strcmpi(UserResp,
'yes');
```

CC 18-14

(a)
```
if MTime<PersBest
    fprintf('New personal best\n')
    PersBest=MTime;
end
```
(b)
```
if TruckWt>MaxWt
    fprintf('Reduce the load in the
             truck.\n')
else
    fprintf('Load Acceptable.\n')
end
```

CC 18-15

```
if TW <= 0
    phase='solid';
elseif TW >= 100
    phase='vapor';
else
    phase='liquid';
end
```

CC 18-16

```
function[x,y]=SumItUp(a,b,c)
if a + b > 100
    fprintf('The inputs are %0.2f, %0.2f,
            and %0.2f',a,b,c)
else
    fprintf('The inputs are %0.0f, %0.0f,
            and %0.0f',a,b,c)
end
x=a+b;
y=b+c;
```

CC 18-17

Choices (a) and (c).

CC 18-18

```
%{
Comprehension Check 18-18

Program description: Classify mass-velocity
pair as possible or not possible to achieve
given the classification diagram.
Input
    Mass - Mass of object [kg]
    Vel - Velocity of object [m/s]
```

```
       MData, VData - data points defining the
                boundary between the two cases
Output
    Class - Either 'possible' or
              'not possible'
%}

% Housekeeping
clear
clc

% Get data
Mass=input('Enter the mass of the object
           [kg]: ');
Vel=input('Enter the velocity of the object
           [m/s]: ');

% Specify data for boundary line
MData=[5 20 40];
VData=[10 5 3.54];

% Determine model parameters for boundary
% line
MP=polyfit(log10(MData),log10(VData),1);
m=MP(1);
b=10^MP(2);

% Classify point
VBound=b*Mass^m;  % velocity on boundary
                  % corresponding to Mass
if Vel<=VBound    % Below or on boundary
                  % line?
    Class='possible';
else
    Class='not possible';
end

% Print results to screen
fprintf('It is %s to throw a %.1f kg object
        at a velocity of %0.1f m/s.\n',
        Class, Mass, Vel)
```

CC 18-19

```
Phase = menu ('Choose type of water sample',
             'Solid','Liquid','Gas');
switch Phase
    case 1
        fprintf('For a solid, water must be
            at a temperature less than 0
            degrees Celsius.\n')
    case 2
        fprintf('For a liquid, water must be
            at a temperature between 0 and
            100 degrees Celsius.\n')
```

```
    case 3
        fprintf('For a gas, water must be
            at a temperature more than 100
            degrees Celsius.\n')
    otherwise
        fprintf('You did not make a proper
            selection.\n')
end
```

CC 18-20

```
Flower = input('Type the name of your
               favorite flower: ','s');
switch Flower
    case 'Rose'
        fprintf('Many roses have thorns.\n')
    case 'Lily'
        fprintf('Lilies are popular at
            Easter.\n')
    case 'Tulip'
        fprintf('The Dutch produce about
            3 billion tulip bulbs each
            year.\n')
    case 'Peony'
        fprintf('Peonies abort their blooms
            if the weather is too hot.\n')
    otherwise
        fprintf('%s is not in the database of
            this program.\n',Flower)
end
```

CC 18-21

```
% The following assumes w and m have
% already been defined.
if m==0
    error('Mass=0! Program terminated.');
elseif m<0
    warning('Mass is negative. Using the
            absolute value instead.')
    m=-m;
end
g=w/m;
fprintf('The gravity is %0.1f m/s^2.\n',g);
```

CC 18-22

NOTE: This solution assumes we will allow multiplication of a matrix by a scalar. To exclude this case, additional code would need to be added since the try-catch structure would not flag this as an error.

```
M1=input('Please enter a matrix.\n');
M2=input('Enter another matrix with the
         same dimensions.\n');
try
   MatProd=M1.*M2;
```

```
catch MultError
    error('Matrices entered incorrectly.
          \n%s',MultError.message)
end
```

CHAPTER 19

CC 19-1

Known:
- Lowest power of 5 in product is 5
- Highest power of 5 in product is 50

Unknown:
- Product of all integer powers of 5 between the lowest and highest powers

Assumptions:
- The result will not exceed the maximum representable number in the computer. Note that by the time the sequence gets to $x = 17$, the product is approximately one googol (10^{100}).

Algorithm:
1. Set product = 1
2. Set x = 5
3. If x is less than or equal to 50
   a. Multiply product by 5^x
   b. Add 1 to x
   c. Return to Step 3 and ask the question again
4. End the process

CC 19-2

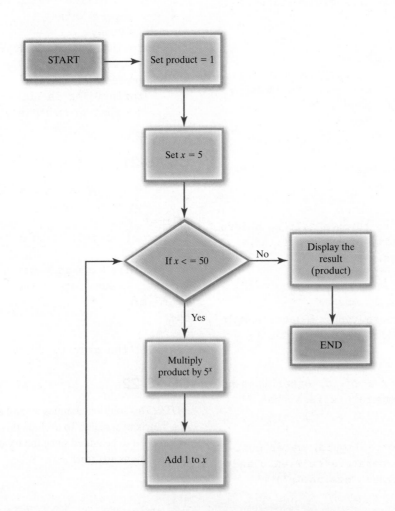

CC 19-3

The arrow leaving the box labeled "Warn the user the input is invalid" should go up to the box labeled "Input the lower bound of the sequence" instead of down to the "END" box.

CC 19-4

```
%{
Comprehension Check 19-4
%}

% Initialization
Prod=1; % To hold the running product
MaxProd=1E6; % Maximum product allowed
NumEl=length(V2); % Number of elements in V2
E=1; % E will point to consecutive
     % elements, starting with the first

% Calculate running product
while E<=NumEl && Prod<MaxProd
    Prod=Prod*V2(E); % Multiply by current
                     % element of V2
    E=E+1; % Step to next element of V2
end
```

CC 19-5

```
%{
Comprehension Check 19-5
%}

% Initialization
Repeat='yes'; % Define Repeat to enter loop
              % the first time

% Calculate running product
while strcmpi(Repeat,'yes')
    LVel=input('Enter launch velocity
             [m/s]: ');
    LAng=input('Enter launch angle
             [degrees]: ');
    Dist=CalcDist(LVel,LAng);
    fprintf('The projectile traveled %.1f
             meters.\n',Dist)

    % Ask user if they wish to enter another
    % set of data
    Repeat=input('Do you wish to calculate a
             new distance (Yes or No)? ','s');
end
```

CC 19-6

```
%{
Comprehension Check 19-6
%}

% First attempt to enter a positive value
Val=input('Enter a positive number: ');

% Ask the user to try again if the value is
% negative
while Val<0
    Val=input('You entered a negative value.
             Try again: ');
end
```

CC 19-7

```
%{
Comprehension Check 19-7
%}

% First attempt to enter a positive integer
% divisible by 7
V7=input('Enter a positive multiple of
         seven: ');

% Ask the user to try again if the value
% not a positive multiple of seven
while V7<=0 || rem(V7,7)~=0
    warning('The number entered is
            either not positive or
            not a multiple of 7.')
    V7=input('Please try again: ');
end
```

CC 19-8

```
%{
Comprehension Check 19-8

The initial estimate will be the midpoint
between 0 and pi/2.
With each iteration, either the low or high
bound will be updated to the most recent
estimate, depending on whether it is too
low or too high. Each successive iteration
is the midpoint between the narrowed
bounds. Eventually, the estimate will be
within the desired accuracy.
%}

% Housekeeping
clear
clc
```

```
% Get target value. Assume user enters a
% valid value.
Target=input('Enter a target value between
            0 and 1: ');

% Initialization
LoX=0; % Initial low X
HiX=pi/2; % Initial High X
Error=1; % Initialization to get into loop

% Iteratively guess solution
while Error>0.000002 % aiming for <= 0.0002%
    X=(LoX+HiX)/2; %get midpoint
    EstVal=exp(-X/10)*cos(X); % Calculate
                    % function with estimate
    Error=abs(EstVal-Target)/Target;
    % Calculate error
    if EstVal>Target % Update High or Low X
        LoX=X;
    else
        HiX=X;
    end
end
fprintf('X=%.8f yields %.8f.\n',X,EstVal)
% Display results
```

CC 19-9

```
%{
Comprehension Check 19-8
%}

% Housekeeping
clear
clc

% Get target value. Assume user enters a
% valid value.
V=input('Enter positive number for square
        root, negative to terminate. \n');

% As long as entered value is positive,
% take square root
while V>=0
    SRV=sqrt(V);
    fprintf('You entered %.0f. The square
            root of %.0f is %.3f.\n',V,V,SRV)
    V=input('Enter positive number for
            square root, negative to
            terminate. \n');
end
fprintf('You entered %.2f. Program
        terminated.\n\n',V)
```

CC 19-10

```
%{
Comprehension Check 19-10
Program to enter a positive value between
10 and 100
%}

% Housekeeping
clear
clc

% Initial input attempt
P=input('Enter a value between 10 and 100: ');

% Initialize counter for number of attempts
Tries=1;

while (P<10||P>100)&&Tries<10 % Try again
            % if wrong and more attempts left
    warning('Invalid Input - try again')
    P=input('Enter a value between 10 and
            100: ');
    Tries=Tries+1; % Increment attempt
                    % counter
end
```

CC 19-11

```
%{
Comprehension Check 19-11
Display all even numbers from 2 to 20
%}

% Housekeeping
clear
clc

% Print desired values
for N=2:2:20
    fprintf ('%.0f\n',N)
end
```

CC 19-12

```
%{
Comprehension Check 19-12
Display all multiples of 5 from 5 to 50
%}

% Housekeeping
clear
clc
```

```
% Print desired values
for N=5:5:50
    fprintf('%.0f\n',N)
end
```

CC 19-13

```
%{
Comprehension Check 19-12
Display all odd numbers from 13 to -11
%}

% Housekeeping
clear
clc

% Print desired values
for N=13:-2:-11
    fprintf('%.0f\n',N)
end
```

CC 19-14

```
% Find each positive value in Vals, and add
% to running sum
PosSum=0; % Initialize sum
for N=1:length(Vals)
    if Vals(N)>0
        PosSum=PosSum+Vals(N);
    end
end
```

CC 19-15

```
% Comprehension Check 19-15

% Housekeeping
clear; clc

% Necessary Initialization
D=0;
G=-1;
I=1;

% Print header. Spaces used to fine-tune
% alignment over columns.
fprintf(' k\t A\t B\t  C\t D\t E\t F\t G
        \t    H\t    I\t J\n')

% Create table
for k=1:10
```

```
% Determine next value in each sequence
A=2*k;
B=A-1;
C=k^2;
D=D+k;
E=7+k/4;
F=11-k;
G=-1*G;
H=2^k;
I=I*k;
J=1/k;
% Print next row of table
fprintf('%2.0f\t%2.0f\t%2.0f\t%3.0f
        \t%2.0f\t%.2f\t%2.0f\t%2.0f
        \t%4.0f%9.0f\t%.2f\n',
        k,A,B,C,D,E,F,G,H,I,J)
end
```

CC 19-16

```
% Initialization
[rows,cols]=size(M1);
PosNums=zeros(1,cols);  % Set initial count
                       % for each column to zero

for C=1:cols
    for R=1:rows
        if M1(R,C)>=0
            PosNums(C)=PosNums(C)+1;
        end
    end
end
```

CC 19-17

NOTE: There are many possible approaches to this problem. We show only one here.

```
% Initialization
[rows,cols]=size(M2);
PosCounter=zeros(rows,1);  % Counter for
                % positive values in each row
NegRows=0;
for R=1:rows  % Step through rows
    % Inner loop will store number of
    % non-negative values in PosCounter
    for C=1:cols
        if M2(R,C)>=0
            PosCounter(R)=PosCounter(R)+1;
        end
    end
end
```

```
% Elements of PosCounter now contain the
% number of non-negatives in the
% corresponding rows of M2

NegRows=sum(PosCounter==0)
%{
EXPLANATION: PosCounter==0 is 1 for each 0
(all negative row) in
PosCounter. These ones are then added up
and placed in NegRows, giving the
total number of all negative rows.
%}
```

NOTE: On the power of MATLAB:
The above problem can be implemented without for loops using only a single line of code.
```
NegRows= sum(0==sum((M2>=0)'));
```

CC 19-18

```
Up=1    Down=10
Up=2    Down=8
Up=3    Down=6
Program Terminated>>
```

NOTE: The command prompt (>>) appears on the same line as "Program Terminated". This could be "fixed" by including \n in the termination print statement.

CC 19-19

```
55
35
15
R Negative>>
```

NOTE: The command prompt (>>) appears on the same line as "R Negative". This could be "fixed" by including \n in the "R Negative" print statement.

INDEX

A

ABCs of evaluating information, 100
ABET design approach, 74
Abscissa, 361, 415, 503
Absolute address, 299, 301
Absolute pressure, 219
Absolute temperature scales, 216
Acceleration, 202, 375, 381
AES (American Engineering System), 171, 203
Aggregation, estimation by, 129–130
Agricultural engineering (AgE), 14
Algorithms, 619–627, 732–737
  defined, 619
  graphical, 619, 623–624
  iterative, 792–794
  loops, 788–794
  scope of, 619–621
  written, 619
American Engineering System (AES),
  171, 203
Amperes (A), 237
Ampersand operator (&), 306, 307
Analogy, estimation by, 129
AND logical operator, 311, 736, 738
Approximation, 147
Archimedes, 164–165
Arithmetic sequences, 803–805
Assignment operator, 554
Atmospheric pressure, 219
Atomic mass unit (amu), 211
Atomic weight, 211–212
Average, 506
Avogadro's number, 211
`axis` command, MATLAB, 686

B

Backslash character (\), 693
Bar graph, 371
Bin, 504
Binary behavior, 311, 731
Bingham plastics, 420
Bioengineering (BioE), 14
Biomedical engineering (BME), 14
Biomimetics, 68
Biosystems engineering (BE), 14–15
Block comment, 631
Boltzmann constant, 212
Boolean logic, 311, 731
Brainstorming, in engineering design process,
  67–68
Breakeven analysis, 383–384
Breakeven point, 384
Breakpoints, 651
British thermal unit (BTU), 227
Buckingham Pi method, 279

C

Calculator notations, 141
Calorie (cal), 227
Capacitance, 241–242, 430
Capacitors, 429–431, 439
Career
  choosing, 8–9
    engineering, 9–11
    specific engineering field, 14–22
  gathering information on, 24–26
  sample paths and possible majors, 10
  student opportunities, 26–37
  websites, 25–26

Cell arrays, 587–596
  defined, 588–590
  extracting data from, 590–593
  functions used with, 594–596
Cell references, 298–301
  absolute address, 299, 301
  mixed addressing, 301
  relative cell address, 300
Celsius scale, 215
Centipoise (cP), 420
Central tendency, 506
Character strings, 585–587
  apostrophe, 586
  combining text strings, 586
  defined, 586
  functions used with, 586–587
  indexing with, 586
Charge (Q), 237
Chemical engineering (ChE), 15
Civil engineering (CE), 15–16
`close all` command, MATLAB, 685–686
Coefficient of determination (R^2 value),
  470–471
Colon operator, 561
Column graph, 371
Command Window, MATLAB, 629–630
Commenting, in MATLAB, 631
Communication
  basic presentation skills, 91–94
  sample presentations, 94–97
  technical. *See* Technical communications
  technical writing skills, 98–101
Comprehensive Assessment of Team-Member
    Effectiveness (CATME), 76
Computer engineering (CpE), 16
Concatenate, 306
CONCAT function, 307
Conditional formatting, 324–327
Conditional statements
  compound, 317
  defined, 311
  in MATLAB, 637, 746–750
  nested, 314–317, 748–749
  sample, 311–312
  single, 312, 747–748
Conductance (G), 239
Conduction, 272
Constants
  fundamental, 269, 270
  material, 270, 271
  Napier, 436
  in problem-solving, 153–154
  spring, 418
  time, 439
  with units, 269–272
  universal gravitational, 270
Control variable, 80
Conversions
  involving equations, 187–189
  involving new units, 178–179
  procedure, 171–174
  steps, 174–178
  temperature, 217
  unit, 135, 625–626
Cooperative education, 26–30
Coulomb (C), 237
Coulomb's law, 237
Counter variable, 792–793
COUNTIF function, 306
Cumulative distribution function, 515–517

D

Dalton (Da), 211
Data types, MATLAB, 636–637
Data validation, in Excel, 319–324
Debugging, in MATLAB, 650–652
Decay function, 436, 439
Decay rate, 439
Degrees Rankine (°R), 216
Delimiters, 560
Density, 206–210, 271
Dependent variable, 80, 361
Derivative, 375
Derived dimensions, 179–183
Derived unit, 180
  of radian, 183
  in SI system, 181, 202
Design process, for engineering programs, 61–84
  brainstorming in, 67–68
  experimental design, 79–82
  project management, 84
  project timeline and dynamics, 82–83
  prototyping and testing, 70
  sustainability, 70–72
  working in teams, 73–79
Development team, 84
Differential pressure, 220
Dimensional analysis, 275–278
Dimensionless numbers
  common, 272–274
  dimensional analysis, 275–278
Dimensionless quantities, 275
Dimensions, 167–189
  defined, 167, 269
  of energy, 227
  metric system, 168–170
  of power, 229
`disp` function, 680
Distribution, data, 505, 509–515
  cumulative, 515–517
  mean, shift in, 512
  normal, 509–510
  skewed data, 513
  variance, decrease in, 512
Documentation, in programming, 628–629
Dot operator, 563
Dynamic viscosity (μ), 420

E

Editor/Debugger window, MATLAB, 630
Efficiency, 231–235
Elasticity, 421
Electrical engineering (EE), 17
Electric charge, 236–237
Electric current (I), 237, 419
Electric power, 240
Elementary charge, 212
Element-wise operator, 563
E-mail, 102–103
Empty element, 569
Energy, 226–230
  defined, 226
  dimensions of, 227
  kinetic, 226
  potential, 226
  thermal, 226
  work, 226
Energy of translational motion, 226
Engineering
  agricultural, 14

Engineering (*continued*)
  bioengineering, 14
  biomedical, 14
  biosystems, 14–15
  as career, 6–7
  chemical, 15
  civil, 15–16
  computer, 16
  defined, 4, 5
  electrical, 17
  environmental, 18–19
  industrial, 19
  materials, 20
  mechanical, 21
  nuclear, 22
  reverse, 13
  technology, 22–24
Engineering notation, 140, 168
Engineer's Creed, 52–53
Environmental engineering (EnvE), 18–19
Equations, 154
  conversion of, 187–189
  laws
    addition, 183–185
    division, 185–186
    multiplication, 185–186
    subtraction, 183–185
  manipulation, 155
Error (s)
  in Excel, 303–310
  formula coding, 628
  formula derivation, 628
  messages
    in Excel, 303–304
    in MATLAB, 758–759
  programming, 628
  runtime, 628
  syntax, 628
Estimation, 124–141
  by aggregation, 129–130
  by analogy, 129
  hints for, 127–128
  significant figures, 131–135
  by upper and lower bounds, 130
  using modeling, 130–131
Ethical decision making
  alternative courses of action from different
    perspectives, 47–49
  determining issues and stakeholders, 46–47
Ethics, 45–53
  Engineer's Creed, 52–53
  fundamental canons, 52–53
  plagiarism, 51–52
  social responsibility, 53
Euler's number, 436
Excel®
  absolute address, 299, 301
  cell references, 298–301
  CONCAT function, 307
  conditional formatting, 324–327
  conditional statements. *See* Conditional
    statements
  converting scales to log in, 480–481
  COUNTIF function, 306
  data validation in, 319–324
  error messages in, 303–304
  filtering data, 330–331
  functions in, 302–310
  graphs in, 369–371
  IFERROR function, 304–305
  input/output, 700–705
  limitations of, 482–486
  lookup function in, 319–324
  mathematical functions in, 302
  miscellaneous functions in, 303
  mixed addressing, 301
  polynomial models, 471–472
  relational operators, 311
  relative address, 300

  ROUND function, 305–306
  ROUNDUP function, 306
  sorting, 327–330
  statistical functions in, 303
  statistics in, 523–528
  SUBTOTAL function, 331–333
  trigonometric functions in, 302
  worksheet, 294
Experiential learning, 26
Experimental data, proper plot containing
  trendlines, 413–414
Experimental design, 79–82
  measurements, 80
  PERIOD parameters, 80–82
Exponential constant "e," 436
Exponential models, 412, 435–440
  decay function, 436, 439
  derivation of, 476
  exponential constant "e," 436
  growth function, 436, 437
Exponential notation, 141, 180

F
Fahrenheit scale, 215
Faraday constant, 212
Farads (*F*), 241, 430
Feedback loop, 789
Fermi problems, 125
Fixed costs, 383
Flowcharts
  actions, 623–624
  decision making in, 733–736
  defined, 619
  rules for creating, 623
Flow rate
  mass, 373
  volumetric, 373
Fluid flow, 419–421
Force (*F*), 202–203
for loop, 801–812
  arithmetic sequences, 803–805
  ending, 809–812
  execution process, 802–803
  index, manipulating, 806
  syntax, 802
  using index variable as array index, 807–809
  in variable recursion, 806
Formula coding errors, 628
Formula derivation errors, 628
Formula weight, 212
Fourier's law, 272
fprintf function, 680–683
Full logarithmic/log–log, 474
Functions
  creation guidelines, 638–639
  in Excel, 302–310
  MATLAB, 637–642
  structure and use, 639–640
  testing, 640–642
Fundamental Canons of the Code of Ethics, 52–53
Fundamental constants, 269, 270
Fundamental dimensions. *See* Dimensions

G
Gas constant, 223
Gas pressure, 219, 223–225
Gauge pressure, 219
Global variables, 642–644
Graphical algorithms, 619, 623–624
Graphs/Graphing
  drawing using logarithmic axes, 690–692
  in Excel, 369–371
  graphical solutions, 382–388
  interpretation of, 372–376
  line shapes, 376–381
  logarithmic, 474–478
  proper plots, 362–368
  terminology, 361

Gravity (*g*), 205, 271
grid command, MATLAB, 686
Growth function, 436, 437
Growth rate, 437

H
Henrys (*H*), 243, 412, 430
Histograms, 503–506
HLOOKUP function, 319–320
Hooke's law, 418
Horizontal gridlines, 362
Horsepower (hp), 229
Hydrostatic pressure, 219, 220–221

I
Ideal gas law, 223, 270, 432
Ideal gas law constant (*R*), 270
Ideal problems, 148
IFERROR function, 304–305
Independent variable, 80, 361
Index variable, 792–793, 803
  as array index, 807–809
Inductance (*L*), 243, 430
Inductor, 412
Inductors, 429–431
Industrial engineering (IE), 19
Infinite feedback loop, 790
Input
  energy, 231–235
  Excel, 700–705
  MATLAB, 637, 669–679
  menu-driven, 674–679
inputdlg function, 678–679
Instrumentation error, 80
Integral, 375
Intercept (*c*), 415
Internship, 30–32
Interpolation, 216, 362
Interpretation, of graphs, 372–376
Irrational number, 436

J
JERK (unit), 381
Joule, 227
Joule's first law, 389

K
Katal (unit), 178
Kelvin scale, 216
Kinematic viscosity, 420
Kinetic energy (KE), 226

L
Laminar flow, 284
Law (s)
  of arguments, 638
  Coulomb's, 237
  of equations, 183–186
  Fourier's, 272
  Hooke's, 418
  ideal gas, 223, 270, 432
  Joule's first law, 389
  Moore's, 437
  Newton's second law. *See* Newton's second law
  Ohm's, 239, 367, 419, 426–427
  Pascal's, 220
  Per, 185
  Plus, 184
  Unit, 184
  Universal Gravitation, 270
Law of Arguments, 274
Law of Universal Gravitation, 270
Leadership in Energy and Environmental Design
  (LEED), 65–66
legend command, MATLAB, 686

Linear fit, 472
Linear models, 412, 414–422
Linear relationships, 417–431
Line plot, 370
Line shapes, 376–381
Local variables, 642–644
Logarithmic axes, 469
    drawing graphs using, 690–692
    proper plot of experimental data, 479–480
Logarithmic graphs, 474–478
Logic, 310–319
    Boolean, 311, 731
    conditional statements. *See* Conditional
        statements
    equivalent forms of, 749–750
Logical expression, 311, 736
Logical negation, 744
Logical operators, 736–739
    AND, 311, 736, 738
    NOT, 311, 737
    OR, 311, 736, 738–739
Logical variables, 739–746
    calculations using, 743
    functions associated with, 745–746
    scalar, 739–740
Loops
    `for`, 801–812
    algorithms, 788–794
    feedback, 789
    infinite feedback, 790
    `while`, 794–801
Loss (negative profit), 384
Lower limit estimation, 130

M

Mach Number (*Ma*), 273
Mac OS, 635, 700
    statistics on, 527–528
Mantissa, 132, 169
Mars Climate Orbiter (MCO) spacecraft, 166
Mass
    atomic, 211–212
    molar, 212
    molecular, 212
    of object, 205
    unit of, 203
Mass flow rate, 373
Material constants, 270, 271
Materials engineering, 20
Mathematical functions, in Excel, 302
Mathematical model, 465–486
    deriving, 646–650
    logarithmic graphs, 474–478
    trendline type, selection of, 466–473
Mathematical symbols, 183
MATLAB, 627
    `axis` command, 686
    character strings, 585–587
    classification diagram, 750–754
    `close all` command, 685–686
    Command Prompt, 551
    Command Window, 629–630
    commenting in, 631
    conditional statements in, 637, 746–750
    data types, 636–637
    debugging, 650–652
    Editor/Debugger window, 630
    error messages, 758–759
    figure windows in, 684–685
    functions, 637–642
    global variables, 642–644
    `grid` command, 686
    input, 637, 669–679
    interface, 550–552, 629–630
    `legend` command, 686
    local variables, 642–644
    logical variables, 739–746
    looping, 637

mathematical inequality in, 737
matrices. *See* Matrices, MATLAB
M-files, 631
    numeric display format in, 559
    numeric functions, 558–559
    numeric types, 556–557
    order of search, 630–631
    output, 637, 680–684
    plotting, 684–694
    `polyfit` command, 695–700
    `polyfit` function, 646–650
    predefined constants, 557–558
    program structure and use, 631–636
    relational and logical operators, 736–739
    scalars, 557
    statistics in, 528–534
    subplots in, 692–694
    `switch` statement, 755–758
    `title` command, 686
    trendlines, 695–700
    `try-catch` statement, 760–761
    variables, 552–556
    vectors. *See* Vectors, MATLAB
    warning messages, 759
    `xlabel` command, 686
    `xlsinfo` function, 700–701
    `xlsread` function, 700–703
    `xlswrite` function, 704–705
    `ylabel` command, 686
Matrices, MATLAB, 572–585
    calculations with
        addition/subtraction, 577–578
        division, 581
        multiplication, 578–579
        powers, 579–580
    cell arrays. *See* Cell arrays
    creation using indexing, 573
        colon operator for, 574
    defined, 572
        from another matrix, 574
    functions used with, 582–585
    individual elements, deleting, 576
    logical, 740–742
    row/column
        accessing, 574–575
        deleting, 575
    structure arrays. *See* Structure arrays
    transpose, 576–577
Mean, 506
Measurements
    accuracy, 127, 137
    experimental design, 79–82
Mechanical engineering (ME), 21
Median, 506–507
Memo (one-page limit), 104–105
Menu-driven input, 674–679
    `menu` function, 674–675
Metric system, 168–170
M-files, 631, 652
Microfarads (*μF*), 439
Microsoft Excel. *See* Excel®
Mixed addressing, 301
Modeling, estimation using, 130–131
Model(s)
    defined, 412
    exponential, 412, 435–440
    linear, 414–422
    power functions, 432–435
Modulus of elasticity, 421
Modulus of elasticity (*E*), 421
Molar mass, 212
Mole (mol), 211
Molecular mass, 212
Moore's law, 437

N

Napier constant, 436
National Academy of Engineering (NAE), 4, 5

National Academy of Engineering (NAE) Grand
    Challenges, 11–12, 536
Negatively skewed data, 513
Nelson rules, 519–521
Nested conditional statements, 314–317
Newton (N), 203
Newtonian fluids, 420
Newton's first law, 417
Newton's law of viscosity, 420
Newton's second law, 180
    and elastic materials, 421–422
    generalized, 412, 417
Normalized plot, 515
Notation, 139
    calculator, 141
    engineering *vs.* scientific, 140, 168
    exponential, 141, 180
NOT logical operator, 311, 737
NSPE Code of Ethics, 52–53
Nuclear engineering (NucE), 22
Number (s)
    Avogadro's, 211
    dimensionless, 272–278
    irrational, 436
    Mach, 273
Numeric types, MATLAB, 556–557

O

Objectives, problem-solving, 151
Observations, problem-solving, 151–153
Occam's razor, 472
Ohm (Ω), 419
Ohm's law, 239, 367, 419, 426–427
Order of search, MATLAB, 630–631
Orders of magnitude, 127, 231–235
Ordinate, 361
OR logical operator, 311, 736, 738–739
Output
    energy, 231–235
    Excel, 700–705
    MATLAB, 637, 680–684

P

Pairwise comparisons, 69
Parametric problems, 148
Pascal (*Pa*), 218
Pascal's law, 220
PERIOD parameters, 80–82
Per law, 185
Pi (*π*), 273
Pie graph, 371
Plagiarism, 51–52
Plotting, MATLAB, 684–694
Plus law, 184
Poiseuille's equation, 433
`polyfit` command, MATLAB, 695–700
    for exponential relationships, 698–700
    for linear relationships, 695–697
    for power relationships, 697–698
`polyfit` function, 646–650
Polynomial models, 471–472
Portrait orientation, graphs, 364
Positively skewed data, 513
Poster presentation, 107–110
Potential energy (PE), 226
Power, 229–230
    dimensions, 229
Power law model, 412
    derivation of, 475
Precision, 137
Prefix, SI, 168–169
Pressure, 218–225
    absolute, 219
    atmospheric, 219
    differential, 220
    dimensions of, 218
    gas, 219, 223–225

Pressure (*continued*)
  gauge, 219
  hydrostatic, 219, 220–221
  total, 219, 221–222
  vacuum, 219
Prior art, research on, 68
Probability, 502
Problems
  RAPID process, 146–149
  solving. *See* Problem-solving
Problem-solving
  representing final results, 155
  SOLVEM approach. See SOLVEM
Product owner, 84
Profit, 384
Programming, 627
  documentation, 628–629
  program testing, 628
Project management, 84
Prototyping, 70

Q

Quality control chart, 518
questdlg function, 675–678

R

Radian, 183
Random error, 80
Range restriction, 513
Rankine scale, 216
Ratio, as problem type, 146
Rayleigh's method, 278–285
Reasonableness, 135–139
Rectilinear, 474
References, in technical communications, 100–101
Relational expression, 311, 736
Relational operators, 311, 736
Relative cell address, 300
Repeatability, 137
Resistance *(R)*, 238–239, 419
Resistors
  in parallel, 426–428
  in series, 426
Revenue, 384
Reverse engineering, 13
Reynolds number, 283–284
Rheology, 420
ROUND function, 305–306
ROUNDUP function, 306
Runtime errors, 628

S

Scalars, 557
Scatter plot, 370
Scientific method, 79
Scientific notation, 140, 168
Scrum master, 84
Selling price, 384
Semilogarithmic/semilog, 474
Shear/strain rate, 419
Shear stress (τ), 419
Shockley equation, 382
Short-cut operator, 307
Short report (two to four pages), 106
Significant, meaning of, 132
Significant figures, 131–135
SI prefix, 168–169
SI system (Le Système International d'Unités), 168–170. *See also* Units
  derived units in, 181, 202
  official rules, 170
Sketch, of problem, 149–150
Skewed data, 513
Slope, line, 373, 415
SOLVEM, 149–158
  avoiding common mistakes, 155–156

  examples of, 156–158
  final results, 155
Specific gravity (SG), 208–210, 273
Specific heat, 271–272
Specific weight, 206
Spring constant, 418
Springs, 418
  in parallel, 422–423
  in series, 423–424
sprintf function, 683–684
Standard deviation, 507
Statistical functions, in Excel, 303
Statistical process control (SPC), 518–522
Statistics, 502–534
  defined, 502
  distributions. *See* Distribution, data
  in Excel, 523–528
  histograms, 503–506
  in MATLAB, 528–534
  statistical behavior, 506–509
  statistical process control, 518–522
Stokes (St), 420
Strain (ε), 422
Stress (σ), 422
Structure arrays, 596–601
  defined, 596–598
  extracting data from, 598–599
  functions used with, 599–601
Study abroad, 34–37
subplot command, MATLAB, 692–694
SUBTOTAL function, 331–333
Sustainability, of engineering design, 70–72
switch statement, 755–758
Syntax errors, 628
Systematic error, 80

T

Technical communications, 102–107
  e-mail, 102–103
  memo (one-page limit), 104–105
  poster presentation, 107–110
  short report (two to four pages), 106
Technical writing, 98–101
Temperature, 214–217
  conversions, 217
  values, calculating, 215–216
Thermal energy, 226
Time constant, 439
title command, MATLAB, 686
Total pressure, 219, 221–222
Transition region, 284
Transpose (´) operator, 561
Trendlines, MATLAB, 695–700
Trigonometric functions, in Excel, 302
try-catch statement, 760–761
Turbulent flow, 284

U

Unit conversions, 135, 171–174
United States Customary System (USCS), 171
Unit law, 184
Units, 167, 269
  AES (American Engineering System), 171, 203
  constants with, 269–272
  conversion procedure for, 171–174
    involving equations, 187–189
    involving new units, 178–179
    steps, 174–178
  derived dimensions and, 179–183
  non-SI, 171
  of radian, 183
  SI system. *See* SI system (Le Système International d'Unités)
  USCS (United States Customary System), 171
Universal gravitational constant *(G)*, 270

Upper limit estimation, 130
USCS (United States Customary System), 171

V

Vacuum pressure, 219
Variable costs, 384
Variable recursion, 806
Variables
  control, 80
  counter, 792–793
  dependent, 80, 361
  global, 642–644
  independent, 80, 361
  index, 792–793, 803
  local, 642–644
  MATLAB, 552–556
  plotting, 685
  in problem-solving, 153–154
  saving and restoring, 555–556
Variance, 507
Vectors, MATLAB, 560–571
  calculations, 570–571
    addition/subtraction, 562
    combined operations, 564
    multiplication/division, 562–563
    powers, 563
  defined, 560
  deleting values from, 569
  functions used with, 565–566
  indexing, 567–568
    linear sequences as, 569–570
  individual values of, 568
  linear sequences, 561
  notes on creating, 566–567
  transpose (´) operator, 561
Velocity, 180
Velocity gradient, 419
Vertical axis, 374
Vertical gridlines, 362
VLOOKUP function, 319–320, 324
Voltage *(V)*, 238, 240, 419, 429
Volumetric flow rate, 373

W

Warning messages, MATLAB, 759
Watt *(W)*, 229
Weight *(w)*, of object, 205–206
Weighted benefit analysis, 69
while loop, 794–801
  application, 796–800
  ending, 812–813
  initializing, 794–796
  syntax, 794
  tracking number of iterations in, 800–801
  "until" logic in, 800
Work (*W*), 226
Worksheet, 294
Workspace, 551
Written algorithms, 619, 621–623
  decision making in, 732–733

X

xlabel command, MATLAB, 686
xlsfinfo function, 700–701
xlsread function, 700–703
xlswrite function, 704–705

Y

Yield stress (τ_0), 420
ylabel command, MATLAB, 686
Young's modulus, 421

Z

Zero number, 132

Equation Tables

Geometric Formulas

Rectangle Area = a b Perimeter = 2 a + 2 b	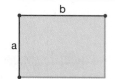	**Rectangular Parallelepiped** Volume = a b c Surface Area = 2 (a b + a c + b c)	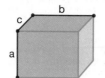
Circle Area = πr^2 Perimeter = $2 \pi r$		**Sphere** Volume = $\frac{4}{3} \pi r^3$ Surface Area = $4 \pi r^2$	
Triangle Area = $\frac{1}{2} b H$		**Right Circular Cone** Volume = $\frac{1}{3} \pi r^2 H$	
Torus (doughnut) Volume = $2 \pi^2 R r^2$	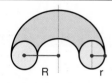	**Right Circular Cylinder** Volume = $\pi r^2 H$ Lateral Surface Area = $2 \pi r H$	

Physical Constants [Value and Units]

c	speed of light in a vacuum	$3 \times 10^8 \frac{m}{s}$
	speed of sound in air [20 °C]	$343.59 \frac{m}{s}$
e	Euler number (base of natural logarithm)	$2.71828\ldots$
	elementary charge of an electron	1.602×10^{-19} C
F	Faraday's constant	$9.65 \times 10^4 \frac{C}{mol}$
φ	golden ratio	$1.61803\ldots$
g	acceleration due to gravity	$9.8 \frac{m}{s^2}$ on Earth
G	gravitational constant	$6.67 \times 10^{-11} \frac{N\,m^2}{kg^2}$
h	Planck constant	6.62×10^{-34} J s
k	Boltzmann constant	$1.38065 \times 10^{-23} \frac{J}{K}$
N_A	Avogadro's number	$6.022 \times 10^{23} \frac{1}{mol}$
π	ratio of circle circumference to diameter	$3.14159\ldots$
R	ideal gas constant	$0.08206 \frac{atm\,L}{mol\,K} = 8314 \frac{Pa\,L}{mol\,K}$
σ	Stefan-Boltzmann constant	$5.67 \times 10^{-8} \frac{W}{m^2 K^4}$

Equations

Newton's Second Law [8.1]	Weight [8.2]	Density [8.3]
$F = m\,a$	$w = m\,g$	$\rho = \dfrac{m}{V}$

Specific Gravity [8.3]	Specific Weight [8.3]	Molecular Weight [8.4]
$SG = \dfrac{\rho_{object}}{\rho_{water}}$	$\gamma = \dfrac{w}{V}$	$MW = \dfrac{m}{n}$

Temperature: °F to °C [8.5]	Temperature: °C to K [8.5]	Temperature: °F to °R [8.5]
$\dfrac{T[°F] - 32}{180} = \dfrac{T[°C] - 0}{100}$	$T[K] = T[°C] + 273$	$T[°R] = T[°F] + 460$

Pressure [8.6]	Pascal's Law of Hydrostatic [8.6]	Pressure: Total [8.6]
$P = \dfrac{F}{A}$	$P_{hydro} = \rho\,g\,H$	$P_{total} = P_{hydro} + P_{surface}$

Ideal Gas Law [8.7]	Energy: Work [8.8]	Energy: Potential [8.8]
$P\,V = n\,R\,T$	$W = F\,\Delta x$	$PE = m\,g\,\Delta H$

Energy: Kinetic, translational [8.8]	Energy: Kinetic, rotational [8.8]	Energy: Kinetic, total [8.8]
$KE_T = \dfrac{1}{2}\,m\,(v_f^2 - v_i^2)$	$KE_R = \dfrac{1}{2}\,I\,(\omega_f^2 - \omega_i^2)$	$KE = KE_T + KE_R$

Energy: Thermal [8.8]	Power [8.9]	Efficiency [8.10]
$Q = m\,C_p\,\Delta T$	$P = \dfrac{E}{t}$	$\eta = \dfrac{P_{out}}{P_{in}}$

Coulomb's Law of Charge [8.11]	Current, related to charge [8.11]	Ohm's Law of Resistance [8.11; 12.1]
$|F| = k_e \dfrac{|Q_1 Q_2|}{r^2}$	$Q = I\,t$	$V = I\,R$

Conductance [8.11]	Joule's First Law of Power [8.11]	Capacitance, related to charge [8.11]
$G = \dfrac{1}{R}$	$P = V\,I = \dfrac{V^2}{R} = I^2 R$	$Q = C\,V$

Energy: Capacitor [8.11]	Inductance [8.11]	Energy: Inductor [8.11]
$E_c = \dfrac{1}{2}\,C\,V^2$	$V = L\dfrac{dI}{dt}$	$E_L = \dfrac{1}{2}\,L\,I^2$

MATLAB Graphing Properties

Most MATLAB graphing functions have the ability to apply special properties for customization of the appearance of a graph. This is a partial list of common property names and values available. In most cases, the property name and property values are placed inside of single quotes; exceptions are noted below. All functions are case sensitive, but properties names and parameters are not case sensitive. For example, the command Plot will not work in place of plot; for color 'black' and 'Black' and 'b' and 'B' all work; 'MarkerSize', 'MARKERSIZE', and 'markersize' all work.

Line Style		Marker Style				Color	
-	solid	*	Asterisk	.	Point	k	Black
- -	dashed	o	Circle	v	Triangle: Downward-point	b	Blue
- .	dash-dot	x	Cross	<	Triangle: Left-point	c	Cyan
:	dotted	+	Plus sign	>	Triangle: Right-point	g	Green
				^	Triangle: Upward-point	m	Magenta
		d / diamond	Diamond	h / hexagram	Hexagram (6-pointed star)	r	Red
		s / square	Square	p / pentagram	Pentagram (5-pointed star)	y	Yellow
						w	White

The Marker Styles as noted and Color properties can be expressed as either the abbreviation (d) or the full word (diamond).

Legend Properties: `legend`

All legend properties except `location` must be adjusted using the `set` command, not directly in the legend function. First, a legend name (a handle) is defined by `LegendName = legend(…)`. The `LegendName` is used with `set` to adjust the legend display properties.

Property Name	Purpose: To Change . . .	Values, in Single Quotes	Default
LineWidth	Width of line (in units of points) bordering the legend	Numerical values without single quotes	0.5 points
EdgeColor	Border color around legend box	See COLOR list	black
Color	Background color of legend box	See COLOR list	white
TextColor	Text color of legend entries	See COLOR list	black
FontSize FontAngle FontWeight	Attribute of text legend entries	See related entries under Text Properties	
Location	Location of legend in FIGURE window NW N NE W E SW S SE	North | South | East | West NorthEast | NorthWest SouthEast | SouthWest	Inside plot box
		__Outside (__ is a directional operator above)	Outside plot box
		Best	Inside plot box with least data conflict
		BestOutside	Outside plot box in least unused space

NOTE: The use of `Color` as a property in MATLAB is inconsistent. In a legend, `Color` refers to background color, whereas in most text related objects it refers to font color and `BackgroundColor` is for the background color. In a legend, the font color is assigned using `TextColor`.

Plotting Properties: `plot`

Property Name	Purpose: To Change . . .	Values, in Single Quotes	Default
`LineWidth`	Width of line (in units of points)	Numerical values without single quotes	0.5 points
`MarkerEdgeColor`	Border color of marker	See `COLOR` list	Default marker color
`MarkerFaceColor`	Fill color of marker *Works for markers: o d s h p ∨ < > ∧*	See `COLOR` list	Default marker color
`MarkerSize`	Size of marker (in units of points)	Numerical values without single quotes	6 points

Text Properties: `text, xlabel, ylabel, title`

Property Name	Purpose: To Change . . .	Values, Place in Single Quotes			Default
`BackgroundColor`	Background color of the textbox	See `COLOR` list			
		`[X,Y,Z]` Custom: Amount of red (X), green (Y), and blue (Z); each value is between 0 and 1			Transparent
`Color`	Text color in the textbox	See `Color` list			`black`
`EdgeColor`	Border color around the textbox	See `Color` list			`none`
The following commands are only enabled when the `EdgeColor` property is changed to a visible color.					
`LineStyle`	Line style bordering the textbox	See `LINE STYLE` list			-
`LineWidth`	Line width (in unit of points) of line bordering the textbox	Numerical values without single quotes			0.5 points
`Margin`	Space (in unit of points) between text in textbox and line bordering textbox	Numerical values without single quotes			2.0 points
`FontAngle`	Font angle	`normal`	`italic`		`normal`
`FontSize`	Font size (in units of points)	Numerical values without single quotes			10.0 points
`FontWeight`	Font weight	`normal`	`bold`	`demi`	`normal`
`HorizontalAlignment`	Alignment of textbox at (x,y) location	`left`	`center`	`right`	`left`
`VerticalAlignment`		`top`	`middle`	`bottom`	`middle`
		`baseline`	`cap`		

Special Characters

These special characters are available for use in graphs in the title, legend, axis labels and text boxes.

Greek Letters						Math Symbols				Just for Fun	
α	\alpha	ε	\epsilon	π	\pi	≈	\approx	±	\pm	♥	\heartsuit
β	\beta	η	\eta	ρ	\rho	≤	\leq	°	\circ	←	\leftarrow
γ	\gamma	θ	\theta	σ	\sigma	≥	\geq	∂	\partial	⇐	\Leftarrow
Γ	\Gamma	λ	\lambda	Σ	\Sigma	≠	\neq	∫	\int	→	\rightarrow
δ	\delta	μ	\mu	τ	\tau	∞	\infty	÷	\div	⇒	\Rightarrow
Δ	\Delta	ν	\nu	Ω	\Omega					©	\copyright